STRENGTH OF MATERIALS

STRENGTH OF MATERIALS

DR. SADHU SINGH

Formerly, **Professor & Head**, Mechanical Engineering Department
& **Dean**, Faculty of Engineering & Technology
Govind Ballabh Pant University of Agriculture & Technology
Pantnagar - 263 145 (Uttarakhand)

Formerly, **Director** (Colleges)
Punjab Technical University, Jalandhar

KHANNA BOOK PUBLISHING CO. (P) LTD.

Publisher of Engineering and Computer Books

4C/4344, Ansari Road, Darya Ganj, New Delhi-110002
Phone : 011-23244447-48 **Mobile:** +91-9910909320
E-mail : contact@khannabooks.com
Website : www.khannabooks.com

Dear Readers,

To prevent the piracy, this book is secured with HIGH SECURITY HOLOGRAM on the front title cover. In case you don't find the hologram on the front cover title, please write us to at contact@khannabooks.com or whatsapp us at +91-99109 09320 and avail special gift voucher for yourself.

Specimen of Hologram on front Cover title:

Moreover, there is a SPECIAL DISCOUNT COUPON for you with EVERY HOLOGRAM.

How to avail this SPECIAL DISCOUNT:

Step 1: Scratch the hologram
Step 2: Under the scratch area, your "coupon code" is available
Step 3: Logon to www.khannabooks.com
Step 4: Use your "coupon code" in the shopping cart and get your copy at a special discount
Step 5: Enjoy your reading!

Thanks,

Khanna Publishing House

Price: ₹ 425.00

STRENGTH OF MATERIALS

Dr. Sadhu Singh

ISBN: 978-93-81068-80-9

Edition: 2017
Reprint: 2021

Published by:

KHANNA BOOK PUBLISHING CO. (P) LTD.
4C/4344, Ansari Road, Darya Ganj, New Delhi-110 002
Phone: 011-23244447-48 **Mobile:** +91-9910909320
E-mail: contact@khannabooks.com

Printed in India by: SPS Enterprises, Delhi

PREFACE

STRENGTH OF MATERIALS is a Text Book for B.E. / B. Tech. students of U.P. Technical University, Lucknow and Uttrakhand Technical University, Dehradun. The book has been written as per the new syllabi of these Universities.

The book contains fourteen chapters on Compound stresses and strains; Theories of failure; Three dimensional stresses and strains; Stresses in beams; Deflection of beams; Fixed and continuous beams; Torsion of shafts; Columns and struts; Helical and leaf springs; Thin cylinders and spheres; Thick cylinders; Rotating discs and cylinders; Curved beams; and Unsymmetrical bending.

The book contains a large number of solved and unsolved problems. The special features of the book are: Summary, Short answer type questions and end of chapters Exercises.

The help provided by Mr. Mukul Seth, and Puneet Khanna during the preparation of the manuscript and otherwise is highly appreciated and acknowledged.

My grand children, Kanupreet Kaur and Amitoej Singh have been a source of great inspiration during the preparation of the manuscript. They kept my mood cheerful, which helped in improving the text. Mrs. Narinderpal Kaur, my daughter-in-law, took proper care during writing of the manuscript and otherwise. I have all appreciations for them.

It is helped that the book in its present form shall be quite useful to the students and teachers. Further suggestions for the improvement of the book are invited by the publishers and the author.

Dr. Sadhu Singh

CONTENTS

CHAPTER-3 :THREE DIMENSIONAL STRESSES AND STRAINS 98

CHAPTER-4: STRESSES IN BEAMS 150

Chapter-1

Compound Stresses and Strains

1.1 INTRODUCTION

In this chapter, we shall study the stresses and strains on an oblique plane under uniaxial, biaxial and complex stresses and strains. The determination of principal stresses and elastic constants shall also be discussed.

1.2 STRESSES ON AN OBLIQUE PLANE

(a)Uniaxial Loading: Consider a uniaxial tensile stress σ acting on plane *BC* of a bar of uniform area of cross-section, as shown in Fig. 1.1 (*a*). We are interested to determine the normal and shear stresses on plane *AC*, whose normal is inclined at an angle θ with the line of action of σ. The forces acting at a point on plane *AC* are shown in Fig. 1.1 (b)

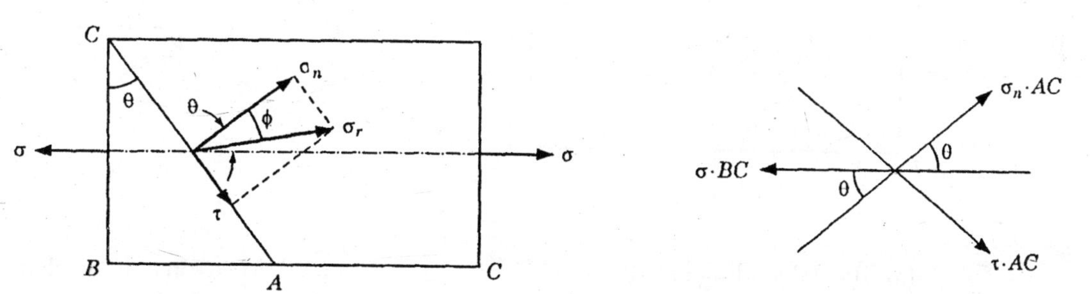

(a) Uniaxial loading on a bar (b) Forces on oblique plane

Fig. 1.1 Stresses on an oblique plane due to uniaxial loading.

Resolving the forces perpendicular and parallel to the plane *AC*, we have

$$\sigma_n \times AC = \sigma \times BC \cdot \cos\theta$$

$$\sigma_n = \sigma \cdot \frac{BC}{AC} . \cos\theta = \sigma \cos^2\theta = \frac{1}{2}\sigma(1+\cos 2\theta) \quad \text{...(1.1)}$$

$$\tau \times AC = \sigma \times BC.\sin\theta$$

$$\tau = \sigma \frac{BC}{AC}\sin\theta = \sigma\cos\theta.\sin = \frac{1}{2}\sigma\sin 2\theta \quad \text{... (1.2)}$$

At $\theta = 45°$ and $135°$, $\tau_{max} = \frac{1}{2}\sigma$

Resultant stress, $\sigma_r = \sqrt{\sigma_n^2 + \tau^2} = \frac{1}{2}\sigma\sqrt{(1+\cos 2\theta)^2 + \sin^2 2\theta}$

$$= \frac{1}{2}\sigma\sqrt{1+\cos^2 2\theta + 2\cos 2\theta + \sin^2 2\theta}$$

$$= \frac{1}{2}\sigma\sqrt{2+2\cos 2\theta} = \frac{1}{2}\sigma\sqrt{2\times 2\cos^2\theta} = \sigma\cos\theta \quad \text{... (1.3)}$$

$$\tan\phi = \frac{\tau}{\sigma_n} = \tan\theta$$

or $\phi = \theta$ ... (1.4)

(b) Biaxial Loading: Now consider the case of biaxial loading, as shown in Fig. 1.2. Resolving the forces perpendicular and parallel to the plane AC, we have

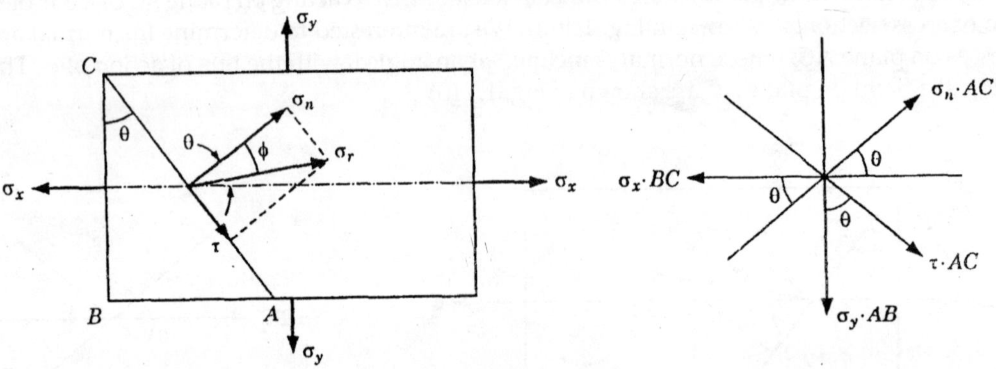

(a) Biaxial loading on a bar (b) Forces on oblique plane

Fig. 1.2 Stresses on an oblique plane due to biaxial loading.

$$\sigma_n \times AC = \sigma_x \cdot BC.\cos\theta + \sigma_y \cdot AB\sin\theta$$

$$\sigma_n = \sigma_x \frac{BC}{AC}\cdot\cos\theta + \sigma_y \cdot \frac{AB}{AC}\sin\theta$$

$$= \sigma_x\cos^2\theta + \sigma_y\sin^2\theta$$

$$= \frac{\sigma_x}{2}(1+\cos 2\theta)+\frac{\sigma_y}{2}(1-\cos 2\theta)$$

$$= \frac{1}{2}(\sigma_x+\sigma_y)+\frac{1}{2}(\sigma_x-\sigma_y)\cos 2\theta \quad ...(1.5)$$

$$\tau . AC = \sigma_x BC \sin\theta - \sigma_y AB.\cos\theta$$

$$\tau = \sigma_x \cos\theta \sin\theta - \sigma_y \sin\theta \cos\theta = \frac{1}{2}(\sigma_x-\sigma_y)\sin 2\theta \quad ...(1.6)$$

Maximum shear stress,
$$\tau_{max} = \frac{1}{2}(\sigma_x-\sigma_y) \text{ at } \theta = 45° \quad ...(1.7)$$

Resultant stress,
$$\sigma_r = \sqrt{\sigma_n^2+\tau^2} + \sqrt{\sigma_x^2\cos^2\theta+\sigma_y^2\sin^2\theta} \quad ...(1.8)$$

$$\tan\phi = \frac{\tau}{\sigma_n} = \frac{(\sigma_x-\sigma_y)\sin\theta\cos\theta}{\sigma_x\cos^2\theta+\sigma_y\sin^2\theta} = \frac{\sigma_x-\sigma_y}{\sigma_x\cot\theta+\sigma_y\tan\theta} \quad ...(1.9)$$

For ϕ to be maximum,
$$\frac{d(\tan\phi)}{d\theta} = 0$$

or
$$\frac{d}{d\theta}(\sigma_x\cot\theta+\sigma_y\tan\theta) = 0$$

$$-\sigma_x \text{cosec}^2\theta+\sigma_y\sec^2\theta = 0$$

$$\tan\theta = \sqrt{\frac{\sigma_x}{\sigma_y}} \quad ...(1.10)$$

$$\tan\phi_{max} = \frac{\sigma_x-\sigma_y}{2\sqrt{\sigma_x\sigma_y}} \quad ...(1.11)$$

1.3 SHEAR STRESSES

Consider a two-dimensional lamina shown in Fig. 1.3 (a) subjected to shear stresses τ_{xy} and τ_{yx}. Because shear stresses are complementary in nature, therefore, $\tau_{xy} = \tau_{yx}$. To prove the shear stresses to be complementary in nature, let $AB = DC = dx$ and $BC = AD = dy$. Considering unit thickness of lamina *ABCD* and taking moments about *A*, we have

$$\tau_{xy} \times BC \times AB = \tau_{yx} \times DC \times AD$$

or
$$\tau_{xy} \times dy \times dx = \tau_{yx} \times dx \times dy$$

$\therefore$
$$\tau_{xy} = \tau_{yx} \quad ...(1.12)$$

Considering the force normal to the diagonal AC shown in Fig. 1.3 (b), we have

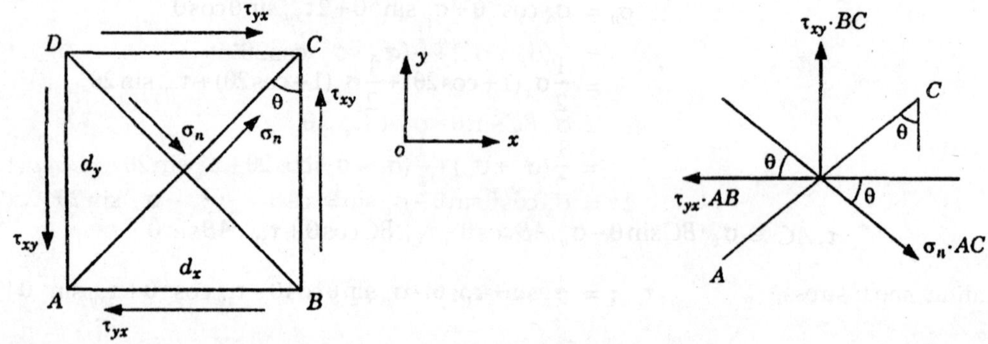

(a) Two-dimensional lamina under shear stresses. (b) Force on diagonal AC.

Fig. 1.3 Shear Stresses

$$\sigma_n AC = \tau_{xy} BC\sin\theta + \tau_{xy} AB\cos\theta$$

$$\sigma_n = \tau_{xy}\sin\theta\cos\theta + \tau_{xy}\sin\theta\cos\theta = \tau_{xy}\sin 2\theta$$

For $\theta = 45°, \sigma_n = \tau_{xy}$

The normal stress σ_n is compressive on diagonal AC and is equal to the shear stress. Similarly on diagonal $BD, \sigma_n = \tau_{xy}$ and is tensile in nature.

1.4 COMPOUND STRESSES

Consider the compoud stresses acting on a body, as shown in Fig. 1.4 (a). The forces on AC are shown in Fig. 1.4 (b). Resolving the forces perpendicular and parallel to the plane AC, we have

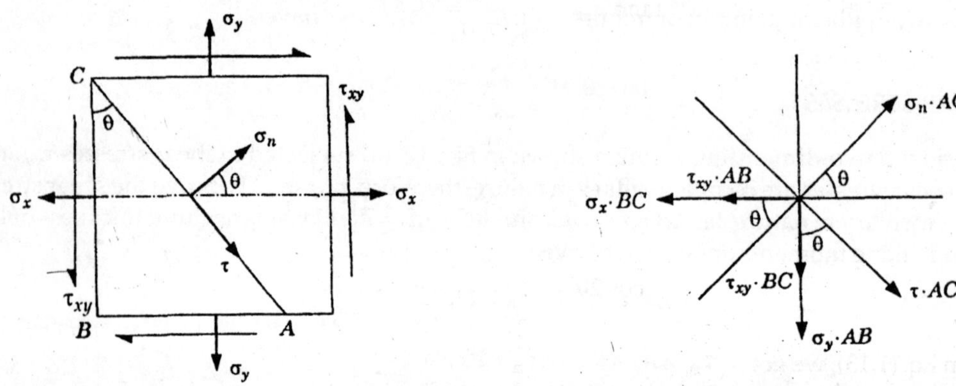

(a) Compound stresses on a body (b) Force on oblique plane

Fig. 1.4 Compound stresses

$$\sigma_n AC = \sigma_x BC.\cos\theta + \sigma_y .AB\sin\theta + \tau_{xy}.BC.\sin\theta + \tau_{xy}AB\cos\theta$$

$$\sigma_n = \sigma_x \cos^2\theta + \sigma_y \sin^2\theta + 2\tau_{xy}\sin\theta\cos\theta$$

$$= \frac{1}{2}\sigma_x(1+\cos 2\theta) + \frac{1}{2}\sigma_y(1-\cos 2\theta) + \tau_{xy}\sin 2\theta$$

$$= \frac{1}{2}(\sigma_x + \sigma_y) + \frac{1}{2}(\sigma_x - \sigma_y)\cos 2\theta + \tau_{xy}\sin 2\theta \qquad \text{... (1.13)}$$

$$\tau . AC = \sigma_x \cdot BC\sin\theta - \sigma_y .AB\cos\theta - \tau_{xy}.BC\cos\theta + \tau_{xy}.AB\sin\theta$$

$$\tau = \sigma_x \sin\theta\cos\theta - \sigma_y \sin\theta\cos\theta - \tau_{xy}\cos^2\theta + \tau_{xy}\sin^2\theta$$

$$= \frac{1}{2}(\sigma_x - \sigma_y)\sin 2\theta - \tau_{xy}\cos 2\theta \qquad \text{... (1.14)}$$

For σ_n to have a maximum or minimum value,

$$\frac{d\sigma_n}{d\theta} = 0$$

$$-(\sigma_x - \sigma_y)\sin 2\theta + 2\tau_{xy}\cos 2\theta = 0$$

$$\tan 2\theta = \frac{2\tau_{xy}}{\sigma_x - \sigma_y} \qquad \text{... (1.15)}$$

1.5 PRINCIPAL STRESSES AND PRINCIPAL PLANES

At any point within a stressed body, no matter how complex the state of stress may be, there always exist three mutually perpendicular planes on each of which the resultant stress is normal stress. These mutually perpendicular planes are called principal planes, and the resultant normal stresses acting on them are called principal stresses.

For σ_n to be either maximum or minimum, from Eq. (1.15), we have

$$\tan 2\theta = \frac{2\tau_{xy}}{\sigma_x - \sigma_y}$$

$$\therefore \quad \sin 2\theta = \frac{2\tau_{xy}}{\pm\sqrt{(\sigma_x - \sigma_y)^2 + 4\tau_{xy}^2}}$$

$$\cos 2\theta = \frac{\sigma_x - \sigma_y}{\pm\sqrt{(\sigma_x - \sigma_y)^2 + 4\tau_{xy}^2}}$$

From Eq. (1.13), we get

$$\sigma_n = \frac{1}{2}(\sigma_x + \sigma_y) + \frac{1}{2}(\sigma_x - \sigma_y)\frac{\sigma_x - \sigma_y}{\pm\sqrt{(\sigma_x - \sigma_y)^2 + 4\tau_{xy}^2}} + \tau_{xy} \times \frac{2\tau_{xy}}{\pm\sqrt{(\sigma_x - \sigma_y)^2 + 4\tau_{xy}^2}}$$

$$= \frac{1}{2}(\sigma_x + \sigma_y) - \frac{1}{2}\left[\frac{(\sigma_x - \sigma)^2 + 4\tau_{xy}^2}{\pm\sqrt{(\sigma_x - \sigma_y)^2 + 4\tau_{xy}^2}}\right]$$

$$\sigma_{1,2} = \frac{1}{2}(\sigma_x + \sigma_y) \pm \frac{1}{2}[(\sigma_x - \sigma_y)^2 + 4\tau_{xy}^2]^{1/2} \quad \text{... (1.16)}$$

Eq. (1.16) gives two principal stresses and Eq. (1.15) gives their orientation. For shear stress τ to be maximum or minimum,

$$\frac{d\tau}{d\theta} = 0$$

Using Eq. (1.14), we have

$$(\sigma_x - \sigma_y)\cos 2\theta + 2\tau_{xy}\sin 2\theta = 0$$

$$\tan 2\theta = \frac{-(\sigma_x - \sigma_y)}{2\tau_{xy}} \quad \text{... (1.17)}$$

The planes of maximum shear stress are θ_3 and $\theta_4 = \theta_3 + 90°$. These planes are inclined at 45° to the principal planes.

$$\tau_{max} = \frac{1}{2}(\sigma_x - \sigma_y)\frac{(\sigma_x - \sigma_y)}{\pm\sqrt{(\sigma_x - \sigma_y)^2 + 4\tau_{xy}^2}} + \tau_{xy} \times \frac{2\tau_{xy}}{\pm\sqrt{(\sigma_x - \sigma_y)^2 + 4\tau_{xy}^2}}$$

$$= \pm\frac{1}{2}\sqrt{(\sigma_x - \sigma_y)^2 + 4\tau_{xy}^2} \quad \text{... (1.18)}$$

$$\tau_{max} = \frac{1}{2}(\sigma_1 - \sigma_2) \quad \text{... (1.19)}$$

1.6 MOHR'S CIRCLE

Mohr's circle is a circular diagram to determine the normal, shear and resultant stresses on an oblique plane when a body is subjected to biaxial or compound state of stress. The values of principal stresses and principal planes can also be determined from this diagram. The radius of the circle gives the maximum shear stress.

(a) Biaxial stresses (Fig. 1.5)

Form Eqs. (1.5) and (1.6), we have

$$\sigma_n = \frac{1}{2}(\sigma_x + \sigma_y) + \frac{1}{2}(\sigma_x - \sigma_y)\cos^2\theta$$

$$\tau = \frac{1}{2}(\sigma_x - \sigma_y)\sin 2\theta$$

$$\left(\sigma_n - \frac{\sigma_x + \sigma_y}{2}\right)^2 + \tau^2 = \left(\frac{\sigma_x - \sigma_y}{2}\right)^2 \qquad \text{... (1.20)}$$

Eq. (1.20) represents a circle in the $\sigma_n - \tau$ plane. The centre of the circle is $\left(\frac{\sigma_x + \sigma_y}{2}, 0\right)$ and radius $\frac{\sigma_x - \sigma_y}{2}$.

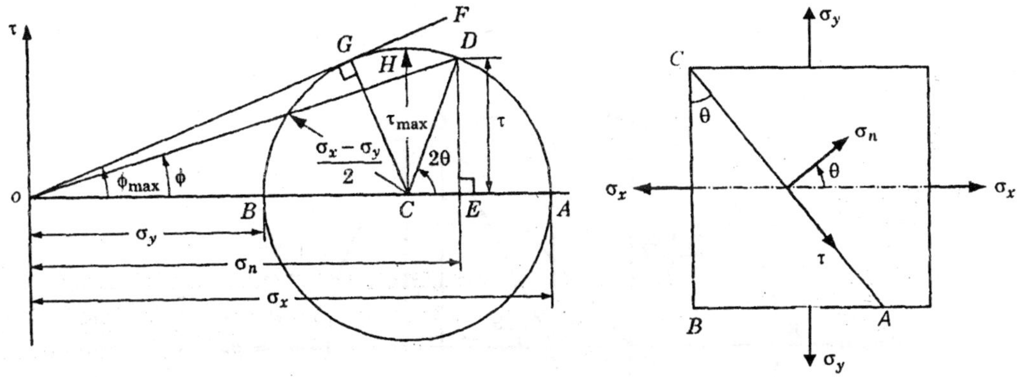

Fig. 1.5 Mohr's stress circle for biaxial stresses when both stresses are tensile.

Construction:

(*i*) $\sigma_x > \sigma_y$

1. On a convenient scale, take $OA = \sigma_x$, $OB = \sigma_y$, $AC = CB = \frac{\sigma_x - \sigma_y}{2}$.
2. With C as centre and *CA* as radius, draw the circle.
3. Make $\angle ACD = 2\theta$ anticlockwise.
4. Drop $DE \perp OA$ and join *OD*. Then $OE = \sigma_n$, $DE = \tau$ and $OD = \sigma_r$, $\angle AOD = \phi$.
5. Draw *OF* tangential to the circle, touching the circle at *G*. Join *CG*. Then $\angle AOF = \phi_{max}$, the maximum obliquity of the resultant stress.
6. Draw $HC \perp OA$. Then $HC = \tau_{max}$.

Proof:

$$DE = CD \sin 2\theta = \frac{(\sigma_x - \sigma_y)}{2} \sin 2\theta = \tau$$

$$OE = OC + CE = \frac{\sigma_x + \sigma_y}{2} + \frac{(\sigma_x - \sigma_y)}{2} \cos 2\theta = \sigma_n$$

$$OD^2 = OE^2 + DE^2$$

$$\sigma_r^2 = \sigma_n^2 + \tau^2$$

$$\tan\phi = \frac{DE}{OE} = \frac{\tau}{\sigma_n}$$

$$\sin\phi_{max} = \frac{CG}{OC} = \frac{\sigma_x - \sigma_y}{\sigma_x + \sigma_y}$$

(ii) σ_y is compressive.

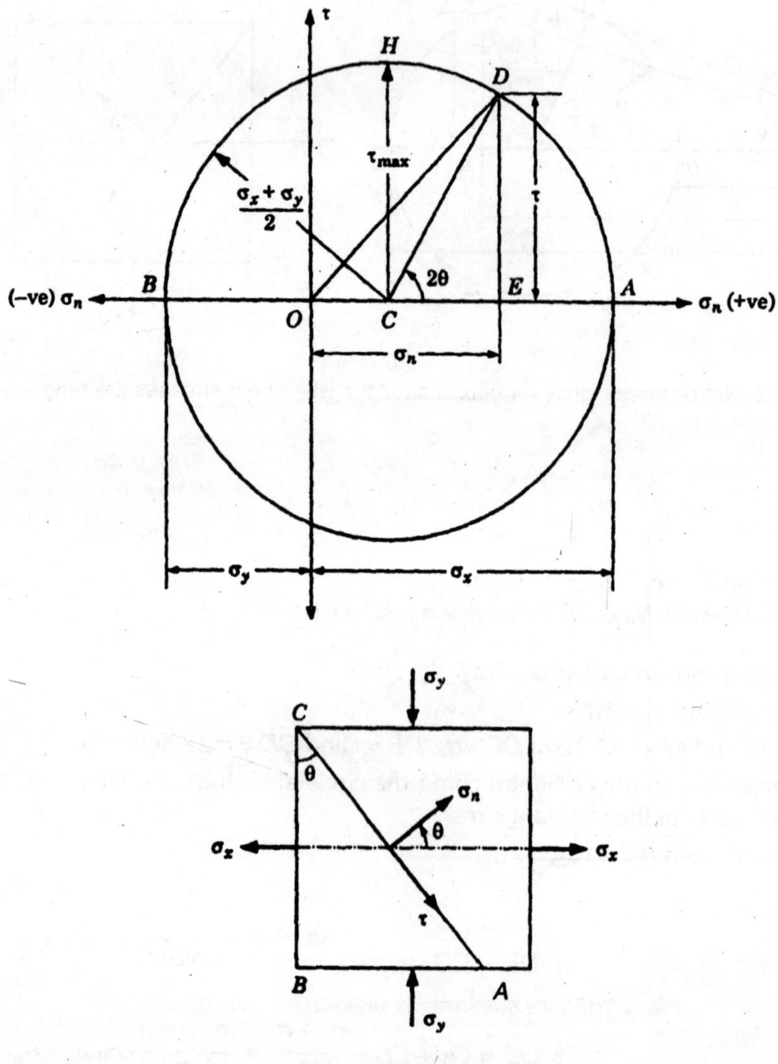

Fig. 1.6 Mohr's stress circle for biaxial stresses when σ_y is compressive

In this case, take a_y along the -ve a_n axis. Then draw the Mohr's circle as explained in *(i)*. *(iii)* a_x is compressive

The Mohr's circle is shown in Fig. 1.7.

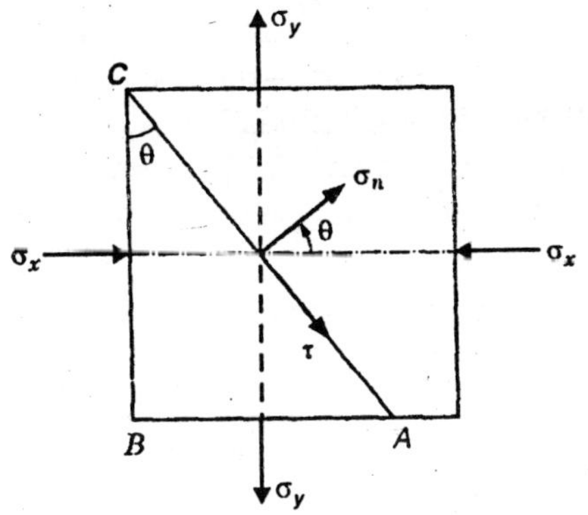

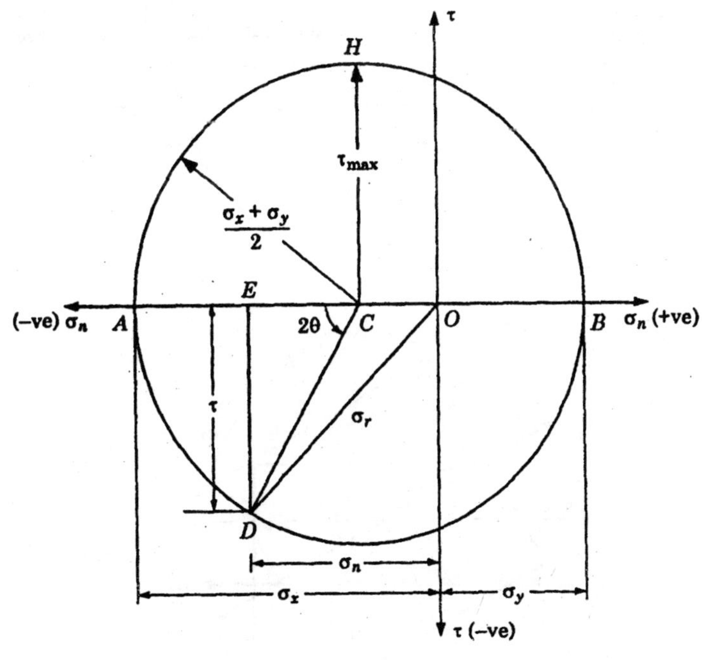

Fig. 1.7 Mohr's stress circle when σ_x is compressive.

(b) Compound Stresses

Stress conventions:

1. Normal stresses are plotted as horizontal coordinates, with tensile stresses (positive) plotted to the right of origin and compressive stresses (negative) plotted to the left.
2. Shearing stresses are plotted as vertical coordinates, with those tending to produce a clockwise rotation of the stress element plotted above the σ_n-axis, and those tending to produce a counterclockwise rotation of the stress element plotted below the σ_n - axis. Consider the stress system shown in Fig. 1.8 (a). The Mohr's circle is drawn in Fig. 1.8 (b) following the above conventions.

(*i*) σ_x and σ_y are both tensile ($\sigma_x > \sigma_y$) and τ_{xy} positive.

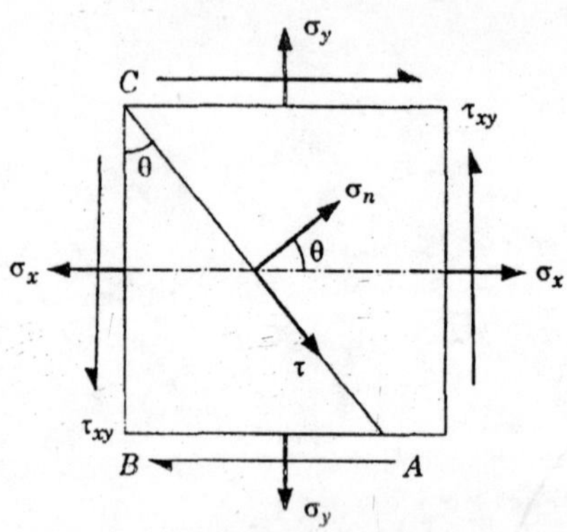

(a) Compound stresses on a body

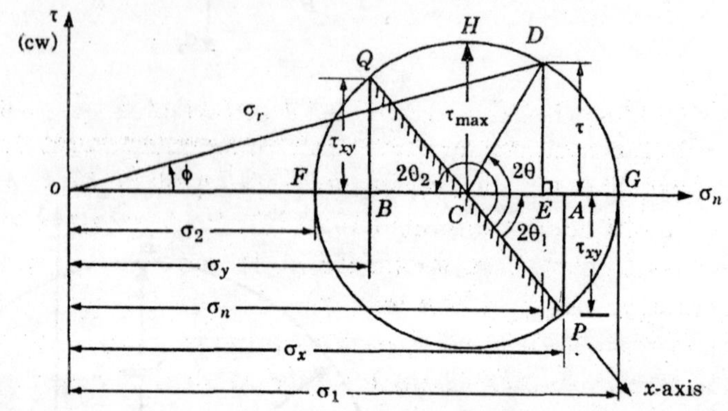

(b) Mohr's circle

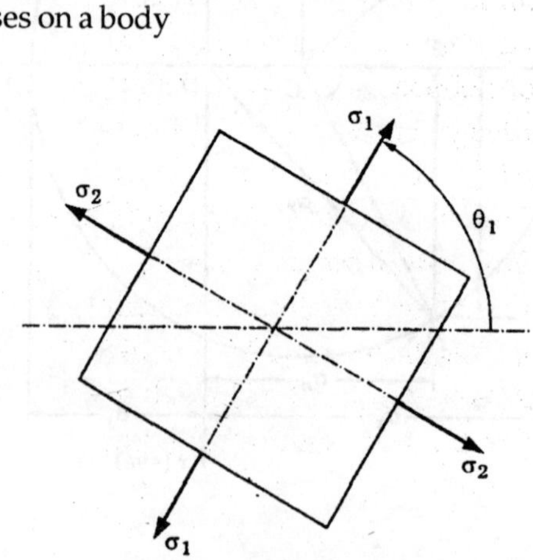

(c) Orientation of principal planes

Fig. 1.8 Mohr's stress circle for compound stresses.

Construction:

1. On a convenient scale, take $OA = \sigma_x$, $OB = \sigma_y$, $AP = BQ \perp OA$ and $= \tau_{xy}$.
2. Join PQ to intersect OA at C.
3. With CP as radius, draw the circle. Then $OG = \sigma_1$ and $OF = \sigma_2$.
4. Make $\angle PCD = 2\theta$. Drop $DE \perp OA$ and join OD. Then $OE = \sigma_n$, $DE = \tau$ and $OD = \sigma_r$. $\angle AOD = \phi$.
5. $\angle PCG = 2\theta_1$ and $PCF = 2\theta_2$

The orientation of principal planes is shown in Fig. 1.8. (*c*)

Proof:

$$AB = \sigma_x - \sigma_y$$

$$AC = CB = \frac{\sigma_x - \sigma_y}{2}$$

$$OC = OB + BC = \sigma_y + \frac{\sigma_x - \sigma_y}{2} = \frac{\sigma_x + \sigma_y}{2}$$

$$OE = OC + CE = OC + CD\cos 2(\theta - \theta_1)$$

$$= OC + CD[\cos 2\theta . \cos 2\theta_1 + \sin 2\theta . \sin 2\theta_1]$$

$$= OC + CD\left[\cos 2\theta \frac{AC}{CP} + \sin 2\theta \frac{AP}{CP}\right]$$

$$= OC + AC . \cos 2\theta + AP . \sin 2\theta \qquad [\because CD = CP]$$

$$\sigma_n = \frac{\sigma_x + \sigma_y}{2} + \frac{\sigma_x - \sigma_y}{2}\cos 2\theta + \tau_{xy}\sin 2\theta$$

$$DE = CD\sin 2(\theta - \theta 1)$$

$$= CD(\sin 2\theta \cos 2\theta_1 - \cos 2\theta \cdot \sin 2\theta_1)$$

$$= CD\left(\sin 2\theta \cdot \frac{AC}{CP} - \cos 2\theta \cdot \frac{AP}{CP}\right)$$

$$= AC\sin 2\theta - AP\cos 2\theta$$

$$\tau = \frac{\sigma_x - \sigma_y}{2}\sin 2\theta - \tau_{xy}\cos 2\theta$$

$$CH = \tau_{max} = CP = \sqrt{AC^2 + AP^2} = \sqrt{\left(\frac{\sigma_x - \sigma_y}{2}\right)^2 + \tau_{xy}^2}$$

$$= \frac{1}{2}\sqrt{(\sigma_x - \sigma_y)^2 + 4\tau_{xy}^2}$$

$$\sigma_1 = OG = OC + CG = OC + CP$$

$$= \frac{\sigma_x + \sigma_y}{2} + \frac{1}{2}\sqrt{(\sigma_x - \sigma_y)^2 + 4x_{xy}^2}$$

$$\sigma_2 = OF = OC - FC = OC - CP$$

$$= \frac{\sigma_x + \sigma_y}{2} - \frac{1}{2}\sqrt{(\sigma_x - \sigma_y)^2 + 4\tau_{xy}^2}$$

$$\sigma_2 = OD = \sqrt{OE^2 + DE^2} = \sqrt{\sigma_n^2 + \tau^2}$$

(ii) σ_x and σ_y both positive ($\sigma_x > \sigma_y$) and τ_{xy} negative.

Consider the stress system shown in Fig. 1.9 *(a)*. The Mohr's circle is drawn in Fig. 1.9 *(b)*.

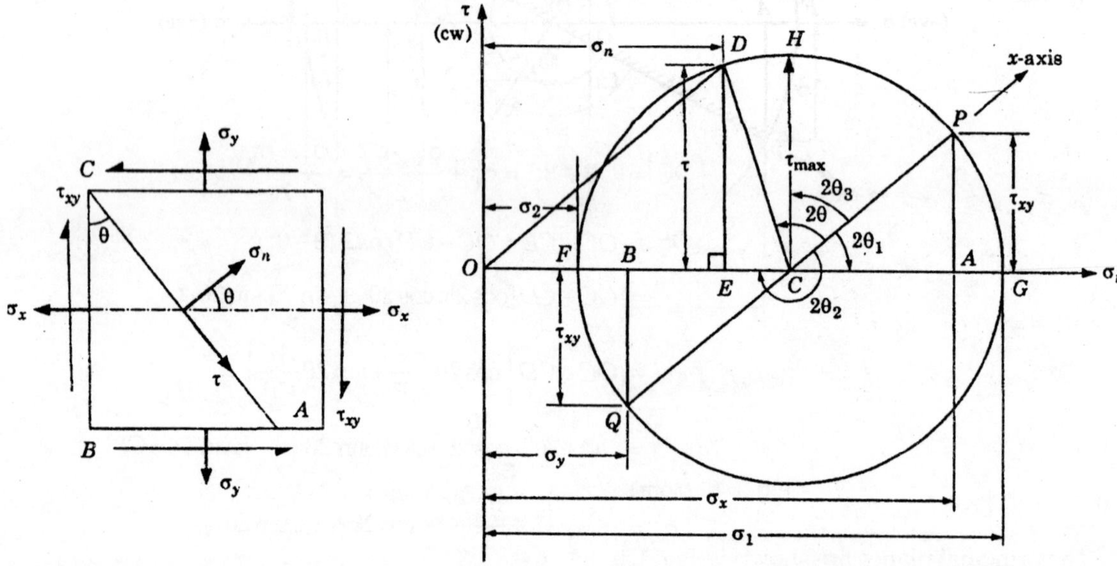

(a) (b)

Fig. 1.9 Mohr's stress circle for –ve τ_{xy}.

(iii) σ_x is compressive The Mohr's circle is shown Fig. 1.10 *(b)*

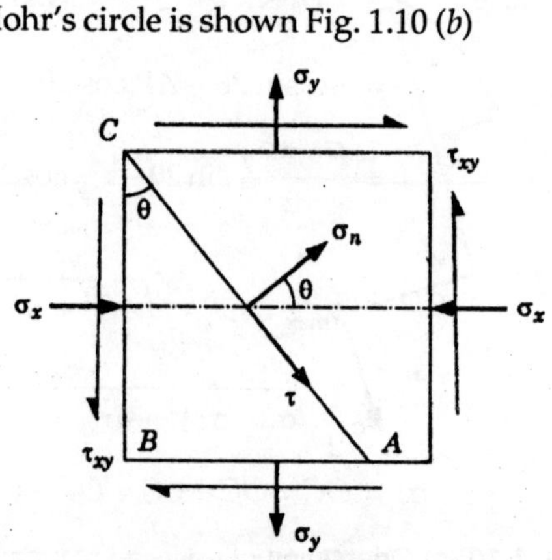

(a)

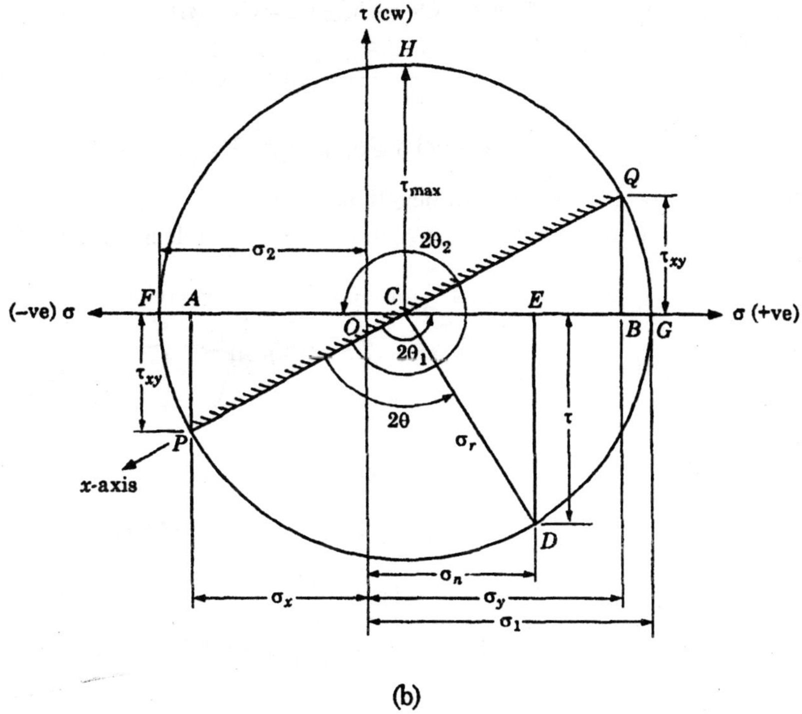

(b)

Fig. 1.10 Mohr's circle for σ_x compressive.

The principal planes are shown in Fig. 1.10 (*c*).

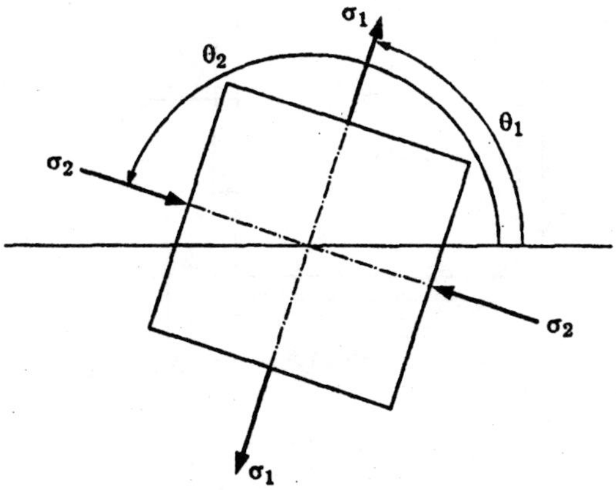

Fig. 1.10 (c) Orientation of principal planes.

Example 1.1 *The principal stresses at a point in a piece of material are 90 MPa tensile and 60 MPa compressive. Find the intensity and direction of the stress across a plane the normal of which is inclined at 30° to the axis of 90 MPa stress.*

Solution:

Given: $\sigma_x = 90 \text{ MPa}, \sigma_y = -60 \text{ MPa}, \theta = 30°$

$$\sigma_n = \frac{1}{2}(\sigma_x + \sigma_y) + \frac{1}{2}(\sigma_x - \sigma_y)\cos 2\theta$$

$$= \frac{1}{2}(90 - 60) + \frac{1}{2}(90 + 60)\cos 60°$$

$$= 15 + 37.5 = 52.5 \text{ MPa}$$

$$\tau = \frac{1}{2}(\sigma_x - \sigma_y)\sin 2\theta = \frac{1}{2}(90 + 60)\sin 60°$$

$$= 64.95 \text{ MPa}$$

$$\sigma_r = \sqrt{\sigma_n^2 + \tau^2} = \sqrt{(52.5)^2 + (64.95)^2}$$

$$= 83.52 \text{ MPa}$$

$$\tan\phi = \frac{\tau}{\sigma_n} = \frac{64.95}{52.5} = 1.23714, \phi = 51.05°$$

Example 1.2 *At a point in a piece of material the intensity of resultant stress on a certain plane is 50 MPa inclined at 30° to normal to that plane. The plane normal to this has a resultant stress whose normal component is 30 MPa. Find (a) the principal planes and principal stresses, and (b) the plane of maximum shear and its intensity.*

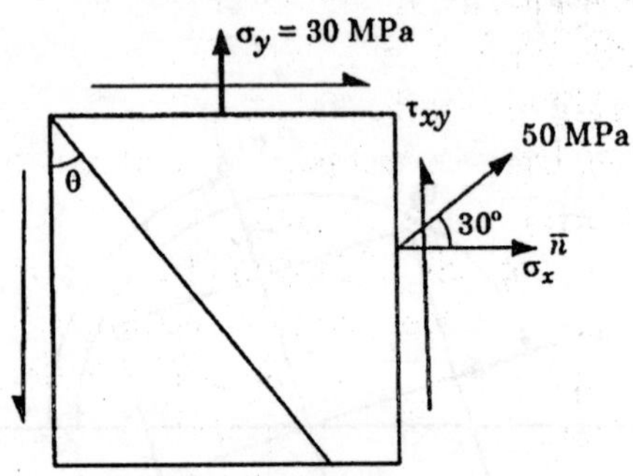

Fig. 1.11

Solution:

The stress system is shown in Fig. 1.11

$$\sigma_x = 50 \cos 30°$$

$$= 43.3 \text{ MPa}$$

$$\tau_{xy} = 50 \sin 30°$$

$$= 25 \text{ MPa}$$

$$\sigma_y = 30 \text{ MPa}$$

(*a*) Principal stresses:

$$\sigma_{1,2} = \frac{1}{2}(\sigma_x + \sigma_y) \pm \frac{1}{2}\sqrt{(\sigma_x - \sigma_y)^2 + 4\tau_{xy}^2}$$

$$= \frac{1}{2}(43.3 + 30) \pm \frac{1}{2}\sqrt{(43.3 - 30)^2 + 4 \times 25^2}$$

$$= 62.52 \text{ MPa}, 10.76 \text{ MPa}$$

Principal Planes are given by,

$$\tan 2\theta = \frac{2\tau_{xy}}{\sigma_x - \sigma_y} = \frac{2 \times 25}{43.3 - 30} = 3.7594$$

$$2\theta = 75.1^\circ$$

$$\theta_1 = 37.55^\circ$$

$$\theta_2 = 90^\circ + \theta_1 = 127.55^\circ$$

$$\tau_{max} = \frac{\sigma_1 - \sigma_2}{2} = \frac{62.52 - 10.76}{2} = 25.88 \text{ MPa}$$

Planes of maximum shear stress are given by

$$\theta_3 = \theta_1 + 45^\circ = 37.55^\circ + 45^\circ = 82.55^\circ$$

$$\theta_4 = \theta_3 + 90^\circ = 172.55^\circ$$

Example 1.3 *At a point in a material subjected to two direct stresses on planes at right angles, the resultant stress on a plane A is 80 MPa inclined at 30° to the normal, and on plane B is 40 MPa inclined at 45° to the normal. Find the principal stresses and show the position of the two planes A and B relative to the principal stresses.*

Solution:

(A) Analytical Method

The planes *A* and *B* are shown in Fig. 1.12. (a).

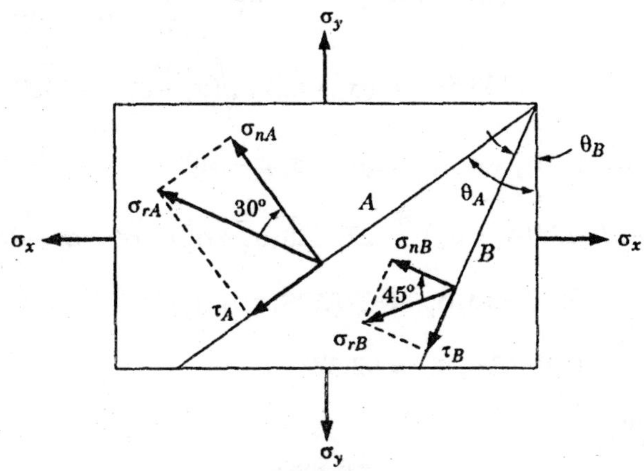

Fig.1.12(a)

Plane A:

$$\sigma_{nA} = \sigma_{rA}\cos\phi_A = 80\cos 30^\circ$$
$$= 69.282 \text{ MPa}$$
$$\tau_A = \sigma_{rA}\sin\phi_A = 80\sin 30^\circ$$
$$= 40 \text{ MPa}$$
$$\sigma_{nA} = \frac{1}{2}(\sigma_x+\sigma_y)+\frac{1}{2}(\sigma_x-\sigma_y)\cos 2\theta_A$$

$$\therefore \quad 138.564 = (\sigma_x+\sigma_y)+(\sigma_x-\sigma_y)\cos 2\theta_A \quad \text{... (1)}$$

$$\tau_A = \frac{1}{2}(\sigma_x-\sigma_y)\sin 2\theta_A$$

$$\therefore \quad 80 = (\sigma_x-\sigma_y)\sin 2\theta_A \quad \text{... (2)}$$

Plane B:

$$\sigma_{nB} = \sigma_{rB}\cos\phi_B = 40\cos 45^\circ$$
$$= 28.284 \text{ MPa}$$
$$\tau_B = \sigma_{rB}\sin\phi_B = 40\sin 45^\circ$$
$$= 28.284 \text{ MPa}$$

$$\therefore \quad 56.568 = (\sigma_x+\sigma_y)+(\sigma_x-\sigma_y)\cos 2\theta_B \quad \text{... (3)}$$

$$56.568 = (\sigma_x-\sigma_y)\sin 2\theta_B$$

From Eq. (2), we have

$$\sin 2\theta_A = \frac{80}{\sigma_x-\sigma_y}$$

$$\cos 2\theta_A = \sqrt{1-\sin^2 2\theta_A} = \sqrt{1-\frac{6400}{(\sigma_x-\sigma y)^2}}$$

Eq. (1) becomes

$$138.564 = (\sigma_x+\sigma_y)+\sqrt{(\sigma_x-\sigma_y)^2-6400}$$

or

$$[138.564-(\sigma_x+\sigma_y)]^2 = (\sigma_x-\sigma_y)^2-6400$$

$$19200+\sigma_x^2+\sigma_y^2+277.128(\sigma_x+\sigma_y) = 277.128(\sigma_x+\sigma_y)=\sigma_x^2+\sigma_y^2-2\sigma_x\sigma_y-6400$$

$$25600+4\sigma_x\sigma_y = 277.128(\sigma_x+\sigma_y)$$

$$\sigma_x+\sigma_y-0.01443\,\sigma_x\sigma_y = 92.376$$

From Eq. (4), we have

$$\sin 2\theta_B = \frac{56.568}{\sigma_x-\sigma_y}$$

$$\cos 2\theta_B = \sqrt{1-\sin^2 2\theta_B} = \sqrt{1-\left(\frac{56.568}{\sigma_x - \sigma_y}\right)^2}$$

$$= (\sigma_x + \sigma_y) + \sqrt{(\sigma_x - \sigma_y)^2 - 3200}$$

$$[56.568 - (\sigma_x + \sigma_y)]^2 = (\sigma_x - \sigma_y)^2 - 3200$$

$$3200 + \sigma_x^2 + \sigma_y^2 + 2\sigma_x\sigma_y - 1113.136\,(\sigma_x + \sigma_y) = \sigma_x^2 + \sigma_y^2 - 2\sigma_x\sigma_y - 3200$$

$$6400 + 4\sigma_x\sigma_y = 113.136(\sigma_x + \sigma_y)$$

$$\sigma_x + \sigma_y - 0.03535\sigma_x\sigma_y = 56.569 \quad \text{...(6)}$$

Subtracting Eq. (6) From (5), we get

$$0.02092\,\sigma_x\,\sigma_y = 35.807$$

$$\sigma_x\,\sigma_y = 1711.6 \quad \text{...(7)}$$

From Eq. (5), we get

$$\sigma_x + \sigma_y = 92.376 + 0.01443 \times \times 1711.6 = 117.075 \quad \text{...(8)}$$

Now

$$138.564 = 117.075 + \sqrt{(\sigma_x - \sigma_y)^2 - 6400}$$

$$(\sigma_x - \sigma_y)^2 - 6400) = (138.564 - 117.075)^2 = 461.794$$

$$\sigma_x - \sigma_y = 82.836 \quad \text{...(9)}$$

Adding Eqs. (8) and (9), we get

$$\sigma_x = 99.955 \simeq 100 \text{ MPa}$$

$$\sigma_y = 17 \text{ MPa}$$

$$\sin 2\theta_A = \frac{80}{82.836} = 0.96576$$

$$\sigma_A = 37.482°$$

$$\sin 2\theta_B = \frac{56.568}{82.836} = 0.68289$$

$$\theta_B = 21.535°$$

Graphical Method:

1. Draw $OA = 8$ cm $= 80$ MPa at 30° to σ_n – axis and $OB = 4$ cm $= 40$ MPa at 45°, as shown in Fig. 1.12 (b).
   $\sigma_x = OD = 10$ cm $= 100$ MPa
   $\sigma_y = OE = 1.7$ cm 17 MPa
2. Join AB and draw the right bisector of AB to intersect σ_n – axis at C.
3. With C as centre and $CA = CB$ as radius draw the circle. Join CA and CB.

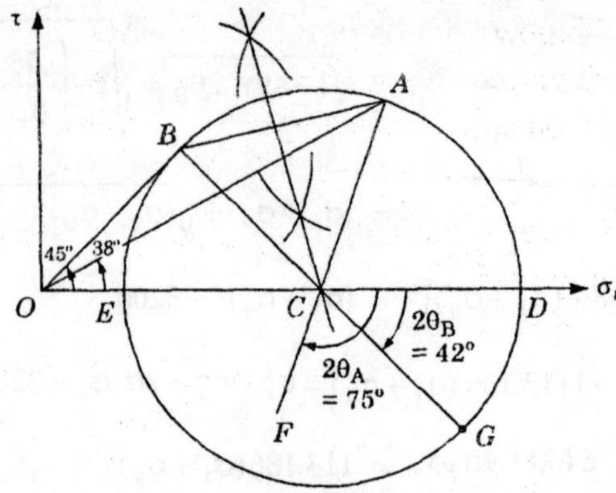

Fig. 1.12(b)

4. $OD = \sigma_x = 10$ cm = 100 MPa, $OE = \sigma_y = 1.7$ cm 17 MPa.
$\angle DCF = 2\theta_A = 75°$ so that $\theta_A = 37.5°$
$\angle DCG = 2\theta_B = 42°$ so that $\theta_A = 21°$

Example 1.4 *The principal stresses at a point in a material are 120 MPa and 60 MPa. Find the magnitude and direction of stress on a plane inclined at 30° to the direction of 60 MPa stress. Find alsc the plane on which the resultant stress is most oblique and its value.*

Solution:

Given: $\sigma_x = 120 \text{ MPa}, \sigma_y = 60 \text{ MPa}, \theta = 90° - 30° = 60°$

$$\sigma_n = \frac{1}{2}(\sigma_x + \sigma_y) + \frac{1}{2}(\sigma_x - \sigma_y)\cos 2\theta$$

$$= \frac{1}{2}(120+60) + \frac{1}{2}(120-60)\cos 120°$$

$$= 90 - 15 = 75 \text{ MPa}$$

$$\tau = \frac{1}{2}(\sigma_x - \sigma_y)\sin 2\theta = \frac{1}{2}(120-60)\sin 120°$$

$$= 25.98 \text{ MPa}$$

$$\sigma_r = (\sigma_n^2 + \tau^2)^{1/2} = [75^2 + (25.98)^2]^{1/2} = 79.37 \text{ MPa}$$

$$\tan\phi = \frac{\tau}{\sigma_n} = \frac{25.98}{75} = 0.3464$$

$$\phi = 19.1°$$

$$\tan\phi_{max} = \frac{\sigma_x - \sigma_y}{2\sqrt{\sigma_x \sigma_y}} = \frac{120-60}{2\sqrt{120 \times 60}} = 0.35355$$

$$\phi_{max} = 19.47°$$

$$\tan\theta = \sqrt{\frac{\sigma_x}{\sigma_y}} = \sqrt{\frac{120}{60}} = \sqrt{2}, \theta = 54.73°$$

$$(\sigma_r)_{max} = [\sigma_x^2 \cos^2\theta + \sigma_y^2 \sin^2\theta]^{1/2}$$

$$= \sqrt{4801+2400} = 84.85 \text{ MPa}$$

Example 1.5 *The state of stress at a point is given by:*

$\sigma_x = 60$ MPa, $\sigma_y = -30$ MPa, $\tau_{xy} = 30$ MPa (ccw)

Calculate (a) principal stresses, (b) principal planes, (c) maximum shear stress, show the orientation of principal planes and planes of maximum shear stress. Use Mohr's circle.

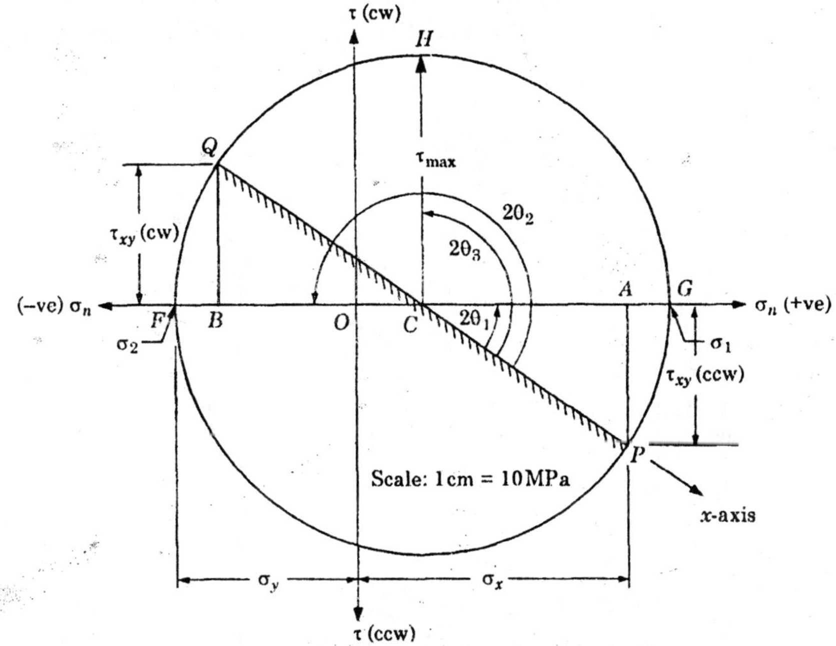

(a) Mohr's stress circle

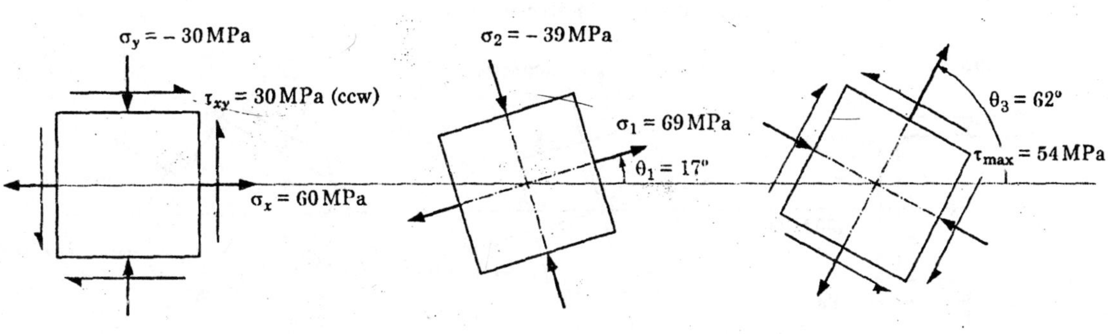

(b) Stress system (c) Principal stresses and planes (d) Maximum shear stresses

Fig. 1.13

Solution:

Draw the Mohr's circle to a scale of 1 cm = 10 MPa as shown in Fig. 1.13 (a).

$$\sigma_1 = OG = 6.9 \text{ cm} = 69 \text{ MPa}$$
$$\sigma_2 = OF = 3.9 \text{ cm} = -39 \text{ MPa}$$
$$\theta_1 = 17°, \theta_2 = 107° \text{ (ccw)}$$
$$\tau_{max} = CH = 5.4 \text{ cm} = 54 \text{ MPa}$$
$$\theta_3 = 62°, \theta_4 = 152° \text{ (ccw)}$$

The orientation of principal streses and principal planes is shown in Fig 1.13 (b) to (d).

Example 1.6 *At a point in a load carrying member, the state of stress is as given below:*

$$\sigma_x = 400 \text{ MPa}, \ \sigma_y = -300 \text{ MPa}, \ \tau_{xy} = 200 \text{ MPa (cw)}$$

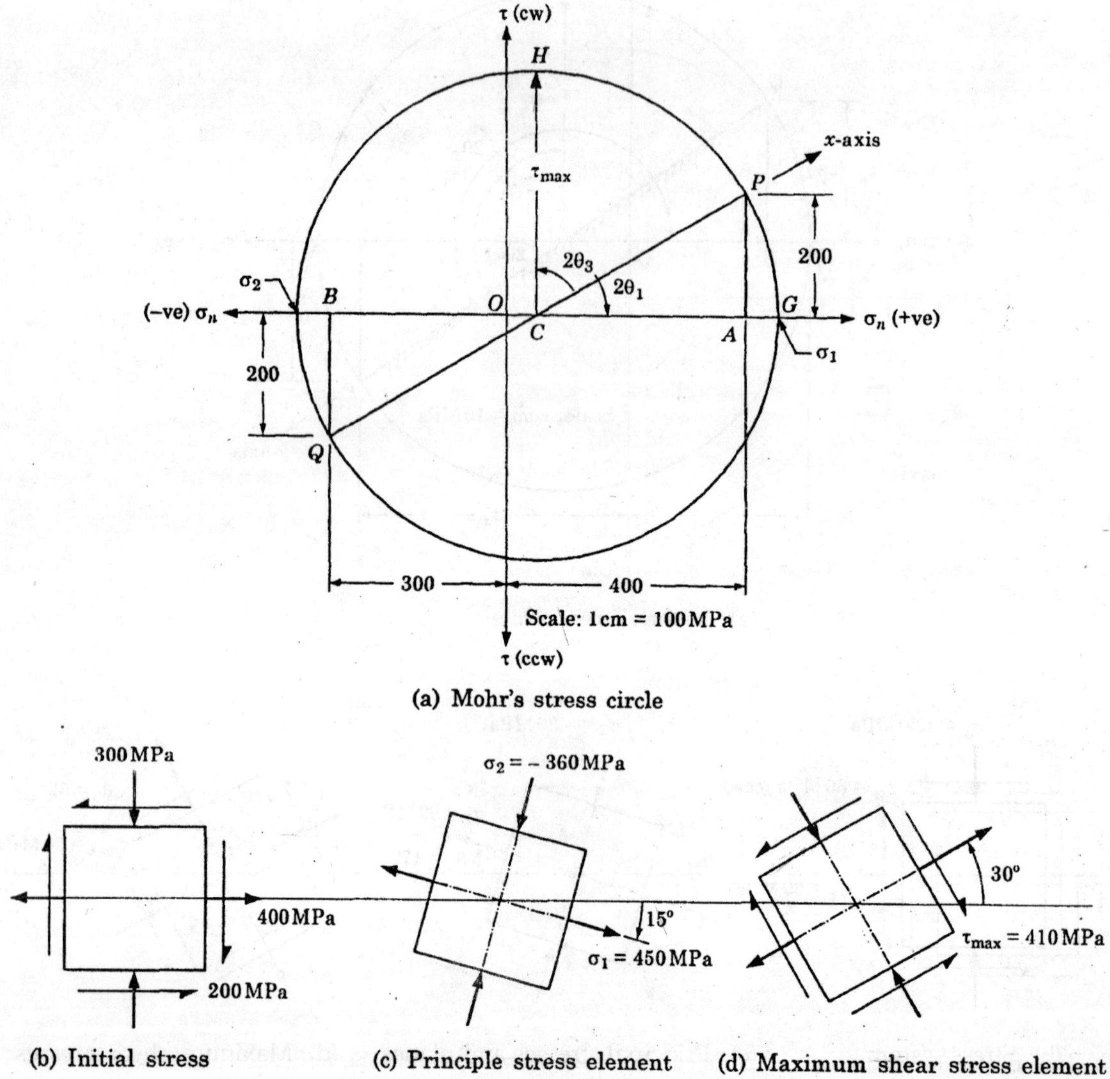

(a) Mohr's stress circle

(b) Initial stress (c) Principle stress element (d) Maximum shear stress element

Fig. 1.14

Draw the Mohr's circle and find the principal stresses and principal planes. Also draw the principal and maximum shear stress elements.

Solution:

The Mohr's circle has been drawn in Fig. 1.14 (a) to a scale of 1 cm = 100 MPa. The stress elements are shown in Figs. 1.14 (b) to (d).

$$\sigma_1 = OG = 4.5 \text{ cm} = 450 \text{ MPa}, \sigma_2 = OF = -3.6 \text{ cm} = -360 \text{ MPa}$$
$$2\theta_1 = 30°, \theta_1 = 15°, \theta_2 = 105° \text{ (ccw)}$$
$$r_{max} = CH = 4.1 \text{ cm} = 410 \text{ MPa}, \ 2\theta_3 = 60° \ \theta_3 = 30° = 120° \text{ (ccw)}$$

Example 1.7 *The state of stress at a point is given by:*

$$\sigma_x = -120 \text{ MPa}, \sigma_y = 180 \text{ MPa}, \tau_{xy} = 80 \text{ MPa (ccw)}$$

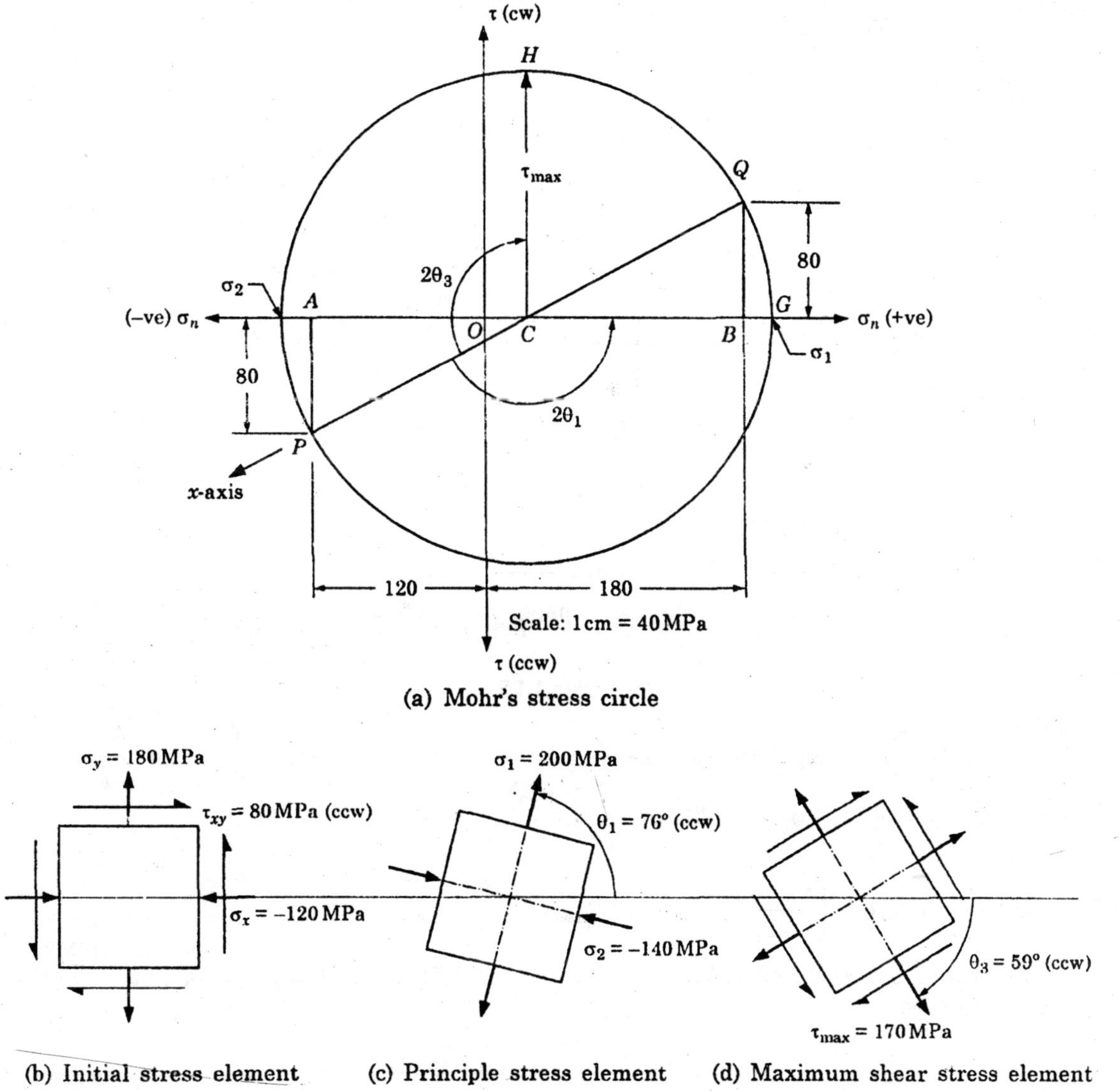

Fig. 1.15

Draw the Mohr's circle and find the principal stress, principal planes and maximum shear stress. Show the results on stress element.

Solution:

Draw the Mohr's circle as shown in Fig. 1.15 (a) to a scale of 1 cm = 40 MPa. The stress elements are shown in Fig. 1.15 (*b*) to (*d*)

$$\sigma_1 = OG = 5 \text{ cm} = 200 \text{ MPa}, \sigma_2 = OF = 3.5 \text{ cm} = 140 \text{ MPa}$$
$$\theta_1 = 76°, \theta_2 = 166° \text{ (ccw)}$$
$$\tau_{max} = CH = 4.3 \text{ cm} = 172 \text{ MPa}$$
$$\theta_3 = 59°, \theta_4 = 149° \text{ (ccw)}$$

Example 1.8 *The principal stresses at a point in a stressed body are 20 MPa and 8 MPa. Determine the maximum obliquity of the resultant stress.*

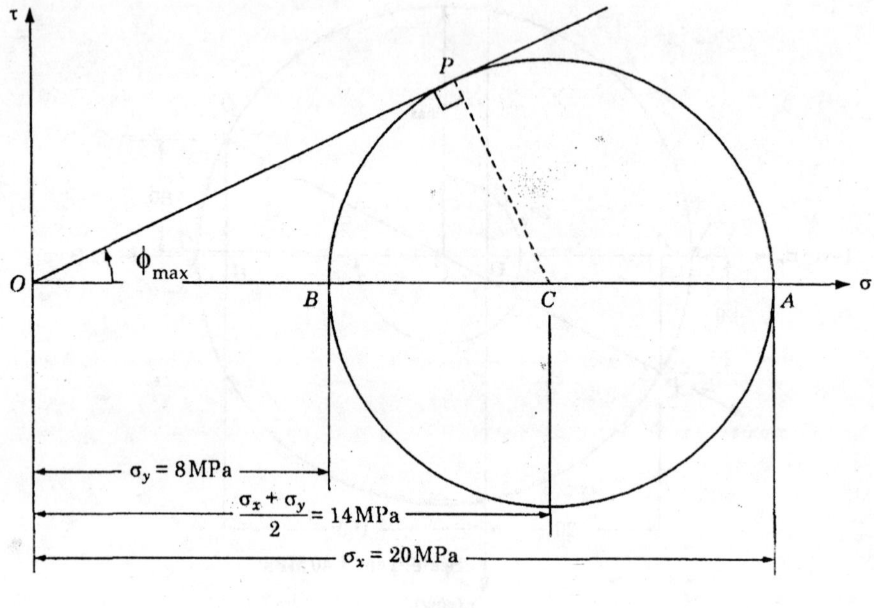

Fig. 1.16

Solution:

Draw Mohr's circle (Fig. 1.16) with $OA = \sigma_x = 20$ MPa, $OB = \sigma_y = 8$ MPa to a scale of 1 cm = 2 MPa.

Radius of circle is $\frac{1}{2}(\sigma_x - \sigma_y) = \frac{1}{2}(20 - 8) = 6$ MPa

Centre of circle, *C* is $\left(\frac{\sigma_x + \sigma_y}{2}, 0\right) = (14, 0)$

Draw a line from *O* tangent to the circle touching it at P. Then $\angle POC = \phi_{max}$ = maximum obliquity of resultant stress.

$$\sin \phi_{max} = \frac{CP}{OC} = \frac{6}{14} = 0.42857$$
$$\phi_{max} = 25.38°$$

By measurement, $\phi_{max} = 25°$

Example 1.9 *The state of stress at a point is given by:*

$$\sigma_x = 400 \text{ MPa}, \sigma_y = 200 \text{ MPa}, \tau_{xy} = 100 \text{ MPa (ccw)}$$

Calculate the normal, shear and resultant stresses on a plane making an angle of 30° with the x—axis.

Solution:

Refer to Fig. 1.17 (*a*).

$$\sigma_n = \frac{1}{2}(\sigma_x + \sigma_y) + \frac{1}{2}(\sigma_x - \sigma_y)\cos 2\theta + \tau_{xy}\sin 2\theta$$

$$= \frac{1}{2}(400 + 200) + \frac{1}{2}(400 - 200)\cos 60° + 100\sin 60°$$

$$= 436.6 \text{ MPa}$$

$$\tau = \frac{1}{2}(\sigma_x - \sigma_y)\sin 2\theta - \tau_{xy}\cos 2\theta$$

$$= \frac{1}{2}(400 - 200)\sin 60° - 100\cos 60°$$

$$= 36.6 \text{ MPa}$$

$$\text{Resultant stress, } \sigma_r = \sqrt{\sigma_n^2 + \tau^2}$$

$$= \sqrt{(436.6)^2 + (36.6)^2}$$

$$= 438.13 \text{ MPa}$$

$$\tan\phi = \frac{\tau}{\sigma_n} = \frac{36.6}{436.6} = 0.08383$$

$$\phi = 4.79°$$

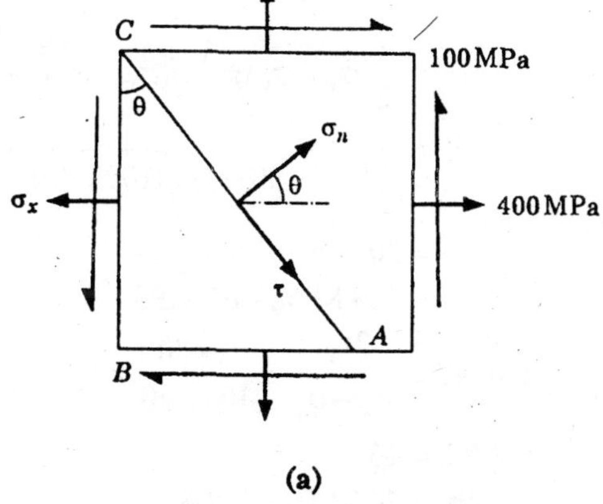

(a)

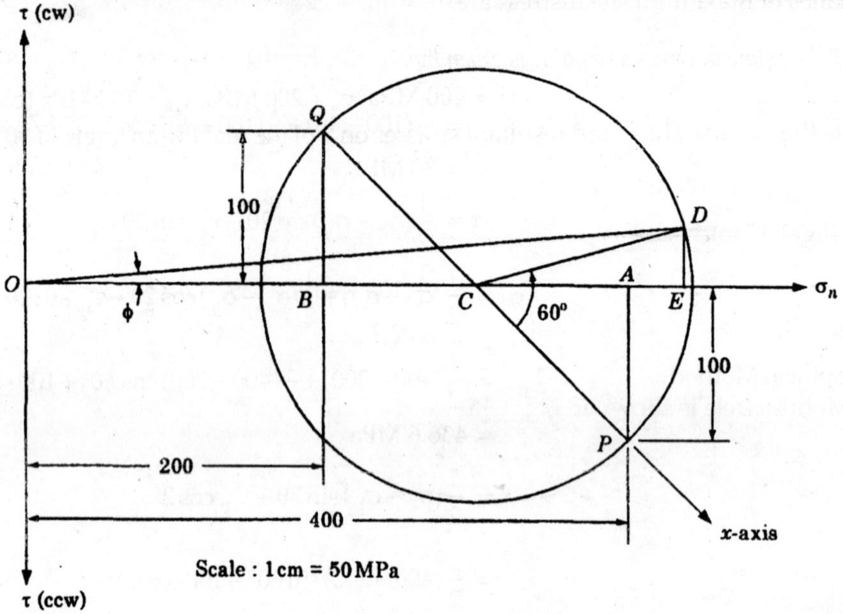

Fig. 1.17

Draw the Mohr's circle as shown in Fig. 1.17 (*b*) to a scale of 1 cm = 50 MPa.

$$\sigma_n = OE = 8.13 \text{ cm} = 436.5 \text{ MPa}$$

$$\tau = DE = 0.7 \text{ cm} = 35 \text{ MPa}$$

$$\phi = 5°$$

Example 1.10 *The normal stresses on two mutually perpendicular planes are* **100** *MPa (tensile) and 40 MPa (compressive) together with shear stress of 70 MPa. Determine graphically or otherwise, (a) the magnitude and direction of principal stresses, and (b) the magnitude of normal and shear stresses on a plane on which the shear stress is maximum.*

Solution:

Given: $\sigma_x = 100 \text{ MPa}, \sigma_y = -40 \text{ MPa}, \tau_{xy} = 70 \text{ MPa}$

(*i*) Analytical Method

(a)
$$\sigma_{1,2} = \frac{1}{2}(\sigma_x + \sigma_y) \pm \frac{1}{2}\sqrt{(\sigma_x - \sigma_y)^2 + 4\tau_{xy}^2}$$

$$= \frac{1}{2}(100 - 40) \pm \frac{1}{2}\sqrt{(100 + 40)^2 + 4 \times 70^2}$$

$$= 30 \pm 99$$

$$= 129 \text{ MPa}, -69 \text{ MPa}$$

$$\tan 2\theta = \frac{2\tau_{xy}}{\sigma_x - \sigma_y} = \frac{2 \times 70}{100 + 40} = 1$$

$$2\theta = 45°$$

$$\theta_1 = 22.5°, \theta_2 = 112.5°$$

(b) Planes of maximum shear stress are: $\theta_3 = 45° + 22.5° = 67.5°$, $\theta_4 = 157.5°$

$$\sigma_n = \frac{1}{2}(\sigma_x + \sigma_y) + \frac{1}{2}(\sigma_x - \sigma_y)\cos 2\theta + \tau_{xy}\sin 2\theta$$

$$= \frac{1}{2}(100 - 40) + \frac{1}{2}(100 + 40)\cos 135° + 70\sin 135°$$

$$= 30 \text{ MPa}$$

$$\tau = \frac{1}{2}(\sigma_x - \sigma_y)\sin 2\theta - \tau_{xy}\sin 2\theta$$

$$= \frac{1}{2}(100 + 40)\sin 135° - 70\cos 135°$$

$$= 99 \text{ MPa}$$

(ii) Graphical Method.

The Mohr's circle is shown in Fig. 1.18.

$$\sigma_1 = OG = 6.5 \text{ cm} = 130 \text{ MPa}$$

$$\sigma_2 = OF = -3.5 = -70 \text{ MPa}$$

$$2\theta_1 = 45°, \theta_1 = 22.5°, \theta_2 = 112.5°$$

$$\tau_{max} = CH = 4.95 = 99 \text{ MPa}$$

$$2\theta_3 = 135°, \theta_3 = 67.5°, \theta_4 = 157.5°$$

On the plane of maximum shear stress,

$$\sigma_n = OC = 1.5 \text{ cm} = 30 \text{ MPa}$$

$$\tau = \tau_{max} = CH = 99 \text{ MPa}$$

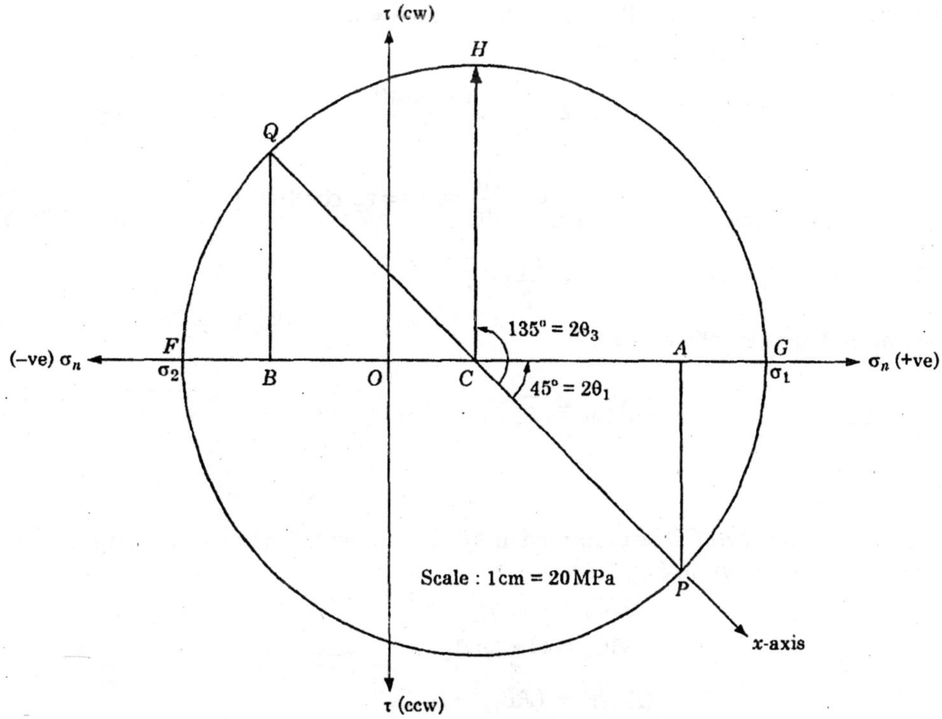

Fig. 1.18

1.7 NORMAL AND SHEAR STRAINS ON AN OBLIQUE PLANE

1.7.1 Uniaxial Strains

Let the rectangular body $ABCD$ be extended to $AB_1C_1D_1$ under uniaxial strain ε_x as shown in Fig. 1.19. Let ε_n be the normal strain in the diagonal AC. Hence

$$AC_1 = (1 + \varepsilon_n)$$

Now

$$(AC_1)^2 = (AB_1)^2 + (B_1C_1)^2$$

$$= (AB)^2 (1 + \varepsilon_x)^2 + (BC)^2$$

$$(AC)^2 (1+2\varepsilon_n) = (AB)^2 2(1 + 2\varepsilon_x) + (BC)^2$$

(Neglecting $\varepsilon_x^2, \varepsilon_n^2$, being small)

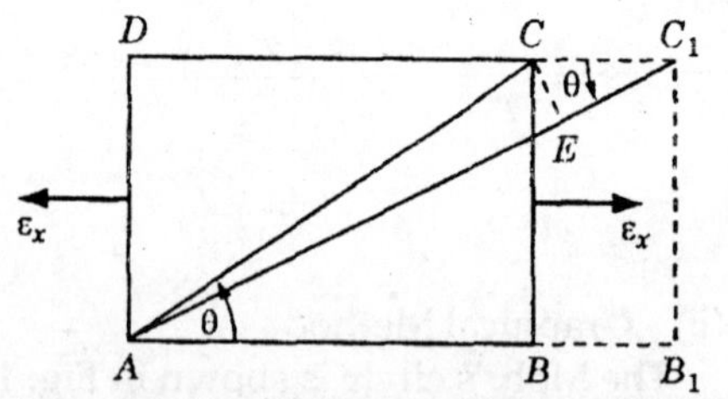

Fig. 1.19 Strains on an oblique plane under uniaxial strain

$$1 + 2\varepsilon_n = \left(\frac{AB}{AC}\right)^2 (1+2\varepsilon_x) + \left(\frac{BC}{AC}\right)^2$$

$$= \cos^2\theta\,(1+2\varepsilon_x) + \sin^2\theta = 1 + 2\varepsilon_x \cos^2\theta$$

∴

$$\varepsilon_n = \varepsilon_x \cos 2\theta$$

$$= \varepsilon_x \left(\frac{1+\cos 2\theta}{2}\right) \qquad \text{... (1.21)}$$

ε_n is maximum when θ is 0° or 180° and is zero when $\theta = 90°$

Let ε_s be the shear strain in AC.

∴

$$\varepsilon_s = \frac{CE}{AC} = \frac{CC_1 \sin\theta}{AC}$$

$$= \varepsilon_x \cdot \frac{AB}{AC} \cdot \sin\theta = \varepsilon_x \cdot \cos\theta \cdot \sin\theta$$

$$= \frac{\varepsilon_x}{2} \sin 2\theta \qquad \text{... (1.22)}$$

ε_s is maximum when $2\theta = 90°$ or $\theta = 45°$

$$(\varepsilon_s)_{max} = \frac{\varepsilon_x}{2} \qquad \text{... (1.23)}$$

1.7.2 Shear Strain

Let the rectangular body $ABCD$ be deformed to $AB_lC_lD_l$ under the action of a shearing force so that the shear strain is γ_{xy} as shown in Fig. 1.20

Now

$$CC_1 = BC.\ \gamma_{xy}$$

$$AC_1 = (1+\varepsilon_n)\, AC$$

$$(AC_1)^2 = (AB_1)^2 + (B_1C_1)^2$$

$$= (AB + CC_1)^2 + (BC)^2$$

$$(AC)^2(1+\varepsilon_n)^2 = (AB+BC\cdot\gamma_{xy})^2+(BC)^2$$

Negleting small quantities, we get

$$1+2\varepsilon_n = \left(\frac{AB}{AC}\right)^2+2\cdot\frac{AB}{AC}\cdot\frac{BC}{AC}\cdot\gamma_{xy}+\left(\frac{BC}{AC}\right)^2$$

$$= \cos^2\theta+2.\cos\theta.\sin\theta.\gamma_{xy}+\sin^2\theta$$

$$= 1+\gamma_{xy}.\sin 2\theta$$

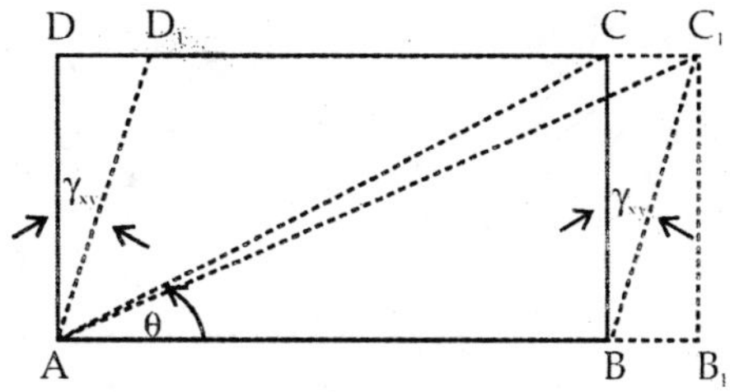

Fig. 1.20: Normal strain under shear strain.

$$\varepsilon_n = \frac{\gamma_{xy}}{2}\sin 2\theta \qquad \text{... (1.24)}$$

ε_n is maximum when $\theta = 45°$ and

$$(\varepsilon_n)_{max} = \frac{\gamma_{xy}}{2} \qquad \text{... (1.25)}$$

1.7.3 Biaxial Strains

Consider a rectangular plate *ABCD* acted upon by strains ε_x and ε_y as shown in Fig. 1.21. After deformation, the rectangle takes up the shape $A_1B_1C_1D_1$.

Normal strain in *OC*

$$\varepsilon_n = \frac{HC_1}{OC}=\frac{HE+EC_1}{OC}=\frac{GG+EC_1}{OC}=\frac{CF\cos\theta+C_1F\sin\theta}{AC/2}$$

$$= \frac{(1/2)\,DC.\varepsilon_x.\cos\theta+(1/2)BC.\varepsilon_y.\sin\theta}{AC/2}=\varepsilon_x\cdot\frac{DC}{AC}\cos\theta+\varepsilon_y\cdot\frac{BC}{AC}\cdot\sin\theta$$

$$= \varepsilon_x.\cos^2\theta+\varepsilon_y.\sin^2\theta$$

$$= \frac{1}{2}(\varepsilon_x+\varepsilon_y)+\frac{1}{2}(\varepsilon_x-\varepsilon_y)\cos 2\theta \qquad \text{... (1.26)}$$

Shear strain in OC,

$$\varepsilon_s = \frac{CH}{OC}-\frac{CG-HG}{OC}=\frac{CG-EF}{OC}$$

$$= \frac{CF\sin\theta-C_1F\cos\theta}{OC}$$

$$= \frac{(1/2)DC.\varepsilon_x.\sin\theta-(1/2)BC.\varepsilon_y.\cos\theta}{AC/2}$$

$$= \varepsilon_x.\frac{DC}{AC}.\sin\theta-\varepsilon_y.\frac{BC}{AC}.\cos\theta$$

$$\varepsilon_s = (\varepsilon_x - \varepsilon_y)\sin\theta\cos\theta$$

$$\varepsilon_s = \frac{1}{2}(\varepsilon_x - \varepsilon_y)\sin 2\theta \quad \text{... (1.27)}$$

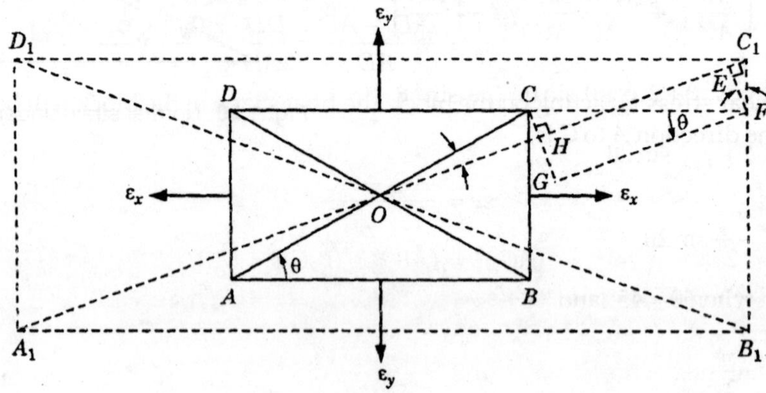

Fig. 1.21 Strains on an oblique plane under biaxial strains

Now $\varepsilon_s = \frac{\gamma}{2}$, the angular unit shear. For shear strain to be maximum, $\sin 2\theta = 1.0$ or $2\theta = 90°$. Hence $\theta = 45°$ and

$$(\varepsilon_s)_{max} = \frac{1}{2}(\varepsilon_x - \varepsilon_y) \quad \text{... (1.28)}$$

Resultant strain, $\varepsilon_r = \sqrt{\varepsilon_n^2 + \varepsilon_s^2} = \sqrt{(\varepsilon_x\cos^2\theta + \varepsilon_y\sin^2\theta)^2 + (\varepsilon_x - \varepsilon_y)^2\sin^2\theta.\cos^2\theta}$

$$= [\varepsilon_x^2\cos^4\theta + \varepsilon_y^2\sin^4\theta + 2\varepsilon_x\varepsilon_y\sin^2\theta.\cos^2\theta + (\varepsilon_x^2 + \varepsilon_y^2 - 2\varepsilon_x\varepsilon_y)\sin^2\theta.\cos^2\theta]^{1/2}$$

$$\varepsilon_r = \sqrt{\varepsilon_x^2\cos^2\theta + \varepsilon_y^2\sin^2\theta} \quad \text{... (1.29)}$$

1.7.4 Biaxial Normal Strains Combined with Shear Strains

Consider a block *ABCD* subjected to direct stresses σ_x and σ_y and shear stress τ_{xy} as shown in Fig. 1.22 (a). Let ε_x be the total normal strain in *AB*, then form Fig. 1.22 (b), we have

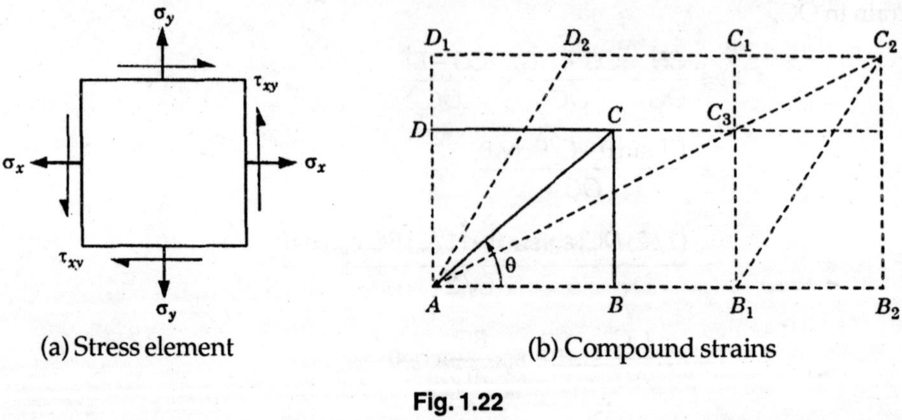

(a) Stress element (b) Compound strains

Fig. 1.22

$$\varepsilon_x = \frac{AB_1 - AB}{AB} = \frac{BB_1}{AB} = \frac{\sigma_x}{E} - \nu\frac{\sigma_y}{E}$$

and ε_y be the total normal strain in AD, i.e.

$$\varepsilon_y = \frac{AD_1 - AD}{AD} = \frac{DD_1}{AD} = \frac{\sigma_y}{E} - \nu\frac{\sigma_x}{E}$$

Due to the shear stress τ_{xy} acting on the block, the final shape of the block will be $AB_1C_2D_2$. The direct strain in the direction A to C is,

$$\varepsilon_n = \frac{AC_2 - AC}{AC}$$

Now $(AC_2)^2 = (AB_2)^2 + (B_2C_2)^2 = (AB_1 + B_1B_2)^2 + (AD_1)^2$

$= (AB_1 + C_1C_2)^2 + (AD_1)^2 = [AB\,.\,(1 + \varepsilon_x) + \gamma_{xy}\,(1 + \varepsilon_y)\,AD]^2 + (AD)^2\,(1 + \varepsilon_y)^2$

$= (AB)^2\,(1 + \varepsilon_x)^2 + \gamma_{xy}{}^2\,(1 + \varepsilon_y)^2 (AD)^2 + 2\gamma_{xy}\,(1 + \varepsilon_x)\,(1 + \varepsilon_y)\,AB\,.\,AD + (AD)^2\,(1 + \varepsilon_y)^2$

Simplifying and neglecting small quantities, we get

$$(AC_2)^2 = (AB)^2\,(1 + 2\varepsilon_x) + 2\gamma_{xy}\,.\,AB.\,AD + (AD)^2\,(1 + 2\varepsilon_y)$$

$$= 2(AB)^2\,.\,\varepsilon_x + 2\gamma_{xy}\,.\,AB\,.\,AD + [(AB)^2 + 2(AD)^2] + 2(AD)^2.\varepsilon_y$$

$$= 2(AB)^2\,.\varepsilon_x + 2\gamma_{xy}.\,AB\,.\,AD + (AC)^2 + 2(AD)^2.\,\varepsilon_y$$

Now $(AC_2)^2 = (AC)^2\,(1 + \varepsilon_n)^2 = (AC)^2\,(1 + 2\varepsilon_n)$

$\therefore$

$$1 + 2\varepsilon_n = 2\varepsilon_x\left(\frac{AB}{AC}\right)^2 + 2\gamma_{xy}\cdot\frac{AB}{AC}\cdot\frac{AD}{AC} + 1 + 2\varepsilon_y\cdot\left(\frac{AD}{AC}\right)^2$$

$$\varepsilon_n = \varepsilon_x.\cos^2\theta + \gamma_{xy}\cos\theta.\sin\theta + \varepsilon_y.\sin^2\theta$$

or

$$\varepsilon_n = \varepsilon_x.\cos^2\theta + \varepsilon_y\cdot\sin^2\theta + \gamma_{xy}\sin\theta.\cos\theta$$

or

$$\varepsilon_n = \frac{1}{2}(\varepsilon_x + \varepsilon_y) + \frac{1}{2}(\varepsilon_x - \varepsilon_y)\cos 2\theta + \frac{\gamma_{xy}}{2}\sin 2\theta \qquad \text{... (1.30)}$$

The maximum and minimum values of ε_n are given when $\frac{d\varepsilon_n}{d\theta} = 0$

$$-\frac{1}{2}(\varepsilon_x - \varepsilon_y)\cdot 2\sin 2\theta + \frac{1}{2}\gamma_{xy}.2\cos 2\theta = 0$$

$$\tan 2\theta = \frac{\gamma_{xy}}{\varepsilon_x - \varepsilon_y} \qquad \text{... (1.31)}$$

$$\sin 2\theta = \frac{\gamma_{xy}}{\pm\sqrt{\varepsilon_x - \varepsilon_y)^2 + \gamma_{xy}^2}}$$

$$\cos 2\theta = \frac{\varepsilon_x - \varepsilon_y}{\pm\sqrt{(\varepsilon_x - \varepsilon_y)^2 + \gamma_{xy}^2}}$$

Substituting in Eq. (1.30), we get

$$\varepsilon_{1,2} = \frac{1}{2}(\varepsilon_x + \varepsilon_y) \pm \sqrt{(\varepsilon_x - \varepsilon_y)^2 + \gamma_{xy}^2} \qquad \text{... (1.32)}$$

as the two principal strains.

In order to determine the shearing strain, which is defined as the change in angle between two mutually perpendicular planes, the relative displacement of two points C and E will be considered as shown in Fig. 1.23.

Now $\qquad \alpha \cong \tan\alpha$ for small α.

$$= \frac{CF}{AC} = \frac{CM + MF}{AC}$$

$$= \frac{C_1H - C_1L + CM}{AC}$$

$$= \frac{C_1C_2 \sin\theta - C_1C_3 \cos\theta + CC_3 \sin\theta}{AC}$$

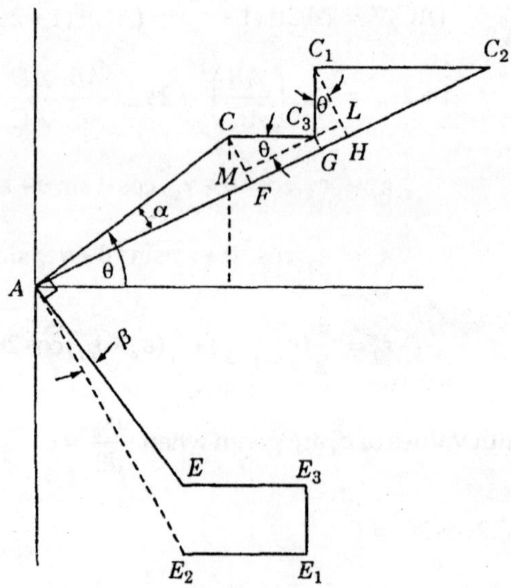

Fig. 1.23

$$= \frac{\gamma_{xy}(1+\varepsilon_y)BC\sin\theta - \varepsilon_y.BC.\cos\theta + \varepsilon_x.AB.\sin\theta}{AC}$$

$$= \gamma_{xy}\sin^2\theta - \varepsilon_y \sin\theta \times \cos\theta + \varepsilon_x \cos\theta \sin\theta$$

$$\alpha = \gamma_{xy} \sin^2 \theta + (\varepsilon_x - \varepsilon_y) \sin\theta \cos\theta$$

Similarly by putting $\theta = -(90° - \theta)$, we get

$$\beta = \gamma_{xy} \sin^2\theta - (\varepsilon_x - \varepsilon_y) \sin\theta \cos\theta$$

Shear strain $\gamma = -\beta + \alpha$

$$2\varepsilon_s = \gamma = 2(\varepsilon_x - \varepsilon_y)\sin\theta \cos\theta - \gamma_{xy} \cos 2\theta$$

$$\therefore \quad \varepsilon_s = \frac{1}{2}(\varepsilon_x - \varepsilon_y)\sin 2\theta - \frac{\gamma_{xy}}{2}\cos 2\theta \qquad \text{... (1.33)}$$

For shearing strain to be maximum or minimum,

$$\frac{d\varepsilon_s}{d\theta} = 0$$

$$\therefore \quad (\varepsilon_x - \varepsilon_y)\cos 2\theta + \gamma_{xy}.\sin 2\theta = 0$$

$$\tan 2\theta = -\left(\frac{\varepsilon_x - \varepsilon_y}{\gamma_{xy}}\right) \qquad \text{... (1.34)}$$

Substituting Eq. (1.34) in Eq. (1.33), we get

$$(\varepsilon_s)_{max} = \left(\frac{\gamma}{2}\right)_{max} = \pm[(\varepsilon_x - \varepsilon_y)^2 + \gamma_{xy}^2]^{1/2} \qquad \text{... (1.35)}$$

1.8 MOHR'S CIRCLE OF STRAIN

Mohr's strain circle is a circular diagram to determine the normal, shear and resultant strains at a point in a body when subjected to biaxial or compound state of strain. The principal strains, principal planes and maximum shear strain can also be determine form this circle

1.8.1 Biaxial Strains

We have from Eqs. (1.26) and (1.27)

$$\varepsilon_n = \frac{1}{2}(\varepsilon_x + \varepsilon_y) + \frac{1}{2}(\varepsilon_x - \varepsilon_y)\cos 2\theta$$

$$\varepsilon_s = \frac{1}{2}(\varepsilon_x - \varepsilon_y)\sin 2\theta$$

$$\therefore \quad \left(\varepsilon_n - \frac{\varepsilon_x + \varepsilon_y}{2}\right)^2 + \varepsilon_s^2 = \left(\frac{\varepsilon_x - \varepsilon_y}{2}\right)^2 \qquad \text{... (1.36)}$$

Eq. (1.36) represents the equation of a circle whose centre is at $\left(\frac{\varepsilon_x + \varepsilon_y}{2}, 0\right)$ in the $\varepsilon_n - \frac{\gamma}{2}$ coordinate system and with radius of the circle equal to $\left(\frac{\varepsilon_x - \varepsilon_y}{2}\right)$. This leads us to represent the state of strain at a point by a circular diagram.

Construction:

(a) Like strains

Take $OA = \varepsilon_x$, $OB = \varepsilon_y$. Let C be the centre of AB. With C as centre and CA as radius, describe the circle as shown in Fig. 1.24. Make $\angle ACD = 2\theta$. Drop $DE \perp BA$. Join OD. Then $OE = \varepsilon_n$, $DE = \varepsilon_s$, $OD = \varepsilon_r$ and $CF = (\varepsilon_s)_{max}$.

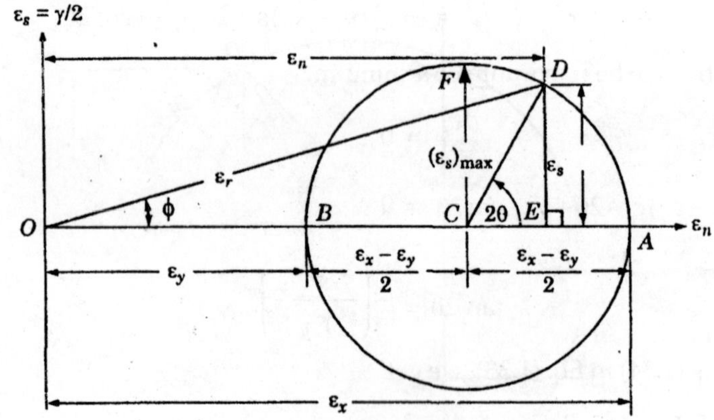

Fig. 1.24 Mohr's strain circle for like biaxial strains.

Proof:

$$OC = OB + BC$$

$$= \varepsilon_y + \frac{\varepsilon_x - \varepsilon_y}{2} = \frac{1}{2}(\varepsilon_x + \varepsilon_y)$$

$$CA = \frac{1}{2}(\varepsilon_x - \varepsilon_y)$$

$$OE = OC + CE = OC + CD\cos 2\theta$$

$$\varepsilon_n = \frac{1}{2}(\varepsilon_x + \varepsilon_y) + \frac{1}{2}(\varepsilon_x - \varepsilon_y)\cos 2\theta$$

$$\varepsilon_s = DE = CD\sin 2\theta$$

$$= \frac{1}{2}(\varepsilon_x - \varepsilon_y)\sin 2\theta$$

Maximum shear strain,

$$(\varepsilon_s)_{max} = CF = \frac{1}{2}(\varepsilon_x - \varepsilon_y)$$

Resultant strain, $\varepsilon_r = OD = \sqrt{OE^2 + DE^2}$

(b) Unlike strains

Let ε_y be compressive and ε_x tensile in nature.

Draw $OA = \varepsilon_x$, $OB = \varepsilon_y$. Let C be the centre of AB. With C as centre and CA as radius, describe the circle. Make $\angle ACD = 2\theta$. Drop $DE \perp CA$. Join OD. Then $OE = \varepsilon_n$, $DE = \varepsilon_s$, $OD = \varepsilon_r$ as shown in Fig. 1.25.

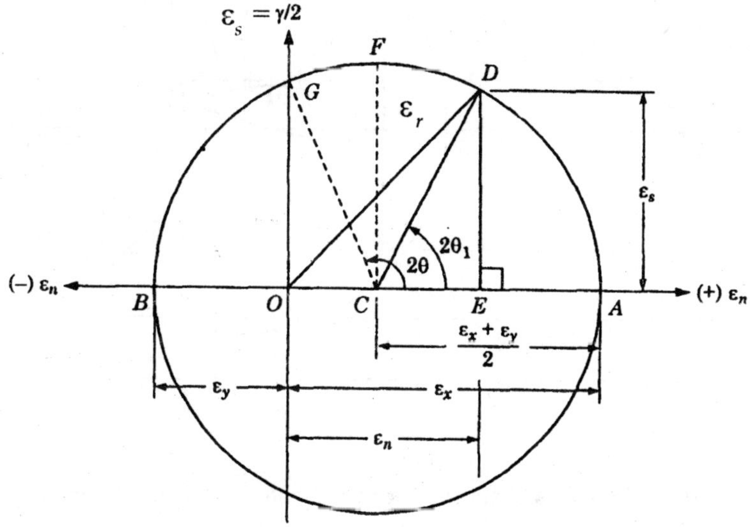

Fig. 1.25 Mohr's strain circle for unlike biaxial strains.

Beyond $\angle ACG = 2\theta_1$, the normal strain will become compressive.

1.8.2 Normal Strains Combined with Shear Strains

In this case, we have form Eqs (1.30) and (1.33)

$$\varepsilon_n = \frac{1}{2}(\varepsilon_x + \varepsilon_y) + \frac{1}{2}(\varepsilon_x - \varepsilon_y)\cos 2\theta + \frac{1}{2}\gamma_{xy}\sin 2\theta$$

$$\varepsilon_s = \frac{1}{2}(\varepsilon_x - \varepsilon_y)\sin 2\theta - \frac{1}{2}\gamma_{xy}\cos 2\theta$$

$$\left(\varepsilon_n - \frac{\varepsilon_x + \varepsilon_y}{2}\right)^2 + \varepsilon_s^2$$

$$\therefore \quad = \left[\frac{1}{2}\{(\varepsilon_x - \varepsilon_y)^2 + \gamma_{xy}^2\}^{1/2}\right]^2 \quad \text{... (1.37)}$$

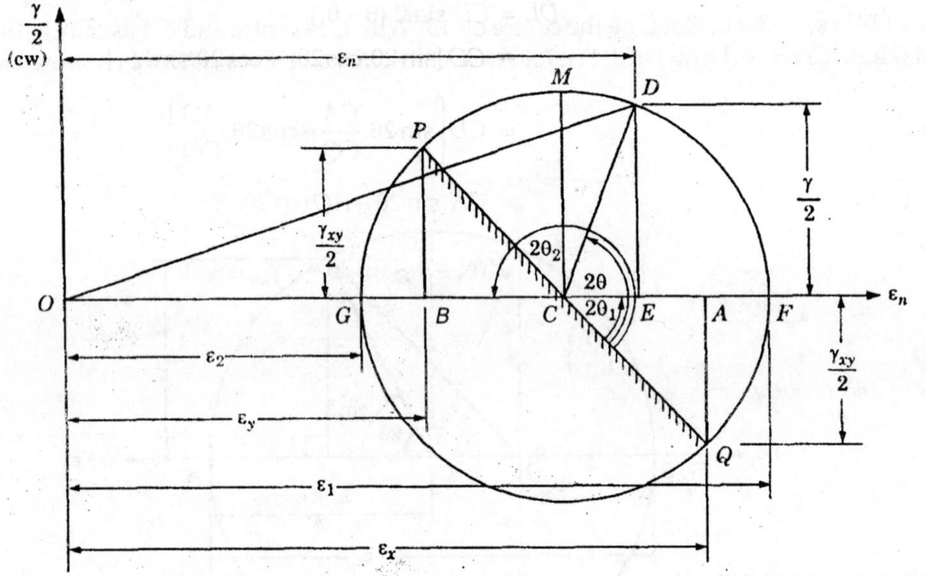

Fig. 1.26 Mohr's strain circle for combined normal and shear strains.

Eq. (1.37) represents a circle whose centre is at $\left(\dfrac{\varepsilon_x+\varepsilon_y}{2},0\right)$ and radius $\dfrac{1}{2}[(\varepsilon_x-\varepsilon_y)^2+\gamma_{xy}^2]^{1/2}$ in the $\varepsilon_n-\varepsilon_s$ plane.

Construction: Take $OA=\varepsilon_x$, $OB=\varepsilon_y$. At A draw $AQ\perp OA$ and at B draw $BP\perp OA$ so that $AQ=BP=\dfrac{1}{2}\gamma_{xy}$. Join PQ. With C as centre, describe a circle with CQ as radius. Make $\angle QCD=2\theta$.

Drop $DE\perp OA$. Join OD. Then $OE=\varepsilon_n$, $DE=\dfrac{\gamma}{2}$, $OD=\varepsilon_r$.

$\angle QCF=2\theta_1, \angle QCG=2\theta_2, OG=\varepsilon_2, OF=\varepsilon_1, CM=\left(\dfrac{\gamma}{2}\right)_{max}$, as shown in Fig. 1.26.

Proof:

$$\begin{aligned}OE &= OC+CE=OC+CD\cos 2(\theta-\theta_1)\\ &= OC+CD[\cos 2\theta.\cos 2\theta_1+\sin 2\theta.\sin 2\theta_1]\\ &= OC+CD\left[\cos 2\theta.\frac{CA}{CQ}+\sin 2\theta.\frac{AQ}{CQ}\right]\\ &= OC+CD.\cos 2\theta+AQ.\sin 2\theta \qquad [\because\ CD=CQ]\end{aligned}$$

$$\varepsilon_n = \frac{1}{2}(\varepsilon_x + \varepsilon_y)\frac{1}{2}(\varepsilon_x - \varepsilon_y)\cos 2\theta + \frac{1}{2}\gamma_{xy}\sin 2\theta$$

$$DE = CD \sin 2(\theta - \theta_1)$$

$$= CD[\sin 2\theta . \cos 2\theta_1 - \cos 2\theta . \sin 2\theta_1]$$

$$= CD\left[\sin 2\theta . \frac{CA}{CQ} - \cos 2\theta . \frac{AQ}{CQ}\right]$$

$$= CA . \sin 2\theta - AQ \cos 2\theta.$$

$$= (\varepsilon_x - \varepsilon_y)\sin 2\theta - \frac{1}{2}\gamma_{xy}\cos 2\theta$$

$$\left(\frac{\gamma}{2}\right)_{max} = CM = \sqrt{(AC)^2 + (AQ)^2} = \pm\frac{1}{2}\sqrt{(\varepsilon_x - \varepsilon y)^2 + \gamma_{xy}^2}$$

$$\tan 2\theta_1 = \frac{AQ}{AC} = \frac{\gamma_{xy}}{\varepsilon_x - \varepsilon_y}$$

$$\theta_2 = \theta_1 + 90°$$

1.9 PRINCIPAL STRAINS

Consider a body subjected to normal strains ε_x, ε_y and shearing strain γ_{xy} in the two-dimensional system. Then it can bc shown that on a plane inclined at an angle θ with ε_x.

Normal strain,

$$\varepsilon_n = \frac{1}{2}(\varepsilon_x + \varepsilon_y) + \frac{1}{2}(\varepsilon_x - \varepsilon_y)\cos 2\theta + \frac{\gamma_{xy}}{2}\sin 2\theta$$

The principal strains are given by,

$$\varepsilon_{1,2} = \frac{1}{2}(\varepsilon_x + \varepsilon_y) \pm \frac{1}{2}\sqrt{(\varepsilon_x - \varepsilon_y)^2 + \gamma_{xy}^2}$$

The princple planes are given by,

$$\tan 2\theta = \frac{\gamma_{xy}}{\varepsilon_x - \varepsilon_y}$$

Maximum Shearing strain;

$$\left(\frac{\gamma}{2}\right)_{max} = \sqrt{(\varepsilon_x - \varepsilon_y)^2 + \gamma_{xy}^2} = \frac{1}{2}(\varepsilon_1 - \varepsilon_2)$$

In a three-dimensional state of stress, the principal strains are:

$$\left.\begin{aligned} \varepsilon_1 &= \frac{1}{E}[\sigma_1 - \nu(\sigma_2 + \sigma_3)] \\ \varepsilon_2 &= \frac{1}{E}[\sigma_2 - \nu(\sigma_3 + \sigma_1)] \\ \varepsilon_3 &= \frac{1}{E}[\sigma_3 - \nu(\sigma_1 + \sigma_2)] \end{aligned}\right\} \quad \text{...(1.38)}$$

1.10 RELATIONSHIP BETWEEN ELASTIC CONSTANTS

Consider a cube of unit size acted upon by simple shear stress τ_{xy}, as shown in Fig. 1.27 on its faces AB and DC.

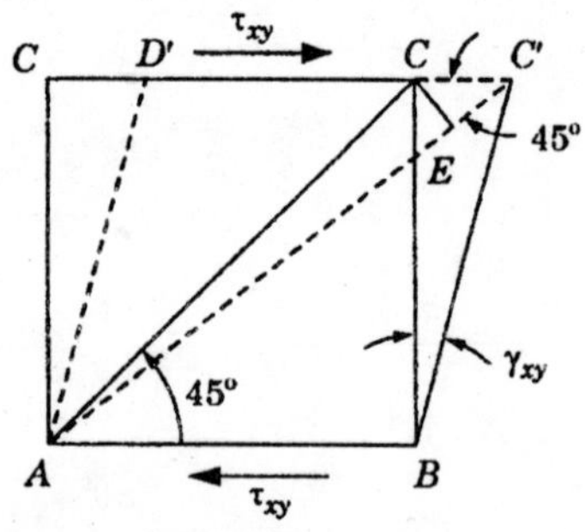

Fig. 1.27 Simple shear of a cubic element

Shear strain, $$\varepsilon_{xy} = \frac{CC'}{BC} = \frac{\tau_{xy}}{G}$$

From C drop $CE \perp AC'$, so that $AC = AE$

Elongation of diagonal $AC = EC'$

$$= CC'\cos 45° = \frac{CC'}{\sqrt{2}}$$

Now $$AC = \sqrt{2}\, AB$$

Now if the cube is subjected to simple shear, then the normal stress on the diagonal of the cube is equal to the shear stress, *i.e.*

$$\sigma_n = \tau_{xy}$$

as shown in Fig. 1.28. This can be shown as explained below.

For the ΔACB, resolving forces along the normal, we have

$$\sigma_n\, AC = \tau_{xy} \cdot AB\cos 45° + \tau_{xy} \cdot BC\sin 45°$$

$$\sigma_n = \tau_{xy} \cdot \frac{AB}{AC}\cos 45° + \tau_{xy} \cdot \frac{BC}{AC}\sin 45°$$

$$= \tau_{xy}\cos^2 45° + \tau_{xy}\sin^2 45°$$

$$= \frac{\tau_{xy}}{2} + \frac{\tau_{xy}}{2} = \tau_{xy}$$

$\therefore$ Shear strain in the diagonal, $$\gamma = \frac{1}{2}\frac{\sigma_n}{G}$$

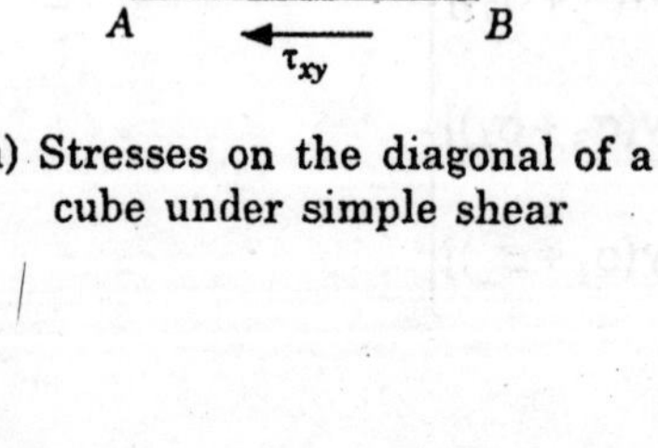

(a) Stresses on the diagonal of a cube under simple shear

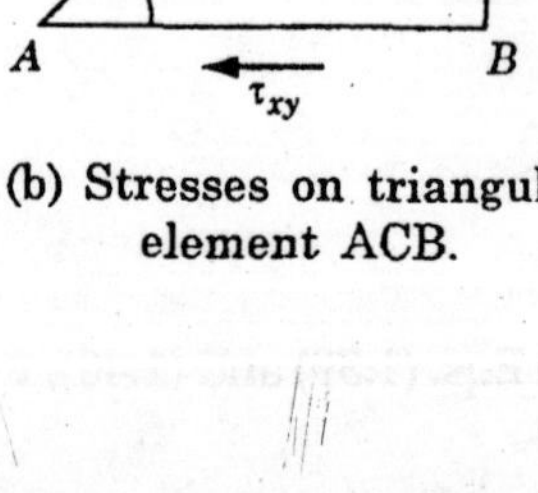

(b) Stresses on triangular element ACB.

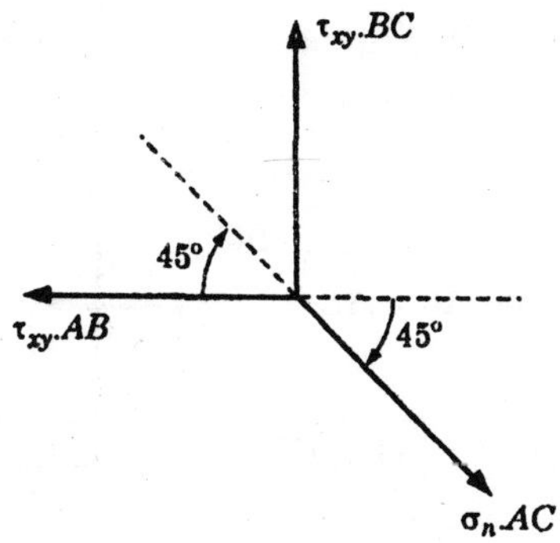

(c) Forces on triangular element ACB.

Fig. 1.28

If σ_n acted alone in the direction of the diagonal AC, then the resultant strain would be $\frac{\sigma_n}{E}$, but there is an equal and opposite normal stress in the direction of the diagonal BD also. The strain caused by this in the direction $AC = -\nu\frac{\sigma_n}{E}$

Total strain in the direction of diagonal $AC = \frac{\sigma_n}{E} - \left(-\frac{\nu\sigma_n}{E}\right) = \frac{\sigma_n}{E}(1+\nu)$

$$\therefore \qquad \frac{1}{2}\frac{\sigma_n}{G} = \frac{\sigma_n}{E}(1+\nu)$$

or

$$G = \frac{E}{2(1+\nu)} \qquad \text{... (1.39)}$$

Now consider cube under an equal compressive stress acting on all its faces, as shown in Fig. 1.29.

Than, volume strain, $\varepsilon_v = \frac{\sigma_n}{K}$

where K = bulk modulus

Total strain along the edge $= \frac{\sigma_n}{E}(1-2\nu)$

$$\varepsilon_v = 3\frac{\sigma_n}{E}(1-2\nu)$$

$$\frac{\sigma_n}{K} = 3\frac{\sigma_n}{E}(1-2\nu)$$

$$E = 3K(1-2\nu) \qquad \text{... (1.40)}$$

From Eqs. (1.39) and (1.40), we have

$$2G(1+\nu) = 3K(1-2\nu)$$

$$2G+2G\nu = 3K-6\nu K$$

$$\nu(2G+6K) = 3K-2G$$

$$\nu = \frac{3K-2G}{2G+6K} \quad \text{... (1.41)}$$

From Eq. (3), we have

$$\nu = \frac{E-2G}{2G} \quad \text{... (1.42)}$$

$$\therefore \quad \frac{E-2G}{2G} = \frac{3K-2G}{2G+6K}$$

$$E = \frac{9KG}{G+3K} \quad \text{... (1.43)}$$

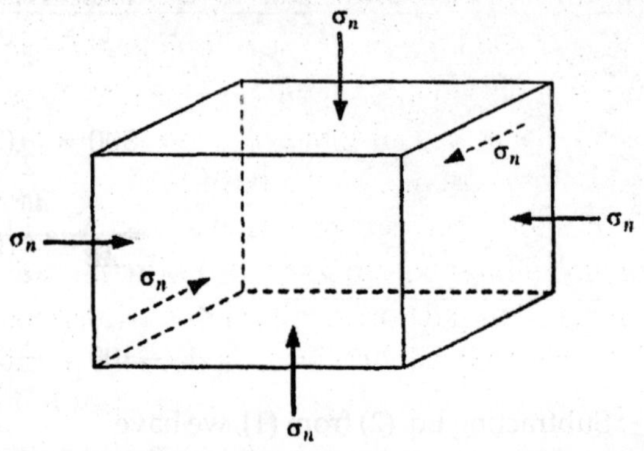

Fig. 1.29 Cubic element under hydrostatic stress.

Example 1.11 *A bar of 20 mm diameter is subjected to a pull of 20 kN. The measured extension over a gauge length of 200 mm is 0.1 mm and the change in diameter is 0.0035 mm. Calculate the value of modulus of elasticity, bulk modulus, and Poisson's ratio.*

Solution:

Given: $d = 20$ mm, $P = 20$ kN, $l = 200$ mm, $\delta l = 0.1$ mm, $\delta d = 0.0035$ mm

$$\text{Area of cross-section of bar, } A = \frac{\pi}{4}d^2 = \frac{\pi}{4}(20)^2 = 314.16 \text{ mm}^2$$

$$\text{Modulus of elasticity, } E = \frac{Pl}{A\delta l} = \frac{20\times10^3\times200}{314.16\times0.1} = 127\times10^3 \text{ N/mm}^2$$

$$\text{Longitudinal strain, } \varepsilon_l = \frac{\delta l}{l} = \frac{0.1}{200} = 5\times10^{-4}$$

$$\text{Lateral strain, } \varepsilon_d = \frac{\delta d}{d} = \frac{0.0035}{20} = 1.75\times10^{-4}$$

$$\text{Poisson's ratio, } \nu = \frac{\varepsilon_d}{\varepsilon_l} = \frac{1.75}{5} = 0.35$$

$$\text{Bulk modulus, } K = \frac{E}{3(1-2\nu)} = \frac{127\times10^3}{3(1-0.7)} = 141.1\times10^3 N/mm^2$$

Example 1.12 A material is subjected to two mutually perpendicular linear strains together with a shear strain. One of the linear strains is 250 x 10^{-6} tensile. Determine the magnitude of the other linear strain and shear strain if the principal strains are – 100 × 10^{-6} and + 300 × 10^{-6}.

Solution:

Given: $\varepsilon_x = 250\times10^{-6}, \varepsilon_1 = 300\times10^{-6}, \varepsilon_2 = -100\times10^{-6}$

$$\varepsilon_y = ?\ \gamma_{xy} = ?$$

$$\varepsilon_1 = \frac{1}{2}(\varepsilon_x + \varepsilon_y) + \frac{1}{2}\sqrt{(\varepsilon_x - \varepsilon_y)^2 + \gamma_{xy}^2}$$

$$300 = \frac{1}{2}(250 + \varepsilon_y) + \frac{1}{2}\sqrt{(250 - \varepsilon_y)^2 + \gamma_{xy}^2} \quad \text{... (1)}$$

$$\varepsilon_2 = \frac{1}{2}(\varepsilon_x + \varepsilon_y) - \frac{1}{2}\sqrt{(\varepsilon_x - \varepsilon_y)^2 + \gamma_{xy}^2}$$

$$-100 = \frac{1}{2}(250 + \varepsilon_y) - \frac{1}{2}\sqrt{(250 - \varepsilon_y)^2 + \gamma_{xy}^2} \quad \text{... (2)}$$

Subtracting Eq. (2) from (1), we have

$$400 = \sqrt{(250 - \varepsilon_y)^2 + \gamma_{xy}^2} \quad \text{... (3)}$$

Substituting Eq. (3) in (1), we get

$$300 = \frac{1}{2}(250 + \varepsilon_y) + 200$$

$$600 = 650 + \varepsilon_y$$

$$\varepsilon_y = 50\ \mu \text{ strain or } - 0.00005$$

Substituting the value of ε_y in Eq. (3), we have

$$400 = \sqrt{(250 + 50)^2 + \gamma_{xy}^2}$$

$$= \sqrt{(300)^2 + \gamma_{xy}^2}$$

Squaring both sides, we have

$$400^2 = 300^2 + \gamma_{xy}{}^2$$

$$\gamma_{xy} = 264.6\ \mu - \text{strain or } 0.0002646$$

Example 1.13 *A circle of 100 mm diameter is scribed on an aluminium plate before it is loaded as shown in Fig. 1.30. After stressing the circle deforms into an ellipse. Determine the lengths of major and minor axes of the ellipse and their orientation.*

For aluminium $E = 10$ GPa and $\nu = 0.32$.

Solution:

Given: $\sigma_x = 80$ MPa, $\sigma_y = 60$ MPa, $\tau_{xy} = 25$ MPa, $E = 70$ GPa, $v = 0.32$

Principal stresses in aluminium plate,

$$\sigma_{1,2} = \frac{1}{2}(\sigma_x + \sigma_y) \pm \sqrt{(\sigma_x - \sigma_y)^2 + 4\tau_{xy}^2}$$

$$= \frac{1}{2}(80 + 60) \pm \frac{1}{2}\sqrt{(80 - 60)^2 + 4 \times 25^2}$$

$$= 70{\cdot}26 \pm 96.926$$

$$= 96.926 \text{ MPa}, 43.074 \text{ MPa}$$

$$\tan 2\theta = \frac{2\tau_{xy}}{\sigma_x - \sigma_y} = \frac{2 \times 25}{80 - 60} = 2.5$$

$$\theta_1 = 34°, \theta_2 = 124°$$

Principal strains are:

$$\varepsilon_1 = \frac{1}{E}[\sigma_1 - \nu \sigma_2]$$

$$= \frac{1}{70 \times 10^3}[96.926 - 0.32 \times 43.074]$$

$$= 1.188 \times 10\text{–}3$$

Increase in diameter of circle, $\delta d_1 = \varepsilon_1 d = 1.188 \times 10^{-3} \times 100 = 0.1188$ mm
Major diameter of ellipse = $d + \delta d_{1'} = 100.1188$ mm

$$\varepsilon_2 = \frac{1}{E}[\sigma_2 - \nu \sigma_1]$$

$$= \frac{1}{70 \times 10^3}[43.074 - 0.32 \times 96.926]$$

$$= 1.7225 + 10^{-4}$$

Increase in diameter of circle perpendicular to previous diameter,

$$\delta d_2 = \varepsilon_2 d = 1.7225 \times 10^{-4} \times 100 = 0.017225 \text{ mm.}$$

Minor diameter of ellipse $= d + \delta d_2 = 100.0172$ mm
The deformed shape of hole is shown in Fig. 1.30

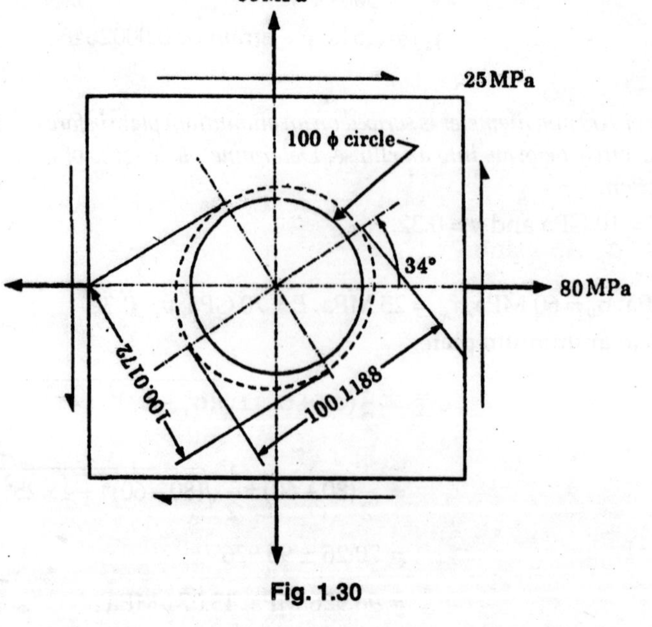

Fig. 1.30

Example1.14 *A point in a strained material is subjected to a tensile stress of 65 N/mm² and a compressive stress of 45 N/mm², acting on two mutually perpendicular planes and a shear stress of 10 N/mm², are acting on these planes. Find the normal stress, tangential stress and resultant stress on a plane inclined at 30° with the plane of the compressive strees.*

Solution:

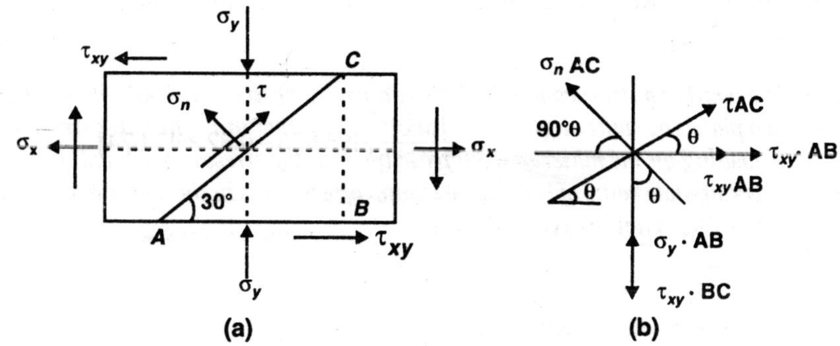

Fig. 1.31

Given: $\sigma_x = 65\ \text{N/mm}^2, \sigma_y = -45\ \text{N/mm}^2, \tau_{xy} = 10\text{N/mm}^2,$

$$\sigma_n \cdot AC = \tau_{xy} \cdot BC \times \cos\theta + \tau_{xy} \cdot AB \times \sin\theta + \sigma_x . BC \times \sin\theta - \sigma_y . AB \times \cos\theta$$

$$\sigma_n = \tau_{xy} \sin\theta\cos\theta + \tau_{xy} \cos\theta\sin\theta + \sigma_x \sin^2\theta - \sigma_y \cos^2\theta$$

$$= \tau_{xy} \sin 2\theta + \frac{1}{2}\sigma_x(1-\cos 2\theta) - \frac{1}{2}\sigma_y(1+\cos 2\theta)$$

$$= \frac{1}{2}(\sigma_x - \sigma_y) - \frac{1}{2}(\sigma_x + \sigma_y)\cos 2\theta + \tau_{xy} \sin 2\theta$$

$$= \frac{1}{2}(65-45) - \frac{1}{2}(65+45)\cos 60° + 10\sin 60°$$

$$= 10 - 27.5 + 8.66 = -8.84\ \text{N/mm}^2$$

$$\tau . AC = \tau_{xy} . BC \times \sin\theta - \sigma_y . AB \times \sin\theta - \tau_{xy} . AB \times \cos\theta - \sigma_x . BC \times \cos\theta$$

$$\tau = \tau_{xy} \sin^2\theta - \sigma_y \cos\theta . \sin\theta - \tau_{xy} \cos^2\theta - \sigma_x \sin\theta \cos\theta$$

$$= \frac{-1}{2}(\sigma_x + \sigma_y)\sin 2\theta + \tau_{xy}(\sin^2\theta - \cos^2\theta)$$

$$= -\frac{1}{2}(\sigma_x + \sigma_y)\sin 2\theta + \tau_{xy}\left[\frac{1-\cos 2\theta}{2} - \frac{1+\cos 2\theta}{2}\right]$$

$$= -\frac{1}{2}(\sigma_x + \sigma_y)\sin 2\theta - \frac{1}{2}\tau_{xy} \cos 2\theta$$

$$= -\frac{1}{2}(65+45)\sin 60° - \frac{1}{2}\times 10\cos 60°$$

$$= -50.13\ \text{N/mm}^2$$

$$\text{Resultant stress} = \sqrt{(-8.84)^2 + (-50.13)^2} = 50.9\ \text{N/mm}^2$$

Example 1.15 *At a point in a strained material, there are normal stresses of 30N/mm², tension and 20 N/mm², compression on two planes at right angles to one another, together with shearing stresses of 15N/mm² on the same planes. If the loading on the material is increased so that the stresses reach values of K times those given, find the maximum permissible value of K if the maximum direct stress in the material is not to exceed 80 N/mm², and maximum shear stress is not to exceed 50N/mm².*

Solution:

Given: $\sigma_x = +30\text{N/mm}^2, \sigma_y = -20\text{N/mm}^2\ \tau_{xy} = 15N/\text{mm}^2,$

$$(\sigma_1)_{max} = 80\text{N/mm}^2, \tau_{max} = 50\ \text{N/mm}^2$$

$$\sigma_{1,2} = \frac{1}{2}(\sigma_x + \sigma_y) \pm \frac{1}{2}\sqrt{(\sigma_x - \sigma_y)^2 + 4\tau_{xy}^2}$$

$$= \frac{1}{2}(30-20) \pm \frac{1}{2}\sqrt{(30+20)^2 + 4\times 15^2}$$

$$= 5 \pm 29.15$$

$$\sigma_1 = 34.15\ \text{N/mm}^2, \quad \sigma_2 = -24.14\ \text{N/mm}^2$$

$$\tau_{max} = \frac{1}{2}(\sigma_1 - \sigma_2) = 29.15\ \text{N/mm}^2$$

$$K\sigma_1 = (\sigma_1)_{max}$$

$$K = \frac{80}{34.15} = 2.34$$

$$K\tau_{max} = 50$$

$$K = \frac{50}{29.15} = 1.71$$

$$\therefore \quad K = 1.71$$

Example 1.16 *The stresses in the three principal direction are +65 MN/mm², +20 MN/m² and –85 MN/m². Find the principal strain. Take ν = 0.3 and E = 200 GN/m².*

Solution:

Given: $\sigma_1 = 65\text{MPa}, \sigma_2 = 20\text{MPa}, \sigma_3 = -85\ \text{MPa}, v = 0.3, E = 200\ \text{GPa}$

$$\varepsilon_1 = \frac{1}{E}[\sigma_1 - \nu(\sigma_2 + \sigma_3)]$$

$$= \frac{10^6}{200\times 10^9}[65-0.3\,(20-85)] = 422.5\times 10^{-6}$$

$$\varepsilon_2 = \frac{1}{E}[\sigma_2 - \nu\,(\sigma_3+\sigma_1)]$$

$$= \frac{10^6}{200\times 10^9}[20-0.3\,(-85+65)] = 130\times 10^{-6}$$

$$\varepsilon_3 = \frac{1}{E}[\sigma_3 = \nu(\sigma_1+\sigma_2)]$$

$$= \frac{10^6}{200\times 10^9}[-85-0.3(65+20)] = -552.5\times 10^{-6}$$

Example 1.17 *A flywheel weighing 500kg is mounted on shaft 75 mm diameter and midway between bearing 0.6m apart. If the shaft is transmitting 30 kW at 360 r.p.m, calculate the principal stresses and the maximum shear stress at the ends of a vertical dimeter in a plane close to the flywheel.*

Solution:

Given: $M_f = 500$ kg, $d_s = 75$mm, $l = 0.6m$, $P = 30$ kW, $N = 360$ r.pm.

$$P = \frac{2\pi NT}{60\times 10^3},\ T = \frac{30\times 60\times 10^3}{2\pi\times 360} = 795.78\ \text{N.m}$$

Maximum B.M. due to flywheel weight, $M = \dfrac{M_f g\times l}{4}$

$$= \frac{500\times 9.81\times 0.6}{4} = 735.75\ \text{N.m}$$

Torsional shear stress, $\tau = \dfrac{16T}{\pi d_s^3} = \dfrac{16\times 795.78\times 10^3}{\pi\times 75^3} = 9.6\ \text{N/mm}^2$

Bonding strees, $\sigma = \dfrac{32\,M}{\pi d_s^3} = \dfrac{32\times 735.75\times 10^3}{\pi\times 75^3} = 17.76\ \text{N/mm}^2$

Principal stresses, $\sigma_{1,2} = \dfrac{\sigma}{2}\pm\dfrac{1}{2}\sqrt{\sigma^2+4\tau^2}$

$$= \frac{17.76}{2}\pm\frac{1}{2}\sqrt{(17.76)^2+4\times(9.6)^2}$$

$$= 8.88\pm 13.07$$

$$= 21.95\ \text{N/mm}^2, -4.19\ \text{N/mm}^2$$

Maximum shear stress, $\tau_{max} = \dfrac{1}{2}(\sigma_1-\sigma_2) = 13.07\ \text{N/mm}^2$

Example 1.18 *If, at a point in a material, the minimum and maximum principle stresses are 30 MPa and 90 MPa, both tension, respectively, find the shear stress and normal stress on a plane through this point making an angle of tan⁻¹ (0.25) with the plane on which the maximum principle stress occurs. Also find the value of the maximum shear stresse and associated normal stress. Show all the states of stress on properly oriented elements.*

Solution:

$\sigma_{max} = 90$ MPa, $\sigma_{min} = 30$ MPa, $\tan^{-1} 0.25 = 14°$

At $\theta = 14°$ (see Fig. 2.32)

$$\theta_n = \frac{1}{2}(\sigma_{max} + \sigma_{min}) + \frac{1}{2}(\sigma_{max} - \sigma_{min})\cos 2\theta$$

$$= \frac{1}{2}(90+30) + \frac{1}{2}(90-30)\cos 28°$$

$$\tau = \frac{1}{2}(\sigma_{max} - \sigma_{min})\sin 2\theta$$

$$= \frac{1}{2}(90-30)\sin 28° = 14.1 \text{ MPa}$$

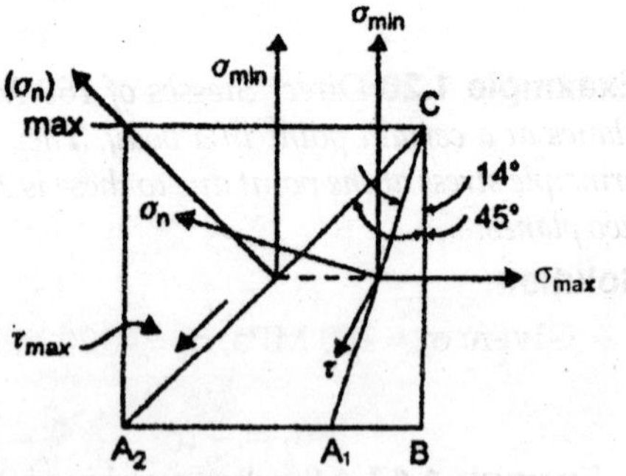

Fig. 2.32

Shear stress is maximum when sin 2θ = 1 or sin 2θ = 90° or θ = 45°

$$\tau_{max} = \frac{1}{2}(90-30) = 30 \text{ MPa}$$

Then $$(\sigma_n)_{max} = \frac{1}{2}(\sigma_x + \sigma_y) = \frac{1}{2}(90+30) = 60 \text{ MPa}$$

Example 1.19 *The state of stress is shown in Fig. 1.33. Determine the magnitude of the principle stresses and maximum shearing strees.*

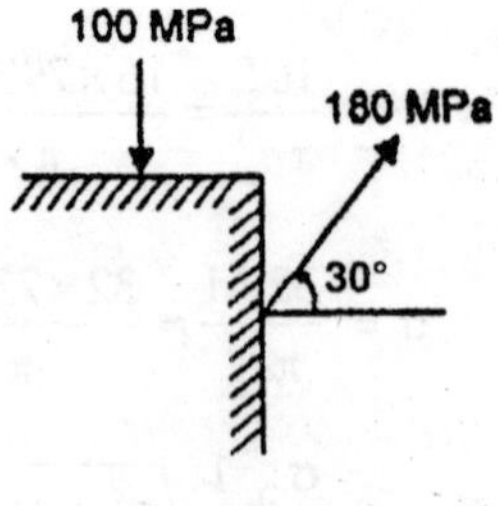

Fig: 1.33

Solution:

$\sigma_x = 1.80 \cos 30° = 155.9$, MPa, $\sigma_y = -100$ MPa, $\tau_{xy} = 180 \sin 30° = 90$ MPa

Principal stresses, $$\sigma_{1,2} = \frac{1}{2}(\sigma_x + \sigma_y) \pm \frac{1}{2}\sqrt{(\sigma_x - \sigma_y)^2 + 4\tau_{xy}^2}$$

$$= \frac{1}{2}(155.9 - 100) \pm \frac{1}{2}\sqrt{(155.9+100)^2 + 4 \times (90)^2}$$

$$= 27.95 \pm 156.43 = 184.38 \text{ MPa}, -128.48 \text{ MPa}$$

Maximum shear streass, $\tau_{max} = \frac{1}{2}(\sigma_1 - \sigma_2)$

$$= \frac{1}{2}(184.38 + 128.48) = 156.43 \text{ MPa}$$

Exaxmple 1.20 *Direct Stesses of 160 MPa tensile and 120 MPa compressive exist on two perpendicular planes at a certain point in a body. They are also accompained by shear stesses on the planes. The greatest principle stress at the point due to these is 200 MPa. (i) What must be magintude of the shearing stresses on the two planes.*

Solution:

Given: $\sigma_x = 160$ MPa, $\sigma_y = -120$ MPa, $\sigma_1 = 200$ MPa

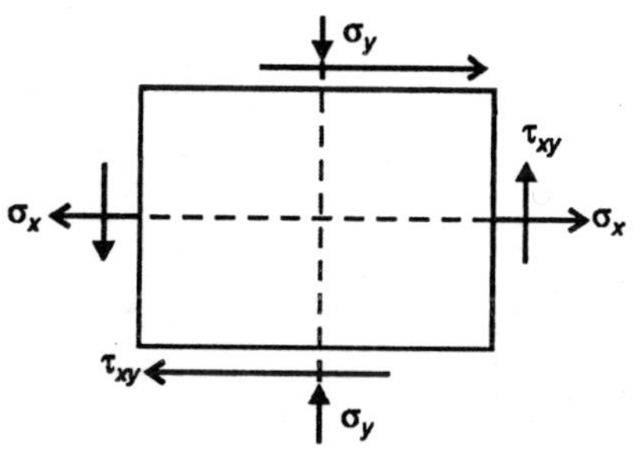

Fig. 1.34

$$\sigma_1 = \frac{1}{2}(\sigma_x + \sigma_y) + \frac{1}{2}\sqrt{(\sigma_x - \sigma_y)^2 + 4\tau_{xy}^2}$$

$$200 = \frac{1}{2}(160 - 20) + \frac{1}{2}\sqrt{(160 + 120)^2 + 4\tau_{xy}^2}$$

$$= 20 + \frac{1}{2}\sqrt{280^2 + 4\tau_{xy}^2}$$

$$(180 \times 2)^2 = 280^2 + 4\tau^2 xy$$

$$\tau_{xy} = 113.137 \text{ MPa}$$

Example 1.21 *An element in a strained material is subjected to normal stresses of 60 MPa and – 40 MPa on two mutually perpendiucular planes together with a shear stress of 40 MPa. Determine the principal stresses and maximum shear stress together with their planes.*

Solution:

Given: $\sigma_x = 60$ MPa, $\sigma_y = -40$ MPa, $\tau_{xy} = 40$ MPa

$$\sigma_{1,2} = \frac{1}{2}(\sigma_x + \sigma_y) \pm \frac{1}{2}\sqrt{(\sigma_x - \sigma_y)^2 + 4\tau^2 xy}$$

$$= \frac{1}{2}(60 - 40) \pm \frac{1}{2}\sqrt{(60 + 40)^2 + 4 \times 40^2} = 10 \pm 64.03$$

$$= 74.03 \text{ MPa}, -54.03 \text{ MPa}$$

$$\tau_{max} = \frac{1}{2}(\sigma_1 - \sigma_2) = \frac{1}{2}(74.03 + 54.03) = 64.03 \text{ MPa}$$

$$\tan 2\theta = \frac{2\tau_{xy}}{\sigma_x - \sigma_y} = \frac{2 \times 40}{60 + 40} = \frac{80}{100} = 0.8$$

$$2\theta = 38.66$$

$$\theta_1 = 19.33°,\ \theta_1 = 109.33°$$

Planes of maximum shear stress are:

$$\theta_3 = 64.33°, \theta_4 = 154.33°$$

Example 1.22 *The state of stress at a point in a stressed material is given by: σ_x = 20MPa, σ_y = 10 MPa and σ_{xy} = 25 MPa. Compute the principal stresses in the material.*

Solution:

Given: $\sigma_x = 20$ MPa, $\sigma_y = 10$ MPa, $\tau_{xy} = 25$ MPa

$$\text{Principal stresses, } \sigma_{1,2} = \frac{1}{2}(\sigma_x + \sigma_y) \pm \frac{1}{2}\sqrt{(\sigma_x - \sigma_y)^2 + 4\tau_{xy}^2}$$

$$= \frac{1}{2}(20+10) \pm \frac{1}{2}\sqrt{(20-10)^2 + 4 \times 25^2}$$

$$= 15 \pm 25.495$$

$$= 40.495 \text{ MPa}, -10.495 \text{ MPa}.$$

Example 1.23 *The state of stress on an element is shown Fig. 1.35. Determine (a) Principle stresses and their orientation (b) Maximum/minimum shear stressess, their orientiations along with the associated normal stresses (c) Values of normal and shear stresses on planes indicated. Show all the results on properly oriented planes.*

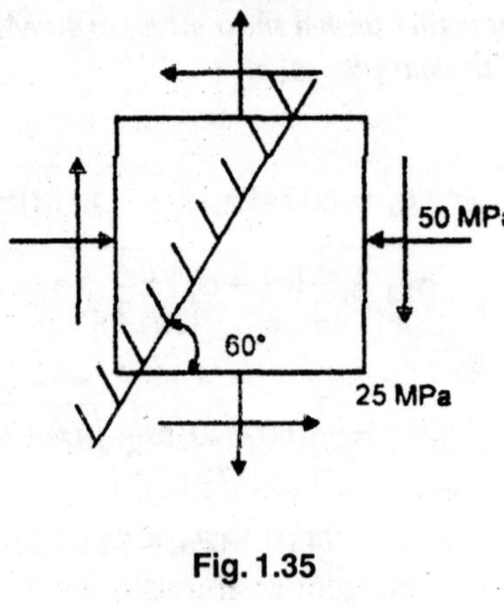

Fig. 1.35

Solution:

$$\sigma_x = -50 \text{ MPa}, \quad \sigma_y = 100 \text{ MPa}, \quad \tau_{xy} = -25 \text{ MPa}$$

(a)
$$\sigma_{1,2} = \frac{1}{2}(\sigma_x + \sigma_y) \pm \frac{1}{2}\sqrt{(\sigma_x - \sigma_y)^2 + 4\tau_{xy}^2}$$

$$= \frac{1}{2}(-50+100) \pm \frac{1}{2}\sqrt{(-50-100)^2 + 4\times 25^2}$$

$$= 25 \pm 79.057$$

$$\sigma_{1,2} = 104.057 \text{ MPa}, -54.057 \text{ MPa}$$

$$\tan 2\theta = \frac{-2\tau_{xy}}{\sigma_x - \sigma_y} = \frac{-2\times(-25)}{-50+100} = \frac{50}{50} = 1$$

$$2\theta = 45°, \theta_1 = 22.5°, \theta_2 = 112.5°$$

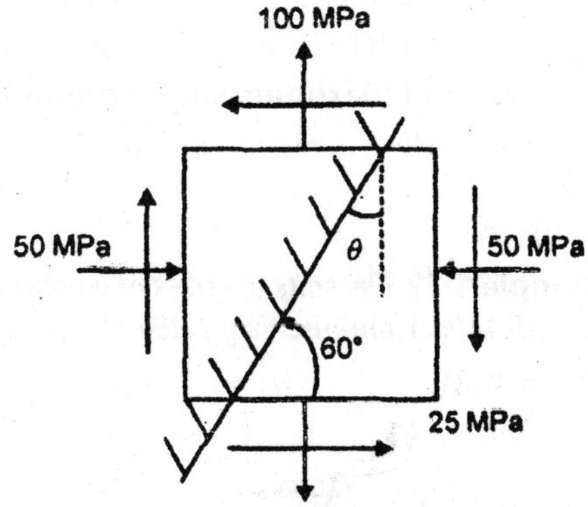

Fig. 1.36

(b) $$\tau_{max} = \frac{1}{2}(\sigma_1 - \sigma_2) = \pm 79.057 \text{ MPa}$$

$$\theta_3 = 22.5° + 45° = 67.5°, \theta_4 = 157.5°$$

(c) Here $\theta = 90° - 30°$ (see Fig. 2.36)

$$\sigma_n = \frac{1}{2}\sigma_x(1+\cos 2\theta) + \frac{1}{2}\sigma_y(1-\cos 2\theta)\tau_{xy}\sin 2\theta$$

$$= \frac{1}{2}(-50)(1+\cos 60°) + \frac{1}{2}\times 100(1-\cos 60°) + 25\sin 60°$$

$$= -37.5 + 25 + 21.65 = 9.15 \text{ MPa}$$

$$\tau = \frac{1}{2}(\sigma_x - \sigma_y)\sin 2\theta + \tau_{xy}\cos 2\theta = \frac{1}{2}(-50-100)\sin 60° - 25\cos 60°$$

$$= -57.027 - 12.5 = -69.527 \text{ MPa}$$

The orientation of planes is shown in the Fig. 1.37.

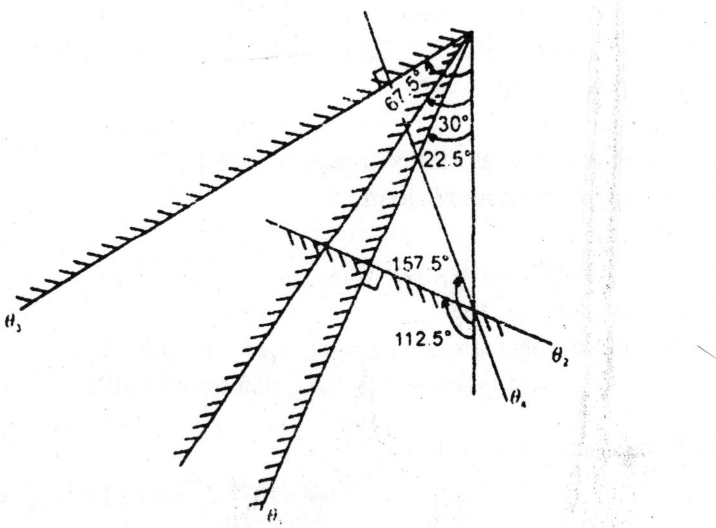

Fig. 1.37

Example: 1.24 *Direct stresses of 160MPa tensile and 120 MPa compressive exist on two perpendicular planes at a certain point in a body. They are also accompained by shear stresses on the planes. The greatest principle stress at the point due to these is 200 MPa.*

(a) *What must be the magnitude of the shearing stresses on the two-planes?*

(b) *What will be the maximum shearing stress and the associated normal stress at the point?*

Use the Mohr's circle approach. Show the results on properly oriented elements.

Solution:

$\sigma_x = +160$ MPa, $\sigma_y = -120$ MPa, $\sigma_1 = 200$ MPa. (see Fig. 1.38)

Mohr's circle (see Fig. 1.39)

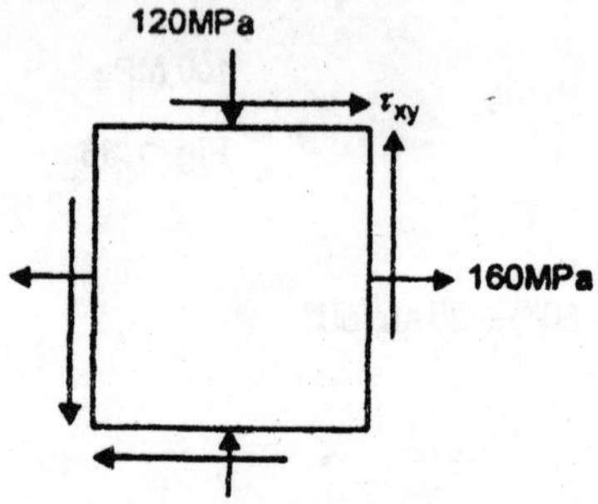

Fig. 1.38

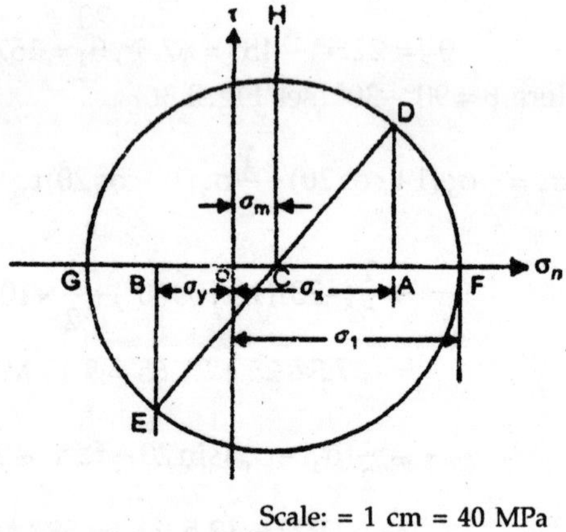

Fig. 1.39

1. 1. Draw $\sigma_n - \tau$ axes. Select scale of 1 cm = 40 MPa.
2. Cutt off $OA = \sigma_x = 4$ cm and $OB = 3$cm, as shown in the Fig. 1.39
3. Locate the centre C of AB.
4. Cutt off, $OF = \sigma_1 = 5$ cm.
5. With C as centre and CF as radius draw the circle.
6. Erect perpendicular at A and B. Joint DE.
7. Measure AD, then

   Shear stress, $\tau_{xy} = AD = 3$ cm $= 120$ MPa.
8. Erect perpendicular at C, then

$$\text{Maximum shear stress, } \tau_{max} = CH = 4.5 \text{ cm} = 180 \text{ MPa.}$$
$$\text{Normal stress, } \tau_n = OC = 0.5 \text{ cm} = 20 \text{ MPa}$$

Example: 1.25 *The strains on an element are:*

$$\varepsilon_x = -8\times10^{-4}, \varepsilon_y = -2\times10^{-4}, \gamma_{xy} = -8\times10^{-4}$$

Determine analytically as well as graphically the principal strains and their direction. What are the corresponding principal stresses and in which direction do these act (E = 200 GPa and ν = 0.3)

Solution:

Given: $\varepsilon_x = -8 \times 10^{-4}$, $\varepsilon_y = -2 \times 10^{-4}$, $\gamma_{xy} = -8 \times 10^{-4}$, E = 200 GPa, ν = 0.3

$$\text{Principal strains, } \varepsilon_{1,2} = \frac{\varepsilon_x + \varepsilon_y}{2} \pm \frac{1}{2}\sqrt{(\varepsilon_x - \varepsilon_y)^2 + \gamma_{xy}^2}$$

$$= \frac{1}{2}(-8-2)\times 10^{-4} \pm \frac{1}{2}\times 10^{-4}\sqrt{(-8+2)^2 + (-8)^2}$$

$$= -5\times 10^{-4} \pm 5\times 10^{-4}$$

$$= \text{zero}, -10 \times 10^{-4}$$

$$\tan 2\theta = \frac{\gamma_{xy}}{\varepsilon_x - \varepsilon_y} = \frac{-8}{-8+2} = \frac{4}{3}$$

$$2\theta = 53.13°$$

$$\theta_1 = 26.565°$$

$$\theta_2 = 116.565°$$

$$\varepsilon_1 = \frac{1}{E}[\sigma_1 - \nu\sigma_2]$$

$$\varepsilon_2 = \frac{1}{E}[\sigma_2 - \nu\sigma_1]$$

$$0 = \frac{1}{E}[\sigma_1 - \nu\sigma_2]$$

$$\therefore \quad \sigma_1 - \nu\sigma_2 = 0 \quad \text{...(1)}$$

$$-10 \times 10^{-4} \times 200 \times 10^3 = \sigma_2 - \nu\sigma_1$$

$$\sigma_2 - \nu(\nu\sigma_2) = -200 \quad \text{...(2)}$$

$$\therefore \quad \sigma_2 = \frac{-200}{1-(0.3)^2} = -219.78 \text{ N/mm}^2$$

$$\sigma_1 = \nu\sigma_2$$

$$= 0.3 \times (-219.78)$$

$$= -65.934 \text{ N/mm}^2$$

Direction of principle stresses are same as those of principal strains.
Graphical construction for principal strains is shown in Fig. 1.40

Scale 1 cm = 1×10^{-4}

OA = 2 cm

$AP = BQ$
$= 4$ cm
$2\theta_1 = 54°$
$\theta_1 = 27°$
$\theta_2 = 117°$
$\varepsilon_1 = 0$
$\varepsilon_2 = OD = 10$ cm
$= -10 \times 10^{-4}$

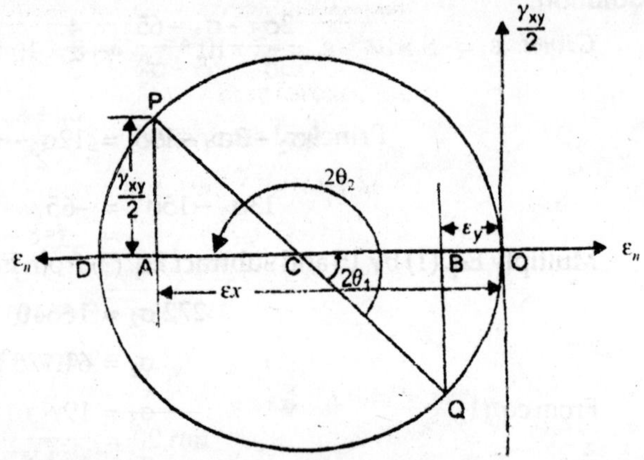

Fig. 1.40

Example 1. 26 *A piece of elastic material is subjected to 3 mutually perpendicular tensile stresses. The strains in the three directions are in the ratio of 3:4:5. Find the magntitude of these stresses if the greatest stress is 65 MPa. Take Poisson's ratio as 1/3.*

Solution:

Let the three principal stresses be $\sigma_1 > \sigma_2 > \sigma_3$.

Then $\sigma_1 = 65$ MPa. Also $\nu = \frac{1}{3}$ and $\varepsilon_1 : \varepsilon_2 : \varepsilon_3 :: 5 : 4 : 3$

Now
$$\varepsilon_1 = \frac{1}{E}[\sigma_1 - \nu(\sigma_2 + \sigma_3)]$$

$$\varepsilon_2 = \frac{1}{E}[\sigma_2 - \nu(\sigma_3 + \sigma_1)]$$

$$\varepsilon_3 = \frac{1}{E}[\sigma_3 - \nu(\sigma_1 + \sigma_2)]$$

$$\frac{\varepsilon_1}{\varepsilon_2} = \frac{\sigma_1 - \nu(\sigma_2 + \sigma_3)}{\sigma_2 - \nu(\sigma_3 + \sigma_1)} = \frac{65 - \frac{1}{3}(\sigma_2 + \sigma_3)}{\sigma_2 - \frac{1}{3}(\sigma_3 + 65)} = \frac{5}{4}$$

$$\frac{195 - \sigma_2 - \sigma_3}{3\sigma_2 - \sigma_3 - 65} = \frac{5}{4}$$

$$780 - 4\sigma_2 - 4\sigma_3 = 15\sigma_2 - 5\sigma_3 - 325$$

$$19\sigma_2 - \sigma_3 = 1105 \quad \text{... (1)}$$

$$\frac{\varepsilon_2}{\varepsilon_3} = \frac{\sigma_2 - \nu(\sigma_3 + \sigma_1)}{\sigma_3 - \nu(\sigma_1 + \sigma_2)} = \frac{\sigma_2 - \frac{1}{3}(\sigma_3 + 65)}{\sigma_3 - \frac{1}{3}(65 + \sigma_2)} = \frac{4}{3}$$

$$\frac{3\sigma_2 - \sigma_3 - 65}{3\sigma_3 - 65 - \sigma_2} = \frac{4}{3}$$

$$9\sigma_2 - 3\sigma_3 - 195 = 12\sigma_3 - 260 - 4\sigma_2 \quad \text{... (2)}$$

$$13\sigma_2 - 15\sigma_3 = -65$$

Multiply Eq. (1) by 15 and subtract Eq. (2) from Eq. (1), we get

$$272\,\sigma_2 = 16640$$

$$\sigma_2 = 61.176 \text{ MPa}$$

From eq. (1), $\sigma_3 = 19 \times 61.176 - 1105 = 57.344$ MPa

$\therefore$ $\sigma_1 = 65$ MPa, $\sigma_2 = 61.176$ MPa, $\sigma_3 = 57.344$ MPa.

Example 1.27 (i) *Draw and describe Mohr's stress circle choosing suitable coordinate system and sign convention for pure 1–dmensional tension case and pure shear case.*

(ii) *If, at a point within a a material, the minimum and maximum principal stresses are 20 N/mm² and 100 N/mm², both tension, find the shear stress and normal stress on a plane through this point making an angle of 30° with the plane on which the maximum principal stress acts.*

Solution:

(*i*) Mohr's circle for 1-D tension case (see Fig. 1.41)

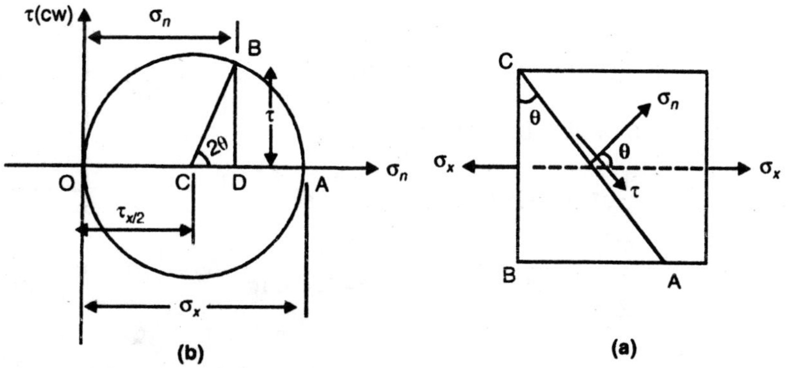

Fig. 1.41 Mohr's circle for 1-dimensional tension case.

1. Draw σ_n and τ axes with σ_n tensile stress to the right and τ (cw) upwards.
2. Cutt off $OA = \sigma_x$.
3. Locate the centre C of OA so that $OC = \frac{1}{2}\sigma_x$.
4. With C as centre and OC as radius, draw the circle.

5. Draw $\angle ACB = 2\theta$ (ccw). Draw $BD \perp$ on OA. Then $OD = \sigma_n$ and $DB = \tau$ on the plane inclined at angle θ with the plane on which σ_x acts.

Mohr' circle for pure shear case (see Fig. 1.42)

1. Draw σ_n and τ axes with σ_n tensile stress to the right and τ(cw) upwards.
2. With O as centre and τ_{xy} as radius, draw the circle.
3. Draw $\angle BOA = 2\theta$ (ccw). Draw $AB \perp \sigma_n$ axis. Then $OB = \sigma_n$ and $AB = \tau$.

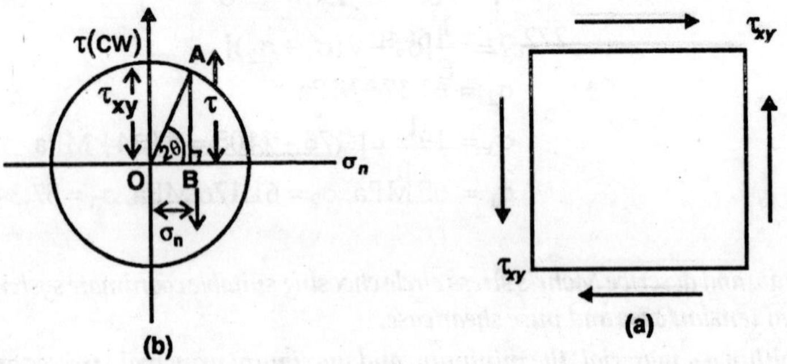

Fig. 1.42 Mohr's circle for pure shear case.

(ii) Given: $\sigma_1 = 100\ \text{N/mm}^2, \sigma_2 = 20\ \text{N/mm}^2, \theta = 30°$

Normal stress, (see. Fig. 1.43)

$$\sigma_n = \frac{1}{2}(\sigma_1+\sigma_2)+\frac{1}{2}(\sigma_1-\sigma_2)\cos 2\theta$$

$$= \frac{1}{2}(100+20)+\frac{1}{2}(100-20)\cos 60°$$

$$= 60 + 20 = 80\ \text{N/mm}^2$$

Shear stress,
$$\tau = \frac{1}{2}(\sigma_1-\sigma_2)\sin 2\theta$$

$$= \frac{1}{2}(100-20)\sin 60° = 34.64\ \text{N/mm}^2$$

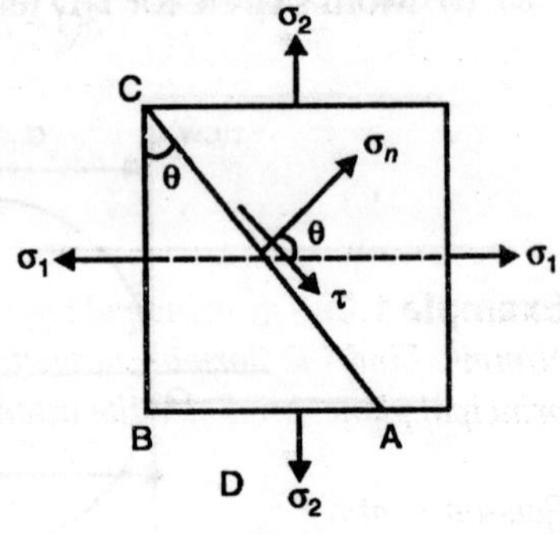

Fig. 1.43

Example 1.28 *A piece of material is subjected to three mutually perpendicular stresses of 70, 56 and 84 MPa. If E = 200 GPa. ν = 0.28, determine (a) principle strains (b) shear modulus and (c) bulk modulus.*

Solution:

Given: $\sigma_1 = 84$ MPa, $\sigma_2 = 70$ MPa, $\sigma_3 = 56$ MPa, $\nu = 0.28$, $E = 200$ 6Pa.

(*a*) Principal strains:

$$\varepsilon_1 = \frac{1}{e}[\sigma_1 - \nu(\sigma_2+\sigma_3)]$$

$$= \frac{1}{200\times 10^3}[84 - 0.28\,(70+56)$$

$$= 243.6 \times 10^{-6}$$

$$\varepsilon_2 = \frac{1}{E}[\sigma_2 - \nu(\sigma_3 + \sigma_1)]$$

$$= \frac{1}{2\times10^5}[70 - 0.28(56+84)]$$

$$= 154 \times 10^{-6}$$

$$\varepsilon_3 = \frac{1}{E}[\sigma_3 - \nu(\sigma_1 + \sigma_2)]$$

$$= \frac{1}{2\times10^5}[56 - 0.28(84+70)]$$

$$= 64.4 \times 10^{-6}$$

(*b*) Shear modulus, $G = \dfrac{E}{2(1+2\nu)}$

$$= \frac{2\times10^5}{2(1+0.28)} = 78.125 \text{ kN/mm}^2$$

(*c*) Bulk modulus, $K = \dfrac{E}{3(1-2\nu)}$

$$= \frac{2\times10^5}{3(1-2\times0.28)} = 151.51 \text{ kN/mm}^2$$

Example 1.29 *The principal tensile stresses at a point across two perpendicular planes are 80 N/mm² and 40 N/mm². Find the normal, tangential and resultant stresses, its obliquity on a plane at 20° with the major principal plane. Find also the intensity of stress which acting alone can produce same maximum strain. Take Poisson's ratio* $= \frac{1}{2}$.

Solution:

Given: $\sigma_x = 80 \text{ N/mm}^2, \sigma_y = 40 \text{ N/mm}^2, \theta = 20°, \nu = 0.25$

Refer to Fig. 1.44.

Normal stress,
$$\sigma_n = \frac{1}{2}(\sigma_x + \sigma_y) + \frac{1}{2}(\sigma_x - \sigma_y)\cos 2\theta$$

$$= \frac{1}{2}(80+40) + \frac{1}{2}(80-40)\cos 40° = 75.32 \text{ N/mm}^2$$

Shear stress,
$$\tau = \frac{1}{2}(\sigma_x - \sigma_y)\sin 2\theta$$

$$= \frac{1}{2}(80 - 40)\sin 40°$$

$$= 12.85\ \text{N/mm}^2$$

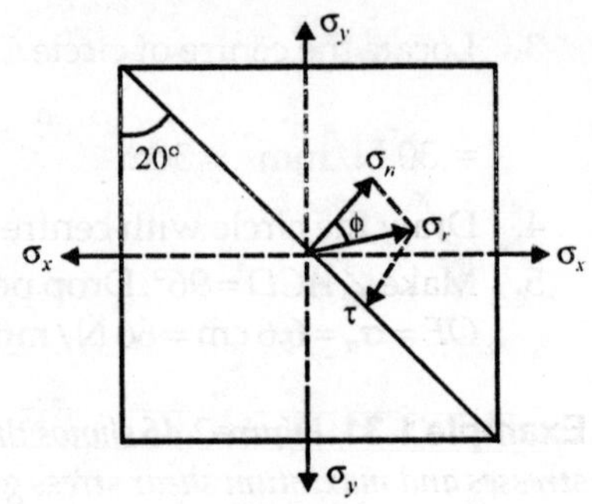

Fig. 1.44

Resultant stress, $\sigma_n = \sqrt{\sigma_n^2 + \tau^2}$

$$= \sqrt{(75.32)^2 + (12.85)^2} = 76.41\ \text{N/mm}^2$$

Angle of obliquity, $\tan\phi = \frac{\tau}{\sigma_n} = \frac{12.85}{75.32} = 0.1706$

$$\phi = 9.68°$$

Maximum strain, $\varepsilon_1 = \frac{1}{E}[\sigma_x - \nu\sigma_y] = \frac{\sigma}{E}$

$$\therefore \quad \sigma = \sigma_x - \nu\sigma_y = 80 - 0.25 \times 40 = 70\text{N/mm}^2$$

Example 1.30 *At a certain point in a strained material the principal stresses are 100 N/mm² and 40 N/mm² both tensile. Find the normal tangential and resultant stresses across a plane through the point at 48° to the major principal plane using Mohr's circle of stress.*

Solution:

Given: $\sigma_x = 100\ \text{N/mm}^2$, $\sigma_y = 40\ \text{N/mm}^2$, $\theta = 48°$

Mohr's circle construction: (see Fig. 1.45)

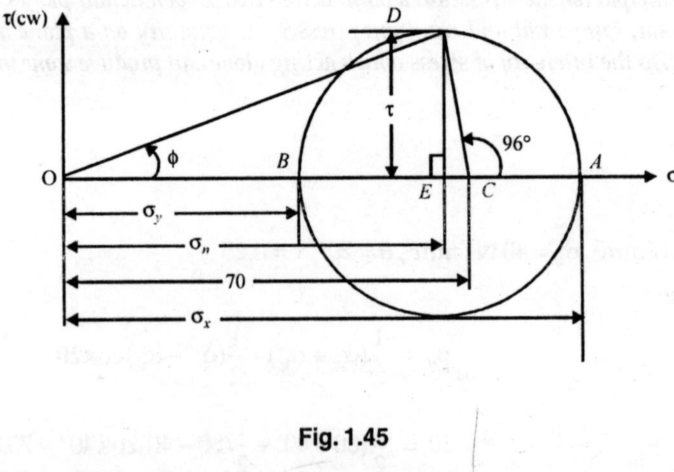

Fig. 1.45

1. Draw σ_n axes Choose a scale of 1 cm = 10 N/mm².
2. Cut off $OA = 100\ \text{N/mm}^2 = 10$ cm and $OB = 40\ \text{N/mm}^2 = 4$ cm.

3. Locate the centre of circle C at $\frac{1}{2}(100+40) = 70\ N/mm^2 = 7cm$. Radius of circle = $\frac{1}{2}(100-40)$ $= 30\ N/mm^2 = 3cm$
4. Draw the circle with centre (7, 0) and radius 3 cm.
5. Make $\angle ACD = 96°$. Drop perpendicular DE on OA and join OD. Then by measurment, we have $OE = \sigma_n = 6.6$ cm $= 66\ N/mm^2$ $DE = \tau = 3$ cm $= 30\ N/mm^2$, $OD = \sigma_r = 7.2$ cm $= 72\ N/mm^2$, $\phi = 24°$

Example 1.31 *Figure 2.46 shows the state of stress at a point. Find the location of principal planes, principal stresses and maximum shear stress graphically or otherwise,*

Solution:

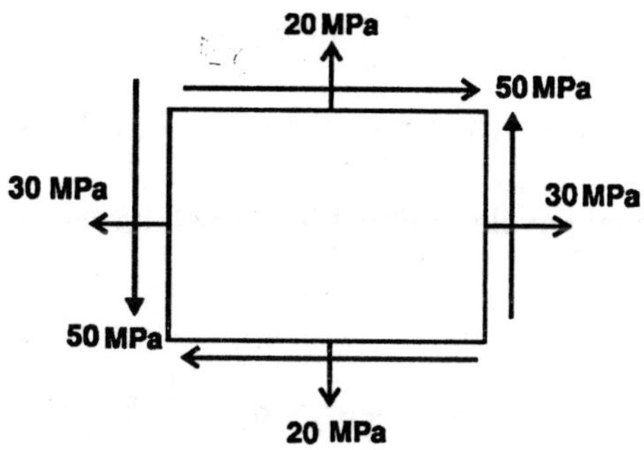

Fig. 1.46

Given: $\sigma_x = 30$ MPa, $\sigma_y = 20$ MPa, $\tau_{xy} = 50$ MPa,

Principal stresses:

$$\sigma_{1,2} = \frac{1}{2}(\sigma_x + \sigma_y) \pm \frac{1}{2}\sqrt{(\sigma_x - \sigma_y)^2 + 4\tau_{xy}^2}$$

$$= \frac{1}{2}(30+20) \pm \frac{1}{2}\sqrt{(30-20)^2 + 4\times 50^2}$$

$$= 25 \pm 50.25$$

$$\sigma_1 = 75.25 \text{ MPa}, \ \sigma_2 = -25.25 \text{ MPa}$$

Principal planes:

$$\tan 2\theta = \frac{2\tau_{xy}}{\sigma_x - \sigma_y} = \frac{2\times 50}{30-20} = 10$$

$$2\theta = 84.29°$$

$$\theta_1 = 42.245°, \theta_2 = \theta_1 + 90° = 132°.245°$$

Maximum shear stress,

$$\tau_{max} = \frac{1}{2}(\sigma_1 - \sigma_2)$$

$$= \frac{1}{2}(75.25+25.25)$$

$$= 50.25\ \text{N/mm}^2$$

Summary

1. Stresses on an oblique plane.

$$\sigma_n = \frac{\sigma}{2}(1+\cos 2\theta)$$

$$\tau = \frac{\sigma}{2}\sin 2\theta$$

Maximum shear stress occurs on a plane inclined at 45° with the line of action of uniaxial stress σ and is half of it.

2. A material under uniaxial tensile shall fail in shear if its shear strength is less then half of its tensile strength.
3. Biaxial loading.

$$\sigma_n = \frac{1}{2}(\sigma_x+\sigma_y)+\frac{1}{2}(\sigma_x-\sigma_y)\cos 2\theta$$

$$\tau = \frac{1}{2}(\sigma_x-\sigma_y)\sin 2\theta$$

$$\tau_{max} = \frac{1}{2}(\sigma_x-\sigma_y)$$

$$\sigma_r = \sqrt{\sigma_x^2\cos^2\theta+\sigma_y^2\sin^2\theta}$$

$$\tan\phi = \frac{\sigma_x-\sigma_y}{\sigma_x\cot\theta+\sigma_y\tan\theta} = \frac{\tau}{\sigma_n}$$

For ϕ to be maximum,

$$\tan\phi = \sqrt{\frac{\sigma_x}{\sigma_y}}$$

$$\tan\phi_{max} = \frac{\sigma_x-\sigma_y}{2\sqrt{\sigma_x\sigma_y}}$$

4. Shear stresses are complementary in nature, i.e. $\tau_{xy} = \tau_{yx}$
5. A two-dimensional lamina under shear develops tensile and compressive stresses on the diagonals which are equal in magnitude to the shear stress.
6. Compound stresses.

$$\sigma_n = \frac{1}{2}(\sigma_x+\sigma_y)+\frac{1}{2}(\sigma_x-\sigma_y)\cos 2\theta+\tau_{xy}\sin 2\theta$$

$$\tau = \frac{1}{2}(\sigma_x-\sigma_y)\sin 2\theta-\tau_{xy}\cos 2\theta$$

For σ_n to be maximum or minimum,

$$\tan 2\theta = \frac{2\tau_{xy}}{\sigma_x-\sigma_y}$$

7. Principal stresses are the maximum or minimum values of normal stresses at a point on the plane. These planes are called the principal planes.

$$\sigma_{1,2} = \frac{1}{2}(\sigma_x+\sigma_y)\pm\frac{1}{2}\sqrt{(\sigma_x-\sigma_y)^2+4\tau_{xy}^2}$$

$$\tau_{max} = \frac{1}{2}(\sigma_1-\sigma_2)$$

Principal planes are given by,

$$\tan 2\theta_1 = \frac{2\tau_{xy}}{\sigma_x-\sigma_y}$$

$$\theta_2 = 90° + \theta_1$$

For shear stress to be maximum,

$$\tan 2\theta_3 = \frac{-(\sigma_x-\sigma_y)}{2\tau_{xy}}$$

$$\theta_4 = 90° + \theta_3$$

The planes of maximum shear and principal planes are separated by 45°.

8. Mohr's circle.

(a) Biaxial stresses centre of circle: $\left(\frac{\sigma_x+\sigma_y}{2},0\right)$

Radius of circle: $\frac{1}{2}(\sigma_x-\sigma_y)$

(b) Compound stresses:

Centre of circle: $\left(\frac{\sigma_x+\sigma_y}{2},0\right)$

Radius of circle: $\sqrt{\left(\frac{\sigma_x-\sigma_y}{2}\right)^2+\tau_{xy}^2}$

(*i*) Normal stresses are plotted as horizontal coordinates and shear stresses as vertical coordinates.

(*ii*) Shear stresses producing clockwise rotation of stress element are plotted along the positive shear stress axis and those producing counter-clockwise rotation are plotted along the negative shear stress axis.

(*iii*) The resultant stress shall have maximum obliquity when a line is drawn tangential to the circle from the origin.

9. Strains on an oblique plane.

Normal strain, $$\varepsilon_n = \frac{\varepsilon_x}{2}(1+\cos 2\theta)$$

Shear strain, $$\varepsilon_n = \frac{\varepsilon_x}{2}\sin 2\theta$$

10. Biaxial strains.

$$\varepsilon_n = \frac{1}{2}(\varepsilon_x+\varepsilon_y)+\frac{1}{2}(\varepsilon_x-\varepsilon_y)\cos 2\theta$$

$$\varepsilon_s = \frac{1}{2}(\varepsilon_x-\varepsilon_y)\sin 2\theta$$

$$\varepsilon_r = \sqrt{\varepsilon_x^2\cos^2\theta+\varepsilon_y^2\sin^2\theta}$$

$$(\varepsilon_s)_{max} = \frac{1}{2}(\varepsilon_x-\varepsilon_y) \text{ when } \theta = 45° \text{ or } 135°$$

$$\varepsilon_s = \frac{\gamma_{xy}}{2}$$

11. Compound strains.

$$\varepsilon_n = \frac{1}{2}(\varepsilon_n+\varepsilon_y)+\frac{1}{2}(\varepsilon_x-\varepsilon_y)\cos 2\theta+\frac{1}{2}\gamma_{xy}\sin 2\theta$$

$$\varepsilon_s = \frac{1}{2}(\varepsilon_x-\varepsilon_y)\sin 2\theta-\frac{1}{2}\gamma_{xy}\cos 2\theta$$

For normal strain to be maximum or minimum,

$$\tan 2\theta = \frac{\gamma_{xy}}{\varepsilon_x-\varepsilon_y}$$

For shear strain to be maximum or minimum,

$$\tan 2\theta_3 = -\left(\frac{\varepsilon_x-\varepsilon_y}{\gamma_{xy}}\right)$$

12. Principal strains.

$$\varepsilon_{1,2} = \frac{1}{2}(\varepsilon_x + \varepsilon_y) \pm \frac{1}{2}\sqrt{(\varepsilon_x - \varepsilon_y)^2 + \gamma_{xy}^2}$$

Principal planes are given by,

$$\tan 2\theta_1 = \frac{\gamma_{xy}}{\varepsilon_x - \varepsilon_y}$$

$$\theta_2 = 90° + \theta_1$$

Maximum shear strain, $(\varepsilon_s)_{max} = \frac{1}{2}(\varepsilon_1 - \varepsilon_2)$

13. Mohr's strain circle can be drawn as the stress circle with ε_n along horizontal axis and $\frac{\gamma}{2}(cw)$ along vertical (+ *ve*) axis.

14. Principle strains under three-dimensional stresses.

$$\varepsilon_1 = \frac{1}{E}[\sigma_1 - \nu(\sigma_2 + \sigma_3)]$$

$$\varepsilon_2 = \frac{1}{E}[\sigma_1 - \nu\,(\sigma_3 + \sigma_1)]$$

$$\varepsilon_3 = \frac{1}{E}[\sigma_3 - \nu(\sigma_3 - \nu(\sigma_1 + \sigma_2)]$$

15. Relationship between elastic constants.

$$G = \frac{E}{2(1+\nu)}$$

$$E = 3K(1-2v)$$

$$E = \frac{9KG}{G+3K}$$

$$v = \frac{3K-2G}{2G+6K}$$

16. Principal stresses from principal strains.

$$\sigma_1 = \frac{E}{(1+\nu)(1-2\nu)}[(1-\nu)\varepsilon_1 + \nu(\varepsilon_2 + \varepsilon_3)]$$

$$\sigma_2 = \frac{E}{(1+\nu)(1-2\nu)}[(1-\nu)\varepsilon_2 + \nu(\varepsilon_3 + \varepsilon_1)]$$

$$\sigma_3 = \frac{E}{(1+\nu)(1-2\nu)}[(1-\nu)\varepsilon_3 + \nu(\varepsilon_1 + \varepsilon_2)]$$

Short Answer Type Questions

1. Define Principal stresses.
2. What are principal planes?
3. For the case of biaxial stresses (both tensile) what is the centre and radius of the Mohr's circle?
4. If the uniaxial tensile stress in a bar is σ, then what is the maximum shear stress and its plane?
5. In a biaxial state of stress (both tensile), what is the maximum angle of obliquity of the resultant stress?
6. If E = 210 GPa and G = 80 GPa, then what is Poisson's ratio?
7. If the bulk modulus for a material is infinitely large, then what is the Poisson's ratio?
8. What shall be the shape of a cicular hole after deformation in a plate subjected to compound state of stress system?

EXERCISES

1. The biaxial stresses in a material are + 100 MPa and - 50 MPa. Determine the normal, shear and resultant stresses on a plane whose normal is inclined at 30° to the axis of first stress.
2. At a certain point in a material the resultant stress on a plane is 80 MPa tensile inclined at 30° to the normal to the plane. The stress on a plane at right angles to this plane has a normal tensile stress 50 MPa. Find (a) the resultant stress on second plane, (b) principal stresses, and (c) principal planes.
3. The state of stress at a point is given by:
$$\sigma_x = 90 \text{ MPa}, \sigma_y = -50 \text{ MPa}, \tau_{xy} = 40 \text{ MPa}$$
Determine (a) the planes of maximum shear stress, and (b) resultant stress on these planes.
4. The resultant stress at a point on plane A is + 90 MPa having an obliquity of 15°. On another plane B, the resultant stress is + 50 MPa having an obliquity of 35°. Determine (a) the angle between the planes A and B, and (b) the maximum obliquity the resultant stress can have with any plane at the point.
5. The state of stress at a point is given by:
$$\sigma_x = 140 \text{ MPa}, \sigma_y = 100 \text{ MPa}, \tau_{xy} = 60 \text{ MPa}$$
Calculate (a) the principal stresses, and (b) the principal planes.
6. The state of stress at a point is given by:
$$\sigma_x = 300 \text{ MPa}, \sigma_y = -200 \text{ MPa}, \tau_{xy} = 100 \text{ MPa (cw)}$$
Draw the Mohr's circle and determine (a) the principal stresses and principal planes, (b) normal and shear stresses on a plane inclined at 40° with σ_x.
7. The state of stress at a point is given by:
$$\sigma_x = -150 \text{ MPa}, \sigma_y\ 250 \text{ MPa}, \tau_{xy} = 100 \text{ MPa (ccw)}$$
Draw the Mohr's circle and determine (a) the principal stresses and principal planes, (b) the maximum shear stress and its planes, and (c) draw the orientation of principal and maximum shear stress planes on the stress element.
8. A rectangular block 250 mm x 100 mm x 50 mm is subjected to the following state of stress:
$$\sigma_x = 100 \text{ MPa}, \sigma_y = -60 \text{ MPa}, \sigma_z = 80 \text{ MPa}$$
If ν = 0.25 and E = 200 GPa, determine its change in volume.

9. A square bar of 20 mm side and 200 mm long is subjected to a compressive load of 200 kN applied in the direction of its length. If all the strains in the other two directions are prevented, determine the contraction in length.
10. A bar is subjected to the following stresses:

$$\sigma_x = 50 \text{ MPa}, \sigma_y = -40 \text{ MPa}, \sigma_z = 70 \text{ MPa}$$

If $\nu = 0.25$ and $E = 200$ GPa, calculate (a) the principal strains, (b) volumetric strain, (c) shear modulus, and (d) bulk modulus.
11. The following strains were measured on pressure vessel:

$$\varepsilon_x = 500 \times 10^{-6}, \varepsilon_y = -100 \times 10^{-6}$$

and at 45° with x-axis, $\varepsilon_z = -200 \times 10^{-6}$ Calculate (a) the principal strains, and (b) principal stresses.

Take $\nu = 0.3$ and $E = 210$ GPa.
12. The state of strain at a point is given by:

$$\varepsilon_x = 400 \times 10^{-6}, \varepsilon_y = -200 \times 10^{-6}, \gamma_{xy} = 250 \times 10^{-6}$$

Calculate (a) the principal strains, and (b) principal planes.
13. The principal strains at a point are given by:

$$\varepsilon_1 = 800 \times 10^{-6}, \varepsilon_2 = -600 \times 10^{-6}, \varepsilon_3 = 400 \times 10^{-6}$$

If $\nu = 0.25$ and E - 200 GPa, calculate the principal stresses.
14. A circle 50 mm in diameter is scribed on a mild steel plate before it is subjected to the following state of stress:

$$\sigma_x = 200 \text{ MPa}, \sigma_y = 100 \text{ MPa}, \tau_{xy} = 80 \text{ MPa}.$$

Calculate the length of the major and minor axes of the ellipse after deformation. Take $\nu = 0.28$ and $E = 210$ GPa.

Chapter-2

Theories of Failure

2.1 STRAIN ENERGY

When an elastic material is loaded, it deforms. To deform the material work has to be done. If the load applied is such that the material remains in the elastic stage then the work of deformation is stored as elastic strain energy.

The strain energy stored per unit volume of material is called *resilience*. The maximum strain energy which can be stored by a material is called *proof resilience*. The proof resilience per unit volume is called *modulus of resilience*.

2.2 EXPRESSIONS FOR STRAIN ENERGY

(a) *Axial loading*. Consider a bar of length L, area of cross-section A, subjected to an axial load P, applied gradually as shown in Fig.2.1 (a).

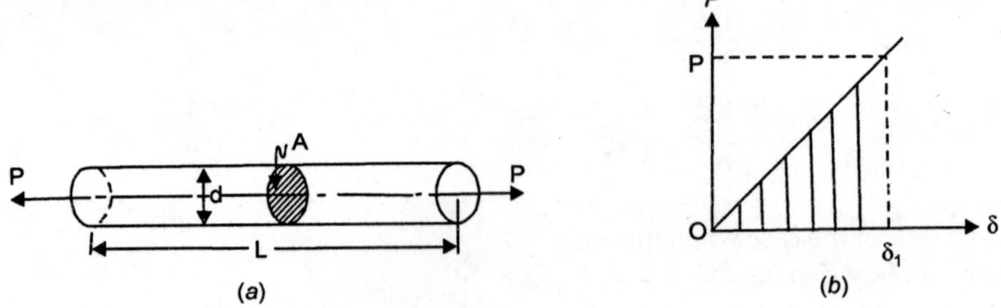

(a) (b)

Fig. 2.1 Bar subjected to axial loading.

Let δ = Extension of the bar

E = Modulus of elasticity

Then $$\delta = \frac{PL}{AE} = \frac{\sigma L}{E}$$

where $$\sigma = \frac{P}{A}$$

Work done in straining the bar [Fig. 2.1 (b)],

$$W = \frac{1}{2} \times \text{Load} \times \text{Extension} = \frac{1}{2}P\delta = \frac{1}{2}P \times \frac{PL}{AE}$$

$$= \frac{1}{2} \times \sigma \times \sigma \times \frac{AL}{E} = \frac{1}{2}\frac{\sigma^2}{E} \times V$$

where V = Volume of the bar = $\frac{\pi}{4}d^2L$

But work done in straining the bar

= Strain energy U stored in the bar

$$\therefore \quad U = \frac{1}{2}\frac{\sigma^2}{E} \times V \quad \text{...(2.1)}$$

Strain energy per unit volume

$$u = \frac{1}{2}\frac{\sigma^2}{E} \quad \text{...(2.2)}$$

If $\sigma = \sigma_{pr}$, the limit of proportionality for the material, then

$$\text{Proof resilience} = \frac{1}{2}\frac{\sigma^2_{pr}}{E} \times V \quad \text{...(2.3)}$$

$$\therefore \text{Modulus of resilience} = \frac{1}{2}\frac{\sigma^2_{pr}}{E} \quad \text{...(2.4)}$$

Also $\sigma = E\varepsilon$

where $\varepsilon = \delta/L$

$$\therefore \quad U = \frac{1}{2}\frac{E^2\varepsilon^2}{E}V = \frac{1}{2}E\varepsilon^2V$$

or

$$U = \frac{1}{2} \times \sigma \times \frac{\sigma}{E} \times V = \frac{1}{2}\sigma\varepsilon V$$

= ½ Stress × Strain × Volume ...(2.5)

Now we consider the case when the area of cross-section of the bar is not uniform as shown in Fig.2.2.

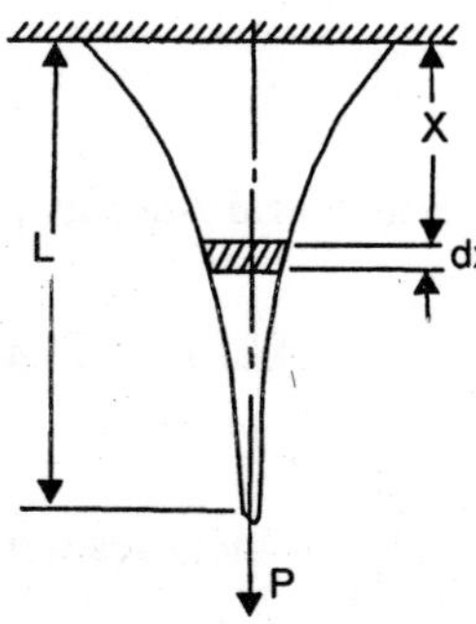

Fig. 2.2 Bar of varying cross-section under tension.

Consider a section at a distance x from the top of the bar.

Let dx be the length of the bar between x and $x + dx$.

Then

$$dU = \frac{1}{2}\frac{\sigma^2}{E} \times A\,dx$$

where A is the area of cross-section of the bar at section x.

Now

$$\sigma = \frac{P}{A}$$

$$dU = \frac{1}{2}\frac{P^2}{AE}dx$$

Total strain energy,

$$\int dU = \frac{1}{2}\int_0^L \frac{P^2}{AE}dx$$

$$U = \frac{1}{2}\frac{P^2}{E}\int_0^L \frac{dx}{A} \quad ...(2.6)$$

(b) *Shear resilience.* Consider a block of length l, width b and height h, rigidly fixed at the base and subjected to a shearing force F as shown in Fig. 2.3

Shearing stress, $\tau = \dfrac{F}{bl}$

Deflection in the direction of F,

$$\delta = \phi h$$

or

$$\delta = \frac{\tau}{G}\cdot h$$

Shear strain energy

$$U = \frac{1}{2}F\delta = \frac{1}{2}\cdot \tau bl \cdot \frac{\tau}{G}h$$

$$= \frac{1}{2}\frac{\tau^2}{G}\cdot blh = \frac{\tau^2}{2G}\cdot Volume \quad ...(2.7)$$

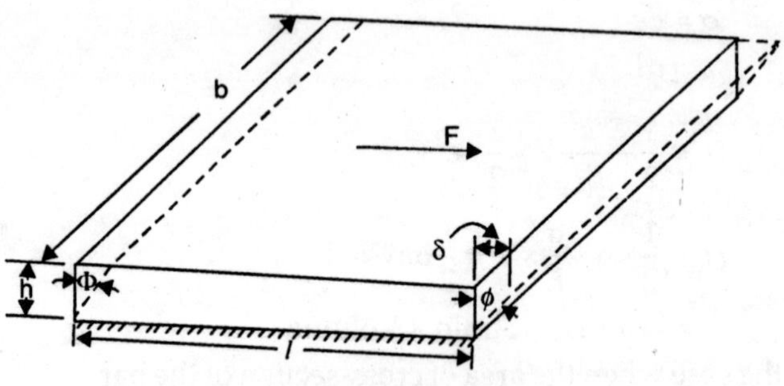

Fig. 2.3 Rectangular block subjected to a shear force.

∴ Shear strain energy per unit volume, *i.e.*

$$\text{Shear resilience} = \frac{\tau^2}{2G} \quad ...(2.8)$$

∴ $$\text{Proof shear resilience} = \frac{\tau^2_{pr}}{2G}\times V \quad ...(2.9)$$

where τ_{pr} = Proof shear stress,

Modulus of shear resilience

$$= \frac{\tau^2_{pr}}{2G} \quad ...(2.10)$$

(c) *Complementary strain energy*. The complementary work performed by a gradually applied force P undergoing a deflection δ is,

$$W_c = \int_0^{P1} \delta\, d\, P \quad ...(2.11)$$

where δ is the component of the total deflection in the direction of P. The complementary work is illustrated by the shaded area as shown in Fig. 2.4. The complementary strain energy of the system U_c, by definition, is equal to the complementary work performed W_c.

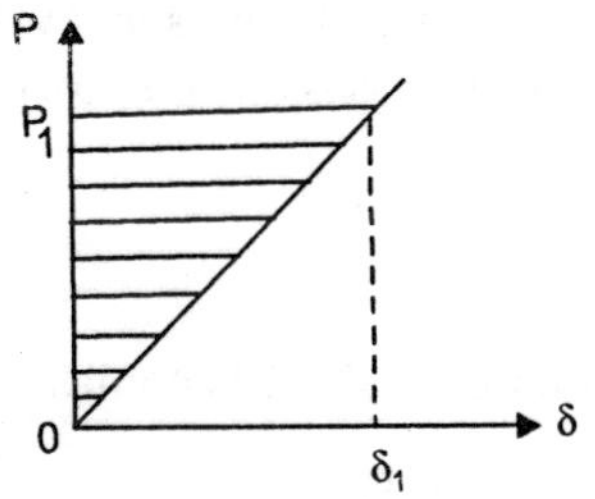

Fig. 2.4

$$U_c = \int_0^{P1} \delta \cdot d\, P \quad ...(2.12)$$

For a linear material $U = U_c$ ...(2.13)

(d) *Elastic strain energy under three-dimensional principal stresses*. Let σ_1, σ_2, σ_3 be the principal stresses acting on a body at a point.

Work done per unit volume by σ_1

$$= \frac{\sigma_1^2}{2E}$$

Work done per unit volume reduced due to lateral strain produced by σ_2 and σ_3 along the σ_1 direction

$$= \frac{\sigma_1}{2}\left\{\frac{\nu}{E}(\sigma_2 + \sigma_3)\right\}$$

$\therefore$ Net work done by σ_1 per unit volume,

$$W_1 = \frac{\sigma_1^2}{2E} - \frac{\nu}{2E}(\sigma_1\sigma_2 + \sigma_1\sigma_3)$$

Similarly net work done by σ_2 and σ_3 per unit volume is,

$$W_2 = \frac{\sigma_2^2}{2E} - \frac{\nu}{2E}(\sigma_2\sigma_3 + \sigma_2\sigma_1)$$

$$W_3 = \frac{\sigma_3^2}{2E} - \frac{v}{2E}(\sigma_3\sigma_1 + \sigma_3\sigma_2)$$

Total work done per unit volume under the influence of all the stresses becomes,

$$W = W_1 + W_2 + W_3$$

$$u = \frac{1}{2E}\left[\sigma_1^2 + \sigma_2^2 + \sigma_3^2 - 2v(\sigma_1\sigma_2 + \sigma_2\sigma_3 + \sigma_3\sigma_1)\right] \quad ...(2.14)$$

Let $\sigma_3 = 0$, then

$$u = \frac{1}{2E}\left(\sigma_1^2 + \sigma_2^2 - 2v\sigma_1\sigma_2\right) \quad ...(2.15)$$

(e) ***Strain energy under pure shear.*** In the case of pure shear we know that across the planes inclined at 45° to the planes of pure shear, normal stresses act as shown in Fig. 2.5.

We know that

$$\sigma_1 = \tau, \sigma_2 = -\tau$$

and $\sigma_3 = 0$

Strain energy per unit volume under pure shear becomes :

$$u = \frac{1}{2E}\left(\sigma_1^2 + \sigma_2^2 - 2v\sigma_1\sigma_2\right) = \frac{1}{2E}\left(\tau^2 + \tau^2 + 2v\tau^2\right)$$

$$= \frac{1}{E}\left(\tau^2 + v\tau^2\right) = \frac{\tau^2}{E}(1+v)$$

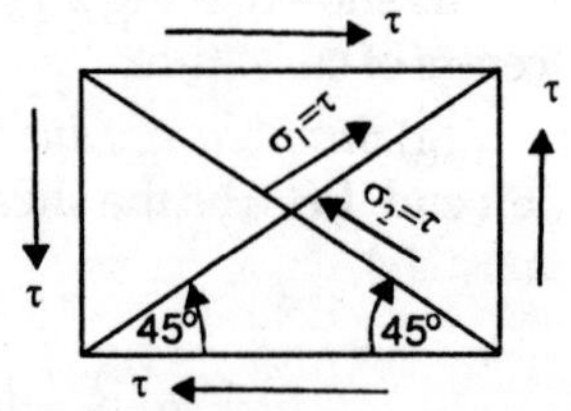

Fig. 2.5 Element under Pure shear.

Now $G = \dfrac{E}{2(1+v)}$

$$u = \frac{\tau^2}{2G} \quad ...(2.16)$$

(*f*) ***Strain energy due to bending.*** Consider a small length dx of the beam at the distance x from left end as shown in Fig. 2.6 (*a*). Let σ be the bending stress on an elementary area dA of the beam at distance y from the neutral axis.

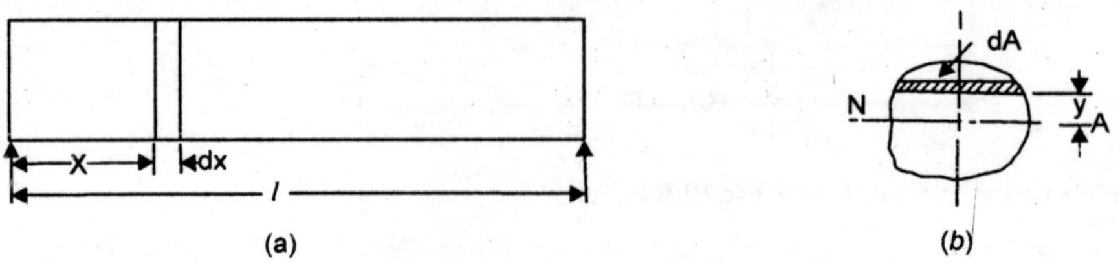

Fig. 2.6 Beam under bending.

Then $$\sigma = \frac{My}{I}$$

The strain energy stored in-the elementary strip due to the bending stress is,

$$dU = \int \frac{\sigma^2}{2E} \times \text{Volume} = \int \frac{M^2 y^2}{2EI^2} \times dx.dA$$

$$= \frac{M^2 dx}{2EI^2} \int y^2 dA = \frac{M^2 dx}{2EI^2} \cdot I = \frac{M^2 dx}{2EI}$$

The flexural strain energy stored in the entire beam is,

$$U = \int_0^l \frac{M^2 dx}{2EI} = \frac{M^2 l}{2EI}$$

$$= \frac{1}{2} M\phi = \frac{\sigma^2}{6E} \times \text{Volume} \qquad ...(2.17)$$

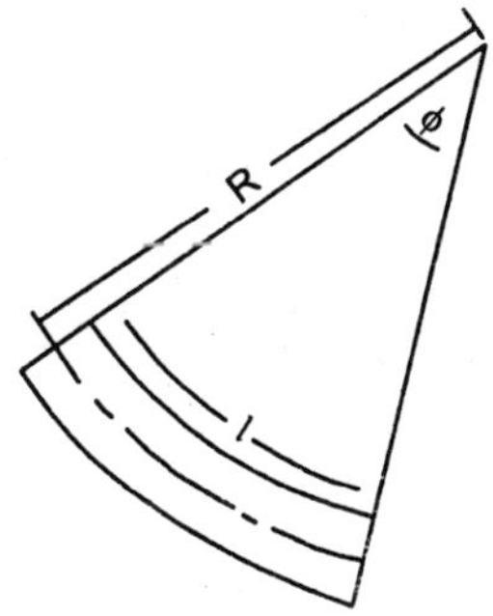

Fig. 2.7

where $\phi = \frac{Ml}{EI} = \frac{l}{R}$

as shown in Fig 2.7, is the angle substented by the bent beam at the centre of curvature.

(g) *Strain energy due to horizontal shear.* Consider a small length dx of the beam at a distance x from left end. Let F be the shear force at this cross-section. The shear stress at a distance y from the neutral axis is

$$\tau = \frac{FA\bar{y}}{Ib}$$

where b = width of the beam at a distance y from the neutral axis

$\bar{y}$ = distance of the centroid of the area above the distance y from the neutral axis as shown in Fig. 2.8.

Fig. 2.8 Horizontal shear in a beam.

Volume of the elementary strip = $bdy.\, dx$

The strain energy stored in the elementary strip due to shear stress τ is :

$$dU = \int \frac{\tau^2}{2G} \cdot bdydx = \int \left(\frac{FA\bar{y}}{Ib}\right)^2 \cdot \frac{1}{2G} bdydx = \left(\frac{1+\nu}{E}\right) \int \left(\frac{FA\bar{y}}{Ib}\right)^2 dxdA$$

$$U = \left(\frac{1+\nu}{EI^2}\right)\int_0^l F^2 dx \iint \frac{(A\bar{y})^2}{b^2} dA \quad \text{...(2.18)}$$

(h) *Strain energy due to torsion.* Consider a shaft of length l subjected to a torque T as shown in Fig. 2.9. Consider a section of the shaft at a distance x from left end. Let τ be the shear stress acting on the elementary area dA at a distance r from the centre. Then the strain energy stored in the elementary strip due to torsion, known as the trosinoal strain energy is given by,

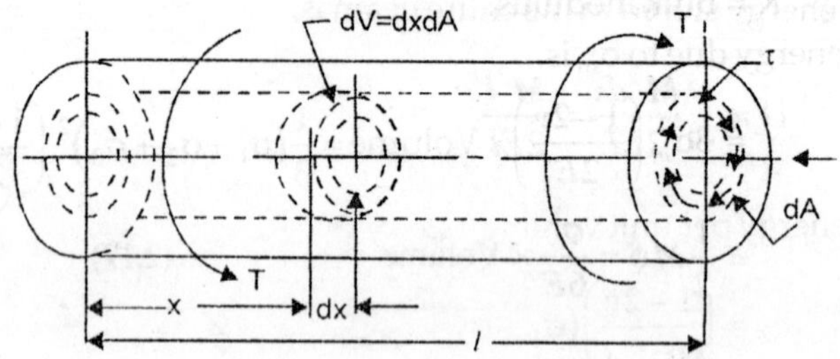

Fig. 2.9 Hollow circular shaft subjected to torsion.

$$dU = \int \frac{\tau^2}{2G} \times \text{Volume} = \int \left(\frac{1+\nu}{E}\right)\left(\frac{Tr}{J}\right)^2 \times \text{Volume}$$

$$= \left(\frac{1+\nu}{E}\right)\int \left(\frac{Tr}{J}\right)^2 dx\, dA = \left(\frac{1+\nu}{E}\right)\frac{T^2 dx}{J^2}\int r^2\, dA$$

$$= \left(\frac{1+\nu}{E}\right)\frac{T^2 dx}{J^2} \cdot J = \left(\frac{1+\nu}{E}\right)\frac{T^2 dx}{J}$$

∴ Total strain energy stored in the shaft of uniform cross-section

$$U = \left(\frac{1+\nu}{E}\right)\int_0^l \frac{T^2 dx}{J} = \left(\frac{1+\nu}{E}\right)\frac{T^2 l}{J} = \frac{T^2 l}{2GJ} \quad \text{...(2.19)}$$

Now $$\theta = \frac{Tl}{GJ}$$

$$U = \frac{1}{2}T\theta \quad \text{...(2.20)}$$

(i) *Shear strain energy due to principal stresses.*

Shear strain energy = Total strain energy – Strain energy due to volumetric strain.

Total strain energy per unit volume due to principal stresses $\sigma_1, \sigma_2, \sigma_3$ is given by (Eq. 2.14)

$$u_1 = \frac{1}{2E}\left[\sigma_1^2 + \sigma_2^2 + \sigma_3^2 - 2\nu(\sigma_1\sigma_2 + \sigma_2\sigma_3 + \sigma_3\sigma_1)\right]$$

Mean stress $\sigma_m = \frac{1}{3}(\sigma_1 + \sigma_2 + \sigma_3)$

If $\sigma_1 = \sigma_2 = \sigma_3 = \sigma$ then

$$u_1 = \frac{1}{2E}\left[3\sigma^2 - 2\nu(3\sigma^2)\right] = \frac{3\sigma^2}{2E}(1-2\nu) = \frac{\sigma^2}{2K} \quad ...(2.21)$$

where K = Bulk modulus.

Volumetric strain energy due to σ_m is

$$= 3\sigma_m^2\left(\frac{1-2\nu}{2E}\right) \times \text{Volume} = \frac{1}{3}(\sigma_1 + \sigma_2 + \sigma_3)^2\left(\frac{1-2\nu}{2E}\right) \times \text{Volume}$$

Volumetric strain energy per unit volume

$$u_2 = \left(\frac{1-2\nu}{6E}\right)(\sigma_1 + \sigma_2 + \sigma_3)^2 \quad ...(2.22)$$

∴ Shear strain energy per unit volume

$$u = u_1 - u_2$$

$$= \frac{1}{2E}\left[\sigma_1^2 + \sigma_2^2 + \sigma_3^2 - 2\nu(\sigma_1\sigma_2 + \sigma_2\sigma_3 + \sigma_3\sigma_1)\right]$$

$$-\left(\frac{1-2\nu}{6E}\right)\left[\sigma_1^2 + \sigma_2^2 + \sigma_3^2 - 2\nu(\sigma_1\sigma_2 + \sigma_2\sigma_3 + \sigma_3\sigma_1)\right]$$

$$= \left(\frac{1+\nu}{6E}\right)\left[(\sigma_1 - \sigma_2)^2 + (\sigma_2 - \sigma_3)^2 + (\sigma_3 - \sigma_1)^2\right]$$

$$= \frac{1}{12G}\left[(\sigma_1 - \sigma_2)^2 + (\sigma_2 - \sigma_3)^2 + (\sigma_3 - \sigma_1)^2\right] \quad ...(2.23)$$

If $\sigma_2 = \sigma_3 = 0$, then

$$u = \frac{\sigma_1^2}{6G} \quad ...(2.24)$$

2.3 STRESSES DUE TO VARIOUS TYPES OF LOADING

(a) *Gradually applied load.* Suppose that a load P is applied gradually to a material. When the load P has been applied to the material, the material deforms and stress σ is introduced in the material.

Potential energy lost by the load

= Strain energy stored *in* the material.

Referring to Fig. 2.10, we get

$$\therefore \quad \frac{1}{2}Px = \frac{1}{2}\sigma A x$$

$$\therefore \quad \sigma = \frac{P}{A} \quad ...(2.25)$$

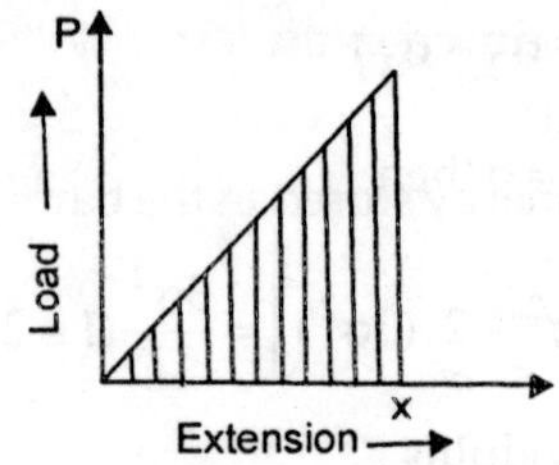

Fig. 2.10 Gradually applied load.

(b) *Suddenly applied load.* Let a tensile load P be applied to a material suddenly and produce an extension x. The work done will be given by Px. Suppose P_1 to be a load which increases gradually from zero and produces the same extension x, the work done now is given by $\frac{1}{2} P_1 x$. (See Fig. 2.11)

Since the bar is strained an equal amount in each case, the strain energy is the same in each case, and consequently the applied energies will be equal,

$$\therefore \quad \frac{1}{2} P_1 x = Px$$

$$P_1 = 2P \qquad \text{...(2.26)}$$

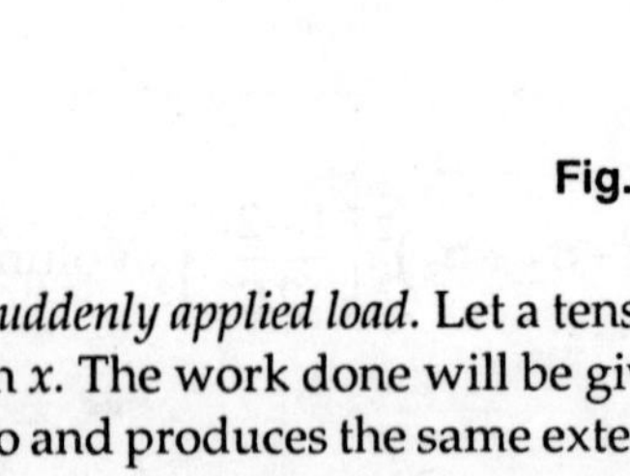

Fig. 2.11 Suddenly applied load.

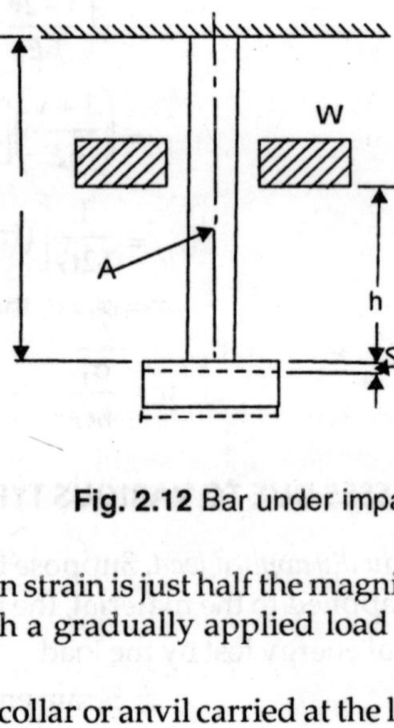

Fig. 2.12 Bar under impact.

Thus a suddenly applied load required to produce a given strain is just half the magnitude of that required, for the same strain, when gradually applied. Such a gradually applied load is called an *equivalent load.*

(c) *Impact or shock load.* Let the weight W fall freely on to a collar or anvil carried at the lower end of a bar of area of cross-section A, length l, fixed at the upper end, through a height h, as shown in Fig. 2.12. The following assumptions are made :

(i) The weight W falls freely.

(ii) The support is rigid,

(iii)There is no loss of energy on striking the anvil or collar due to rebound. This assumption is not true in practice.

Let x = extension of the bar

σ = stress induced in the bar

Potential energy lost by the weight

= Strain energy stored in the bar

$$\therefore \quad W(h+x) = \frac{1}{2}\,\sigma A x$$

Now
$$x = \frac{\sigma l}{E}$$

$$W\left(h+\frac{\sigma l}{E}\right) = \frac{1}{2}\cdot\sigma\cdot A\cdot\frac{\sigma l}{E}$$

or
$$\sigma^2\left(\frac{Al}{2E}\right)-\sigma\left(\frac{WI}{E}\right)-Wh=0$$

$$\therefore \quad \sigma = \frac{\dfrac{Wl}{E}\pm\sqrt{\left(\dfrac{Wl}{E}\right)^2+4\left(\dfrac{Al}{2E}\right)Wh}}{\dfrac{Al}{E}} = \frac{\dfrac{WI}{E}\pm\sqrt{\left(\dfrac{Wl}{E}\right)^2+2\left(\dfrac{Al}{E}\right)\times Wh}}{\dfrac{Al}{E}}$$

$$= \frac{W}{A}+\sqrt{\left(\frac{W}{A}\right)^2+\frac{2E}{Al}\times Wh} = \frac{W}{A}\left[1+\sqrt{1+\frac{2AEh}{Wl}}\right]$$

Maximum stress,
$$\sigma_{max} = \frac{W}{A}\left[1+\sqrt{1+\frac{2AEh}{Wl}}\right] \qquad \text{...(2.27)}$$

If $h = 0$, then

$$\sigma_{max} = \frac{2W}{A} \qquad \text{...(2.28)}$$

(d) *Impact on a simply supported beam.* Let a weight W fall on a simply supported beam of length L at the mid-span through a height h as shown in Fig. 2.13. Let δ be the deflection produced in the beam. Then the work done by W is $W(h+\delta)$, provided the resultant stress produced in the beam is within the elastic limit.

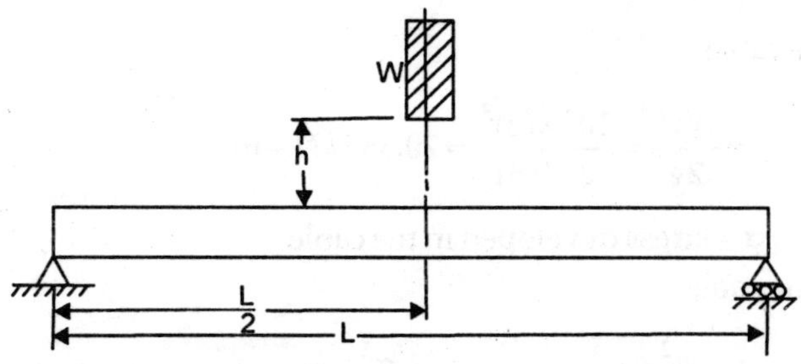

Fig. 2.13 Simply supported beam under the impact loading

If W_1 is the equivalent static load at mid-span to produce the deflection δ, then work done by W_1 is ½W_1δ. The strain energy produced in the beam is the same in each case.

$$\therefore \quad \frac{1}{2}W_1\delta = W(h+\delta)$$

Now
$$\delta = \frac{1}{48}\frac{W_1L^3}{EI}$$

$$\therefore \quad \frac{W_1^2L^3}{96EI} = W\left(h+\frac{W_1L^3}{48EI}\right)$$

$$W_1^2\frac{L^3}{96\,EI} - W_1\frac{WL^3}{48EI} - Wh = 0$$

$$W_1^2 - 2\,WW_1 - \frac{96WHEI}{l^3} = 0$$

$$W_1 = W + \sqrt{W^2 + \frac{96WhEI}{L^3}} \qquad \text{...(2.29)}$$

If $h = 0$, then $W_1 = 2W$ ...(2.30)

Maximum bending moment,

$$M_{max} = \frac{W_1L}{4}$$

Bending stress,

$$\sigma_b = \frac{M_{max}\,y}{I}$$

Example 2.1 *A trolly of weight 100 kN is descending a slope with a uniform velocity of 2 m/s at the end of a steel cable which is wound round a drum. When the length of the cable laid out is 600 m, emergency brakes are suddenly applied and the wagon is brought to a halt. If the cross-sectional area of the cable is 6 sq. cm and E = 200 GPa, find the stress developed in the cable. What would be the stress if the length of the cable laid out is 240 m?*

Solution.

Kinetic energy in the cable

$$= \frac{Wv^2}{2g} = \frac{10^5\times(2)^3}{2\times 9.81} = 20.387\ kN-\text{m}$$

Let σ = stress developed in the cable.

∴ Strain energy in the cable

$$= \frac{\sigma^2}{2E}\times\text{Volume} = \frac{\sigma^2}{2\times200\times10^9}\times6\times600\times10^{-4}$$

$= 9 \times 10^{-13} \times \sigma^2$ N . m

Neglecting losses, due to suddenly stopping, the whole of the kinetic energy is absorbed by the cable.

$$9 \times 10^{-13} \sigma^2 = 20.387 \times 10^3$$

$$\sigma = 150.51 \text{ MPa}$$

When length of the cable is 240 m, then

$$\frac{\sigma^2}{2 \times 200 \times 10^9} \times 6 \times 600 \times 10^{-4} = 20.387 \times 10^3$$

$$\sigma = 237{\cdot}97 \text{ MPa}$$

Example 2.2 *A vertical tie rod rigidly fixed at the top end consists of a steel rod 2.5 m long and 20 mm diameter is encased throughout in a brass tube 20 mm internal diameter and 30 mm external diameter. The rod and the casing are fixed together at both ends. The compound bar is suddenly loaded in tension by a weight of 10 kN falling freely through 3 mm before being arrested by the tie. Calculate the maximum stresses in steel and brass.*

$$E_s = 200 \text{ GPa}, E_b = 100 \text{ GPa}$$

Solution.

Let x = extension of the bar in m. (Fig. 2.14)

Then strain

$$\varepsilon_s = \varepsilon_b = \frac{x}{2{\cdot}5}$$

$$\sigma_s = E_s \cdot \varepsilon_s = 200 \times 10^9 \times \frac{x}{2{\cdot}5}$$

$$= 80x \times 10^9 \text{ Pa}$$

$$\sigma_b = E_b \times \varepsilon_b = 100 \times 10^9 \times \frac{x}{2{\cdot}5}$$

$$= 40x \times 10^9 \text{ Pa}$$

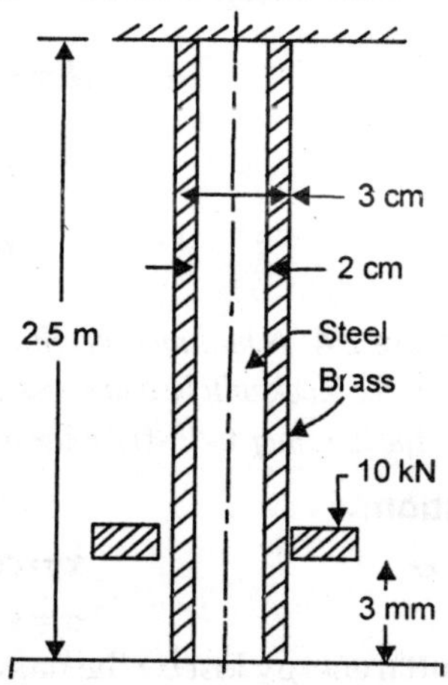

Fig. 2.14

Potential energy lost by weight

$$= 10^4 (3 \times 10^{-3} + x) \text{ N-m} \quad \text{...(a)}$$

Strain energy stored in the bar

$$= \frac{\sigma_s^2}{2E_s} \times \frac{\pi}{4} \times 2^2 \times 2.5 \times 10^{-4}$$

$$+ \frac{\sigma_b^2}{2E_b} \times \frac{\pi}{4}(9-4) 2.5 \times 10^{-4}$$

$$= \frac{(80x \times 10^9)^2}{2 \times 200 \times 10^9} \times \pi \times 2{\cdot}5 \times 10^{-4} + \frac{(40x \times 10^9)^2}{2 \times 100 \times 10^9} \times \frac{\pi \times 5 \times 2{\cdot}5 \times 10^{-4}}{4}$$

$$= 12{\cdot}567 \times 10^6 \times x^2 + 7{\cdot}854 \times 10^6 \times x^2 = 20{\cdot}421 \times 10^6 \times x^2 \text{ N-m} \quad \text{...(b)}$$

Equating Eqs. (*a*) and (*b*), we get

or $20{\cdot}421 \times 10^6 \times x^2 = 10^4 (3 \times 10^{-3} + x)$

or $2042{\cdot}1 \times x^2 = 3 \times 10^{-3} + x$

or $2042{\cdot}1\, x^2 - x - 3 \times 10^{-3} = 0$

$$x = \sqrt{\frac{1+24{\cdot}505}{2\times 2042{\cdot}1}} = 0{\cdot}0014814\,\text{m}$$

$$\sigma_s = 80 \times 0{\cdot}0014814 \times 10^9 = 118{\cdot}51\text{ MPa}$$

$$\sigma_b = 40 \times 0{\cdot}0014814 \times 10^9 = 59{\cdot}256\text{ MPa}$$

Example 2.3 *A flat ribbon of spring steel 3·2 mm wide and 0·5 mm thick is wound around a cylinder 50 cm in diameter. Find the maximum stress in the steel ribbon and the energy in bending stored per metre length of the ribbon. Take E = 220 GPa.*

Solution.

Now $$\frac{\sigma}{y} = \frac{E}{R}$$

$$\sigma = \frac{Ey}{R} = \frac{220\times 10^9 \times 0{\cdot}025}{25} = 220\text{ MPa}$$

Strain energy stored per metre length,

$$U = \frac{\sigma^2}{6E} \times \text{volume}$$

$$= \frac{\left(220\times 10^6\right)^2}{6\times 220\times 10^9} \times 3{\cdot}2 \times 0{\cdot}5 \times 1\times 10^{-6} = 58{\cdot}67 \times 10^{-3}\text{ N·m}$$

Example 2.4 *A hammer weighing 100 N falls 1·25 m on a 5 cm cube iron block before coming to rest. Find the distance the block will be compressed and the instantaneous stress induced in it. Also determine the velocity with which the hammer will strike the block. E = 200 GPa.*

Solution.

Let x = compression of the block in m

σ = stress induced in the block, N/m²

Potential energy lost by the hammer

$$= 100\,(1{\cdot}25 + x)\text{ N-m} \qquad \text{... (a)}$$

Strain energy stored in the block

$$U = \frac{\sigma^2}{2E} \times \text{volume}$$

$$U = \frac{\sigma^2}{2\times 200\times 10^9} \times 125 \times 10^{-6} = 31{\cdot}25 \times 10^{-17} \times \sigma^2 \qquad \text{...(b)}$$

And $$x = \frac{\sigma l}{E} = \frac{\sigma\times 5\times 10^{-2}}{200\times 10^9} = 2{\cdot}5 \times 10^{-13} \times \sigma\text{ m}$$

Equating Eqs. (*a*) and (*b*), we get

$$100(1{\cdot}25+2{\cdot}5\times 100^{-13}\times\sigma) = 31{\cdot}25\times 10^{-17}\times\sigma^2$$

$$\sigma^2 - 8\times 10^4\times\sigma - 4\times 10^{17} = 0$$

$$\sigma = \frac{[8\times 10^4 + 10^4\sqrt{64+16\times 10^9}\,]}{2} = 632{\cdot}5 \text{ MPa}$$

$$x = 2{\cdot}5\times 10^{-13}\times 632{\cdot}5\times 10^6 = 158{\cdot}125\times 10^{-6}\text{ m}$$

Let v = Velocity of the hammer.

Then
$$W(h+x) = \frac{1}{2}\cdot\frac{W}{g}\cdot v^2$$

$$100\,(1{\cdot}25 + 158{\cdot}125\times 10^{-6}) = \frac{1}{2}\times\frac{100}{9.81}\times v^2$$

$$v^2 = \frac{100\times 2\times 9.81\times 1.250158}{100} = 24{\cdot}528$$

$$v = 4{\cdot}953 \text{ m/s}$$

Example 2.5 A bar 3 *m long, 3 cm diameter hangs vertically and has a collar securely attached at the lower end. Find the maximum stress induced, when*

(a) a *weight of* 2500 *N falls* 12·5 *cm on the collar, and*
(b) *a weight of* 25 *kN falls* 1·25 *cm on the collar.*

$$E = 200 \text{ GPa}$$

Solution.

Work done on the bar = strain energy stored in the bar

Let x = extension of the bar, m

σ = stress induced in the bar, Pa

$$W(h+x) = \frac{\sigma^2}{2E}\times\text{Volume of bar}$$

$$\text{(a) } 2500\,(12{\cdot}5\times 10^{-2}+x) = \frac{\sigma^2}{2\times 200\times 10^9}\times\frac{\pi}{4}\times 25\times 10^{-4}\times 3$$

$$x = \frac{\sigma l}{E} = \frac{\sigma\times 3}{200\times 10^9}\text{ m}$$

$$\therefore\quad 2500\left(0{\cdot}125+\frac{3\sigma}{200\times 10^9}\right) = 14{\cdot}7262\times 10^{-15}\times\sigma$$

$$0{\cdot}125 + 1{\cdot}5\times 10^{-11}\times\sigma = 5{\cdot}89\times 10^{-18}\times\sigma^2$$

$$5{\cdot}89\times 10^{-18}\times\sigma^2 - 1{\cdot}5\times 10^{-11}\times\sigma - 0{\cdot}125 = 0$$

Solving for σ, we get $\sigma = 146{\cdot}95$ MPa

(b) $25 \times 10^3 \left(1{\cdot}25\times10^{-2}+\dfrac{3\sigma}{200\times10^9}\right)=14{\cdot}7262\times10^{-15}\times\sigma^2$

$0{\cdot}0125 + 1{\cdot}5 \times 10^{-11} \times \sigma = 5{\cdot}89 \times 10^{-19} \times \sigma^2$

or $\quad 5{\cdot}89 \times 10^{-19} \times \sigma^2 - 1{\cdot}5 \times 10^{11} \times \sigma - 0{\cdot}0125 = 0$

Solving for σ, we get $\sigma = 158{\cdot}96$ MPa

Example 2.6 *A bar 1·2 cm internal diameter and 2 cm external diameter gets stretched by 0·3 cm under a steady load of 8 kN.*

What stress would be produced in the bar by a weight of 800 N which falls 8 cm before commencing the stretching of the rod which is initially unstressed? E = 200 GPa.

Solution.

Under a steady load,

$$\sigma = \frac{W}{A} = \frac{xE}{l}$$

$$\frac{l}{A} = \frac{xE}{W} = \frac{0{\cdot}3\times10^{-2}\times200\times10^9}{8\times10^3} = 7{\cdot}5\times10^4\ \text{m}^{-1}$$

$$l = 7{\cdot}5\times10^{-4}\times\frac{\pi}{4}[2^2-(1{\cdot}2)^2]\times10^{-4} = 15{\cdot}08\ \text{m}$$

Under impact loading,

$$\sigma = \frac{W}{A}\left[1+\sqrt{1+\frac{2AEh}{Wl}}\right]$$

$$= \frac{800\times4}{\pi\times2{\cdot}56\times10^{-4}}\times\left[1+\sqrt{1+\frac{2\times\pi\times2{\cdot}56\times10^{-4}\times200\times10^9\times8\times10^{-2}}{4\times800\times15{\cdot}08}}\right]$$

$$= 95{\cdot}927\ \text{MPa}$$

Example 2.7 *A bar 120 cm long is 2 cm² in cross-sectional area for 60 cm of its length and 1·5 cm² for the remaining 60 cm. If a load of 150 N falls on the collar which is provided at one end of the rod, the other being fixed, from a height of 5 cm: find the maximum stress induced in the bar. E = 200 GPa.*

Solution.

Let x be the total extension of the bar shown in Fig. 2.15 and W_1 equivalent gradually applied load to produce the same stress as due to the impact load W.

Now $$x = \frac{W_1 l_1}{A_1 E} + \frac{W_1 l_2}{A_2 E}$$

$$= W_1\left[\frac{60\times10^{-2}}{2\times10^{-4}\times200\times10^9}+\frac{60\times10^{-2}}{1{\cdot}5\times10^{-4}\times200\times10^9}\right]$$

$$= 0{\cdot}35\times10^{-7}\times W_1\ \text{m}$$

Also $$W(h+x) = \frac{1}{2} W_1 x$$

$$150\,(0{\cdot}05 + 0{\cdot}35 \times 10^{-7} \times W_1)$$

$$= \frac{1}{2} W_1 \times 0{\cdot}35 \times 10^{-7} \times W_1$$

$$W_1^2 - 300W_1 - 4{\cdot}2857 \times 10^8 = 0$$

$$W_1 = \frac{300 + 10^2\sqrt{9 + 17{\cdot}1428 \times 10^4}}{2}$$

$$= \frac{300 + 41405}{2} = 20852{\cdot}5\ N$$

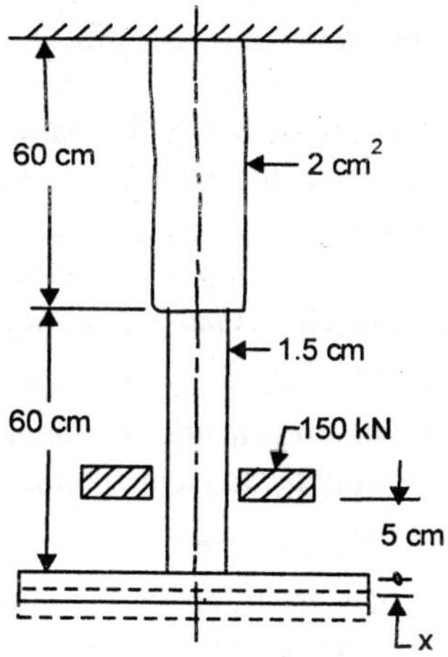

Fig. 2.15

Maximum stress will occur in smaller cross-section

$$\therefore \quad \sigma_{max} = \frac{W_1}{1{\cdot}5} = \frac{20852{\cdot}5}{1{\cdot}5 \times 10^{-4}} = 139{\cdot}02 \text{ MPa}$$

Example 2.8 *A static concentrated load of 8 kN applied at the mid span of a simply supported beam produces a maximum deflection of 0·5 cm and a maximum bending stress of 12 MPa. Calculate the instantaneous stress produced in the beam when a load of 5 kN is allowed to fall through a height of 2·5 cm at the mid-span of the beam.*

Solution.

For a simply supported beam carrying a concentrated load at mid-span, the central deflection,

$$\delta = \frac{Wl^3}{48EI}$$

$$0{\cdot}5 \times 10^{-2} = \frac{800 \times l^3}{48EI}$$

$$\frac{l^3}{EI} = \frac{48 \times 0{\cdot}5 \times 10^{-2}}{8000} = 3 \times 10^{-5}$$

Now equivalent static load,

$$W_1 = W + \sqrt{W^2 + \frac{96\,W\,h\,EI}{l^3}} = 5 + \sqrt{(5)^2 + \frac{96 \times 5 \times 2{\cdot}5 \times 10^{-2}}{3 \times 10^{-5} \times 10^3}}$$

$$= 5 + \sqrt{25 + 400} = 5 + 20{\cdot}6155 = 25{\cdot}6155 \text{ kN}$$

$$\therefore \quad \text{Stress produced} = \frac{12}{8} \times 25{\cdot}6155 = 38{\cdot}423 \text{ MPa}$$

2.4 THEORIES OF ELASTIC FAILURE

A material is said to have failed when it is stressed beyond the elastic limit and permanent deformation occurs. Failure of a material represents either direct separation of particles from each other' (brittle fracture) or slipping of particles (ductile fracture or yielding) accompanied by considerable plastic deformations. Failure does not necessarily mean rupture. If the material is subjected to a single type of stress (*i.e.* axial, bending or torsional), then it is very easy to anticipate when the failure is likely to occur. If, however the material is subjected to a complex stress system then it is not easy to predict the failure of the material straight away. In order to do this, the following theories of failure have been put forward:

(i) The maximum principal stress theory.
(ii) The maximum principal strain theory.
(iii) Maximum shear stress theory.
(iv) Maximum strain energy theory.
(v) Maximum shear strain energy theory.
(vi) Mohr's theory.

Let $\sigma_1, \sigma_2, \sigma_3$ be the three principal stresses at a point in a body taken in descending order and σ is the magnitude of the direct stress in simple tension at the elastic limit.

(i) *The maximum principal stress theory or Rankine Theory.* According to this theory failure of the material is assumed to have taken place under a state of complex stresses when the value of the maximum principal stress σ_1 reaches, a value equal to that of the elastic limit stress σ as found in a simple tensile test, *i.e.*

$$\sigma_1 = \sigma \quad \text{... (2.31)}$$

This theory has been found to predict the failure behaviour of brittle materials, like cast iron quite satisfactorily. The elastic limit stress in tension and compression is assumed to be the same.

(ii) *The maximum principal strain theory or St. Venant's theory.* According to this theory, failure of the material is deemed to have taken place when the maximum principal strain reaches a value calculated from a simple tensile test. Thus

$$\frac{1}{E} = [\sigma_1 - \nu(\sigma_2 + \sigma_3)] = \frac{\sigma}{E}$$

or $$\sigma = \sigma_1 - \nu\,(\sigma_2 + \sigma_3) \quad \text{...(2.32)}$$

when $$\sigma_3 = 0$$

$$\sigma = \sigma_1 - \nu\sigma_2 \quad \text{...(2.33)}$$

This theory over estimates the behaviour of ductile materials.

(iii) *Maximum shear stress theory or Guest-Coulomb's theory or Tresca's theory.* According to this theory the failure of the material is deemed to have taken place when the maximum shear stress exceeds the maximum shear stress in a simple tensile test, *i.e.*

$$\frac{\sigma_1 - \sigma_3}{2} = \frac{\sigma}{2}$$

$$\sigma_1 - \sigma_3 = \sigma \quad \text{...(2.34)}$$

This theory has been found to give quite satisfactory results for ductile materials. There is a serious drawback in this theory, because the effect of the intermediate stress is not taken into account at all.

(iv) *Maximum strain energy theory or Beltrami-Haigh's theory.* According to this theory failure is assumed to take place when the total strain energy exceeds the strain energy determined from a simple tensile test. Thus

$$\frac{1}{2E}\left[\sigma_1^2+\sigma_2^2+\sigma_3^2-2\nu(\sigma_1\sigma_2+\sigma_2\sigma_3+\sigma_3\sigma_1)\right]=\frac{\sigma^2}{2E}$$

$$\sigma_1^2+\sigma_2^2+\sigma_3^2-2\nu(\sigma_1\sigma_2+\sigma_2\sigma_3+\sigma_3\sigma_1)=\sigma^2 \qquad ...(2.35)$$

This theory has been found to give good results for ductile materials.

(v) *Maximum shear strain energy or Distortion energy theory or Mises-Hencky theory.* According to this theory failure is assumed to take place when the maximum shear-strain energy exceeds the shear strain energy in a simple tensile test, *i.e.*

$$\frac{1}{12G}\left[(\sigma_1-\sigma_2)^2+(\sigma_2-\sigma_3)^2+(\sigma_3-\sigma_1)^2\right]=\frac{\sigma^2}{6G}$$

or

$$(\sigma_1-\sigma_2)^2+(\sigma_2-\sigma_3)^2+(\sigma_3-\sigma_1)^2=2\sigma^2 \qquad ...(2.36)$$

This theory gives very satisfactory results for ductile materials.

(vi) *Mohr's theory.* Mohr has proposed a theory to predict the failure of brittle materials based on his stress circle.

In Fig. 2.16 the circle on diameter *OA* is that of pure tension, the circle on diameter OB that for pure compression and the circle with centre *O* and diameter *CD* is that for pure shear. An envelope to these circles then represents the failure envelope according to the Mohr theory. A failure condition is then indicated when a stress circle for a particular complex stress condition is found to cut the envelope. As a close approximation to this Mohr suggested that only the pure tension and pure compression failure circles need be drawn with *OA* and *OB* equal to the yield strengths σ_{yt} and σ_{yc} in tension and compression respectively. Common tangents to these circles may then be used as the failure envelope. Circles drawn tangent to this envelope then represent the condition of failure at the point of tangency.

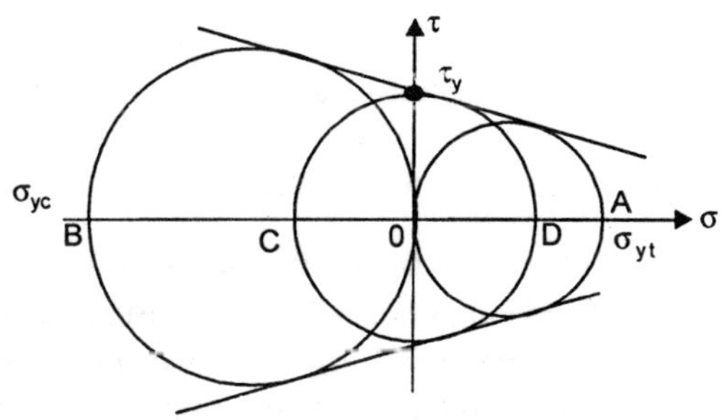

Fig.2.16 Mohr's theory on $\sigma-\tau$ axes

An expression for Mohr's theory may be derived as follows with principal stresses σ_1 and σ_2. From the geometry of the Fig. 2.17,

$$\frac{AC}{AB}=\frac{HC}{GB}$$

Now $AC = AO + CO = AO + (CK - OK)$

$$=\frac{1}{2}\sigma_{yt}+\frac{1}{2}(\sigma_1+\sigma_2)-\sigma_1$$

$$=\frac{1}{2}(\sigma_{yt}+\sigma_2-\sigma_1)$$

$$AB = OA + OB = \frac{1}{2}(\sigma_{yt}+\sigma_{yc})$$

$$HC = FC - HF = FC - AD$$

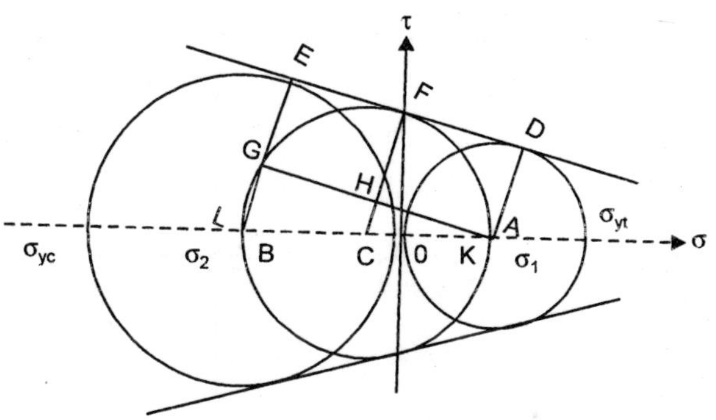

Fig. 2.17

$$= \frac{1}{2}(\sigma_1 + \sigma_2) - \frac{1}{2}\sigma_{yt}$$

$$= \frac{1}{2}(\sigma_1 + \sigma_2 - \sigma_{yt})$$

$$GB = BE - GE = BE - AD$$

$$= \frac{1}{2}\sigma_{yc} - \frac{1}{2}\sigma_{yt} - \frac{1}{2}(\sigma_{yc} - \sigma_{yt})$$

$$\frac{\frac{1}{2}\left(\sigma_{yt} + \sigma_2 - \sigma_1\right)}{\frac{1}{2}\left(\sigma_{yt} + \sigma_{yc}\right)} = \frac{\frac{1}{2}\left(\sigma_1 + \sigma_2 - \sigma_{yt}\right)}{\frac{1}{2}\left(\sigma_{yc} - \sigma_{yt}\right)}$$

Simplifying, we get

$$\frac{\sigma_1}{\sigma_{yt}} + \frac{\sigma_2}{\sigma_{yc}} = 1 \qquad \text{... (2.37)}$$

which is the modified Mohr's shear stress theory for brittle materials.

2.4.1 Graphical Representation.

The theories of failure may be represented graphically as explained below:

(i) *Maximum principal stress theory.* This theory states that failure will occur when σ_1 or $\sigma_2 = \sigma_t$ or σ_c. Assuming $\sigma_t = \sigma_c = \sigma$, these conditions are represented graphically on σ_1, σ_2 coordinates as shown in Fig. 2.18. Any point having coordinates (σ_1, σ_2) falling outside the square will indicate failure.

(ii) *Maximum shear stress theory.* According to this theory $\sigma_1 = \sigma$ when $\sigma_2 = 0$ and $\sigma_2 = \sigma$ when $\sigma_1 = 0$. When σ_2 is tensile and σ_1 is compressive, *i.e.* in the second quadrant. this theory becomes

$$\sigma_2 + \sigma_1 = \sigma$$

and in the fourth quadrant $\quad \sigma_1 + \sigma_2 = \sigma$

These two equations represent straight lines.

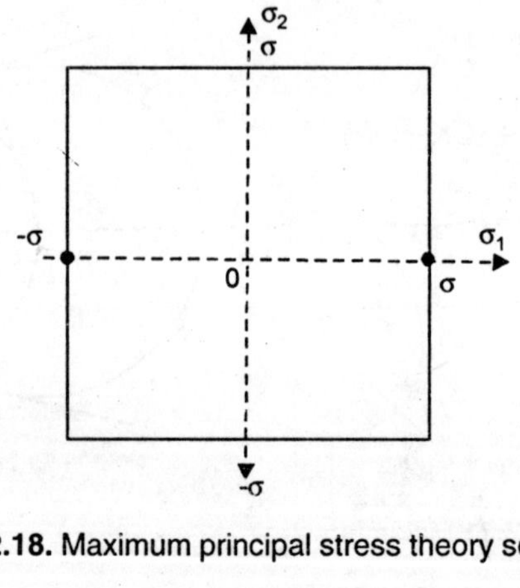

Fig. 2.18. Maximum principal stress theory square.

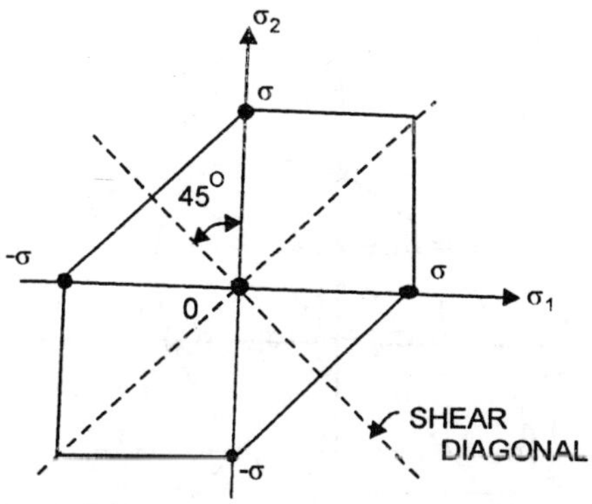

Fig. 2.19 Maximum shear stress theory hexagon.

In the third quadrant both the stresses are negative. Hence the curve is again a square as in the first quadrant. This failure theory is shown in Fig. 2·19.

(iii) *Maximum principal strain theory.* For yielding in tension this theory states that

$$\sigma_1 - \nu\sigma_2 = \sigma$$

and for compressive yield, with σ_2 compressive

$$\sigma_2 - \nu\sigma_1 = \sigma$$

These equations when plotted produce the rhomboid failure envelope as shown in Fig. 2·20

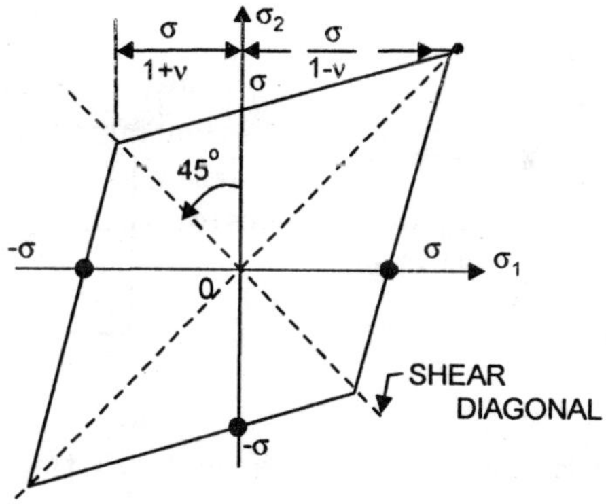

Fig. 2.20 Maximum principal strain theory rhomboid.

(iv) *Maximum strain energy theory.*

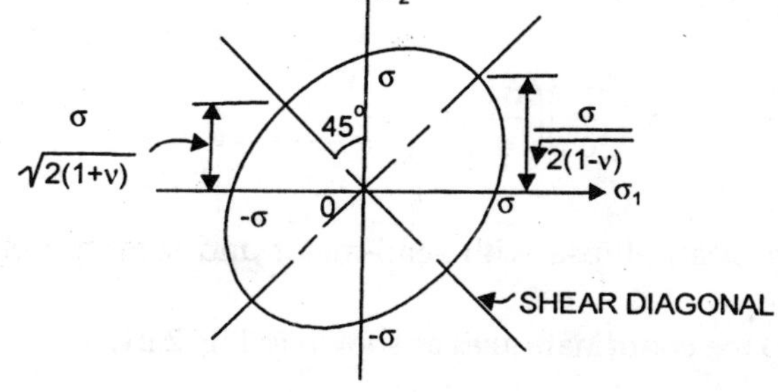

Fig. 2.21 Maximum strain energy theory ellipse.

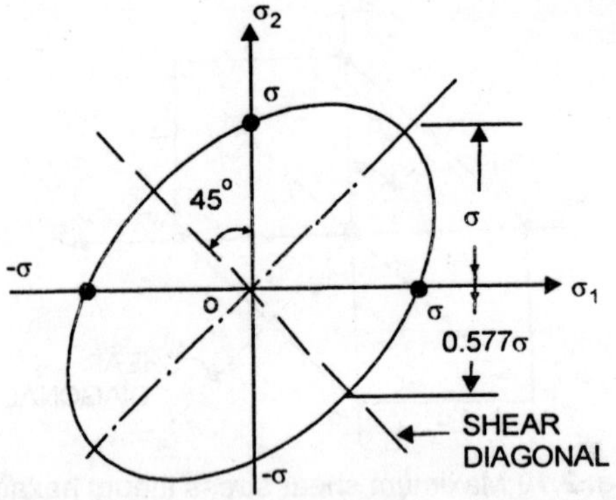

Fig. 2.22 Maximum shear strain energy theory ellipse.

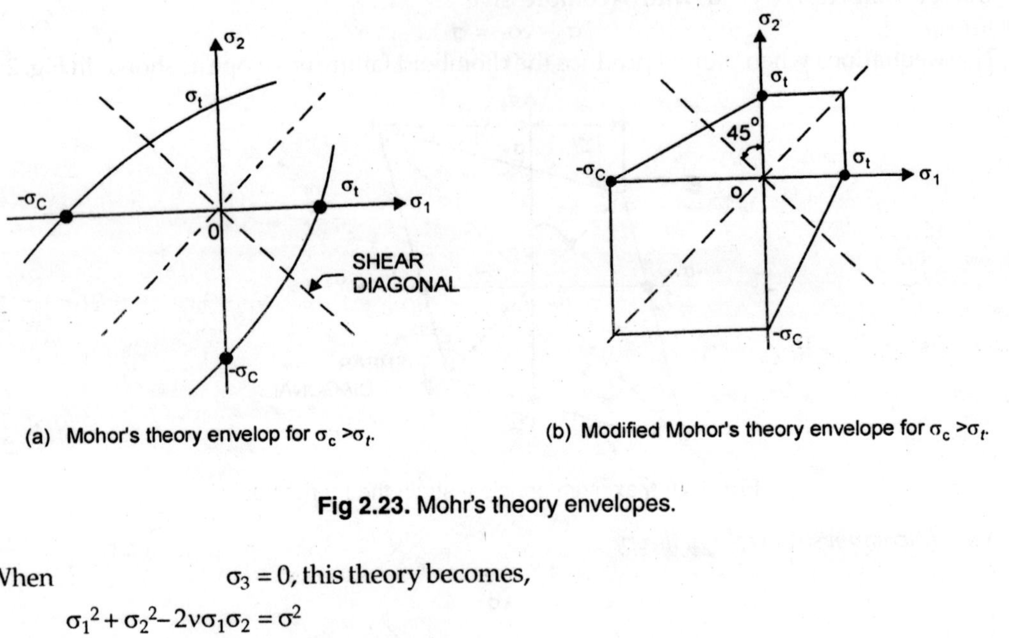

(a) Mohor's theory envelop for $\sigma_c > \sigma_t$.

(b) Modified Mohor's theory envelope for $\sigma_c > \sigma_t$.

Fig 2.23. Mohr's theory envelopes.

When $\sigma_3 = 0$, this theory becomes,

$$\sigma_1^2 + \sigma_2^2 - 2\nu\sigma_1\sigma_2 = \sigma^2$$

or
$$\left(\frac{\sigma_1}{\sigma}\right)^2 + \left(\frac{\sigma_2}{\sigma}\right)^2 - 2\nu\left(\frac{\sigma_1}{\sigma}\right)\left(\frac{\sigma_2}{\sigma}\right) = 1$$

This is the equation of an ellipse with semi-major and semi-minor axes $\frac{\sigma}{\sqrt{(1-\nu)}}, \frac{\sigma}{\sqrt{(1+\nu)}}$ respectively each at45° to the coordinate axes as shown in Fig. 2.21.

(v) *Maximum shear strain energy theory.*

When $\sigma_3 = 0$, this theory becomes,

or
$$\left(\frac{\sigma_1}{\sigma}\right)^2 + \left(\frac{\sigma_2}{\sigma}\right)^2 - \left(\frac{\sigma_1}{\sigma}\right)\left(\frac{\sigma_2}{\sigma}\right) = 0$$

This is the equation of an ellipse with semi-major and semi- minor axes $\sqrt{2}\sigma$, and $\sqrt{2/3}\sigma$ respectively at 45° to the coordinate axes as shown in Fig. 2.22.

(vi) *Mohr's theory.* Based on the results of pure tension, pure compression and pure shear the Mohr's theory envelope is shown in Fig, 2.23 (a). In its modified form, based on pure tension and pure compression, the failure envelope is shown in Fig. 2.23 (b).

Example 2.9 *The principal stresses in a material are* 50 *MPa,* 40 *MPa and* – 30 *MPa. Calculate:* (a) *total strain energy,* (b) *volumetric strain energy, and* (c) *shear strain energy. If the material yields at* 100 *MPa, calculate the factor of safety on the total strain energy criterion.*

$$E = 200\ GPa,\ \nu = 0{\cdot}30.$$

Solution.

(*a*) Total strain energy per unit volume is :

$$\frac{1}{2E}\left[\sigma_1^2 + \sigma_2^2 + \sigma_3^2 - 2\nu(\sigma_1\sigma_2 + \sigma_2\sigma_3 + \sigma_3\sigma_1)\right]$$

$$= \frac{10^{14}}{[illegible]}[25 + 16 \ [illegible] \ 15)]$$

$$= \frac{10^3}{4}[50 + 4{\cdot}2] = \frac{54{\cdot}2 \times 10^3}{4} = 13{\cdot}55 \text{ kN-m/m}^3$$

(b) Volumetric strain energy per unit volume is :

$$\frac{1}{3}(\sigma_1 + \sigma_2 + \sigma_3)^2\left(\frac{1-2\nu}{2E}\right) = \frac{(1-0{\cdot}6)\times 10^{14}}{6 \times 200 \times 10^9}[5+4-3]^2$$

$$= \frac{0{\cdot}4 \times 36 \times 10^3}{12} = 1{\cdot}2 \text{kN-m/m}^3.$$

(c) Shear strain energy per unit volume is :

$$u = \frac{1}{12G}\left[(\sigma_1 - \sigma_2)^2 + (\sigma_2 - \sigma_3)^2 + (\sigma_3 - \sigma_1)^2\right]$$

Now
$$G = \frac{E}{2(1+\nu)} = \frac{200 \times 10^9}{2 \times 1{\cdot}3} = 76{\cdot}923 \text{ GPa}$$

$\therefore$
$$u = \frac{10^{14}}{12 \times 76{\cdot}923 \times 10^9}\left[(5-4)^2 + (4+3)^2 + (-3-5)^2\right]$$

$$= \frac{10^{14}}{12 \times 76{\cdot}923 \times 10^9}[1 + 49 + 64]$$

$$= 12{\cdot}35 \text{kN-m/m}^3$$

Now strain energy per unit volume under unaxial loading is :

$$= \frac{\sigma^2}{2E} = \frac{(100)^2 \times 10^{12}}{2 \times 200 \times 10^9}$$

$$= 25 \text{ kN-m/m}^3$$

$$\therefore \quad \text{Factory of safety} = \frac{25}{13{\cdot}55} = 1{\cdot}845$$

Example 2.10 *For a brittle material the tensile and compressive strengths are 400 MN/ m² and 1200 MN/m² respectively: (a) If the maximum value of the tensile principal stress is to be limited to one-quarter of the tensile strength, determine the maximum value and nature of the other principal stress using Mohr's modified theory of failure.*

(b) *What would be the values of the principal stresses associated with a maximum shear stress of 500 MN/m² according to Mohr's modified theory?*

(a) Maximum principal stress,

$$\sigma_1 = \frac{400}{4} = 100 \text{ N/m}^2$$

According to modified Mohr's theory :

$$\frac{\sigma_1}{\sigma_t} + \frac{\sigma_2}{\sigma_c} = 1$$

$$\frac{100}{400} = \frac{\sigma_2}{-1200} = 1$$

$$\sigma_2 = -900 \text{ MN/m}^2$$

(b) At any point the radius of the circle represents the maximum shear stress value. Therefore in the $\sigma - \tau$ diagram shown in Fig, 2.24, draw a line at a distance of 500 MN/ m² parallel to the Mohr's failure envelope, cutting the σ axis at C. Draw a circle with C as centre and tangential to the envelopes. Then

$$\sigma_1 = 0{\cdot}4 \text{ cm} = 80 \text{ MN/m}^2$$

$$\sigma_2 = 4{\cdot}8 \text{ cm} = -960 \text{ MN/m}^2$$

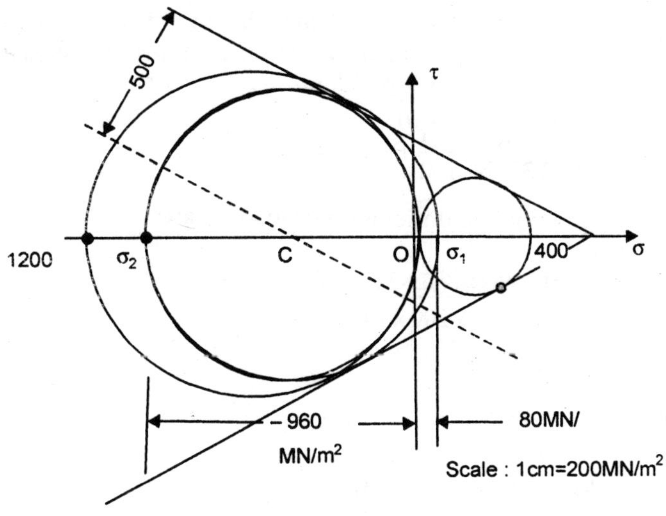

Fig. 2.24

2.5 CASTIGLIANO'S THEOREM

This theorem states that the rate of change of the strain energy with respect to statically independent force gives the component deflection of this force in the direction of the force.

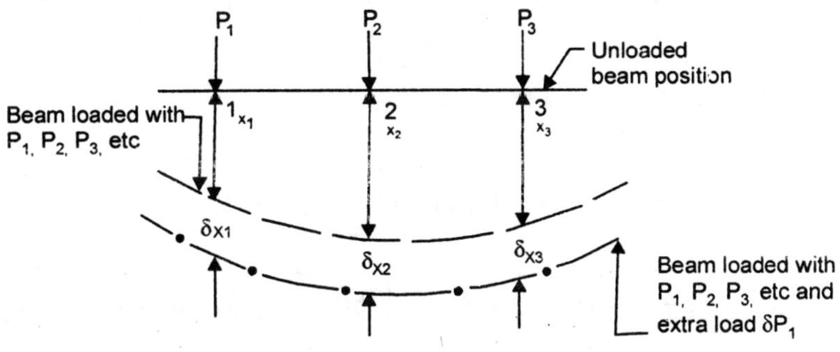

Fig. 2.25. Beam subjected to a system of concentrated lodes.

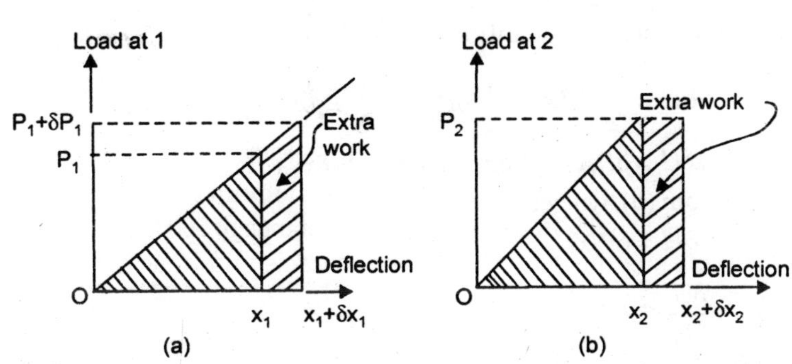

Fig. 2.26 Load-deflection curves for positions 1 and 2.

In order to prove the theorem, consider the beam shown in Fig. 2.25. with forces P_1, P_2, P_3, etc. acting at points 1, 2, 3, etc. If x_1, x_2, x_3, etc., are the deflections in the direction of the forces then the total strain energy of the system is equal to the work done.

$$U = \frac{1}{2} P_1 x_1 + \frac{1}{2} P_2 x_2 + \frac{1}{2} P_3 x_3 + \ldots \qquad \ldots(a)$$

If one of the loads, P_1, is now increased by an amount δP_1 the changes in deflections will be δx_1, δx_2, δx_3 etc. as shown in Fig. 2·25.

Extra work done at 1, as shown in Fig. 2.26 (a) is

$$= \left(P_1 + \frac{1}{2}\delta P_1\right) \delta x_1$$

Extra work done at 2, as shown in Fig. 2.26 (b) is:

$$= P_2 \delta x_2$$

Similarly extra work done at 3 etc. is $P_3 \delta x_3$, etc.

Increase in strain energy = Total extra work done

$$\therefore \qquad \delta U = P_1 \delta x_1 + \frac{1}{2} \delta P_1 \delta x_1 + P_2 \delta x_2 + P_3 \delta x_3 + \ldots$$

Neglecting the product of small quantities, we get

$$\delta U = P_1 \delta x_1 + P_2 \delta x_2 + P_3 \delta x_3 + \ldots \qquad \ldots(b)$$

If the loads $P_1 + \delta P_1$, P_2, P_3, etc., were applied gradually from zero the total strain energy would be,

$$U + \delta U = \Sigma \frac{1}{2} \times \text{load} \times \text{deflection}$$

$$= \frac{1}{2}(P_1 + \delta P_1)(x_1 + \delta x_1) + \frac{1}{2} P_2 (x_2 + \delta x_2) + \frac{1}{2} P_3 (x_3 + \delta x_3) + \ldots\ldots$$

$$= \frac{1}{2} P_1 x_1 + \frac{1}{2} P_1 \delta x_1 + \frac{1}{2} \delta P_1 x_1 + \frac{1}{2} \delta P_1 \delta x_1 + \frac{1}{2} P_2 x_2 + \frac{1}{2} P_2 \delta x_2$$

$$+ \frac{1}{2} P_3 x_3 + \frac{1}{2} P_3 \delta x_3 + \ldots \qquad \ldots(c)$$

Neglecting the product of small quantities and subtracting Eq. (a) from (c), we get

$$\delta U = \frac{1}{2} P_1 \delta x_1 + \frac{1}{2} \delta P_1 x_1 + \frac{1}{2} P_2 \delta x_2 + \frac{1}{2} P_3 \delta x_3 + \ldots$$

or

$$2\delta U = \delta P_1 x_1 + P_1 \delta x_1 + P_2 \delta x_2 + P_3 \delta x_3 + \ldots \qquad \ldots(d)$$

Subtracting Eq. (b) from (d), we get

$$\delta U = \delta P_1 x_1$$

or

$$\frac{\delta U}{\delta P_1} = x_1$$

or in the limit

$$\frac{\partial U}{\partial P_1} = x_1$$

i.e., the partial differential of the strain energy U with respect to P_1 gives the deflection under P_1 in the direction of P_1. Similarly

$$\frac{\partial U}{\partial P_2} = x_2$$

and $$\frac{\partial U}{\partial P_3} = x_3, \text{ etc.}$$

or in general, $$\frac{\partial U}{\partial P_i} = x_i \qquad ...(2.38)$$

which is the Castigliano's theorem.

Castigliano's theorem can also be applied to determine angular rotations under the action of bending moment or torque. Proceeding as above it can be shown that the angle of rotation

$$\theta_i = \frac{\partial U}{\partial M_i} \qquad ...(2.39)$$

where M is the couple applied.

(a) *Deflection under axial load.*

Strain energy under axial load P,

$$U = \int \frac{1}{2} \cdot \frac{P^2 dx}{AE}$$

$\therefore$ $$\delta = \frac{\partial U}{\partial P} = \int \frac{P dx}{AE} \qquad ...(2.40)$$

(b) *Deflection under bending.*

Strain energy under bending moment M,

$$U = \int \frac{M^2 dx}{2EI}$$

$$\delta = \frac{\partial U}{\partial P} = \frac{\partial U}{\partial M} \times \frac{\partial M}{\partial P} = \int \frac{M}{EI} dx \times \frac{\partial M}{\partial P} = \int \frac{M}{EI} \frac{\partial M}{\partial P} dx \qquad ...(2.41)$$

(c) *Rotation under torsion*

Strain energy under torque T,

$$U = \int \frac{T^2 dx}{2GJ}$$

$\therefore$ $$\theta = \frac{\partial U}{\partial T} = \int \frac{T dx}{GJ} \qquad ...(2.42)$$

(d) *Rotation under bending.*

Now $$U = \int \frac{M^2 dx}{2EI}$$

$$\therefore \qquad \phi = \frac{\partial U}{\partial M} = \int \frac{Mdx}{2EI} \qquad ...(2.43)$$

(e) *Deflection under torsion.*

Now $$U = \int \frac{T^2 dx}{2GJ}$$

$$\therefore \qquad \delta = \frac{\partial U}{\partial P} = \frac{\partial U}{\partial T} \times \frac{\partial T}{\partial P} = \int \frac{T}{GJ} dx \frac{\partial T}{\partial P} = \int \frac{T}{GJ} \frac{\partial T}{\partial P} dx \qquad ...(2.44)$$

(f) *Deflection under shear.*

Strain energy under shear force F,

$$U = \int \frac{F^2 dx}{2AG} \qquad ...(2.45)$$

$$\therefore \qquad \delta = \frac{\partial U}{\partial F} = \int \frac{Fdx}{AG}$$

(g) *Deflection under horizontal shear.*

Now $$U = \left(\frac{1+\nu}{EI^2}\right) \int F^2 dx \iint \frac{(A\bar{y})^2}{b^2} dA$$

where F is a function of P.

$$\therefore \qquad \delta = \frac{\partial U}{\partial P} \qquad ...(2.46)$$

Example. 2.11 *Using Castigliano's theorem, obtain the deflection under a single concentrated load applied to a simply supported beam shown in Fig. 2·27.*

$EI = 2\ MNm^2$.

Solution.

Let the load at C be denoted by P kN.

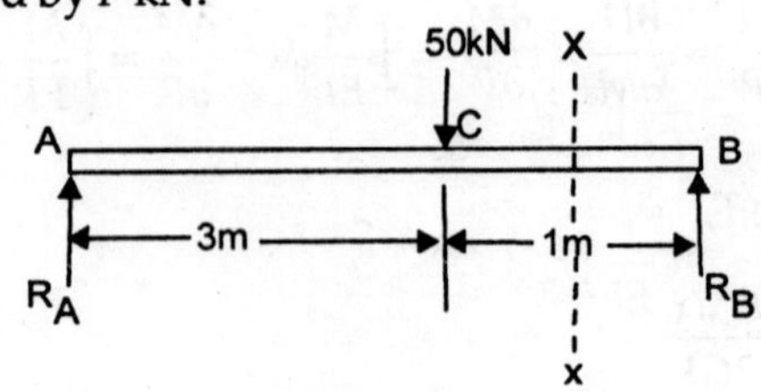

Fig. 2.7

$$R_A = \frac{P}{4}\ \text{kN}$$

$$M_x = R_A \times x - P(x-3) = \frac{Px}{4} - P(x-3)$$

$$\frac{\partial M}{\partial P} = \frac{x}{4} - (x-3)$$

$$\delta = \int \frac{M}{EI}\frac{\partial M}{\partial P}dx$$

$$= \frac{1}{EI}\int_0^3 \frac{Px}{4} \times \frac{x}{4}dx + \frac{1}{EI}\int_3^4 \left\{\frac{Px}{4} - P(x-3)\right\} \times \left\{\frac{x}{4} - (x-3)\right\}dx$$

$$= \frac{P}{16EI}\left|\frac{x^3}{3}\right|_0^3 + \frac{9P}{16EI}\left|16x + \frac{x^3}{3} - 4x^2\right|_3^4$$

$$= \frac{9P}{16EI} + \frac{9P}{16EI}[16 + 12{\cdot}33 - 28] = \frac{9P}{16EI}\ \frac{3P}{16EI} = \frac{3P}{4EI}$$

$$\delta = \frac{3\times 50\times 10^3}{4\times 2\times 10^6} = 1{\cdot}875 \times 10^{-2}\text{ m} = 18{\cdot}75\text{ mm}$$

Example 2.12 *Using Castigliano's theorem, obtain the deflection at the centre of a beam carrying u.d.l. of 20 kN /m over the whole span. The beam is simply supported over a span of 3 m and EI = 2·5 MN-m².*

Solution.

Let P be the dummy load at the mid-span. Then taking moments about the right hand support.

$$R_A \times 3 = 20 \times 3 \times 1{\cdot}5 + P \times 1\text{–}5 = 90 + 1{\cdot}5P$$

$$R_A = (30 + 0{\cdot}5P)\text{ kN}$$

$$M_x = (30 + 0{\cdot}5P)\,x - 10x^2 - P(x - 1{\cdot}5)$$

$$\frac{\partial M_x}{\partial P} = 0{\cdot}5x - (x - 1{\cdot}5)$$

$$\delta = \int\frac{M}{EI}\frac{\partial M}{\partial P}dx = \frac{1}{EI}\int_0^{1.5}\{(30+0.5P)x - 10x^2\}\,0.5x\,dx$$

$$+\frac{1}{EI}\int_0^{1.5}\{(30+0.5P)x - 10x^2 - P(x-1.5)\}\times\{0.5x - (x-1.5)\}dx$$

$$= \frac{1}{EI}\int_0^{1.5}\{(15+0.25P)x^2 - 5x^3\}dx$$

$$+\frac{1}{EI}\int_{1.5}^{3}\left[(45-1.5P)x + (0.25P-30)x^2 + 5x^3 + 2.25P\right]dx$$

$$= \frac{1}{EI}\left\{\left|(15+0.25P)\frac{x^3}{3} - \frac{5}{4}x^4\right|\right\}_0^{1.5} + \frac{1}{EI}\left|(45-1.5P)\frac{x^2}{2}\right.$$

$$\left. + (0.25P-30)\frac{x^3}{3} + \frac{5}{4}x^4 + 2.25Px\right|_{1.5}^{3}\Bigg]$$

Putting $\quad P = 0$, we get

$$\delta = \frac{1}{EI}[5\times1.5 - 1.25\times1.5^4] + \frac{1}{EI}[22.5\times6.75 - 10\times23.625 + 1.25\times75.9375]$$

$$= \frac{1}{EI}(16.875 - 6.328) + \frac{1}{EI}(151.875 - 236.25 + 94.9218)$$

$$= \frac{2.0938}{EI} = \frac{21.0938\times10^3}{2.5\times10^6} = 8.4375\times10^{-3}\,\text{m} = 8.4375\text{ mm}$$

Example 2.13 *A bent wire is loaded as shown in Fig. 2.28. Determine the translation and rotation of end A about the x, y and z axes by using energy method.*

Solution.

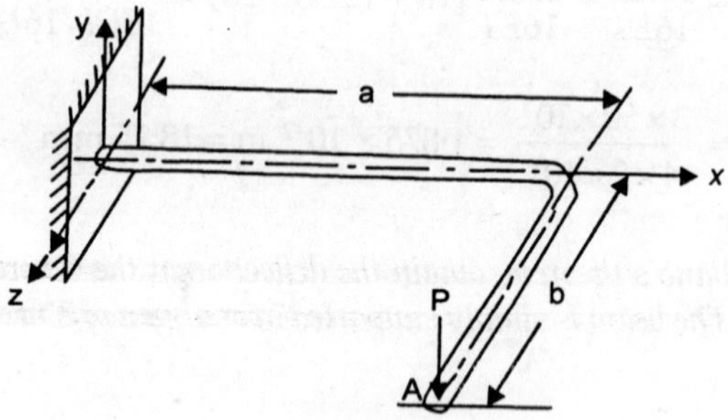

Fig. 2.28

$$\delta_x = \delta_z = 0$$

$$\delta_y = \int_0^b \frac{Px^2}{EI}dx + \int_0^b \frac{Px^2}{EI}dx + \int_0^b \frac{Pb^2}{GJ}x\,dx = \frac{Pb^3}{3EI} + \frac{Pa^3}{3EI} + \frac{Pa^2b}{2GJ}$$

$$\theta_x = \int_0^b \frac{Pb}{2EI}dx + \int_0^a \frac{Pb^2}{GJ}dx = \frac{Pb^2}{2EI} + \frac{Pab}{GJ}$$

$$\theta_y = 0$$

$$\theta_z = -\int_0^a \frac{Pa}{2EI}dx = -\frac{Pa^2}{2EI}$$

Example 2.14 *Find the central deflection of the uniform bend shown in Fig. 2.29 by Castigliano's theorem.*

Solution.

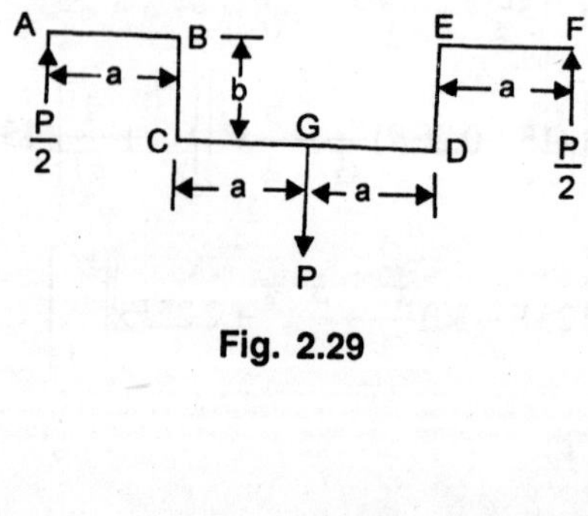

Fig. 2.29

$$U_{ABCG} = \int_0^a \frac{\left[(P/2)x\right]^2 dx}{2EI} + \int_0^b \frac{\left[(P/2)a\right]^2 dx}{2EI} + \int_0^a \left\{\frac{P}{2}(x+a)\right\}^2 \frac{dx}{2EI}$$

$$= \frac{P^2a^3}{24EI} + \frac{P^2a^3b}{8EI} + \frac{P^2}{24EI}\left|(x+a)^3\right|_0^a$$

$$= \frac{P^2a^3}{24EI} + \frac{P^2a^2b}{8EI} + \frac{7P^2a^3}{24EI}\left[8a^3 - a^3\right]$$

$$= \frac{P^2a^3}{24EI} + \frac{P^2a^2b}{8EI} + \frac{7P^2a^3}{24EI} = \frac{P^2a^2}{24EI}\left[a+3b+7a\right]$$

Total strain energy, $U = 2U_{ABCG} = \dfrac{P^2a^3}{12EI}(8a+3b) = \dfrac{P^2a^2}{24EI}(8a+3b)$

Vertical deflection of $G = \dfrac{\partial U}{\partial P} = \dfrac{Pa^2}{6EI}(8a+3b)$

Example 2.15 *Find strain energy stored by the structure shown in Fig. 2.30 and hence determine the vertical deflection of end A. Assume that the member is of uniform cross-section throughout.*

Solution.

Section AB:

$$M_x = Px$$

$$U_{AB} = \int_0^a \frac{M_x^2\,dx}{2EI} = \int_0^a \frac{P^2x^2dx}{2EI} = \frac{P^2a^3}{6EI}$$

Section BC:

$$M_y = Pa$$

$$U_{BC} = \int_0^l \frac{M_y^2\,dy}{2EI} = \int_0^l \frac{P^2a^2dy}{2EI} = \frac{P^2a^2l}{6EI}$$

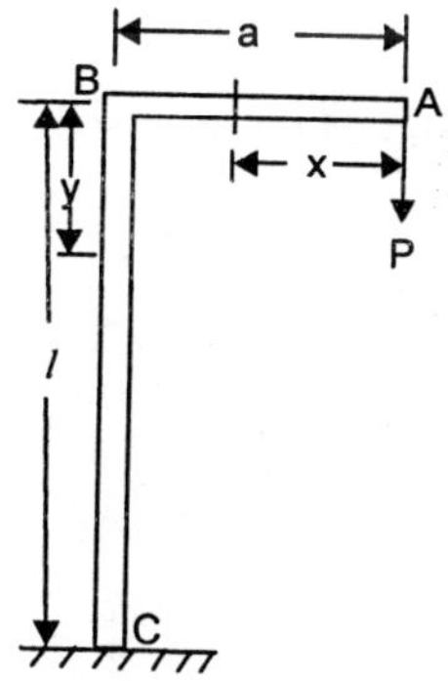

Fig. 2·30

Total strain energy,

$$U = U_{AB} + U_{BC}$$

$$= \frac{P^2a^3}{6EI} + \frac{P^2a^2l}{2EI} = \frac{P^2a^2}{6EI}(a+3l)$$

Let δ be the deflection of end A. Then

Work done by P = Total strain energy stored

$$\frac{1}{2}P\delta = \frac{P^2a^2}{6EI}(a+3l)$$

$$\delta = \frac{Pa^2}{3EI}(a+3l)$$

Example 2.16 *A vertical steel rod 2 m long is securely held at its upper end and a weight sliding freely on the rod falls on to a stop at the lower end of the rod. When the weight falls from a height of 40 mm the maximum stress produced in the rod is 160 MPa. Determine the stress if the load had been applied (a) gradually (b) to fall from a height of 60 mm. Take E = 200 GPa.*

Solution.

(a) Extension of rod, $\delta l = \frac{\sigma l}{E} = \frac{160\times10^6\times2}{200\times10^9} = 1.6$ mm

Let W = falling weight

Potential energy lost by weight = strain energy gained by rod

$$W(h+\delta l) = \frac{1}{2}\frac{\sigma^2}{E}Al$$

$$W(40+1.6) = \frac{1}{2}\times\frac{(160)^2}{2\times10^5}\times A\times2\times10^3$$

$$41.6\,W = 128A$$

$$\frac{W}{A} = 3.077\text{ N/mm}^2$$

(b) Maximum stress produced undrer falling weight,

$$\sigma = \frac{W}{A}\left[1+\sqrt{1+\frac{2AEh}{Wl}}\right]$$

$$= 3.077\left[1+\sqrt{1+\frac{2\times2\times10^5\times60}{3.077\times2\times10^3}}\right]$$

$$= 195.26\text{ N/mm}^2$$

Example 2.17. *A steel bar 60 mm diameter is bent to the shape shown in Fig. 2.31 and the lower end is firmly fixed in the ground in a vertical position. A load of 800 N is applied at the free end. Calculate the vertical deflection of the free end. E = 200 GN/m².*

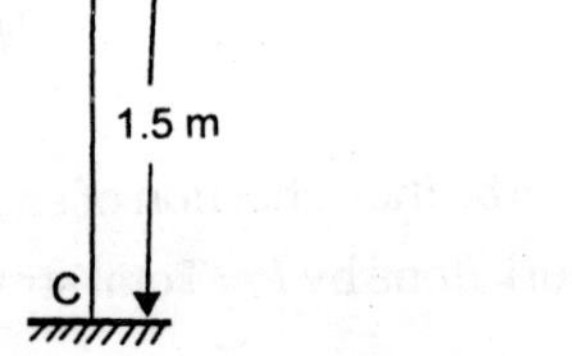

Fig. 2·31

Solution.

$$D = 60\text{ mm}, R = 1\text{ m}, P = 800\text{ N}$$

$$I = \frac{\pi D^4}{64} = \frac{\pi\times60^4}{64} = 63.62\text{ cm}^4$$

AB:
$$M = P \times 1 \sin\theta$$
$$\frac{\partial M}{\partial P} = \sin\theta$$

BC:
$$M = 1 \times P$$
$$\frac{\partial M}{\partial P} = 1$$

$$(\delta_A)_V = \frac{1}{EI}\int M\frac{\partial M}{\partial P}ds,\ ds = R\,d\theta = \theta$$
$$= \frac{1}{EI}\left[\int_0^{\pi/2} P\sin\theta.\sin\theta.d\theta + \int_0^{1.5} P dx\right]$$
$$= \frac{1}{EI}\left[\frac{P}{2}\int_0^{\pi/2}(1-\cos 2\theta)d\theta + \int_0^{1.5} P dx\right] = \frac{1}{EI}\left[\frac{P}{2}\left|\theta + \frac{\sin\theta}{2}\right|_0^{\pi/2} + P|x|_0^{1.5}\right]$$
$$= \frac{1}{EI}\left[\frac{P}{2}\left(\frac{\pi}{2}\right) + 1.5P\right] = \frac{800\times 2.2854}{200\times 10^9\times 63.62\times 10^{-8}} = 14.37 \text{ mm}$$

Problem 2.18. *A steel ring shown in Fig. 2.32 is formed in a circular arc subtending an angle of 270° at the centre. The lower end is rigidly fixed and a vertical force of 5 N is applied at the free end.*

If the section of the spring is 12 mm. wide and 3 mm thick, calculate the vertical and horizontal displacements of the free end. E = 200 GNIm³.

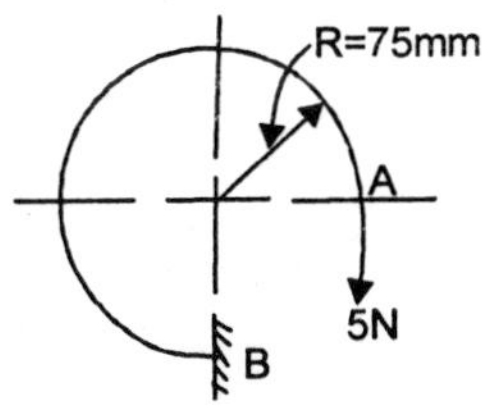

Fig. 2.32

Solution.

Apply a fictitious horizontal force Q *at* A.

$$M = PR(1-\cos\theta) + QR\sin\theta$$
$$\frac{\partial M}{\partial P} = R(1-\cos\theta)$$
$$\frac{\partial M}{\partial Q} = R\sin\theta$$
$$I = \frac{1.2\times(0.3)^3}{12} = 0.0027\text{cm}^4$$
$$(\delta_A)_V = \frac{1}{EI}\int M\frac{\partial M}{\partial P}ds = \frac{1}{EI}\int_0^{3\pi/2} PR(1-\cos\theta).R(1-\cos\theta).R\,d\theta$$
$$= \frac{PR^3}{EI}\int_0^{3\pi/2}(1+\cos^2\theta - 2\cos\theta)d\theta = \frac{PR^3}{EI}\int_0^{3\pi/2}\left[1+\frac{1+\cos 2\theta}{2} - 2\cos\theta\right]d\theta$$
$$= \frac{PR^3}{EI}\left[\theta + \frac{1}{2}\left(\frac{\sin 2\theta}{2}\right) - 2\sin\theta\right]_0^{3\pi/2} = \frac{PR^3}{EI}\left[\frac{3\pi}{2} + \frac{1}{2}\left(\frac{3\pi}{2}\right) + 2\right]$$

$$= \frac{9.0686 \times 5 \times (7.5)^3 \times 10^{-6}}{200 \times 10^9 \times 0.0027 \times 10^{-8}} = 3.54 \text{ mm}$$

$$(\delta_A)_H = \frac{1}{EI} \int M \frac{\partial M}{\partial Q} ds = \frac{PR^3}{EI} \int_0^{3\pi/2} \left(\sin\theta - \frac{\sin 2\theta}{2} \right) d\theta$$

$$= \frac{PR^3}{EI} \left[-\cos\theta + \frac{\cos 2\theta}{4} \right]_0^{3\pi/2}$$

$$= \frac{PR^3}{EI} \left[-\frac{1}{4} + 1 - \frac{1}{4} \right] = \frac{PR^3}{2EI} = \frac{5 \times (7.5)^3 \times 10^{-6}}{2 \times 200 \times 10^9 \times 0.0027 \times 10^{-8}} = 0.195 \text{ mm}$$

Example 2.19 *A steel tube having outside and inside diameters of 100 mm and 60 mm espectively is bent into the form of a quadrant of 2 m radius as shown in Fig. 2.33. One end is rigidly attached to a horizontal base plate to which a tangent to that end is perpendicular, and the free end supports a load of 1 kN. Determine the vertical and horizontal deflections of the free end under this load using Castigliano's theorem. Given E = 200 GPa.*

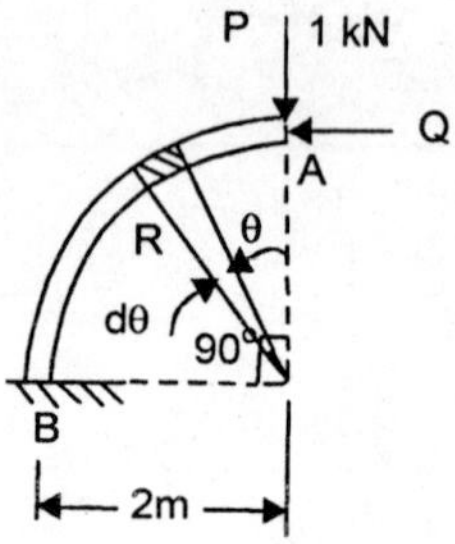

Fig. 2.33

Solution.

$$I = \frac{\pi}{64} (100^4 - 60^4) = 4272566 \text{ mm}^4$$

Vertical deflection:

$$M = PR \sin\theta$$

$$U = \int_0^{\pi/2} \frac{M^2 ds}{2EI} = \int_0^{\pi/2} \frac{P^2R^2 \sin^2\theta \times R d\theta}{2EI} = \frac{P^2R^3}{2EI} \int_0^{\pi/2} \sin^2\theta . d\theta$$

$$\delta_V = \frac{\partial U}{\partial P} \frac{PR^3}{2EI} \int_0^{\pi/2} (1 - \cos 2\theta) d\theta = \frac{PR^3}{2EI} \left| \theta - \frac{\sin 2\theta}{\theta} \right|_0^{\pi/2}$$

$$= \frac{PR^3}{2EI} \times \frac{\pi}{2} = \frac{10^3 \times 8 \times \pi}{4 \times 200 \times 10^9 \times 4272566 \times 10^{-12}} = 7.353 \text{ mm}$$

Horizontal deflection:

$$M = PR \sin\theta - QR(1 - \cos\theta)$$

$$U = \frac{1}{2EI} \int_0^{\pi/2} [PR \sin\theta - QR(1 - \cos\theta)]^2 R d\theta$$

$$= \frac{R^3}{2EI} \int_0^{\pi/2} [P \sin\theta - Q(1 - \cos\theta)]^2 d\theta$$

$$\delta_H = \frac{\partial U}{\partial Q} = \frac{R^3}{EI} \int_0^{\pi/2} [P \sin\theta - Q(1 - \cos\theta)](-1 + \cos\theta) d\theta$$

Put $Q = 0$

$$\delta_H = \frac{R^3}{EI}\int_0^{\pi/2} -P\sin\theta(1-\cos\theta)d\theta = -\frac{PR^3}{EI}\int_0^{\pi/2}\left(\sin\theta - \frac{1}{2}\sin\theta\right)d\theta$$

$$= -\frac{PR^3}{EI}\left|-\cos\theta + \frac{\cos 2\theta}{4}\right|_0^{2\pi/2} = -\frac{PR^3}{EI}\left(-\frac{1}{4}+1\frac{1}{4}\right)$$

$$= -\frac{PR^3}{EI} = -\frac{1000\times 8}{2\times 200\times 10^9\times 4272566\times 10^{-12}} = -4.681 \text{ mm}$$

SUMMARY

1. The strain energy is the work done on the material.
2. Strain energy stored per unit volume is called resilience.
3. Expressions for strain energy.
   (i) Axial loading:

$$U = \frac{1}{2}\frac{\sigma^2}{E}V$$

   (ii) Shear loading

$$U = \frac{1}{2}\frac{\tau^2}{G}V$$

   (iii) Three - dimensional principal stresses.

$$U = \frac{1}{2E}\left[\sigma_1^2 + \sigma_2^2 + \sigma_3^2 - 2\nu(\sigma_1\sigma_2 + \sigma_2\sigma_3 + \sigma_3\sigma_1)\right]$$

   (iv) Bending loading

$$U = \frac{M^2 l}{2EI} = \frac{1}{2}M\phi,\ \phi = \frac{Ml}{EI} = \frac{l}{R}$$

   (v) Horizonted shear.

$$U = \frac{(1+\nu)}{EI^2}\int_0^l F^2 dx \iint_A \frac{(A\bar{y})^2}{b^2}dA$$

   (vi) Torsion loding

$$U = \frac{1}{2}\frac{T^2 l}{GJ} = \frac{1}{2}T\theta$$

   (vii) Shear strain energy due to principal stresses.

$$U = \frac{1}{12G}\left[(\sigma_1-\sigma_2)^2 + (\sigma_2-\sigma_3)^2 + (\sigma_3-\sigma_1)^2\right]$$

4. Sudden load = 2 × Gradual load
5. Impact load on a collar fitted to a bar.

$$\sigma = \frac{W}{A}\left[1+\sqrt{1+\frac{2AEh}{Wl}}\right]$$

6. Impact on a simply suprroted beam.

Equivalent static load, $W_1 = W + \sqrt{W^2 + \frac{96W\,h\,EI}{L^3}}$

Bending stress, $\sigma_b = \frac{W_1\,L}{4} \times \frac{y}{I}$

7. Theories of failure
   (i) Maximmum principal stress theory, $\sigma_1 = \sigma$
   (ii) Maximmum principal strain theory, $\sigma_1 - \nu\sigma_2 = \sigma$
   (iii) Maximmum shear stress theory, $\sigma_1 - \sigma_3 = \sigma$
   (iv) Maximmum strain energy theory, $\sigma_1^2 + \sigma_2^2 + \sigma_3^2 - 2\nu(\sigma_1\sigma_2 + \sigma_3 + \sigma_3\sigma_1) = \sigma^2$
   (v) Maximmum shear strain energy theory or distortion energy theory,

$$(\sigma_1 - \sigma_2)^2 + (\sigma_2 - \sigma_3)^2 + (\sigma_3 - \sigma_1)^2 = 2\sigma^2$$

   (vi) Mohr's theory,

$$\frac{\sigma_1}{\sigma_{yt}} + \frac{\sigma_2}{\sigma_{yc}} = 1$$

8. Castigliano's theorem.
   (i) Deflection under axial load, $\delta = \frac{\partial U}{\partial P} = \int \frac{P}{AE} dx$
   (ii) Deflection under bending, $\delta = \int \frac{M}{EI} \cdot \frac{\partial M}{\partial P} . dx$
   (iii) Rotation under torsion, $\theta = \int \frac{Tdx}{GJ}$
   (iv) Rotation under bending, $\phi = \int \frac{Mdx}{2EI}$
   (v) Deflection under torsion, $\delta = \int \frac{T}{GJ} \cdot \frac{\partial T}{\partial P} dx$
   (vi) Deflection under shear, $\delta = \int \frac{F\,dx}{AG}$

Short Answer Type Questions

1. Deffferentiate between gradual load, sudden load and impact load.
2. What is the necessity of theries of faillure?
3. Name the theories of failure sutiable for ductile materials.
4. Name the theories of failure sutiable for brittle materials.
5. State Castigliano's theorem.
6. Define strain energy, resilience and proof resilience.

EXERXISES

1. Compare the maximum shear stress theory and the octahedral shear stress theory of failure in uniaxial tension and plain strain conditions.
2. What is a failure theory ? Discuss its importance. Enumerate various failure theories and mention their fields of applications.
3. What are the various theories of failure *when* a material is subjected to stress ? Discuss any two of them. How does Mohr's theory of failure differs from others ?
4. Explain the salient features of maximum dislocation energy theory of elastic failure and discuss how it compares with maximum strain energy and maximum shear stress theories.
5. A rod of constant flexural rigidity *EI* is formed into a planar member as shown in Fig. 2.34. Find the horizontal and vertical deflections of point A, using energy method. (Consider *only* flexural effect). $\left[\textbf{Ans.}\ \frac{10\,Pa^3}{3EI}\cdot\frac{6.57\,Pa^3}{El}\right]$

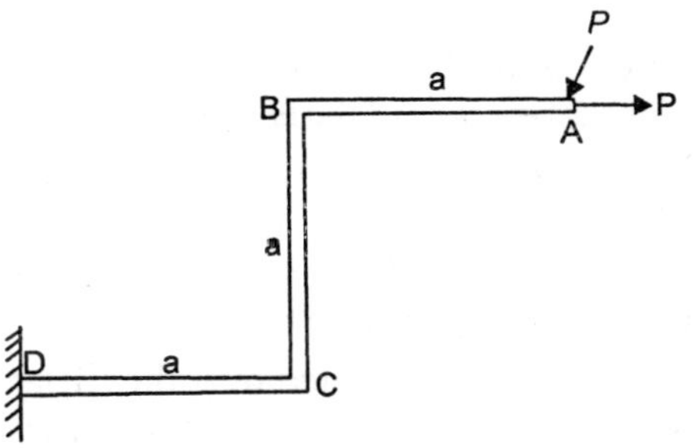

Fig. 2.34

6. man weighing 80 kg jumps onto a diving board, as shown in Fig. 2.35, from a height of 0·6 m. What maximum bending stress will this cause in the board ? The diving board is 50 × 300 mm² cross-section, and its *E* = 12 GPa. [**Ans.** 59·3 MPa]

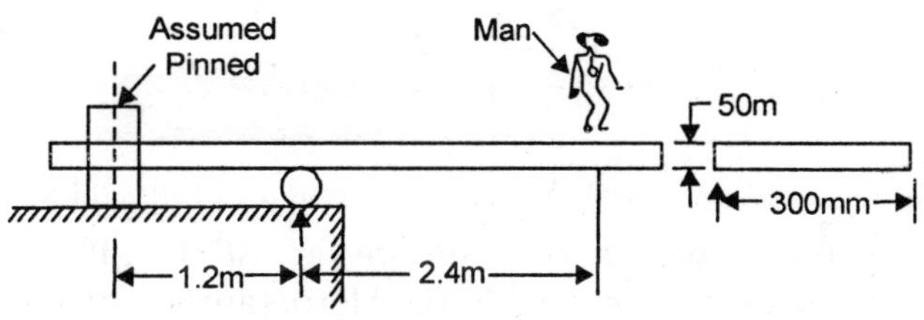

Fig. 2.35

Chapter-3

Three Dimensional Stresses and Strains

3.1 FORCE DISTRIBUTION

The forces acting on a body can be distinguished as follows :

(a) Internal forces.

(b) External forces.

(a) Internal forces: The internal mechanical state of a body is described by the stresses and strains at all points in it. The internal forces are the reactive forces which are set-up due to external applied forces. And within elastic limits, the internal forces are numerically equal to the external forces.

(b) External forces: The state of stress and strain in a body arises due to external influences. The external forces acting on a body can be divided into two types.

(i) *Surface forces*–which are the forces distributed over the surface or boundary of the body and acting from the surrounding medium, like atmospheric pressure, hydraulic pressure or contact pressure exerted by one body on another. An exterior surface is any surface comprising a part of the bounding surface of the body and includes for example, the surface surrounding an internal cavity. The surface forces acting on the external surface of the body are described in terms of forces per unit area and, as such, are called applied stresses. Their components along the x, y and z axes of reference are represented by S_x, S_y and S_z, such that

$$\left.\begin{aligned} S_x &= \underset{\Delta A \to 0}{\text{Limit}} \frac{\Delta F_x}{\Delta A} \\ S_y &= \underset{\Delta A \to 0}{\text{Limit}} \frac{\Delta F_y}{\Delta A} \\ S_x &= \underset{\Delta A \to 0}{\text{Limit}} \frac{\Delta F_z}{\Delta A} \end{aligned}\right\} = 0 \qquad \text{...(3.1)}$$

where ΔF_x, ΔF_y and ΔF_z are the components of the surface force ΔF along the x, y and z co-ordinate axes respectively and acting on elementary area ΔA. Surface forces are also called the contact forces.

An internal surface is created only by an imaginary cut through the body and lies, prior to such a cut, entirely inside the boundaries of the body. In order to distinguish between surface force distribution at actual and imaginary boundaries, we call the surface force distribution on the actual boundary, the traction force and on the imaginary boundary, the surface force.

(ii) *Body forces*—which act throughout the body and are not produced by physical contact with other bodies. Body force intensities are designated as forces per unit volume or per unit mass. For example, gravitational forces electromagnetic forces, centrifugal forces and inertia forces are body forces. They are represented by B_x, B_y and B_z, such that

$$\left.\begin{aligned} B_x &= \underset{\Delta V \to 0}{\text{Limit}} \frac{\Delta F_x}{\Delta V} \text{ or } \underset{\Delta m \to 0}{\text{Limit}} \frac{\Delta F_x}{\Delta m} \\ B_y &= \underset{\Delta V \to 0}{\text{Limit}} \frac{\Delta F_y}{\Delta V} \text{ or } \underset{\Delta m \to 0}{\text{Limit}} \frac{\Delta F_y}{\Delta m} \\ B_z &= \underset{\Delta V \to 0}{\text{Limit}} \frac{\Delta F_z}{\Delta V} \text{ or } \underset{\Delta m \to 0}{\text{Limit}} \frac{\Delta F_z}{\Delta m} \end{aligned}\right\} \quad \dots(3.2)$$

where ΔV and Δm are the elementary volume or elementary mass respectively.

3.2 DEFINITION OF STRESS

Consider a body as shown in Fig.3.1, which is in static equilibrium under the action of the system of external forces $F_1, F_2, \dots, F_n$. Suppose now that the body is cut into two parts 1 and 2 by an imaginary plane AB. Then along the plane AB there are forces between the particles in part 1 and those in part 2, that tend to hold the body together. F_{21} is the sum of all the reactions of the particles in part 1 on the particles in part 2. For the body to remain in one piece and for its equilibrium to be maintained, F_{21} must be opposed by an equal and opposite force F_{12}, which is equal to the sum of all the particle forces in part 2 on the particles in part 1.

Thus $\qquad F_{21} = -F_{12}$

or $\qquad F_{21} + F_{12} = 0$

Therefore, when a body is in equilibrium under the action of external forces, any arbitrary portion of that body must be in equilibrium under the action of the external and internal forces on that portion.

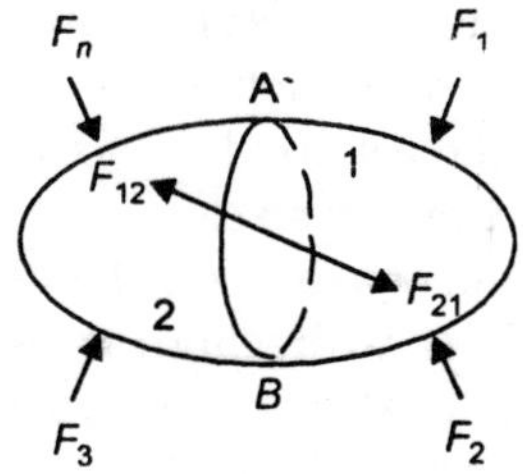

Fig. 3.1 Internal forces in a continuous body.

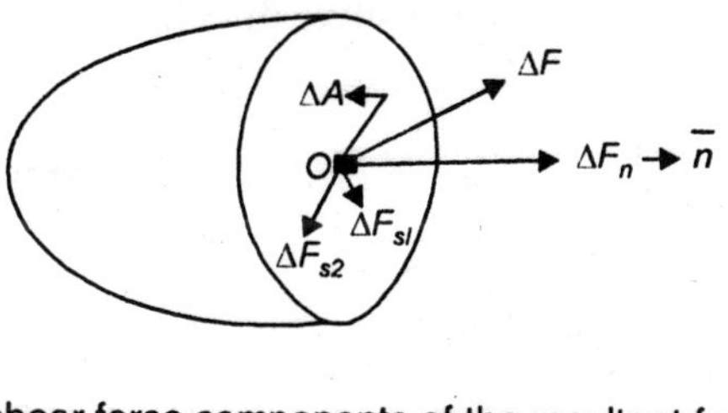

Fig. 3.2 Normal and shear force components of the resultant force ΔF acting on an elementary area ΔA at a point O.

Consider now an element of area ΔA on this section, as shown in Fig. 3.2. Let the resultant internal forces acting on this area be ΔF, such that $\Sigma \Delta A . \Delta F = F_{21}$. The component of this force in the direction of unit normal $\bar{n}$ is ΔF_n. And the other two components at right angles to each other in the plane of the section are ΔF_{S_1} and ΔF_{S_2}. Then the normal component of stress at point O *is* defined as:

$$\sigma = \underset{\Delta A \to 0}{\text{Limit}} \frac{\Delta F_n}{\Delta A} \quad \text{...(3.3)}$$

and the shear components of stress are:

$$\left.\begin{aligned} \tau_{s_1} &= \underset{\Delta A \to 0}{\text{Limit}} \frac{\Delta F_{s_1}}{\Delta A} \\ \tau_{s_2} &= \underset{\Delta A \to 0}{\text{Limit}} \frac{\Delta F_{s_2}}{\Delta A} \end{aligned}\right\} \quad \text{...(3.4)}$$

Therefore, on a plane at a point there are three stress components, one normal to the plane and the other two tangential to the plane. It should be remembered that if a different plane is passed through the same point in the body, then the stresses will be different. Further, we have taken only two directions mutually perpendicular to each other in the section to determine only two components of shear stress. However, there are infinite number of directions along which shear stresses may be determined.

Since in addition to magnitude and direction, a plane of reference must also be specified for the stress, therefore, we can say that stress is not a vector but a tensor quantity. A tensor quantity is a more general quantity than a vector quantity since it needs more than three components for its complete definition (a vector needs only three quantities for its complete definition). For the stress tensor, six components must be known, three of these are the vector quantities, direction and magnitude, and the other three components are those necessary to define a plane of reference to which the stress is referred.

3.3 STRESS NOTATIONS

In order to understand clearly the state of stress at a point, double subscript system representing stress components is utilised. The first subscript denotes the direction of the outward drawn normal on the plane on which the stress acts, and the second subscript denotes the direction towards which the stress acts. Therefore, on the positive x-face of an elementary parallelopiped, the stresses are τ_{xx}, τ_{xy} and τ_{xz} in the Cartesian co-ordinate system of axes. Out of which τ_{xx} is the normal stress, because the first subscripted x, denotes the direction (*i.e.*, x-direction) of the outward drawn normal on the plane-x and the second subscripted x, denotes the direction (*i.e.*, x-direction) along which the stress acts. Whereas τ_{xy} and τ_{xz} are the shear stresses, because the subscripted x gives the direction of the outward drawn normal, *i.e.*, the x-direction and the subscripts y and z give the directions (*i.e.*, y and z-directions) along which the stresses are measured. The stress components on a plane will be all positive if the outward drawn normal on that plane is in the positive direction of the co-ordinate axes.

3.4 STRESS TENSOR AT A POINT

The state of stress at a point in the Cartesian co-ordinates can be represented by the following nine components of stress known as the stress tensor:

$$\tau_{ij} = \begin{bmatrix} \tau_{xx} & \tau_{xy} & \tau_{xz} \\ \tau_{yx} & \tau_{yy} & \tau_{yz} \\ \tau_{zx} & \tau_{zy} & \tau_{zz} \end{bmatrix}$$

The first subscript in τ_{ij} can be considered to identify a row of the array of terms whereas the second subscript can be considered to identify a column of the array. The principal or leading diagonal terms are all normal stresses and all off-diagonal terms are shear stresses. The normal stresses can also be written as $\sigma_{xx}, \sigma_{yy}, \sigma_{zz}$ or $\sigma_x, \sigma_y, \sigma_z$ instead of $\tau_{xx}, \tau_{yy}, \tau_{zz}$ or τ_x, τ_y, τ_z. Throughout this book we shall write the normal stresses as σ_x, σ_y and σ_z. Therefore, the stress tensor may be written as

$$\begin{bmatrix} \sigma_x & \tau_{xy} & \tau_{xz} \\ \tau_{yx} & \sigma_y & \tau_{yz} \\ \tau_{zx} & \tau_{zy} & \sigma_z \end{bmatrix}$$

All the stress components have been shown on a parallelopiped in Fig. 3.3. It can be noted that there are nine components of stress in total for a general state of stress.

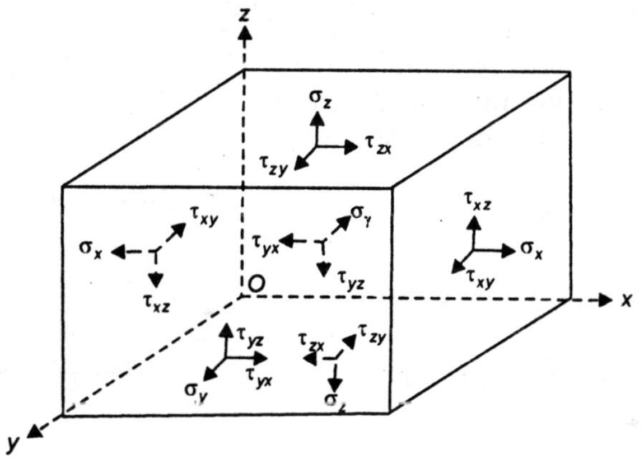

Fig.3.3 Components of stress tensor acting on the faces of a rectangular parallelopiped.

In the same manner the stress tensors in cylindrical (r, θ, z) and spherical (r, θ, ϕ) co-ordinates may be written as follows :

$$\begin{bmatrix} \sigma_r & \sigma_{r\theta} & \tau_{rz} \\ \tau_{\theta r} & \sigma_\theta & \tau_{\theta z} \\ \tau_{zr} & \tau_{z\theta} & \sigma_z \end{bmatrix} \text{ and } \begin{bmatrix} \sigma_r & \sigma_{r\theta} & \tau_{r\phi} \\ \tau_{\theta r} & \sigma_\theta & \tau_{\theta\phi} \\ \tau_{\phi r} & \tau_{\phi\theta} & \sigma_\phi \end{bmatrix}$$

3.5 STRESS GRADIENT

The variation of stress with distance is called stress gradient. If the stress tensor at a point is known then the stress components in the neighbourhood of the point can be known by expanding the stress components by Taylor series expansion. Thus if σ_x is **the** stress at a point then the stress at a point at an infinitely small distance δx will be given by

$$\sigma_x + \frac{\partial}{\partial x}\sigma_x \cdot \delta_x + \text{higher order terms.}$$

Neglecting higher order terms, we get $\sigma_x + \dfrac{\partial \sigma_x}{\partial x}\delta x$

In general, it can be written as $\tau_{ij} + \frac{\partial \tau_{ij}}{\partial x_j} \cdot \delta x_j$.

Taylor series expansion of a function $f(x)$ at $(x + h)$, where h is an infinitely small distance, is given by

$$f(x+h) = f(x) + hf'(x) + \frac{h^2}{2!} f''(x) + \frac{h^3}{3!} f'''(x) + \ldots$$

where dashes represent differentiation with respect to x.

3.6 NATURE OF STRESS TENSOR

Consider an infinitesimal parallelopiped of sides dx, dy and dz, as shown in Fig. 3.4. The shear stress τ_{xy} and τ_{yx} are shown acting on its faces perpendicular to the x and y-axes. By expressing the condition of equilibrium of moments, $\Sigma M_z = 0$ about point O, we get

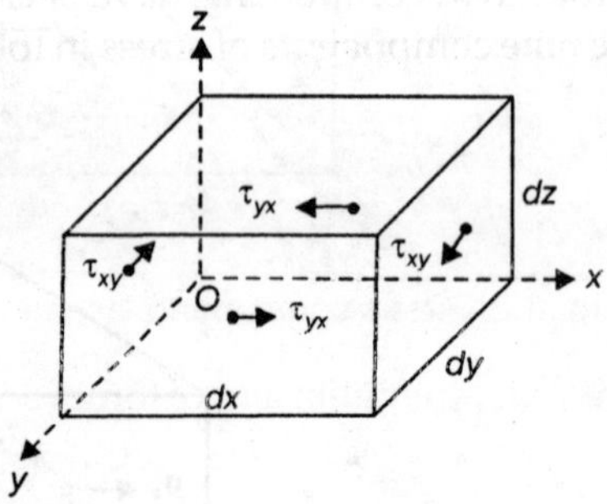

Fig. 3.4 Moment equilibrium of the stress component acting on the faces of a rectangular parallelopiped.

$$(\tau_{xy}\, dydz)\, dx - (\tau_{yx} \cdot dxdy)\, dz = 0$$

$$\therefore \quad \tau_{xy} = \tau_{yx}$$

Similarly by considering the equilibrium of moments

$$\Sigma M_x = 0, \Sigma M_y = 0, \text{ we get}$$

$$\tau_{yz} = \tau_{zy}$$

$$\tau_{xz} = \tau_{zx}$$

Therefore, the shear stresses are complimentary in nature and hence the stress tensor is symmetric in nature.

$$\tau_{ij} = \tau_{ji}; i \neq j \text{ and } i, j = x, y, z \qquad \ldots (3.5)$$

Hence, the stress tensor in Cartesian co-ordinates may be written as:

$$\begin{bmatrix} \sigma_x & \tau_{xy} & \tau_{xz} \\ \tau_{xy} & \sigma_y & \tau_{yz} \\ \tau_{xz} & \tau_{yz} & \sigma_z \end{bmatrix}$$

Therefore, the components of stress tensor reduces from 9 to 6 independent components.

If we consider stress gradients in the parallelopiped, then the higher order terms like $\frac{\partial \tau_{xy}}{\partial x} dx$ etc., contribute only higher order terms in dx, dy, dz to the moments. Also, the acceleration effects, being proportional to the mass moment of inertia of the element, are of higher degree in dx, dy, dz, and hence do not contribute to the moments.

3.7 EQUILIBRIUM EQUATIONS IN CARTESIAN CO-ORDINATES

Consider an infinitesimal rectangular parallelopiped, as shown in Fig. 3.5, having sides dx, dy and dz under the action of surface forces, inertia forces and body forces.

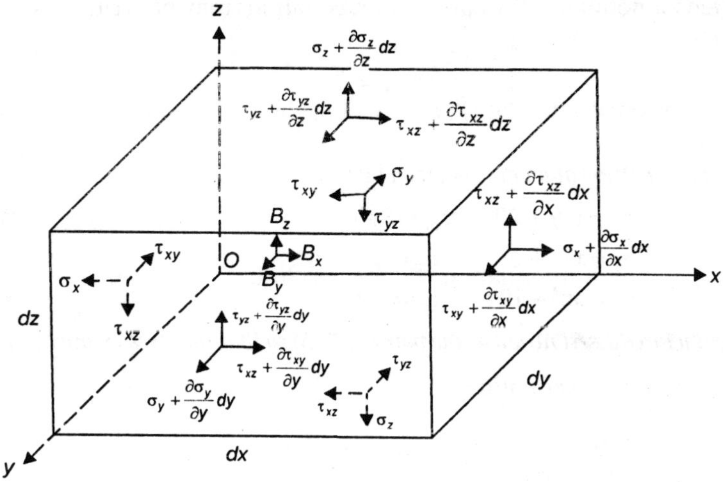

Fig. 3.5 Stress components acting on the faces of an elementary rectangular parallelopiped.

Consider the equilibrium of forces in the x-direction. By using Newton's second law of motion, we get

$$\left(\sigma_x + \frac{\partial \sigma_x}{\partial x} \cdot dx\right) dy\, dz - \sigma_x . dydz + \left(\tau_{xy} + \frac{\partial \tau_{xy}}{\partial y} dy\right) dxdz - \tau_{xy}\, dxdz$$

$$+ \left(\tau_{xz} + \frac{\partial \tau_{xz}}{\partial z} dz\right) dxdy - \tau_{xz}\, dxdy \ + B_x . dxdydz = \rho dxdydz . \bar{a}_x$$

where B_x = component of the body force in the x-direction, expressed per unit volume.

$\bar{a}_x$ = acceleration in the x-direction.

ρ = density of the material of the parallelopiped.

Simplifying, we get $\frac{\partial \sigma_x}{\partial x} + \frac{\partial \tau_{xy}}{\partial y} + \frac{\partial \tau_{xz}}{\partial z} + B_x = \rho \bar{a}_x$

Similarly considering the equilibrium of forces in the y- and z directions, we get respectively

$$\frac{\partial \tau_{xy}}{\partial x} + \frac{\partial \sigma_y}{\partial y} + \frac{\partial \tau_{yz}}{\partial z} + B_y = \rho \bar{a}_y$$

$$\frac{\partial \tau_{xz}}{\partial x} + \frac{\partial \tau_{yz}}{\partial y} + \frac{\partial \sigma_z}{\partial z} + B_z = \rho \bar{a}_z$$

In the absence of inertia and body forces, we get the equilibrium equations as:

$$\left.\begin{aligned} \frac{\partial \sigma_x}{\partial x} + \frac{\partial \tau_{xy}}{\partial y} + \frac{\partial \tau_{xz}}{\partial z} = 0 \\ \frac{\partial \tau_{xy}}{\partial x} + \frac{\partial \sigma_y}{\partial y} + \frac{\partial \tau_{yz}}{\partial z} = 0 \\ \frac{\partial \tau_{xz}}{\partial x} + \frac{\partial \tau_{yz}}{\partial y} + \frac{\partial \sigma_z}{\partial z} = 0 \end{aligned}\right\} \quad \text{...(3.6)}$$

In the form of tensor notations, the equilibrium equations may be written as :

$$\frac{\partial \tau_{ij}}{\partial x_j} + B_i = 0 \quad \text{... (3.7)}$$

Example 3.1 *The state of stress at a point is given by :*

$$\sigma_x = x^2y + 20 \; ; \quad \tau_{xy} = 3x^2y$$
$$\sigma_y = x^3z + y^2 \; ; \quad \tau_{yz} = yz$$
$$\sigma_z = yz^2 + 10 \; ; \quad \tau_{xz} = xz.$$

Determine the body force distribution at the point (1,2,3) so that the stresses are in equilibrium.

Solution. Using the equilibrium equation,

$$\frac{\partial \sigma_x}{\partial x} + \frac{\partial \tau_{xy}}{\partial y} + \frac{\partial \tau_{xz}}{\partial z} + B_x = 0$$

we get $\quad 2xy + 3x^2 + x + B_x = 0$

$\therefore \quad B_x = -(2xy + 3x^2 + x)$

At the point (1, 2, 3), B_x becomes

$$B_x = -(2 \times 1 \times 2 + 3 \times 1 + 1)$$
$$= -(4 + 3 + 1) = -8$$

Similarly, using the other two equilibrium equations, we get

$$B_y = -(6xy + 2y + y) = -18$$
$$B_z = -(2z + 2yz) = -18$$

$\therefore$ The body force distribution required for equilibrium becomes :

$$B = -8\hat{i} - 18\hat{j} - 18\hat{k}$$

Example 3.2 *The stress components at a point in a body subjected to two dimensional state of stress are given by :*

$$\sigma_x = 2x^2y + 3xy$$
$$\sigma_y = 2x^3 + 5xy^2$$
$$\tau_{xy} = 4x^2y^2$$

Determine whether the equations of equilibrium are satisfied at the point (– 2, 3) or not.

Solution. The equations of equilibrium in the absence of body forces in two-dimensions are given by:

$$\frac{\partial \sigma_x}{\partial x} + \frac{\partial \tau_{xy}}{\partial y} = 0$$

$$\frac{\partial \tau_{xy}}{\partial x} + \frac{\partial \sigma_y}{\partial y} = 0$$

Substituting the stress components, we get

$$(4xy + 3y) + 8x^2y$$

and $\quad 8xy^2 + 10xy$

At the point (– 2, 3), we get

$$4(-2)(3) + 3 \times 3 + 8 \times 4 \times 3 = -24 + 9 + 96 = 81$$

and $$8(-2)(9) + 10(-2)3 = -144 - 60 = -204$$

Therefore we find that the equations of equilibrium are not satisfied.

3.8 PRINCIPAL STRESSES AND PRINCIPAL PLANES

Consider an infinitesimal tetrahedron having three faces with known stresses on the reference planes, as shown in Fig. 3.6.

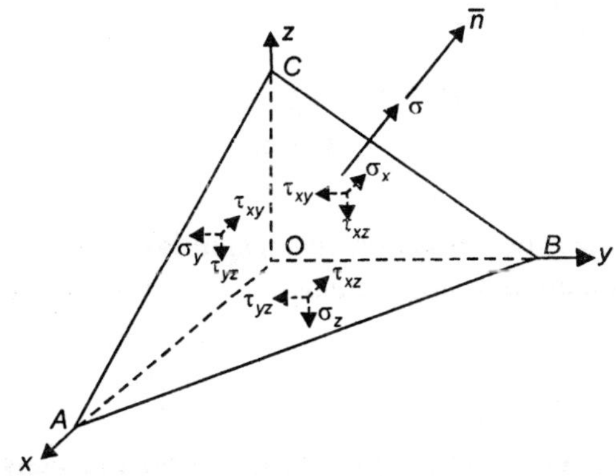

Fig. 3.6 Stress components acting on the faces of an elementary tetrahedron.

Assume that the inclined face *ABC* is a principal plane. The principal stress σ on this plane is along the normal $\bar{n}$ to this plane having direction cosines a_{nx}, a_{ny} and a_{nz}.

Applying Newton's second law of motion in the *z*-direction and neglecting body and inertia forces, we get

$$\sigma.ABC.a_{nz} - \sigma_z.AOB - \tau_{yz}.AOC - \tau_{xz}.BOC = 0$$

Dividing throughout by area ABC, we get

$$\sigma.a_{nz} - \sigma_z.a_{nz} - \tau_{yz}.a_{ny} - \tau_{xz}a_{nx} = 0$$

$$\tau_{xz}.a_{nx} + \tau_{yz}.a_{ny} + (\sigma_z - \sigma)a_{nz} = 0$$

Similarly, by considering the equilibrium of forces in the *x* and *y* directions, we get

$$(\sigma_x - \sigma)a_{nx} + \tau_{xy}a_{ny} + \tau_{xz}a_{nz} = 0$$

$$\tau_{xy}.a_{nx} + (\sigma_y - \sigma)a_{ny} + \tau_{yz}a_{nz} = 0$$

Therefore, the three simultaneous homogeneons equations can be written as :

$$\begin{bmatrix} \sigma_x - \sigma & \tau_{xy} & \tau_{xz} \\ \tau_{xy} & \sigma_y - \sigma & \tau_{yz} \\ \tau_{xz} & \tau_{yz} & \sigma_z - \sigma \end{bmatrix} \begin{bmatrix} \alpha_{nx} \\ a_{ny} \\ a_{nz} \end{bmatrix} = \begin{bmatrix} 0 \\ 0 \\ 0 \end{bmatrix}$$

By using Cramer's rule, we get:

$$a_{nx} = \frac{\begin{vmatrix} 0 & \tau_{xy} & \tau_{xz} \\ 0 & \sigma_y - \sigma & \tau_{yz} \\ 0 & \tau_{yz} & \sigma_z - \sigma \end{vmatrix}}{\begin{vmatrix} \sigma_x - \sigma & \tau_{xy} & \tau_{xz} \\ \tau_{xy} & \sigma_y - \sigma & \tau_{yz} \\ \tau_{xz} & \tau_{yz} & \sigma_z - \sigma \end{vmatrix}}$$

a_{nx} will be zero, as will the other direction cosines, unless the denominator in the preceeding equation is zero so as to permit an indeterminate result. But all the direction cosines cannot be zero, because

$$a_{nx}^2 + a_{ny}^2 + a_{nz}^2 = 1$$

Thus a necessary condition required for the solution of this problem is:

$$D = \begin{vmatrix} \sigma_x - \sigma & \tau_{xy} & \tau_{xz} \\ \tau_{xy} & \sigma_y - \sigma & \tau_{yz} \\ \tau_{xz} & \tau_{yz} & \sigma_z - \sigma \end{vmatrix} = 0$$

Expanding the determinant, we get:

$$\sigma^3 - (\sigma_x + \sigma_y + \sigma_z)\sigma^2 + (\sigma_x\sigma_y + \sigma_y\sigma_z + \sigma_x\sigma_z - \tau_{xy}^2 - \tau_{yz}^2 - \tau_{xz}^2)\sigma$$

$$-(\sigma_x\sigma_y\sigma_z - \sigma_x\tau_{yz}^2 - \sigma_y\tau_{xz}^2 - \sigma_z\tau_{xy}^2 + 2\tau_{xy}\tau_{yz}\tau_{zx}) = 0 \qquad \text{... (3.8)}$$

This is a cubic equation and must have three distinct roots, giving three principal stresses. Further it may be borne in mind that principal stresses depend only on the state of stress at a point in a body and do not depend on the orientation of the co-ordinate axes.

3.8.1 Stress Invariants

The combination of stresses at a point which do not change with the orientation of the co-ordinate axes are called stress invariants. Therefore, we define,

$$\sigma_x + \sigma_y + \sigma_z = I_1 = \text{First invariant of stress} \qquad \text{... (3.9 a)}$$

$$\sigma_x\sigma_y + \sigma_y\sigma_z + \sigma_x\sigma_z - \tau_{xy}^2 - \tau_{yz}^2 - \tau_{xz}^2 = I_2$$

$$= \text{Second invariant of stress} \qquad \text{... (3.9 b)}$$

$$\sigma_x\sigma_y\sigma_z - \sigma_x\tau_{yz}^2 - \sigma_y\tau_{xz}^2 - \sigma_z\tau_{xy}^2 + 2\tau_{xy}\tau_{yz}\tau_{xz} = I_3$$

$$= \text{Third invariant of stress} \qquad \text{... (3.9 c)}$$

It can be noticed that

$$I_1 = \sigma_x + \sigma_y + \sigma_z = \text{trace of the stress tensor}$$

$$I_2 = \begin{vmatrix} \sigma_x & \tau_{xy} \\ \tau_{xy} & \sigma_y \end{vmatrix} + \begin{vmatrix} \sigma_y & \tau_{yz} \\ \tau_{yz} & \sigma_z \end{vmatrix} + \begin{vmatrix} \sigma_x & \tau_{xz} \\ \tau_{xz} & \sigma_z \end{vmatrix}$$

$$I_3 = \begin{vmatrix} \sigma_x & \tau_x & \tau_{xz} \\ \tau_{xy} & \sigma_y & \tau_{yz} \\ \tau_{xz} & \tau_{yz} & \sigma_z \end{vmatrix}$$

$$= \text{value of the determinant of the stress tensor.}$$

Therefore, the cubic equation can be written as

$$\sigma^3 - I_1\sigma^2 + I_2\sigma - I_3 = 0 \qquad \text{... (3.10)}$$

The roots of this equation can be determined by hit and trial method. This is more time consuming and cumbersome. Therefore, it should always be preferred to determine the roots by a numerical technique. For the above equation, Newton-Raphson method may be used to determine the roots. This method can give the roots with a sufficiently high accuracy in three or four iterations and the convergence is also very rapid.

3.8.2 Solution of Cubic Equation

1. First Method

The cubic equation is :

$$\sigma^3 - I_1\sigma^2 + I_2\sigma - I_3 = 0$$

Put

$$\sigma = r\cos\theta + \frac{I_1}{3}$$

$$r^3\cos^3\theta + \frac{I_1^3}{27} + I_1 r^2\cos^2\theta + \frac{I_1^2}{3} r\cos\theta - I_1 r^2\cos^2\theta$$

$$-\frac{I_1^3}{9} - \frac{2}{3} I_1^2 r\cos\theta + I_2 r\cos\theta + \frac{I_1 I_2}{3} - I_3 = 0$$

$$r^3\cos^3\theta - \left(\frac{I_1^2}{3} - I_2\right) r\cos\theta - \left(\frac{2}{27} I_1^3 - \frac{I_1 I_2}{3} + I_3\right) = 0$$

$$\cos^3\theta - \frac{1}{r^2}\left(\frac{I_1^2}{3} - I_2\right)\cos\theta - \frac{1}{r^3}\left(\frac{2}{27}I_1^3 - \frac{I_1 I_2}{3} + I_3\right) = 0 \qquad \text{... (a)}$$

Now $\qquad \cos 3\theta = 4\cos^3\theta - 3\cos\theta$

or

$$\cos^3\theta - \frac{3}{4}\cos\theta - \frac{1}{4}3\theta = 0 \qquad \text{... (b)}$$

Comparing Eqs. (*a*) and (*b*), we have

$$\frac{1}{r^2}\left(\frac{I_1^2}{3} - I_2\right) = \frac{3}{4}$$

$$r = \left[\frac{4}{3}\left(\frac{I_1^2}{3} - I_2\right)\right]^{1/2} \qquad \text{... (c)}$$

and

$$\frac{1}{r^3}\left(\frac{2}{27}I_1^3 - \frac{I_1 I_2}{3} + I_3\right) = \frac{1}{4}\cos 3\theta$$

or

$$\cos 3\theta = \frac{4}{r^3}\left(\frac{2}{27}I_1^3 - \frac{I_1 I_2}{3} + I_3\right)$$

$$\theta_1 = \frac{1}{3}\cos^{-1}\left[\frac{4}{r^3}\left(\frac{2}{27}I_1^3 - \frac{I_1 I_2}{3} + I_3\right)\right] \qquad \text{... (d)}$$

$$\theta_2 = 120° - \theta_1$$

$$\theta_3 = 120° + \theta_1$$

Then

$$\sigma_1 = r\cos\theta_1 + \frac{I_1}{3}$$

$$\sigma_2 = r\cos\theta_2 + \frac{I_1}{3}$$

$$\sigma_3 = r\cos\theta_3 + \frac{I_1}{3}$$

2. Second Method

Consider the cubic equation :

$$y^3 + py^2 + qy + r = 0$$

Putting $\qquad y = x - \frac{p}{3}$, we get:

$$x^3 + \left(q - \frac{p^2}{3}\right)x + \left(\frac{2}{27}p^3 - \frac{pq}{3} + r\right) = 0$$

or $$x^3 + ax + b = 0$$

where $$a = q - \frac{p^2}{3}$$

$$b = \frac{2}{27}p^3 - \frac{pq}{3} + r$$

Put $$\cos\theta = \frac{-b}{2\left(-\frac{a^3}{27}\right)^{1/2}}$$

$$g = 2\left(-\frac{a}{3}\right)^{1/2}$$

Then $$y_1 = g\cos\frac{\theta}{3} - \frac{p}{3}$$

$$y_2 = g\cos\left(120° + \frac{\theta}{3}\right) - \frac{p}{3}$$

$$y_3 = g\cos\left(240° + \frac{\theta}{3}\right) - \frac{p}{3}$$

Here $$p = -I_1, \quad q = I_2, r = -I_3$$

$$y_1 = \sigma_1, \quad y_2 = \sigma_2, \text{ and } y_3 = \sigma_3$$

3. Newton-Raphson Method

Let $$f(\sigma) = \sigma^3 - I_1\sigma^2 + I_2\sigma - I_3$$

If σ_i is the approximate root chosen for the above equation, then the better approximated root will be given by,

$$\sigma_{i+1} = \sigma_i - \frac{f(\sigma_i)}{f'(\sigma_i)}$$

where $i = 0, 1, 2, \ldots$ and $f'(\sigma_i)$ is the first derivative of $f(\sigma_i)$.

The iteration process may be continued till the difference between the two successive roots is not appreciable as demanded by the accuracy.

Thus, the three principal stresses σ_1, σ_1, σ_3 can be determined. Each principal stress when substituted back will give three direction cosines. For σ_1 to be substituted, let

$$A_1 = \begin{bmatrix} \sigma_y - \sigma_1 & \tau_{yz} \\ \tau_{yz} & \sigma_z - \sigma_1 \end{bmatrix}$$

= cofactor of first term of determinant D.

$$B_1 = -\begin{bmatrix} \tau_{xy} & \tau_{yz} \\ \tau_{xz} & \sigma_z - \sigma_1 \end{bmatrix}$$

= cofactor of second term of determinant D.

$$C_1 = \begin{bmatrix} \tau_{xy} & \sigma_y - \sigma_1 \\ \tau_{xz} & \tau_{yz} \end{bmatrix}$$

= cofactor of third term of determinant D.

Then the non-trivial (non-zero) solution is:

$$\frac{a_{nx_1}}{A_1} = \frac{a_{ny_1}}{B_1} = \frac{a_{nz_1}}{C_1} = K_1$$

where $K_1 = \text{constant} \neq 0$

$$K_1 = \frac{\sqrt{a_{nx_1}^2 + a_{ny_1}^2 + a_{nz_1}^2}}{\sqrt{A_1^2 + B_1^2 + C_1^2}} = \pm \frac{1}{\sqrt{A_1^2 + B_1^2 + C_1^2}}$$

Therefore, the direction cosines for principal stress σ_1 become:

$$a_{nx_1} = \frac{A_1}{\sqrt{A_1^2 + B_1^2 + C_1^2}}$$

$$a_{ny_1} = \frac{B_1}{\sqrt{A_1^2 + B_1^2 + C_1^2}}$$

$$a_{nz_1} = \frac{C_1}{\sqrt{A_1^2 + B_1^2 + C_1^2}}$$

Similarly, direction cosines for other principal stresses can be determined.

Discussion

(i) *Case 1.* If the three roots have distinct values, *i.e.*, $\sigma_1 \neq \sigma_2 \neq \sigma_3$, then the three corresponding principal stresses occur on a single set of mutually perpendicular principal planes.

(ii) *Case* 2. If two of the three roots are equal (say $\sigma_1 \neq \sigma_2 = \sigma_3$) then all planes perpendicular to the principal planes on which σ_1, acts are principal planes acted on by the principal stresses $\sigma_2 = \sigma_3$.

(iii) *Case* 3. If the three roots are equal, *i.e*, $\sigma_1 = \sigma_2 = \sigma_3$ then every plane is a principal plane. This type of state of stress is called Hydrostatic.

If the three principal stresses are distinct then all the nine direction cosines for the three principal stresses can be determined. However, when two of the principal stresses are equal to each other or there is a hydrostatic state of stress, then all the nine direction cosines cannot be determined directly. For such a case, the following procedure may be followed.

(iv) Let $\sigma_1 \neq \sigma_2 = \sigma_3$.

Let the direction cosines for the three principal stresses, σ_1, σ_2 and σ_3 be given by $a_{nx_1}, a_{ny_1}, a_{nz_1}$; $a_{nx_2}, a_{ny_2}, a_{nz_2}$; $a_{nx_3}, a_{ny_3}, a_{nz_3}$ respectively. Then direction cosines for σ_1 and or σ_2 can be determined directly by the procedure as outlined above. In order to determine the direction cosines for σ_3, the following equations may be used.

$$a_{nx3}^2 + a_{ny3}^2 + a_{nz3}^2 = 1$$

$$a_{nx1}.a_{nx3} + a_{ny1}.a_{ny3} + a_{nz1}.a_{nz3} = 0$$

$$a_{nx2}.a_{nx3} + a_{ny2}.a_{ny3} + a_{nz2}.a_{nz3} = 0$$

From these three equations, we can determine a_{nx3}, a_{ny3}, a_{nz3}.

(v) Let $\sigma_1 = \sigma_2 = \sigma_3$.

For such a state of stress, every direction is a principal direction. Since we are limiting our study to the three-coordinate system, therefore, we are interested only to determine three directions which are mutually perpendicular to each other. To determine these three directions the following procedure may be adopted.

Let the direction cosines for σ_1 be a_{nx1}, a_{ny1}, a_{nz1}. Then any two other directions (say X_2 and X_3) can be assumed, which are at right angles to the direction of σ_1 by using the orthogonality condition. Then in order to ensure that X_2 and X_3' are also perpendicular to each other, the following linear relation must hold good.

$$(X_2 + kX'_3) \perp X_2$$

where k is a constant and can be determined by using orthogonality condition. Then the X_3 direction which is perpendicular to the direction of σ_1 and direction X_2 is

$$X_3 = X_2 + kX_3'$$

Hence the three mutually perpendicular directions can be determined.

Example 3.3 *The state of stress at a point is given by the following array of terms :*

$$\begin{bmatrix} 9 & 6 & 3 \\ 6 & 5 & 2 \\ 3 & 2 & 4 \end{bmatrix} MPa$$

Determine the principal stresses and principal directions.

Solution. Principal stresses are the roots of the cubic equation :

$$\sigma^3 - I_1\sigma^2 + I_2\sigma - I_3 = 0$$

wher

$$I_1 = \sigma_x + \sigma_y + \sigma_z = 9 + 5 + 4 = 18$$

$$I_2 = \sigma_x\sigma_y + \sigma_y\sigma_z + \sigma_x\sigma_z - \tau_{xy}^2 - \tau_{yz}^2 - \tau_{xz}^2$$
$$= 9 \times 5 + 5 \times 4 + 9 \times 4 - 6^2 - 2^2 - 3^2$$
$$= 45 + 20 + 36 - 36 - 4 - 9 = 52$$

$$I_3 = \sigma_x\sigma_y\sigma_z - \sigma_x\tau_{yz}^2 - \sigma_y\tau_{xz}^2 - \sigma_z\tau_{xy}^2 + 2\tau_{xy}\tau_{yz}\tau_{xz}$$
$$= 9 \times 5 \times 4 - 9 \times 4 - 5 \times 9 - 4 \times 36 + 2 \times 6 \times 2 \times 3$$
$$= 180 - 36 - 45 - 144 + 72$$
$$= 27$$

$\therefore$ The cubic equation becomes

$\sigma^3 - 18\sigma^2 + 52\sigma - 27 = 0$

We shall solve this equation by Newton Raphson method.

Let the approximate root be assumed = 3. Then we have from,

$$\text{First approximation} = 3 - \frac{f(3)}{f'(3)}$$

$$= 3 - \frac{27 - 162 + 156 - 27}{27 - 108 + 52} = 3 - \frac{6}{29}$$

$$= 3 - 0.207 = 2.793$$

Second approximation

$$= 2.793 - \frac{f(2.793)}{f'(2.793)}$$

$$= 2.793 - \frac{0.387}{25.148}$$

$$= 2.793\text{-}0.01538 = 2.777$$

Third approximation

$$= 2.777 - \frac{f(2.777)}{f'(2.777)}$$

$$= 2.777 - \frac{0.004}{-24.837}$$

$$= 2.777 + 0.00016 = 2.776$$

Therefore, we take one root as 2.776. Dividing the cubic equation by ($\sigma - 2.776$), we get

$$\sigma^2 - 15.224\sigma + 9.74$$

The roots of this quadratic equation are 14.554 and 0.669. Hence the three principal stresses are

$$\sigma_1 = 14.554 \text{ MPa} ; \sigma_2 = 2.776 \text{ MPa and } \sigma_3 = 0.669 \text{ MPa}$$

Now we determine the principal directions for σ_1 stress. For σ_1 to be substituted, the determinant D becomes

$$\begin{vmatrix} 9-14.554 & 6 & 3 \\ 6 & 5-14.554 & 2 \\ 3 & 2 & 4-14.554 \end{vmatrix} = \begin{vmatrix} -5.554 & 6 & 3 \\ 6 & -9.554 & 2 \\ 3 & 2 & -10.554 \end{vmatrix}$$

$$A_1 = \begin{vmatrix} -9.554 & 2 \\ 2 & -10.554 \end{vmatrix} = 100.83 - 4 = 96.83$$

$$B_1 = -\begin{vmatrix} 6 & 2 \\ 3 & -10.554 \end{vmatrix} = -(-63.324 - 6) = 69.324$$

$$C_1 = \begin{vmatrix} 6 & -9.55 \\ 3 & 2 \end{vmatrix} = 12 + 28.662 = 40.662$$

$$\sqrt{A_1^2+B_1^2+C_1^2} = \sqrt{(96.83)^2+(69.324)^2+(40.662)^2}$$
$$= \sqrt{9376+4805.8+1653.4}$$
$$= \sqrt{15835.2} = 125.83$$

$$a_{nx_1} = \frac{A_1}{\sqrt{A_1^2+B_1^2+C_1^2}} = \frac{96.83}{125.83} = 0.769$$

$$a_{ny_1} = \frac{B_1}{\sqrt{A_1^2+B_1^2+C_1^2}} = \frac{69.324}{125.83} = 0.550$$

$$a_{nz_1} = \frac{C_1}{\sqrt{A_1^2+B_1^2+C_1^2}} = \frac{40.662}{125.840} = 0.325$$

Similary,

$$a_{nx_2} = -0.226; \quad a_{nx_3} = 0.596$$
$$a_{ny_2} = -0.177; \quad a_{ny_3} = -0.800$$
$$a_{nz_2} = 0.944; \quad a_{nz_3} = 0.057$$

Example 3.4 *The components of stress at a point are given by;* $\sigma_x = 4, \sigma_y = 6, \sigma_z = 5, \tau_{xy} = 2, \tau_{yz} = 1, \tau_{zx} = 3$ *MPa. Determine the principal stresses and principal planes.*

Solution. Now

$$I_1 = \sigma_x + \sigma_y + \sigma_z$$
$$= 4 + 6 + 5 = 15$$

$$I_2 = \sigma_x\sigma_y + \sigma_y\sigma_z + \sigma_z\sigma_x - \tau_{xy}^2 - \tau_{yz}^2 - \tau_{zx}^2$$
$$= 4 \times 6 + 6 \times 5 + 4 \times 5 - 4 - 1 - 9$$
$$= 24 + 30 + 20 - 4 - 1 - 9$$
$$= 60$$

$$I_3 = -(\sigma_x\tau_{yz}^2 + \sigma_y\tau_{2x}^2 + \sigma_z\tau_{xy}^2 - \sigma_x\sigma_y\sigma_z - 2\tau_{xy}\tau_{yz}\tau_{zx})$$
$$= -(4 \times 1 + 6 \times 9 + 5 \times 4 - 120 - 12)$$
$$= +54$$

∴ Cubic equation becomes

$$\sigma^3 - 15\sigma^2 + 60\sigma - 54 = 0$$

Now

$$\cos 3\theta = 4\cos^3\theta - 3\cos\theta$$

∴

$$\cos^3\theta - \frac{3}{4}\cos\theta - \frac{1}{4}\cos 3\theta = 0 \qquad \text{... (a)}$$

Put

$$\sigma = r\cos\theta + \frac{I_1}{3} = r\cos\theta + 5$$

∴ The cubic equation becomes:

$$r^3\cos^3\theta - 15r^3\cos^2\theta + 60r\cos\theta - 54 + 125 + 15r^2\cos^2\theta + 75r\cos\theta - 375 - 150r\cos\theta + 300 = 0$$

$$r^3\cos^3\theta - 15r\cos\theta - 4 = 0$$

$$\cos^3\theta - \frac{15}{r^2}\cos\theta - \frac{4}{r^3} = 0 \quad \text{... (b)}$$

Hence, Eqs. (*a*) and (*b*) are indentical if

$$\frac{15}{r^2} = \frac{3}{4}, \quad i.e., \quad r = \sqrt{20} = 4.47$$

and
$$\frac{4}{r^3} = \frac{\cos 3\theta}{4}$$

or
$$\cos 3\theta = \frac{16}{(4.47)^3} = \frac{16}{89.4} = 0.179$$

∴
$$\theta_1 = 26.6°, \theta_2 = 93.4°, \theta_3 = 146.6°$$

∴
$$r_1\cos\theta_1 = 4.47\cos 26.6° = 4.00$$
$$r_2\cos\theta_2 = 4.47\cos 93.4° = -0.265$$
$$r_3\cos\theta_3 = 4.47\cos 146.6° = -3.731$$

$$\sigma_1 = r_1\cos\theta_1 + 5 = 4 + 5 = 9.0\,\text{MPa}$$

∴
$$\sigma_1 = r_2\cos\theta_2 + 5 = -0.265 + 5 = 4.735\text{ MPa}$$
$$\sigma_3 = r_3\cos\theta_3 + 5 = -3.731 + 5 = 1.268\text{ MPa}$$

Example 3.5 *Calculate the principal stresses for the stress components given below :*

$$\begin{bmatrix} 3 & 5 & 6 \\ 5 & 4 & 2 \\ 6 & 2 & 5 \end{bmatrix} \text{MPa.}$$

Solution.
$$I_1 = 3 + 4 + 5 = 12$$
$$I_2 = 12 + 20 + 15 - 25 - 4 - 36 = -18$$
$$I_3 = 3 \times 16 - 5 \times 13 + 6 \times (-14) = -101$$
$$f(\sigma) = \sigma^3 - 12\sigma^2 - 18\sigma + 101$$

Comparing with cubic equation

$$y^3 + py^2 + qy + r = 0$$

we have
$$p = -12, q = -18, r = 101$$

$$a = q - \frac{p^2}{3} = -18 - 48 = -66$$

$$b = \frac{2}{27}p^3 - \frac{pq}{3} + r = \frac{2}{27}(-1728) - \frac{1}{3}(-12)(-18) + 101$$
$$= -128 - 72 + 101 = -99$$

$$\cos\theta = \frac{-b}{2\left(-\frac{a^3}{27}\right)^{1/2}} = \frac{99}{2(103.189)} = 0.4797$$

$$\theta = 61.334°$$

$$g = 2\sqrt{\frac{66}{3}} = 9.381$$

$$\sigma_1 = y_1 = g\cos\frac{\theta}{3} - \frac{p}{3}$$

$$= 9.381 \cos 20.445° + 4 = 12.79 \text{ MPa}$$

$$\sigma_2 = y_2 = g\cos\left(120° + \frac{\theta}{3}\right) - \frac{p}{3}$$

$$= 9.381 \cos 140.445° + 4 = -3.23 \text{ MPa}$$

$$\sigma_3 = y_3 = g\cos\left(240° + \frac{\theta}{5}\right) - \frac{p}{3}$$

$$= 9.381 \cos 260.445° + 4 = 2.44 \text{ MPa}$$

3.9 RIGID BODY DISPLACEMENTS

The simplest type of displacement vector is $\bar{u}$ = constant. This type of motion defines a translation of a medium. A second group of rigid motions is that of rotations. These two groups of translations and rotations combine to form the group of all rigid displacements, since any rigid displacement may be composed of a rotation about an axis in a fixed direction, plus a translation, that depends on the choice of this axis.

From the displacement components at a single point, it is not possible to tell whether the displacement is due to distortion of the body or to movement of the body without distortion. Distortion or strain implies a change in displacement from one point to the other, and can be associated with derivatives of displacements with respect to x_j, *i.e.*, $\frac{\partial u_i}{\partial x_j}$. In order to eliminate the effect of rigid body rotation, we write

$$\frac{\partial u_i}{\partial x_j} = \frac{1}{2}\left(\frac{\partial u_i}{\partial x_j} + \frac{\partial u_j}{\partial x_i}\right) + \frac{1}{2}\left(\frac{\partial u_i}{\partial x_j} - \frac{\partial u_j}{\partial x_i}\right)$$

$$= \varepsilon_{ij} + \omega_{ij} \qquad \text{...(3.11)}$$

= Strain matrix + Rotation matrix.

The components of strain matrix are

$$\varepsilon_{ij} = \begin{bmatrix} \varepsilon_{xx} & \varepsilon_{xy} & \varepsilon_{xz} \\ \varepsilon_{yx} & \varepsilon_{yy} & \varepsilon_{yz} \\ \varepsilon_{zx} & \varepsilon_{zy} & \varepsilon_{zz} \end{bmatrix} = \begin{bmatrix} \varepsilon_x & \frac{\gamma_{xy}}{2} & \frac{\gamma_{xz}}{2} \\ \frac{\gamma_{xy}}{2} & \varepsilon_y & \frac{\gamma_{yz}}{2} \\ \frac{\gamma_{xz}}{2} & \frac{\gamma_{yz}}{2} & \varepsilon_z \end{bmatrix} \qquad \text{....(3.12 a)}$$

$$\varepsilon_{ij} = \begin{bmatrix} \dfrac{\partial u_x}{\partial x} & \dfrac{1}{2}\left(\dfrac{\partial u_x}{\partial y}+\dfrac{\partial u_y}{\partial x}\right) & \dfrac{1}{2}\left(\dfrac{\partial u_x}{\partial z}+\dfrac{\partial u_z}{\partial x}\right) \\ \dfrac{1}{2}\left(\dfrac{\partial u_y}{\partial x}+\dfrac{\partial u_x}{\partial y}\right) & \dfrac{\partial u_y}{\partial y} & \dfrac{1}{2}\left(\dfrac{\partial u_y}{\partial z}+\dfrac{\partial u_z}{\partial y}\right) \\ \dfrac{1}{2}\left(\dfrac{\partial u_z}{\partial x}+\dfrac{\partial u_x}{\partial z}\right) & \dfrac{1}{2}\left(\dfrac{\partial u_z}{\partial y}+\dfrac{\partial u_y}{\partial z}\right) & \dfrac{\partial u_z}{\partial z} \end{bmatrix} \quad \text{... (3.12 b)}$$

We find from Eq. (3.12 b) that

$$\varepsilon_{ij} = \varepsilon_{ji}, i \neq j, i, j = x, y, z \quad \text{... (3.13)}$$

Therefore the strain tensor is symmetric

The components of rotation matrix are

$$\omega_{ij} = \begin{bmatrix} \omega_{xx} & \omega_{xy} & \omega_{xz} \\ \omega_{yz} & \omega_{yy} & \omega_{yz} \\ \omega_{zx} & \omega_{zy} & \omega_{zz} \end{bmatrix}$$

$$\begin{bmatrix} 0 & \dfrac{1}{2}\left(\dfrac{\partial u_x}{\partial y}-\dfrac{\partial u_y}{\partial x}\right) & \dfrac{1}{2}\left(\dfrac{\partial u_x}{\partial z}-\dfrac{\partial u_z}{\partial x}\right) \\ \dfrac{1}{2}\left(\dfrac{\partial u_y}{\partial x}-\dfrac{\partial u_x}{\partial y}\right) & 0 & \dfrac{1}{2}\left(\dfrac{\partial u_y}{\partial z}-\dfrac{\partial u_z}{\partial y}\right) \\ \dfrac{1}{2}\left(\dfrac{\partial u_z}{\partial x}-\dfrac{\partial u_x}{\partial z}\right) & \dfrac{1}{2}\left(\dfrac{\partial u_z}{\partial y}-\dfrac{\partial u_y}{\partial z}\right) & 0 \end{bmatrix} \quad \text{...(3.14)}$$

We find from Eq. (3.14) that

$$\omega_{ij} = -\omega_{ji}, i \neq j, i, j = x, y, z \quad \text{.... (3.15)}$$

Therefore the rotation tensor is skew-symmetric. If $\delta\phi$ is the angle of rotation having rotation components $\delta\phi_x$, $\delta\phi_y$ and $\delta\phi_z$, then we have

$$\mathrm{d\,f} = \delta\phi_x \hat{i} + \delta\phi_y \hat{j} + \delta\phi_z \hat{k} \quad \text{... (3.15a)}$$

where
$$\left.\begin{aligned} \delta\phi_x &= \omega_{zy} \\ \delta\phi_y &= \omega_{xz} \\ \delta\phi_z &= \omega_{yx} \end{aligned}\right\}$$

3.10 ROTATION OF A VOLUME ELEMENT

Consider an infinitesimal volume element surrounding any particle P of a continuous media. Under a deformation, this infinitesimal volume is altered not only in position, but also in dimensions and shape. We define the "rotation of the volume element" to be the mean value of the rotations experienced by the set of infinitesimal line elements emanating from point *P*. Accordingly, in our present discussion, we are concerned with only rotations of line elements.

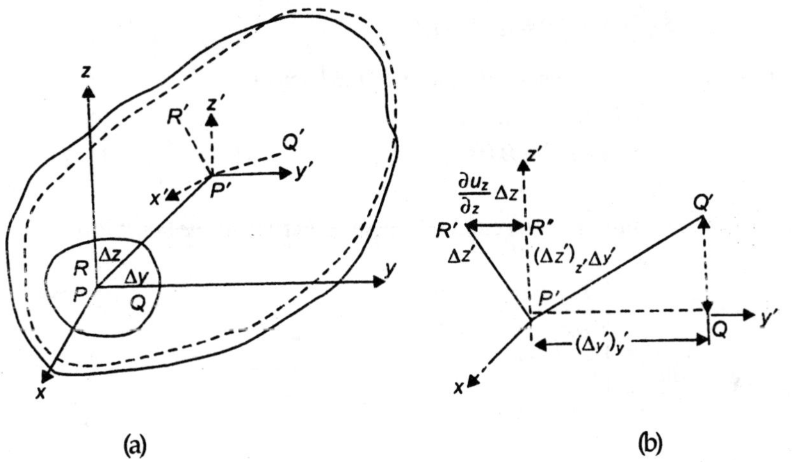

Fig. 3.7 Rotation of a volume element.

Let an element of a body at point P undergoing infinitesimal deformation move to a point P', as shown in Fig. 3.7. Thus, there is a translation $\bar{u}$ of the element in addition to rigid body rotation. The rotation of Δy about the x-axis can be expressed as :

$$\delta\phi_x = \frac{\left(u_z + \frac{\partial u_z}{\partial y}\Delta y\right) - u_z}{(\Delta y')_{y'} \simeq \Delta y}$$

where $(\Delta y')_{y'}$ = projection of $\Delta y'$ on y'-axis and for small deformations $\cong \Delta y$.

$$\therefore \quad \delta\phi_x = \frac{\partial u_z}{\partial y}$$

The rotation of Δz about the x-axis is :

$$\delta\phi_x = \frac{-\left(u_y + \frac{\partial u_y}{\partial z}\cdot\Delta z\right) + u_y}{(\Delta z')_{z'} \cong \Delta z}$$

$$\delta\phi_x = -\frac{\partial u_y}{\partial z}$$

$\therefore$ Average rotation about x-axis is :

$$\delta\phi_x = \frac{1}{2}\left(\frac{\partial u_z}{\partial y} - \frac{\partial u_y}{\partial z}\right)$$

$$= \omega_{zy} \qquad \text{... (3.16 a)}$$

Similarly

$$\delta\phi_y = \omega_{xz} = \frac{1}{2}\left(\frac{\partial u_x}{\partial z} - \frac{\partial u_z}{\partial x}\right) \qquad \text{... (3.16 b)}$$

$$\delta\phi_z = \omega_{yx} = \frac{1}{2}\left(\frac{\partial u_y}{\partial x} - \frac{\partial u_x}{\partial y}\right) \qquad \text{... (3.16 c)}$$

Therefore the rotation vector becomes

$$\overline{\delta\phi} = \delta\phi_x\hat{i} + \delta\phi_y\hat{j} + \delta\phi_z\hat{k} \qquad \text{... (3.17 a)}$$

$$\overline{\delta\phi} = \omega_{zy}\hat{i}+\omega_{xz}\hat{j}+\omega_{yx}\hat{k} \quad \text{... (3.17 b)}$$

Example 3.6 *The displacement components in a strained-body are :*

$$u_x = 0.01xy + 0.02y^2$$
$$u_y = 0.02x^2 + 0.01z^3y$$
$$u_z = 0.0lxy^2 + 0.05z^2$$

Determine the strain matrix, rotation matrix and the angle of rotation at the point P(3, 2,-5).

Solution. Now
$$\frac{\partial u_i}{\partial x_j} = \begin{bmatrix} \frac{\partial u_x}{\partial x} & \frac{\partial u_x}{\partial y} & \frac{\partial u_x}{\partial z} \\ \frac{\partial u_y}{\partial x} & \frac{\partial u_y}{\partial y} & \frac{\partial u_y}{\partial z} \\ \frac{\partial u_z}{\partial x} & \frac{\partial u_z}{\partial y} & \frac{\partial u_z}{\partial z} \end{bmatrix}$$

$$\frac{\partial u_i}{\partial x_j} = \begin{bmatrix} 0.01y & 0.01x+0.04y & 0 \\ 0.04x & 0.01z^3 & 0.03z^2y \\ 0.01y^2 & 0.02xy & 0.10\,z \end{bmatrix}$$

$$\left[\frac{\partial u_i}{\partial x_j}\right]_{P(3,2,-5)} = \begin{bmatrix} 0.02 & 0.11 & 0 \\ 0.12 & -1.25 & 1.50 \\ 0.04 & 0.12 & -0.50 \end{bmatrix}$$

Now
$$\varepsilon_{ij} = \frac{1}{2}\left(\frac{\partial u_i}{\partial x_j}+\frac{\partial u_j}{\partial x_i}\right)$$

∴ Strain matrix becomes:

$$\varepsilon_{ij} = \begin{bmatrix} 0.02 & 0.115 & 0.02 \\ 0.115 & -1.25 & 0.81 \\ 0.02 & 0.81 & -0.50 \end{bmatrix}$$

Rotation matrix,

$$\omega_{ij} = \frac{1}{2}\left(\frac{\partial u_i}{\partial x_j}-\frac{\partial u_j}{\partial x_i}\right)$$

∴
$$\omega_{ij} = \begin{bmatrix} 0 & -0.005 & -0.02 \\ 0.005 & 0 & 0.69 \\ 0.02 & -0.69 & 0 \end{bmatrix}$$

Also
$$\overline{\delta\phi} = \omega_{zy}\hat{i}+\omega_{xz}\hat{j}+\omega_{yz}\hat{k}$$
$$= -0.69\hat{i}-0.02\hat{j}+0.005\hat{k}$$

3.11 PRINCIPAL STRAINS

The principal strains are the eigen values of the strain matrix ε_{ij}. Let ε_i, $i = 1, 2, 3$ be the eigen-values of the strain matrix ε_{ij}, then the characteristic equation to determine the eigen-values is the determinant $|\varepsilon_{ij} - \varepsilon_i I| = 0$, where I is the unit matrix.

$$\begin{vmatrix} \varepsilon_x - \varepsilon_i & \frac{\gamma_{xy}}{2} & \frac{\gamma_{xz}}{2} \\ \frac{\gamma_{xy}}{2} & \varepsilon_y - \varepsilon_i & \frac{\gamma_{yz}}{2} \\ \frac{\gamma_{xz}}{2} & \frac{\gamma_{yz}}{2} & \varepsilon_z - \varepsilon_i \end{vmatrix} = 0$$

Solving the determinant, we get

$$\varepsilon_i^3 - (\varepsilon_x + \varepsilon_y + \varepsilon_z)\varepsilon_i^2 + \left(\varepsilon_x \varepsilon_y + \varepsilon_y \varepsilon_z + \varepsilon_x \varepsilon_z - \frac{\gamma_{xy}^2}{4} - \frac{\gamma_{yz}^2}{4} - \frac{\gamma_{xz}^2}{4}\right)\varepsilon_i$$

$$-\left(\varepsilon_x \varepsilon_y \varepsilon_z - \varepsilon_x \cdot \frac{\gamma_{yz}^2}{4} - \varepsilon_y \cdot \frac{\gamma_{xz}^2}{4} - \varepsilon_z \cdot \frac{\gamma_{xy}^2}{4} + \frac{\gamma_{xy} \cdot \gamma_{yz} \cdot \gamma_{xz}}{4}\right) = 0 \qquad \text{... (3.18)}$$

We define

$$J_1 = \varepsilon_x + \varepsilon_y + \varepsilon_z = \text{first invarian of strain} \qquad \text{.... (3.19 a)}$$

$$J_2 = \varepsilon_x \varepsilon_y + \varepsilon_y \varepsilon_z + \varepsilon_z \varepsilon_x - \frac{\gamma_{xy}^2}{4} - \frac{\gamma_{yz}^2}{4} - \frac{\gamma_{xz}^2}{4}$$

$$= \text{second invariant of strain} \qquad \text{.... (3.19 b)}$$

$$J_3 = \begin{vmatrix} \varepsilon_x & \frac{\gamma_{xy}}{2} & \frac{\gamma_{xz}}{2} \\ \frac{\gamma_{xy}}{2} & \varepsilon_y & \frac{\gamma_{yz}}{2} \\ \frac{\gamma_{xz}}{2} & \frac{\gamma_{yz}}{2} & \varepsilon_z \end{vmatrix}$$

$$= \text{third invariant of strain} \qquad \text{... (3.19 c)}$$

∴ Eq. (3.18) may be written as :

$$\varepsilon_i^3 - J_1 \varepsilon_i^2 + J_2 \varepsilon_i - J_3 = 0 \qquad \text{... (3.20)}$$

The same procedure as outlined for principal stresses may be followed to determine the principal strains and principal directions also.

Example 3.7 *At a point in a body the components of strain tensor are :*

$$\varepsilon_x = 0.01, \quad \varepsilon_y = -0.005, \quad \varepsilon_z = 0.005, \varepsilon_{xy} = 0.03,$$
$$\varepsilon_{yz} = 0.01, \quad \varepsilon_{xz} = -0.08$$

Determine the principal strains and principal strain directions.

Solution.

Given : The strain tensor is

$$\varepsilon_{ij} = \begin{bmatrix} \varepsilon_x & \frac{\gamma_{xy}}{2} & \frac{\gamma_{xz}}{2} \\ \frac{\gamma_{xy}}{2} & \varepsilon_y & \frac{\gamma_{yz}}{2} \\ \frac{\gamma_{xz}}{2} & \frac{\gamma_{yz}}{2} & \varepsilon_z \end{bmatrix} = \begin{bmatrix} 0.01 & 0.015 & -0.004 \\ 0.015 & -0.005 & 0.005 \\ -0.004 & 0.005 & 0.005 \end{bmatrix}$$

The principal strains are the roots of the equation :

$$\varepsilon^3 - J_1\varepsilon^2 + J_2\varepsilon - J_3 = 0$$

where $J_1 = \varepsilon_x + \varepsilon_y = \varepsilon_z$

$$= 0.01 - 0.005 + 0.005 = 0.01$$

$$J_2 = \varepsilon_x\varepsilon_y + \varepsilon_y\varepsilon_z + \varepsilon_x\varepsilon_z - \frac{\gamma_{xy}^2}{4} - \frac{\gamma_{yz}^2}{4} - \frac{\gamma_{xz}^2}{4}$$

$$= 0.01 \times (-0.005) + (-0.005)(0.005) + 0.01 \times 0.005 - \frac{1}{4}(0.030)^2 - \frac{1}{4}(0.01)^2 - \frac{1}{4}(-0.008)^2$$

$$= -0.00005 - 0.000025 + 0.0005 - 0.000225 - 0.000025 - 0.000016$$

$$= -0.000291$$

$$J_3 = 0.01[-0.005 \times 0.005 - 0.005 \times 0.005] - 0.015\,[0.015 \times 0.005 + 0.005 \times 0.004] - 0.004\,[0.015 \times 0.005 - 0.005 \times 0.004] = 0.01\,[-0.000025 - 0.000025] - 0.015\,[0.000075 - 0.000020] - 0.004[0.000074 - 0.00020]$$

$$= -0.0000003 - 0.000001425 - 0.000000220$$

$$= -0.000002145$$

$$\therefore \quad \varepsilon^3 - 0.01\,\varepsilon - 0.000291\,\varepsilon + 0.000002145 = 0$$

We shall solve this equation by Newton-Raphson method. Let the approximate root of this equation be = 0.01.

$\therefore$ First approximation gives :

$$= 0.01 - \frac{f(0.01)}{f'(0.01)}$$

$$= 0.01 - \frac{-0.765 \times 10^{-3}}{-0.0191 \times 10^{-2}}$$

$$= 0.01 - 0.004 = 0.006$$

Second approximation gives

$$= 0{\cdot}006 - \frac{f(0.006)}{f'(0.06)}$$

$$= 0.006 - \frac{0.225 \times 10^{-6}}{-303 \times 10^{-6}}$$

$$= 0.006 + 0.00084$$
$$= 0.00684$$

We take one root as 0.00684. Dividing the cubic equation by $(\varepsilon - 0.00684)$, we get

$$\varepsilon^2 - 0.00316\varepsilon - 269.4 \times 10^{-6} = 0$$

Its roots are

$$= \frac{0.00316 \pm \sqrt{(-0.00316)^2 - 4(-269.4 \times 10^{-6})}}{2}$$

$$= \frac{0.00316 \pm 10^{-3}\sqrt{1087.55}}{2}$$

$$= \frac{0.00316 \pm 10^{-3} \times 33}{2}$$

$$= 0.01808, \ -0.01492$$

$$\therefore \quad e_1 = 0.01808; \ \varepsilon_2 = 0.00684, \ \varepsilon_3 = -0.01592$$

Principal directions for ε_1 can be determined as follows :

$$|\varepsilon_{ij} - \varepsilon_1 I| = \begin{vmatrix} -0.00808 & 0.015 & -0.004 \\ 0.015 & -0.022308 & 0.005 \\ -0.004 & 0.005 & -0.01308 \end{vmatrix}$$

$$\therefore \quad A_1 = 0.02308 \times 0.01308 - 0.005 \times 0.005$$
$$= 0.00030188 - 0.000025$$
$$= 0.00027688$$
$$B_1 = -[-0.015 \times 0.01308 + 0.005 \times 0.004]$$
$$= -[-0.0001962 + 0.000020] = 0.0001762$$
$$C_1 = 0.015 \times 0.005 - 0.02308 \times 0.004$$
$$= 0.000075 - 0.00009232$$
$$= -0.00001730$$

$$K_1 = \sqrt{A_1^2 + B_1^2 + C_1^2} = 10^{-3}\sqrt{0.07666 + 0.03104 + 0.000299}$$
$$= 10^{-3}\sqrt{0.0107999}$$
$$= 10^{-3} \times 0.32863$$

$$\therefore \quad a_{nx_1} = \frac{A_1}{K_1} = \frac{0.27688 \times 10^{-3}}{0.32863 \times 10^{-3}} = 0.8425$$

$$a_{ny_1} = \frac{B_1}{K_1} = \frac{0.1762 \times 10^{-3}}{0.32863 \times 10^{-3}} = 0.5361$$

$$a_{nz_1} = \frac{C_1}{K_1} = \frac{-0.01730 \times 10^{-3}}{0.32863 \times 10^{-3}} = -0.05264$$

Similarly principal directions for $\varepsilon_2 = 0.00684$ can be deter-as follows :

$$|\varepsilon_{ij} - \varepsilon_2 I| = \begin{vmatrix} -0.00316 & 0.015 & -0.004 \\ 0.015 & -0.01184 & 0.005 \\ -0.004 & 0.005 & -0.00184 \end{vmatrix}$$

$$\begin{aligned} A_2 &= 0.01184 \times 0.00184 - 0.005 \times 0.005 \\ &= 10^{-3}\ (0.0217856\text{-}0.025) \\ &= 10^{-3}\ (-0.0032144) \\ B_2 &= -[-0.015 \times 0{,}00184 + 0.005 \times 0.004] \\ &= -[-0.0276 + 0.020]\ 10^{-3} \\ &= 0.0076 \times 10^{-3} \\ C_2 &= 0.015 \times 0.005\text{-}0.01184 \times 0.004 \\ &= 10^{-3}\ (0.075 - 0.04736) \\ &= 10^{-3}\ (0.02764) \\ K_2 &= \sqrt{A_2^2 + B_2^2 + C_2^2} = 10^{-3}\sqrt{0.00001033 + 0.00005776 + 0.00076397} \\ &= 10^{-3}\sqrt{0.0008319} \\ &= 10^{-3} \times 0.02884 \end{aligned}$$

$\therefore$

$$a_{nx_2} = \frac{A_2}{K_2} = \frac{-0.0032144 \times 10^{-3}}{0.2884 \times 10^{-3}} = -0.1114$$

$$a_{ny_2} = \frac{B_2}{K_2} = \frac{0.0076 \times 10^{-3}}{0.02884 \times 10^{-3}} = 0.026352$$

$$a_{nz_2} = \frac{C_2}{K_2} = \frac{0.02764 \times 10^{-3}}{0.02884 \times 10^{-3}} = 0.95839$$

And for $\varepsilon_3 = -0.01492$, we have

$$|\varepsilon_{ij} - \varepsilon_3 I| = \begin{vmatrix} -0.014392 & 0.015 & -0.004 \\ 0.015 & -0.00992 & 0.005 \\ -0.004 & 0.005 & -0.01992 \end{vmatrix}$$

$\therefore$

$$\begin{aligned} A_3 &= 0.00992 \times 0.01992 - 0.005 \times 0.005 \\ &= 10^{-3}\ (0.1976 - 0.025) \\ &= 0.1726 \times 10^{-3} \\ B_2 &= -[0.015 \times 0.01992 + 0.005 \times 0.004] \\ &= 10^{-3}\ [0.2988 - 0.020] \\ &= -0.2788 \times 10^{-3} \\ C_3 &= 0.015 \times 0.005 + 0.00992 \times 0.004 \\ &= 10^{-3}\ [0.075 + 0.3968] \\ &= 0.11468 \times 10^{-3} \\ K_3 &= \sqrt{A_2^3 + B_3^2 + C_3^2} \end{aligned}$$

$$= 10^{-3}\sqrt{(0.1726)^2 + (0.2788)^2 + (0.11468)^2}$$
$$= 10^{-3}\sqrt{0.02979 + 0.07779 + 0.01315}$$
$$= 10^{-3}\sqrt{0.120669}$$
$$= 0.34737 \times 10^{-3}$$

$$a_{nx_3} = \frac{A_3}{K_3} = \frac{0.1726 \times 10^{-3}}{0.34737 \times 10^{-3}} = 0.49687$$

$$a_{ny_3} = \frac{B_3}{K_3} = \frac{-0.2788 \times 10^{-3}}{0.34737 \times 10^{-3}} = -0.80260$$

$$a_{nz_3} = \frac{C_3}{K_3} = \frac{0.11468 \times 10^{-3}}{0.34737 \times 10^{-3}} = 0.33013$$

3.12 COMPONENTS OF STRAIN IN RECTANGULAR CARTESIAN COORDINATES

Consider two lines OA and OB along the axes ox and oy in the undeformed geometry of a body. After deformation the points O, A and B pass to new positions O', A' and B' and the right angle AOB changes to $A'O'B'$, as shown in Fig. 3.8. The displacements of a point will be represented by three rectangular components u_x, u_y, u_z parallel to the axes x, y, z respectively. The components of the displacement of points A and B are then given by:

$$(u_x)_A = u_x + \frac{\partial u_x}{\partial x}\cdot dx$$

$$(u_y)_A = u_y + \frac{\partial u_y}{\partial x}\cdot dx$$

$$(u_x)_B = u_x + \frac{\partial u_x}{\partial y}\cdot dy$$

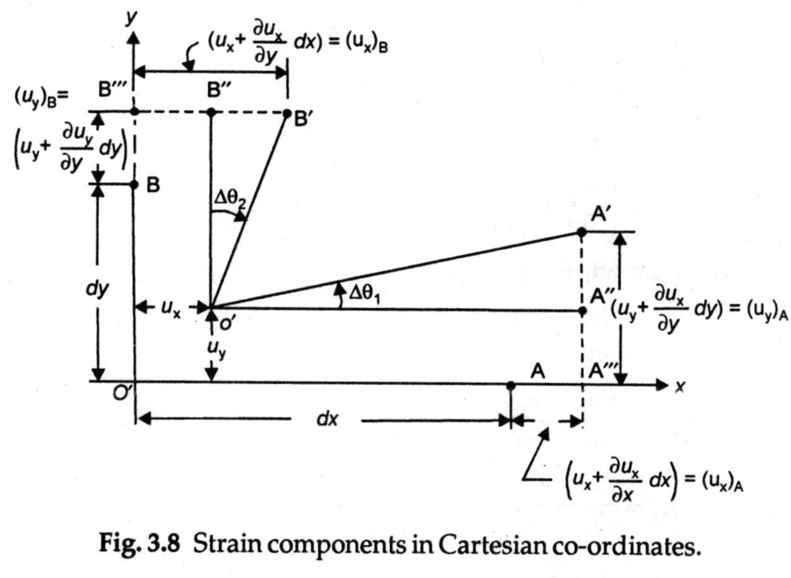

Fig. 3.8 Strain components in Cartesian co-ordinates.

$$(u_y)_B = u_y + \frac{\partial u_y}{\partial y} \cdot dy$$

$$\therefore \qquad O'A'' = dx + (u_x)_A - u_x$$

$$= dx + u_x + \frac{\partial u_x}{\partial x} \cdot dx - u_x$$

$$= dx + \frac{\partial u_x}{\partial x} dx$$

Assuming that $\angle A'O'A'' = \Delta\theta_1$ is very small, its cosine may be taken as unity.

$$\therefore \qquad = O'A' - O'A''$$

Hence, change in length of OA becomes

$$= O'A' - OA$$

$$= dx + \frac{\partial u_x}{\partial x} \cdot dx - dx$$

$$= \frac{\partial u_x}{\partial x} dx$$

$\therefore$ Strain,
$$\varepsilon_x = \frac{O'A' - OA}{OA} = \frac{\partial u_x}{\partial x} \qquad \text{... (3.21 a)}$$

Similarly,
$$\varepsilon_y = \frac{\partial u_y}{\partial y} \qquad \text{... (3.21 b)}$$

and
$$\varepsilon_z = \frac{\partial u_z}{\partial z} \qquad \text{... (3.21 c)}$$

Further
$$\tan\Delta\theta_1 = \Delta\theta_1$$

$$= \left(\frac{\partial u_y}{\partial x} dx\right)\frac{1}{dx} = \frac{\partial u_y}{\partial x}$$

$$\tan \Delta\theta_2 \simeq \Delta\theta_2$$

$$= \left(\frac{\partial u_x}{\partial y} dy\right)\frac{1}{dy} = \frac{\partial u_z}{\partial y}$$

Similarly,
$$\gamma_{xy} = \Delta\theta_1 + \Delta\theta_2$$

$$= \frac{\partial u_x}{\partial y} + \frac{\partial u_y}{\partial x} \qquad \text{... (3.22 } a\text{)}$$

Similarly,
$$\gamma_{yz} = \frac{\partial u_y}{\partial z} + \frac{\partial u_z}{\partial y} \qquad \text{.... (3.22 } b\text{)}$$

and
$$\gamma_{xz} = \frac{\partial u_x}{\partial z} + \frac{\partial u_z}{\partial x} \qquad \text{... (3.22 } c\text{)}$$

3.13 COMPATIBILITY EQUATIONS IN CARTESIAN CO-ORDINATES

We have the strain tensor given by:

$$\varepsilon_{ij} = \frac{1}{2}\left(\frac{\partial u_i}{\partial x_j} + \frac{\partial u_j}{\partial x_i}\right)$$

Therefore, for a given displacement field, the strain components can be easily determined by using the above relations. However, if the strain components are known and it is desired to know the displacement field, then integration of the six partial differential equations must be carried out. In order to ensure a single-valued, continuous solution for the displacement components, certain restrictions have to be imposed on the strain components. These resulting equations are termed the compatibility equations

Now
$$\varepsilon_x = \frac{\partial u_x}{\partial x} \quad \text{...(1)}$$

$$\varepsilon_y = \frac{\partial u_y}{\partial y} \quad \text{...(2)}$$

$$\varepsilon_z = \frac{\partial u_z}{\partial z} \quad \text{...(3)}$$

$$\gamma_{xy} = \frac{\partial u_y}{\partial x} + \frac{\partial u_x}{\partial y} \quad \text{...(4)}$$

$$\gamma_{xz} = \frac{\partial u_x}{\partial z} + \frac{\partial u_z}{\partial x} \quad \text{...(5)}$$

$$\gamma_{yz} = \frac{\partial u_y}{\partial z} + \frac{\partial u_z}{\partial y} \quad \text{...(6)}$$

Differentiating Eq. (1) w.r.t. y and Eq. (2). w.r.t. x twice, we get

$$\frac{\partial^2 \varepsilon_x}{\partial y^2} = \frac{\partial^3 u_x}{\partial y^2 \partial x} \quad \text{...(7)}$$

$$\frac{\partial^2 \varepsilon_y}{\partial x^2} = \frac{\partial^3 u_y}{\partial x^2 \partial y} \quad \text{...(8)}$$

Adding Eqs. (7) and (8), we get

$$\frac{\partial^2 \varepsilon_x}{\partial y^2} + \frac{\partial^2 \varepsilon_y}{\partial x^2} = \frac{\partial^3 u_x}{\partial y^2 dx} + \frac{\partial^3 u_y}{\partial x^2 \partial y} \quad \text{...(9)}$$

Taking the derivative of Eq. (4) w.r.t. x and y together, we get

$$\frac{\partial^2 \gamma_{xy}}{\partial x \, \partial y} = \frac{\partial^3 u_y}{\partial x^2 \partial y} + \frac{\partial^3 u_x}{\partial y^2 \partial x} \quad \text{...(10)}$$

From Eqs. (9) and (10), we get

$$\frac{\partial^2 \varepsilon_x}{\partial y^2} + \frac{\partial^2 \varepsilon_y}{\partial x^2} = \frac{\partial^2 \gamma_{xy}}{\partial x \partial y} \quad \text{... (11)}$$

Similarly, we can get

$$\frac{\partial^2 \varepsilon_y}{\partial z^2} + \frac{\partial^2 \varepsilon_z}{\partial y^2} = \frac{\partial^2 \gamma_{yz}}{\partial y \partial z} \quad \text{... (12)}$$

$$\frac{\partial^2 \varepsilon_z}{\partial x^2} + \frac{\partial^2 \varepsilon_x}{\partial z^2} = \frac{\partial^2 \gamma_{xz}}{\partial x \partial z} \quad \text{... (13)}$$

Now take the mixed derivative of Eq. (1) w.r.t. z and y.

$$\therefore \quad \frac{\partial^2 \varepsilon_x}{\partial z \partial y} = \frac{\partial^3 u_x}{\partial x\, \partial y\, \partial z} \quad \text{... (14)}$$

And taking the partial derivative of Eq.(4) w.r.t. z and x, we get

$$\frac{\partial^2 \gamma_{xy}}{\partial x \partial z} = \frac{\partial^3 u_x}{\partial x\, \partial y\, \partial z} + \frac{\partial^3 u_y}{\partial z\, \partial x^2} \quad \text{... (15)}$$

Also take the partial derivative of Eq. (5) w.r.t. y and x, thus

$$\frac{\partial^2 \gamma_{xz}}{\partial x\, \partial y} = \frac{\partial^3 u_x}{\partial x\, \partial y\, \partial z} + \frac{\partial^3 u_z}{\partial x^2\, \partial y} \quad \text{... (16)}$$

And take the partial derivative of Eq. (6) w.r.t. x twice.

$$\frac{\partial^2 \gamma_{yz}}{\partial x^2} = \frac{\partial^3 u_z}{\partial x^2 \partial y} + \frac{\partial^3 u_y}{\partial x^2 \partial z} \quad \text{... (17)}$$

Add Eqs. (15) and (16) and subtract Eq. (17) from the result. Thus, we get

$$-\frac{\partial^2 \gamma_{yz}}{\partial x^2} + \frac{\partial^2 \gamma_{xz}}{\partial x\, \partial y} + \frac{\partial^2 \gamma_{xy}}{\partial x\, \partial z} = \frac{2\partial^3 u_x}{\partial x\, \partial y\, \partial z} \quad \text{... (18)}$$

By using Eq. (14), we get

$$\frac{2\partial^2 \varepsilon_x}{\partial y\, \partial z} = \frac{\partial}{\partial x}\left[-\frac{\partial \gamma_{yz}}{\partial x} + \frac{\partial \gamma_{xz}}{\partial y} + \frac{\partial \gamma_{xy}}{\partial z}\right] \quad \text{... (19)}$$

Similarly, we can get

$$\frac{2\partial^2 \varepsilon_y}{\partial x\, \partial z} = \frac{\partial}{\partial y}\left[-\frac{\partial \gamma_{zx}}{\partial y} + \frac{\partial \gamma_{yx}}{\partial z} + \frac{\partial \gamma_{yz}}{\partial x}\right] \quad \text{... (20)}$$

$$\frac{2\partial^2\varepsilon_z}{\partial x\,\partial y} = \frac{\partial}{\partial z}\left[-\frac{\partial\gamma_{xy}}{\partial z}+\frac{\partial\gamma_{yx}}{\partial x}+\frac{\partial\gamma_{zx}}{\partial y}\right] \quad ...(21)$$

Eqs. (11) to (13) and (19) to (21)are the six compatibility equations.

3.14 GENERALISED HOOKE'S LAW

The most general linear relationship which connects stresses to trains is known as the generalised Hooke's law. It can be expressed as:

$$\varepsilon_{ij} = c_{ljkl}\tau_{kl}$$

for an elastic anisotropic media.

where c_{ijkl} = coefficients of deformation

Solving for stresses, we get

$$\tau_{kl} = [c_{ijkl}]^{-1}\varepsilon_{ij}$$

or $$\tau_{ij} = C_{ijkl}\,\varepsilon_{kl}$$

where C_{ijkl} = moduli of elasticity or elastic constants.

These are 81 elastic constants for the most general case.

(i) If there is symmetry in the last two indices, i.e.,

$$C_{ijkl} = C_{ijlk}$$

then the number of independent elastic constants reduces by 27 to 54.

(ii) If there is symmetry in the first two indices also, i.e.

$$C_{ijkl} = C_{jilk}$$

then the number of independent elastic constants further reduces by 18 to 36.

$\therefore$ $$\tau_{ij} = c_{ij}\,\varepsilon_{ij}$$

(iii) If an elastic potential exists when the variation of the body under deformation occurs isothermally or adiabatically then the number of independent elastic constants in the most general case of anisotropy is reduced to 21. This happens on account of the matrix of elastic constants to be symmetric.

(iv) A body is said to be isotropic if its elastic properties are the same in all directions. In order to determine the number of independent elastic constants for various crystallographic form of materials, the symmetry of the crystal structure is taken into accout.

1. **Monoclinic materials :** If the coordinate system is rotated about the *z*-axis by 180°, then there is elastic symmetry about the plane *yz* and *xz*. This reduces the number of independent elastic constants by 8 to 13.
2. **Orthotropic materials :** If the coordinate system is rotated about the x-axis by 180°, the number of independent elastic constants reduces further by 4 to 9. Such like materials exhibit symmetry about the three diagonal planes.
3. **Tetragonal materials :** A tetragonal material is an orthotropic material which has the same properties along two of the axes but different properties along the third axis. This reduces the number of independent elastic constants to 6.
4. **Cubic materials :** In cubic materials, the mechanical properties are same along three orthagonal

directions. This reduces the total number of elastic constants to 3.

5. **Isotropic materials :** These materials possess elastic properties which are independent of the orientation of the axes. This reduces the number of independent elastic constants to 2.

Therefore, the stress-strain relationships may be expressed as :

$$\begin{Bmatrix} \sigma_x \\ \sigma_y \\ \sigma_z \\ \tau_{xy} \\ \tau_{yz} \\ \tau_{xz} \end{Bmatrix} = \begin{bmatrix} 2G+\lambda & \lambda & \lambda & 0 & 0 & 0 \\ \lambda & 2G+\lambda & \lambda & 0 & 0 & 0 \\ \lambda & \lambda & 2G+\lambda & 0 & 0 & 0 \\ 0 & 0 & 0 & G & 0 & 0 \\ 0 & 0 & 0 & 0 & G & 0 \\ 0 & 0 & 0 & 0 & 0 & G \end{bmatrix} \begin{Bmatrix} \varepsilon_x \\ \varepsilon_y \\ \varepsilon_z \\ \gamma_{xy} \\ \gamma_{yz} \\ \gamma_{xz} \end{Bmatrix} \quad \text{...(3.23)}$$

$$\therefore \quad \left.\begin{aligned} \sigma_x &= (2G+\lambda)\varepsilon_x + \lambda(\varepsilon_y + \varepsilon_z) \\ \sigma_y &= (2G+\lambda)\varepsilon_x + \lambda(\varepsilon_z + \varepsilon_x) \\ \sigma_z &= (2G+\lambda)\varepsilon_z + \lambda(\varepsilon_x + \varepsilon_y) \\ \tau_{xy} &= G\,\gamma_{xy} \\ \tau_{yz} &= G\,\gamma_{yz} \\ \tau_{xz} &= G\,\gamma_{xz} \end{aligned}\right\} \quad \text{....(3.24)}$$

Expressing strains in terms of stresses, we get

$$\left.\begin{aligned} \varepsilon_x &= \frac{\lambda+G}{G(3\lambda+2G)}\sigma_x - \frac{\lambda}{2G(3\lambda+2G)}(\sigma_y+\sigma_z) \\ \varepsilon_y &= \frac{\lambda+G}{G(3\lambda+2G)}\sigma_x - \frac{\lambda}{2G(3\lambda+2G)}(\sigma_z+\sigma_x) \\ \varepsilon_z &= \frac{\lambda+G}{G(3\lambda+2G)}\sigma_z - \frac{\lambda}{2G(3\lambda+2G)}(\sigma_x+\sigma_y) \\ \gamma_{xy} &= \frac{\tau_{xy}}{G} \\ \gamma_{yz} &= \frac{\tau_{yz}}{G} \\ \gamma_{xz} &= \frac{\tau_{xz}}{G} \end{aligned}\right\} \quad \text{... (3.25)}$$

In a. simple tensile test, we have

$$\varepsilon_x = \frac{\sigma_x}{E} = \frac{\lambda+G}{G(3\lambda+2G)}\sigma_x$$

$$\therefore \qquad E = \frac{G(3\lambda+2G)}{\lambda+G} \qquad \text{... (3.26)}$$

where $\qquad E$ = modulus of elasticity.

Also

$$\varepsilon_y = -\nu\varepsilon_x = -\nu\frac{\sigma_x}{E}$$

$$\varepsilon_z = -\nu\varepsilon_x = -\nu\frac{\sigma_x}{E}$$

where $\qquad \nu$ = Poisson's ratio

For $\qquad \sigma_y = \sigma_z = 0$, we get

$$\frac{-\lambda}{2G(3\lambda+2G)}\sigma_x = -\frac{\nu}{E}\sigma_x$$

$$\therefore \qquad \frac{\nu}{E} = \frac{\lambda}{2G(3\lambda+2G)} \qquad \text{... (3.27)}$$

Substituting the value of E from Eq. (3.26), we get

$$\frac{\nu(G+\lambda)}{G(3\lambda+2G)} = \frac{\lambda}{2G(3\lambda+2G)}$$

$$\therefore \qquad 2\nu(G+\lambda) = \lambda$$

$$\nu = \frac{\lambda}{2(2G+\lambda)} \qquad \text{... (3.28)}$$

Solving for λ from Eqs. (3.26) and (3.27), we get

$$\lambda = \frac{G(2G-E)}{(E-3G)} = \frac{4G^2\nu}{E-6G\nu} \qquad \text{... (3.29)}$$

$$\therefore \qquad G = \frac{E}{2(1+\nu)} \qquad \text{... (3.30)}$$

For a hydrostatic state of stress, *i.e.*, all round compression p,

$$\sigma_x = \sigma_y = \sigma_z = -p$$

$$\therefore \qquad \varepsilon_z + \varepsilon_y + \varepsilon_z = \frac{-3(1-2\nu)p}{E}$$

$$\therefore \qquad -p = \frac{E(\varepsilon_x+\varepsilon_y+\varepsilon_z)}{3(1-2\nu)}$$

$$= \left(\lambda + \frac{2G}{3}\right)(\varepsilon_x+\varepsilon_y+\varepsilon_z)$$

$$= K(\varepsilon_x+\varepsilon_y+\varepsilon_z)$$

$$\therefore \qquad K = \lambda + \frac{2G}{3} \qquad \text{... (3.31)}$$

where K = Bulk modulus of elasticity.

Relationships between the various elastic constants have been given in Table 3.1.

Using Eqs. (3.29) and (3.30), Eqs. (3.25) become

$$\left.\begin{aligned}
\varepsilon_x &= \frac{1}{E}[\sigma_x - \nu(\sigma_y + \sigma_z)] \\
\varepsilon_y &= \frac{1}{E}[\sigma_y - \nu(\sigma_x + \sigma_z)] \\
\varepsilon_z &= \frac{1}{E}[\sigma_z - \nu(\sigma_x + \sigma_y)] \\
\gamma_{xy} &= \frac{\tau_{xy}}{G} \\
\gamma_{yz} &= \frac{\tau_{yz}}{G} \\
\gamma_{xz} &= \frac{\tau_{xz}}{G}
\end{aligned}\right\} \quad \text{...(3.32)}$$

Table 3.1 Relationship Between the Various Elastic Constants

	λ, G	K, G	G, ν	E, ν	E, G
$\lambda =$	1	$K - \frac{2G}{E}$	$\frac{2G\nu}{1-2\nu}$	$\frac{\nu E}{(1+\nu)(1-2\nu)}$	$\frac{G(E-2G)}{(3G-E)}$
$G =$	G	G	G	$\frac{E}{2(1+\nu)}$	G
$K =$	$\lambda + \frac{2G}{3}$	K	$\frac{2G(1+\nu)}{3(1-2\nu)}$	$\frac{E}{3(1-2\nu)}$	$\frac{EG}{3(3G-E)}$
$E =$	$\frac{G(3\lambda+2G)}{(\lambda+G)}$	$\frac{9KG}{3K+G}$	$2G(1+\nu)$	E	E
$\nu =$	$\frac{\lambda}{2(\lambda+G)}$	$\frac{3K-2G}{GK+2G}$	ν	ν	$\frac{E}{2G} - 1$

Example 3.8 *The state of strain at a point is given by* $\varepsilon_x = 0.001$, $\varepsilon_y = 0.003$, $\varepsilon_z = \gamma_{xy} = 0$, $\gamma_{xy} = -0.004$, $\gamma_{xz} = 0.001$. *Determine the stress tensor at this point. Take* $E = 210 \times 10^6$ *kPa. and* $\nu = 0.28$. *Also compute Lame's constant.*

Solution. Lame's constant,

$$\lambda = \frac{\nu E}{(1+\nu)(1-2\nu)}$$

$$= \frac{0.28 \times 210 \times 10^6}{(1+0.28)(1-0.56)}$$

$$= \frac{0.28 \times 210 \times 10^6}{1.28 \times 0.44} = 104.5 \times 10^6 \text{ kPa}$$

$$G = \frac{E}{2(1+\nu)}$$

$$= \frac{210 \times 10^6}{2 \times 1.28} = 82 \times 10^6 \text{ kPa}$$

$$\sigma_x = (2G + \lambda)\varepsilon_x + \lambda(\varepsilon_y + \varepsilon_z)$$

$$= (2 \times 82 + 104.5)\, 10^6 \times 0.001 + 104.5 \times 10^6 \times (-0.003)$$

$$= 268.5 \times 10^6 \times 0.001 - 104.5 \times 0.003 \times 10^6$$

$$= (268.5 - 313.5)\, 10^3 = -45 \text{ MPa}$$

$$\sigma_y = (2G + \lambda)\, \varepsilon_y + \lambda\, (\varepsilon_z + \varepsilon_x)$$

$$= 268.5 \times 10^6 \times (-0.003) + 104.5 \times 10^6 \times 0.001$$

$$= -805.5 \times 103 + 104.5 \times 10^3$$

$$= -701.0 \text{ MPa}$$

$$\sigma_z = (2G + \lambda)\varepsilon_z + \lambda(\varepsilon_x + \varepsilon_y)$$

$$= 268.5 \times 10^6 \times 0 + 104.5 \times 10^6\, (0.001 - 0.003)$$

$$= 209.0 \text{ MPa}$$

$$\tau_{xy} = G\gamma_{xy}$$

$$= 82 \times 10^6 \times 0 = 0$$

$$\tau_{yz} = G\gamma_{yz}$$

$$= 82 \times 10^6 \times 0.001 = 82 \text{ MPa}$$

$$\tau_{xz} = G\gamma_{xz}$$

$$= 82 \times 10^6 \times (-0.004)$$

$$= -328 \text{ MPa}$$

Example 3.9 *For a shear modulus of* 80×10^6 *kPa and an elastic modulus of* 200×10^6 *kPa, compute the strain tensor for the following state of stress :*

$$\sigma_x = 100,\ \sigma_y = 50,\ \sigma_z = -200,\ \tau_{xy} = -500,\ \tau_{yz} = 50,\ \tau_{xz} = 0 \text{ MPa}.$$

Solution.

Given:

$$G = \frac{E}{2(1+\nu)}$$

$\therefore$

$$\nu = \frac{E}{2G} - 1 = \frac{200 \times 10^6}{2 \times 80 \times 10^6} - 1$$

$$= 1.25 - 1 = 0.25$$

Now

$$\varepsilon_x = \frac{1}{E}[\sigma_x - \nu(\sigma_y + \sigma_z)]$$

$$= \frac{10^3}{200 \times 10^6}[100 - 0.25(50 - 200)]$$

$$= \frac{10^3}{200 \times 10^6}[100 - 37.5]$$

$$= \frac{137.5 \times 10^3}{200 \times 10^6} 687.5 \times 10^{-6}$$

$$\varepsilon_y = \frac{1}{E}[\sigma_x - \nu(\sigma_z + \sigma_x)]$$

$$= \frac{10^3}{200 \times 10^6}[50 - 0.25(-200 + 100)]$$

$$= \frac{10^3}{200 \times 10^6}[50 + 25]$$

$$= \frac{75}{200} \times 10^{-3} = 375 \times 10^{-6}$$

$$\varepsilon_z = \frac{1}{E}[\sigma_x - \nu(\sigma_x + \sigma_y)]$$

$$= \frac{10^3}{200 \times 10^6}[-200 - 0.25(100 + 50)]$$

$$= \frac{10^3}{200 \times 10^6}[-200 - 37.5]$$

$$= \frac{-237.5 \times 10^3}{200 \times 10^6} = 1187.5 \times 10^{-6}$$

$$\gamma_{xy} = \frac{\tau_{xy}}{G} = \frac{-500 \times 10^3}{80 \times 10^6}$$

$$= 6250 \times 10^{-6}$$

$$\gamma_{yz} = \frac{\tau_{yz}}{G} = \frac{50 \times 10^3}{80 \times 10^6} = 625 \times 10^{-6}$$

$$\gamma_{xz} = \frac{\tau_{xz}}{G} = 0$$

3.15 TWO-DIMENSIONAL PROBLEMS IN CARTESIAN COORDINATES

By neglecting either the stress or strain components in the z-direction two type of problems arise, *i.e.*, the plane stress and plane strain.

3.15.1 Plane Stress

For plane stress, we have

$$\left.\begin{aligned} &\sigma_z = \tau_{yz} = \tau_{zx} = 0 \\ \text{and} \qquad &\sigma_x, \sigma_y, \tau_{xy} = \phi(x, y) \end{aligned}\right\} \qquad \text{... (3.33)}$$

The state of stress in a thin plate which is subjected to forces applied at the boundary, parallel to the plane of the plate and uniformly distributed over the thickness, represents a state of plane stress. Other practical examples are, deep beams, tie bars etc.

The governing equations for plane stress are

(i) Equilibrium equations

$$\left.\begin{aligned} &\frac{\partial \sigma_x}{\partial x} + \frac{\partial \tau_{xy}}{\partial y} + B_x = 0 \\ &\frac{\partial \tau_{xy}}{\partial x} + \frac{\partial \sigma_y}{\partial y} + B_y = 0 \end{aligned}\right\} \qquad \text{... (3.34)}$$

(ii) Strain-displacement relations

$$\left.\begin{aligned} &\varepsilon_x = \frac{\partial u_x}{\partial x}, \ \varepsilon_y = \frac{\partial u_y}{\partial y} \\ &\gamma_{xy} = \frac{\partial u_x}{\partial y} + \frac{\partial u_y}{\partial x} \end{aligned}\right\} \qquad \text{... (3.35)}$$

(iii) Stress-strain relations

$$\left.\begin{aligned} &\varepsilon_x = \frac{1}{E}[\sigma_x - \nu\sigma_y] \\ &\varepsilon_y = \frac{1}{E}[\sigma_y - \nu\sigma_x] \\ &\varepsilon_z = -\frac{\nu}{E}(\sigma_x + \sigma_y) \\ &\gamma_{xy} = \frac{\tau_{xy}}{G}, \gamma_{yz} = \gamma_{zx} = 0 \end{aligned}\right\} \qquad \text{... (3.36)}$$

3.15.2 PLANE STRAIN

Here the length of the member in the z-direction is either very large so that no displacement is possible or the movements along the z-axis are otherwise prevented so that $\varepsilon_z = 0$. Further, if the loading is so arranged that it does not change in the z-direction, then each element will have same loading and hence shear strains $\gamma_{yz} = \gamma_{zx} = 0$. Thus for plane strain we have

$$\left.\begin{aligned} &\varepsilon_z = \gamma_{yz} = \gamma_{zx} = 0 \\ &u_x = u_x(x, y), u_y = u_y(x, y), u_z = 0 \end{aligned}\right\} \qquad \text{... (3.37)}$$

Some important examples of plane strain problems are, a long cylindrical tube under internal pressure, a dam restrained at both ends and a tunnel etc. The governing equations for plane strain problems are :

(i) Equilibrium equations

$$\left.\begin{aligned} \frac{\partial \sigma_x}{\partial x} + \frac{\partial \tau_{xy}}{\partial y} + B_x &= 0 \\ \frac{\partial \tau_{xy}}{\partial x} + \frac{\partial \sigma_y}{\partial y} + B_y &= 0 \\ \frac{\partial \sigma_z}{\partial z} &= 0 \end{aligned}\right\} \quad ...(3.38)$$

(ii) Strain-displacement relations

$$\left.\begin{aligned} \varepsilon_x &= \frac{\partial u_x}{\partial x}, \varepsilon_y = \frac{\partial u_y}{\partial y} \\ \gamma_{xy} &= \frac{\partial u_x}{\partial y} + \frac{\partial u_y}{\partial x} \end{aligned}\right\} \quad ...(3.39)$$

(*iii*) Stress-strain relations

$$\varepsilon_z = \frac{1}{E}[\sigma_z - \nu(\sigma_x + \sigma_y)] = 0$$

$$\because \quad \sigma_z = \nu(\sigma_x + \sigma_y)$$

$$\therefore \quad \left.\begin{aligned} \varepsilon_x &= \frac{1}{E}[\sigma_x - \nu(\sigma_y + \sigma_z)] \\ &= \left(\frac{1+\nu}{E}\right)[(1-\nu)\sigma_x - \nu\sigma_y] \\ \varepsilon_y &= \left(\frac{1+\nu}{E}\right)[(1-\nu)\sigma_y - \nu\sigma_x] \end{aligned}\right\} \quad ...(3.40)$$

$$\gamma_{xy} = \frac{\tau_{xy}}{G}$$

Plane strain equations may be transformed to plane stress by replacing E and ν by E' and ν' respectively, such that

$$E' = \frac{E(1+2\nu)}{(1+\nu)^2}, \nu' = \frac{\nu}{1+\nu} \quad ...(3.41)$$

3.16 AIRY'S STRESS FUNCTION IN CARTESIAN COORDINATES

Equilibrium equations are :

$$\frac{\partial \sigma_x}{\partial x} + \frac{\partial \tau_{xy}}{\partial y} + B_x = 0$$

$$\frac{\partial \tau_{xy}}{\partial x} + \frac{\partial \sigma_y}{\partial y} + B_y = 0$$

Let

$$\left.\begin{aligned} \sigma_x &= \frac{\partial^2 \phi}{\partial y^2} - V \\ \sigma_y &= \frac{\partial^2 \phi}{\partial x^2} - V \\ \tau_{xy} &= \frac{-\partial^2 \phi}{\partial x \, \partial y} \end{aligned}\right\} \quad \text{.... (3.42)}$$

where $\quad B_x = \frac{\partial V}{\partial x}, B_y = \frac{\partial V}{\partial y}$

and ϕ is Airy's stress function.

We immediately observe that Eq. (3.42) satisfy the equilibrium equations identically. Hence, the stress components defined by Airy's stress function satisfy the equilibrium equations. The quantity V is the body force potential.

3.16.1 Polynomials as Airy's Stress Functions

We shall illustrate some of the stress functions by the use of polynomials only.

(i) Polynomial of the first degree

$$\phi_1 = a_1 x + b_1 y$$

$$\therefore \quad \sigma_x = \frac{\partial^2 \phi_1}{\partial y^2} = 0$$

$$\sigma_y = \frac{\partial^2 \phi_1}{\partial x^2} = 0$$

$$\tau_{xy} = -\frac{\partial^2 \phi_1}{\partial x \, \partial y} = 0$$

This stress function gives a stress free body.

(ii) Polynomial of the second degree.

$$\phi_2 = \frac{a_2}{2} x^2 + b_2 xy + \frac{c_2}{2} y^2$$

$$\therefore \quad \sigma_x = \frac{\partial^2 \phi_2}{\partial y^2} = c_2$$

$$\sigma_y = \frac{\partial^2 \phi_2}{\partial x^2} = a_2$$

$$\tau_{xy} = \frac{-\partial^2 \phi_2}{\partial x \partial y} = -b_2$$

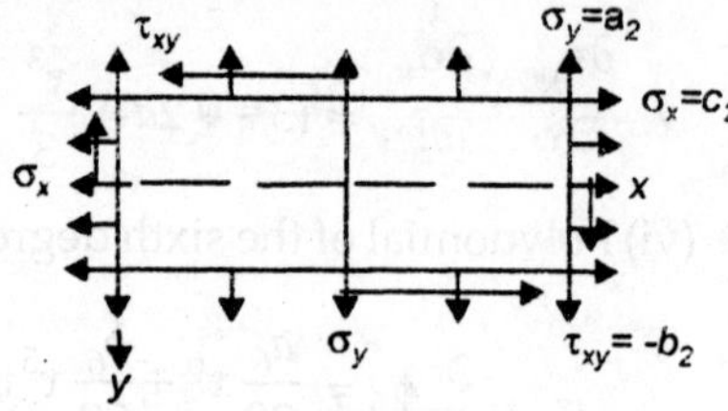

Fig. 3.9 Constant stress field

This state of stress is represented by a constant stress field as shown in Fig. 3.9.

(iii) Polynomial of the third degree.

$$\phi_3 = \frac{a_3}{6}x^3 + \frac{b_3}{2}x^2 y + \frac{c_3}{2}xy^2 + \frac{d_3}{6}y^3$$

$$\therefore \quad \sigma_x = c_3 x + d_3 y$$

$$\sigma_y = a_3 x + b_3 y$$

$$\tau_{xy} = -b_3 x - c_3 y$$

Fig 3.10 Pure bending of a beam.

This stress function gives a linearly varying stress field. If $a_3 = b_3 = c_3 = 0$, then $\sigma_x = d_3 y$. $\sigma_y = \tau_{xy} = 0$ This corresponds to pure bending on the face perpendicular to the x-axis, as shown in Fig. 3.10.

(*iv*) Polynomial of the fourth degree.

$$\phi_4 = \frac{a_4}{12}x^4 + \frac{b^4}{6}x^3 y + \frac{c_4}{2}x^2 y^2 + \frac{d_4}{6}xy^3 + \frac{e_4}{12}y^4$$

This stress function satisfies $\nabla^4 \phi_4 = 0$, if

$$e_4 = -(2c_4 + a_4)$$

$$\sigma_x = c_4 x^2 + d_4 xy - (2c_4 + a_4)\, y^2$$

$$\sigma_y = a_4 x^2 + b_4 xy - c_4 y^2$$

$$\tau_{xy} = -\frac{b_4}{2}x^2 - 2c_4 xy - \frac{d_4}{2}y^2$$

This stress function gives parabolic stress distribution.

(v) Polynomial of the fifth degree.

$$\phi_5 = \frac{a_5}{20}x^5 + \frac{b_5}{12}x^4 y + \frac{c_5}{6}x^3 y^2 + \frac{d_5}{5}x^2 y^3 + \frac{e_5}{12}xy^4 + \frac{f_5}{20}y^5$$

This stress function satisfies $\nabla^4 \phi_5 = 0$, if

$$e_5 = -3a_5 - 2c_5$$

$$b_5 = -2d_5 - 3f_5$$

$$\sigma_x = \frac{c_5}{3}x^3 + d_5 x^2 y - (3a_5 + 2c_5)\, xy^2 + f_5 y^3$$

$$\sigma_y = a_5 x^3 - (3f_5 + 2d_5)\, x^2 y + c_5 xy^2 + \frac{d_5}{3}y^3$$

$$\tau_{xy} = (3f_5 + 2d_5)\frac{x^3}{3} - c_5x^3y - d_5xy^2 + (3a_5 + 2c_5)\frac{y^3}{3}$$

(vi) Polynomial of the sixth degree.

$$\phi_6 = \frac{a_6}{30}x^6 + \frac{b_6}{20}x^5y + \frac{c_6}{12}x^4y^2 + \frac{d_6}{6}x^3y^3 + \frac{e_6}{12}x^2y^4 + \frac{f_6}{20}xy^5 + \frac{g_6}{30}y^6$$

This stress function satisfies $\nabla^4\phi_6 = 0$, if

$$e_6 = -6a_6 - 2c_6$$
$$b_6 = -f_6 - 2d_6$$
$$g_6 = 2a_6 + \frac{c_6}{2}$$

$$\sigma_x = \frac{c_6}{6}x^4 + d_6x^3y - (6a_6 + 2c_6)x^2y^2 + f_6xy^3 + (4a_6 + c_6)\frac{y^4}{2}$$

$$\sigma_y = a_6x^4 - (f_6 + 2d_6)x^3y + x_6xy^2 + d_6xy^3(-3a_6 + c_6)\frac{y^4}{3}$$

$$\tau_{xy} = (f_6 + 2d_6)\frac{x^4}{4} - \frac{2}{3}c_6x^3y - \frac{3}{2}d_6x^2y^2 + \frac{4}{3}(3a_6 + c_6)xy^3 - \frac{f_6}{4}y^4 - (4\alpha_6 + c_5)\frac{y^5}{10}$$

3.17 APPLICATIONS TO SIMPLE PROBLEMS

3.17.1 Pure Tension

Consider a rectangular plate under simple tension, as shown in Fig. 3.11

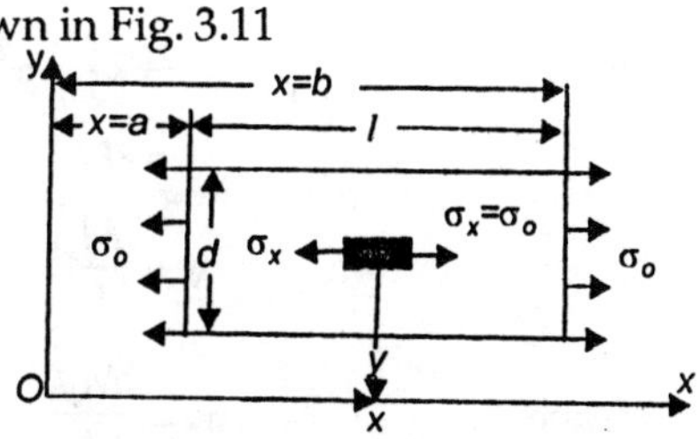

Fig.3.11 Plate under pure tension.

Let $\phi(x, y) = \frac{c}{2}y^2$

$\therefore$ $\nabla^4\phi = 0$

Hence $\sigma_x = c = \sigma_0$

$\sigma_y = 0$

$\tau_{xy} = 0$

3.17.2 Pure Shear

Consider a member under pure shear, as shown in Fig. 3.12

Let

$$\phi(x,y) = -cxy$$

which satisfies

$\nabla^4\phi = 0$. Hence

$\sigma_x = 0,\ \sigma_y = 0,\ \tau_{xy} = c = \tau.$

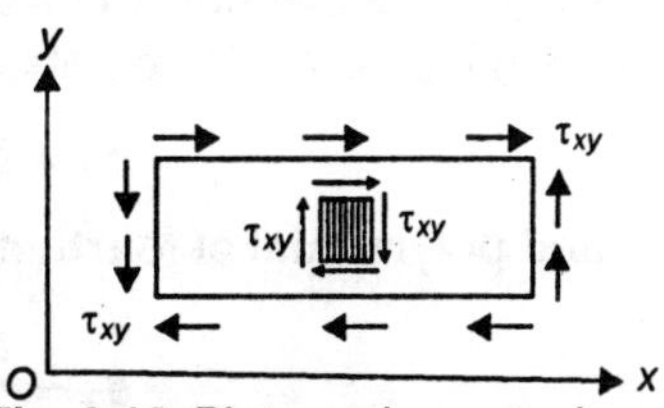

Fig. 3.12 Plate under pure shear.

3.17.3 Pure Bending

Consider a beam under pure bending, as shown in Fig. 3.13. We take

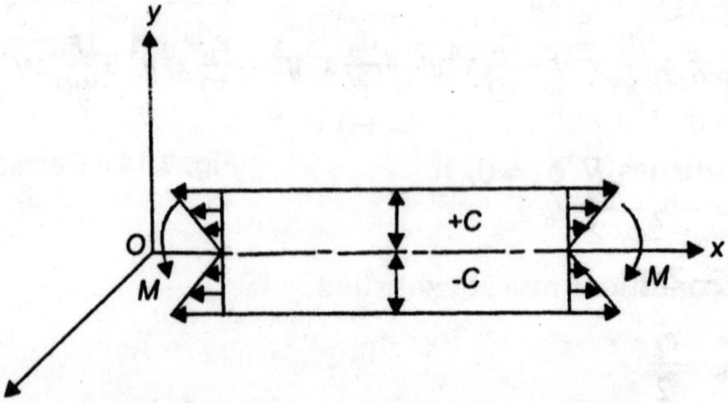

Fig. 3.13 Beam under pure bending.

$$\phi(x, y) = \frac{c}{6} y^3$$

Which satisfies $\nabla^4 \phi = 0$

Hence $$\sigma_x = cy = ky = \frac{My}{I}$$

$$\sigma_x = 0, \quad \tau_{xy} = 0$$

Now bending moment,

$$M = \int_{-c}^{+c} \sigma_x y \cdot dA$$

$$= k \int_{-c}^{+c} y^2 dA$$

$$= kI$$

$\therefore$ $$k = \frac{M}{I}$$

Hence $$\sigma_x = \frac{My}{I}$$

3.17.4 Cantilever Beam Loaded at the End

Consider a cantilever beam of narrow rectangular cross-section carrying a load P at the end as shown in Fig. 3.14. Using polynomial of second degree with $a_2 = c_2 = 0$, we have

$$\phi_2 = b_2 xy \qquad \text{... (1)}$$

and polynomial of fourth degree with $a_4 = b_4 = c_4 = e_4 = 0$

$$\phi_4 = \frac{d_4}{6} xy^3 \qquad \text{... (2)}$$

Adding both the stress functions, we get

$$\phi = \phi_2 + \phi_4$$

$$= b_2 xy + \frac{d_4}{6} xy^3$$

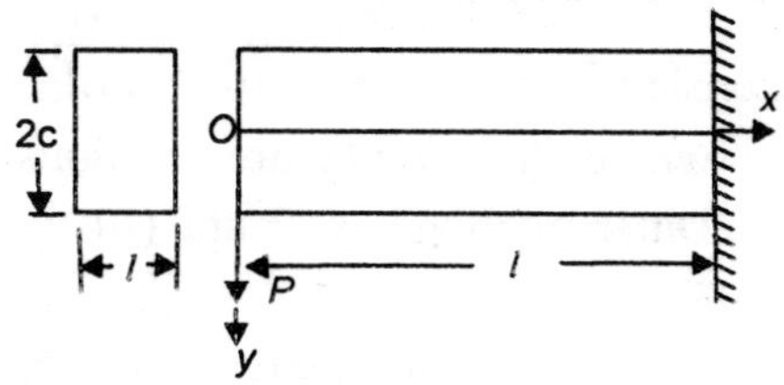

Fig. 3.14 Cantilever having end load.

$$\therefore \quad \sigma_x = d_4 xy \quad \text{... (3)}$$

$$\sigma_y = 0 \quad \text{... (4)}$$

$$\tau_{xy} = -b_2 - \frac{d_4}{2} y_6$$

The following boundary conditions must be satisfied

(i) $(\tau_{xy})_{y=\pm c} = 0$ gives $d_4 = -\frac{2d_2}{c^2}$

(ii) $\int_{-c}^{+c} \tau_{xy} dy = -P$

$$\int_{-c}^{+c} \left(-b_2 - \frac{b_2}{c^2} y^2 \right) dy = -P$$

$$\therefore \quad b_2 = \frac{3}{4} \frac{P}{c}$$

$\therefore$ Eqs. (4) become

$$\therefore \quad \left. \begin{aligned} \sigma_x &= -\frac{3}{2} \frac{P}{c^3} xy = \frac{My}{I} \\ \sigma_y &= 0 \\ \tau_{xy} &= \frac{3P}{4c}\left(1 - \frac{y^2}{c^2}\right) = -\frac{P}{I}\left(\frac{c^2 - y^2}{2}\right) \end{aligned} \right\} \quad \text{... (5)}$$

where $I = \frac{2}{3} c^3, M = -Px$

To determine the displacements, we have

$$\varepsilon_x = \frac{\partial u_x}{\partial x} = \frac{\sigma_x}{E} = -\frac{Pxy}{EI} \quad \text{... (6)}$$

$$\varepsilon_y = \frac{\partial u_y}{\partial y} = -\nu \frac{\sigma_x}{E} = \frac{\nu Pxy}{EI} \quad \text{... (7)}$$

$$\gamma_{xy} = \frac{\partial u_x}{\partial y} + \frac{\partial u_y}{\partial x} = \frac{\tau_{xy}}{G} = -\frac{P}{2IG}(c^2 - y^2) \quad \text{... (8)}$$

Integrating Eqs. (6) and (7), we get

$$u_x = \frac{Px^2 y}{2EI} + f_1(y) \quad \text{... (9)}$$

$$u_y = \frac{\nu Pxy^2}{2EI} + f_2(x) \qquad \text{... (10)}$$

where $f_1(y)$ and $f_2(x)$ are functions of y and x only respectively.

Differentiating Eqs. (9) and (10) w.r.t. y and x respectively and adding and comparing with Eq. (8), we get

$$\frac{\partial u_x}{\partial y} + \frac{\partial u_y}{\partial x} = \frac{Px^2}{2EI} + \frac{df_1(y)}{dy} + \frac{\nu Py^2}{2EI} + \frac{df_2(x)}{dx} = -\frac{P}{2IG}(c^2 - y^2)$$

Rearranging, we find

$$\left[-\frac{Px^2}{2EI} + \frac{df_2(x)}{dx}\right] + \left[\frac{\nu Py^2}{2EI} + \frac{Py^2}{2IG} + \frac{df_1(y)}{dy}\right] = -\frac{Pc^2}{2IG}$$

$\therefore \qquad F_1(x) + F_2(y) = k = \text{constant}$

$\therefore$ $F_1(x)$ and $F_2(y)$ are both constants equal to k_1 and k_2 respectively.

Let
$$-\frac{Px^2}{2EI} - \frac{Py^2}{2IG} + \frac{df_2(x)}{dx} = k_1$$

$$\frac{\nu Py^2}{2EI} - \frac{Py^2}{2IG} + \frac{df_1(y)}{dy} = k_2$$

or
$$\frac{df_2(x)}{dx} = k_1 + \frac{Px^2}{2EI}$$

$$\frac{df_2(x)}{dy} = k_2 - \frac{\nu Py^2}{2EI} + \frac{Py^2}{2IG}$$

Integrating, we get

$$f_2(x) = k_1 x + \frac{Px^3}{6EI} + k_3$$

$$f_1(y) = k_2 y - \frac{\nu Py^3}{6EI} + \frac{Py^3}{6IG} + k_3$$

Substituting in Eqs. (9) and (10), we get

$$u_x = -\frac{Px^2y}{2EI} - \frac{\nu Py^3}{6IE} + \frac{Py^2}{6GI} + K_2 y + k_4 \qquad \text{... (11)}$$

$$u_y = \frac{\nu Pxy^2}{2EI} + \frac{Px^3}{6EI} + k_1 x + k_3 \qquad \text{... (12)}$$

The constants k_1, k_2, k_3, k_4 are to be determined from the boundary conditions

$$\left.\begin{aligned} u_x = u_y &= 0 \\ \frac{\partial u_y}{\partial x} &= 0 \end{aligned}\right\} \text{for } x = l \text{ and } y = 0$$

$$\therefore \qquad k_4 = 0 \quad \text{and } k_1 l + k_3 = -\frac{Pl^3}{6EI}$$

Now
$$\frac{\partial u_y}{\partial x} = \frac{\nu P y^2}{2EI} + \frac{Px^2}{2EI} + k_1$$

$$\therefore \qquad k_1 = -\frac{Pl^2}{2EI}$$

$$k_3 = -\frac{Pl^3}{6EI} - \frac{Pl^3}{2EI} = \frac{Pl^3}{3EI}$$

Also
$$k_1 + k_2 = -\frac{Pc^2}{2IG}$$

$$\therefore \qquad k_2 = -\frac{Pc^2}{2IG} + \frac{Pl^2}{2EI}$$

$$\therefore \qquad u_x = -\frac{Px^2 y}{2EI} - \frac{\nu P y^2}{6EI} + \frac{Py^2}{6IG} - \frac{Pc^2 y}{2IG} + \frac{Pl^2 y}{2EI} \qquad \text{...(13)}$$

$$u_y = \frac{\nu Pxy^2}{2EI} + \frac{Px^3}{6EI} - \frac{Pl^2 x}{2EI} + \frac{Pl^3}{3EI} \qquad \text{...(14)}$$

The equation of the deflection curve is

$$(u_y)_{y=0} = \frac{Px^3}{6EI} - \frac{Pl^2 x}{2EI} + \frac{Pl^3}{3EI} \qquad \text{...(15)}$$

At $x = 0$

$$(u_y)_{\substack{y=0\\x=0}} = \frac{Pl^3}{3EI} \qquad \text{...(16)}$$

Example 3.10 *A cantilever beam of 2m length is of rectangular cross-section 80 mm × 120 mm. It is subjected to an end load of 1 kN. Calculate the stress distribution at the fixed end.*

Solution.

Given:
$$M = Pl = 1 \times 2 = 2 \text{ kN.m}$$

$$I = \frac{80 \times 120^3}{12} \times 10^{-12} = 1152 \times 10^{-8} \text{m}^4$$

$$\sigma_x = \frac{My}{I}$$

$$= \frac{2 \times 10^3}{1152 \times 10^{-8}} \times y \times 10^{-3} = \frac{y}{5.76} \text{MPa}$$

$$\sigma_y = 0$$

$$\tau_{xy} = -\frac{P}{I}\left(\frac{c^2 - y^2}{2}\right)$$

Here $c = 60$ mm

$$\tau_{xy} = -\frac{1-10^3}{2\times1152\times10^{-8}}(3600 - y^2)\times 10^{-6}$$

$$= -\left(\frac{3600 - y^3}{23040}\right) \text{MPa}$$

y, mm	0	10	20	30	40	50	6
σ_x MPA	0	1.736	3.472	5.208	6.944	8.680	10.416
τ_{xy} MPa	– 0.15665	–0.15191	–0.1389	–0.11718	– 0.0868	– 0.04774	0

3.17.5 BENDING OF A BEAM BY UNIFROM LOAD

Let us consider a beam of rectangular cross-section having unit width, supported at the ends, be bent by a uniformly distributed load of intensity q, as shown in Fig. 3.15

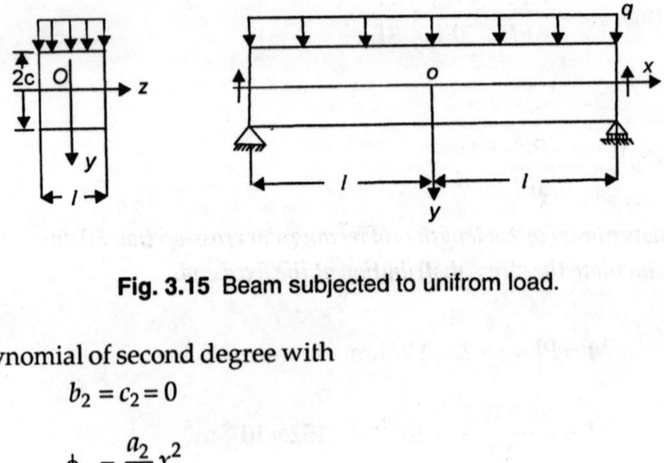

Fig. 3.15 Beam subjected to unifrom load.

Consider a polynomial of second degree with

$$b_2 = c_2 = 0$$

$$\phi_2 = \frac{a_2}{2}x^2$$

a polynomial o f thrid degree with $a_3 = a_3 = 0$

$$\phi_3 = \frac{b_3}{6}x^2 y + \frac{d_3}{6}y^3$$

and a polynomial of fifth degree with $a_5 = b_5 = c_5 = e_5 = 0$

$$\phi_5 = \frac{d_5}{6}x^2y^3 - \frac{d_5}{30}y^5 \qquad \left[\because\ f_5 = -\frac{2}{3}d_5\right]$$

$$\therefore \quad \phi = \phi_2 + \phi_3 + \phi_5$$

$$= \frac{a_2}{2}x^2 + \frac{d_3}{2}x^2y + \frac{d_3}{6}y^3 + \frac{d_5}{2}x^2y^3 - \frac{d_5}{30}y^5 \qquad ...(1)$$

$$\sigma_x = d_3y + d_5\left(x^2y - \frac{2}{3}y^3\right) \qquad ...(2)$$

$$\sigma_y = a_2 + b_3y + \frac{d_5}{3}y^3 \qquad ...(3)$$

$$\tau_{xy} = -b_3x - d_5xy^2 \qquad ...(4)$$

The following boundary conditions must be satisfied

(*i*) $(\tau_{xy})_{y=\pm c} = 0$

(*ii*) $(\sigma_y)_{y=+c} = 0$

(*iii*) $(\sigma_y)_{y=c} = -q$

(*iv*) $\int_{-c}^{+c}(\sigma_x)_{x=\pm l}\,dy = 0$

(*v*) $\int_{-c}^{+c}(\tau_{xy})_{x=\pm l}\,dy = \pm ql$

(*vi*) $\int_{-c}^{+c}(\sigma_x)_{x=\pm l}\,y\,dy = 0$

The first three conditions when substiuted in Eqs. (3) and (4) give

$$-b_3 - d_5c^2 = 0$$

$$a_2 + b_3c + \frac{d_5}{3}c^3 = 0$$

$$a_2 - b_3c - \frac{d_5}{3}c^3 = -q$$

which gives on solving $\quad a_2 = \frac{q}{2}, b_3 = \frac{3q}{4c}, d_5 = -\frac{3q}{4c^3}$

From condition (*vi*), we have

$$\int_{-c}^{+c}\left[d_3y + d_5\left(x^2y - \frac{2}{3}y^3\right)\right]y\,dy = 0$$

which gives $\quad d_3 = -d_5\left(l^2 - \frac{2}{5}c^3\right)$

$$= \frac{3q}{4c}\left(\frac{l_2}{c^2} - \frac{2}{5}\right)$$

$$\therefore \qquad \sigma_x = \frac{3q}{4c}\left(\frac{l_2}{c^2}-\frac{2}{5}\right)y - \frac{3q}{4c^3}\left(x^2y-\frac{2}{3}y^3\right)$$

$$\sigma_y = -\frac{q}{2}+\frac{3q}{4c}y-\frac{q}{4c^3}y^3$$

$$\tau_{xy} = -\frac{3q}{4c}x+\frac{3q}{4c^3}xy^2$$

Now

$$I = \frac{2}{3}c^3$$

$$\left.\begin{aligned}\sigma_x &= \frac{q}{2I}(l^2-x^2)y+\frac{q}{I}\left(\frac{y^3}{3}-\frac{c^2y}{5}\right)\\ \sigma_y &= -\frac{q}{2I}\left(\frac{y^3}{3}-c^2y+\frac{2}{3}c^3\right)\\ \tau_{xy} &= -\frac{q}{2I}x(c^2-y^2)\end{aligned}\right\} \qquad \text{...(5)}$$

The corresponding results from strenght of materials are

$$\left.\begin{aligned}\sigma_x &= \frac{My}{I}=\frac{q}{2I}(l^2-x^2)y\\ \sigma_y &= 0\\ \tau_{xy} &= -\frac{q}{2I}(c^2-y^2)x\end{aligned}\right\} \qquad \text{...(6)}$$

The comparison of both the results have been shown in Fig. 3.16.

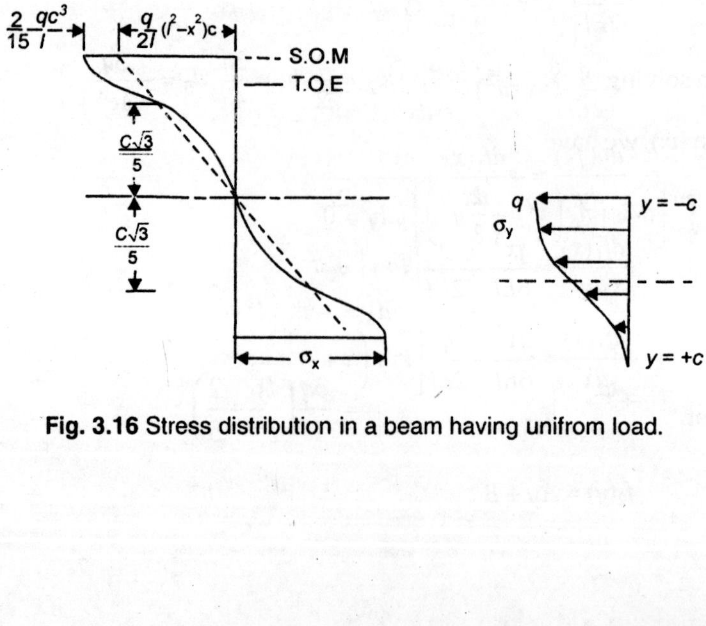

Fig. 3.16 Stress distribution in a beam having unifrom load.

Now $$\varepsilon_x = \frac{\partial u_x}{\partial x} = \frac{1}{E}[\sigma_x - \nu\sigma_y]$$

$$= \frac{q}{2EI}\left[\left\{(l^2 - x^2)y + \frac{2}{3}y^3 - \frac{2}{5}c^2y\right\} + \nu\left\{\frac{1}{3}y^3 - c^2y + \frac{2}{3}c^3\right\}\right] \quad \text{... (7)}$$

$$\varepsilon_y = \frac{\partial u_x}{\partial y} = \frac{1}{E}(\sigma_y - \nu\sigma_x)$$

$$= \frac{-q}{2EI}\left[\left(\frac{1}{3}y^3 - c^2 + \frac{2}{3}c^3\right) + \nu\left\{(l^2 - x^2)y + \left(\frac{2}{3}y^3 - \frac{2}{5}c^2y\right)\right\}\right] \quad \text{... (8)}$$

$$\gamma_{xy} = \frac{\partial u_x}{\partial y} + \frac{\partial u_y}{\partial x} = \frac{2(1+\nu)}{E}\tau_{xy} = \frac{\tau_{xy}}{G}$$

$$= \frac{q}{2IG}(c^2 - y^2)x \quad \text{... (9)}$$

Intergrating Eqs. (7) and (8), we get

$$u_x = \frac{q}{2EI}\left[\left(l^2x - \frac{x^3}{3}\right)y + x\left(\frac{2}{3}y^3 - \frac{2}{5}c^2y\right) + \nu x\left(\frac{y^3}{3} - c^2y + \frac{2}{3}c^3\right)\right] + f_2(y) \quad \text{... (10)}$$

$$u_y = -\frac{q}{2EI}\left[\left(\frac{y^4}{12} - \frac{c^2y^2}{2} + \frac{2}{3}c^3y\right) + \nu\left\{(l^2 - x^2)\frac{y^2}{2} + \left(\frac{y^4}{6} - \frac{1}{5}c^2y^2\right)\right\}\right] + f_1(x) \quad \text{...(11)}$$

where $f_1(x)$ and $f_2(y)$ are function of x and y only respectively.
Substituting the values of u_x and u_y in Eq.(9), we get

$$\frac{q}{2EI}\left[\left(l^2x - \frac{x^3}{3}\right) + x\left(2y^2 - \frac{2}{5}c^2\right) + \nu x(y^2 - c^2)\right] + \frac{df_2(y)}{dy}$$

$$-\frac{q}{2EI}(-\nu xy^2) + \frac{df_1(x)}{dx} = \frac{-q}{2IG}(c^2 - y^2)x$$

$$\therefore \quad \frac{dy(f_2)}{dy} = -\frac{df_1(x)}{dx} + \frac{qx^3}{6EI} - \frac{q}{2EI}\left[l^2 + \left(\nu + \frac{8}{5}\right)c^2\right]x = A$$

$$\frac{df_1(x)}{dx} = \frac{qx^3}{6EI} - \frac{q}{2EI}\left[l^2 + \left(\nu + \frac{8}{5}\right)c^2\right]x - A$$

$$\frac{df_1(x)}{dx} = \frac{qx^3}{6EI} - \frac{q}{2EI}\left[l^2 + \left(\nu + \frac{8}{5}\right)c^2\right]x = A$$

where A is a constant.
Integrating, we get

$$f_2(y) = Ay + B$$

$$f_1(x) = \frac{qx^4}{24EI} - \frac{q}{4EI}\left[l^2 + \left(\nu + \frac{8}{5}\right)c^2\right]x^2 - Ax - C$$

where B and C are another constants.

SUMMARY

1. The stress tensor is symmetric in nature.
2. Equilibrium equtions in Cartesian coorsinates.

$$\frac{\partial \sigma_x}{\partial x} + \frac{\partial \tau_{xy}}{\partial y} + \frac{\partial \tau_{xz}}{\partial z} + B_x = 0$$

$$\frac{\partial \tau_{xy}}{\partial_x} + \frac{\partial \sigma_y}{\partial y} + \frac{\partial \tau_{yz}}{\partial z} + B_z = 0$$

$$\frac{\partial \tau_{xz}}{\partial x} + \frac{\partial \tau_{yz}}{\partial y} + \frac{\partial \sigma_z}{\partial z} + B_z = 0$$

3. Principal stresses are the roots of cubic equation,

$$\sigma^3 - I_1\sigma^2 + I_2\sigma - I_3 = 0$$

where $I_1 = \sigma_x + \sigma_y + \sigma_z$ = first invariant of stress

$I_2 = \sigma_x\sigma_y + \sigma_y\sigma_z + \sigma_z\sigma_x - \tau^2{}_{xy} - \tau^2{}_{yz} - \tau^2{}_{zx}$ = second invariant of stress

$I_3 = |\tau_{ij}|$ = third invariant of stress

4. $\frac{\partial u_i}{\partial x_j} = \varepsilon_{ij} + \omega_{ij}$ = Strain matrix + Rotation matrix

$$\varepsilon_{ij} = \varepsilon_{ji} = \text{and } \omega_{ij} = -\omega_{ji},\ i \neq j$$

5. Principal strains are the roots of cubic eqution

$$\varepsilon^3 - J_1\varepsilon^2 + J^2\varepsilon - J_3 = 0$$

where $J_1 = \varepsilon_x + \varepsilon_y + \varepsilon_z$ = first invariant of strain

$J_2 = \varepsilon_x\varepsilon_y + \varepsilon_y\varepsilon_z + \varepsilon_z\varepsilon_x - \frac{1}{4}(\gamma_{xy}^2 + \gamma_{yz}^2 + \gamma_{zx}^2)$ = second invariant of strain

$J_3 = |\varepsilon_{ij}|$ = third invariant of strain

6. Strain component is Catersian coordinates.

$$\varepsilon_x = \frac{\partial u_x}{\partial x}, \varepsilon_y = \frac{\partial u_y}{\partial y}, \varepsilon_z = \frac{\partial u_z}{\partial z}, \gamma_{xy} = \frac{\partial u_x}{\partial y} + \frac{\partial u_y}{\partial x}, \gamma_{yz} = \frac{\partial u_y}{\partial z} + \frac{\partial u_z}{\partial y}, \gamma_{xz} = \frac{\partial u_x}{\partial z} + \frac{\partial u_z}{\partial x}$$

7. Generalised Hooke's law.

(Type of material)	(Number of elastic constants)
(i) Most general case	81
(ii) $C_{ijkl} = C_{ijlk}$	54
(iii) $C_{ijkl} = C_{jilk}$	36
(*iv*) $C_{ij} = C_{ij}$, anisotropy	21
(v) Monoclinic	13
(vi) Orthotropic	9
(vii) Tetragonal	6
(viii) Cubic	3
(ix) Isotropic	2 (λ and G)

8. Airy' stress function.

$$\sigma_x = \frac{\partial^2\phi}{\partial y^2}, \sigma_y = \frac{\partial^2\phi}{\partial x^2}, \tau_{xy} = -\frac{\partial^2\phi}{\partial x \partial y}$$

Short Answer Type Questions

1. Differentiate between surface force and body force.
2. What is the nature of stress tensor?
3. What are stress invariants?
4. What is the relationship between ε_{ij} and γ_{ij}?
5. Differentiate beteeen simple shear and pure shear.
6. Differentiate between plane stress and plane strain.
7. What are Airy's stress functions?

EXERCISES

1. (a) Define, body force, surface force and give examples.
   (b) Differentiate clearly between internal and external forces.
2. The components of stress at a point are given by :

$$\sigma_x = 3xy^2z + 2x, \quad \tau_{xy} = 0$$

$$\sigma_y = 5xyz + 3y \quad \tau_{yz} = \tau_{xz} = 3xy^2 + 2xy$$

$$\sigma_z = x^2y + y^2z.$$

Determine whether these components of stress satisfy the equilibrium equations or not at the point $(1, -1, 2)$. If not then find the body force. **[Ans.** No $\overline{B} = -11\hat{i} - 16\hat{j} + 5\hat{k}$]

3. The state of stress at a point for a given reference axes xyz is given by the following array of terms:

$$\begin{bmatrix} 150 & 80 & -60 \\ 80 & -120 & 50 \\ -60 & 50 & 80 \end{bmatrix} \text{kN/m}^2.$$

(a) Determine the stress invariants. **[Ans.** (*a*) $I_1 = 110$, $I_2 = -28100$, $I_3 = -2,375,000$**]**

4. The components of stress at a point are :

$$\sigma_x = 1, \sigma_y - 2, \sigma_z = 2, \tau_{xy} = 2, \tau_{yz} = 3, \tau_{xz} = 3\,\text{kPa}.$$

Determine:

The principal stresses at the point. **[Ans.** (*a*) $\sigma_1 = 6.13$, $\sigma_2 = -1.425$, $\sigma_3 = -3.650$ kPa.**]**

5. It is desired to drill an oil hole through a member subjected to stress field given by : $\sigma_x = 1050$, $\sigma_y = -350$, $\sigma_z = 700$, $\tau_{xy} = -210$, $\tau_{yz} = 0$, $\tau_{xz} = 70$ kN/m^2. This oil hole should be drilled in a line parallel to the line of maximum normal tension. Indicate on a one cm cube where the centre line of the oil hole should be located. **[Ans.** 1083.51 kN/m^2, 8.625°, 98.32°, 87.77°**]**
6. Derive equilibrium equations in three-dimensional Cartesian co-ordinate system. State the assumptions which you make in the derivation.
7. The stress tensor at a point is defined by the array :

$$\begin{bmatrix} 3 & 5 & 8 \\ 5 & 1 & 0 \\ 8 & 0 & 2 \end{bmatrix} \text{MN/m}^2$$

Determine the stress invariants, principal stresses and principal directions.

8. The state of stress at a point is given by the following array :

$$\begin{bmatrix} 20 & -40 & -60 \\ -40 & 40 & 20 \\ -60 & 20 & -20 \end{bmatrix} \text{MPa.}$$

Determine the three principal stresses and associated principal directions.

$$\textbf{Ans.} \begin{bmatrix} 97.39\text{MPa}, & 6.44\text{MPa}, & -63.83\text{MPa}; \\ a_{nx_1} = 0.6574 & a_{nx_2} = 0.4488 & a_{nx_3} = 0.6053 \\ a_{ny_1} = -0.6116 & a_{ny_2} = 0.7871 & a_{ny_3} = 0.0807 \\ anz_1 = 0.4402 & anz_2 = 0.4232 & anz_3 = 0.7919 \end{bmatrix}$$

9. The state of stress at a point is given by :

$$\sigma_x = 70\text{ MPa}, \sigma_y = 10\text{ MPa}, \sigma_z = -20\text{ MPa},$$
$$\tau_{xy} = -40\text{ MPa}, \tau_{yz} = 20\text{ MPa} = \tau_{xz}.$$

Determine the principal stresses, maximum shear stress and the maximum principal stress direction.

10. For what body forces will the following stress field describe a state of equilibrium ?

$$\sigma_x = -2x^2 + 3y^2 - 5z$$
$$\sigma_y = -2y^2$$

$$\sigma_z = -3x + y + 3z - 5$$
$$\tau_{xy} = z + 4xy - 7$$
$$\tau_{xz} = -3x + y + 1$$
$$\tau_{yz} = 0$$

11. Given the polynomial
$$\phi = C_1x^2 + C_2xy + C_3y^2.$$
If used as an Airy stress function, what boundary problem it may be considered a solution of in plane stress ; in plane strain ?

12. Using the following polynomial as an Airy stress function
$$\phi = C_1x^3 + C_2x^2y + C_3xy^2 + C_4y^3$$
evaluate the constants so that you have a solution to the beam under pure bending. Take the lengths of the beam as L and width equa to h.

13. Show that for a simply supported beam of length L, depth h and unit thickness, loaded by a concentrated load P at the mid-span, the stress function satisfying the loading condition is
$$\phi = \frac{C_1}{6}x^3 + C_2xy$$
where C_1 and C_2 are constants. The positive direction of y being upwards and $x = 0$ at mid-span. Also determine the stress distribution.

Chapter-4

Stresses in Beams

4.1 DEFINITIONS

When a beam is subjected to transverse loads it causes deflection in the beam which produces bending. When the loads producing bending lie in the centroidal plane such that bending is not accompanied by torsion then the bending is said to be **simple bending.** When the beam is subjected to such a system of loads so that the shear force in the beam is zero then the beam is said to be subjected to **pure bending.** Bending is quite a common phenomenon. Beams and plates used in civil engineering construction, transmission shafting, tall structures like chimneys and distillation columns subjected to wind load, etc. are some of the common examples of bending. We shall study the tensile, compressive and shear stresses produced in beams due to simple bending. Some of the terms associated with simple bending of beams are explained below.

Consider a beam subjected to bending as shown in Fig. 4.1.

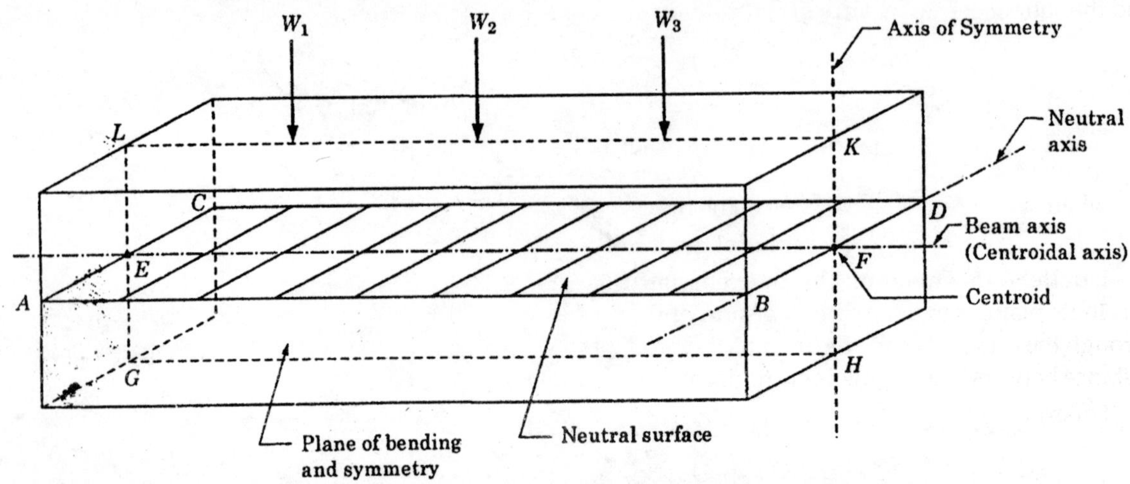

Fig. 4.1 Terms relating to bending of beam

Axis of Beam (or Centroidal axis): It is a line passing through centroids of all cross-sections of the beam. Line *EF* is the axis of beam.

Axis of symmetry: It is a line about which the area of cross-section of the beam is symmetrical. For example, circular, rectangular and square sections have two axes of symmetry, where as T-and I-sections have only one axis of symmetry. *AC, BD, GL* and *HK* are the axes of symmetry.

Plane of Symmetry: It is a plane passing through the axis of symmetry. *GHKL* is the plane of symmetry.

Plane of Bending: It is a plane in which bending takes place. For vertical loading, the plane of symmetry and plane of bending coincide. *GHKL* is the plane of bending.

Natural Surface: It is a surface on which the bending stress is zero and the beam fibres do not undergo any tension or compression. ABCD is the neutral surface.

Neutral axis: It is a line where the neutral surface intersects with the area of cross-section of the beam. *AC* and *BD* are the neutral axes.

Centroid: The centroid of a plane area is that point at which the whole of the area of the plane may be assumed to be concentrated. If a plane area is subdivided into small areas A_1, A_2 ..., and the perpendicular distances of the centroids of these areas from two chosen coordinate axes x and y are y_1, y_2 ..., and x_1, x_2 ..., respectively, then the coordinates of the centroid $(\overline{x}, \overline{y})$ of the whole area are given by:

$$\overline{x} = \frac{A_i x_1 + A_2 x_2 + \ldots}{A_1 + A_2 + \ldots} = \frac{\Sigma A_i x_i}{A}, A = \Sigma A_i \quad \ldots (4.1a)$$

$$\overline{y} = \frac{A_1 y_1 + A_2 y_2 + \ldots}{A_1 + A_2 + \ldots} = \frac{\Sigma A_i y_i}{A} \quad \ldots (4.1b)$$

Moment of Inertia: The moment of inertia or the second moment of area of a plane figure about a given axis is the sum of the product of the elementary areas into which plane figure may be divided and the square of the distance of centroids of elementary areas from the chosen axis. Thus

$$I_x = \Sigma y^2 \cdot dA$$

and

$$I_y = \Sigma x^2 \cdot dA \quad \ldots (4.2)$$

Moment of inertia, $I = AK^2$

where A = area of the figure

K = radius of gyration

Parallel Axis Theorem: This theorem states that the moment of inertia of a plane figure about any axis in its plane is equal to the total sum of the moment of inertia of the figure about a parallel axis through the centroid of the figure plus the product of the total area of the figure and the square of the distance between the parallel centroidal axis and the axis of reference.

For example,

$$\left.\begin{aligned} I_{xx} &= I_{\overline{x}\overline{x}} + A\overline{y}^2 \\ I_{yy} &= I_{\overline{y}\overline{y}} + A\overline{x}^2 \end{aligned}\right\} \quad \ldots (4.3)$$

where $\overline{x} - \overline{x}$ and $\overline{y} - \overline{y}$ are the centroidal axes.

4.2 THEORY OF SIMPLE BENDING

The theory of simple bending was developed by Bernoulli. The following assumptions are made in this theory:

1. The material of the beam is assumed to be homogeneous, perfectly elastic and isotropic. Homogeneous material means that the composition is uniform throughout. Perfectly elastic means, Hooke's law is obeyed and no permanent set is formed on unloading. Isotropic means that the elastic constants like E, G, K are same in all directions in the beam.
2. All transverse sections of the beam, which are planes before bending, remain plane after bending.
3. The radius of curvature of the beam before bending is very large in comparison to its transverse dimensions. This implies that the beam is initially straight.
4. The resultant pull or push across any transverse section of the beam is zero. This implies that total tensile force is equal to the total compressive force in the beam cross-section.
5. The Young's modulus of elasticity is same in tension and compression.
6. The stresses are within the proportional limit.

Consider the simple bending of a beam as shown in Fig. 4.2 (a). In the unloaded state, let GH be the fibres at a distance y from the centroidal axis KL, its length being determined by two transverse parallel planes AD and BC. Fig. 4.2 (*b*) shows the beam after bending, when loaded, and the planes AD and BC assume the positions A_1D_1 and B_1C_1 respectively. The planes A_1D_1 and B_1C_1 intersect at point O, the centre of curvature, and are inclined at an angle θ. Let R be the radius of curvature of the centroidal fibres at E_1F_1. Then

$$G_1H_1 = (R+y)\theta \text{ and } E_1F_1 = R\theta$$

$$\therefore \quad \frac{G_1H_1}{E_1F_1} = \frac{(R+y)\theta}{R\theta} = \frac{R+y}{R} + \frac{G_1H_1}{EF}$$

Strain in the fibres at GH,

$$\varepsilon = \frac{G_1H_1 - GH}{GH} = \frac{G_1H_1 - EF}{EF} = \frac{G_1H_1}{EF} - 1 = \frac{R+y}{R} - 1 = \frac{y}{R}$$

Let $\quad \sigma$ = bending stress in the fibres at GH

Then $$\sigma = E\varepsilon = \frac{E_y}{R}$$

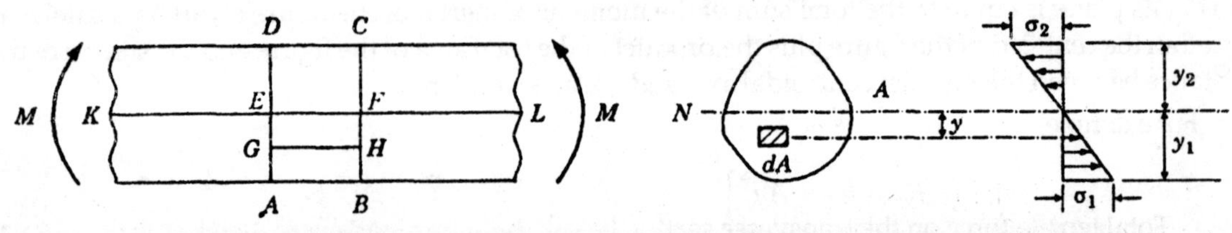

(a) Initially straight beam (c) Force on elementary beam area (d) Stress distribution

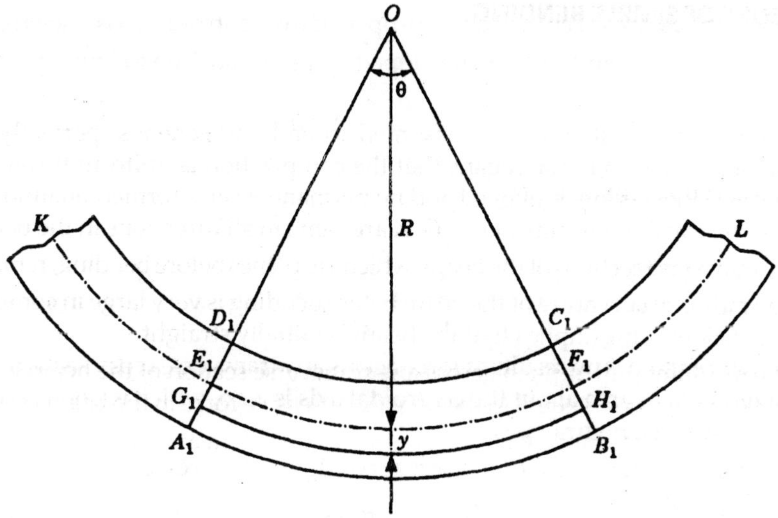

(b) Bent beam

Fig. 4.2 Simple bending of a beam

or $$\frac{\sigma}{y} = \frac{E}{R} \quad \text{...(1)}$$

For a given load on a beam,

$$\frac{E}{R} = \text{const.} = k \text{ (say)}$$

$\therefore$ $$\sigma = ky$$

Therefore, the bending stress at any point in the cross-section is proportional to its distance from the centroidal axis.

Now consider an elementary area dA at a distance y from the centroidal axis, as shown in Fig. 4.2 (c). Force acting on dA is,

$$dF = \sigma \cdot dA$$

But $$\frac{\sigma}{y} = \frac{\sigma_1}{y_1}$$

or $$\sigma = \sigma_1 \frac{y}{y_1}$$

$\therefore$ $$dF = \sigma_1 \frac{y}{y_1} \cdot dA$$

Total tensile force on the transverse section below the centroidal axis is then Fig. 4.2 (*d*),

$$F_1 = \Sigma \sigma_1 \frac{y}{y_1} dA = \frac{\sigma_1}{y_1} \Sigma y . dA$$

If the elementary area is chosen on the upper side of centroidal axis, then total compressive force will be,

$$F_2 = \frac{\sigma_2}{y_2} \Sigma y \cdot dA$$

For equilibrium of the beam,

$$F_1 = F_2$$

$$\therefore \qquad \frac{\sigma_1}{y_1} = \frac{\sigma_2}{y_2}$$

Further, since there is no resultant force across any transverse cross-section therefore, $\Sigma y \cdot dA = 0$, *i.e.* the first moment of area about the centroidal axis is zero, which is possible only if the centroidal axis coincides with the neutral axis.

Now moment of the force acting on dA about the neutral axis is,

$$dM = \frac{\sigma_1}{y_1} y^2 \, dA$$

The total moment of all the forces acting on various elements composing the cross-section forms a couple which is equal to the bending moment M. This total moment is called the *moment of resistance*.

$$\therefore \qquad M = \frac{\sigma_1}{y_1} \Sigma y^2 \cdot dA$$

$$= \frac{\sigma_1}{y_1} I$$

$$\text{or} \qquad \frac{M}{I} = \frac{\sigma_1}{y_1} = \frac{\sigma}{y} \qquad \text{...(2)}$$

where $I = \Sigma y^2 \cdot dA$ = moment of inertia of the cross-section about the neutral axis.

From Eqs. (1) and (2), we get

$$\frac{M}{I} = \frac{\sigma}{y} = \frac{E}{R} \qquad \text{...(4.4)}$$

Eq. (4.4) represents the bending formula. The implications of the bending formula are:

- The variation of bending stress (tensile or compressive) through the beam cross-section is linear,
- The bending stress is directly proportional to the distance of the beam fibres from the neutral axis. Therefore, maximum stress occurs on the outermost fibres of the beam.
- The neutral axis coincides with the centroidal axis.

4.3 SECTION MODULUS

To correlate the bending tensile and compressive stresses with the direct tensile and compressive stresses, a quantity called section modulus is defined. This is analogous to the area of cross-section.

As direct stress, $\sigma_d = \frac{P}{A}$, similarly bending stress, $\sigma_b = \frac{M}{z}$, where z = section modulus.

$$\frac{M}{I} = \frac{\sigma}{y}$$

$$\sigma_{max} = \frac{M_{y_{max}}}{I} = \frac{M}{\left(\dfrac{I}{y_{max}}\right)} = \frac{M}{z} \qquad \text{...(4.5)}$$

where $z = \dfrac{I}{y_{max}}$ is called the "section modulus".

The S.I units of z are mm^3 or m^3.

$$M = \sigma_{max} \times z$$

Those sections which have higher value of section modulus will develop lower value of bending stress for the same value of bending moment applied and same area of cross-section.

4.4 CALCULATION OF BENDING STRESSES

4.4.1 Circular Section [Fig. 4.3 (a)]

For a solid circular section beam of diameter d,

$$z = \frac{I}{y_{max}} = \frac{\pi d^4}{64} \times \frac{2}{d} = \frac{\pi d^3}{32}$$

$$\sigma = \frac{32M}{\pi d^3}$$

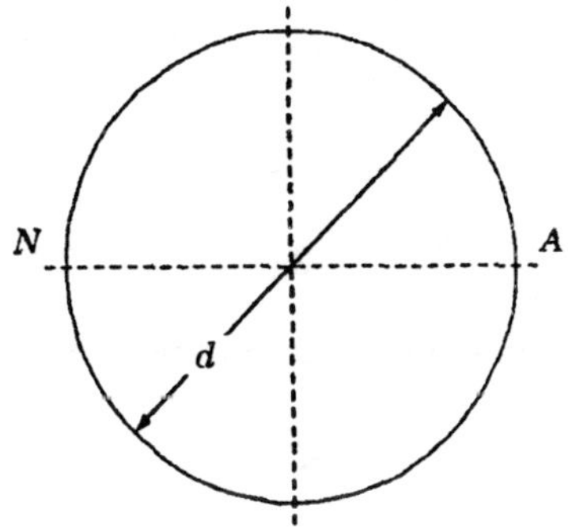

Fig. 4.3 (a) Circular section

For a hollow circular section beam of outer diameter d_0 and inner diameter d_i we have

$$z = \frac{\pi}{64}(d_0^4 - d_i^4) \times \frac{2}{d_0} = \frac{\pi}{32} d_0^3 \left[1 - \left(\frac{d_i}{d_0}\right)^4\right]$$

If $k = \dfrac{d_i}{d_0}$, then $\sigma = \dfrac{32M}{\pi(1-k^4)d_0^3}$

4.4.2 Rectangular Section [Fig. 4.3 (b)]

For a rectangular section of width b and depth h, we have

$$z = \frac{I_{NA}}{y_{max}} = \frac{6h^3}{12} \times \frac{2}{h} = \frac{6h^2}{6}$$

$$\sigma = \pm \frac{6M}{6h^2}$$

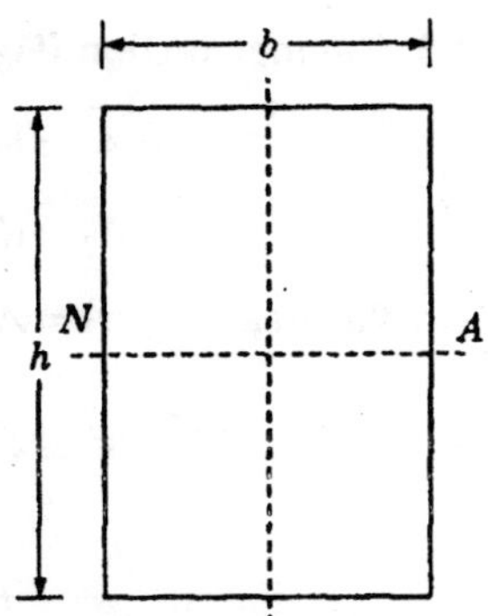

Fig. 4.3 (b) Rectangular section

4.4.3 I-Section

Consider a symmetrical I-section as shown in Fig. 4.3 (*c*). For bending about the neutral-axis, we have

$$I_{NA} = \frac{b(h+2t)^3}{12} - \frac{(b-t_1)h^3}{12}$$

$$y_{max} = \frac{h}{2}+t;\ t = \frac{I_{NA}}{y_{max}};\ \sigma = \pm\frac{M}{z}$$

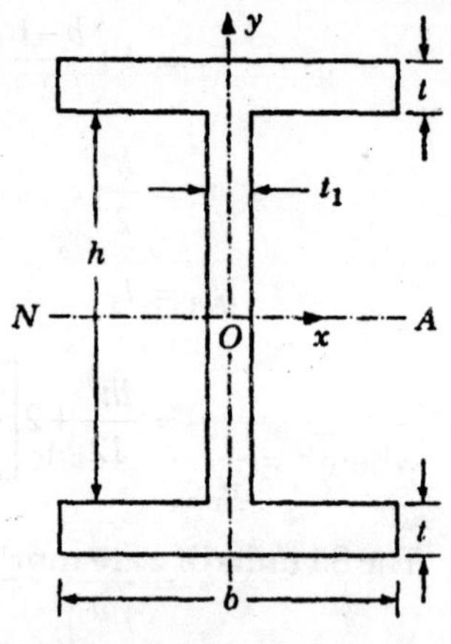

Fig. 4.3 (c) I-section

4.4.4 T-Section

The T-section [Fig. 4.3 (d)] has y-axis as the axis of symmetry.

$$\bar{y} = \frac{\frac{1}{2}b_1t_1^2 + b_2t_2\left(t_1+\frac{b_2}{2}\right)}{(b_1t_1+b_2t_2)}$$

$$I_{NA} = \frac{b_1t_1^3}{12} + b_1t_1\left(\bar{y}-\frac{t_1}{2}\right)^2 + \frac{t_2b_2^3}{12} + t_2b_2\left(t_1+\frac{b_2}{2}-\bar{y}\right)^2$$

Compressive bending stress,

$$\sigma_c = \frac{M\bar{y}}{I_{NA}}$$

Tensile bending stress,

$$\sigma_t = \frac{M(b_2+t_1-\bar{y})}{I_{NA}}$$

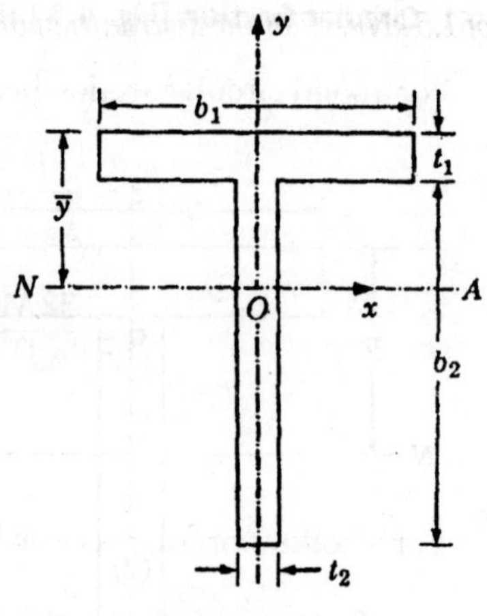

Fig. 4.3(d) T-section

4.4.5 Channel Section [Fig. 4.3 (e)]

$$A_1 = ht$$

$$A_2 = (b-t)t$$

Total area, $A = A_1 + 2A_2 = ht + 2(b-t)t$

$$\bar{x} = \frac{A_1\bar{x}_1 + 2A_2\bar{x}_2}{A}$$

$$\bar{x}_1 = \frac{t}{2}$$

$$\bar{x}_2 = t + \frac{b-t}{2}$$

$$\bar{y} = \frac{h}{2}$$

$$I_{NA} = I_{\bar{x}}$$

$$= \frac{th^3}{12} + 2\left[\frac{(b-t)t^3}{12} + (b-t)t\left\{\frac{h}{2} - \frac{t}{2}\right\}^2\right]$$

$$\frac{M}{I} = \frac{\sigma}{\left(\frac{h}{2}\right)} \Rightarrow \sigma = \frac{Mh}{2I}$$

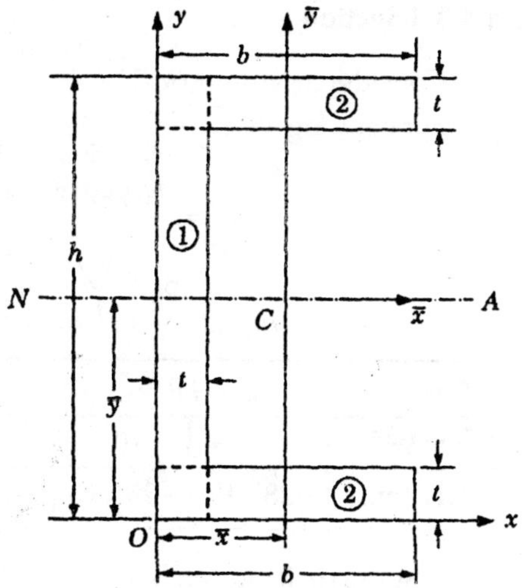

Fig .4.3 (e) Channel section

Example 4.1 *The cross-section of a T-section beam is shown in Fig. 4.4 (a). It is subjected to a bending moment of 1.5 kNm. Plot the distribution of stress due to bending in the beam.*

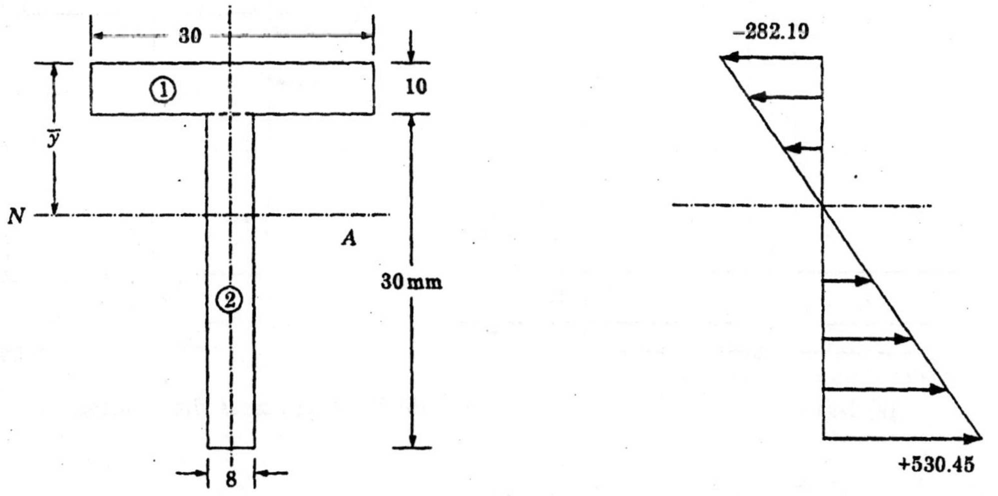

(a) T-section

(b) Bending stress distribution

Fig. 4.4

Solution:

$$\bar{y} = \frac{30 \times 10 \times 5 + 30 \times 8 \times 25}{30 \times 10 + 30 \times 8} = \frac{7500}{540} = 13.89 \text{ mm}$$

$$I_{NA} = \left[\frac{30\times10^3}{12} + 300\times(13.89-5)^2\right] + \left[\frac{8\times30^3}{12} + 240\times(25-13.89)^2\right]$$

$$= [2500 + 23709.6] + [18000 + 29623.7]$$

$$= 73833.3 \text{ mm}^4$$

$$\sigma = \frac{My}{I} = \frac{1.5\times10^6\times y}{73833.3} = 20.316y \text{ N/mm}^2$$

Fibres	*Top Fibres*			*Bottom Fibres*				
y, mm	13.89	10	5	5	10	15	20	26.11
σ, N/mm²	−282.19	−203.16	−101.58	101.58	203.16	304.74	406.32	530.45

The stress distribution is shown in Fig. 4.4 (*b*).

Example 4.2 *The cross-section of an I-beam is shown in Fig. 4.5. The bending moment at the section is 20 kN.m. Plot the distribution of bending stress in the beam.*

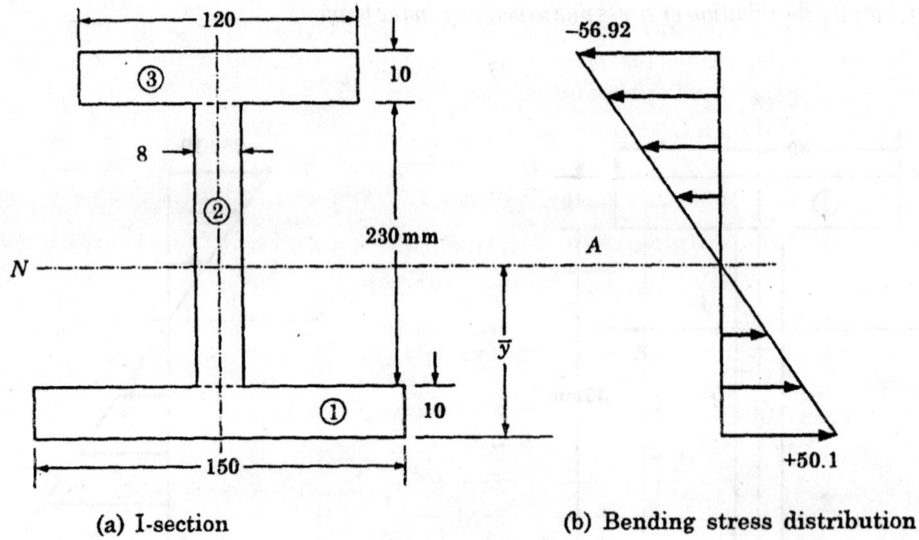

Fig. 4.5

Solution:

$A_1 = 150 \times 10 = 1500 \text{ mm}^2$
$A_2 = 230 \times 8 = 1840 \text{ mm}^2$
$A_3 = 120 \times 10 = 1200 \text{ mm}^2$
$A = 4540 \text{ mm}^2$

$$\bar{y} = \frac{1500\times5 + 1840\times125 + 1200\times245}{4540} = \frac{531500}{4540} = 117\text{mm}$$

$$I_{NA} = \left[\frac{150\times10^3}{12}+1500\times(117-5)^2\right]+\left[\frac{8\times230^3}{12}+1840\times(125-117)^2\right]$$

$$+\left[\frac{120\times10^3}{12}+1200\times(245-117)^2\right]$$

= [12500 + 118 × 10³] + [81113333 + 117760] + [10000 + 196608 ×10²]

= 464727393 mm⁴

$$\sigma = \frac{My}{I} = \frac{020\times10^6\times y}{46728393} = 0.428\,y\ \text{N/mm}^2$$

Fibres	*Top Fibres*			*Bottom Fibres*			
y, mm	133	100	50	0	50	100	117
σ, N/mm²	- 56.92	- 42.8	- 21.4	0	21.4	42.8	50.1

The stress distribution is shown in Fig. 4.5 (*b*).

Example 4.3 *A cast iron channel carries water as shown in Fig. 4.6. It is supported at two points 12 m apart. Density of water is 1000 kg/m³ and that of cast iron 7000 kg/m³. Calculate the depth of water in the channel if the tensile and compressive stresses in bending are not to exceed 20 MPa and 50 MPa respectively.*

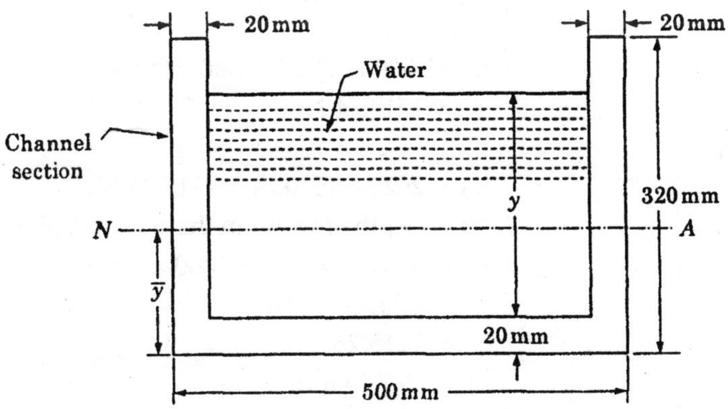

Fig. 4.6 Channel section containing water.

Solution:

$$\bar{y} = \frac{500\times20\times10+2\times300\times20\times170}{500\times20+2\times300\times20} = 97.3\text{ mm}$$

$$I_{NA} = \frac{500\times20^3}{12}+500\times20(97.3-10)^2$$

$$+2\left[\frac{20\times30^3}{12}+20\times300(170-97.3)^2\right]\times10^{-12}$$

$$= 23 \times 10^{-5}\ m^4$$

Moment of resistance in compression,

$$M_c = \frac{\sigma_C I_{NA}}{y_c} = \frac{50\times23\times10^{-5}\times10^6}{(320-93.3)\times10^{-3}} = 51.639\,kN.m$$

Moment of resistance in tension,

$$M_t = \frac{\sigma_t I_{NA}}{y_t} = \frac{20\times10^6\times23\times10^{-5}}{93.3\times10^{-3}} = 47.276\ kN.m$$

Safe bending moment = 47.276 kN.m

$\therefore$

$$\sigma_t = 20\ MPa$$

$$\sigma_c = \frac{20\times222.7}{97.3} = 45.776\ MPa$$

Let w = Total weight of channel and water per metre length, kN/m

$$M = \frac{wl^2}{8} = \frac{w\times12^2}{8} = 18w\ kN.m$$

$\therefore$

$$18w = 47.276$$

$$w = 2.626\ kN/m$$

Weight of channel per meter,

$$w_1 = (500 \times 20 + 2 \times 300 \times 20) \times 10^{-6} \times 7000 \times 9.81$$

$$= 1510.74\ N$$

Weight of water,

$$w_2 = 2626 - 1510.74 = 1115.26\ N/m$$

Let x = depth of water in the channel, mm

Then

$$w_2 = 460 \times x \times 10^{-6} \times 1000 \times 9.81$$

$$= 4.5126\,x\ N/m$$

$$= 1115.26$$

$$x = 247.14\ mm$$

Example 4.4 *A cantilever beam with triangular cross-section as shown in Fig. 4.7 carries a uniformly distributed load of intensity 2 kN/m over its entire length. If the depth h = 2b, determine the minimum dimensions allowing tensile or compressive stress to be 25 MPa.*

Solution:

$$M_{max} = \frac{wl^2}{2} = \frac{2\times2^2}{2} = 4\,kN.m$$

$$I_{NA} = \frac{bh^3}{36} = \frac{h^4}{72}$$

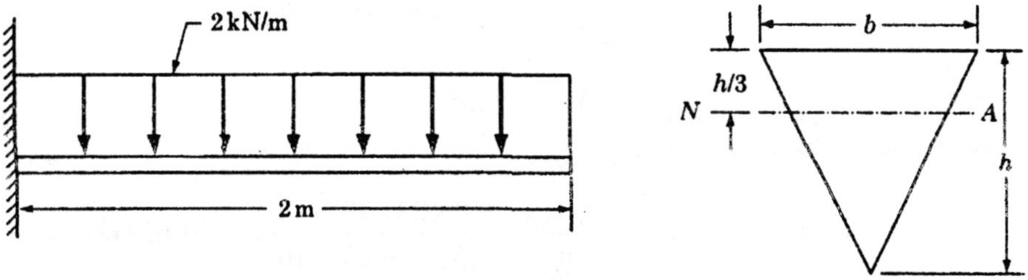

Fig. 4.7

$$\sigma_t = \frac{4\times10^3\times\frac{h}{3}}{\frac{h^4}{72}} = \frac{96\times10^3}{h^3}\,\text{N/m}^2$$

$$\sigma_c = \frac{4\times10^3\times\frac{2h}{3}}{\frac{h^4}{72}} = \frac{192\times10^3}{h^3}\,\text{N/m}^2$$

For minimum dimensions of cross-section,

$$\frac{96\times10^3}{h^3} = 25\times10^6$$

$$h^3 = 3.84\times10^{-3}$$

$$h = 0.1566\text{ m}$$

$$b = \frac{h}{2} = 0.0783\text{ m}$$

4.5 COMPOSITE OR FLITCHED BEAMS

A composite or flitched beam is made of two or more different materials and rigidly connected together. Such beams are used when one material, if used alone, requires a large cross-sectional area.

Consider a beam made of timber and steel. Let t and s be the subscripts for timber and steel respectively. At the common surface,

Strain, $\varepsilon_s = \varepsilon_t$

or $$\frac{\sigma_s}{E_s} = \frac{\sigma_t}{E_t} \qquad \text{...(4.6 a)}$$

or $$\frac{\sigma_s}{\sigma_t} = \frac{E_s}{E_t} \qquad \text{...(4.6 b)}$$

Let y_s, y_t = distance of the farthermost fibres from the neutral axis of steel and timber respectively.

Now $\varepsilon \propto \sigma$

and $\sigma \propto y$

$\therefore$ $\varepsilon \propto y$

Hence
$$\left(\frac{\varepsilon_s}{\varepsilon_t}\right)_{max} = \frac{y_s}{y_t}$$

or
$$\frac{(\sigma_s)_{max}}{E_s} \times \frac{E_t}{(\sigma_t)_{max}} = \frac{y_s}{y_t}$$

or
$$\frac{(\sigma_s)_{max}}{(\sigma_t)_{max}} = \frac{y_s}{y_t} \times \frac{E_s}{E_t} \quad \text{... (4.7)}$$

Now
$$\frac{M_s}{I_s} = \frac{\sigma_s}{y_s}$$

$$M_s = \frac{(\sigma_s)_{max} I_s}{y_s}$$

and
$$M_t = \frac{(\sigma_t)_{max} I_t}{y_t}$$

Total moment of resistance,

$$M = M_s + M_t$$

$$= \frac{(\sigma_s)_{max} I_s}{y_s} + \frac{(\sigma_t)_{max} I_t}{y_t}$$

$$= \frac{(\sigma_t)_{max}}{y_t} \cdot \frac{E_s}{E_t} \cdot I_s + \frac{(\sigma_t)_{max} \cdot I_t}{y_t}$$

$$= \frac{(\sigma_t)_{max}}{y_t}\left[\frac{E_s}{E_t}.I_s + I_t\right] \quad \text{... (4.8)}$$

Similarly
$$M = \frac{(\sigma_s)_{max}}{y_s}\left[I_s + \frac{E_t}{E_s} \cdot I_t\right] \quad \text{... (4.9)}$$

Let
$$\frac{E_s}{E_t} = m, \text{ moduli ratio}$$

$\therefore$
$$M = \frac{(\sigma_t)_{max}}{y_t}[mI_s + I_t] \quad \text{... (4.10)}$$

and
$$M = \frac{(\sigma_s)_{max}}{y_s}\left[I_s + \frac{I_t}{m}\right] \quad \text{... (4.11)}$$

$I_t + mI_s$ = equivalent moment of inertia of the section as if the beam is made of timber.

$I_s + \frac{I_t}{m}$ = equivalent moment of inertia of the section as if the beam is made of steel.

Example 4.5 *A timber beam 20 cm wide and 30 cm deep is reinforced by a steel plate 20 cm wide, 2 cm thick and bolted to its bottom surface. Calculate the moment of resistance if safe stresses in timber and steel are 10 MPa and 150 MPa respectively. Take $E_s = 20\ E_t$.*

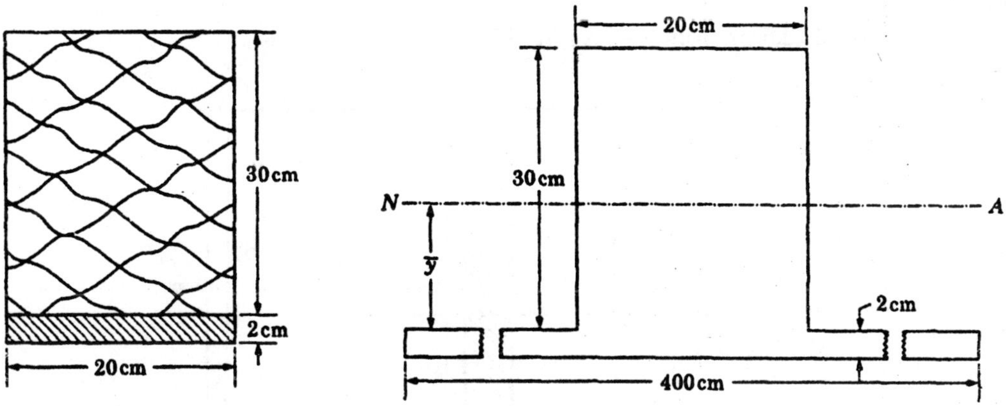

(a) Flitched beam

(b) Equivalent timber beam

Fig. 4.8

Solution:

The composite beam is shown in Fig. 4.8 (*a*)

$$\text{Equivalent (timber) area of steel} = \frac{E_s}{E_t} \times A_s = 20 \times 20 \times 2 = 800\,\text{cm}^2$$

$$\text{Equivalent (timber) width of steel strip} = \frac{800}{2} = 400\,\text{cm}$$

The equivalent timber beam is shown in Fig. 4.8 (*b*).

$$\bar{y} = \frac{20 \times 30 \times (15+2) + 400 \times 2 \times 1}{20 \times 30 + 400 \times 2} = \frac{11000}{1400} = 7.857\,\text{cm}$$

$$I_{NA} = \frac{20 \times 30^3}{12} + 20 \times 30(17 - 7.857)^2 + \frac{400 \times 2^3}{12} + 400 \times 2(7.857 - 1)^2$$

$$= 133038\ \text{cm}^4$$

$$(\sigma_t)_{max} = 10\ \text{MPa}$$

$$(\sigma_s)_{max} = \frac{10 \times 7.857 \times 20}{(32 - 7.857)} = 65.087\ \text{MPa} < 150\text{MPa}$$

Moment of resistance of beam,

$$M = \frac{10\times10^6\times133038\times10^{-8}}{(32-7.857)\times10^{-2}} = 55.104\,\text{kN.m}$$

Example 4.6 *A flitched beam is made of two timber joists each 10 cm wide by 30 cm deep with a steel plate 2 cm thick and 25 cm deep placed symmetrically and firmly fixed to them. Calculate the moment of resistance of beam when the maximum stress in timber reaches 9 MPa. Also find the maximum uniformly distributed load which the beam can carry on a simply supported span of 6 cm. $E_s = 200$ GPa and $E_t = 10$ GPa.*

Solution:

Stress in timber at 12.5 cm from N—A (Fig. 4.9)

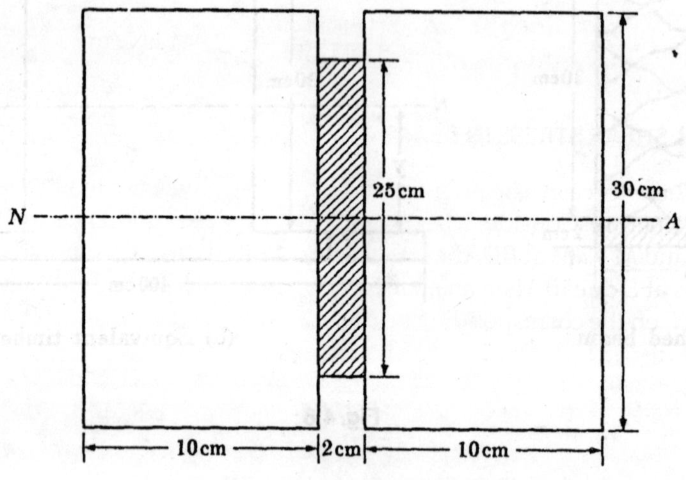

Fig. 4.9

$$\sigma_t = (\sigma_t)_{\max}\times\frac{12.5}{15} = \frac{12.5}{3}(\sigma_t)_{\max}$$

$$\varepsilon_t = \frac{\sigma_t}{E_t} = \frac{2.5}{3}\cdot\frac{(\sigma_t)_{\max}}{E_t} = \varepsilon_s = \frac{(\sigma_s)_{\max}}{E_s}$$

$$\therefore \quad (\sigma_s)_{\max} = \frac{2.5}{3}(\sigma_t)_{\max}\times\frac{E_s}{E_t}$$

$$= \frac{2.5}{3}\times\frac{200}{10}\times(\sigma_t)_{\max}$$

$$= \frac{50}{3}\times9 = 150\,\text{MPa}$$

Moment of resistance of timber,

$$M_t = \frac{(\sigma_t)_{\max}}{y_t}\times I_t$$

$$= \frac{9\times10^6}{0.15}\times2\times\frac{10\times30^3}{12}\times10^{-8} = 27\text{ kN.m}$$

Moment of resistance of steel,

$$M_s = \frac{(\sigma_s)_{max}}{y_s}\times I_s = \frac{150\times10^6}{0.125}\times\frac{2\times25^3\times10^{-8}}{12} = 31.25\text{ kN.m}$$

Total moment of resistance,

$$M = M_t + M_s = 27 + 31.25 = 58.25\text{ kN.m}$$

Now
$$M = \frac{wl^2}{8}$$

$$w = \frac{58.25\times8}{6^2} = 12.944\text{ kN/m}$$

4.6 HORIZONTAL SHEAR STRESS IN BEAMS

The vertical shear force in a beam subjected to bending gives rise to horizontal shear force due to its complementary nature. Consider a beam of uniform cross-section subjected to bending moment M at the section AC and $M + dM$ at BD, the two sections being dx apart, as shown in Fig. 4.10 (a). Let σ be the bending stress at E due to M on an elementary area of width b and thickness dy, as shown in Fig. 4.10 (b), and σ_1 at F on the corresponding area of the cross-section. Then

$$\sigma = \frac{My}{I} \quad\text{and}\quad \sigma_1 = \frac{(M+dM)y}{I}$$

Longitudinal pull at $E = \sigma\cdot dA = \sigma.bdy = \frac{M}{I}.bydy$

Longitudinal pull at $F = \sigma_1\, dA = \sigma_1\, bdy = \frac{(M+dM)}{I}.bydy$

Resultant longitudinal pull [see Fig. 4.10 (c)]

$$= \frac{[(M+dM)-M]}{I}\cdot bydy = \frac{dM}{I}\cdot bydy$$

Total force on the elementary length dx and cross-shaded area above y is

$$= \int_y^{h/2}\frac{dM}{I}bydy = \frac{dM}{I}\int_y^{h/2} by\cdot dy$$

This resultant pull is resisted by the shearing force on the longitudinal section at EF, whose area is $b\,.dx$. Let τ be the shear stress on this area. Then for the equilibrium of $AEFB$, we have

$$\tau\cdot b\cdot dx = \frac{dM}{I}\int_y^{h/2} by.dy$$

$$\tau = \frac{dM}{dx} \cdot \frac{1}{Ib} \int_{y}^{h/2} by \cdot dy$$

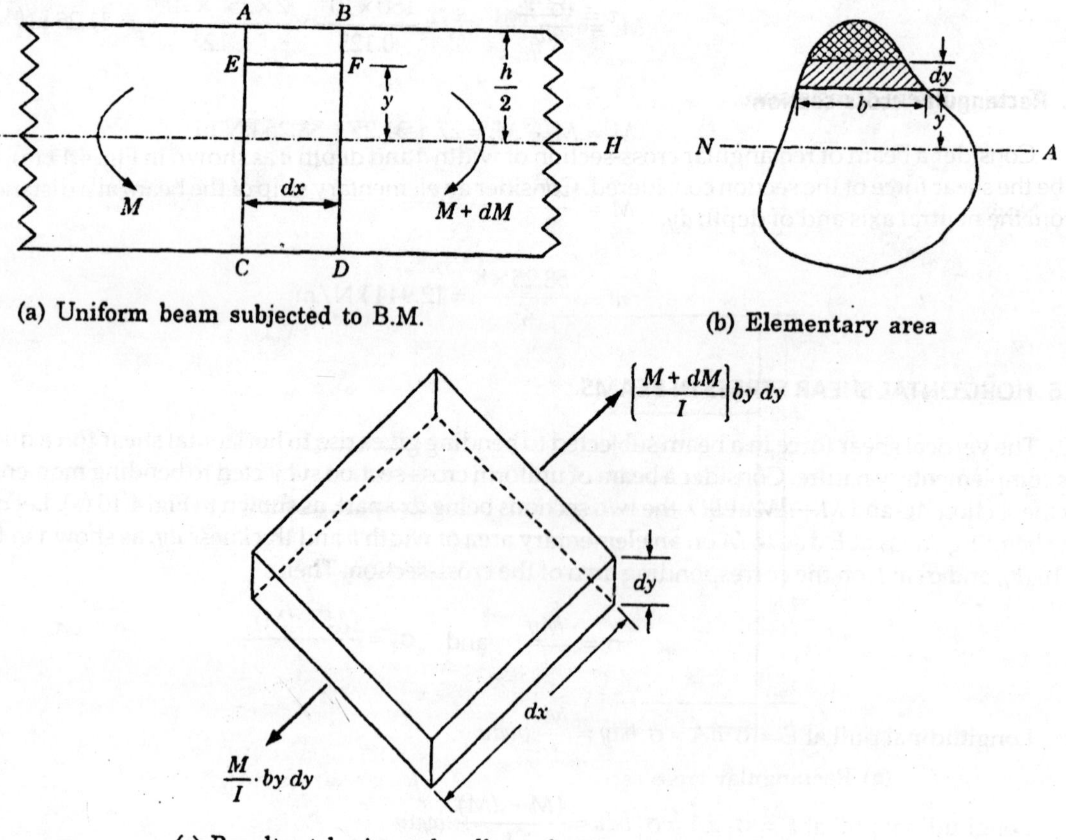

(a) Uniform beam subjected to B.M.

(b) Elementary area

(c) Resultant horizontal pull on elementary area of beam.

Fig. 4.10

But $\dfrac{dM}{dx} = F$, the shear force at the section.

Now $\int_{y}^{h/2} by.dy$ = Sum of the moments of the areas of all the small strips comprising the cross-shaded area, about the neutral axis.

If A = cross-shaded area above the section considered. and $\bar{y}$ = distance of centroid of area A from the neutral axis.

Then $$\int_{y}^{h/2} by.dy = A\bar{y}$$

$$\therefore \qquad \tau = \frac{FA\bar{y}}{Ib} \qquad \text{... (4.12)}$$

1. Rectangular cross-section

Consider a beam of rectangular cross-section of width *b* and depth *h* as shown in Fig. 4.11 (a). Let *F* be the shear force of the section considered. Consider an elementary strip of the beam at a distance *y* from the neutral axis and of depth *dy*.

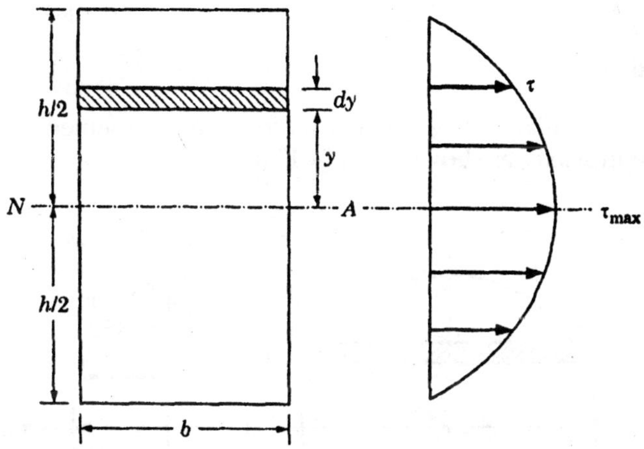

(a) Rectangular cross-section (b) Shear stress distribution

Fig. 4.11

$$\tau = \frac{FA\bar{y}}{Ib} = \frac{F}{Ib}\int_{y}^{h/2} bydy$$

$$= \frac{F}{I}\int_{y}^{h/2} ydy = \frac{F}{I}\left|\frac{y^2}{2}\right|_{y}^{h/2}$$

$$= \frac{F}{2I}\left[\frac{h^2}{4} - y^2\right]$$

$$I = \frac{bh^3}{12}$$

$$\tau = \frac{6F}{bh^3}\left[\frac{h^2}{4} - y^2\right] \qquad ...(4.13)$$

This represents the equation of a parabola. Thus the horizontal shear stress distribution in a beam of rectangular cross-section is parabolic in nature, as shown in Fig. 4.11 (*b*).

At $y = \pm\frac{h}{2}$, $\tau = 0$

At $y = 0$ $\tau_{max} = \frac{3}{2}\frac{F}{bh} = \frac{3}{2}\tau_{mean}$

$$\frac{\tau_{max}}{\tau_{mean}} = \frac{3}{2} \qquad ...(4.14)$$

2. Circular Cross-section

Consider a beam of circular cross-section of radius r and an elementary strip of depth dy at a distance y from the neutral axis, as shown in Fig. 4.12 (a).

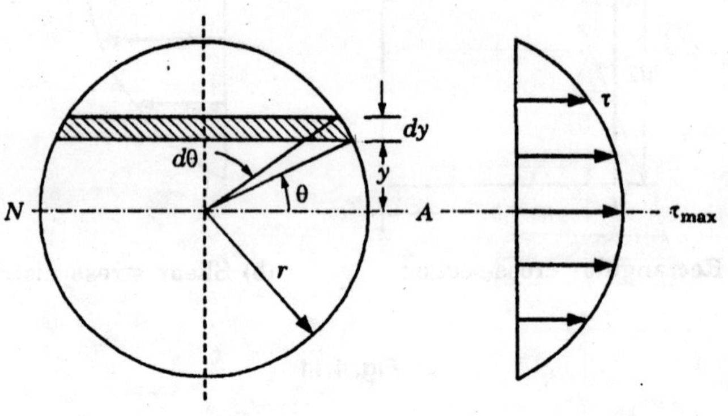

(a) Circular cross-section (b) Shear stress distribution

Fig. 4.12

$$\tau = \frac{F}{Ib}\int_y^r bydy$$

$$b = 2r\cos\theta, y = r\sin\theta, dy = r\cos\theta\, d\theta$$

$$\tau = \frac{F}{2Ir\cos}\int_\theta^{\pi/2} 2r\cos\theta .\, r\sin\theta \cdot r\cos\theta\, d\theta$$

$$= \frac{F}{2Ir\cos\theta}\int_{\theta}^{\pi/2}\cos^2\theta.\sin\theta\, d\theta = \frac{Fr^2}{I\cos\theta}\left|\frac{-\cos^3\theta}{3}\right|_{\theta}^{\pi/2}$$

At $\theta = \frac{\pi}{2}, \tau = 0$

and at $\theta = 0,\ \tau_{max} = \frac{4F}{3\pi r^2}$

The variation of shear stress is parabolic in nature, as shown in Fig. 4.12 (*b*).

$$\tau_{mean} = \frac{F}{\pi r^2}$$

$$\therefore \quad \frac{\tau_{max}}{\tau_{mean}} = \frac{4}{3} \qquad ...(4.15)$$

3. Triangular cross-section

Consider a triangular cross-section beam of base *d* and height *h*. Consider an elementary strip of width *b* and depth *dy* at a distance *y* from the apex *O* of the triangle as shown in Fig. 4.13 (a).

Then

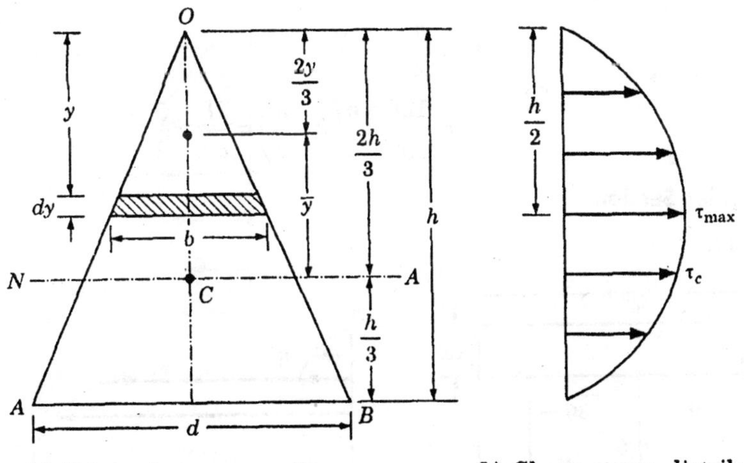

(a) Triangular cross-section (b) Shear stress distribution

Fig. 4.13

$$\frac{b}{y} = \frac{d}{h}$$

or $$b = \frac{d}{h}y$$

$$A = \frac{1}{2}by = \frac{1}{2}\times\frac{d}{h}y^2$$

$$\bar{y} = \frac{2h}{3} - \frac{2}{3}y = \frac{2}{3}(h-y)$$

$$\tau = \frac{FA\bar{y}}{Ib} = \frac{F \times \frac{1}{2}\frac{dy^2}{h} \times \frac{2}{3}(h-y)}{\frac{dh^3}{36} \times \frac{dy}{h}} = \frac{12Fy(h-y)}{dh^3}$$

At $y = 0$, $\tau = 0$. Also $\tau = 0$ at $y = h$

At $y = \frac{2h}{3}$, $\tau_c = \frac{12F}{dh^3} \times \frac{2h}{3} \times \frac{2h}{3}\left(h - \frac{2h}{3}\right) = \frac{8}{3}\frac{F}{dh}$

The variation of shear stress is parabolic in nature, as shown in Fig. 4.13 (*b*).

$$\tau_{mean} = \frac{2F}{dh}$$

$$\tau_{max} / T_{mean} = \frac{4}{3} \quad ...(4.16)$$

For τ to be maximum

$$\frac{d\tau}{dy} = 0$$

$$\frac{12F}{dh^3}[h-2y] = 0$$

$$\therefore \quad y = \frac{h}{2}$$

$$\tau_{max} = \frac{12F}{dh^3} \times \frac{h}{2}\left(h - \frac{h}{2}\right) = \frac{3F}{dh} = \frac{3}{2}\tau_{mean} \quad ...(4.17)$$

4. Hollow Rectangular Section

Refer to Fig. 4.14.

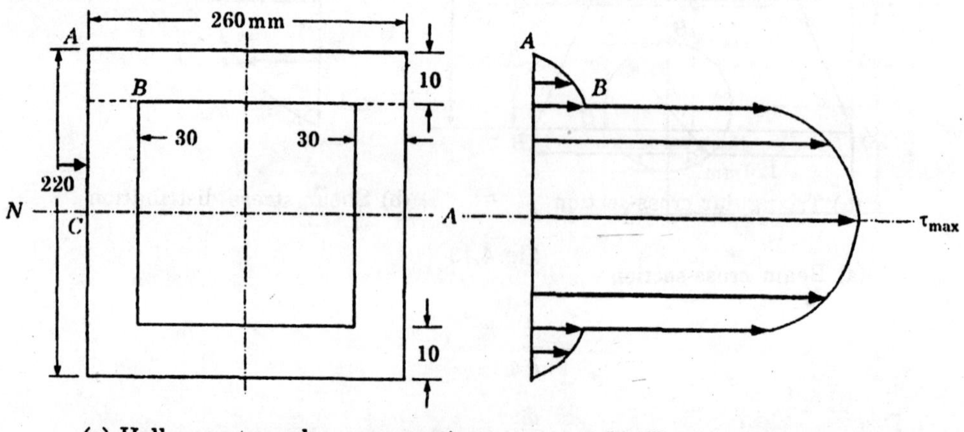

(a) Hollow rectangular cross-section (b) Shear stress distribution

Fig. 4.14

$$I = \frac{260 \times 220^3}{12} - \frac{200 \times 200^3}{12} = 97373333 \text{ mm}^4$$

$$\tau = \frac{FA\bar{y}}{Ib}$$

$$\tau_A = 0$$

$$\tau_B = \frac{F \times (260 \times 10) \times (110 - 5)}{I \times 260} = F \times 1.0783 \times 10^{-5} \text{N/mm}^2$$

Shear stress at N – A:

$$A = 260 \times 10 + 2 \times 100 \times 30 = 8600 \text{ m}^2$$

$$\bar{y} = \frac{260 \times 10 \times 105 + 2 \times 100 \times 30 \times 50}{260 \times 10 + 2 \times 100 \times 30}$$

$$= \frac{5733000}{8600} = 66.28 \text{mm}$$

$$\tau_{max} = \tau_c = \frac{FA\bar{y}}{Ib} = \frac{F \times 8600 \times 66.628}{1 \times 2 \times 30}$$

$$= F \times 9.8076 \times 10^{-5} \text{ N/mm}^2$$

5. Rectangular section with a coaxial Hole

Refer to Fig. 4.15.

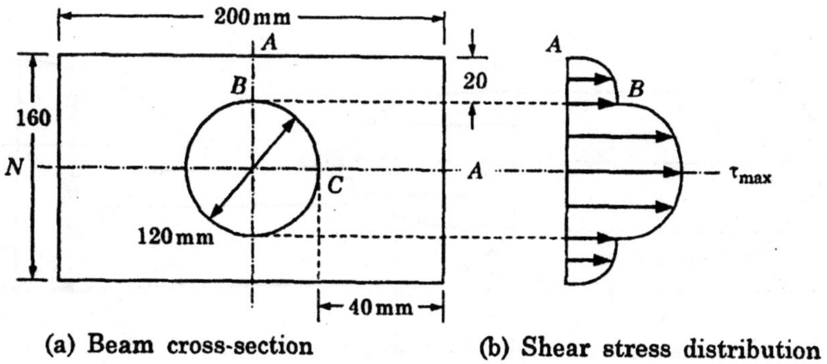

(a) Beam cross-section (b) Shear stress distribution

Fig. 4.15

$$I = \frac{200 \times 160^3}{12} - \frac{\pi \times 120^4}{64} = 58087906 \text{mm}^4$$

$$\tau_A = \frac{FA\bar{y}}{Ib}$$

$$\tau_A = 0$$

$$\tau_B = \frac{F \times (200 \times 20) \times (80 - 10)}{I \times 200}$$

$$= \frac{F \times 200 \times 20 \times 70}{58087906 \times 200} = 2.41 \times 10^{-5} \text{F N/mm}^2$$

Shear stress τ_c at N–A:

$$A_1 = 200 \times 80 = 16000 \text{ mm}^2$$

$$A_2 = \frac{\pi \times 120^2}{8} = 5655 \text{ mm}^2$$

$$A = A_1 - A_2 = 10345 \text{ mm}^2$$

$$\bar{y} = \left(1600 \times 40 - 5655 \times 4 \times \frac{4 \times 60}{3\pi}\right) \times \frac{1}{10345} = 47.95 \text{ mm}$$

$$b = 200 - 120 = 80 \text{ mm}$$

$$\tau_{max} = \tau_c = \frac{F \times 10345 \times 47.95}{58087906 \times 80} = 1.067 \times 10^{-4} \text{ FN/mm}^2$$

6. Circular Section with a Coaxial Hole

Refer to Fig. 4.16.

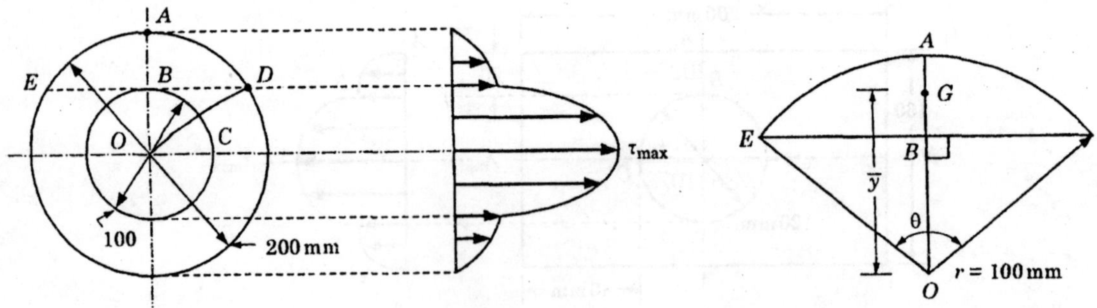

(a) Hollow circular cross-section (d) Shear stress distribution (b) C.G. of area EBDAE

Fig. 4.16

$$\tau = \frac{FA\bar{y}}{Ib}, \quad \tau_A = 0$$

Now $BD = \sqrt{OD^2 - OB^2} = \sqrt{100^2 - 50^2} = 83.6\,\text{mm}$

$b = DE = 2BD = 173.2\text{ mm}$

Area EBDAE, $A = \frac{1}{2}r^2(\theta^c - \sin\theta)$

(SeeFig. 5.16 (b))

$$\cos\frac{\theta}{2} = \frac{OB}{OD} = \frac{50}{100} = 0.5$$

$$\frac{\theta}{2} = 60°$$

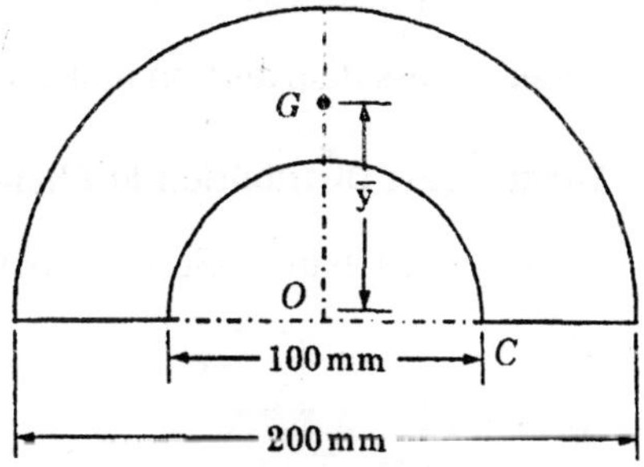

Fig. 4.16 (c) C.G. of semi-circular area

$$\theta = 120° \text{ or } \frac{2\pi}{3} \text{ radians}$$

$$A = \frac{1}{2}\times 100^2\left(\frac{2\pi}{3} - \sin 120°\right) = 6142\text{mm}^2$$

$$\bar{y} = \frac{2r}{3}\left[\frac{\sin^2\frac{\theta}{2}}{\frac{\theta^c}{2} - \sin\frac{\theta}{2}\cos\frac{\theta}{2}}\right]$$

$$= \frac{2\times 100}{3}\left[\frac{\sin^2 60°}{\frac{\pi}{3} - 60°\cos 60°}\right] = 81.4\,\text{mm}$$

$$I = \frac{\pi}{64}(200^4 - 100^4) = 73631078\text{ mm}^4$$

$$\tau_B = \frac{F\times 6142\times 81.4}{73631078\times 173.2} = 3.92\times 10^{-5}\text{F N/mm}^2$$

$$A_1 = \frac{\pi}{8}\times 200^2 = 15708\,\text{mm}^2 \text{ (see Fig. 5.16 (c))}$$

$$A_2 = \frac{\pi}{8}\times 100^2 = 3927\text{mm}^2$$

$$A = A_1 - A_2 = 11781\text{ mm}^2$$

$$\bar{y} = \left(15708\times\frac{4\times 100}{3\pi} - 3927\times\frac{4\times 50}{3\pi}\right)\times\frac{1}{11781} = 49.5\text{mm}$$

$$b = 200 - 100 = 100\text{ mm}$$

$$\tau_{max} = \tau_c = \frac{F \times 11781 \times 49.5}{73631078 \times 100} = 7.92 \times 10^{-5} F\ N/mm^2$$

Shear stress distribution is shown in Fig. 4.16 (*d*)

7. Shear Stress Distribution In Channel Section Beam

Consider a Channel-Section beam as shown in Fig. 4.17(*a*).

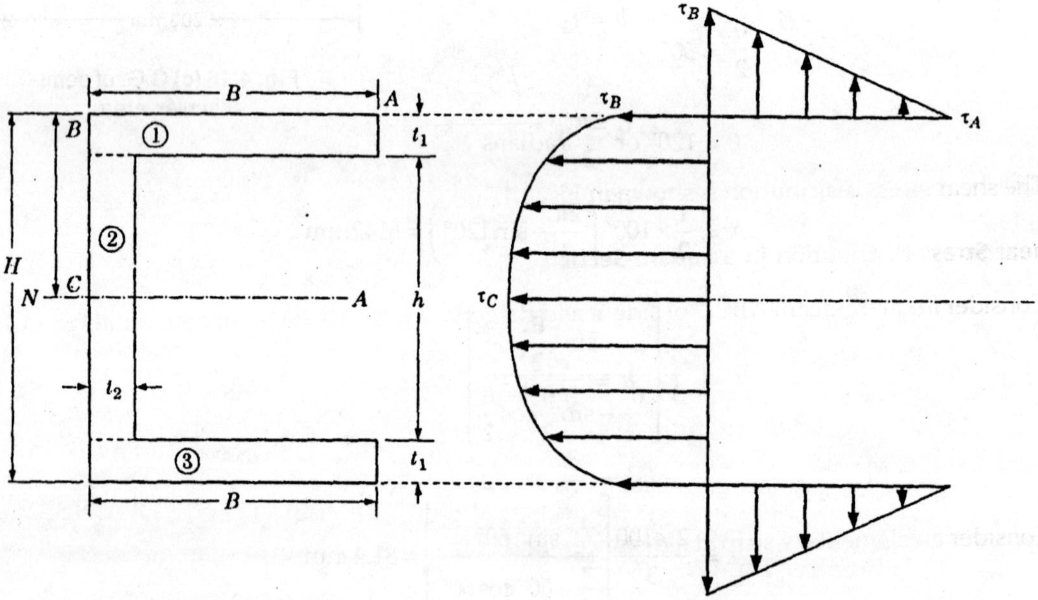

(a) Channel section

(b) Shear stress distribution

Fig. 4.17

$$\bar{x} = \frac{H}{2}$$

$$A_1 = A_3 = Bt_1$$

$$A_2 = ht_2,\ h = H - 2t_1$$

$$I_{NA} = 2\left[\frac{Bt_1^3}{12} + A_1\left(\bar{x} - \frac{t_1}{2}\right)^2\right] + \frac{t_2h^3}{12}$$

Point A: $\tau_A = 0$

Point B: $A = Bt_1$

$$\bar{y} = \bar{x} - \frac{1}{2}t_1$$

$$b = B$$

$$\tau_B = \frac{F \times Bt_1 \times \left(\bar{x} - \frac{1}{2}t_1\right)}{IB}$$

Point C :

$$y = \frac{1}{2}h \times t_2, \bar{x} = \frac{h}{2}$$

$$\bar{y} = \frac{h}{4} = \frac{\bar{x}}{2}$$

$$b = t_2$$

$$\tau_c = \tau_B + \frac{F \times \frac{1}{2}ht_2 \times \frac{h}{4}}{It_2}$$

The shear stress distribution is shown in Fig. 4.17 (*b*).

8. Shear Stress Distribution in a Square-Section

Consider a square-beam ABCD of side '*a* and diagonal *d* with *AC* diagonal horizontal Fig. (4.18).

$$h = \frac{a}{\sqrt{2}}, d = \sqrt{2}\,a$$

$$I_{NA} = \frac{dh^3}{6} = \sqrt{2}a \times \frac{a^3}{2\sqrt{2}} \times \frac{1}{6} = \frac{a^4}{12}$$

Consider an elementary strip at a distance *y* and of thickness *dy* from the natural axis N–A.

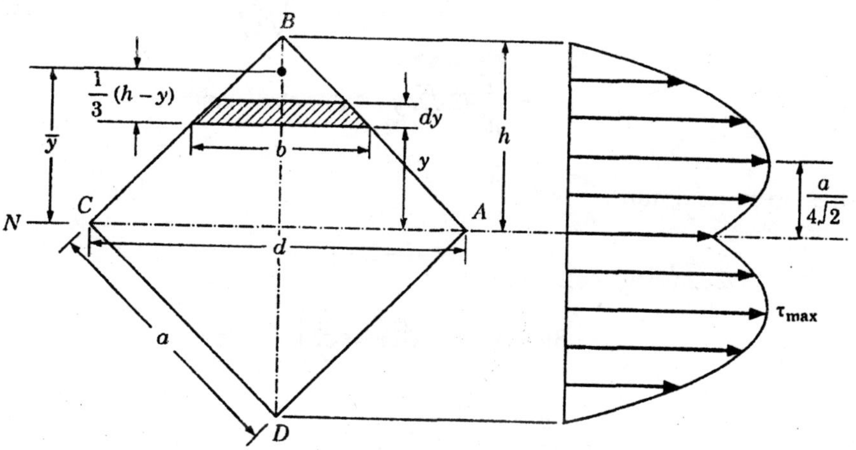

(a) Square section (b) Shear stress distribution

Fig. 4.18

Width of strip, $$b = \frac{d(h-y)}{h} = \frac{\sqrt{2}a\left(\frac{a}{\sqrt{2}}-y\right)}{\frac{a}{\sqrt{2}}}$$

$$= 2\sqrt{a} - 2y$$

$$A = \frac{1}{2}b\times(h-y) = \frac{1}{2}(\sqrt{2a}-2y)\left(\frac{a}{\sqrt{2}}-y\right) = \left(\frac{a}{\sqrt{2}}-y\right)^2$$

$$\bar{y} = y+\frac{1}{3}(h-y) = \frac{2}{3}y+\frac{1}{3}\times\frac{a}{\sqrt{2}}$$

Horizontal shear stress, $$\tau = \frac{FA\bar{y}}{Ib}$$

$$= \frac{F\left(\frac{a}{\sqrt{2}}-y\right)^2\times\frac{1}{3}\left(2y+\frac{a}{\sqrt{2}}\right)}{\frac{a^4}{12}\times(\sqrt{2a}-2y)}$$

$$= \frac{F(\sqrt{2a}=2y)^2\left(2y+\frac{a}{\sqrt{2}}\right)}{a^4(\sqrt{2a}-2y)}$$

$$= \frac{F}{a^4}(\sqrt{2a}-2y)\left(2y+\frac{a}{\sqrt{2}}\right)$$

$$= \frac{F}{a^4}(2\sqrt{2}ay-4y^2+a^2-\sqrt{2}ay)$$

For shear stress to be maximum, $\frac{d\tau}{dy} = 0$

$$2\sqrt{2a} = 8y - \sqrt{2a} = 0$$

$$8y = (2\sqrt{2}-\sqrt{2})a = \sqrt{2a}$$

$$y = \frac{\sqrt{2a}}{8} = \frac{a}{4\sqrt{2}}$$

$$\tau_{max} = \frac{F}{a^4}\left(2\sqrt{2a}\times\frac{a}{4\sqrt{2}}-4\times\frac{a^2}{32}+a^2-\sqrt{2a}\times\frac{a}{4\sqrt{2}}\right)$$

$$= \frac{F}{a^4}\left(\times\frac{a^2}{2} - \frac{a^2}{8} + a^2 - \frac{a^2}{4}\right)$$

$$= \frac{9Fa^2}{8a^4} = \frac{9F}{8a^2}$$

y	0	$\frac{a}{4\sqrt{2}} = \frac{h}{4}$	$\frac{a}{2\sqrt{2}} = \frac{h}{2}$	$\frac{a}{\sqrt{2}} = h$
τ	$\frac{F}{a^2}$	$\frac{9}{8}\frac{F}{a^2}$	$\frac{F}{a^2}$	0

The distribution of shear stress is parabolic in nature as shown in Fig. 5.18 (b).

9. Shear Stress Distribution in I-Section

The *I*-section is shown in Fig. 5.19 (a).

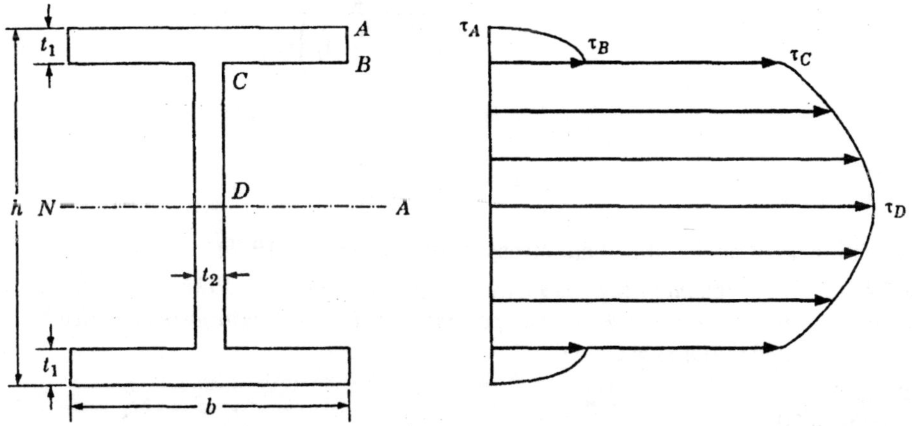

(a) I-beam section

(b) Shear stress distribution

Fig. 4.19

$$\tau = \frac{FA\bar{y}}{Ib}$$

$$I = \frac{bh^3}{12} - \frac{(b-t_2)(h-2t_1)^3}{12}$$

Top flange: $\tau_A = 0$

Area of top flange, $A = bt_1$

$$\bar{y} = \frac{h}{2} - \frac{t_1}{2} = \frac{1}{2}(h - t_1)$$

$$b = b$$

$$\tau_B = \frac{F \times bt_1 \times \frac{1}{2}(h - t_1)}{Ib}$$

Intersection of top flange and web:

$$\tau_C = \tau_B \times \frac{b}{t^2}$$

Neutral axis-point D:

$$A = bt_1 + \left(\frac{h}{2} - t_1\right) \times t_2$$

$$\left[bt_1 + \left(\frac{h}{2} - t_1\right) \times t_2\right]\bar{y} = bt_1 \times \left(\frac{h}{2} - \frac{t_1}{2}\right) + \left(\frac{h}{2} - t_1\right)t_2 \times \frac{1}{2}\left(\frac{h}{2} - t_1\right)$$

$$= \frac{bt_1(h - t_1) + \frac{1}{2}\left(\frac{h}{2} - t_1\right)^2 t_2}{bt_1 + \left(\frac{h}{2} - t_1\right)t_2}$$

$$b = t_2$$

$$\tau_D = \frac{FA\bar{y}}{It_2}$$

The distribution of horizontal shear stress is shown in Fig. 4.19 (b)

Example 4.7 *A 150 mm x 60 mm I-beam is subjected to a shearing force of 15 kN. Determine the distribution of horizontal shear stress in the beam. Find what percentage of the total shear force is carried by the web. Web thickness = 4 mm, flange thickness = 6 mm.*

Solution:

Refer to Fig. 4.20.

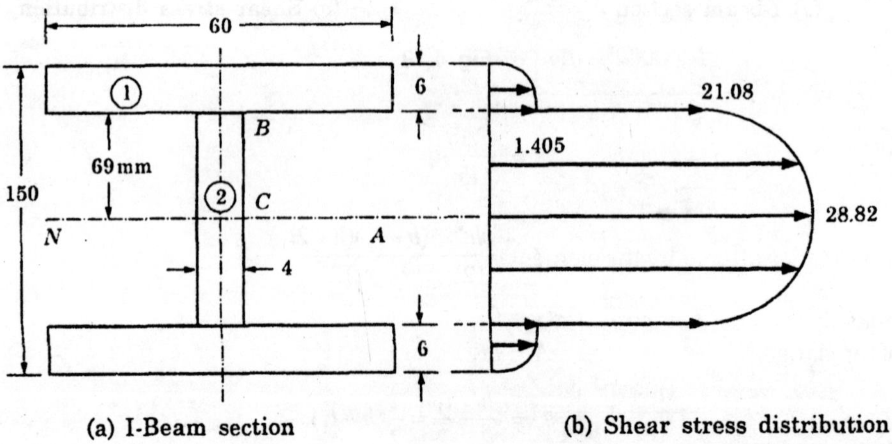

Fig. 4.20

$$I = \frac{60\times150^3}{12} - \frac{56\times138^3}{12} = 4610664 \text{ mm}^4$$

$$\tau = \frac{FA\bar{y}}{Ib}$$

$$\tau_A = 0$$

$$(\tau_B)_{\text{flange}} = \frac{15\times10\times(60\times6)\times(75-3)}{4610664\times60} = 1.405 \text{ N/mm}^2$$

Shear stress at *C*:

$$A_1 = 60 \times 6 = 360 \text{ mm}^2$$
$$A_2 = 69 \times 4 = 276 \text{ mm}^2$$
$$A = 636 \text{ mm}^2$$

$$\bar{y} = \frac{360\times72+276\times34.5}{636} = 55.72 \text{ mm}$$

$$b = 4 \text{ mm}$$

$$\tau_c = \frac{15\times10^3+636\times55.72}{4610664\times4} = 28.82 \text{ N/mm}^2$$

The shear stress distribution is shown is Fig. 4.20 (b). Shear force taken by the web = total shear force- shear force taken by the flange.

Consider elementary strip of thickness dy in the flange at a distance y from the *N–A* (Fig. 4.21).

$$A = 60\,(75-y)$$

$$\bar{y} = y + \frac{75-y}{2} = \frac{75+y}{2}$$

$$b = 60 \text{ mm}$$

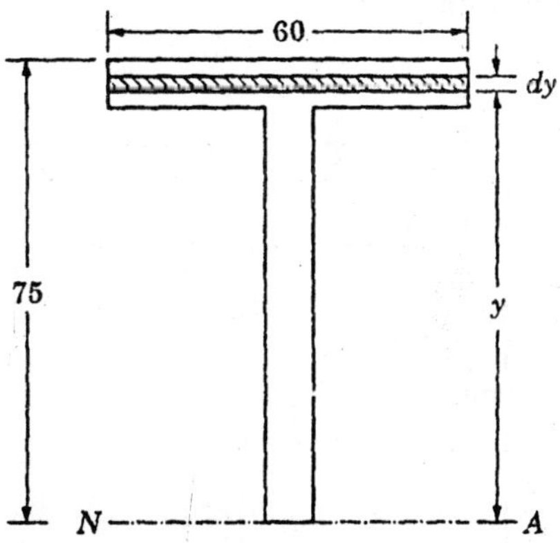

Fig. 4.21

$$\tau = \frac{F\times60(75-y)\times\left(\frac{75+y}{2}\right)}{60I}$$

$$= \frac{F}{2I}(75^2-y^2)$$

Shear resistance offered by the elementary strip of flange = $\tau.dA = \tau\times60\times dy$

$$= \frac{F}{21}(5625-y^2)\times60\times dy$$

$$= \frac{15\times10^3}{20+4610664}\times(5362-y^2)\times60\times dy$$

$$= 0.0976\,(5625-y^2)dy$$

Total resistance of one flange $= \int_{69}^{75} 0.0976(5625 - y^2)\,dy$

$$= 0.09756 \left| 5625y - \frac{y^3}{3} \right|_{69}^{75}$$

$$= 0.0976 \left[5625(75-69) - \frac{1}{3}(45^3 - 69^3) \right]$$

$$= 256.4\text{N}$$

Shear resistance of two flanges

$$= 2 \times 256.4 = 512.8 \text{ N} = 0.5128 \text{ kN}$$

Shear resistance of web $= 15 - 0.5128 = 14.4872$ kN

Percentage of the total shear force carried by the web

$$= \frac{14.4872}{15} \times 100 = 96.58\%$$

Example 4.8 *Draw the horizontal shear stress distribution in a T-beam shown in Fig. 4.22 and subjected to shear force of 10 kN.*

Solution:

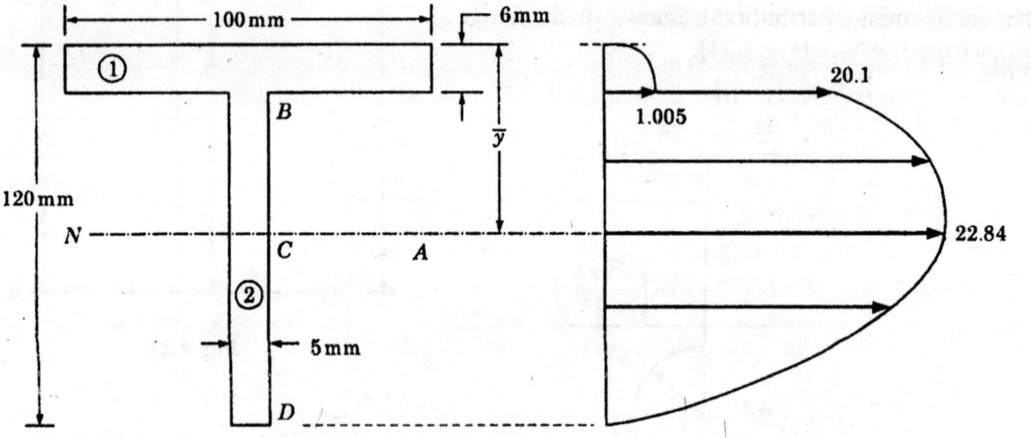

(a) T-beam section (b) Shear stress distribution

Fig. 4.22

$$A_1 = 100 \times 5 = 500 \text{ mm}^2$$

$$A_2 = 115 \times 5 = 575 \text{ mm}^2$$

$$A = A_1 + A_2 = 1075 \text{ mm}^2$$

$$\bar{y} = \frac{500 \times 2.5 + 575 \times (57.5 + 5)}{1075} = 34.593\,\text{mm}$$

$$I_{NA} = \frac{100 \times 5^3}{12} + 500(34.593 - 2.5)^2 + \frac{5 \times 115^3}{12} + 575(62.5 - 34.393)^2$$

$$= 1596593\ \text{mm}^4$$

$$\tau = \frac{FA\bar{y}}{Ib}$$

$$\tau_A = 0$$

$$(\tau_A)_{\text{flange}} = \frac{10 \times 10^3 \times (100 \times 5) \times (34.593 - 2.5)}{1596593 \times 100} = 1.005\,\text{N/mm}^2$$

$$(\tau_B)_{\text{web}} = \frac{1.005 \times 100}{5} = 20.1\ \text{N/mm}^2$$

Shear stress at C:

$$A = 100 \times 5 + (34.593 - 5) \times 5 = 647.965\ \text{mm}^2$$

$$\bar{y} = \left[100 \times 5(34.593 - 2.5) + (34.593) \times 5 \times \left(\frac{34.593 - 5}{2}\right)\right] \times \frac{1}{647.965}$$

$$= 28.143\ \text{mm}$$

$$b = 5\,\text{mm}$$

$$\tau_C = \frac{10 \times 10^3 \times 647.965 \times 28.143}{1596593.5} = 22.84\ \text{N/mm}^2$$

The shear stress distribution is shown in Fig. 4.22(b)

Example 4.9 *A timber beam as shown in Fig. 4.23 is simply supported on a span of 4 m. The weight of the beam is 10 kN/m³ and the permissible stress in timber is 10 MPa. Calculate the safe point load at mid-span.*

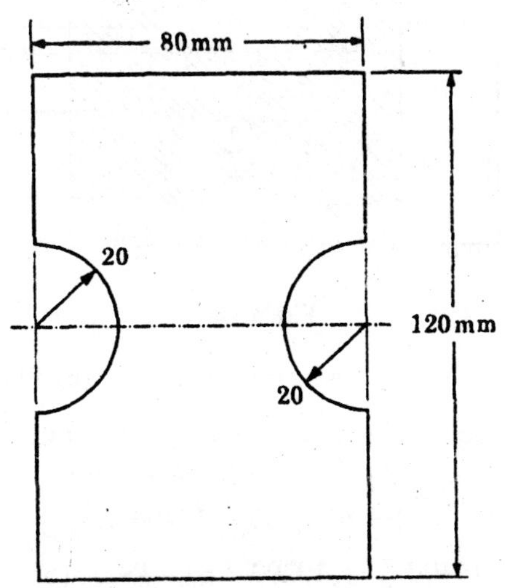

Fig. 4.23

Solution:

$$I = \frac{80\times120^3}{12} - \frac{\pi\times20^4}{64} = 11512146\text{mm}^4$$

Weight of beam, $w = (80 \times 120 - \pi \times 40) \times 10^{-6} \times 10 \times 10^3$

$$= 94.743 \text{ N/m}$$

$$M_{max} = \frac{Wl}{4} + \frac{wl^2}{8}$$

$$= \frac{W\times4}{4} + \frac{94.743\times4^2}{8} = W + 189.487 \text{ N.m}$$

$$z = \frac{I}{y} = \frac{011512146}{60} = 191869 \text{ mm}^3$$

$$M_{max} = \sigma \times z$$

$$W + 189.487 = 10 \times 10^6 \times 191869 \times 10^{-9}$$

$$W = 1729.2 \text{ N}$$

Example 4.10 *A beam of 3 m length is simply supported at the ends and carries a uniformly distributed load of 40 kN/m over its whole span. The cross-section of the beam is shown in Fig. 4.24 Calculate the maximum tensile and compressive stresses in the beam.*

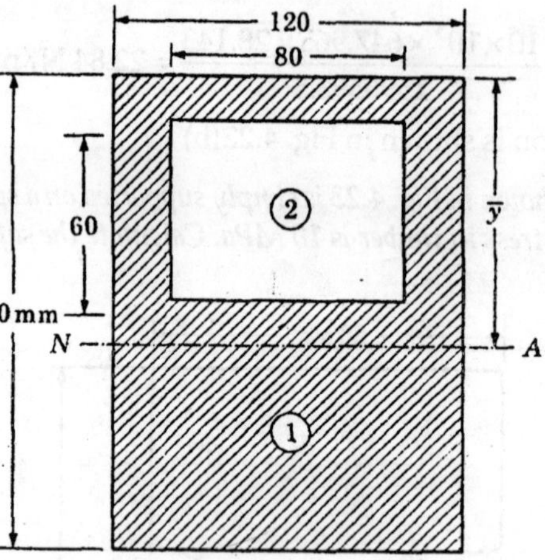

Fig. 4.24

Solution:

$$A_1 = 120 \times 160 = 19200 \text{ mm}^2$$

$$A_2 = 80 \times 60 = 4800 \text{ mm}^2$$

$$A = A_1 - A_2 = 14400 \text{ mm}^2$$

$$14400\ \bar{y} = 19200 \times 80 - 4800 \times 50$$

$$\bar{y} = 90 \text{ mm}$$

$$I_{N\text{-}A} = \frac{120 \times 160^3}{12} + 19200 \times (90 - 80)^2$$

$$= 3376 \times 10^4 \text{ mm}^4$$

$$M_{max} = \frac{wl^2}{8} = \frac{40 \times 10^3 \times 3^2}{8} = 45 \times 10^3 \text{ N.m}$$

$$\text{Compressive stress} = \frac{45 \times 10^3 \times 90 \times 10^3}{3376 \times 10^4} = 119.96 \text{ MPa}$$

$$\text{Tensile stress} = 119.96 \times \frac{70}{90} = 93.3 \text{ MPa}$$

Example 4.11 *A log of timber 300 mm in diameter is available to be used as a beam. Calculate the following:*

(a) Dimensions of the strongest rectangular section that can be cut out of it.

(b) Maximum span to carry a uniformly distributed load of 3 kN/m to limit the bending stress to 10 MPa.

(c) Longest span to use the log as such.

Solution:

(a) Let b = width of section (Fig. 4.25)

h = depth of section.

Diameter of log, d = 300 mm

$$d^2 = b^2 + h^2$$

$$h^2 = 300^2 - b^2$$

$$z = \frac{bh^2}{6} = \frac{b(300^2 - b^2)}{6}$$

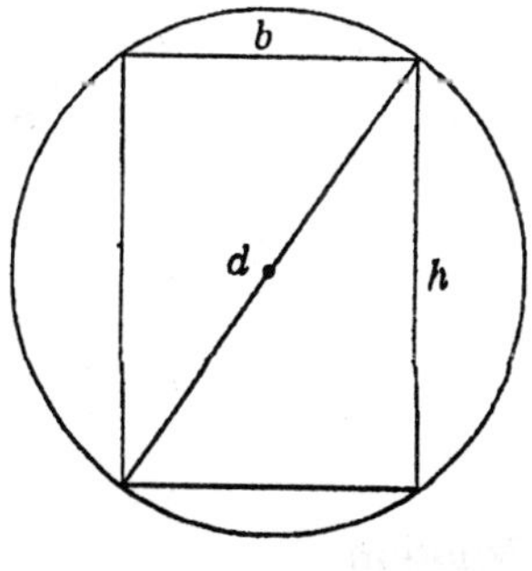

Fig. 5.25

For a strongest section, z should be maximum.

$$\frac{dz}{db} = 0$$

$$300^2 - 3b^2 = 0$$

$$b = 173.2 \text{ mm}$$

$$h^2 = 300^2 - 30000 = 60000$$

$$h = 244.95 \simeq 245 \text{ mm}$$

(b)

$$z = \frac{173.2 \times 60000}{6} = 1732000 \text{ mm}^2$$

$$M = \sigma \times z = 10 \times 10^6 \times 1732000 \times 10^{-9} = 17320 \text{ N.m}$$

$$= \frac{wl^2}{8} = \frac{3 \times 10^3 \times l^2}{8}$$

$$l = 6.796 \text{ m}$$

(c)

$$z = \frac{\pi d^3}{32} = \frac{\pi \times 300^2}{32} = 2650719 \text{ mm}^3$$

$$M = 10 \times 10^6 \times 2650719 \times 10^{-9} = 26507 \text{ N.m}$$

$$= \frac{3\times10^3 \times l^2}{8}$$

$$l = 8.4 \text{ m}$$

Example 4.12 *A beam of I-section shown in Fig. 4.26 is subjected to the bending moment of 12 kN.m. Find the maximum tensile and compressive stresses induced in the beam.*

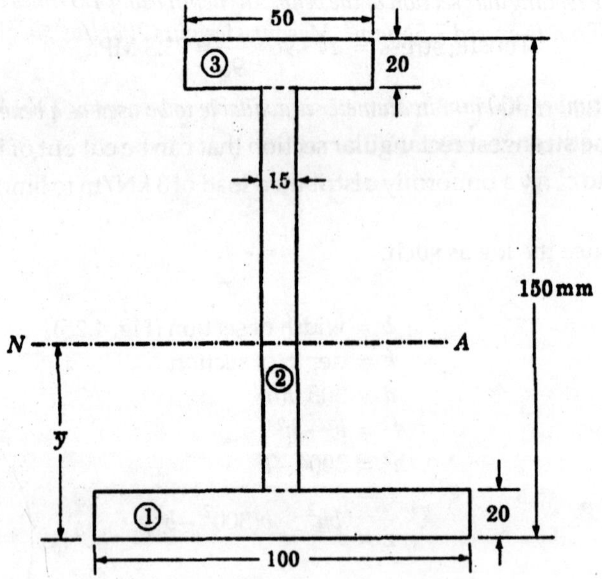

Fig. 4.26 Unequal I-Section beam.

Solution:

$$A_1 = 100 \times 20 = 2000 \text{ mm}^2$$

$$A_2 = 110 \times 15 = 1650 \text{ mm}^2$$

$$A_3 = 50 \times 20 = 1000 \text{ mm}^2$$

$$A = 2000 + 1650 + 1000 = 4650 \text{ mm}^2$$

$$4650\,\bar{y} = 2000 \times 10 + 1650 \times 75 + 1000 \times 140$$

$$\bar{y} = 61 \text{ mm}$$

$$I_{NA} = \left[\frac{100\times20^3}{12} + 200\times(61-10)^2\right]$$

$$+\left[\frac{15\times110^3}{12} + 1650\times(75-61)^2\right]$$

$$+\left[\frac{50\times 20^3}{12}+1000\times(140-61)^2\right]$$

$$= 13530150 \text{ mm4}$$

$$\sigma_t = \frac{12\times 10^6\times 61}{13530150} = 54.1 \text{ MPa}$$

$$\sigma_c = \frac{54.1\times 89}{61} = 78.93 \text{ MPa}$$

Example 4.13 *A wooden beam has to carry a masonry wall over a span of 5 metre. The wall is 300mm thick and 3m high. Design a suitable rectangular section of the beam, the depth being 1.5 times the width, if the maximum bending stress in the beam not to exceed 7.5Nmm². Masonry weights 20kN/m³, neglect self weight of beam.*

Solution:

Given: $L = 5\text{m}, B = 300 \text{ mm}, H = 3 \text{ m}, h = 1.5\text{b}, \sigma = 7.5 \text{ N/mm}^2, \gamma = 20 \text{ kN/m}^3$

Weight per unit length of masonry wall, $w = B \times \text{H} \times 1 \times \gamma$

$$= 0.3 \times 3 \times 20 = 18 \text{ kN/m}$$

$$M_{max} = \frac{wL^2}{8} = \frac{18\times 25}{8} = 56.25 \text{ kN.m}$$

$$z = \frac{bh^2}{6} = \frac{(1.5)^2 b^3}{6} = 0.375\, b^3 \text{mm}^3$$

$$M_{max} = \sigma \times z$$

$$56.25 \times 10^6 = 7.5 \times 0.375\, b^3$$

$$b = 271.4 = 272 \text{ mm}$$

$$h = 1.5 \times 272 = 408 \text{ mm}$$

Example 4.14 *A timber joist of 6 metre span has to carry a load of 15 kN/metre. Find the dimensions of the joist if the maximum permissible stress is limited to 8N / mm². The depth of the joist has to be twice the width.*

Solution:

Given: $l = 6\text{m}, w = 15 \text{ kN/m}, \sigma_{max} = 8\text{N/mm}^2, \text{h} = 2b$

Refer to Fig. 4.27 $\quad I = \frac{bh^3}{12} = \frac{b\times 8b^3}{12} = \frac{2}{3}b^4$

$$M_{max} = \frac{wl^2}{8} = \frac{15\times 6^2}{8} = 67.5 \text{ kN.m}$$

$$\frac{M}{I} = \frac{\sigma}{y}$$

$$\frac{67.5\times 10^3\times 10^3}{\frac{1}{3}b^4} = \frac{8}{\frac{h}{2}=b}$$

$$b^3 = 12656250$$

$$b = 233 \text{ mm}, h = 466 \text{ mm}$$

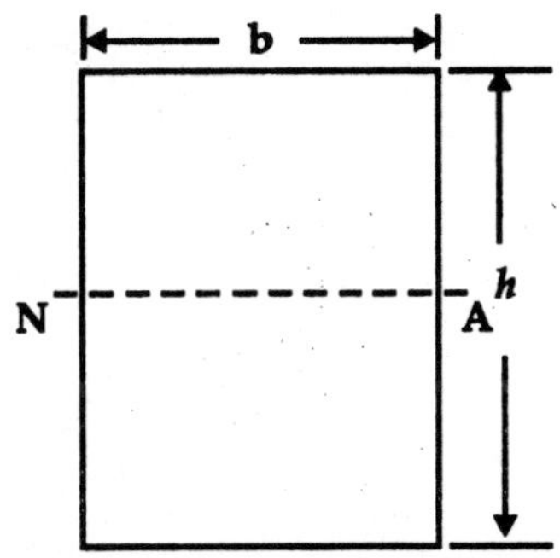

Fig. 4.27

Example 4.15 *A timber beam 100 mm wide and 150 mm deep carries a uniformly distributed load over a span of 2 m. If the permissible stresses are 28 MPa in bending and 2MPa in shear, calculate the maximum UDL, which can be carried by the beam*

Solution:

Given: $b = 100$ mm, $h = 150$ mm, $l = 2$ m, $\sigma = 32$ MPa, $\tau = 2$ MPa

$$M_{max} = \frac{wl^2}{8} = \frac{w \times 2^2}{8} = \frac{w}{2} \text{ N.m}$$

$$z = \frac{bh^2}{6} = \frac{100 \times (150)^2}{6} = 375 \times 10^3 \text{ mm}^3$$

$$M_{max} = \sigma \times z$$

$$\frac{w}{2} \times 10^3 = 32 \times 375 \times 10^3$$

$$w = 24000 \text{ N/m or } 24 \text{ kN/m}$$

Shear force in the beam, $F = \frac{wl}{2} = \frac{w \times 2}{2} = wN$

Maximum shear stress in rectangular beam, $\tau = \frac{3}{2}\tau_{avg}$

$$2 = \frac{3}{2} \times \frac{F}{A} = \frac{3}{2} \times \frac{w}{bh} = \frac{3}{2} \times \frac{w}{100 \times 150}$$

$$w = 20{,}000 \text{ N or } 20 \text{ kN/m}$$

$$\text{Safe } udl = 20 \text{ kN/m}$$

Example 4.16 *A. rectangular beam of 100 mm deep, 50 mm wide and 1.5 m long is simply supported at its ends. Vertical loads of 5 kN each are applied at 0.5 m and 1m from one end. Determine and plot the distribution of longitudinal stress across the section at mid span.*

Solution:

The loaded beam is shown in Fig.4.28 (*a*) The loading is symmetrical. The S.F.D. is shown in Fig. 4.28 (*b*).

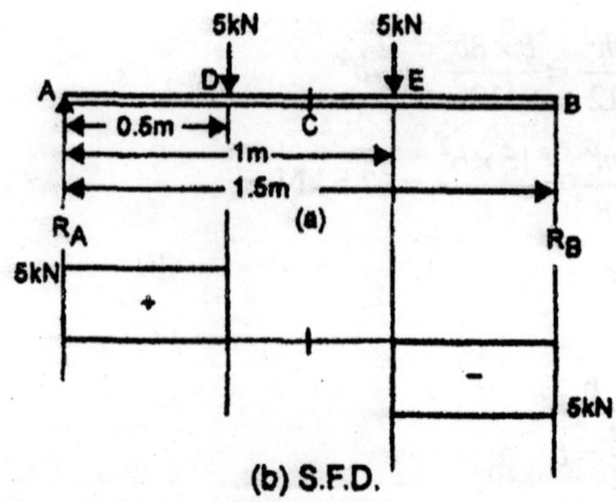

Fig. 4.28

Shear force at the centre C, $F = 0$

Hence, there is no horizontal shear stress at the mid span of the beam.

Example 4.17 *A rectangular beam 100 mm wide and 250 mm deep is subjected to a maximum shear force of 50 kN. Determine the maximum shear stress in the beam.*

Solution:

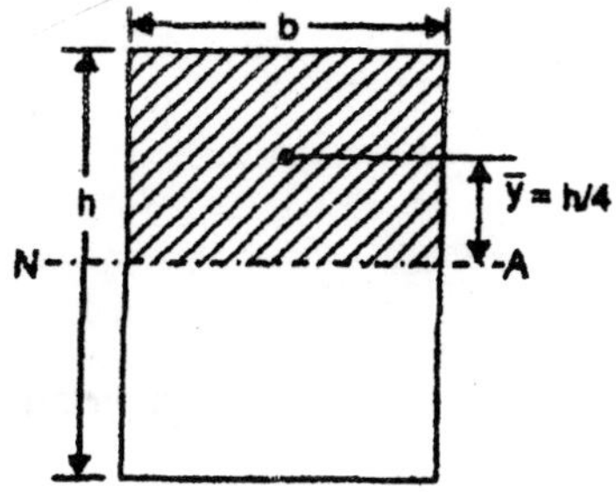

Fig. 5.29

Given: $b = 100$ mm, $h = 250$ mm (Fig. 4.29)

$$I_{NA} = \frac{bh^3}{12} = \frac{100 \times 250^3}{12}$$

$$= 130208333 \text{ mm}^4$$

Shear stress, $$\tau = \frac{FA\bar{y}}{Ib}$$

τ_{max} occurs at the N – A.

$$F = 50 \text{ kN},$$

$$A = b \times \frac{h}{2} = 100 \times 125 = 12500 \text{ mm}^2$$

$$\bar{y} = \frac{h}{4} = \frac{250}{4} = 62.5 \text{ mm}$$

$$\tau_{max} = \frac{50 \times 10^3 \times 12500 \times 62.5}{130208330 \times 100} = 3 \text{ N/mm}^2$$

other wise, $$\tau_{max} = \frac{3}{2}\tau_{mean} = \frac{3}{2}\frac{F}{bh}$$

$$= \frac{3}{2} \times \frac{50 \times 10^3}{100 \times 250} = 3 \text{ N/mm}^2$$

Example 4.18 *A cast iron pipe having a bore of 60 cm and wall thickness 4 cm is simply supported over a span of 4 m. Find the intensity of maximum bending stress when it is full of water. Density of cast iron = 70 kN/m³ and density of water = 10 kN/m³.*

Solution:

Given: $d_i = 60$ cm, $t = 4$ cm, $l = 4$m,

Weight of water per m length of pipe,

$$w_w = \frac{\pi}{4} d_i^2 \times 1 \times (\rho_w \times g)$$

$$= \frac{\pi}{4} \times (0.6)^2 \times 1 \times 10 \times 10^3 = 2827.43 \text{ N/m}$$

Self weight of C.I. pipe per m lengths,

$$w_{C.I.} = \frac{\pi}{4} \times (d_0^2 - d_i^2) \times 1 \times (\rho_{C.I} \times g)$$

$$= \frac{\pi}{4}[(0.68)^2-(0.6)^2]\times 1021\times 70\times 10^3$$

$$= 5629.73 \text{ N/m}$$

Total $$udl, w = w_w + w_{C.I.}$$

$$= 2827.43 + 5629.73$$

$$= 8457.16 \text{ N/m}$$

$$M_{max} = \frac{wl^2}{8} = \frac{8457.16\times 4^2}{8} = 16914.3 \text{ N.m.}$$

$$\text{Section modulus of pipe, } z = \frac{\frac{\pi}{64}(d_0^4 - d_i^4)}{(d_0/2)} = \frac{\pi}{32}\left[\frac{(0.68)^4-(0.6)^4}{0.68}\right] = 0.01216 \text{ m}^3$$

Maximum bending stress in pipe,

$$\sigma = \frac{M_{max}}{z} = \frac{16914.3}{0.01216} = 1.39 \text{ MPa}$$

Example 4.19 *A CI beam of I section Fig. (4.30) is simply supported over a span of 10 m. If the permissible stresses are 100 N/mm² in compression and 40 N/mm² in tension, what UDL can be safely applied on the beam?*

X-Section Details:

Flange width = 200 mm

Overall depth = 400 mm

Thickness of web and flange = 20 mm.

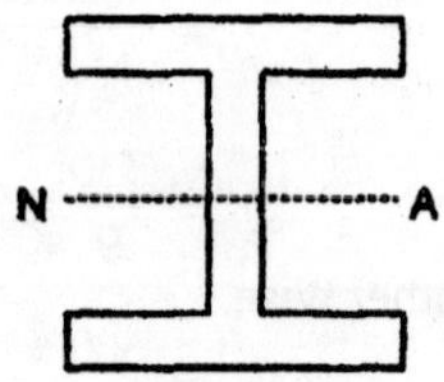

Fig. 4.30

Solution:

For symmetrical I-beam section, (See Fig. 4.31)

$$\bar{y} = 200 \text{ mm}$$

$$I_{NA} = \frac{200\times 400^3}{12} - \frac{180\times 360^3}{12}$$

$$= 366826667 \text{ mm}^4$$

For simply supported beam carrying *udl* of intensity *w*,

$$M_{max} = \frac{wl^2}{8} = \frac{w\times 10^2}{8}$$

$$= 12.5\, w \text{ N.m}$$

$$y_t = y_c = 200 \text{ mm}$$

$\sigma_t < \sigma_c$, therefore σ_t shall be the basis for w_{max}.

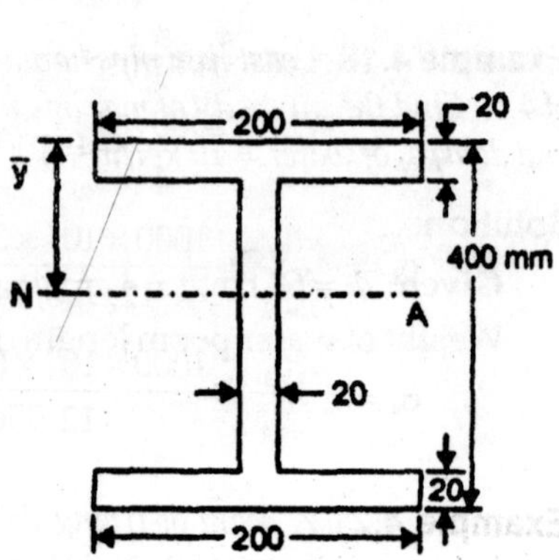

Fig. 4.31

$$\frac{M_{max}}{I_{NA}} = \frac{\sigma_t}{y_t}$$

$$\frac{125\,w\times 10^3}{366826667} = \frac{40}{200}$$

$$N = 5869.2 \text{ N/m}$$

Example 4.20 *A composite beam is formed by fixing a 30 mm × 10 mm steel strip symmetrically at the top of a timber joist of 100 mm depth and 50 mm width. The composite beam is simply supported between supports 2 m apart. If a load of 2 kN is applied at the mid span, determine the stress distribution at that section.* E_{steel} *= 200 GPa and* E_{timber} *= 10 GPa.*

Solution:

$$\text{Equivalent (timber) area of steel strip} = \frac{E_s}{E_t} \times A_s = \frac{200}{10} \times (30 \times 10) = 6000 \text{ mm}^2$$

Equivalent (timber) width of steel strip (Fig. 4.32)

$$= \frac{6000}{10} = 600 \text{mm}$$

$$A_1 = 600 \times 10 = 6000 \text{ mm}^2$$

$$A_2 = 50 \times 100 = 5000 \text{ mm}^2$$

$$A = A_1 + A_2 = 11000 \text{ mm}^2$$

$$\bar{y}_1 = 5 \text{ mm}, \ \bar{y}_2 = 10 + 50 = 60 \text{ mm}$$

$$\bar{y} = \frac{A_1\bar{y}_1 + A_2\bar{y}_2}{A}$$

$$= \frac{6000 \times 5 + 5000 \times 60}{11000}$$

$$= \frac{330}{9} = 36.67 \text{ mm}$$

$$I_{N-A} = \frac{600 \times 10^3}{12} + 6000 \times (36.67 - 5)^2$$

$$+ \frac{50 \times 100^3}{12} + 5000 \times (60 - 36.67)^2$$

$$= 12{,}956{,}045 \text{ mm}^4$$

$$M_{max} = \frac{WL}{4} = \frac{2000 \times 2}{4} = 1000 \text{ N.m}$$

$$\sigma_t = \frac{My_t}{I_{N-A}} = \frac{1000 \times 10^3 \times 36.67}{12{,}956{,}045} = 2.83 \text{ N/mm}^2$$

$$\sigma_s = \frac{My_s}{I_{N-A}} = \frac{1000 \times 10^3 \times (110 - 36.67)}{12{,}956{,}045} = 5.66 \text{ N/mm}^2$$

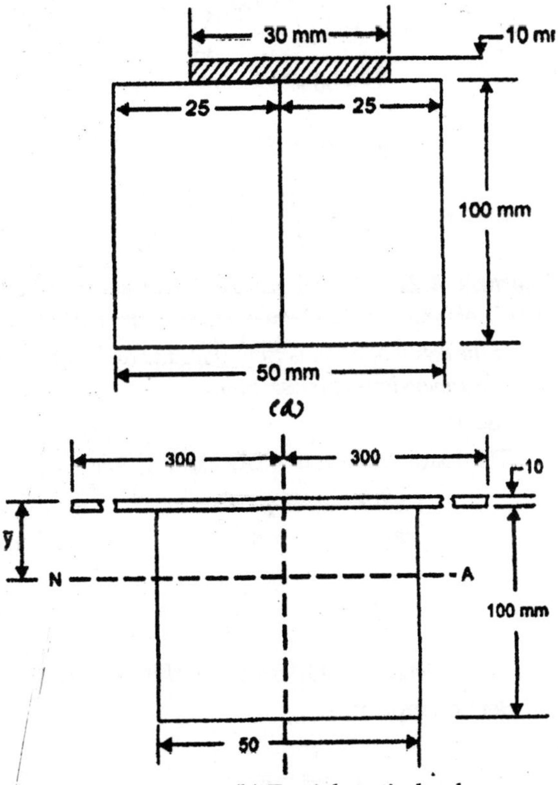

(b) Equialent timber beam

Fig. 4.32

Example 4.21 *A beam of triangular section with 100 mm base width and 250 mm deep is subjected to a maximum shear force of 50 kN. Determine the maximum shear stress in the beam.*

Solution:

Maximum horizontal shear stress occurs at the mid-depth. (See Fig. 4.33)

$$\tau = \frac{FA\bar{y}}{Ib}$$

Now $\frac{b}{100} = \frac{1}{2}$

$b = 50$ mm

$A = \text{area } \Delta \text{ ADE} = \frac{1}{2} \times 50 \times 125$

$= 3125 \text{ mm}^2$

$\bar{y} = \left(125 + \frac{125}{3}\right) - \frac{250}{3} = 83.3 \text{ mm}$

$I = I_{NA} = \frac{100 \times 250^3}{36} = 43402778 \text{ mm}^4$

$F = 50$ kN

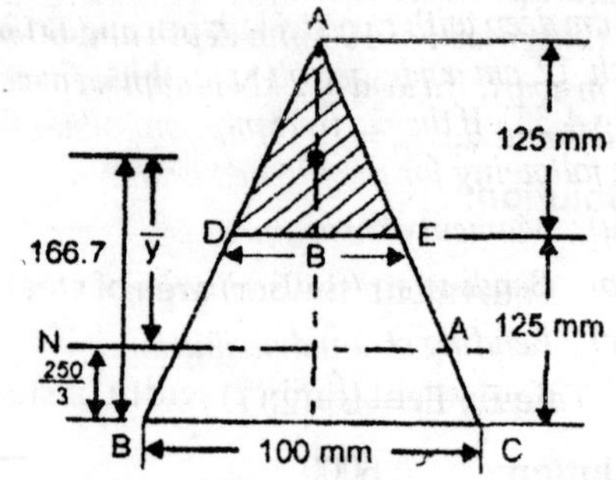

Fig. 4.33

$$\tau_{max} = 50 \times 10^3 \times 3125 \times 83.3 \times \frac{1}{43402778 \times 50} = 6 \text{N/mm}^2$$

$$\text{Otherwise:} \tau_{max} = \frac{3F}{dh} = \frac{3 \times 50 \times 10^3}{100 \times 250} = 6 \text{ N/mm}^2$$

Example 4.22 *A timber beam of rectangular section of length 8 m is simply supported. The beam carries a U.D.L. of intensity 12 kN/m over the entire length and a point load of 10 kN at 3 m from the left support.*

If the depth of beam is two times the width and the stress in the timber is not to exceed 8 N/mm², find the suitable dimensions of the section.

Solution:

The loaded beam is shown in Fig. 4.34 (a)

$h = 2b,$

$\sigma = 8 \text{N/mm}^2$

$$I = \frac{bh^3}{12} = \frac{b \times (2b)^3}{12} = \frac{2}{3}b^4$$

$M_{max} = 111.586$ kN.m (Fig. 4.34 (c))

Section modulus,

$$z = \frac{I}{y} = \frac{I}{b} = \frac{2}{3}b^3$$

$$M_{max} = \sigma \times z$$

$$111.586 \times 10^6 = 8 \times \frac{2}{3}b^3$$

$b = 275.6$ mm

$h = 551.2$ mm

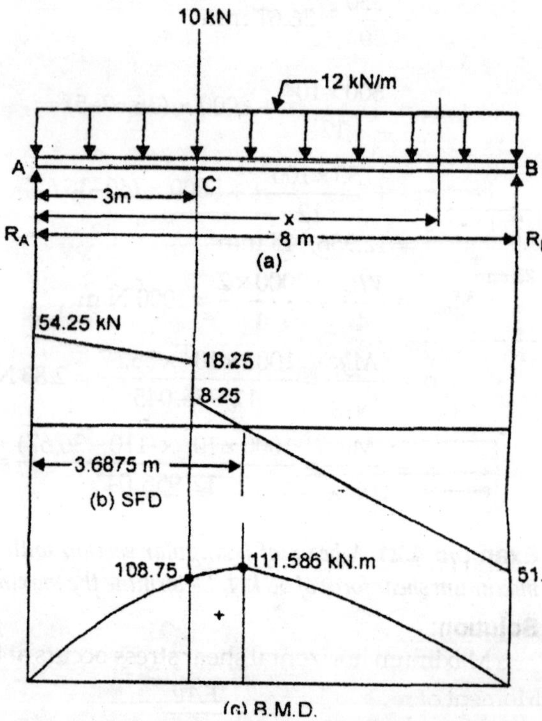

Fig. 4.34

Example 4.23 *A flitched beam consists of a timber beam 12 cm wide and 30 cm deep with two steel plates, one at the top while other at the bottom, each 12 cm wide and 10 cm thick firmly secured to the timber beam (Fig.4.35). If the maximum permissible stress for timber is 8 N/mm², find the following for the flitched beam.*

(a) Moment of resistance.

(b) Bending stress distribution

(c) Bending strain distribution.

Take $E_s/E_t = m = 20$.

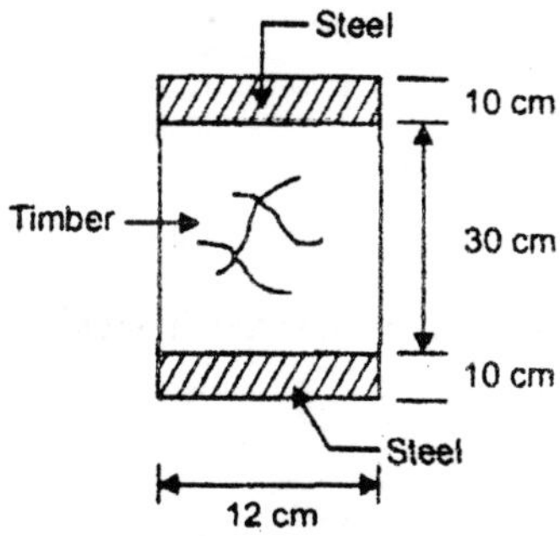

Fig. 4.35

Solution:

$$(\sigma_t)_{max} = 8 \text{N/ mm}^2$$

$$m = \frac{E_s}{E_t} = 20$$

$$\text{Equivalent timber area of steel} = A_s \times \frac{E_s}{E_t}$$

$$= 12 \times 10 \times 20 = 2400 \text{ cm}^2$$

$$\text{Equivalent timber width of steel plate} = \frac{2400}{10} = 240 \text{ cm}$$

The equivalent timber beam is shown in Fig. 4.36.

$$I_{NA} = \frac{240 \times 50^3}{12} - \frac{228 \times 30^3}{12}$$

$$= 198700 \text{ cm}^4$$

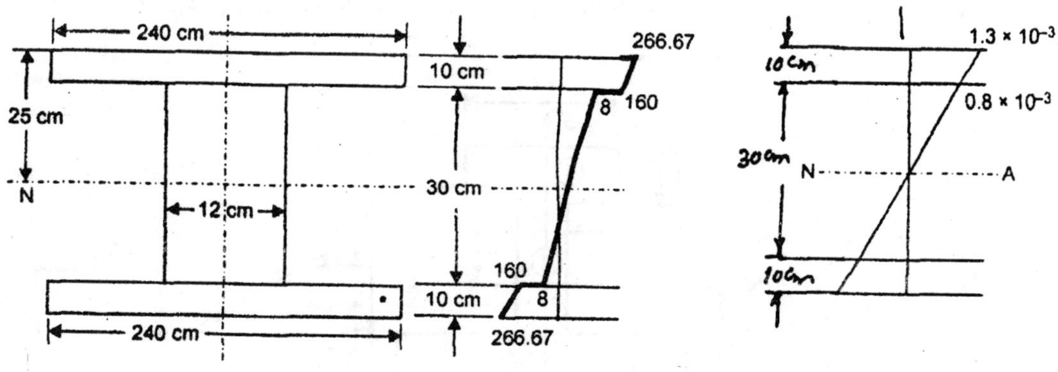

(a) Equivalent timber beam (b) Stress distribution (c) Strain distribution

Fig. 4.36

$$\text{Moment of resistance, M} = \frac{(\sigma_t)_{max} \times I_{NA}}{y_t}$$

$$= \frac{8 \times 1987000 \times 10^4 \times 10^{-6}}{150}$$

$$= 1059.73 \text{ kN.m}$$

Maximum stress in steel, $$(\sigma_s)_{max} = \frac{y_s}{y_t} \times \frac{E_s}{E_t} \times (\sigma_t)_{max}$$

$$= \frac{25}{15} \times 20 \times 8$$

$$= 266.67 \text{ N/mm}^2$$

Stress in steel in contact with timber = 8× 20 = 160 N/mm²

The stress distributin is shown in Fig. 4.36 (*b*)

Let $E_s = 200$ GPa. Then $$E_t = \frac{200}{20} = 10 \text{ GPa}$$

$$(\varepsilon_s)_{max} = \frac{266.67}{200 \times 10^3} = 1.3 \times 10^{-3}$$

$$(\varepsilon_s)_{min} = \frac{160}{200 \times 10^3} = 0.8 \times 10^{-3}$$

$$(\varepsilon_t)_{max} = \frac{8}{10 \times 10^3} = 0.8 \times 10^{-3}$$

The strain distribution is shown in Fig. 4.36 (*b*)

Example 4.24 *(i) Adopting a sign convention and a coordinate system derive the bending stress formula, (ii) Determine the centroid and flexural rigidity for the beam section shown below in Fig. 4.37.*

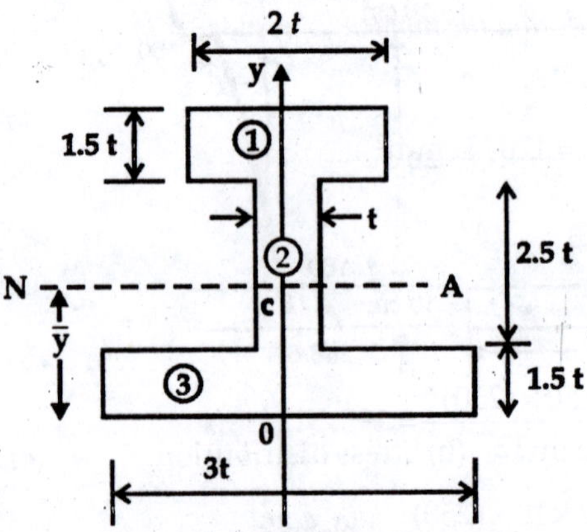

Fig. 4.37

Solution:

(*i*) Refer to art. 4.2

(*ii*) Centroid

$$A_1 = 2t \times 1.5t = 3\,t^2$$
$$A_2 = t \times 2.5t = 2.5\,t^2$$
$$A_3 = 3t \times 1.5t = 4.5\,t^2$$
$$A = A_1 + A_2 + A_3 = (3 + 2.5 + 4.5)t^2 = 10t^2$$
$$\bar{y}_1 = (1.5t + 2.5\,t + 0.7\,t) = 4.7\,5t$$
$$\bar{y}_2 = 1.5\,t + 1.25t = 2.75\,t$$
$$\bar{y}_3 = 0.75\,\text{t}$$
$$\bar{y} = \frac{A_1\bar{y}_1 + A_2\,\bar{y}_2 + A_3\,\bar{y}_3}{A}$$
$$= \frac{(3\times4.5+2.5\times2.75+4.5\times0.75)t^3}{10t^2}$$
$$= 2.375\,t$$
$$I_{N\text{-}A} = \frac{2\times(1.5)^3 t^4}{12} + 3t^2(4.75-2.375)^2 t^2$$
$$= (17.484+1.65+12.726)t^4$$
$$= 31.864\,t^4$$
$$\text{Flexural rigidity} = \text{EI} = 31.864\,t^4E$$

Example 4.25 *A flitched beam consists of a wooden joist 150 mm wide and 250 mm deep strengthened by steel plates 10 mm thick and 200 mm deep, one on either side of the joist. If the stresses in wood and steel are not to exceed 7 N/mm² and 120 N/mm², find the moment of resistance of section of the beam. Take $E_s = 20\,E_w$.*

Solution:

Given: $\sigma_w = 1\ \text{N/mm}^2, \sigma_s = 120\ \text{N/mm}^2$

$$\frac{E_s}{E_w} = 20$$

Refer to Fig. 4.38

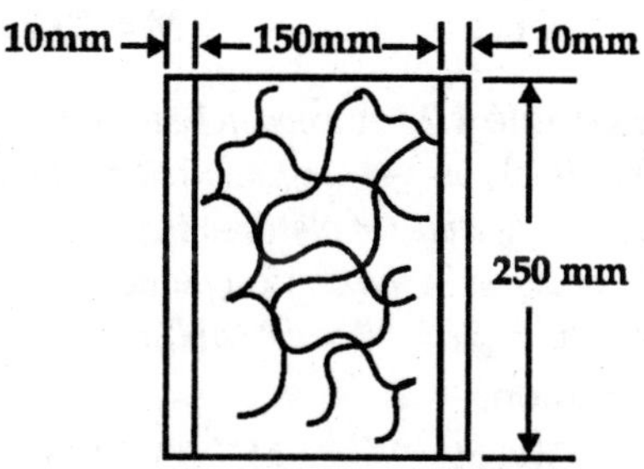

Fig. 4.38

$$I_w = \frac{150\times(250)^3}{12} = 195{,}312\ \text{mm}^4$$

$$I_s = \frac{2\times10\times(250)}{12} = 26{,}041.7\text{mm}^4\ \text{rt}$$

$$y_s = 125\ \text{mm}$$

$$\text{Moment of resistance, } M = \frac{\sigma_s}{y_s}\left[I_s + \frac{E_w}{E_s} \times I_w\right]$$

$$= \frac{120}{125}\left[26{,}041.7 + \frac{1}{20} \times 195{,}312.5\right]$$

$$= 34{,}375 \text{ N.mm}$$

or
$$M = \frac{\sigma_w}{y_w}\left[\frac{E_s}{E_w} \cdot I_s + I_w\right]$$

$$= \frac{7}{125}[20 \times 26{,}041.7 + 195{,}312]$$

$$= 40{,}104 \text{ N.mm}$$

Hence $M = 34{,}375$ N.mm

Example 4.26 *A flitched beam consists of a wooden joints 120 mm wide and 200 mm deep strengfhend by a steel plate 10 mm thick and 180 mm deep, one on either side of the joint. If the stresses in wood and steel are not to exceed 7N/mm² and 120 N/mm², find the moment of resistance of the section of the beam. Take $E_s = 20E_w$. Refer to Fig. 4.39.*

Solution:

$$I_w = \frac{120 \times 200^3}{12} = 8 \times 10^7 \text{ mm}^4$$

$$I_s = \frac{2 \times 10 \times 180^3}{12} = 972 \times 10^4 \text{ mm}^4$$

$$y_s = 90 \text{ mm}$$

$$M = \frac{120}{90}\left[972 \times 10^4 + \frac{1}{20} \times 8 \times 10^7\right]$$

$$= 18{,}293{,}333 \text{ N.mm}$$

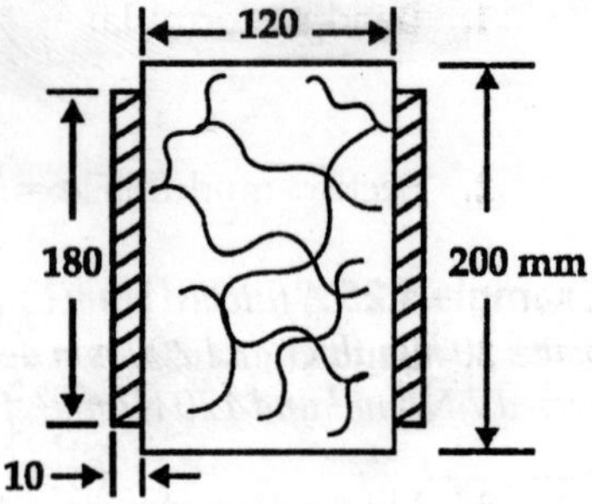

Fig. 4.39

or
$$M = \frac{7}{100}[20 \times 972 \times 10^4 + 8 \times 10^7]$$

$$= 19{,}208{,}000 \text{ N.mm}$$

Hence $M = 18{,}293{,}333$ N.mm

Example 4.27 *A wooden beam 100 mm wide by 300 mm deep is reinforced by two similar steel plates each 12 mm thick, one—being secured to the upper face of the beam and the other to the lower face. Determine the necessary width of plates so that the composite section thus formed should be able to resist a bending moment of 8300 N - m in the vertical plane without the stress in wood exceeding 5.0 MN/m². Take $E_{steel} = 200 \times 10^3$ MN/m² and $E_{wood} = 20 \times 10^3$ MN/m².*

Solution:

Given: $M = 8300$ N.m, $(\sigma_w)_{max} = 5$ MN/m²

$E_s = 200 \times 10^3$ MN/m²

$$m = \frac{E_s}{Ew} = \frac{200}{20} = 10$$

Let b = width of steel plates (Fig. 4.40) Equivalent (timber) moment of inertia,

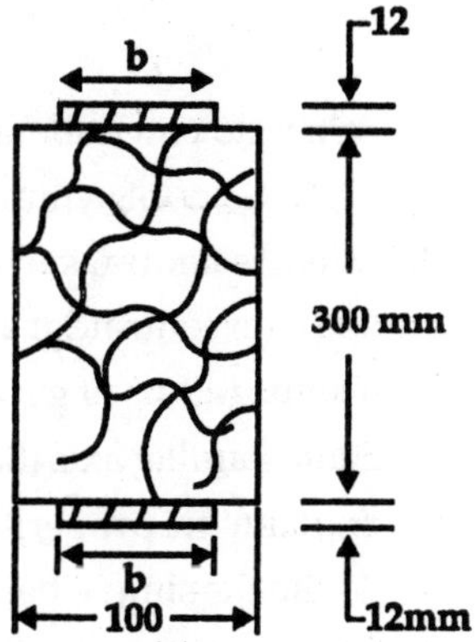

Fig. 4.40

$$I_e = I_w + m\,I_s$$

$$= \frac{100 \times 300^3}{12} + 10\left[\frac{b+12^3}{12} + b \times 12 \times (150+6)^2\right] \times 2$$

$$= 225 \times 10^6 + 5{,}845{,}520 \times b \text{ mm}^4$$

$$M = \frac{(\sigma_w)_{\max} I_e}{y_w} = \frac{5}{150}[225 \times 10^6 + 5{,}843{,}520 \times b]$$

$$= 7.5 \times 10^6 + 194{,}784 \times b = 8300 \times 10^3$$

$$b = 41 \text{ mm}$$

Summary

1. Bending formula: $\frac{M}{I} = \frac{\sigma}{y} = \frac{E}{R}$

2. Section modulus, $z = \frac{I}{y}$

3. Bending stress, $\sigma = \frac{M}{z}$

4. Flitched beams: $\frac{(\sigma_s)_{\max}}{(\sigma_t)_{\max}} = \frac{y_s}{y_t} \times \frac{E_s}{E_t}$

5. Total moment of resistance, $M = M_s + M_t$

$$= \frac{(\sigma_t)_{\max}}{y_t}\left[\frac{E_s}{E_t} \cdot I_s + I_t\right]$$

$$= \frac{(\sigma_s)_{\max}}{y_s}\left[I_s + \frac{E_t}{E_s} \cdot I_t\right]$$

6. Horizontal shear stress, $\tau = \frac{FA\bar{y}}{Ib}$

Short Answer Type Questions

1. What do you understand by simple bending of beams?
2. Define axis of symmetry and plane of symmetry of a beam.
3. What is neutral surface of a beam?
4. Explain centroidal axis and neutral axis of a beam.
5. Define radius of gyration.
6. State parallel axis theorem.
7. Explain the properites of the material to which theory of simple bending is applicable.
8. Write the simple bending formula and give the units of all terms.
9. Define a flitched beam.
10. State the locations for maximum bending stress and maximum horizontal stress in a beam.
11. Define equivalent bending moment.
12. What is section modulus?

EXERCISES

1. An inverted T-beam of 4 m span is simply supported at the ends. Its cross-section is 300 mm × 250 mm × 10 mm. If permissible stresses in tension and compression are 60 MPa and 80 MPa respectively, calculate the safe uniformly distributed load on the whole span.
2. A rolled steel joist 400 mm × 200 mm × 20 mm is freely supported on a span of 4 m. The flanges are strengthened by two 250 mm × 20 mm plates, one riveted to each flange. Calculate (a) the maximum central load the beam can carry if the stress in the plate is not to exceed 125 MPa, (*b*) the minimum length of the strengthening plates required to restrict the maximum stress to 125 MPa.
3. A timber beam 200 mm × 300 mm is to be reinforced with two steel flitches 150 mm-x 10 mm in section. Compare their moments of resistance for the following two cases:
   (*a*) Flitches are attached at top and bottom.
   (*b*) Flitches are attached symmetrically on the sides.
4. Two rectangular bars, one of brass and other of steel, each 50 mm x 10 mm, are placed together to form a beam 50 mm × 20 mm. The composite beam is supported on a span of 1.0 m. The brass bar is on top of steel. Determine the maximum central load which can be applied to the beam, if the bars are:
   (*a*) separate and can bend independently,
   (*b*) firmly secured to each other throughout their length
   Take $E_b = 90$ GPa, $E_s = 200$ GPa, $(\sigma_b)_{max} = 80$ MPa, and $(\sigma_s)_{max} = 120$ MPa..
5. An *I*-section beam 60 mm × 250 mm × 20 mm is subjected to a shear force of 250 kN. Determine the distribution of shear stress in the beam.
6. A T-section beam 150 mm x 100 mm x 10 mm is required to carry a shear force of 200 kN. Plot the distribution of shear stress in the beam.

7. A beam of 120 mm square cross-section is used with a diagonal in a vertical position. The vertical shear force at a section is 2 kN. Plot the distribution of shear stress in the beam.

8. A rectangular cross-section beam has a width of 100 mm. Determine the depth of the beam so that maximum bending stress in the beam does not exceed 35 MPa.

9. A cast iron water main pipe 12 m long, 500 mm inside diameter and 25 mm wall thickness runs full of water. It is supported at its ends. If the density of cast iron and water are 7200 kg/m^3 and 1000 kg/m^3 respectively, calculate the maximum stress induced in cast iron.

10. A uniform heavy girder of length 10 m is to be placed over two supports 6 m apart. Calculate the length of overhang for the girder at both ends so that the maximum bending moment, positive or negative, due to self weight shall be as small as possible.

Chapter-5

Deflection of Beams

5.1 INTRODUCTION

When a beam is subjected to transverse loads, it deflects or sags. The neutral axis, which is initially a straight line, becomes a curved line. This line is called the elastic curve. The deflection of the beam is a vertical distance between a point on the elastic curve and the unloaded neutral axis.

The slope of the beam is the angle between the tangent to the elastic curve at a point on the elastic line and the unloaded neutral axis.

It is very important from structural design point of view to determine the deflection of the beam to have adequate stiffness. A simply supported beam is shown in Fig. 5.1 (a) when unloaded and deflected under concentrated load in Fig. 5.1 (*b*).

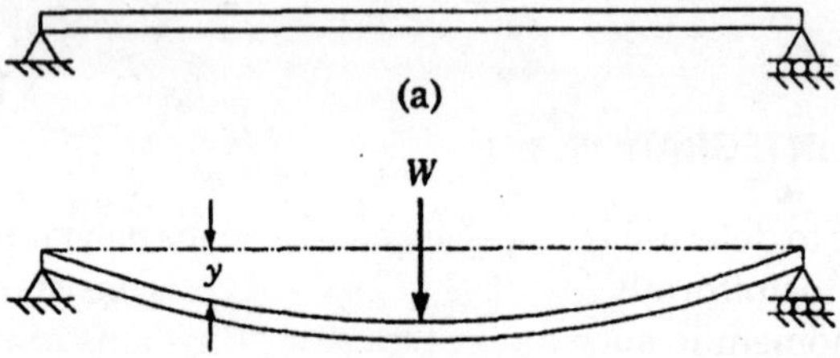

Fig. 5.1 (a) Unloaded, and (b) deflected beam.

5.2 RELATIONSHIP BETWEEN MOMENTS, SLOPE AND DEFLECTION

The curvature of a curve $y = f(x)$ of radius R is given by,

$$\frac{1}{R} = \frac{\frac{d^2y}{dx^2}}{\left[1+\left(\frac{dy}{dx}\right)^2\right]^{3/2}}$$

If the curvature is very small then $\frac{dy}{dx}$ is also small and its square is negligible. Hence

$$\frac{1}{R} = \frac{d^2y}{dx^2} = \frac{M}{EI}$$

where M = B.M., E = Young's modulus of elasticity, and I = moment of inertia. If the curvature is positive then $\frac{M}{EI}$ decreases. Thus

$$\frac{1}{R} = -\frac{M}{EI} = \frac{d^2y}{dx^2}$$

$$\therefore \qquad M = -EI\frac{d^2y}{dx^2} \qquad \text{...(5.1)}$$

Slope,
$$\frac{dy}{dx} = -\int\frac{M}{EI}dx \qquad \text{...(5.2)}$$

Deflection,
$$y = -\iint\frac{M}{EI}dx.dx \qquad \text{...(5.3)}$$

Shear force,
$$F = \frac{dM}{dx} = -EI\frac{d^3y}{dx^3} \qquad \text{...(5.4)}$$

Uniform load,
$$w = -EI\frac{d^4y}{dx^4} \qquad \text{...(5.5)}$$

5.3 METHOD OF DOUBLE INTEGRATION

This method is suitable to determine the deflection and slope of beams subjected to concentrated loads and can be extended to uniformly distributed loads also. It consists of successive integration of expressions for bending moment in such a way that same constants of integration are valid for all portions of the beam even though the law of bending moment differs from portion to portion. The following procedure is adopted to apply this method:

1. Write the general expression for the bending moment covering all the loads just before the right hand support.
2. A bracket is disregarded when for a particular value of x the contents of the bracket becomes negative.
3. The bracket terms must be kept intact and not multiplied out until a numerical substitution is made for x.
4. If there is a uniformly distributed load starting anywhere and extending upto the right end then the general equation for bending moment will hold good for the entire beam. However, if the load does not extend upto the right end, the load may be extended upto the right end, and an equal and opposite load may be added to counteract the effect due to the additional load.
5. A couple 'C at a distance 'a' from left end may be written: $C(x-a)^0$.

5.3.1 Concentrated Load on a Simply Supported Beam

Consider a simply supported beam *AB* of span *l* carrying concentrated load *W* at *C* at a distance '*a*' from left end support as shown in Fig. 5.2.

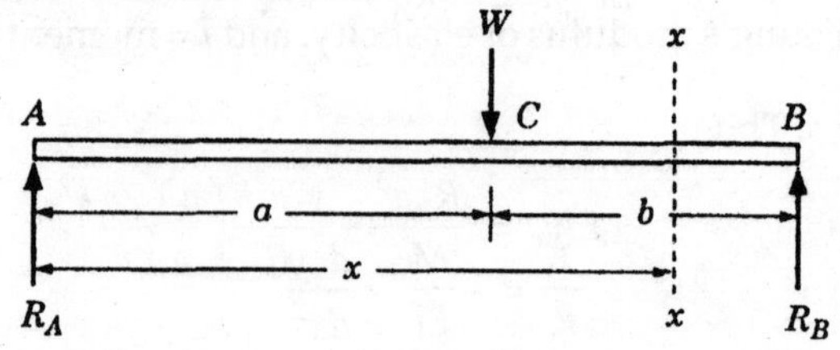

Fig. 5.2 Simply supported beam carrying off-centre concentrated load

$$R_A = \frac{Wb}{l} \text{ and } R_B = \frac{wa}{l}$$

Taking a section $x-x$ at a distance x from left end A, the bending moment there is,

$$M_x = R_A x - W(x-a)$$

$$\therefore \quad EI\frac{d^2y}{dx^2} = -M_x = -R_A x + W(x-a)$$

Integrating, we get

$$EI\frac{dy}{dx} = -R_A\frac{x^2}{2} + \frac{W(x-a)^2}{2} + C_1$$

where C_1= a constant of integration

Integrating again, we get

$$Ely = -R_A\frac{x^3}{6} + \frac{W(x-a)^3}{6} + C_1 x + C_2$$

where C_2 = another constant of integration

The end conditions are :

At $x = 0, y = 0$

$$\therefore \quad C_2 = 0$$

At $x = l, y = 0$

$$0 = -R_A\frac{l^3}{6} + \frac{Wb^3}{6} + C_1 l$$

$$C_1 = \frac{R_A l^2}{6} - \frac{Wb^3}{6l} = \frac{Wb(a+b)}{6} - \frac{Wb^3}{6(a+b)}$$

$$= Wb\left[\frac{a^2+b^2+2ab-b^2}{6(a+b)}\right] = \frac{Wab}{6}\left(\frac{a+2b}{a+b}\right)$$

$$EIy = -R_A\frac{x^3}{6} + \frac{W(x-a)^3}{6} + \frac{Wabx}{6}\left(\frac{a+2b}{a+b}\right) \quad \text{...(1)}$$

$$EI\frac{dy}{dx} = -R_A\frac{x^2}{2} + \frac{W(x-a)^2}{2} + \frac{Wab}{6}\left(\frac{a+2b}{a+b}\right) \quad \text{...(2)}$$

At $x = a$, we have

$$EIy_c = \frac{R_A a^3}{6} + \frac{Wa^2b}{6}\left(\frac{a+2b}{a+b}\right)$$

$$= -\frac{Wba^3}{6(a+b)} + \frac{Wa^2b}{6}\left(\frac{a+2b}{a+b}\right)$$

$$= \frac{Wa^2b}{6(a+b)}[-a+a+2b]$$

$$= \frac{Wa^2b^2}{3(a+b)}$$

$$y_c = \frac{Wa^2b^2}{3EIl} \quad \text{...(5.6a)}$$

where $l = a + b$

At $x = \frac{l}{2}$, we have

$$EIy = \frac{Wb}{l}\times\frac{l^3}{48} + \frac{W}{6}\left(\frac{l}{2}-a\right)^3 + \frac{Wabl}{12}\left(\frac{a+2b}{l}\right)$$

$$= \frac{W}{48l}\left[-bl^3 + 8l\left(\frac{l}{2}-a\right)^3 + 4abl(a+2b)\right]$$

$$= \frac{W}{48l}\left[-l^3(l-a) + 8l\left(\frac{l}{2}-a\right) + 4al(l-a)(2l-a)\right]$$

$$y = \frac{Wa}{48EI}[(3l^2 - 4a^2)] \quad \text{...(5.6b)}$$

If $a = b = \frac{l}{2}$, then

$$(y)_{x=\frac{l}{2}} = \frac{Wl^3}{48EI} \quad \text{...(5.7)}$$

For deflection to be maximum,

$$\frac{dy}{dx} = 0$$

$$-\frac{Wbx^2}{2(a+b)}+\frac{W(x-a)^2}{2}+\frac{Wab}{6}\left(\frac{a+2b}{a+b}\right) = 0$$

For $x > a$, second bracket becomes negative, and is ignored.

$$x^2 = \frac{a(a+2b)}{3}$$

or

$$x = \sqrt{\frac{a(a+2b)}{3}} = \sqrt{\frac{l^2-b^2}{3}}$$

$\therefore$

$$Ely_{\max} = \frac{Wab(a+2b)}{9(a+b)}\left[\frac{a}{3}(a+2b)\right]^{\frac{1}{2}} \quad \text{...(5.8)}$$

At $x = 0$,

$$EI\left(\frac{dy}{dx}\right)_{x=0} = \frac{Wab}{6}\left(\frac{a+2b}{a+b}\right)$$

$\therefore$

$$\theta_A = \frac{Wab}{6EI}\left(\frac{a+2b}{a+b}\right) \quad \text{...(5.9)}$$

For $a = b = \frac{l}{2}$, we get

$$\theta_A = \frac{Wl^2}{16EI} \quad \text{...(5.10)}$$

At $x = l$,

$$EI\left(\frac{dy}{dx}\right)_{x=l} = -\frac{Wbl}{2}+\frac{Wb^2}{2}+\frac{Wab}{6}\left(\frac{a+2b}{a+b}\right)$$

$$= -\frac{Wab}{l}\left(\frac{2a+b}{a+b}\right) \quad \text{...(5.11)}$$

For $a = b = \frac{l}{2}$, we get

$$\theta_B = -\frac{Wl^2}{16EI} \quad \text{...(5.12)}$$

5.3.2 Uniformly Distributed Load on Simply Supported Beam

Consider a beam AB of length l, simply supported at the ends and carrying a uniformly distributed load of intensity w per unit length, as shown in Fig. 5.3. The bending moment at a section x – x is,

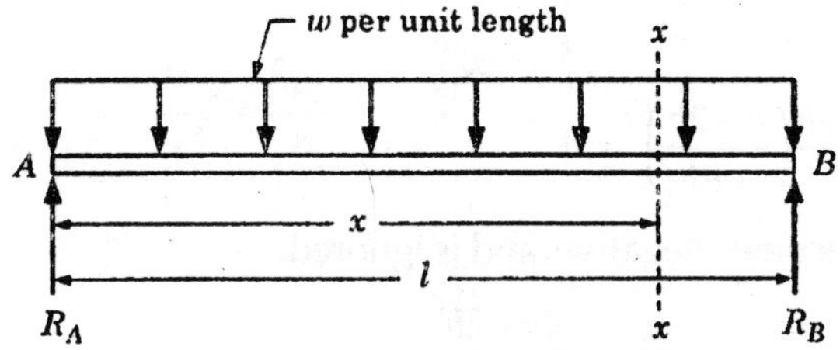

Fig. 5.3 Simply supported beam carrying *udl*.

$$M_x = R_A x - \frac{wx^2}{2}$$

where $R_A = R_B = \dfrac{wl}{2}$

$$EI\frac{d^2y}{dx^2} = -M_x = -\frac{wl}{2}x + \frac{wx^2}{2}$$

Integrating, we get

$$EI\frac{dy}{dx} = -w\frac{lx^2}{4} + \frac{wx^3}{6} + C_1$$

where $C_1 = a$ constant of integration

Integrating again, we get

$$EIy = -wl\frac{x^3}{12} + \frac{wx^4}{24} + C_1x + C_2$$

At $x = 0, y = 0$

$\therefore$ $$C_2 = 0$$

At $x = l, y = 0$

$$0 = -\frac{wl^4}{12} + \frac{wl^4}{24} + C_1 l$$

$$C_1 = \frac{wl^3}{24}$$

$\therefore$ $$EIy = -wl\frac{x^3}{12} + \frac{wx^4}{24} + \frac{wl^3}{24}x \qquad \text{... (5.13)}$$

At $x = \dfrac{l}{2}$, the deflection is maximum.

$$y_{max} = \frac{5}{384}\frac{wl^4}{EI} \quad \text{... (5.14)}$$

Now

$$EI\frac{dy}{dx} = -\frac{wlx^2}{4} + \frac{wx^3}{6} + \frac{wl^3}{24} \quad \text{... (5.15)}$$

At $x = 0$,

$$\theta_A = \frac{wl^3}{24EI} \quad \text{... (5.16)}$$

At $x = l$,

$$\theta_B = -\frac{wl^3}{24EI} \quad \text{... (5.17)}$$

5.3.3 Concentrated Load at the free end of a Cantilever Beam

Consider a cantilever beam *AB* of span *l* carrying concentrated load *W at* the free end, as shown in Fig. 5.4. Bending moment at a distance *x* from the free end is,

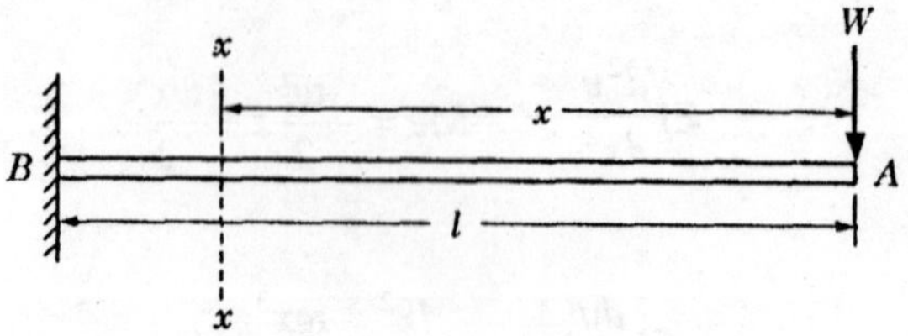

Fig. 5.4 Cantilever beam carrying end load

$$M_x = -Wx$$

being a hogging moment.

$$EI\frac{d^2y}{dx^2} = -M_x = Wx$$

Integrating, we get

$$EI\frac{dy}{dx} = \frac{wx^2}{2} + C_1$$

where C_1 = a constant of integration.
Integrating again, we get

$$Ely = \frac{Wx^3}{6} + C_1x + C_2$$

where C_2 = another constant of integration

At $x = l$, $y = 0$ and $\frac{dy}{dx} = 0$

$$\therefore \qquad C_1 = -\frac{Wl^2}{2} \text{ and } C_2 = \frac{Wl^3}{3}$$

Hence
$$El\frac{dy}{dx} = \frac{Wx^2}{2} - \frac{Wl^2}{2} \qquad \text{... (5.18)}$$

$$Ely = \frac{Wx^3}{6} - \frac{Wl^2x}{2} + \frac{Wl^3}{3} \qquad \text{... (5.19)}$$

At $x = 0$,
$$\theta_A = -\frac{Wl^2}{2EI} \qquad \text{... (5.20)}$$

and
$$y_A = \frac{Wl^3}{3EI} \qquad \text{... (5.21)}$$

5.3.4 Cantilever Beam Carrying Uniformly Distributed Load

Consider a cantilever beam AB of span l carrying uniformly distributed load of intensity w per unit length over the whole span. as shown in Fig. 5.5. The bending moment at a distance x from free end A is,

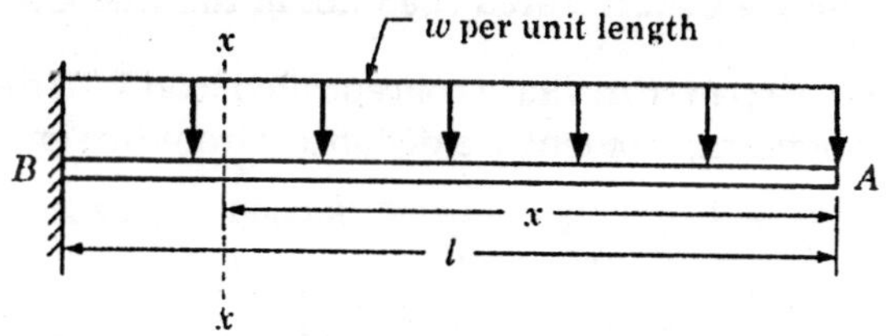

Fig. 5.5 Cantilever beam carrying *udl*

$$M_x = -\frac{wx^2}{2}$$

$$\therefore \qquad EI\frac{d^2y}{dx^2} = -M_x = \frac{wx^2}{2}$$

Integrating, we get

$$EI\frac{dy}{dx} = \frac{wx^3}{6} + C_1$$

Integrating again, we get

$$Ely = \frac{wx^4}{24} + C_1x + C_2$$

At $x = l$, $y = 0$ and $\frac{dy}{dx} = 0$

$$\therefore \quad C_1 = -\frac{wl^3}{6}$$

$$C_2 = \frac{wl^4}{8}$$

$$\therefore \quad EI\frac{dy}{dx} = \frac{wx^3}{6} - \frac{wl^3}{6} = \frac{w}{6}(x^3 - l^3) \quad \text{...(5.22)}$$

$$Ely = \frac{wx^4}{24} - \frac{wl^3x}{6} + \frac{wl^4}{8} \quad \text{...(5.23)}$$

At $x = 0$,

$$\theta_A = -\frac{wl^3}{6EI} \quad \text{...(5.24)}$$

$$y_A = \frac{wl^4}{8EI} \quad \text{...(5.25)}$$

5.3.5 Cantilever Beam Carrying a Concentrated Load not at the free End

Consider a cantilever *AB* of span *l* carrying a concentrated load *W* at distance 'a' from the free end A, as shown in Fig. 5.6. The bending moment at a distance *x* from free end is,

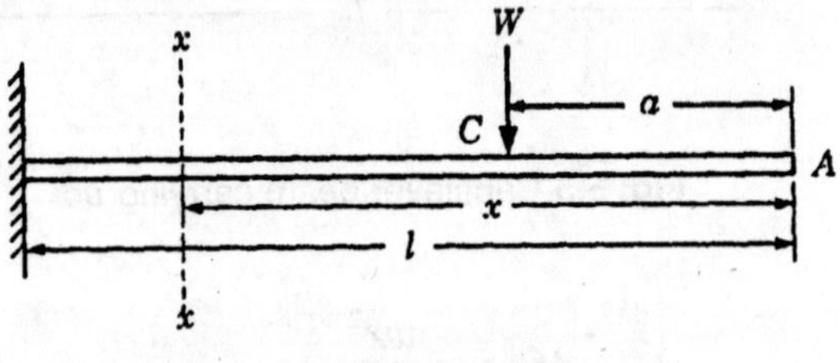

Fig. 5.6 Cantilever beam carrying concentrated load not at the free end.

$$M_x = -W(x-a)$$

$$EI\frac{d^2y}{dx^2} = -M_x = W(x-a)$$

Integrating, we get

$$EI\frac{dy}{dx} = \frac{W(x-a)^2}{2} + C_1$$

Integrating again, we get

$$Ely = \frac{W(x-a)^3}{6} + C_1x + C_2$$

At $x = l, y = 0$ and $\frac{dy}{dx} = 0$

$$C_1 = -\frac{W(l-a)^2}{2}$$

$$0 = \frac{W(l-a)^3}{6} - \frac{W(l-a)^2 l}{2} + C_2$$

$$C_2 = \frac{W(l-a)^3}{6} + \frac{W(l-a)^2 l}{2}$$

$$= \frac{W(l-a)^2}{6}[-(l-a)+3l]$$

$$= \frac{W(l-a)^2}{6}(2l+a)$$

$$EI\frac{dy}{dx} = \frac{W(x-a)^2}{2} - \frac{W(l-a)^2}{a} \quad \text{... (5.26)}$$

$$Ely = \frac{W(x-a)^3}{6} - \frac{W(l-a)^2 x}{2} + \frac{W(l-a)^2}{6}(2l+a) \quad \text{... (5.27)}$$

At $x = a$

$$El\frac{dy}{dx} = -\frac{W(1-a)^2}{2}$$

$$\left(\frac{dy}{dx}\right)_{x=a} = \theta_c = \frac{-W(l-a)^2}{2EI} \quad \text{... (5.28)}$$

$$EI(y)_{x=a} = -\frac{W(l-a)^2 a}{2} + \frac{W(l-a)^2}{6}(2l+a)$$

$$= \frac{W(l-a)^2}{6}[-3a+2l+a]$$

$$y_c = (y)_{x=a} = \frac{W(l-a)^3}{3EI} \quad \text{... (5.29)}$$

At $x = 0$

$$y_A = \frac{W(l-a)^2}{6EI}(2l+a) \quad \text{... (5.30)}$$

Example 5.1 *A beam AB, 15 m long is simply supported at the ends. It carries two concentrated loads of 900 kN and 60 kN at 4m and 10m respectively from the left end. Moment of inertia of the beam about the neutral axis is 15 ×10^{-4} m^4 and E = 200 GPa. Calculate the deflection of the beam under the two loads.*

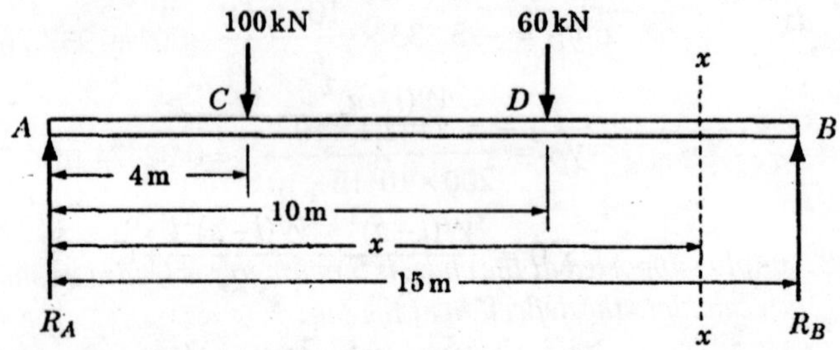

Fig. 5.7

Solution:

The loaded beam is shown in Fig. 5.7.

$$\Sigma M_B = 0, \text{ gives}$$

$$15R_A = 100 \times 11 + 60 \times 5 = 1400$$

$$R_A = 93.333 \text{ kN}$$

At the section $x - x$, the bending moment is,

$$M_x = R_A x - 100\,(x-4) - 60\,(x-10)$$

$$EI\frac{d^2y}{dx^2} = -M_x = -R_A\,x + 100(x-4) + 60(x-10)$$

$$EI\frac{dy}{dx} = -R_A\frac{x^2}{2} + 50\,(x-4)^2 + 30\,(x-10)^2 + C_1$$

$$Ely = -R_A\frac{x^3}{6} + \frac{50}{3}\,(x-4)^3 + 10\,(x-10)^3 + C_1x + C_2$$

At $x = 0$, $y = 0$ so that $C_2 = 0$

At $x = 15$m, $y = 0$,

$$0 = -93.333 \times \frac{15^3}{6} + \frac{50}{3} \times 11^3 + 10 \times 5^3 + 15C_1$$

$$C_1 = 937.76 \text{ kN.m}^3$$

$$Ely = -R_A\frac{x^3}{6} + \frac{50}{3}\,(x-4)^3 + 10\,(x-10)^3 + 1937.76\,x$$

At $x = 4$m,

$$Ely_c = -93.333 \times \frac{4^3}{6} + 1937.76 \times 4 = 6755.51$$

$$y_c = \frac{6755.51 \times 10^3}{200 \times 10^9 \times 15 \times 10^{-4}} = 0.02252 \text{ m}$$

At $x = 10$m,

$$EIy_D = -93.333 \times \frac{10^3}{6} + \frac{50}{3} \times 6^3 + 1937.76 \times 10 - 7422.1$$

$$y_D = \frac{7422.1 \times 10}{200 \times 10^9 15 \times 10^4} = 0.02474 \text{ m}$$

Example 5.2 *A beam AB, simply supported at the ends, is 5 m long. It is loaded as shown in Fig. 5.8 (a). If E = 210 GPa and I = 1 × 10⁻⁴m⁴, calculate the deflection at the centre.*

Solution:

The *udl* has been extended to the end B and an equal and opposite *udl* is applied on the span DB, as shown in Fig. 5.8 (*b*).

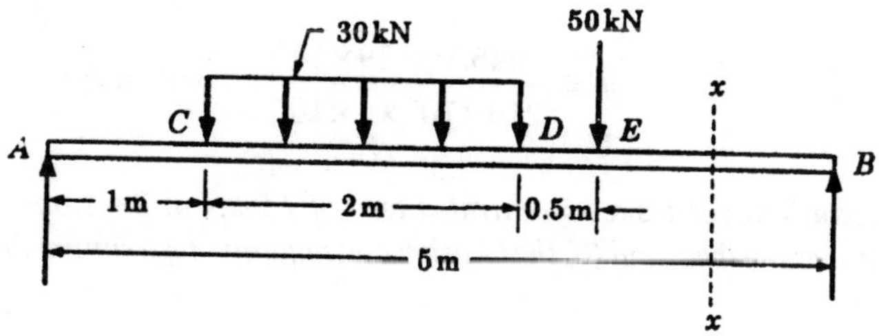

(a)

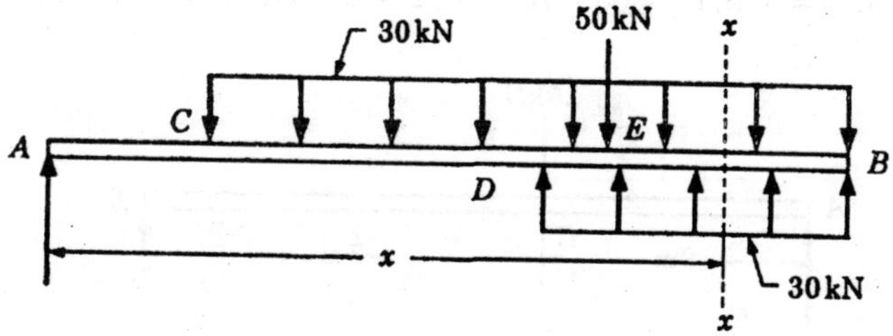

(b)

Fig. 5.8

$$5R_A = 30 \times 2 \times 3 + 50 \times 1.5 = 255$$

$$R_A = 51 \text{ kN}$$

$$M_x = 51x - \frac{30(x-1)^2}{2} + \frac{30(x-3)^2}{2} - 50(x-3.5)$$

$$EI\frac{d^2y}{dx^2} = -M_x = -51x + 15(x-1)^2 - 15(x-3)^2 + 50(x-3.5)$$

$$EI\frac{dy}{dx} = -25.5x^2 + 5(x-1)^3 - 5(x-3)^2 + 25(x-3.5)^2 + C_1$$

$$Ely = -8.5x^3 + 1.25(x-1)^4 - 1.25(x-3)^4 + \frac{25}{3}(x-3.5)^3 + C_1x + C_2$$

At $x = 0$, $y = 0$, so that $C_2 = 0$

$x = 5$ m, $y = 0$,

$$0 = -8.5 \times 5^3 + 1.25 \times 4^4 - 1.25 \times 2^4 + \frac{25}{3} \times 1.5^2 + 5C_1:$$

$$C_1 = 148.75 \text{ kNm}^3$$

$$Ely = -8.5x^3 + 1.25(x-1)^4 - 1.25(x-3)^4 + \frac{25}{3}(x-3.5)^3 + 148.75x$$

At $x = 2.5$m,

$$Ely = -8.5 \times 2.5^3 + 1.25 \times 1.5^4 + 148.75 \times 2.5 = 245.39$$

$$= 245.39$$

$$y = \frac{245.39 \times 39 \times 10^3}{210 \times 10^9 \times 1 \times 10^{-4}} = 0.011658 \text{ m}$$

Example 5.3 *A beam of span 5 m carries a concentrated load of 2.5 kN at 2m from the left hand support. Calculate (a) the deflection under the load, and (b) the location and amount of maximum deflection. Take $I_{xx} = 60 \times (10^{-6}m^4$ and $E = 200$ GPa.*

Solution:

Given: $l = 5$ m, $a = 2$ m, $b = 3$ m, $I_{xx} = 60 \times 10^{-6}$ m⁴, $E = 200$ GPa, W = 2.5 kN

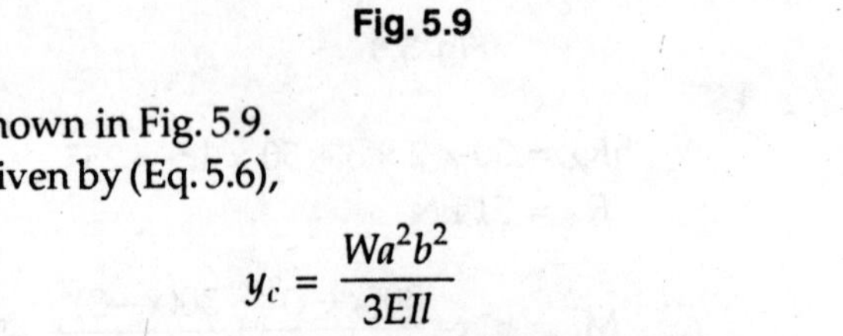

Fig. 5.9

The loaded beam is shown in Fig. 5.9.

(*a*) Deflection at C is given by (Eq. 5.6),

$$y_c = \frac{Wa^2b^2}{3EIl}$$

$$= \frac{25 \times 10^3 \times 2^2 \times 3^2}{3 \times 200 \times 10^9 \times 60 \times 10^{-6} \times 5}$$

$$= 5 \times 10^{-3} \text{ m or } 5 \text{ mm}$$

(*b*) For maximum deflection,

$$x = \sqrt{\frac{a(a+2b)}{3}}$$

$$= \sqrt{\frac{2(2+2\times3)}{3}} = 2.31 \text{ m}$$

$$y_{max} = \frac{Wab(a+2b)}{9EIl}\left[\frac{a}{3}(a+2b)\right]^{1/2}$$

$$= \frac{25\times10^3\times2\times3\times8}{9\times200\times10^9\times60\times10^{-6}\times5}\left[\frac{2}{3}(2+6)\right]^{1/2}$$

$$= 5.132 \times 10^{-3} \text{ m or } 5.132 \text{ mm}$$

Example 5.4 *A beam AB of 4 m span and simply supported at the ends is loaded as shown in Fig. 5.10. Calculate (a) the deflection at midspan, (b) maximum defection, and (c) slope at the end A.*

Take $I = 25 \times 10^{-6}$ m^4, E = 210 G Pa.

Solution:

Taking moments about *B*, we have

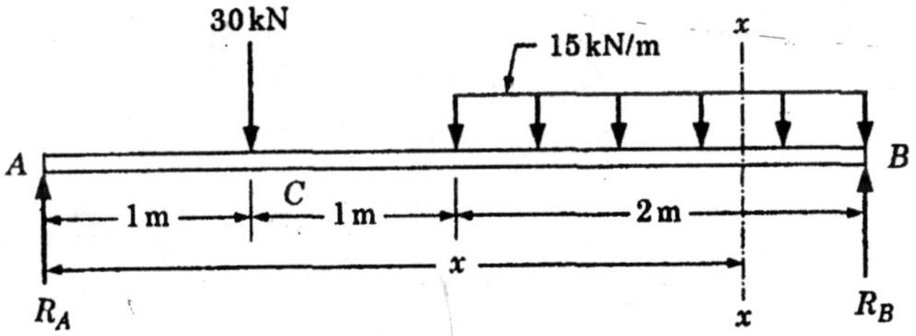

Fig. 5.10

$$R_A \times 4 = 30 \times 3 + 15 \times 2 \times 1$$

$$R_A = \frac{120}{4} = 30 \text{ kN}$$

$$R_B = 60 - 30 = 30 \text{ kN}$$

Taking a section x – *x* near the end *B*,

$$M_x = R_A \times x - 30\,(x-1) - \frac{15}{2}\,(x-2)^2$$

$$= 30\,x - 30\,(x-1) - 7.5\,(x-2)^2$$

$$EI\frac{d^2y}{dx^2} = -M_x$$

$$= -30\,x + 30\,(x-1) + 7.5\,(x-2)^2$$

Integrating, we get

$$EI\frac{dy}{dx} = -15x^2 + 15(x-1)^2 + 2.5(x-2)^3 + C_1$$

Integrating again, we get

$$Ely = -5x^3 + 5(x-1)^3 + 0.625(x-2)^4 + C_1x + C_2$$

At x 0, $y = 0$ and at $x = 4$m, $y = 0$,

$\therefore$

$$C_2 = 0$$

$$0 = -5 \times 64 + 5 \times 27 + 0.625 \times 16 + 4C_1$$

$$C_1 = 43.75$$

$\therefore$

$$Ely = -5x^3 + 5(x-1)^3 + 0.625(x-2)^4 + 43.75x$$

(*a*) At $x = 2$ m,

$$Ely = -5 \times 8 + 5 \times 1 + 0 + 87.5 = 52.5$$

$$y = \frac{52.5 \times 10^3}{210 \times 10^9 \times 25 \times 10-6} = 0.01\text{m or } 10\text{ mm}$$

(*b*) For maximum deflection, $\frac{dy}{dx} = 0$

$$-15x^2 + 15(x-1)^2 + 2.5(x-2)^3 + 43.75 = 0$$

$$-15x^2 + 15x^2 + 15 - 30x + 2.5(x^3 - 8 - 6x^2 + 12x) + 43.75 = 0$$

$$15 - 30x + 2.5x^3 - 20 - 15x^2 + 30x + 43.75 = 0$$

$$2.5x^3 - 15x^2 + 38.75 = 0$$

$$f(x) = x^3 - 6x^2 + 15.5 = 0$$

x, m	2	1.9	1.95	1.955
$f(x)$	–0.5	0.699	0.0998	0.04

We adopt, $x = 1.955$ m

$$Ely_{max} = -5 \times (1.955)^3 + (1.955-1)^3 + 0 + 43.75 \times 1.955$$
$$= 52.5258$$

$$y_{max} = \frac{52.5258 \times 10^3}{210 \times 25 \times 10^{-6}} = 0.01\text{m or } 10\text{ mm}$$

(*c*) Slope at $x = 0$,

$$EI\frac{dy}{dx} = 43.75$$

$$\theta_A = \frac{dy}{dx} = \frac{43.750 \times 10^3}{210 \times 10^9 \times 25 \times 10^{-6}} 8.33 \times 10^{-3} \text{ rad.}$$

Example 5.5 *A simply supported beam is carrying a uniformly distributed triangular load as shown in Fig. 5.11. Calculate (a) deflection at the centre, and (6) slope at the ends.*

Solution.

$$R_A = R_B = \frac{wl}{4}$$

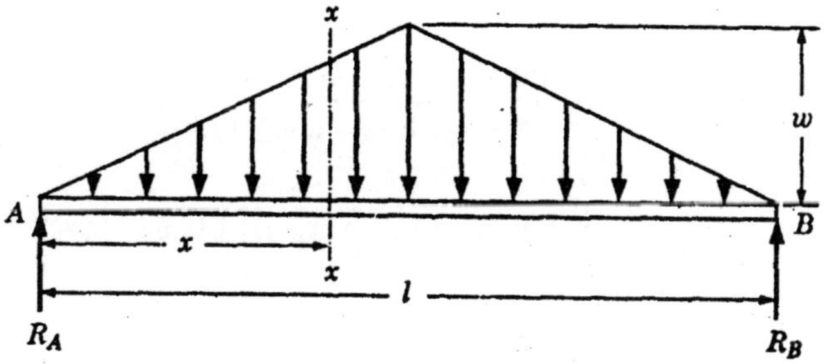

Fig. 5.11

Consider a section $x - x$ at a distance x from A.

Intensity of load at $x - x = \dfrac{2wx}{l}$

$$M_x = R_A \times x - \frac{1}{2} \times \frac{2wx}{l} \times x \times \frac{x}{3}$$

$$= \frac{wl}{4}x - \frac{wx^3}{3l}$$

$$EI\frac{d^2y}{dx^2} = -M_x$$

$$= -\frac{wlx}{4} + \frac{wx^3}{3l}$$

Integrating twice, we have

$$EI\frac{dy}{dx} = -\frac{wlx^2}{8} + \frac{wx^4}{12l} + C_1$$

$$Ely = -\frac{wlx^8}{24} + \frac{wx^5}{60l} + C_1x + C_2$$

At $x = 0, y = 0$ and for $x = \dfrac{l}{2}, \dfrac{dy}{dx} = 0$

$\therefore$

$$C_2 = 0$$

$$0 = \frac{wl^3}{32} + \frac{wl^3}{192} + C_1$$

$$C_1 = \frac{5}{192}wl^3$$

$$EIy = -\frac{wlx^3}{24} + \frac{wx^5}{60l} + \frac{5}{192}wl^3x$$

$$EI\frac{dy}{dx} = -\frac{wlx^2}{8} + \frac{wx^4}{12l} + \frac{5}{192}wl^3$$

(*a*) At $x = \frac{l}{2}$,

$$EIy = -\frac{wl}{192} + \frac{wl^4}{1920} + \frac{5wl^4}{384}$$

$$y = \frac{wl^4}{120\,EI}$$

(*b*) At $x = 0$,

$$\theta_A = \frac{5}{192}\frac{wl^3}{EI}$$

Example 5.6 *An overhanging beam ABC is loaded as shown in Fig. 5.12. Calculate (a) deflection at C, and (b) maximum deflection between A and B. Take E = 200 GPa, I = 24 × 10⁻⁶ m⁴.*

Solution:

Taking moments about *B*,

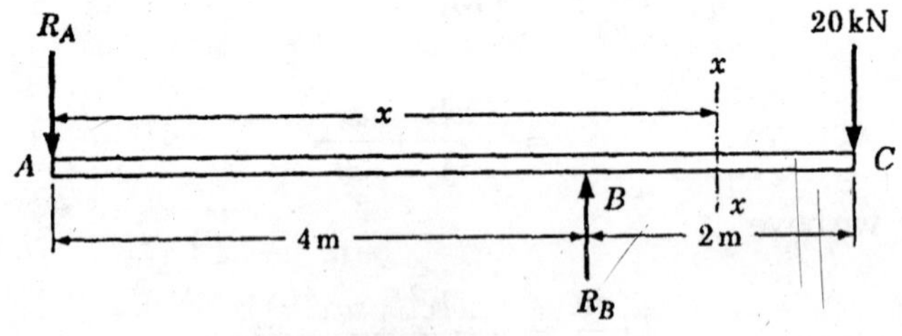

Fig. 5.12

$$R_A \times 4 = 20 \times 2$$

$$R_A = 10\text{ kN}\downarrow, R_B = 30\text{ kN}\uparrow$$

Taking a section x – x at a distance *x* from *A*,

$$M_x = -R_A \times x + R_B(x-4)$$

$$= -10 \times +30(x-4)$$

$$EI\frac{d^2y}{dx} = -M_x$$

$$= 10x - 30(x - 4)$$

Integrating twice, we have

$$EI\frac{dy}{dx} = 5x^2 - 15(x - 4)^2 + C_1$$

$$EIy = \frac{5}{3}x^2 - 5(x - 4)^3 + C_1 x + C_2$$

At $x = 0$ and 4 m, $y = 0$,

$$\therefore \quad C_2 = 0$$

$$0 = \frac{5 \times 64}{3} + 0 + 4C_1$$

$$C_1 = -\frac{80}{3}$$

(*a*) At $x = 6$ m,

$$EIy_c = \frac{5}{3} \times 216 - 5 \times 8 + 6 \times \left(-\frac{80}{3}\right)$$

$$= 160$$

$$y_c = \frac{160 \times 10^3}{200 \times 10^9 \times 24 \times 10^{-6}} = 0.033 \text{ m or 33 mm (downwards)}$$

(*b*) For maximum deflection between *AB*,

$$\frac{dy}{dx} = 0$$

$$\therefore \quad 5x^2 - \frac{80}{3} = 0$$

$$x = -2.31 \text{ m}$$

$$y_{max} = \frac{-41.056 \times 10^3}{200 \times 10^9 \times 24 \times 10^{-6}} - 8.55 \text{ mm (upwards)}$$

5.4 BEAM SUBJECTED TO A COUPLE

5.4.1 Simply Supported Beam Carrying a Couple

Consider a simply supported beam carrying a *ccw* couple C at point *D* as shown in Fig. 5.13. Taking moments about *B*, we have

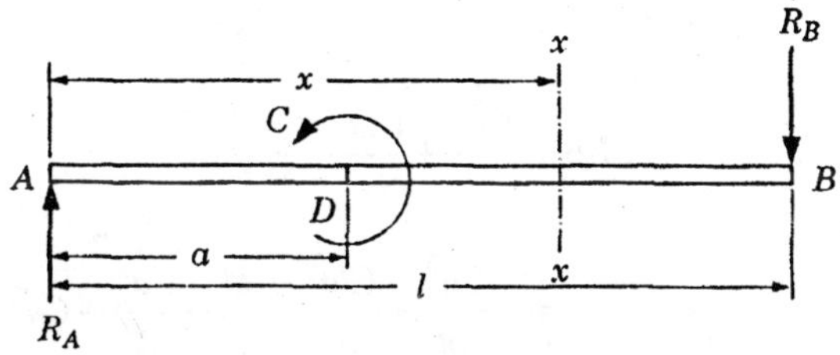

Fig. 5.13 Simply supported beam carrying couple.

$$R_A \times l = C$$

$$R_A = \frac{C}{l} \uparrow$$

$$R_B = \frac{C}{l} \downarrow$$

At a section $x - x$,

$$M_x = R_A \times x - C(x-a)^0$$

$$EI\frac{d^2y}{dx^2} = -M_x$$

$$= -R_A \times x + C(x-a)^0$$

$$= -\frac{Cx}{l} + C(x-a)^0$$

Integrating twice, we have

$$EI\frac{dy}{dx} = -\frac{Cx^2}{2l} + C(x-a) + C_1$$

$$Ely = -\frac{Cx^3}{6l} + \frac{C}{2}(x-a)^2 + C_1 x + C_2$$

$y = 0$ at $x = 0$ and l.

$\therefore$

$$C_2 = 0.$$

$$0 = \frac{Cl^2}{6} + \frac{C}{2}(l-a)^2 + C_1 l$$

$$C_1 = -\frac{Cl^2}{6} + \frac{C}{2l}(l-a)^2$$

$$EI\frac{dy}{dx} = -\frac{Cx^2}{2l} + C(x-a) + \frac{Cl}{6} - \frac{C}{2l}(l-a)^2$$

$$Ely = -\frac{Cx^3}{6l} + \frac{C}{2}(x-a)^2 + \frac{Clx}{6} - \frac{Cx}{2l}(1-a)^2$$

At $x = a$,

$$Ely = \frac{Ca^3}{6l} + \frac{Cla}{6} - \frac{Ca}{2l}(l-a)^2$$

$$= -\frac{Ca^3}{6l} + \frac{Cla}{6} - \frac{Cal}{2} - \frac{Ca^3}{2l} + Ca^2$$

$$= -\frac{2}{3}\frac{Ca^3}{l} - \frac{Cal}{3} + Ca^2 = \frac{aC(2a-l)(l-a)}{3l}$$

$$y_D = \frac{aC(2a-l)(l-a)}{3EIl} \quad ...(5.31)$$

5.4.2 Cantilever Beam Carrying a Couple at Free End

Consider a section at distance x from fixed end as shown in Fig. 5.14.

$$M_x = +C$$

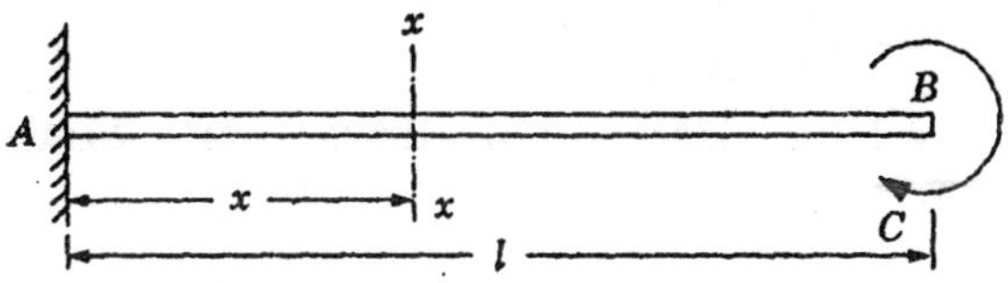

Fig. 5.14 Cantilever beam carrying couple at free end.

$$EI\frac{d^2y}{dx^2} = -M_x = -C$$

$$EI\frac{dy}{dx} = -Cx + C_1$$

At $x = 0$, $\frac{dy}{dx} = 0$, so that, $C_1 = 0$

$$EIy = -\frac{Cx^2}{2} + C_2$$

At $x = l$, $y = 0$, $C_2 = 0$

$$EIy_c = -\frac{Cl^2}{2}$$

$$y_C = -\frac{Cl^2}{2EI} \quad ...(5.32)$$

Example 5.7 *A beam 6 m long is loaded as shown in Fig. 5.15. If $EI = 9 \times 10^4$ kN. m^2, calculate the deflection at the centre of span.*

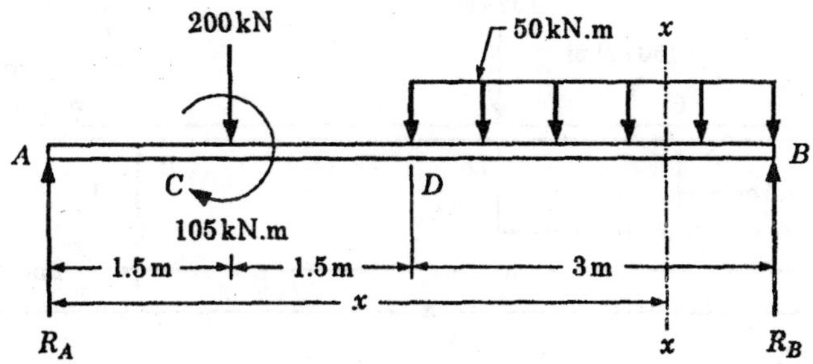

Fig. 5.15

Solution:

$$\Sigma M_B = 0$$
$$R_A \times 6 = 200 \times 4.5 - 105 + 50 \times 3 \times 1.5$$
$$R_A = 170 \text{ kN}$$
$$R_B = 180 \text{ kN}$$

At a section x–x from A,

$$M_x = R_A \times x - 200\,(x-1.5) + 105\,(x-1.5)^0 - \frac{50 \times (x-3)^2}{2}$$
$$= 170\,x - 200\,(x-1.5) + 105\,(x-1.5)^0 - 25\,(x-3)^2$$

$$EI\frac{d^2y}{dx^2} = -M_x$$
$$= 170\,x + 200\,(x-1.5) - 105\,(x-1.5)^0 + 25\,(x-3)^2$$

Integrating twice, we have

$$EI\frac{d^2y}{dx^2} = -85\,x^2 + 100\,(x-1.5)^2 - 105\,(x-1.5) + \frac{25}{3}\,(x-3)^3 + C_1$$

$$Ely = -\frac{85}{3}x^3 + \frac{100}{3}\,(x-1.5)^3 - 52.5\,(x-1.5)^2 + \frac{2.5}{12}\,(x-3)^4 + C_1 x + C_2$$

$y = 0$ at $x = 0$ and 6 m

$\therefore$

$$C_2 = 0$$

$$0 = -\frac{85}{3}\times 6^3 + \frac{100}{3} \times (4.5)^3 - 52.5 \times (4.5)^2 + \frac{25}{12} \times 3^4 + 6\,C_1$$
$$C_1 = 662.8125$$

$$Ely = -\frac{85}{3}x^3 + \frac{100}{3}\,(x-1.5)^3 - 52.5\,(x-1.5)^2 + \frac{25}{3}(x-3)^4 + 662.8125\,x$$

At $x = 3$ m,

$$Ely_D = -\frac{85}{3}\times 3^3 + \frac{100}{3} \times (1.5)^3 - 52.5 \times (1.5)^2 + 662.8125 \times 3$$
$$= 1217.8125$$
$$y_D = \frac{1217.8125}{9\times 10^4} = 0.0135\text{m or } 13.5 \text{ mm}$$

Example 5.8 *A beam is loaded as shown in Fig. 5. 16. Calculate the deflection in the centre of span.* $EI = 20 \times 10^3$ kN m^2.

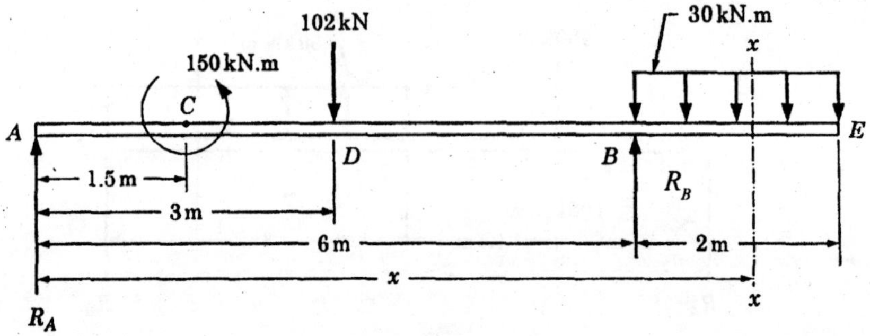

Fig. 5.16

Solution:

$$\Sigma M_B = 0 \text{ gives}$$

$$R_A \times 6 = 150 + 102 \times 5 - 30 \times 2 \times 1$$

$$R_A = 100 \text{ kN}, R_B = 62 \text{ kN}$$

At a section $x - x$ from A,

$$M_x = R_A \times x - 150\,(x - 1.5)^0 - 102\,(x - 3) + 62\,(x - 6) - 15\,(x - 6)^2$$

$$EI\frac{d^2y}{dx^2} = -M_x$$

$$= -100\,x + 150\,(x - 1.5)^0 + 102\,(x - 3) - 62\,(x - 6) + 15\,(x - 6)^2$$

Integrating twice, we have

$$EI\frac{dy}{dx} = -50\,x^2 + 150\,(x - 1.5) + 51\,(x - 3)^2 - 31\,(x - 6)^2 + 5\,(x - 6)^3 + C_1$$

$$EIy = -\frac{50}{3}\,x^3 + 75\,(x - 1.5)^2 + 17\,(x - 3)^3 - \frac{31}{3}\,(x - 6)^3 + 1.25\,(x - 6)^4 + C_1 x + C_2$$

$y = 0$ at $x = 0$ and 6m.

$\therefore$ $$C_2 = 0$$

$$0 = \frac{50}{3} \times 6^3 + 75 \times (4.5)^2 + 17 \times 3^3 + 6C_1$$

$$C_1 = 270.\ 375$$

$$EIy = -\frac{50}{3}x^3 + 75\,(x - 1.5)^2 + 17\,(x - 3)^3 - \frac{31}{3}(x - 6)^3 + 1.25(x - 6)^4 + 270.375x$$

At $x = 3$m,

$$y_D = \frac{1}{20 \times 10^3}\left[-\frac{50}{3} \times 3^3 + 75 \times (1.5)^2 + 270.375 \times 3\right]$$

$$= \frac{1}{20 \times 10^3}[529.875] = 0.265 \text{ m or } 26.5 \text{ mm}$$

5.5 METHOD OF SUPERPOSITION PRINCIPLE

The super position principle states that the total deflection and slope of a point due to a number of loads acting on a beam is equal to the sum of the individual deflections and slopes produced by the various loads considering if other loads were not present. This principle holds good for small values of deflections and slopes.

5.5.1 Simply Supported beam Carrying Concentrated Load and Uniformly Distributed Load

Consider a beam carrying concentrated load W at midspan and udl w/unit length over the whole span as shown in Fig. 5.17. Deflection at midspan due to concentrated load at midspan,

$$y_1 = \frac{Wl^3}{48\,EI}$$

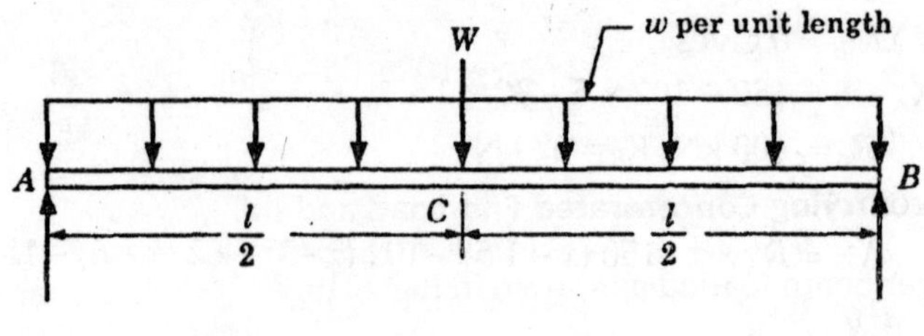

Fig. 5.17

Deflection at midspan due to *udl*,

$$y_2 = \frac{5}{384}\frac{wl^4}{EI}$$

By the principle of superposition, total midspan deflection,

$$y = y_1 + y_2 = \frac{Wl^3}{48\,EI} + \frac{5}{384}\frac{wl^4}{EI}$$

5.5.2 Simply Supported Beam Carrying Two Concentrated Loads

Consider a simply supported beam carrying two concentrated loads W_1 and W_2 as shown in Fig. 5.18.

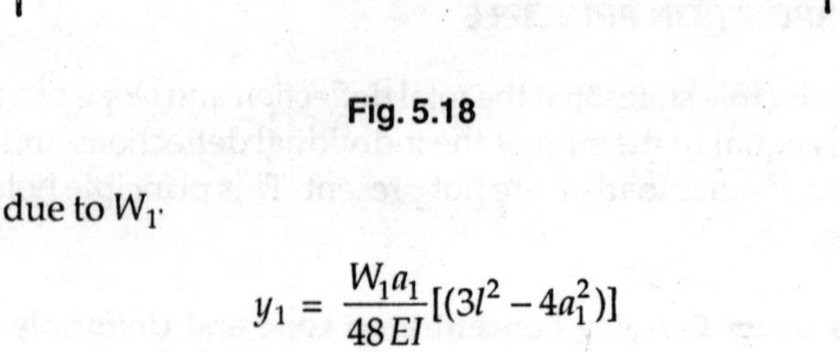

Fig. 5.18

Midspon deflection due to W_1.

$$y_1 = \frac{W_1 a_1}{48\,EI}[(3l^2 - 4a_1^2)]$$

and due to W_2,

$$y_2 = \frac{W_2 a_2}{48\,EI}[(3l^2 - 4a_2^2)]$$

By the principle of superposition,

$$y = y_1 + y_2$$

$$= \frac{1}{48\,EI}[W_1a_1(3l^2 - 4a_1^2) + W_2a_2(3l^2 - 4a_2^2)]$$

5.5.3 Cantilever Beam Carrying Concentrated End Load and udl

Consider a cantiliver beam loaded as shown in Fig. 5.19.

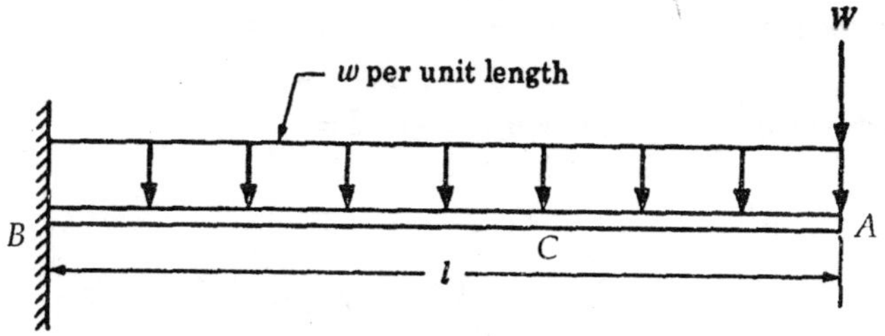

Fig. 5.19

Deflection of free end due to W,

$$y_1 = \frac{Wl^3}{3EI}$$

and due to *udl*,

$$y_2 = \frac{wl^4}{8\,EI}$$

Total deflection of free end, $y = y_1 + y_2$

$$= \frac{Wl^3}{3EI} + \frac{wl^4}{8\,EI}$$

5.5.4 Cantilever Beam Carrying Two Concentrated Loads

Consider a cantilever beam carrying two concentrated loads as shown in Fig. 5.20.

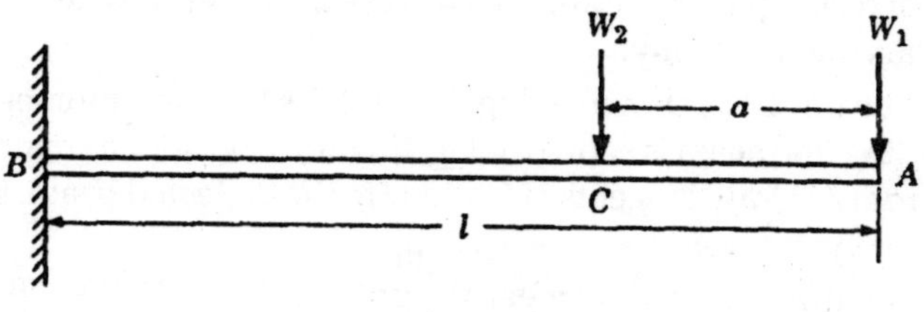

Fig. 5.20

Deflection of A due to W_1, $y_1 = \dfrac{W_1 l^3}{3EI}$

Deflection at A due to W_2, $y_2 = \dfrac{W_2(l-a)^2(2l+a)}{6EI}$

Total deflection at A, $y = y_1 + y_2$

Example 5.9 *A simply supported beam of span 6 m carries a concentrated load 200 kN at mid span and another concentrated load at 4 m from left end support of intensity 150 kN. If EI = 80 × 10³ kN-m², calculate the deflection of beam at midspan by using the principle of superposition.*

Solution:

Given: $W_1 = 200$ kN, $l = 6$ m, $W_2 = 150$ kN, $a = 4$ m, $b = 2$ m,

$$EI = 80 \times 10^3 \text{ kN.m}^2$$

Total midspan deflection,

$$y = \frac{1}{48EI}[W_1 l^3 + W_2 a(3l^2 - 4a^2)]$$

$$= \frac{1}{48 \times 80 \times 10^3}[200 \times 6^3 + 150 \times 4(3 \times 6^2 - 4 \times 4^2)]$$

$$= 0.0181 \text{ m or } 18.1 \text{ mm}$$

Example 5.10 *A cantilever beam of span 5 m carries a concentrated load 100 kN at free end and a udl of intensity 50 kN/m over the whole span. Calculate the free end deflection, if EI = 8 × 10⁵ kN-m². Use principle of superposition.*

Solution:

Given : Total free end deflection,

$$y = \frac{Wl^3}{3EI} + \frac{wl^4}{8EI} = \frac{1}{8 \times 10^5}\left[\frac{100 \times 5^3}{3} + \frac{50 \times 5^4}{8}\right]$$

$$= \frac{1}{8 \times 10^5}[8072.917]$$

$$y = 0.0101 \text{ m or } 10.1 \text{ mm}$$

5.6 MOMENT AREA METHOD

The moment area method is suitable for beams having concentrated loads for which bending moment area consists of rectangles and triangles. However, the method can be used for beams having *udl* also if the load is varying uniformly.

Consider a loaded beam *AB* as shown in Fig. 5.21 (a). Its bending moment diagram is shown in Fig. 5.21 (*b*). After bending, the beam is shown in Fig. 5.21 (c). Consider a small element $CD = ds$ of the beam at a distance x_c from *A*. Draw tangents at *C* and *D* to the deflected beam. to meet the vertical line at *B* at *P* and Q respectively

Also draw normals to these tangents to meet at point *O* so that $\angle COD = d\theta$. In the limiting case for small deflection,

$$ds \simeq dx$$

and
$$\frac{1}{R} = \frac{d\theta}{ds}$$

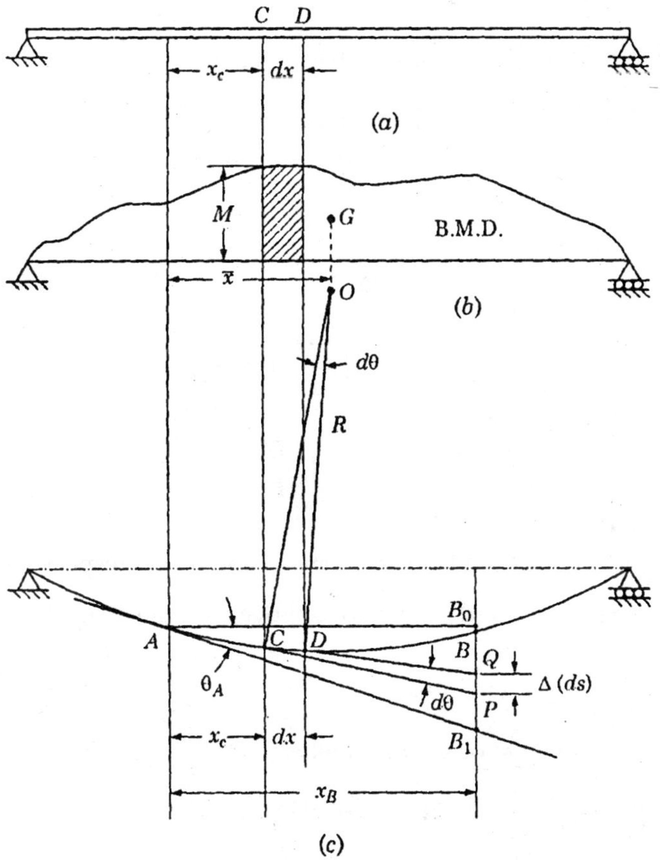

Fig. 5.21 Moment area method.

Also
$$\frac{1}{R} = \frac{M}{EI}$$

∴
$$d\theta = \frac{M}{EI} \times ds \simeq \frac{M \times dx}{EI} = \frac{dA}{EI}$$

where $dA = M \times dx$ = area of bending moment diagram between points C and D.

M = bending moment between C and D

R = radius of bent beam

Thus the angle between the tangents drawn at any two adjacent points on the deflected beam is equal to the area of the bending moment diagram between these two points divided by *EI*.

The total change of slope from *A* to *B* becomes,

$$\int_A^B d\theta = \int_A^B \frac{M}{EI} \times dx = \frac{1}{EI}\int_A^B M \times dx$$

or $\theta_A - \theta_B = \frac{1}{EI}\int_A^B M \times dx$ = area of B.M.D. between *A* to *B*.

Distance of point *B* above the tangent AB_1 at $A = BB_1$

Vertical distance between points C and $D = (x_B - x_c)\, d\theta = \Delta(ds)$

But $$d\theta = \frac{Mdx}{EI}$$

∴ $$(x_B - x_C)\frac{Mdx}{EI} = \Delta(ds)$$

$$BB_1 = \int_{x=x_A}^{x=x_B} \frac{M}{EI}(x_B - x_C)dx$$

$$= \frac{1}{EI}\int_{x=x_A}^{x_B} \frac{M}{EI}(x_B - x_C)dx$$

$BB_1 = \frac{1}{EI}$ [Moment of the B.M.D. between points *A* and *B* with respect to point *B*]

= Distance between the deflected position of *B* above the tangent at A

Now $$BB_0 = y_B = \theta_A \cdot x_B - \int_{x_A}^{x_B} \frac{M}{EI}(x_B - x_C)dx$$

Thus the following theorems should be remembered for the moment area method:

First Theorem: The slope at any point *B* is equal to the area of the $\frac{M}{EI}$ diagram between the point *B* and the point where the slope is zero.

Mathematically, $$\int_A^B d\theta = \int_A^B \frac{M_x}{EI}dx = \frac{\text{Area of B.M.D. between } A \text{ and } B}{EI}$$

or change of slope from *A* to *B* is,

$$\theta_A - \theta_B = \int_A^B \frac{M_x}{EI}dx$$

To determine the absolute value θ_A, the second point *B* must have a tangent in the horizontal direction, *i.e.* $\theta_B = 0$.

Second Theorem: The deflection at any point *B* is equal to the moment about point *B* of the $\frac{M}{EI}$ diagram between the point *B* and the point at which the slope is zero.

Mathematically,

$$\delta_{BA} = \left[\int_{x_A}^{x_B} \frac{M_x}{EI} dx\right] \times \bar{x}_B$$

where $\bar{x}_B$ = distance of the centroid of *B.M.D.* between *A* and *B* from point *B*.

or

$$\delta_A - \delta_B = \left[\int_{x_A}^{x_B} \frac{M_x}{EI} dx\right] \times \bar{x}_B$$

$$= \frac{1}{EI} \text{(Area of B.M.D)} \times \bar{x}_B \qquad \text{... (5.33)}$$

To determine the absolute value of δ_A, the second point *B* must have zero deflection. The centroids of various B.M.D. are given in Table 5.1.

Table 5.1: Centroids and areas of bending moment diagrams.

Loading	*Bending moment diagram*	*Area of BMD*	*Centroid,* $\bar{x}$
1. Concentrated load at mid span on S.S. beam 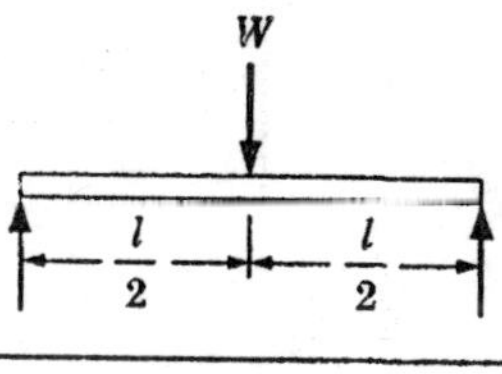	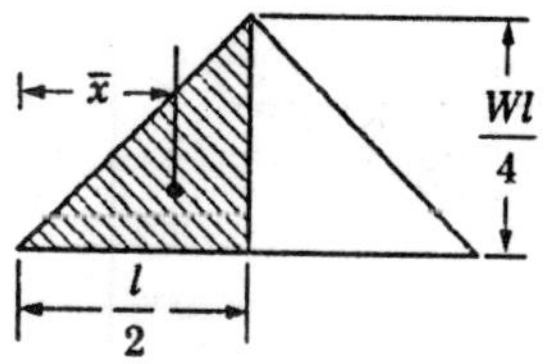	$\frac{Wl^2}{16}$	$\frac{l}{3}$
2. Udl on whole span on S.S. beam 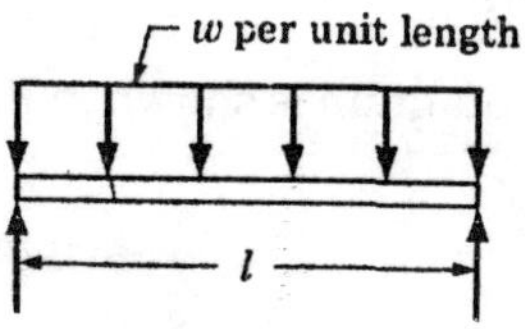	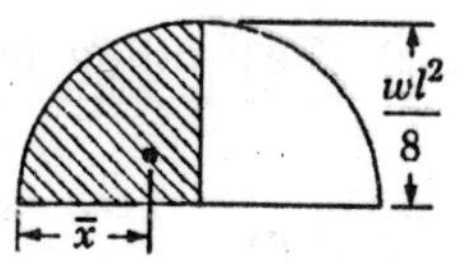	$\frac{wl^3}{24}$	$\frac{5}{16}l$
3. Concentrated load at any point on S.S. beam. 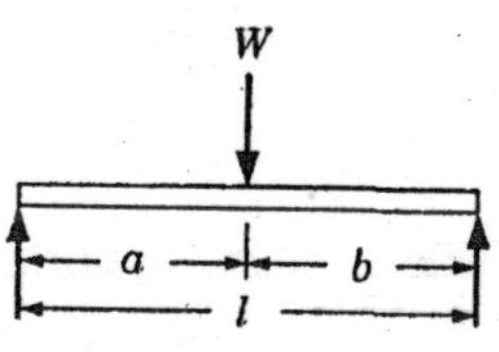	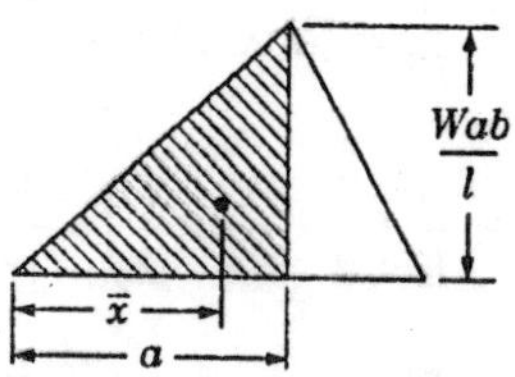	$\frac{1}{2}\frac{Wa^2b}{l}$	$\frac{2}{3}a$

4. Cantilever with concentrated load at free end 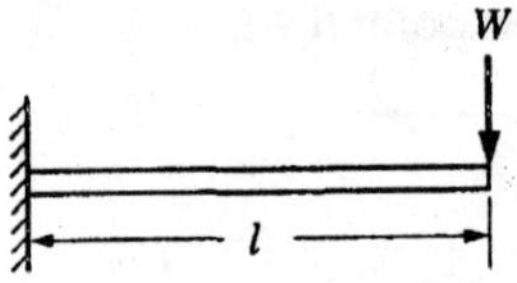	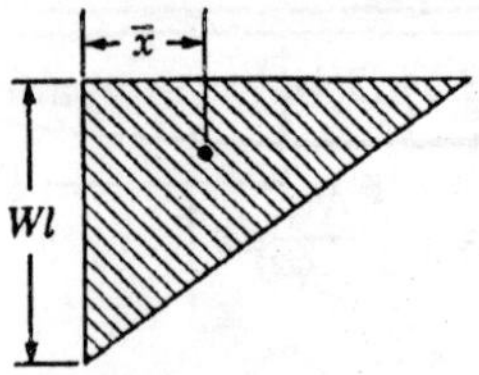	$\frac{1}{2}Wl^2$	$\frac{l}{3}$
5. Cantilever with *udl* over entire span 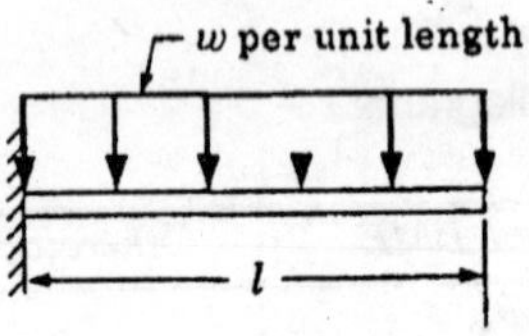	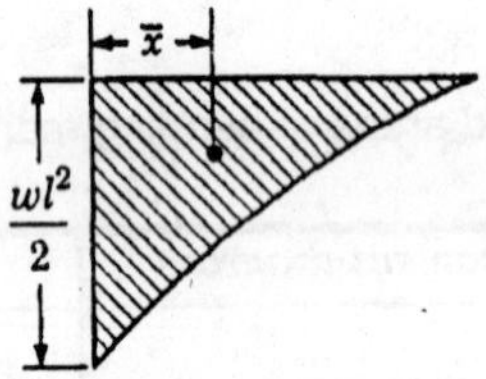	$\frac{wl^3}{6}$	$\frac{l}{4}$
6. Triangular load on cantilever 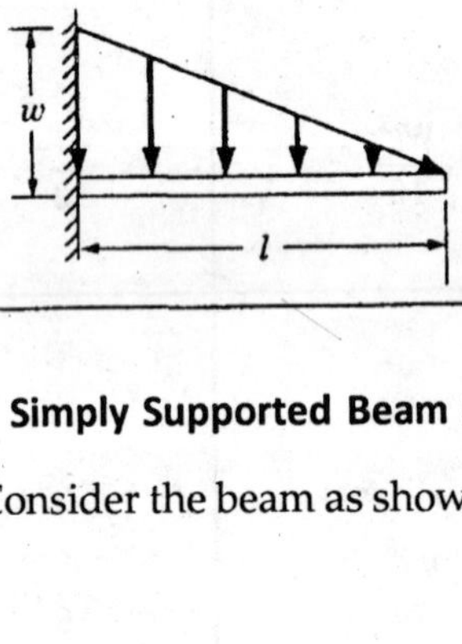	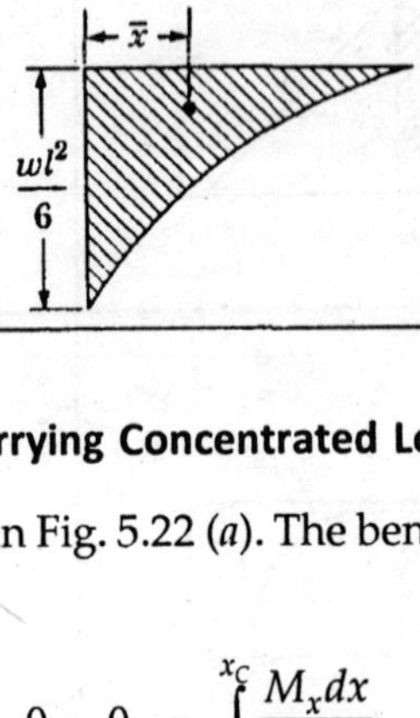	$\frac{wl^3}{24}$	$\frac{l}{5}$

5.6.1 Simply Supported Beam Carrying Concentrated Load at Midspan

Consider the beam as shown in Fig. 5.22 (*a*). The bending moment diagram is shown in Fig.5.22 (*b*).

$$\theta_A - \theta_C = \int_{x_A}^{x_C} \frac{M_x dx}{EI}$$

Here the tangent to the deflection curve at midspan C is horizontal. Hence $\theta_C = 0$

Slope at *A*, $$\theta_A = \frac{1}{EI} = \text{(area of B.M.D. between } A \text{ to C)} = \frac{A}{EI}$$

$$= \frac{1}{EI}\left(\frac{1}{2} \times \frac{Wl}{4} \times \frac{l}{2}\right) = \frac{Wl^2}{16EI}$$

$$\bar{x} = \frac{l}{3}$$

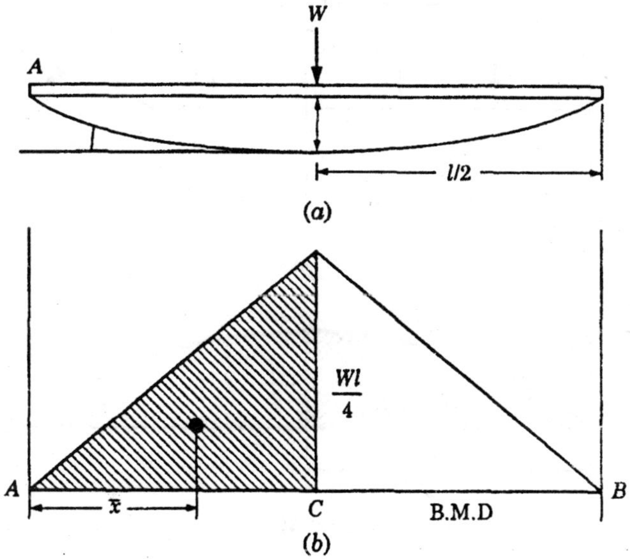

Fig. 5.22

Deflection at C,
$$\delta_C = \frac{A\bar{x}}{EI} = \frac{1}{EI}\left(\frac{1}{2}\times\frac{Wl}{4}\times\frac{l}{2}\times\frac{2}{3}\times\frac{l}{2}\right)$$
$$= \frac{Wl^3}{48\,EI}$$

5.6.2 Simply Supported Beam Carrying Uniformly Distributed Load Over Entire Span

Consider the beam as shown in Fig. 5.23 (*a*). The bending moment diagram is shown in Fig.5.23(*b*) The tangent to the deflection curve at mid span C is horizontal. Hence $\theta_c = 0$

Slope at *A*.
$$\theta_A = \frac{A}{EI}$$
$$= \frac{1}{EI}\left(\frac{2}{3}\times\frac{wl^2}{8}\times\frac{l}{2}\right)$$
$$= \frac{wl^3}{24\,EI}\quad \bar{x} = \frac{5}{16}l$$

Deflection at C,
$$\delta_c = \frac{A\bar{x}}{EI} = \frac{1}{EI}\left(\frac{2}{3}\times\frac{wl^2}{8}\times\frac{l}{2}\times\frac{5}{16}l\right) = \frac{5}{384}\frac{wl^4}{EI}$$

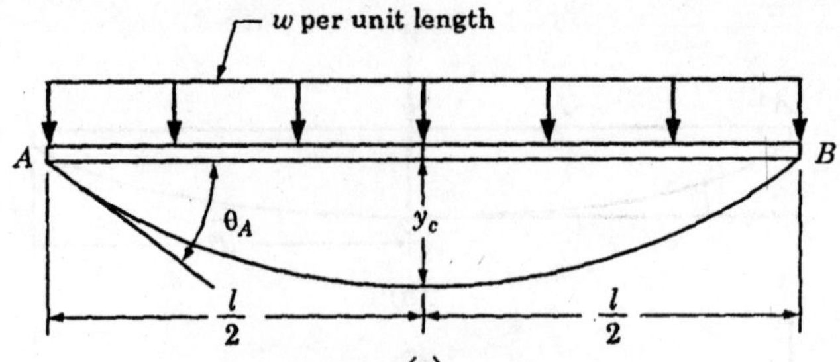

(a)

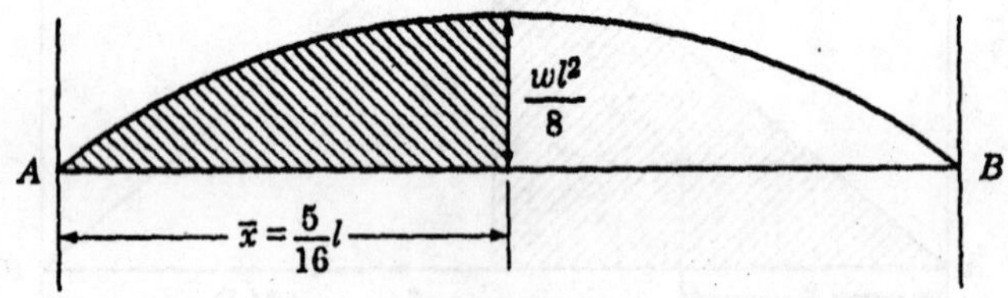

(b) B.M.D.

Fig. 5.23

5.6.3 Cantilever Beam Carrying Concentrated Load at the Free End

Consider the cantiliver beam shown in Fig. 5.24 *(a)* and B.M.D. in Fig. 5.24 *(b)*.

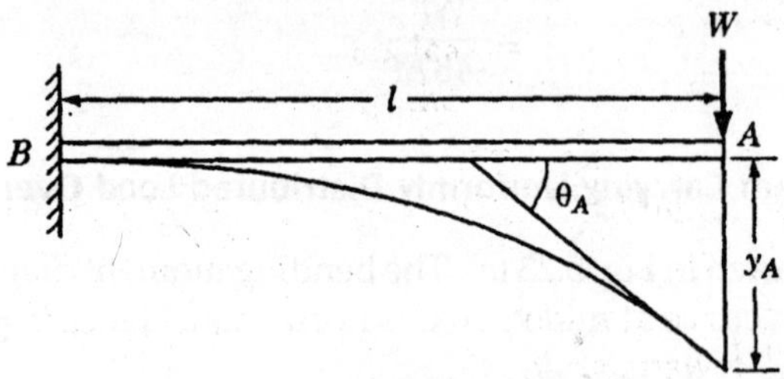

(a)

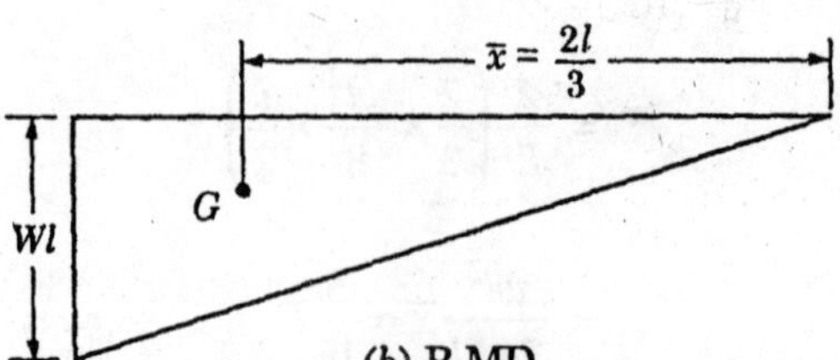

(b) B.MD.

Fig. 5.24

Slope at A,

$$\theta_A = \frac{A}{EI} = -\frac{1}{EI}\left(\frac{1}{2}\times Wl\times l\right)$$

$$= -\frac{Wl^2}{2EI}$$

Deflection at A,

$$\delta_A = \frac{A\bar{x}}{EI} = \frac{1}{EI}\left(\frac{1}{2}\times Wl\times l\times\frac{2}{3}l\right) = \frac{Wl^3}{3EI}$$

5.6.4 Cantilever Beam Carrying udl Over the Entire Span

The cantilever is shown in fig.5.25 (*a*) and B.M.D. in Fig. 5.25 (*b*)

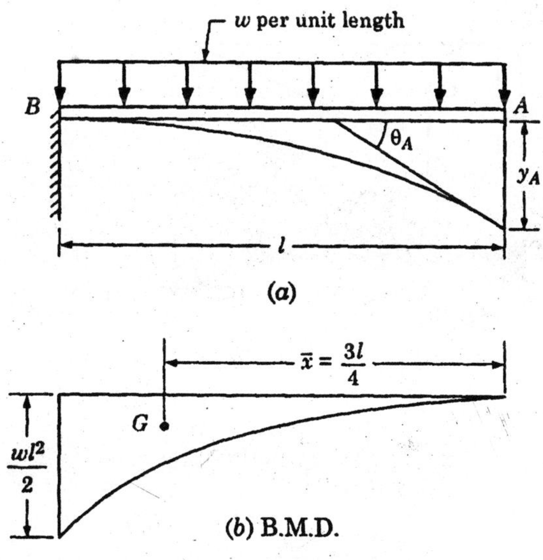

Fig. 5.25

Slope at A,

$$\theta_A = \frac{A}{EI} = -\frac{1}{EI}\left(\frac{1}{3}\times\frac{wl^2}{2}\times l\right)$$

$$= -\frac{wl^3}{6EI}$$

$$\bar{x} = \frac{3}{4}l$$

Deflection at A, $\delta A = \dfrac{A\bar{x}}{EI} = \dfrac{1}{EI}\left(\dfrac{wl^3}{6} \times \dfrac{3}{4}l\right) = \dfrac{wl^4}{8EI}$

5.6.5 Cantilever Beam Carrying Concentrated Load at Any Point

The cantilever beam is shown in Fig. 5.26(a) and B.M.D.in Fig. 5.26.(b).

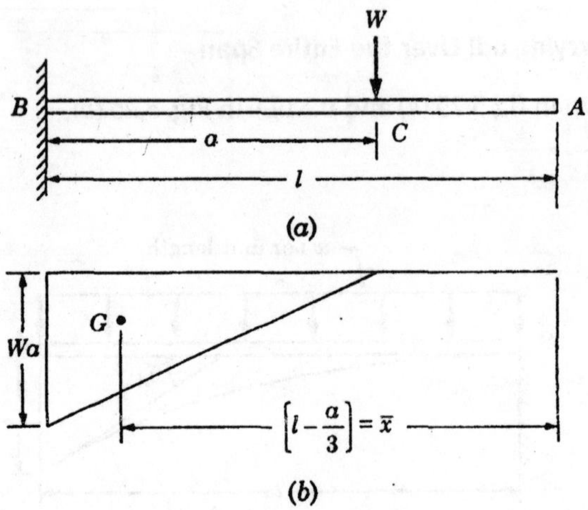

Fig 5.26

$$A = \frac{1}{2} Wa \times a = \frac{Wa^2}{2}$$

$$\theta_A = \frac{Wa^2}{2EI}$$

$$\bar{x} = (l - a) + \frac{2}{3}a = l - \frac{a}{3}$$

$$\delta_A = \frac{A\bar{x}}{EI} = \frac{Wa^2}{2} \times \left(l - \frac{a}{3}\right) \times \frac{1}{EI}$$

$$\delta_C = \frac{Wa^2}{2} \times \frac{2}{3}a \times \frac{1}{EI} = \frac{Wa^3}{3EI}$$

5.6.6 Cantilever Beam Carrying udl on Part of Span

The loaded cantilever is shown in Fig. 5.27 (a) and B.M.D. in Fig. 5.27 (b)

$$M_{max} = \frac{wa^2}{2}$$

$$\bar{x} = (l-a)+\frac{3}{4}a = l-\frac{a}{4}$$

$$A = \frac{1}{3}\times\frac{wa^2}{2}\times a = \frac{wa^3}{6}$$

$$\theta_A = \frac{wa^3}{6EI}$$

$$\delta_A = \frac{wa^3}{6}\times\left(l-\frac{a}{4}\right)\times\frac{1}{EI}$$

$$= \frac{wa^3}{6EI}\left(l-\frac{a}{4}\right)$$

$$= \frac{wa^3}{6EI}\times\frac{3a}{4}$$

$$= \frac{wa^4}{8EI}$$

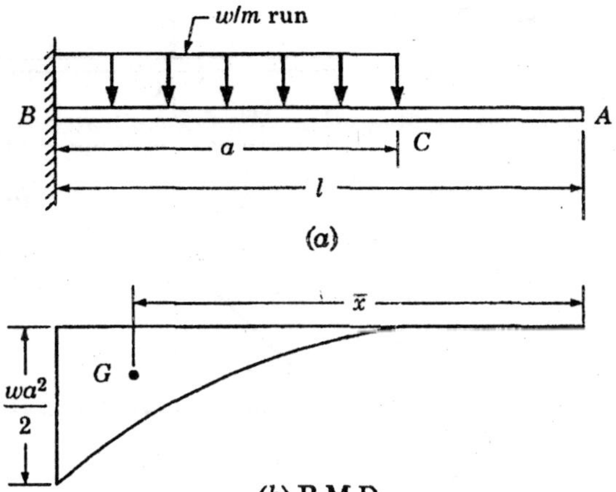

(b) B.M.D.

Fig. 5.27

Example 5.11 *A simply supported beam of 4 m span carries concentrated loads of 15 kN each at 1m from the ends. Calculate (a) slope and deflection under each load, and (b) maximum slope and deflection of beam. Take EI = 1.5. x 10^4 kN.m^2.*

Solution:

Given : The loaded beam and B.M.D. are shown in Fig. 5.28.

(*a*) Area $A_1 = \frac{1}{2} \times 15 \times 1 = 7.5 \text{ kN. m}^2$

$$\theta_{D-B} = \frac{A_1}{EI} = \frac{7.5}{7.5\times10^4} = 5\times10^{-4} \text{ rad}$$

$$A_2 = 15 \times 1 = 15 \text{ kN. m}^2$$

$$A = A_1 + A_2 = 22.5 \; kN. \text{m}^2$$

$$\theta_B = \frac{22.5}{1.5\times10^4} = 15\times10^{-4} \text{ rad}$$

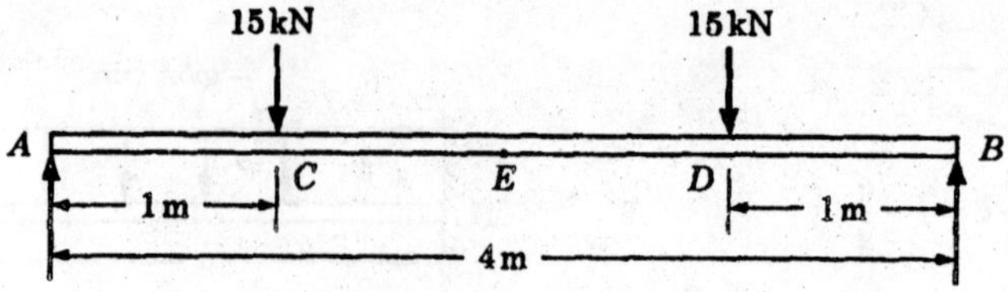

(a)

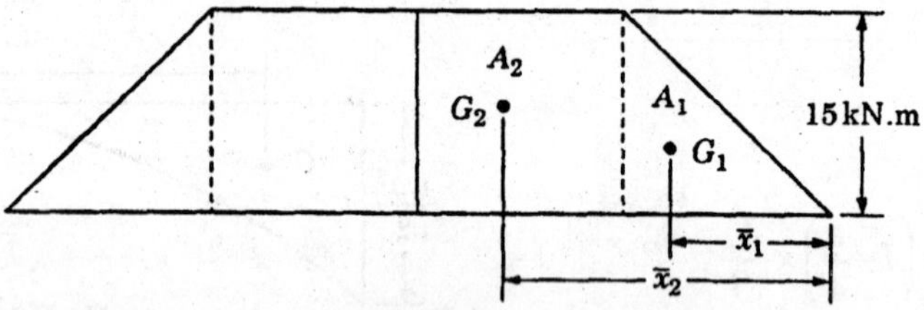

(b) B.M.D.

Fig. 5.28

$$\theta_D = \theta_B - \theta_{D-B} = 15 \times 10^{-4} - 5 \times 10^{-4} = 10^{-3} \text{ rad}$$
$$= \theta_C \text{ (by symmetry)}$$

$$\bar{x}_1 = \frac{2}{3} m, \bar{x}_2 = 1 + 0.5 = 1.5 \text{ m}$$

$$y_D = y_1 + y_2$$

where $y_1 = \dfrac{A_1\bar{x}_1}{EI} = 7.5 \times \dfrac{2}{3} \times \dfrac{1}{1.5 \times 10^4} = 0.333 \times 10^{-3} \text{m}$

$$y_2 = \theta_D \times 1.0 = 10^{-3} \text{ m}$$
$$y_D = (0.333 + 1.0)10^{-3} = 1.333 \text{ mm}$$

(b) $$\theta_{max} = \frac{A}{EI} = \frac{22.5}{1.5 \times 10^4} = 15 \times 10^{-4} \text{ rad}$$

$$y_{max} = \frac{A_1\bar{x}_1 + A_2\bar{x}_2}{EI}$$

$$= \frac{7.5 \times \frac{2}{3} + 15 \times 1.5}{1.5 \times 10^4}$$

$$= \frac{27.5}{1.5 \times 10^4} = 18.33 \times 10^{-4}$$

$$= 1.833 \text{ mm}$$

Example 5.12 *A cantilever beam of length l is loaded only one half of its length from the free end with a udl of w per unit length. Find the deflection at the free end.*

Solution:

The *udl* is extended upto the fixed end and *udl* of same intensity is applied upward for the half length *BC*, which was initially unloaded. The loaded cantilever beam and *B.M.D.* are shown in Fig. 5.29.

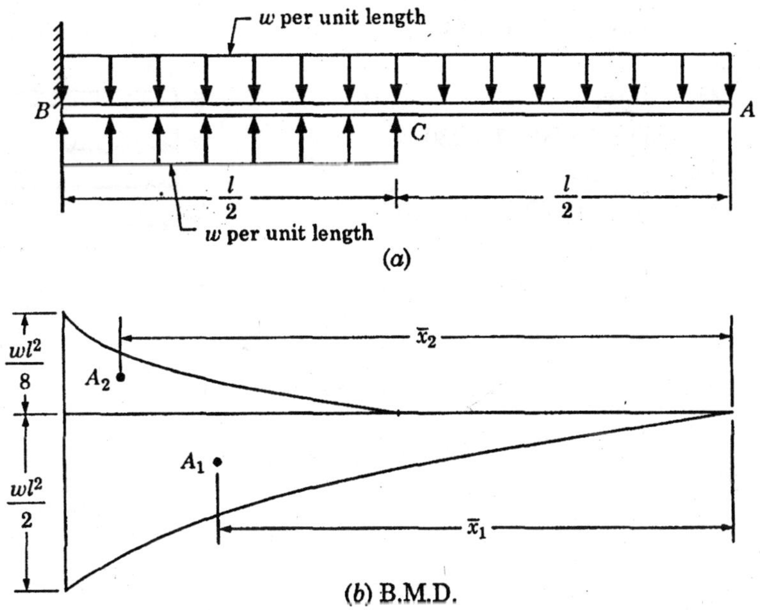

Fig. 5.29

$$\bar{x}_1 = \frac{3}{4}l$$

$$\bar{x}_2 = \frac{l}{2}+\frac{3}{4}\times\frac{l}{2}=\frac{7}{8}l$$

$$A_1 = \frac{1}{3}\times\frac{wl^2}{2}\times l=\frac{wl^3}{6}$$

$$A_2 = \frac{1}{3}\times\frac{wl^2}{8}\times\frac{l}{2}=\frac{wl^3}{48}$$

$$\delta_A = \frac{1}{EI}(A_1\bar{x}_1 - A_2\bar{x}_2)$$

$$= \frac{1}{EI}\left(\frac{wl^3}{6}\times\frac{3}{4}l-\frac{wl^3}{48}\times\frac{7}{8}l\right)$$

$$= \frac{41}{384}\frac{wl^4}{EI}$$

Example 5.13 *A beam, simply supported at ends A and B 4m apart is loaded with two point loads of 60 kN and 50 kN at distance 1 metre and 3 respectively from end A. Determine the position and magnitude of maximum deflection.*

Take $E = 2 \times 10^5\ N/mm^2$ and $I = 8500 cm^4$.

Solution:

Given: $E = 2\times 10^5\ \text{N/mm}^2, I = 8500\ \text{cm}^4$

Refer to Fig. 5.30.

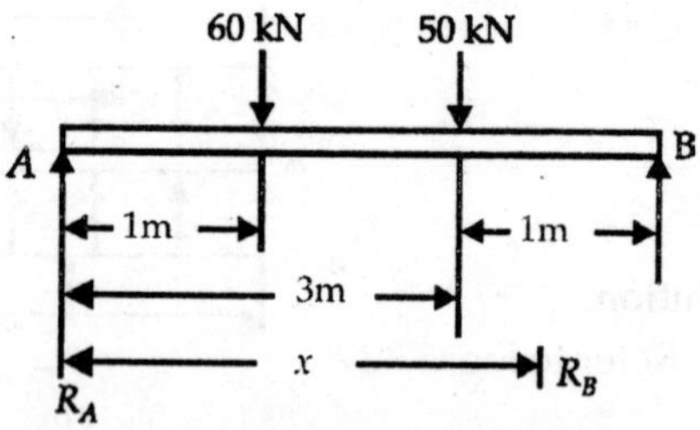

Fig. 5.30

$$\Sigma M_B = 0 \text{ gives,}$$

$$4R_A = 60 \times 3 + 50 \times 1 = 230$$

$$R_A = 57.5 \text{ kN}$$

$$M_x = 57.5 \times x - 60\ (x-1) - 50\ (x-3)$$

$$EI\frac{d^2y}{dx^2} = -M_x = -57.5x + 60\ (x-1) + 50\ (x-3)$$

$$EI\frac{dy}{dx} = -28.75x^2 + 30\ (x-1)^2 + 25\ (x-3)^2 + C_1,$$

$$EIy = -9.58x^3 + 10\ (x-1)^3 + 8.333\ (x-3)^3 + C_1 x + C_2$$

At $x = 0, y = 0$ gives,

$$0 = C_2$$

At $x = 4$ m, $y = 0$ gives,

$$\theta = -9.583 \times 4^3 + 10 \times 3^3 + 8.333 \times 1 + 4C_1$$

$$C_1 = 83.745$$

$$\therefore \quad EIy = -9.583\ x^3 + 10\ (x-1)^3 + 8.333\ (x-3)^3 + 83.745\ x$$

For y to be maximum, $\frac{dy}{dx} = 0$

$$-28.75\ x^2 + 30\ (x^2 + 1 - 2x) + 25\ (x^2 + 9 - 6x) + 83.745 = 0$$

$$26.75x^2 - 210x + 338.\ 745 = 0$$

$$x^2 - 8x + 12.9 = 0$$

$$x = \frac{1}{2}[8 \pm \sqrt{64 - 4 \times 12.9}] = 2.239\text{m}$$

$$y = \frac{1}{EI}[-9.583x^3 + 10(x-1)^3 + 8.333(x-3)^3 + 83.745x]$$

At $x = 2.239$ m

$$y_{max} = \frac{10^{12}}{2 \times 10^5 \times 8500 \times 10^4}[-9.583 \times (2.239)^3 + 10(2.239 - 1)^3 + 83.745 \times 2.239]$$

$$= \frac{10^3}{2 \times 8500} \times 192.856 = 11.3\text{mm}$$

Example 5.14 *A simply supported beam has a flexural rigidity of 24 MN .m² and is loaded as shown in Fig. 5.31 Determine the deflection of mid span.*

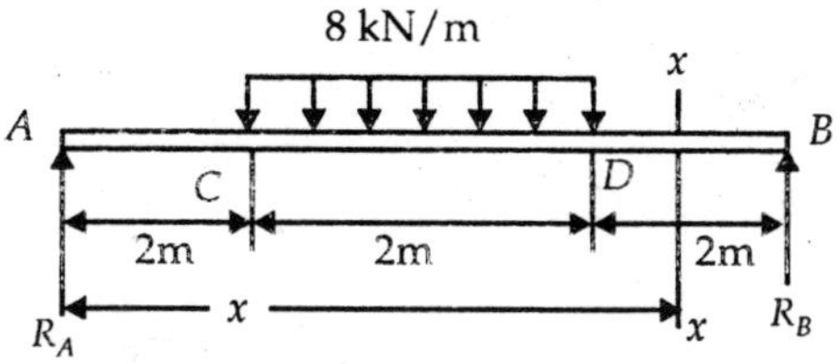

Fig. 5.31

Solution.

Refer to Fig. 5.32

$$\Sigma M_B = 0 \text{ gives,}$$
$$6R_A = 8 \times 2 \times 3 = 48$$
$$R_A = 8 \text{ kN} = R_B$$
$$M_x = 8 \times x - 8\times 2 \times (x-3) = 8x - 16\,(x-3)$$

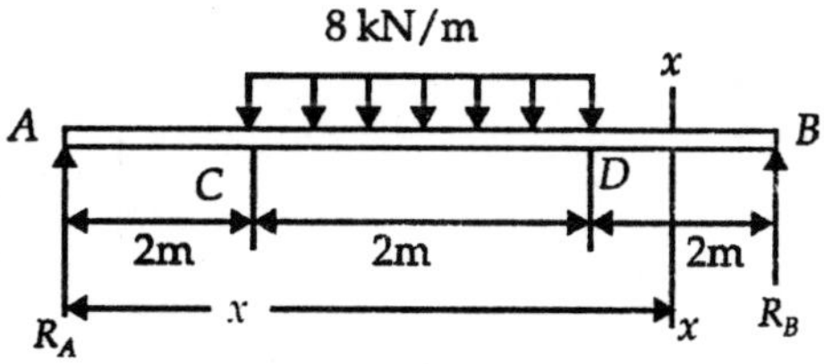

Fig. 5.32

$$EI\frac{d^2y}{dx^2} = -M_x = -8x + 16\,(x-3)$$

$$EI\frac{dy}{dx} = -4x^2 + 8\,(x-3)^2 + C_1$$

$$EIy = \frac{-4}{3}x^3 + \frac{8}{3}(x-3)^3 + C_1x + C_2$$

At $x = 0, y = 0$

$$0 = C_2$$

At $x = 6m, y = 0$

$$0 = -\frac{4}{3}\times 6 + \frac{8}{3}(6-3)^3 + 6C_1$$
$$0 = -288 + 72 + 6C_1$$
$$C_1 = 36$$

$\therefore$
$$EIy = -\frac{4}{3}x^3 + \frac{8}{3}(x-3)^3 + 36x$$

At
$$x = 3\text{ m}$$

$$EIy = \frac{-4}{3}\times 3^3 + 0 + 36\times 3 = 72$$

$$y = \frac{72\times 10^3 \times 10^9}{24\times 16^6 \times 10^6} = 3\text{mm}$$

Example 5.15 *A cantilever beam of uniform EI has a span equal to L. An upward force W acts at the midpoint of the beam and a downward force P acts at the free end. In order that the deflection at the free end is zero, find the relation between P and W.*

Solution:

Refer to Fig. 5.33.

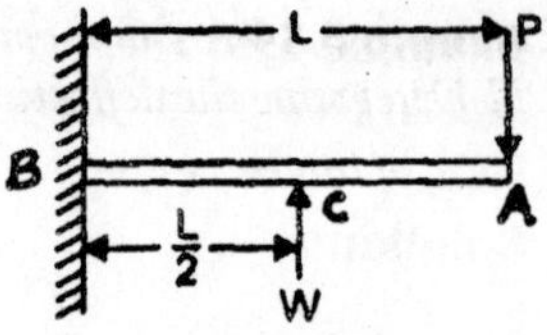

Fig. 5.33

$$\frac{PL^3}{3EI} - \frac{5WL^3}{48EI} = 0$$

$$\frac{P}{W} = \frac{5}{16}$$

Example 5.16 *A cantilevel beam of span 3m is loaded with a concentrated load of 500 N at the free end and a load W at the mid span. Determine the value of W so that the deflection at the end is 10 mm. Take E = 15 GPa, and I = 10^7 mm^4.*

Solution:

Given: d_A = 10mm, E = 15 GPa, I = 10^7 mm^4, L = 3m.

Refer to Fig. 5.34.

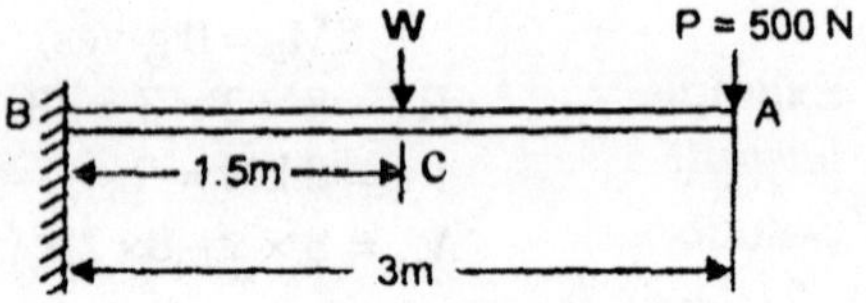

Fig. 5.34

$$\delta_A = \frac{PL^3}{3EI} + \frac{5WL^3}{48EI}$$

$$= \frac{L^3}{3EI}\left[P + \frac{5W}{16}\right]$$

$$10 = \frac{(3\times10^3)^3}{3\times15\times10^3\times10^7}\left[500 + \frac{5W}{16}\right]$$

$$W = -\,1066.7\text{ N}$$

-ve sign indicates that W is applied upwards.

Example 5.17 *If the deflection at the free end of a cantilever of a rectangular cross-section due to certain loading is 0.8 cm, find the deflection of the free end if the depth of the cross-section is doubled keeping the width as same.*

Solution.

$$\delta = \frac{WL^3}{3EI} = 0.3\text{ cm}$$

$$\delta \propto \frac{1}{I}$$

Refer to Fig. 5.35

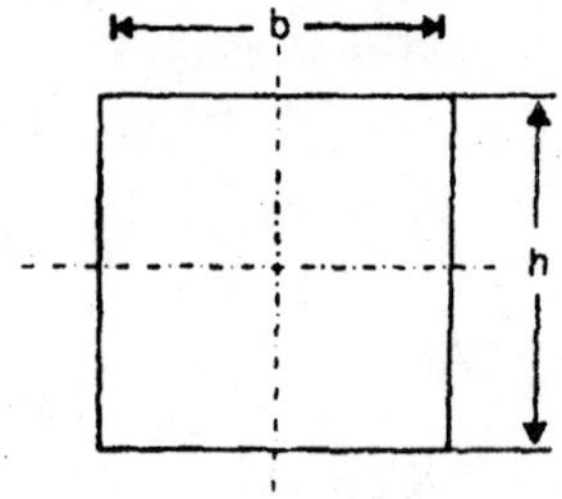

Fig. 5.35

$$\frac{\delta_2}{\delta_1} = \frac{I_1}{I_2} = \frac{\left(\frac{bh^3}{12}\right)}{\frac{b\times(2h)^3}{12}} = \frac{bh^3}{12}\times\frac{12}{8bh^3} = \frac{1}{8}$$

$$\delta_2 = \frac{0.8}{8} = 0.1\text{cm}$$

Example 5.18 *The mid span deflection of a simply supported beam (span = 5 m) loaded with a concentrated load 20 kN at the centre is 2.5 mm. Determine the maximum deflection if the concentrated load is replaced by a UDL of intensity 4 kN/m acting over the whole span of the beam.*

Solution:

Given: $L = 5$ m, $W = 20$ kN, d = 2.5 mm

$$\delta = \frac{WL^3}{48EI}, EI = \frac{WL^3}{48\delta} = \frac{20\times10^3\times5^3}{48\times2.5\times10^{-3}} = 20.833\times10^6 \text{N.m}^2$$

$$\text{Under udl, } \delta = \frac{5}{384}\frac{wL^4}{EI} = \frac{5}{384}\times\frac{5000\times3^4\times10^3}{200\times10^9\times3\times10^6\times10^{-12}}$$

$$= 1.5625 \text{ mm}$$

Example 5.19 *Find the deflection at the centre of a simply supported beam of length 3 m, loaded with a udl of intensity 5kN/m through out. Take E = 200 GPa and I = 3 ×10⁶ mm⁴.*

Solution:

Given L = 3m, w = 5 kN/m, E = 200 GPa, I = 3 × 10^6 mm^4

Deflection at the centre of beam,

$$\delta = \frac{5}{384}\frac{wL^4}{EI} = \frac{5}{384}\times\frac{5000\times3^4\times10^3}{200\times10^9\times3\times10^6\times10^{-12}}$$

$$= 8.79 \text{ mm}$$

Example 5.20 *A beam of 6 m span is simply supported at its ends. The beam is loaded by uniformly distributed load of 50 kN/m which acts over its whole span. Besides this it has two point loads of 26 kN each acting at 1.5 m and 4.5 m from left end support. Use Macaulay's method to determine the deflection under point loads. Assume E - 200 GPa and I = 6.64 × 10^{-4} m^4.*

Solution:

Given: E = 200 GPa, I = 6.64 × 10^{-4} m^4.

The loaded beam is shown in Fig. 5.36.

$\Sigma M_B = 0$ gives

$6 R_A = 50 \times 6 \times 3 + 26 \times 15$

$= 1056$

$R_A = 176$ KN

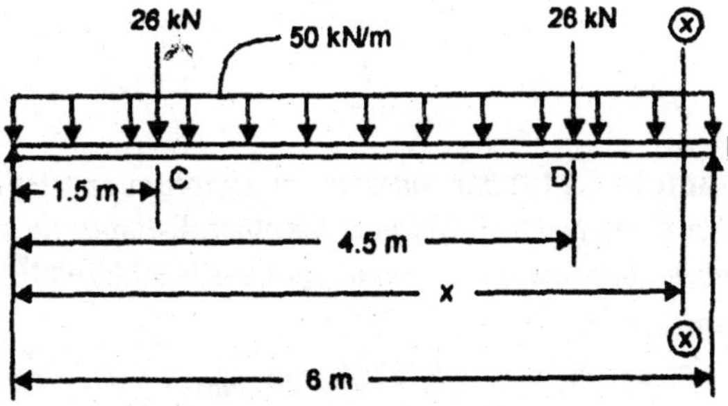

Fig. 5.36

Consider a section $x - x$ at a distance x from left hand support A.

$$M_x = 176\,x - 50\times x\times\frac{x}{2} - 26(x-1.5) - 26(x-4.5)$$

$$= 176x - 25x^2 - 26\,(x-1.5) - 26\,(x-4.5)$$

$$EI\frac{d^2y}{dx^2} = -M_x = -176x + 25x^2 + 26\,(x-1.5) + 26\,(x-4.5)$$

$$EI\frac{dy}{dx} = -88x^2 + \frac{25}{3}x^3 + 13(x-1.5)^2 + 13(x-4.5)^2 + C_1$$

$$Ely = -\frac{88}{3}x^3 + \frac{25}{12}x^4 + \frac{13}{3}(6-15)^3 + \frac{13}{3}(6-45)^3 + C_1x + C_2$$

At $x = 0, y = 0$ and $x = 6m, y = 0$

$$0 = C_2$$

$$0 = -\frac{88}{3}(6)^3 + \frac{25}{12}(6)^4 + \frac{13}{3}(6-15)^3 + \frac{13}{3}(6-4.5)^3 + 6C_1$$

$$= -6336 + 2700 + 394.87 + 14.62 + 6C_1$$

$$C_1 = \frac{3226.51}{6} = 537.75$$

$$\therefore \quad Ely_C = -\frac{88}{3}x^3 + \frac{25}{12}x^4 + \frac{13}{3}(x-15)^3 + \frac{13}{3}(x-4.5)^3 + 537.75x$$

At $x = 1.5$ m,

$$Ely_C = -\frac{88}{3}(1.5)^3 + \frac{25}{12}(1.5)^4 + 0 + 0 + 537.75 \times 1.5$$

$$= -99 + 10.547 + 806.625 = 718.172$$

$$y_C = \frac{718.172 \times 10^3}{200 \times 10^9 \times 6.64 \times 10^{-4}} = 5.4 \times 10^{-3} \text{ m or } 5.4 \text{ mm}$$

At $x = 4.5$m

$$Ely_D = -\frac{88}{3}(4.5)^3 + \frac{25}{12}(4.5)^4 + \frac{13}{3}(4.5-1.5)^3 + 0 + 537.75 \times 4.5$$

$$= -2673 + 854.3 + 117 + 2419.98 = 718.175$$

$$y_D = \frac{718.175 \times 10^3}{200 \times 10^9 \times 6.64 \times 10^{-4}} = 5.4 \times 10^{-3} \text{ m or } 5.4 \text{ mm}$$

Example 5.21 *A horizontal beam, simply supported at its ends, carries a load which varies uniformly from 10 kN/m at one end to 50 kN/m at the other. Estimate the central deflection if the span is 10 m, the section is 400 mm deep, and the maximum bending stress is 90 N/mm². Take E = 205GPa.*

Solution:

Given: L = 10m, h = 400 mm, s_{max} = 90N/mm², E = 205 GPa.

At a distance x from left end A, intensity of loading is (

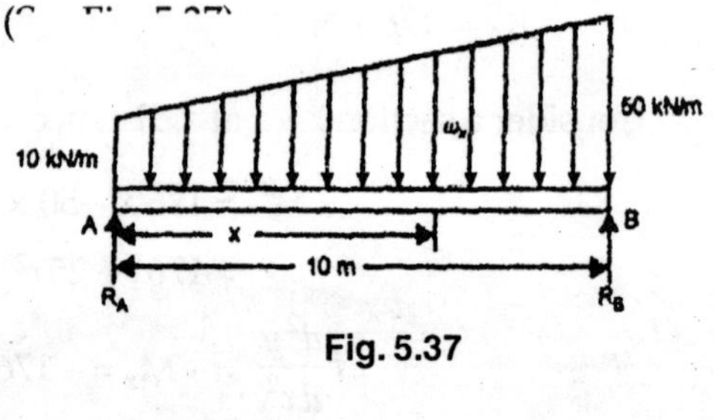

Fig. 5.37

$$w_x = 10 + \frac{(50-10)}{10}x = 10 + 4x$$

$\Sigma M_B = 0$ gives

$$10R_A = 10 \times 10 \times 5 + \frac{1}{2} \times 10 \times 40 \times \frac{10}{3} = 1166.67$$

$$R_A = 116.67 \ kN$$

$$M_x = R_A \times x - 10 \times x \times \frac{x}{2} - \frac{1}{2} \times x \times 4x \times \frac{x}{3}$$

$$= 116.67x - 5x^2 - \frac{2}{3}x^3$$

$$EI\frac{d^2y}{dx^2} = -M_x = -116.67x + 5x^2 - \frac{2}{3}x^3$$

$$EI\frac{dy}{dx} = -58.335x^2 + \frac{5}{3}x^3 + \frac{1}{6}x^4 + C_1$$

$$EIy = -19.445x^3 + \frac{5}{12}x^4 + \frac{1}{30}x^5 + C_1x + C_2$$

At $x = 0, y = 0$ gives $C_2 = 0$

At $x = 10\ m, y = 0$ gives

$$0 = -19445 + \frac{5}{12}\times 10^4 + \frac{1}{30}\times 10^5 + 10C_1$$

$$C_1 = 1194.5$$

At $x = 5\text{m}$

$$EIy = -19.445 \times 5^3 + \frac{5}{12}\times 5^4 + \frac{1}{30}\times 5^5 + \frac{1}{30}\times 5^5 + 1194.5 \times 5$$

$$= 3906$$

$$y = \frac{3906}{EI}$$

For M_x to be maximum, $\frac{dM_x}{dx} = 0$

$$116.67 - 10x - 2x^2 = 0$$

or $x^2 + 5x - 58.335 = 0$

$$x = \frac{1}{2}[-5 \pm \sqrt{25 + 4\times 58.335}] = 5.536\text{m}$$

$$M_{max} = 116.67 \times 5.536 - 5 \times (5.536)^2 - \frac{2}{3}(5.536)^3$$

$$= 379.54\text{ kN.m}$$

$$M_{max} = \sigma \times z$$

$$z = \frac{379.54\times 10^3\times 10^3}{90} = 4.217\times 10^6\text{mm}^3$$

$$\frac{b\times 400^2}{6} = 4.217 \times 10^6$$

$$b = 158\text{ mm}$$

$$I = \frac{bh^3}{12} = \frac{158 \times 400^3}{12} = 842666667 \text{ mm}^4$$

At $$x = 5\text{m}, y = \frac{3906 \times 10^3 \times 10^9}{205 \times 10^3 \times 84266666} = 22.6 \text{mm}$$

Example 5.22 *A simply supported beam of span 8 m is subjected to two concentrated loads of 60 kN and 80 kN at 3 m and 5 m from left end respectively. Determine*

(*a*) slopes at the supports.

(*b*) deflection under the loads.

(*c*) position and magnitude of maximum deflection.

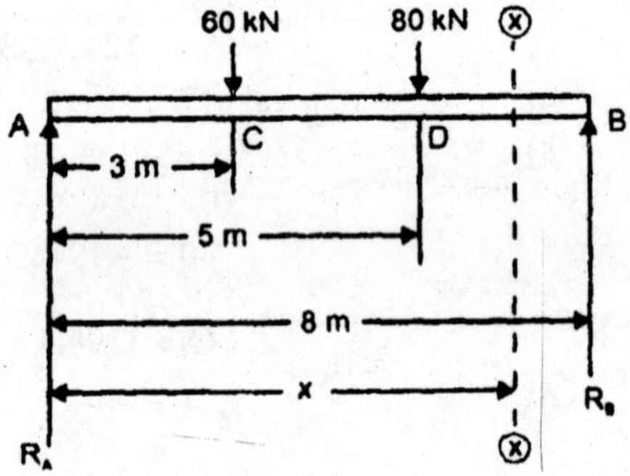

Fig. 5.38

Solution:

Refer to Fig. 5.38

$$R_A \times 8 = 60 \times 5 + 80 \times 3 = 540$$

$$R_A = 67.5 \text{ kN}$$

$$M_x = 67.5\,x - 60\,(x–3) - 80\,(x–5)$$

$$EI\frac{d^2y}{dx^2} = -M_x$$

$$= -67.5x + 60\,(x - 3) + 80\,(x - 5)$$

$$EI\frac{dy}{dx} = -33.75\,x^2 + 30\,(x - 3)^2 + 40\,(x - 5)^2 + C_1$$

$$Ely = -11.25x^3 + 10\,(x - 3)^3 + \frac{40}{3}\,(x - 5)^3 + C_1x + C_2$$

At $x = 0, y = 0$ gives $C_2 = 0$

At $x = 8\ m, y = 0$ gives

$$0 = -11.25 \times 8^3 + 10 \times 5^3 + \frac{40}{3} \times 3^3 + 8C_1$$

$$C_1 = 518.75$$

(*a*) $$EI\frac{dy}{dx} = -33.75\,x^2 + 30\,(x -)^2 + 40\,(x - 5)^2 + 518.75$$

At $x = 0$

$$\left(\frac{dy}{dx}\right)_A = \theta_A = \frac{518.75}{EI}$$

At $x = 8\text{m}$

$$EI\frac{dy}{dx} = -33.75 \times 8^2 + 30 \times 5^2 + 40 \times 3^2 + 518.75 = -531.25$$

$$\left(\frac{dy}{dx}\right)_B = \theta_B = \frac{-531.25}{EI}$$

(b) $EIy = -11.25x^3 + 10(x-5)^3 + \frac{40}{3}(x-5)^3\ 518.75x$

At $\quad x = 3\text{m}$

$$EIy_C = -11.25 \times 3^3 + 518.75 \times 3 = 1252.5$$

$$y_C = \frac{1252.5}{EI}$$

At $\quad x = 5\text{m}$

$$EIy_D = -11.25 \times 5^3 + 10 \times 2^3 + 518.75 \times 5 = 1267.5$$

$$y_D = \frac{1267.5}{EI}$$

(c) For y to be maximum, $\frac{dy}{dx} = 0$

$$-33.75x^2 + 30(x-3)^2 + 40(x-5)^2 + 518.75 = 0$$

$$-33.75x^2 + 30x^2 + 270 - 180x + 40x^2 + 1000 - 400x + 518.75 = 0$$

$$36.25x^2 - 580x + 1788.75 = 0$$

$$x^2 - 16x + 49.345 = 0$$

$$x = \frac{16 \pm \sqrt{256 - 197.38}}{2} = \frac{16 \pm 7.656}{2} = 4.172\text{m} = 4.172\text{ m}$$

$$Ely = -11.25 \times (4.172)^3 + 10(4.172-3)^3 + 0 + 518.75 \times 4.172 = 1363.4$$

$$y_{max} = 1363.4/EI$$

Example 5.23 *An overhanging beam of circular cross-section diameter, d = 101.6 mm, length L = 4.88 m is simply supported symmetrically. The distance between the two supports is 3.05 m. A load P = 18.15 kN acts vertically at each end. Assuming E = 210 GPa, determine the deflection under each load.*

Solution:

Given: $\quad d = 101.6\text{ mm}, E = 210\text{ GPa}$

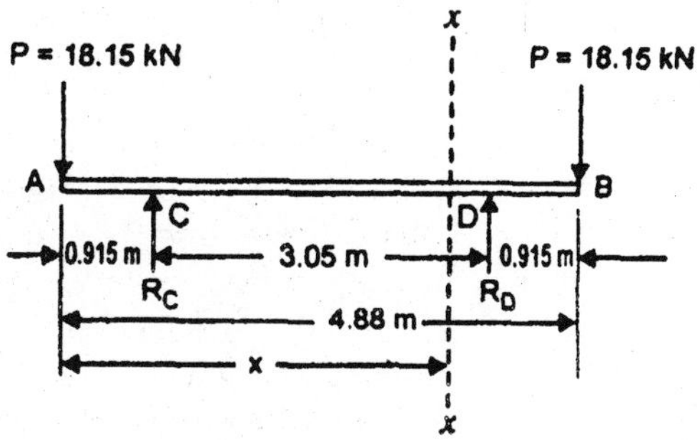

Fig. 5.39

$$\text{Overhang} = \frac{1}{2}(4.88 - 3.05)$$

$$= 0.915\text{ m}$$

Moment of inertia of bar,

$$I = \frac{\pi d^4}{64} = \frac{\pi \times (101.6)^4}{64}$$

$$= 5230518\text{ mm}^4$$

$$EI = 210 \times 10^9 \times 5230518 \times 10^{-12}$$

$$= 1098409\text{ N.m}^2$$

Refer to Fig.5.39

Taking moments about D, we have

$$3.05 \times R_C = 18.15 \times 3.965 - 18.15 \times 0.915$$

$$R_C = 18.15 \text{ kN}$$
$$R_D = 18.15 \text{ kN}$$

At a section x from A,

$$M_x = -Px + P(x-a)$$

where $AC = a = 0.915$ m, L = 3.05 m = CD

$$EI\frac{d^2y}{dx^2} = -M_x = +Px - P(x-a)$$

Integrating twice, we have

$$EI\frac{dy}{dx} = \frac{Px^2}{2} - \frac{P}{2}(x-a)^2 + C_1$$

$$Ely = \frac{Px^3}{6} - \frac{P}{6}(x-a)^3 + C_1x + C_2$$

At $x = a$, $y = 0$ and $x = a + L$, $y = 0$

$$0 = \frac{Pa^3}{6} + C_1a + C_2 \quad \text{... (1)}$$

$$0 = \frac{P}{6}(a+L)^3 - \frac{P}{6}L^3 + C_1(a+L) + C_2 \quad \text{... (2)}$$

Subtracting Eq. (1) from (2), we have

$$\frac{P}{6}[(a+L)^3 - a^3] - \frac{P}{6}L^3 + C_1L = 0$$

$$\frac{P}{6}(a^3 + L^3 + 3a^2L + 3aL^2 - a^3 - L^3) + C_1L = 0$$

$$\frac{P}{2}(a^2L + aL^2) + C_1L = 0$$

$$C_1 = -\frac{P}{2}(a^2 + aL)$$

$$C_2 = -\frac{Pa^3}{6} + \frac{Pa}{2}(a^2 + aL)$$

$$= -\frac{Pa^3}{6} + \frac{Pa^3}{2} + \frac{Pa^2L}{2} = \frac{Pa^3}{3} + \frac{Pa^2L}{2}$$

$$Ely = \frac{Px^3}{6} - \frac{P}{6}(x-a)^{-3} - \frac{Px}{2}(a^2 + aL) + \frac{Pa^3}{3} + \frac{Pa^2L}{2}$$

At $x = 0$,

$$Ely_A = \frac{Pa^3}{3} + \frac{Pa^2L}{2}$$

$$y_A = \frac{P}{EI}\left[\frac{a^3}{3} + \frac{a^2L}{2}\right]$$

$$= \frac{18.15\times10^3}{1098409}\left[\frac{(0.915)^3}{3}+\frac{(0915)^2\times3.05}{2}\right]$$

$$= 0.0253 \text{ m or } 25.3 \text{ mm}$$

The loading is symmetrical.

$\therefore \qquad y_B = y_A = 25.3 \text{ mm}$

Example 5.24 *The rigid bars ABC and CD are supported by pin supports at A and D by a steel rod at B, as shown in Fig. 5.40. There is a roller connection between the bars at C. Compute the vertical deflection of point C caused by the 50 kN load.*

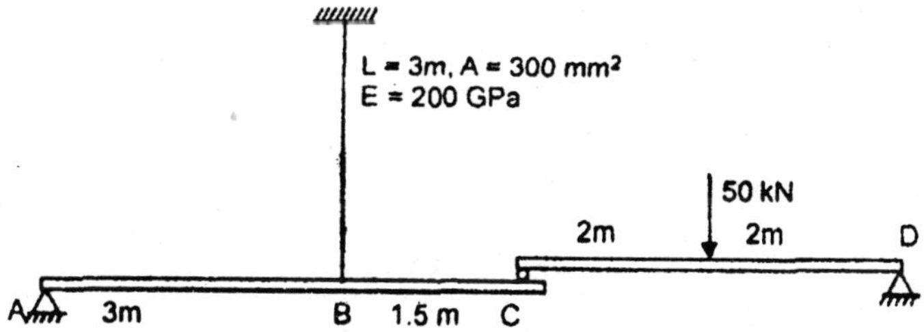

Fig. 5.40

Solution:

Consider free body diagram for bar CD shown in Fig.5.41 (*a*).

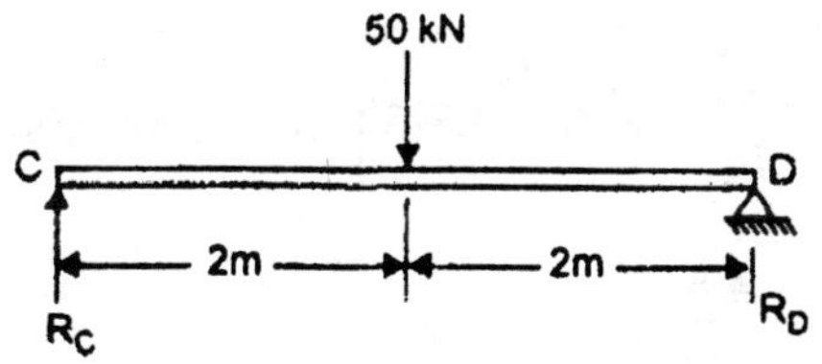

Fig. 5.41 (a)

$$R_C = R_D = \frac{50}{2} = 25 \ kN$$

Let $\qquad W$ = Force in rod

Free body diagram for rod is shown in Fig 5.41 (*b*).

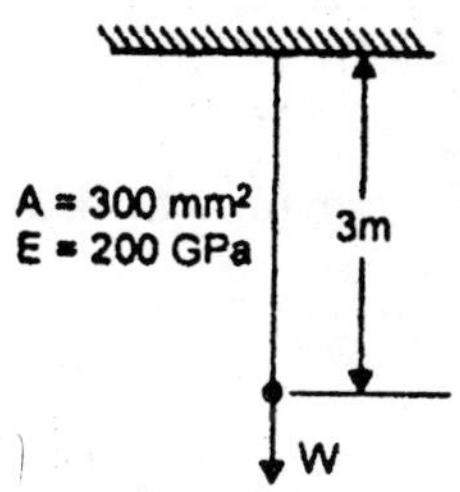

Fig. 5.41 (b)

Free body diagram for bar ABC is shown in Fig. 5.41 (*c*).

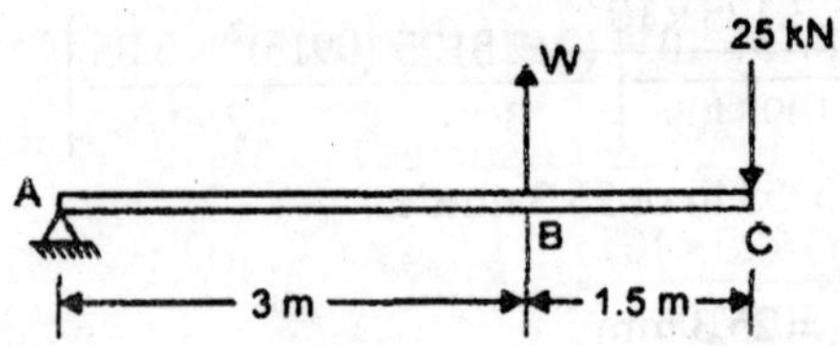

Fig. 5.41 (c)

Taking moments about A, we have

$$3W = 25 \times 4.5$$

$$W = 37.5\text{kN}$$

$$\text{Extension of rod}, \delta_1 = \frac{Wl}{AE} = \frac{37.5\times10^3\times3\times10^3}{300\times10^{-6}\times200\times10^9} = \mathbf{1.875\ mm\ (downwards)}$$

$$\text{Deflection of point B due to load } W,\ \delta_2 = \frac{WL^3}{3EI} = \frac{37.5\times10^3\times3^3}{3\times EI}$$

$$= \frac{337.5\times10^3}{EI}\text{m(upwards)}$$

$$\text{Resultant deflection of point } B, = \delta_2 - \delta_1$$

$$= \left(\frac{337.5\times10^3}{EI}\times10^3 - 1.875\right)\text{ mm (upwards)}$$

$$\text{Deflection of point C, } \delta_3 = \delta_B \times \frac{4.5}{3}$$

$$= \left(\frac{337.5\times10^3}{EI}\times -1.875\right)\times 1.5\text{ mm (upwards)}$$

Deflection of point C due to 25 kN load,

$$\delta_4 = \frac{25\times(4.5)^3\times10^3\times10^3}{3EI}$$

$$= \frac{759.375\times10^6}{EI}\text{ mm (downwards)}$$

Resultant downwards deflection of point C,

$$\delta_C = \delta_4 - \delta_3$$

$$= \frac{759.375\times10^6}{EI} - \frac{337.5\times10^6}{EI}\times1.875\times1.5$$

$$= \frac{25.3.125 \times 10^6}{EI} + 2.8125$$

Let $\quad EI = 50 \times 10^6 \text{ N.m}^2$ (assumed)

$$\delta_C = \frac{253.125 \times 10^6}{50 \times 10^6} + 2.8125$$

$$= 7.875 \text{ mm}$$

Example 5.25 *The simply supported beam ABC (Fig. 5.42) carries the uniformly distributed load between its supports and the concentrated load P at end C. Find the value of P, using the moment area method, for which*

(a) *There is no deflection at C.*

(b) *The deflection curve at B is horizontal.*

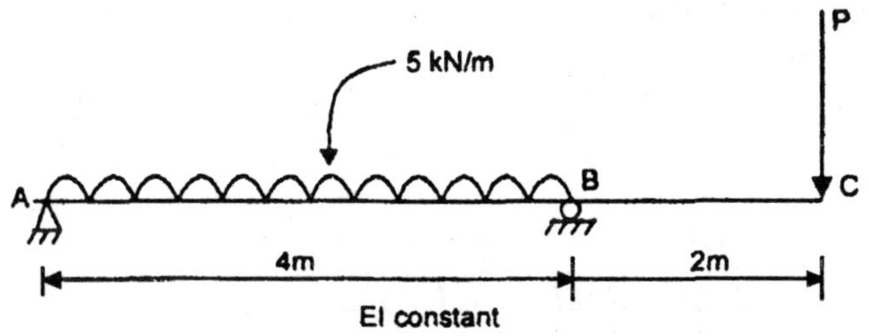

Fig. 5.42

Solution:

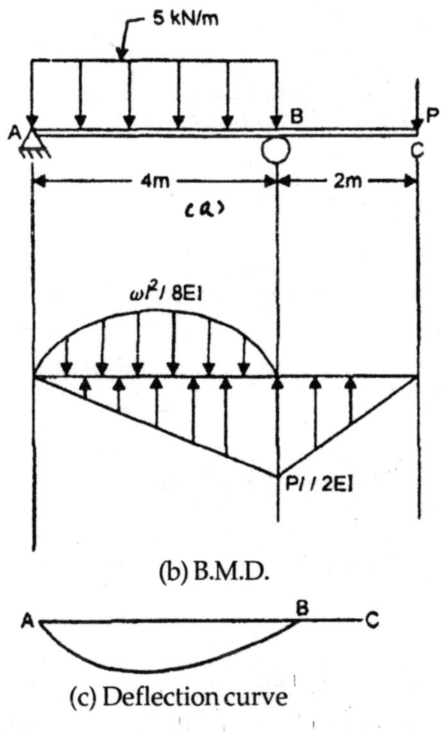

Fig. 5.43

Let $AB = l$ then $BC = l/2$

Refer to Fig. 5.43

Moment of $\frac{M}{EI}$ diagram between A and B about B,

$$Z_B = \frac{wl^2}{8EI} \times \frac{2}{3} l \times \frac{1}{2} - \frac{1}{2} \times \frac{Pl}{2EI} \times l \times \frac{l}{3}$$

$$= \frac{wl^4}{24EI} - \frac{Pl^3}{12EI}$$

Moment of $\frac{M}{EI}$ diagram about C,

$$Z_C = \frac{wl^2}{8EI} - \frac{2l}{3}\left(\frac{l}{2} + \frac{l}{2}\right) - \frac{1}{2} \times \frac{Pl}{2EI} \times l\left(\frac{l}{3} + \frac{l}{2}\right) - \frac{1}{2} \times \frac{Pl}{2EI} \times \frac{l}{2} \times \frac{2}{3} \times \frac{l}{2}$$

$$= \frac{wl^4}{12EI} - \frac{5}{24}\frac{Pl^3}{EI} - \frac{Pl^3}{24EI} = \frac{wl^4}{12EI} - \frac{Pl^3}{4EI}$$

(a) Deflection of C,

$$\delta_C = \frac{2}{3} Z_B - Z_C = \frac{3}{2}\left[\frac{wl^4}{24EI} - \frac{Pl^2}{12EI}\right] - \frac{wl^4}{12EI} - \frac{Pt^3}{4EI} = 0$$

$$= \frac{wl^4}{EI}\left[\frac{1}{16} - \frac{1}{12}\right] - \frac{Pl^3}{EI}\left[\frac{1}{8} - \frac{1}{4}\right] = 0$$

$$\frac{wl^4}{48EI} = \frac{Pl^3}{8EI}$$

$$P = \frac{wl}{6} = \frac{5 \times 4}{6} = \frac{10}{3} \text{kN}$$

(b) Here $\theta_B = 0$

Moment of $\frac{M}{EI}$ diagram between A to B about A,

$$Z_A = \frac{wl^2}{8EI} \times \frac{2l}{3} \times \frac{l}{2} - \frac{1}{2} \times \frac{Pl}{2EI} \times l \times \frac{2l}{3}$$

$$= \frac{wl^4}{24EI} - \frac{Pl^3}{6EI}$$

$$\theta_B = \frac{Z_A}{l} = \frac{wl^3}{24EI} - \frac{Pl^2}{6EI} = 0$$

$$P = \frac{wl}{4} = \frac{5 \times 4}{4} = 5kN$$

5.7 Singularity Function Method

The singularity functions permit the expression of any system of loads (distributed loads, concentrated loads, couples and combinations of them) as an equivalent distributive load. The motivation for the development of the singularity functions resides in the differential relations:

$$\left.\begin{aligned} \frac{dF}{dx} &= w(x) \quad (a) \\ \frac{dM}{dx} &= F(x) \quad (b) \\ \frac{EId^2y}{dx^2} &= M(x) \; (c) \end{aligned}\right\} \qquad \text{.... (5.34)}$$

5.7.1 Macaulay Functions

The Macaulay functions are defined by the relations:

$$(x-a)^n = \begin{cases} 0 & \text{for } x<a \\ (x-a)^n & \text{for } a \le x < \infty, n=0,1,2,.... \end{cases} \qquad \text{... (5.35)}$$

These functions have the following integration property:

$$\int_0^x (\xi-a)^n \, d\xi = \frac{(x-a)^{n+1}}{n+1} \qquad \text{... (5.36)}$$

The equivalent distributive load associated with the various types of loads is:

(1) Distributive load on part of the beam extending upto the right end (Fig.5.44).

Equivalent distributive load,

$$w_e = w_0 (x-a)^0$$
$$(x-a)^0 = 0 \text{ for } x<a \qquad \text{... (7.37)}$$
$$(x-a)^0 = 1 \text{ for } x \ge a$$

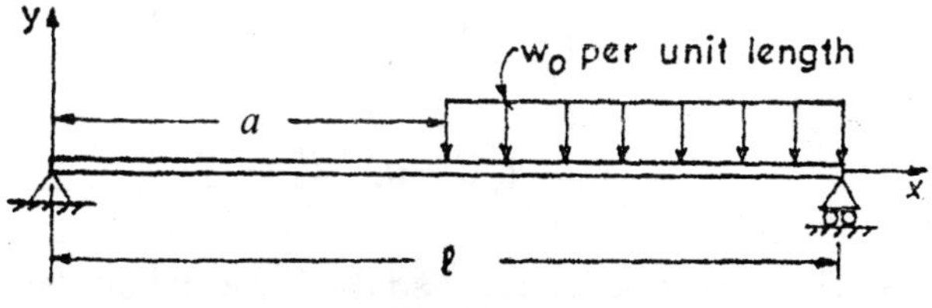

Fig. 5.44

(2) Distributive load on part of the beam starting from the left end (Fig.5.45)

$$w_e = w_0 (x-0)^0 - w_0 (x-a)^0 \qquad \text{.... (5.38)}$$

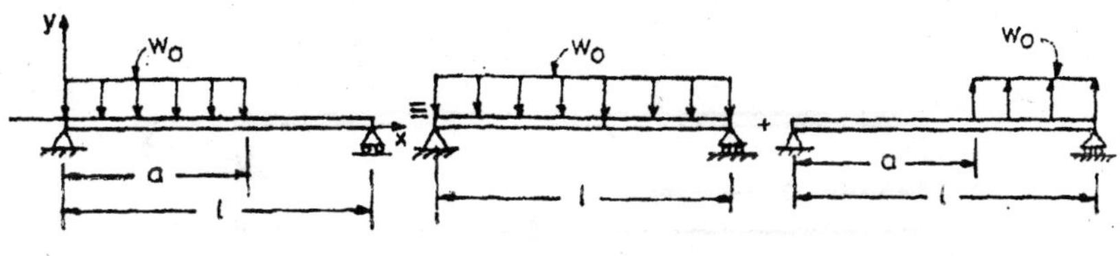

Fig. 5.45

(3) Linearly distributed load. (Fig.5.46)

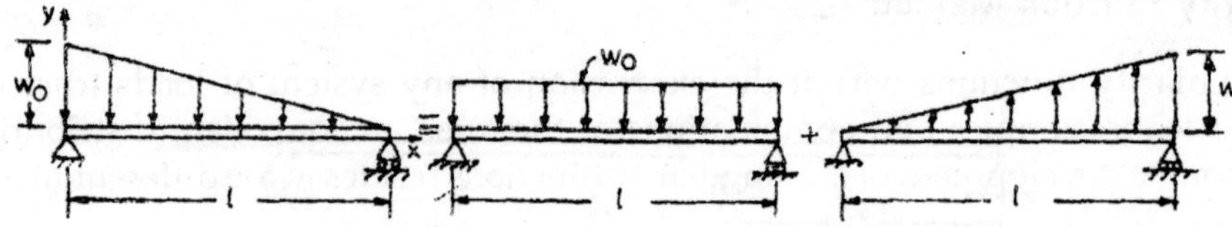

Fig. 5.46

$$w_e = w_0(x-0)^0 - \frac{w_0}{l}(x-0)^1 \qquad \text{... (5.39)}$$

Integration of Macaulay functions:

$$\int_0^x \xi^n (\xi - a)^m d\xi = \frac{(x-a)^{n+m+1}}{n+m+1} + \frac{na(x-a)^{n+m}}{m+n} + \frac{n(n-1)a^2(x-a)^{n+m-1}}{(m+n-1).2!} \qquad \text{... (5.40)}$$

For $n = m = 1$,

$$\int_0^x \xi(\xi - a)^1 d\xi = \frac{(x-a)^3}{3} + \frac{a(x-a)^2}{2} \qquad \text{... (5.41)}$$

For $n = 2, m = 1$,

$$\int_0^x \xi^2 (\xi - a)^1 d\xi = \frac{(x-a)^4}{4} + \frac{2a(x-a)^3}{3} + \frac{a^2(x-a)^2}{2} \qquad \text{... (5.42)}$$

5.7.2 Singularity Functions

To represent a concentrated load as an equivalent distributive load, it is necessary to introduce a new function that has the properties,

$$(x-a)^{-1} = \begin{cases} 0 \; for \; x \neq a \\ \infty \; for \; x = a \end{cases} \qquad \text{... (5.43(a))}$$

and

$$\int_0^x (\xi - a)^{-1} d\xi = < x - a >^0 = \begin{cases} 0 \; for \; x < a \\ 1 \; for \; x > a \end{cases} \qquad \text{... (5.43(b))}$$

This new function is called the Dirac delta or the unit impulse function.

(1) Concentrated load P at $x = a$ (Fig. 5.47)

$$w_c = P(x-a)^{-1} \qquad \text{... (5.44)}$$

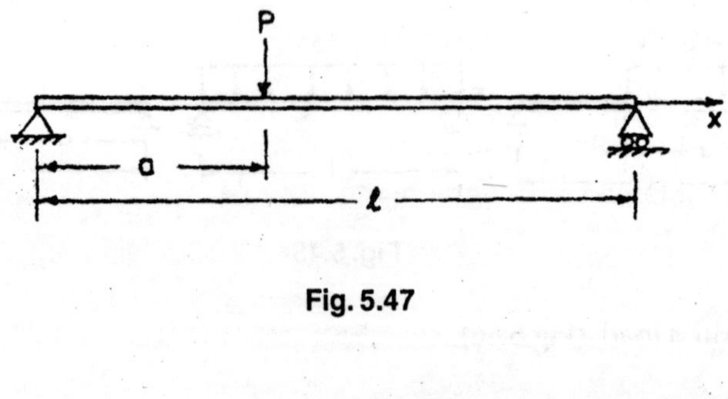

Fig. 5.47

(2) Concentrated couple C at $x = a$ (Fig. 5.48)

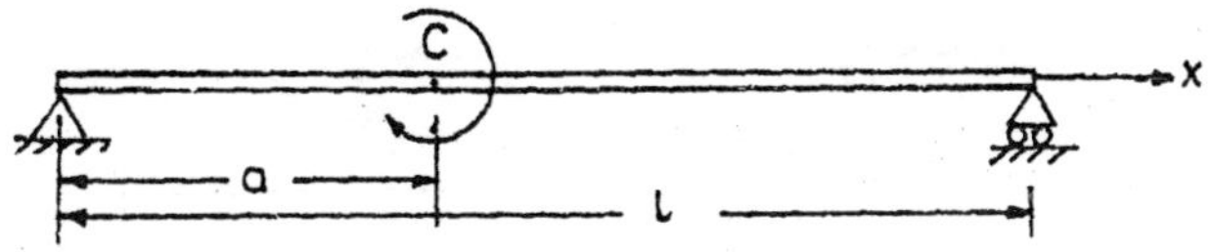

Fig. 5.48

To represent a couple by an equivalent distributive load, we introduce a second singular function that has the properties,

$$(x-a)^{-2} = \begin{cases} 0 \; for \; x \neq a \\ \infty \; for \; x = a \end{cases} \quad \text{... (5.45 a)}$$

and $$\int_0^x (\xi - a)^{-2}\, d\xi = <x-a>^{-1} \quad \text{... (5.45 b)}$$

Hence $$w_e = C <x-a>^{-2} \quad \text{... (5.46)}$$

Example 5.26 *Using Macaulay and singularity functions, determine the shear and bending moment equations, and the slope and deflection equations for the beam shown in Fig. 5.49*

Solution:

Taking moments about B,

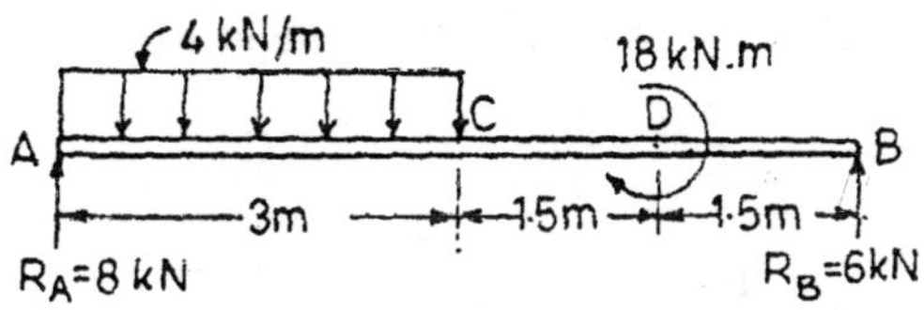

Fig. 5.49

$$R_A \times 6 = 4 \times 3 \times 4.5 - 18$$
$$= 54 - 18 = 36$$
$$R_A = 6 \text{ kN}$$
$$R_B = 4 \times 3 - 6 = 6 \text{ kN}$$

Equivalent distributed load corresponding to all applied loads and reactions is,

$$w_e = 6 <x-0>^{-1} - [4<x-0>^0 - 4<x-3>^0] + 18 <x-4.5>^{-2} + 6 <x-6>^{-1} \quad \text{...(a)}$$

The shear force is obtained by integrating Eq. (a),

$$F(x) = 6 <x-0>^0 - [4 <x-0>^1 - 4 <x-3>^1] + 18 <x-4\text{–}5>^{-1} + 6 <x-6>^0 \quad \text{... (b)}$$

The moment equation is obtained by integrating Eq. (b),

$$M(x) = 6 <x-0>^1 - [2 <x-0>^2 - 2 <x-3>^1] + 18 <x-4.5>^0 + 6 <x-6>^1 \quad \text{... (c)}$$

$$EI\frac{dy}{dx} = 3<x-0>^2 - \left[\frac{2}{3}<x-0>^3 - \frac{2}{3}<x-0>^3\right]$$

$$+18<x-4.5>^1 + 3<x-6>^2 + C_1 \quad \text{...(d)}$$

$$EIy = <x-0>^3 - \left[\frac{1}{6}<x-0>^4 - \frac{1}{6}<x-3>^4\right]$$

$$+9<x-4.5>^2 + <x-6>^3 + C_1 x + C_2 \quad \text{...(e)}$$

At $x = 0, y = 0$ and at $x = 6\ m, y = 0.$

Hence $C_2 = 0$, and

$$6^3 - \left[\frac{1}{6}(6)^4 - \frac{1}{6}(3)^4\right] + 9(1.5)^2 + 0 + 6C_1 = 0$$

$$216 - (216 - 13.5) + 20.25 + 6\,C_1 = 0$$

or

$$C_1 = \frac{33.75}{6} = -5.625$$

For $0 \le x \le 3$ m

$$F(x) = 6 - 4x$$

$$M(x) = 6x - 2x^2$$

or

$$EI\frac{dy}{dx} = 3x^2 - \frac{2}{3}x^3 - 5.625$$

For $3\text{m} \le x \le 4.5\text{m}$

$$F(x) = 6 - 4x + 4(x-3)$$

$$M(x) = 6x - 2x^2 + 2(x-3)^2$$

$$EI\frac{dy}{dx} = 3x^2 - \frac{2}{3}x^3 + \frac{2}{3}(x-3)^3 - 5.625$$

$$EIy = x^3 - \frac{1}{6}x^4 + \frac{1}{6}(x-3)^4 - 5.625x$$

Example 5.27 *Determine the reactions at the roller and the wall for the beam shown in Fig. 5.50 (a).*

Solution.

The equivalent distributive load for the beam shown in Fig. 5.50 (*b*) is

$$w_e = R_A<x-0>^{-1} - M_0<x-l>^{-2}$$

$$- R_B<x-2l>^{-1} - M_B<x-2l>^{-2}$$

$$F(x) = R_A<x-0>^0 - M_0<x-l>^{-1}$$

$$-R_B<x-2l>^0 - M_B<x-2l>^{-1}$$

$$M(x) = R_A<x-0>^1 - M_0<x-l>^0$$

$$-R_B<x-2l>^1 - M_B<x-2l>^0$$

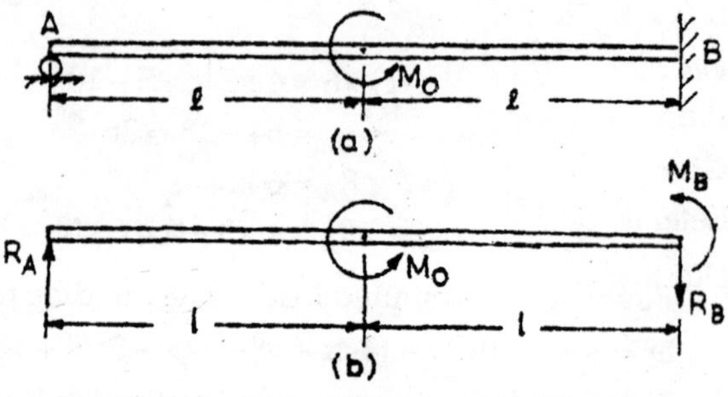

Fig. 5.50

$$EI\frac{dy}{dx} = \frac{R_A < x-0 >^2}{2} - M_0 < x-1 >^1 - \frac{R_B < x-2l >^2}{2} - M_B < x-2l >^1 + C_1$$

$$EIy = \frac{R_A < x-0 >^3}{2} - \frac{M_0 < x-l >^2}{2} - \frac{R_B < x-2l >^3}{6} - \frac{M_B < x-2l >^1}{2}$$

$$+C_1x + C_2$$

Now $\quad y(0) = 0$: Hence $C_2 = 0$.

$$y(2l) = 0: \frac{R_A(2l)^3}{6} - \frac{M_0l^2}{2} + 2C_1l = 0 \qquad \text{...(a)}$$

$$\frac{dy}{dx}(2l) = 0: \frac{R_A(2l)^2}{2} - M_0l + C_1 = 0 \qquad \text{...(b)}$$

Eqs. (*a*) and (*b*) give,

$$C_1 = -\frac{1}{8}M_0l \text{ and } R_A = \frac{9}{16}\frac{M_0}{l}$$

$$F(2l) = 0 : R_A - M_0 < x-l >^{-1} - R_B = 0$$

$$\frac{9}{16}\frac{M_0}{l} - 0 - R_B = 0$$

or

$$R_B = \frac{9}{16}\frac{M_0}{l}$$

$$M(2l) = 0 : 2R_Al - M_0 - M_B = 0$$

$$\frac{9}{8}M_0 - M_0 - M_B = 0$$

$$M_B = \frac{M_0}{8}$$

Example 7.28 *A simply supported beam of length 20 meter has two concentrated loads of 10 kN each and placed at locations which are respectively at 5m end 15 meter from left end. Calculate the deflection under each load and the maximum deflection. Assume E = 2 × 10⁵ N/mm²*

Solution:

Given: $E = 2 \times 10^5$ N/mm²

Refer to Fig. 5.51.

Due to symmetery of loading,

$$R_A = R_B = 10 \text{ kN}$$

$$M_x = 10x - 10(x-5) - 10(x-15)$$

$$EI\frac{d^2y}{dx^2} = -M_x = -10x + 10(x-5) + 10(x-15)$$

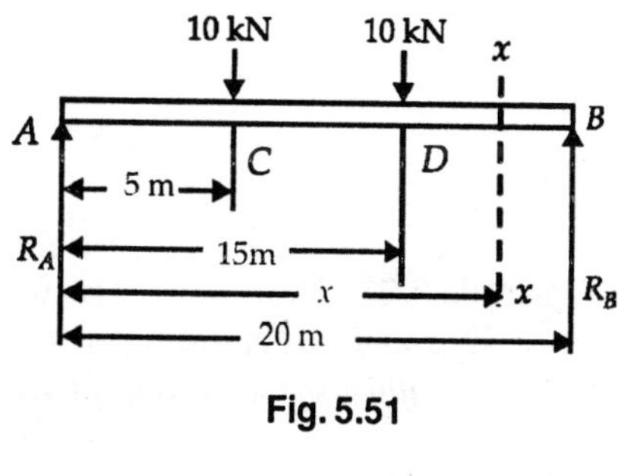

Fig. 5.51

$$EI\frac{dy}{dx} = -5x^2 + 5(x-5)^2 + 5(x-15)^2 + C_1$$

$$EIy = -\frac{5}{3}x^3 + \frac{5}{3}(x-5)^3 + \frac{5}{3}(x-15)^3 + C_1x + C_2$$

At $x = 0, y = 0$ giving $C_2 = 0$

At $x = 20\text{m}, y = 0$

$$0 = -\frac{5}{3}\times 20^3 + \frac{5}{3}\times 15^3 + \frac{5}{3}\times 5^3 + 20C_1$$

$$C_1 = \frac{5}{3\times 20}[20^3 - 15^3 - 5^3] = 375$$

$$EIy = \frac{5}{3}[-x^3 + (x-5)^3 + (x-15)^3] + 375x$$

At $x = 5m,$

$$EIy_1 = \frac{5}{3}[-5^3] + 375\times 5 = 1666.67$$

Let $I = 20 \times 10^8 \text{ mm}^4$

$$y_1 = \frac{1666.67\times 10^{12}}{2\times 10^5\times 20\times 10^8} = 4.17\text{ mm}$$

At $x = 15\text{ m}$

$$EIy_2 = \frac{5}{3}[-15^3 + 10^3] + 375\times 15 = 1666.67$$

$$y_2 = 4.17\text{ mm}$$

Maximum deflection will occur between the loads.

$$-5x^2 + 5(x-5)^2 + 375 = 0$$

$$-x^2 + (x-5)^2 + 75 = 0$$

$$-x^2 + x^2 + 25 - 10x + 75 = 0$$

$$10x = 100$$

$$x = 10\text{ m}$$

$$EIy_{max} = \frac{5}{3}[-10^3 + 5^3] + 375\times 10 = 2291.67$$

$$y_{max} = \frac{2291.67\times 10^{12}}{2\times 10^5\times 20\times 10^8} = 5.73\text{mm}$$

Example 5.29 *A beam section is 10 m long and is simply supported at the ends. It carries concentrated loads of 100 kN and 60 kN at distances of 2 m and 5 m respectively from the left end. Calculate the deflection under each load. Find also the maximum deflection. Take I = 18 × 10⁸ mm⁴ and E = 200 kN/mm².*

Solution:

Given: $I = 18 \times \text{mm}^4$, $E = 200\ \text{kN/mm}^2$

Refer to Fig. 5.52.

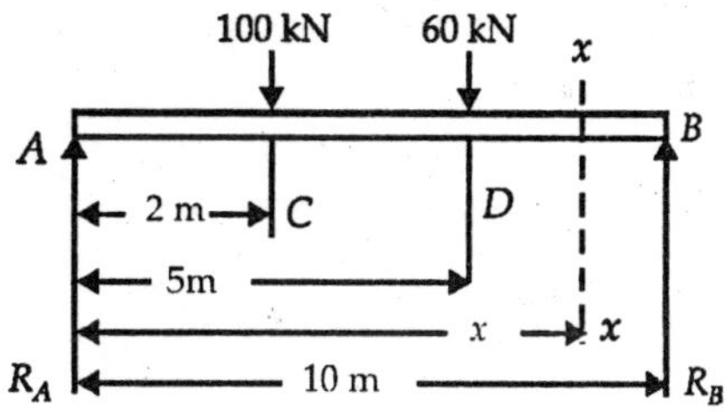

Fig. 5.52

$$\Sigma M_B = 0 \text{ gives,}$$

$$10R_A = 100 \times 8 + 60 \times 5 = 1100$$

$$R_A = 110\ \text{kN}$$

$$M_x = 110x - 100\,(x-2) - 60\,(x-5)$$

$$EI\frac{d^2y}{dx^2} = -M_x = -110x + 100\,(x-2) + 60\,(x-5)$$

$$EI\frac{dy}{dx} = -55x^2 + 50\,(x-2)^2 + 30\,(x-5)^2 + C_1$$

$$EIy = -\frac{55}{3}x^3 + \frac{50}{3}(x-2)^3 + 10(x-5)^2 + C_1x + C_2$$

At $x = 0$, $y = 0$ gives

$$0 = C_2$$

At $x = 10\ m$, $y = 0$ gives

$$0 = -\frac{55}{3}\times 10^3 + \frac{50}{3}\times 8^3 + 10\times 5^3 + 10C_1$$

$$C_1 = 967.5$$

$$EIy = -\frac{55}{3}x^3 + \frac{50}{3}(x-2)^3 + 10(x-5)^3 + 967.5x$$

At $x = 2\text{m}$, $EIy = -\frac{55}{3}\times 2^3 + 967.5\times 5\times 2 = 2 = 1788.3$

$$y_1 = \frac{1788.3\times 10^{12}}{200\times 10^3\times 18\times 10^8} = 4.96\,\text{mm}$$

At $x = 5\text{m}$, $EIy_2 = -\frac{55}{3}\times 5^3 + \frac{50}{3}\times 3^3 + 967.5\times 5 = 2995.8$

$$y_2 = \frac{2995.58\times 6\times 10^{12}}{200\times 10^3\times 18\times 10^8} = 8.32\,\text{mm}$$

Maximum deflection shall occur between the loads.

$$-55x^2 + 50\,(x-2)^2 + 967.5 = 0$$

$$-11x^2 + 10\,(x-2)^2 + 193.5 = 0$$

$$-11x^2 + 11x^2 + 40 - 40x + 193.5 = 0$$

$$-x^2 - 40x + 233.5 = 0$$

$$x = \frac{1}{2}[-40 \pm \sqrt{1600+934}]$$

$$= -20 + 25.17 = 5.17 \text{ m}$$

$$EIy_{max} = -\frac{55}{3}(5.17)^3 + \frac{50}{3}(.3.17)^3 + 10\times(0.17)^3 + 967.5\times 5.17$$

$$= 2999.5$$

$$y_{max} = \frac{2999.5\times 10^{12}}{200\times 10^3 \times 18\times 10^8} = 8.33 \text{mm}$$

Example 5.30 *A horizontal girder of steel having uniform section is 14 metres long and. is simply supported at its ends. It carries a concentrated loads of 120 kN and 80 kN at two points 3 metres and 4.5 metres from the two ends respectively. I = 16 x 10⁸ mm⁴ for girder section and E$_s$ = 210 kN/mm². Calculate the deflection of girder at points under two loads. Find also maximum deflection.*

Solution:

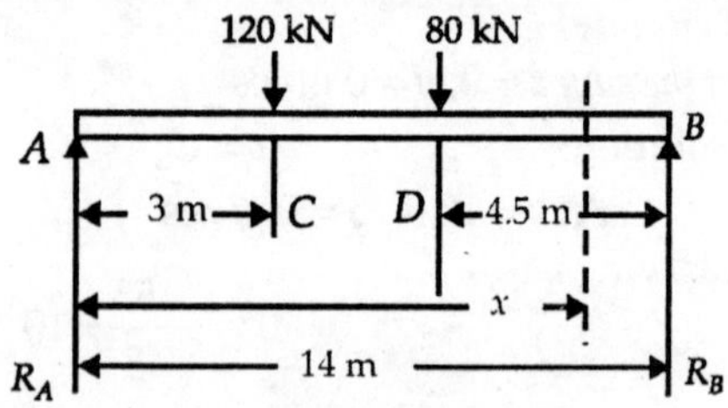

Fig. 5.53

Given: $I = 16 \times 10^8$ mm^4, $E_s = 210$ kN/mm^2

Refer to Fig. 5.53.

$\Sigma M_B = 0$ gives

$$14R_A = 120 \times 11 + 80 \times 4.5 = 1680$$

$$R_A = 120 \text{ kN}$$

$$M_x = 120\,x - 120\,(x-3) - 80\,(x-9.5)$$

$$EI\frac{d^2y}{dx^2} = -M_x = -120x + 120\,(x-3) + 80\,(x-9.5)$$

$$EI\frac{dy}{dx} = -60x^2 + 60\,(x-3)3 + 40\,(x-9.5)^2 + C_1$$

$$EIy = -20x^3 + 20(x-3)^3 + \frac{40}{3}(x-9.5)^3 + C_1x + C_2$$

At $x = 0, y = 0$ gives

$$0 = C_2$$

At $x = 14$ m, $y = 0$

$$0 = -20 \times 14^3 + 20 \times 11^3 + \frac{40}{3} \times (4.5)^3 + 14\,C_1$$

$$C_1 = 1931.8$$

$$EIy = -20x^3 + 20\,(x-3)^3 + \frac{40}{3}\,(x-9.5)^3 + 1931.8x$$

At $x = 3$m

$$EIy_C = -20 \times 3^3 + 1931.8 \times 3 = 5253$$

$$y_C = \frac{5253\times 10^{12}}{210\times 10^3 \times 16\times 10^8} = 15.6 \text{mm}$$

At $x = 9.5$ m

$$EIy_D = -20\,(9.5)^3 + 20 \times (6.5)^3 + 1931.8 \times 9.5 = 6697.1$$

$$y_D = \frac{6697 \times 10^{12}}{210 \times 10^3 \times 16 \times 10^8} = 19.93\,\text{mm}$$

For y_{max}, $-60x^2 + 60\,(x-3)^2 + 1931.8 = 0$

$$-x^2 + (x-3)^2 + 32.197 = 0$$

$$-x^2 + x^2 + 9 - 6x + 32.197 = 0$$

$$6x = 41.197$$

$$x = 6.87$$

$$EIy_{max} = -20 \times (6.87)^3 \times 20 \times (3.87)^3 + 1931.8 \times 6.87$$

$$= 7945.8$$

$$y_{max} = \frac{7945.8 \times 10^{12}}{210 \times 10^3 \times 16 \times 10^8} = 23.65\,\text{mm}$$

Example 5.31 *A simply supported beam carrying a uniformly distributed load over the whole span is proped to its original level at its mid-point with the help of a rigid prop. Calculate the load carried by the prop and each of the supports.*

Solution:

Let R_c = reaction of the prop (Fig. 5.54) Deflection at the mid-point of simply supported beam carrying *udl*,

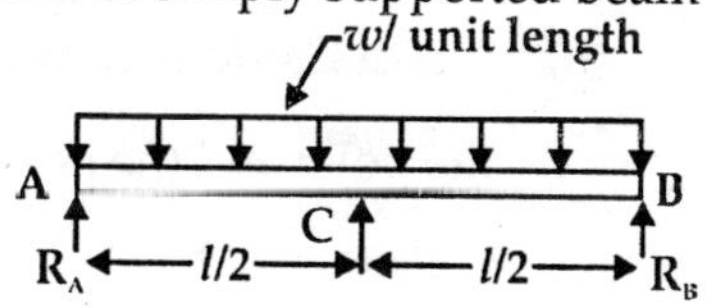

Fig. 5.54

$$\delta = \frac{5}{384}\frac{wl^4}{EI}$$

$$= \frac{R_c l^3}{48EI}$$

$$R_C = \frac{5}{8}wl$$

$$R_A + R_B = wl - \frac{5}{8}wl = \frac{3}{8}wl$$

$$R_A = R_B = \frac{3}{16}wl$$

Summary

1. Bending moment, $M_x = -EI\dfrac{d^2y}{dx^2}$

   Slope, $\dfrac{dy}{dx} = -\int \dfrac{Mx}{EI}dx$

Deflection, $y = \iint \frac{M}{EI} dx\, dx$

2. Double integration method.
   (i) Choose the section $x - x$ nearer to the right end support carrying all the loads.
   (ii) Keep intact the bracket terms until a numerical substitution is made.
   (iii) Disregard the bracket when for a particular value of x the contents of the bracket becomes negative.
   (iv) A couple C at a distance '*a*' from left end support may be written as $C(x-a)^{\circ}$
   (v) If the udl does not extend to the right end then the udl may be extended upto the right end and an equal and opposite udl be added to counteract the effect due to the additional udl.
3. Superposition principle.
   This principle states that the total deflection and slope at a point on the beam due to number of loads is equal to the sum of the individual deflections and slopes produced by the various loads considering if other loads were not present.
4. Moment area method.
   (i) The slope at any point is equal to the area of the $\frac{M}{EI}$ diagram between the point and the point where the slope is zero.
   (ii) The deflection at any point is equal to the moment of the area of the $\frac{M}{EI}$ diagram between the point and the point at which slope is zero about the point under consideration.

Short Answer Type Questions

1. Define slope of a beam.
2. What is a beam?
3. What is Macaulay's method of double integeation?
4. Macaulay's method is suitable for which type of beam loading.
5. How a couple C applied at a distance 'a' from left end of the beam is represented?
6. What is the formula for maximum deflection of a simply supported beam carrying concentrated load at mispan?
7. What is the formula for maximum deflection of a simply supported beam carrying udl over the whole span?
8. What is the maximum deflection of a contilever carrying end load?
9. What is the principle of super position?
10. Explain the moment area method.

EXERCISES

(a) Solve by double integration method.

1. A steel beam 12 m long is simply supported at the ends. It carries concentrated loads of 50 kN and 80 kN at 4 m and 8 m respectively from the left end support. Calculate (a) the deflection at the centre, and (b) maximum deflection.
   Take $E = 200$ GPa, $I = 120 \times 10^{-4}$ m^4.
2. A beam of span 10 m is simply supported at the ends. It carries a uniformly distributed load of 6 kN/m for 5 m from left end and a concentrated load at 7 m from left end of magnitude 20 kN. Calculate the deflection of mid span and maximum deflection.
   Take $E = 210$ GPa and $I = 80 \times 10^{-6}$ m^4.
3. A beam AB of 6 m span and $EI = 8 \times 10^4$ kN.m^2 is subjected to a clockwise couple of 500 kN.m at a distance of 4 m from left end support. Calculate (a) deflection at midspan and (b) maximum deflection.
4. A beam ABC 12 m long is supported at A and B and overhangs by 2 m from B. It carries a concentrated load 100 kN in the middle of AB and udl of intensity 20 kN/m over the entire length BC of overhang. Determine (a) deflection in the middle of AB, and (b) deflection and slope at C.
   Take $E = 200$ GPa, and $I = 10 \times 10^{-6}$ m^4.

(b) Solve by Moment area method.

5. A cantilever beam 4 m long carries a concentrated load 20 kN at free end and udl of 10 kNm over the whole span. If $E = 200$ GPa and $I = 8 \times 10^{-6}$ m^4, calculate the slope and deflection at free end.
6. A simply supported beam of timber 150 mm x 200 mm is 5 m long. It carries a udl of intensity 5 kN/m over the whole span. If E = 10 GPa, calculate the slope at the ends and deflection in the middle.
7. A simply supported beam of 6 m span carries a udl of 20 kN/m on the whole span and a concentrated load of 20 kN at the midspan. Calculate the slope at the ends and maximum deflection.
   Take $E = 200$ GPa and $I = 5 \times 10^3$ m^4.
8. A beam 5 m long and simply supported at the ends carries two concentrated loads 40 kN and 60 kN at 1 m and 3 m from the left end support respectively. It also carries a udl of 15 kN/m on the 2 m length from the right end support. Calculate the central deflection of the beam.
   Take $E = 210$ GPa, $I = 12 \times 10^{-6}$ m^4.

(c) Solve by Macaulay's singularity function method.

9. Calculate the slope at the ends and deflection in the beam loaded as shown in Fig. 5.56
   Take $EI = 3 \times 10^3$ kN.m^2.

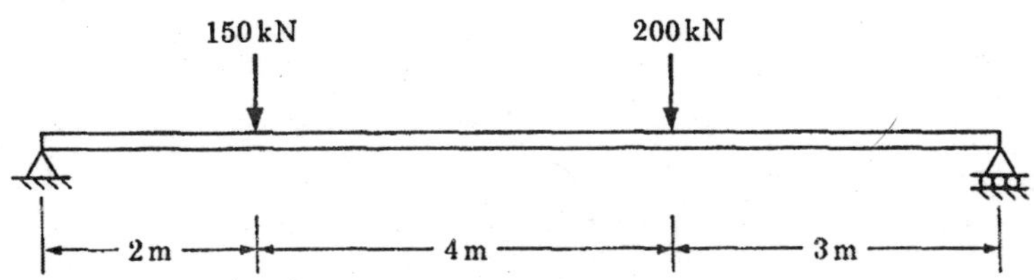

Fig. 5.56

7.10 Calculate the slope at the supports and deflection under the load for the beam shown in Fig. 5.57 $E = 200$ GPa, $I = 15 \times 10^{-6}$ m^4.

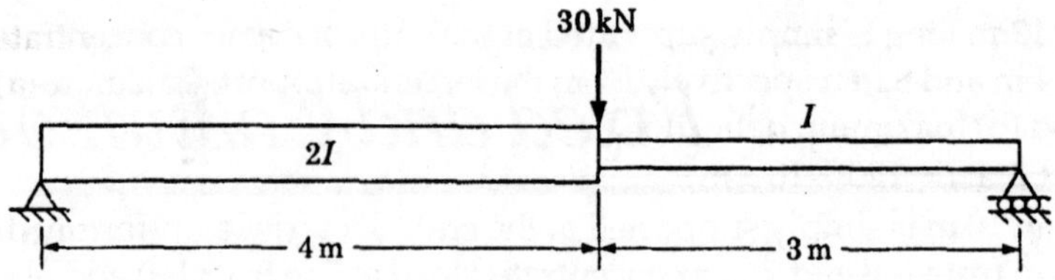

Fig. 5.57

11. Calculate the deflection at the free end of the cantilever beam shown in Fig. 5.58 $E = 200$ GPa, $I = 8 \times 10^{-4}$ m^4.

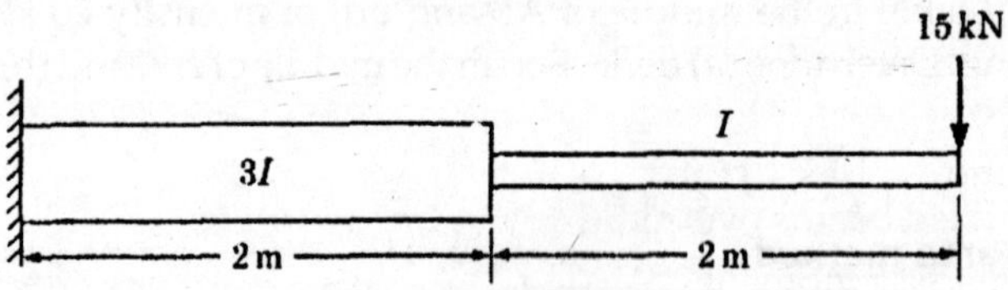

Fig. 5.58

12. Calculate the mid-span deflection fro the beam shown in Fig. 5.59. $E = 210$ GPa, $I = 300 \times 10^{-6}$ m^4.

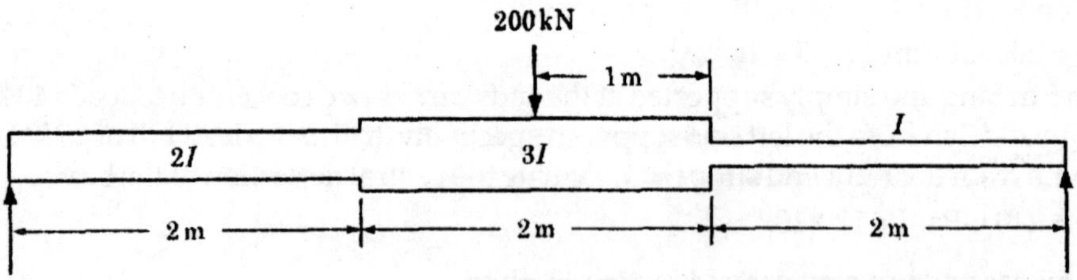

Fig. 5.59

Chapter-6

Fixed and Continuous Beams

6.1 INTRODUCTION

A fixed beam has both of its ends rigidly fixed so that slope at the ends is zero. Such a beam is also called built in or encastre beam. The fixed ends give rise to fixing moments at the ends in addiction to the reactions.

The advantages of fixed beams over simply supported beams are:

1. The maximum bending moment is smaller for the same loads applied.
2. The deflections are also smaller for the same loads applied.
3. They are more stronger and stiffer.

The disadvantages of fixed beams over simply supported beams are:

1. High accurary required in aligning the supported and fixing the ends during erection increases the cost.
2. Small sinking of either support and temperature changes can set up large stresses.
3. The end fixings are normally sensitive to vibrations and fluctuations in bending moments.

We shall use the Macaulay's method and moment area method to determine the deflection of fixed beams.

6.2 DOUBLE INTEGRATION METHOD

6.2.1 Fixed Beam carrying a Central Concentrated Load

Consider a fixed beam *AB* carrying central concentrated load *W* shown in Fig. 6.1 (*a*). Let M_A and M_B be the fixing moments at *A* and *B* respectively. Due to symmetry of loading $M_A = M_B$. The fixing bending moment diagram is shown in Fig. 6.1(b). The free bending moment diagram is shown in Fig.6.1 (*c*) and the resultant B.M.D. in Fig. 6.1(*d*). Equating the areas of fixing and free bending moment diagrams, we have

$$M_A l = \frac{1}{2} \times \frac{Wl}{4} \times l$$

$$M_A = \frac{Wl}{8}$$

and $$M_B = M_A = \frac{Wl}{8} \quad ...(6.1)$$

At mid span, $$M = \frac{Wl}{4} - \frac{Wl}{8} = \frac{Wl}{8} \quad ...(6.2)$$

Due to symmetry, $R_A = R_B = \frac{W}{2}$ ...(6.3)

The two points of inflextion occur at $l/4$ from the ends of beam.

To determine the deflection and slope anywhere, consider a section x-x at a distance x from A. The bending moment at this section is:

$$M_x = R_A \cdot x - W\left(x - \frac{l}{2}\right) - M_A$$

$$\frac{EId^2y}{dx^2} = -M_x$$

$$= -R_A x + W\left(x - \frac{l}{2}\right) + M_A$$

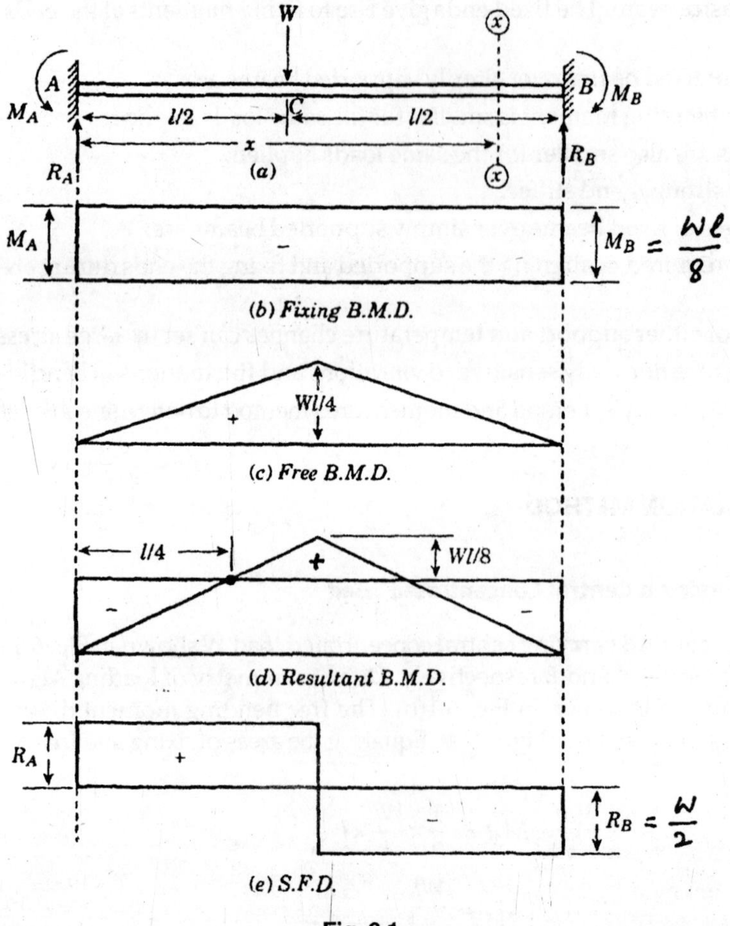

Fig. 6.1

$$EI\frac{dy}{dx} = -R_A + \frac{W\left(x-\frac{l}{2}\right)^2}{2} + M_A x + C_1$$

$$EIy = -R_A\frac{x^3}{6} + \frac{W\left(x-\frac{l}{2}\right)^3}{6} + \frac{M_A x^2}{2} + C_1 x + C_2$$

At $x = 0, y = 0$, giving $C_2 = 0$

At $x = 0, \frac{dy}{dx} = 0$, giving $C_1 = 0$

Hence
$$EI\frac{dy}{dx} = -R_A\frac{x^2}{2} + \frac{W\left(x-\frac{l}{2}\right)^2}{2} + M_A x \quad \text{... (1)}$$

$$EIy = -R_A\frac{x^3}{6} + \frac{W\left(x-\frac{l}{2}\right)^3}{6} + \frac{M_A x^2}{2} \quad \text{... (2)}$$

At $x = l, y = 0$ and $\frac{dy}{dx} = 0$

$$0 = R_A\frac{l^2}{2} + \frac{Wl^2}{8} + M_A l$$

$$0 = -R_A\frac{l^3}{6} + \frac{Wl^3}{48} + \frac{M_A l^2}{2}$$

Simplifying, we get

$$4R_A l - 8M_A = Wl$$
$$8R_A l - 24M_A = Wl$$

Solving, we get

$$M_A = \frac{Wl}{8} \quad \text{and } R_A = \frac{W}{2}$$

Substituting in Eqs. (1) and (2), we get

$$EIy = -\frac{W}{2}\times\frac{x^3}{6} + \frac{W\left(x-\frac{l}{2}\right)^3}{6} + \frac{Wlx^2}{16}$$

$$= -\frac{Wx^3}{12} + \frac{W\left(x-\frac{1}{2}\right)^3}{6} + \frac{Wlx^2}{16}$$

At $x = l/2$, $$EIy = -\frac{Wl^3}{96} + \frac{Wl^2}{64} = \frac{Wl^3}{192}$$

or $$y = \frac{Wl^3}{192EI} \quad \text{... (6.4)}$$

$$EI\frac{dy}{dx} = -\frac{Wx^2}{4} + \frac{W\left(x + \frac{l}{2}\right)^2}{2} + \frac{Wlx}{8}$$

At $x = l/2$, $$EI\frac{dy}{dx} = -\frac{Wl^2}{16} + \frac{Wl^2}{16} = 0 \quad \text{... (6.5)}$$

6.2.2Fixed Beam carrying Concentrated Load off Midspan.

Consider a fixed beam AB carrying concentrated load W at C, as shown in Fig. 6.2 (a). The fixing, free and resultant B.M. diagrams are shown in Figs. 6.2 (*b*) to (*d*). The S.F. diagram is shown in Fig. 6.2 (*e*). At a distance x from A, the bending moment is :

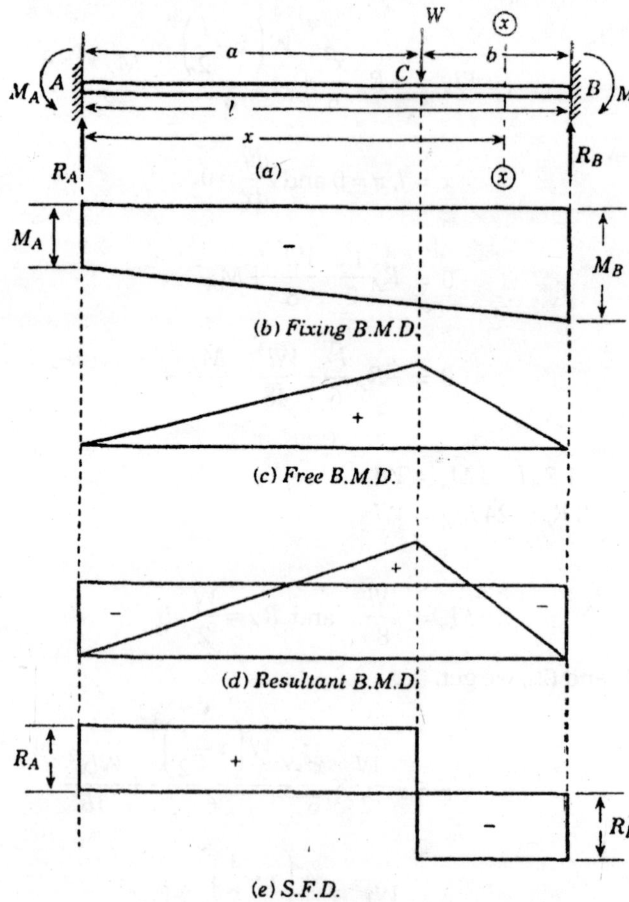

Fig. 6.2

$$M_x = R_A x - W(x-a) - M_A$$

$$EI\frac{d^2y}{dx^2} = -M_x$$

$$= -R_A x + W(x-a) + M_A$$

Integrating twice, we get

$$EI\frac{dy}{dx} = -R_A\frac{x^2}{2} + \frac{W(x-a)^2}{2} + M_A x + C_1$$

$$EIy = -R_A\frac{x^3}{6} + \frac{W(x-a)^3}{6} + M_A\frac{x^2}{2} + C_1 x + C_2$$

At $x = 0$, $y = 0$ and $\frac{dy}{dx} = 0$, which gives $C_1 = 0$ and $C_2 = 0$

At $x = l$, $y = 0$, and $\frac{dy}{dx} = 0$. Thus

$$0 = -R_A\frac{l^2}{2} + \frac{W(l-a)^2}{2} + M_A l$$

$$0 = -R_A\frac{l^3}{6} + \frac{W(l-a)^3}{6} + \frac{M_A l^2}{2}$$

Simplifying, we get

$$R_A l^2 - 2M_A l = W(l-a)^2$$

$$R_A l^3 - 3M_A l^2 = W(l-a)^3$$

Solving, we get

$$M_A = Wa\left(1-\frac{a}{l}\right)^2 = \frac{Wab^2}{l^2} \qquad \text{...(6.6)}$$

$$R_A = W\left(1-\frac{a}{l}\right)^2\left(1+\frac{2a}{l}\right) \qquad \text{...(6.7)}$$

$$M_B = M_A - R_A l + W(l-a)$$

$$= \frac{Wa^2}{l^2}(l-a)$$

$$= \frac{Wa^2 b}{l^2} \qquad \text{...(6.8)}$$

$$R_B = \frac{Wa^2(3l-2a)}{l^3} \qquad \text{...(6.9)}$$

At $$x = a,$$

$$y_c = \frac{1}{EI}\left[-W\left(1-\frac{a}{l}\right)^2\left(1+\frac{2a}{l}\right)\frac{a^3}{6}+Wa\left(1-\frac{a}{l}\right)^2\left(1+\frac{2a}{l}\right)\frac{a^2}{2}\right]$$

$$= \frac{Wa^3}{3EI}\left(1-\frac{a}{l}\right)^3 = \frac{Wa^3b^3}{3EIl^3} \quad \text{...(6.10)}$$

For y to be maximum, $\frac{dy}{dx} = 0$

$$-\frac{W}{l^3}(l-a)^2(l+2a)\frac{x^2}{2}+\frac{Wa}{l^2}(l-a)^2 x = 0$$

$$x = \frac{2al}{l+2a} \quad \text{...(6.11)}$$

$$EI\, y_{max} = -\frac{W}{6l^3}(l-a)^2(l+2a)\left(\frac{2al}{1+2a}\right)^3+\frac{Wa}{2l^2}(l-a)^2\left(\frac{2al}{1+2a}\right)^2$$

$$= \frac{2}{3}\frac{Wa^3b^2}{(3a+b)^2}$$

$$y_{max} = \frac{2}{3}\frac{Wa^3b^2}{(3a+b)^2 EI} \quad \text{...(6.12)}$$

6.2.3 Fixed Beam carrying a Couple

Consider a fixed beam AB carrying a couple M_0 at C, as shown in Fig.6.3. At a distance x from A, the bending moment is:

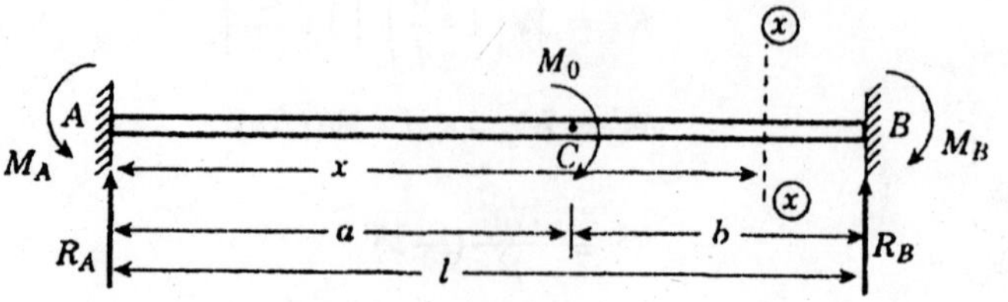

Fig. 6.3

$$M_x = R_A x + M_0 (x-a)^0 - M_A$$

$$EI\frac{d^2y}{dx^2} = -M_x = -R_A x - M_0 (x-a)^0 + M_A$$

Integrating twice, we get

$$EI\frac{dy}{dx} = -R_A\cdot\frac{x^2}{2} - M_0(x-a) + M_Ax + C_1$$

$$EIy = -R_A\cdot\frac{x^3}{6} - \frac{M_0(x-a)^2}{2} + \frac{M_Ax^2}{2} + C_1x + C_2$$

At $x = 0, y = 0$ and $\frac{dy}{dx} = 0$ give, $C_1 = 0, C_2 = 0$

At $x = l, y = 0$ and $\frac{dy}{dx} = 0$

$$0 = -R_A\frac{l^2}{2} - M_0(l-a) + M_Al$$

$$0 = -R_A\frac{l^2}{6} - M_0\frac{(l-a)^2}{2} + M_A\frac{l^2}{2}$$

Solving, we get

$$R_A = -\frac{6M_0\,ab}{l^3}$$

$$M_A = \frac{M_0b(l-3a)}{l^2} \qquad \text{...(6.13)}$$

$$EIy_c = \frac{6M_0\,ab}{l^3}\times\frac{a^3}{6} + \frac{M_0\,b(l-3a)}{l^2}\times\frac{a^2}{2} \qquad \text{...(6.14)}$$

$$= \frac{M_0\,a^4b}{l^3} + \frac{M_0\,a^2b(l-3a)}{2l^2}$$

$$y_c = \frac{M_0\,a^2b^2(l-2a)}{2EI\,l^3} \qquad \text{...(6.15)}$$

For y to be maximum, $\frac{dy}{dx} = 0$

$$\frac{6M_0\,ab}{l^3}\times\frac{x^2}{2} - M_0(x-a) + \frac{M_0b(l-3a)}{l^2}x = 0$$

$$\frac{3M_0\,ab}{l^3}x^2 - M_0(x-a) + \frac{M_0b(l-3a)}{l^2}\,x = 0$$

or $\quad 3ab\,x^2 + bl(l-3a)\,x - (x-a)\,l^3 = 0$

or $\quad 3ab\,x^2 + [bl\,(l-3a) - l^3]\,x + al^3 = 0$

Solving for x, we get

$$x = \frac{l^2}{3b} \qquad \text{...(6.16)}$$

$$y_{max} = \frac{M_0\,a(2l-3a)^2}{54\,EI\,b^2} \qquad \text{...(6.17)}$$

6.2.4 Fixed Beam carrying udl over Whole span

Consider a fixed beam *AB* carrying *udl* of intensity w per unit length as shown in Fig. 6.4 (a). Consider a section x–x at a distance x from A. The bending moment at this section is:

$$M_x = R_A x - \frac{wx^2}{2} - M_A$$

Due to symmetry, $R_A = R_B = \dfrac{wl}{2}$

$$EI\frac{d^2y}{dx^2} = -Mx = -\frac{wl}{2}x + \frac{wx^2}{2} + M_A$$

Integrating twice, we get

$$EI\frac{dy}{dx} = -\frac{wl\,x^2}{4} + \frac{wx^3}{6} + M_A x + C_1$$

$$Ely = -\frac{wlx^3}{12} + \frac{wx^4}{24} + M_A\frac{x^2}{2} + C_1x + C_2$$

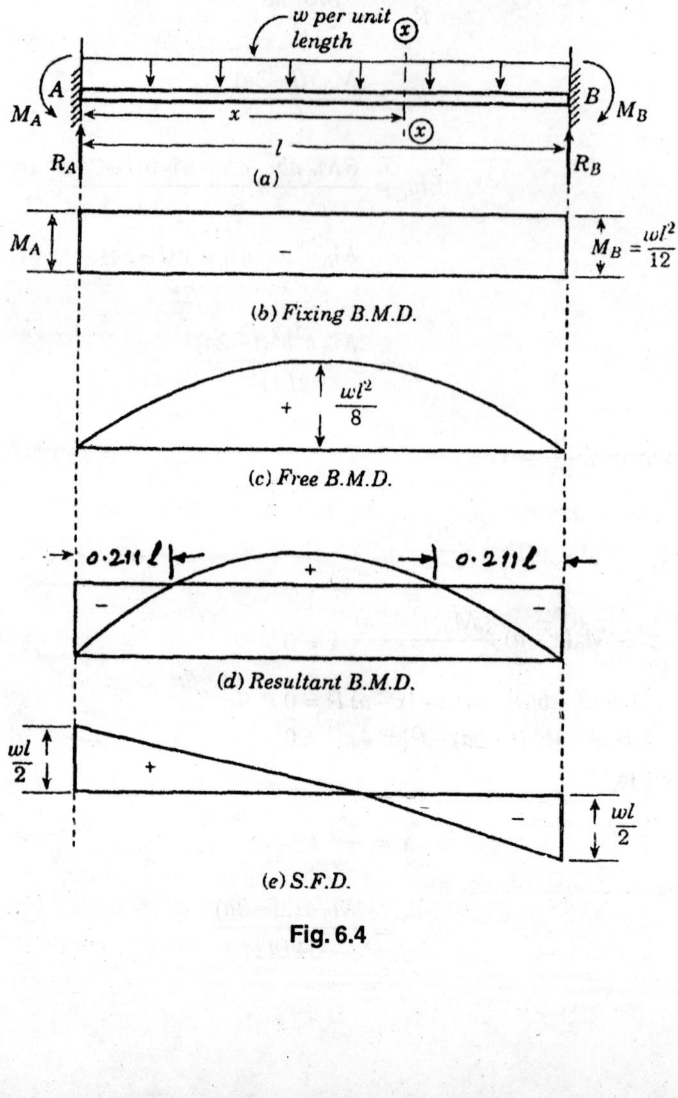

Fig. 6.4

At $x = 0, y = 0$ and $\frac{dy}{dx} = 0 = 0$, which gives

$$C_1 = 0, C_2 = 0$$

At

$$x = l, y = 0$$

$$0 = -\frac{wl^4}{12} + \frac{wl^4}{24} + M_A\frac{l^2}{2}$$

$$M_A = \frac{wl^2}{12} \quad \text{...(6.18)}$$

But

$$M_A = M_B$$

At

$$x = \frac{l}{2}, M = \frac{wl^2}{4} - \frac{wl^2}{8} - \frac{wl^2}{12} = \frac{wl^2}{24}$$

$$y_{max} = \frac{1}{EI}\left[-\frac{wl^4}{96} + \frac{wl^4}{384} + \frac{wl^4}{96}\right] \quad \text{...(6.19)}$$

To determine the points of contraflexure, $M_x = 0$

$$\frac{wlx}{2} - \frac{wx^2}{2} - \frac{wl^2}{12} = 0$$

$$6\,lx - 6x^2 - l^2 = 0$$

$$6x^2 - 6lx + l^2 = 0$$

$$x = \frac{6l \pm \sqrt{36l^2 - 24l^2}}{12} = \frac{6l \pm 2\sqrt{3}l}{12} = \frac{1}{2}l \pm \frac{\sqrt{3}}{6}l$$

$$= 0.211\,l, 0.789\,l \quad \text{...(6.20)}$$

6.2.5 Fixed Beam carrying Triangular Load

Consider a fixed beam AB carrying a triangular load as shown in Fig. 6.5. At a distance x from A, intensity of loading is: $\frac{wx}{l}$.

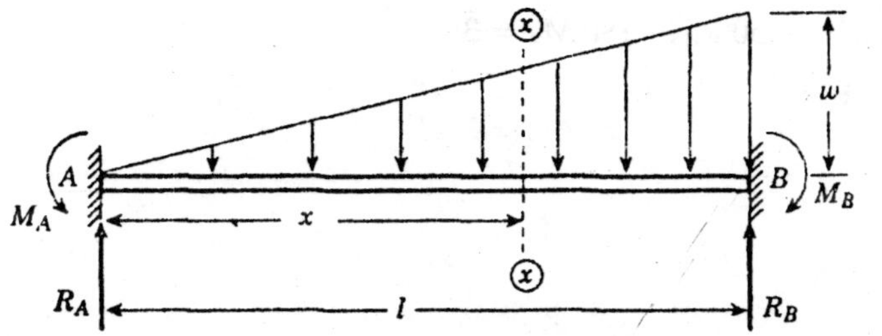

Fig. 6.5

$$M_x = R_A x - \frac{1}{2} \cdot \frac{wx}{l} \times x \times \frac{1}{3} x - M_A$$

$$= R_A x - \frac{wx^3}{6l} - M_A$$

$$EI \frac{d^2 y}{dx^2} = -M_x$$

$$= -R_A x + \frac{wx^3}{6l} + M_A$$

Integrating twice, we get

$$EI \frac{dy}{dx} = -R_A \frac{x^2}{2} + \frac{wx^4}{24l} + M_A x + C_1$$

$$EIy = -R_A \frac{x^3}{6} + \frac{wx^5}{120l} + M_A \frac{x^2}{2} + C_1 x + C_2$$

At $x = 0, y = 0, = \frac{dy}{dx} = 0$, which gives, $C_1 = 0, C_2 = 0$

At $x = l, y = 0, \frac{dy}{dx} = 0$. Hence

$$0 = -R_A \frac{l^2}{2} + \frac{wl^3}{24} + M_A l$$

$$0 = -R_A \frac{l^3}{6} + \frac{wl^4}{120} + M_A \frac{l^2}{2}$$

Simplifying, we get

$$12\, R_A l - 24\, M_A = wl^2 \quad \text{... (1)}$$

$$20\, R_A l - 60\, M_A = wl^2 \quad \text{... (2)}$$

Multiplying Eq. (1) by 5 and (2) by 3, we get

$$60\, R_A l - 120\, M_A = 5\, wl^2$$

$$60\, R_A l - 180\, M_A = 3\, wl^2$$

Subtracting, we get

$$60\, M_A = 2\, wl^2$$

$$M_A = \frac{wl^2}{30} \quad \text{...(6.21)}$$

$$12\, R_A l - 24 \times \frac{wl^2}{30} = wl^2$$

$$R_A = \frac{3}{20}wl, R_B = \frac{17}{20}wl \qquad ...(6.22)$$

$$M_B = \frac{2}{30}wl^2 - \frac{wl^2}{6} - \frac{wl^2}{30} = \frac{wl^2}{20} \qquad ...(6.23)$$

At $$x = \frac{l}{2}, y_C = wl^4/(768\text{ EI}) \qquad ...(6.24)$$

6.3 MOMENT AREA METHOD FOR FIXED BEAMS

Consider a fixed beam *AB* as shown in Fig. 6.6 (a). The fixing, free and resultant B.M. diagrams are shown in Figs. 6.6 (*b*) to (*d*). For a beam of uniform cross-section, the changes in slope is given by:

$$\theta = \frac{A}{EI}$$

where A = area of bending moment diagram.

For a fixed beam, $\theta = 0$ for the whole span. Since the bending stiffness *EI* is not zero, it follows that the area A of the resultant bending moment diagram must be equal to zero.

According to Mohr's theorem,

$$EA(\theta_A - \theta_B) = \Sigma A = A_1 + A_2$$

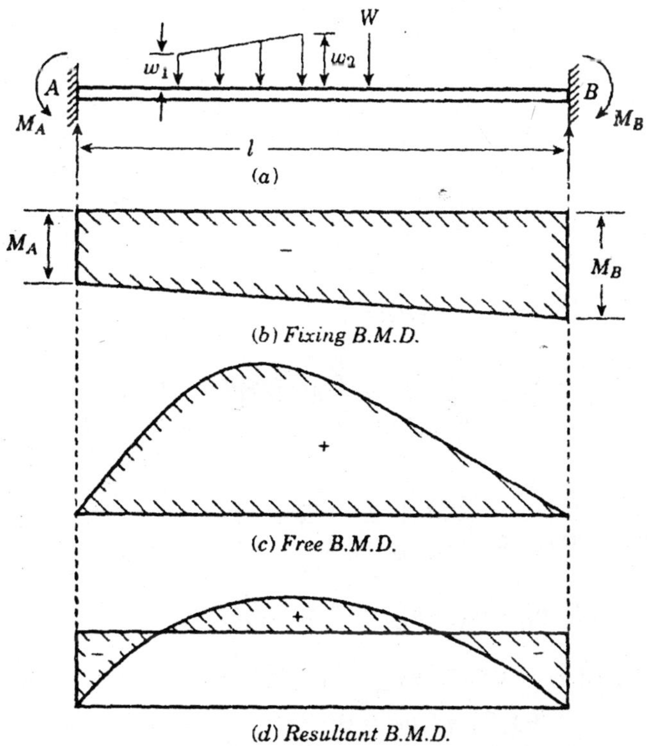

Fig. 6.6

where A_1 = area of fixing B.M. diagram

A_2 = area of free B.M. diagram

Also $EI\,(y_B - y_A) = A\bar{x} = A_1\bar{x}_1 + A_2\bar{x}_2$

where $\bar{x}_1, \bar{x}_2$ = distances of C.G. of areas A_1 and A_2 from end A respectively.

For a fixed beam, $\theta_A - \theta_B = 0$

and $y_B - y_A = 0$

Hence $A_1 + A_2 = 0$

and $A_1\bar{x}_1 + A_2\bar{x}_2 = 0$

or $$M_x = -EI\frac{d^2y}{dx^2} = (M_x)_F + (M_x)_f$$

where $(M_x)_F$ = Fixing B.M. at a distance x from end A

$(M_x)_f$ = Free B.M. at a distance x from end A

Integrating over the whole span, we get

$$\int_0^l -EI\frac{d^2y}{dx^2} = \int_0^l (M_x)_F\,dx + \int_0^l (M_x)_f\,dx$$

$$\left|-EI\frac{dy}{dx}\right|_0^l = EI(\theta_A - \theta_B)$$

$$\int_0^l (M_x)_F\,dx = A_1 \quad \text{and} \quad \int_0^l (M_x)_f\,dx = A_2$$

$\therefore$ $$EI\,(\theta_A - \theta_B) = A_1 + A_2$$

But $\theta_A - \theta_B = 0$

$\therefore$ $A_1 + A_2 = 0$

Also $$\int_0^l -EI\frac{d^2y}{dx^2}x\,dx = \int_0^l (M_x)_F\,x\,dx + \int_0^l (M_x)_f\,x\,dx$$

$$-EI\left[x.\left|\frac{dy}{dx}\right|_A^B - \int_0^l 1\cdot\frac{dy}{dx}dx\right]_0^l = A_1\bar{x}_1 + A_2\bar{x}_2$$

$$-EI\,[x(\theta_B - \theta_A) - y]_0^l = A_1\bar{x}_1 + A_2\bar{x}_2$$

$$EI\,[l\theta_A - y_A) - (l\theta_B - y_B)] = A_1\bar{x}_1 + A_2\bar{x}_2$$

But $\theta_A, \theta_B, y_A, y_B = 0$

$\therefore$ $A_1\bar{x}_1 + A_2\bar{x}_2 = 0$

Now

$$A_1 = \frac{1}{2}(M_A + M_B)l$$

$$A_1\bar{x}_1 = \frac{1}{2}M_A l \times \frac{l}{3} \times \frac{l}{2} M_B l \times \frac{2}{3} l$$

$$= \frac{l^2}{6}[M_A + 2M_B] \quad \text{...(6.25)}$$

6.3.1 Fixed Beam carrying Central Concentrated Load

Consider a fixed beam *AB* carrying concentrated load *W at* mid span C, as shown in Fig. 6.7. Due to symmetry of loading,

$$M_A = M_B$$

Now

$$A_1 = -M_A l$$

$$A_2 = \frac{1}{2} \times \frac{Wl}{4} \times l = \frac{Wl^2}{8}$$

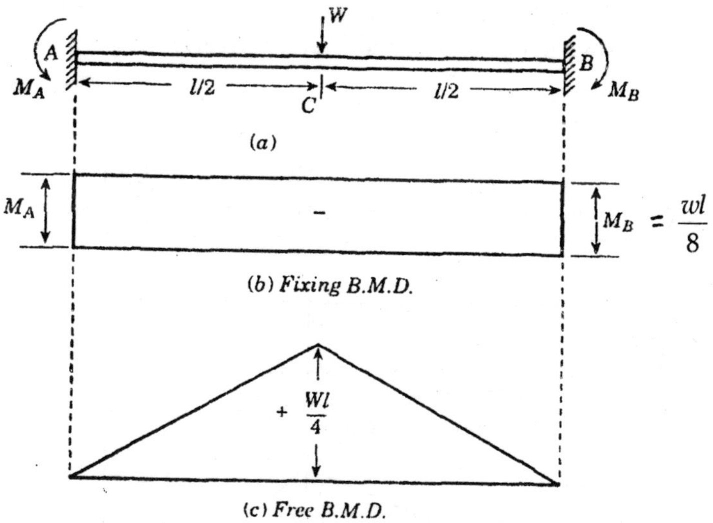

Fig. 6.7

For $A_1 + A_2 = 0$

$$-M_A l + \frac{Wl^2}{8} = 0$$

$$M_A = \frac{Wl}{8}$$

Also
$$R_A = R_B = \frac{W}{2}$$

Deflection at mid span,
$$\delta = \frac{A\bar{x}}{EI} = \frac{1}{EI}(A_1\bar{x}_1 + A_2\bar{x}_2)$$

Considering half the beam.
$$\delta_C = \frac{1}{EI}\left[-\frac{Wl}{8}\times\frac{l}{2}\times\frac{l}{4}+\frac{1}{2}\times\frac{Wl}{4}\times\frac{l}{2}\times\frac{2}{3}\times\frac{l}{2}\right]$$
$$= \frac{1}{EI}\left[-\frac{Wl^3}{64}+\frac{Wl^3}{48}\right]$$
$$= \frac{Wl^3}{192\,EI} \qquad \text{...(6.26)}$$

6.3.2 Fixed Beam carrying Concentrated Load off Mid span.

Consider a fixed beam *AB* carrying concentrated load *W* at C as shown in Fig. 6.8 (*a*). The fixing and free B.M. diagrams are shown in Figs. 6.8 (*b*) and (*c*) respectively.

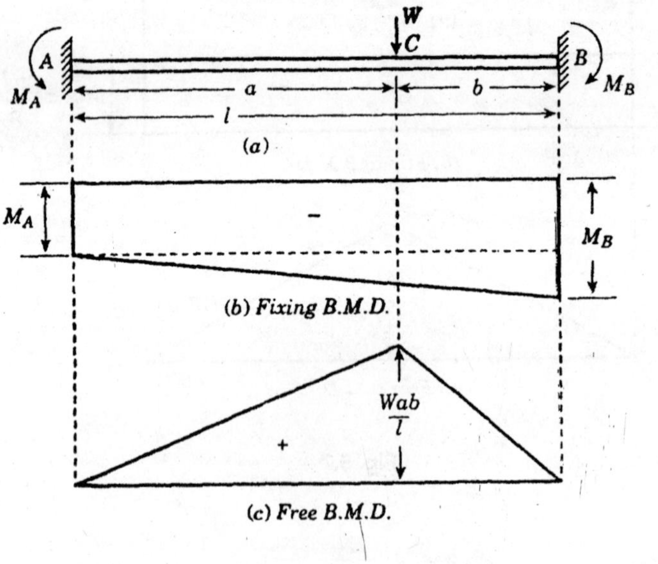

Fig. 6.8

$$A_1 = -\frac{1}{2}(M_A + M_B)l$$

$$A_2 = \frac{1}{2}\times\frac{W\,ab}{l}\times l = \frac{W\,ab}{2}$$

Now $A_1 + A_2 = 0$

$$\therefore \quad M_A + M_B = \frac{Wab}{l} \qquad \text{... (1)}$$

Also

$$A_1\bar{x}_1 + A_2\bar{x}_2 = 0$$

$$A_1\bar{x}_1 = -\left[M_A l\times\frac{l}{2}+\frac{1}{2}(M_B - M_A)l\times\frac{2}{3}l\right]$$

$$= -\left[\frac{1}{6}M_A l^2 + \frac{1}{3}M_B l^2\right]$$

$$A_2\bar{x}_2 = \frac{1}{2}\times\frac{W\,ab}{l}\times a\times\frac{2}{3}a+\frac{1}{2}\times\frac{W\,ab}{l}\times b\times\left(a+\frac{b}{3}\right)$$

$$= \frac{1}{6}W\,ab(2a+b)$$

$$-\left[\frac{1}{6}M_A l^2 + \frac{1}{3}M_B\,l^2\right]+\frac{1}{6}W\,ab(2a+b) = 0$$

$$M_A + 2M_B = W\,ab\,\frac{(2a+b)}{l^2}$$

Solving Eqs. (1) and (2), we get

$$M_A = \frac{W\,ab^2}{l^2} \qquad \text{...(6.27a)}$$

$$M_B = \frac{Wa^2b}{l^2} \qquad \text{...(6.27b)}$$

At C,

$$(M_C)_F = M_A + \frac{(M_B - M_A)a}{l}$$

$$A_1\bar{x}_1 = M_A\times a\times\frac{a}{2}+\frac{1}{2}(M_B - M_A)\frac{a}{l}\times a\times\frac{a}{3}$$

$$= \frac{1}{2}M_A a^2 + \frac{1}{6}(M_B - M_A)\frac{a^3}{l}$$

$$= \frac{Wa^3b^2}{6l^3}(2a+3b)+\frac{W\,a^5b}{6l^3}$$

$$= \frac{W}{6l^3}[2a^4b^2 + 3a^3b^3 + a^5b]$$

$$A_2\bar{x}_2 = \frac{1}{2} \times \frac{Wab}{l} \times a \times \frac{a}{3} = \frac{Wa^3b}{6l}$$

$$= \frac{W}{6l^3}(a^5b + a^3b^3 + 2a^4b^2)$$

$$A\bar{x} = A_1\bar{x}_1 - A_2\bar{x}_2 = \frac{W}{6l^3} \times 2a^3b^3 = \frac{W a^3 b^3}{3l^3}$$

$$\delta_C = \frac{A\bar{x}}{EI} = \frac{W a^3 b^3}{3EIl^3} \quad ...(6.28)$$

6.3.3 FIXED BEAM CARRYING UDL OVER WHOLE SPAN

Consider a fixed beam AB carrying *udl* of intensity w per unit length as shown in Fig. 6.9. Due to symmetry of loading,

$$M_A = M_B$$

$$A_1 = -M_A l$$

$$A_2 = \frac{2}{3} \times \frac{wl^2}{8} \times l = \frac{wl^2}{12}$$

Now $\quad A_1 + A_2 = 0$

$\therefore \quad -M_A + \frac{wl^3}{12} = 0$

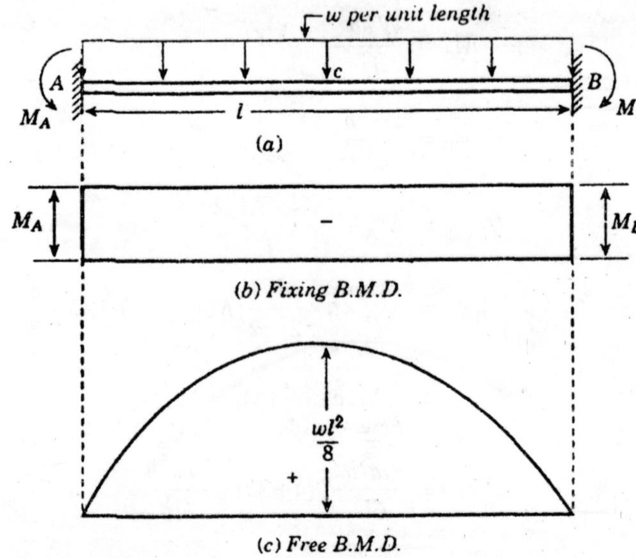

Fig. 6.9

$$M_A = \frac{wl^2}{12} \quad ...(6.29)$$

Considering half the beam,

At $x = \frac{l}{2}$, the deflection is:

$$\delta_C = \frac{A\bar{x}}{EI}$$

$$= \frac{1}{EI}\left[\frac{wl^2}{12}\times\frac{l}{2}\times\frac{l}{4}-\frac{2}{3}\times\frac{wl^2}{8}\times\frac{l}{2}\times\left(\frac{3}{8}\times\frac{l}{2}\right)\right]$$

$$= \frac{1}{EI}\left[\frac{wl^4}{96}-\frac{wl^4}{128}\right]$$

$$= \frac{wl^4}{384\,EI} \quad ...(6.30)$$

Example 6.1 *A built in beam of span 12 m carries a concentrated load of 40 kN acting at 5 m from the right hand support. Calculate the deflection at the mid span. E = 210 GPa, I = 1600 cm⁴.*

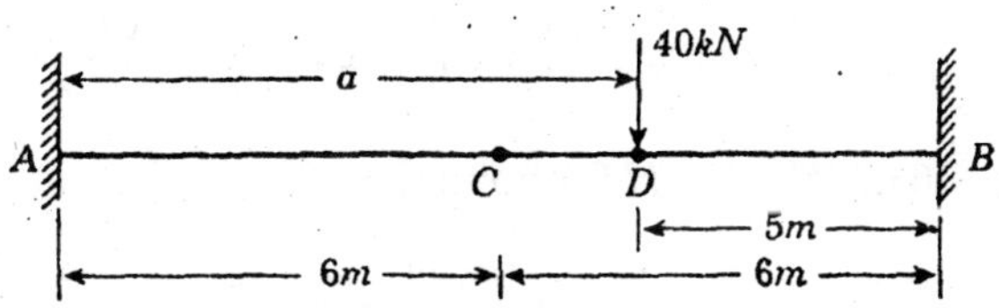

Fig. 6.10

Solution:

From Fig. 6.10, we have

$$Ely = \frac{W}{l^3}(l-a)^2(l+2a)\frac{x^3}{6}+\frac{W(x-a)^3}{6}+\frac{Wa}{l^2}(l-a)^2\frac{x^2}{2}$$

Here $a = 7\text{ m}, l = 12\text{m}, x = 6\text{m}, W = 40\text{ kN}$

$$y_c = \frac{8.16666\times40\times10^3}{210\times10^9\times1600\times10^{-8}} = 97.2\,\text{mm}$$

Example 6.2 *A built-in beam is loaded as shown in Fig. 6.11 Determine the deflection at the mid-span and under the concentrated load, I = 3000 cm⁴, E = 200 GPa.*

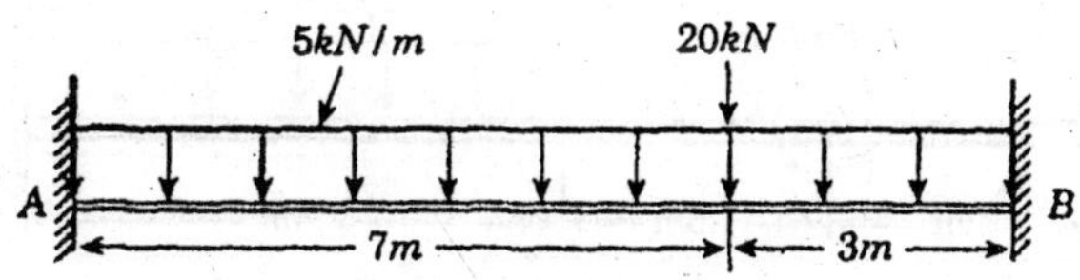

Fig. 6.11

Solution.

For Fig. 6. 11, we have for uniformly distributed load,

$$y = \frac{w}{EI}\left[-\frac{lx^3}{12}+\frac{l^2x^2}{24}+\frac{x^4}{24}\right]$$

At $$x = 5\text{m}, y_c = \frac{wl^4}{384\,EI}$$

Here $$w = 5\text{ kN/m}, l = 10\text{ m}, E = 200\text{ GPa}, I = 3000\text{ cm}^4.$$

$$y_c = 21.7\text{ mm}$$

At $$x = 7\text{m}$$

$$Ely_D = 5\times10^3\left[-\frac{10\times7^3}{12}+\frac{100\times49}{24}+\frac{7^7}{24}\right]$$

$$y_D = \frac{5\times10^3\times18.3767}{200\times10^9\times3000\times10^8} = 15.315\text{ mm}$$

For concentrated load :

$$Ely = -\frac{W}{l^3}(l-a)^2(l+2a)\frac{x^3}{6}+\frac{W(x-a)^3}{6}+\frac{Wa(l-a)^2x^2}{2l^2}$$

Here $x = 5$ m, $a = 7$ m, $l = 10$ m, W = 20 kN for point C.

$$= W\left[-\frac{9\times24\times125}{6\times1000}-\frac{8}{6}+\frac{7\times9\times25}{2\times100}\right]$$

$$y_C = \frac{20\times10^3\times2.042}{200\times10^6\times3000\times10^{-8}} = 6.81\text{mm}$$

$$y_D = \frac{Wa^3(l-a)^3}{3\,El\,l^3} = \frac{20\times10^3\times7^3\times27}{3\times200\times10^9\times3000\times10^{-8}} = 10.29\text{mm}$$

Total deflections :

$$y_c = 21.7 + 6.81 = 28.51\text{ mm}$$

$$y_D = 15.315 + 10.29 = 25.605\text{ mm}$$

Example 6.3 A *Fixed beam is loaded as shown in Fig. 6.12. Determine the maximum deflection, end reactions and fixing moments. E = 200 GPa, I = 3600 cm⁴.*

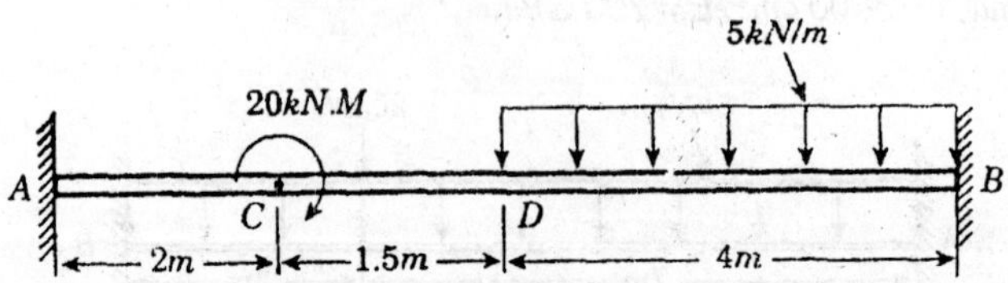

Fig. 6.12

Solution:

$$M_x = R_A \times x + 20\,(x-2)^0 - 2.5\,(x-3.5)^2 - M_A$$

$$EI\frac{d^2y}{dx^2} = -M_x$$

$$= -R_A \times x - 20(x-2)^0 + 2.5\,(x-3.5)^2 + M_A$$

$$EI\frac{dy}{dx} = -R_A\frac{x^2}{2} - 20(x-2) + \frac{2.5}{3}(x-3.5)^2 + M_A x + C_1$$

$$Ely = -R_A\frac{x^3}{6} - 10(x-2)^2 + \frac{2.5}{12}(x-35)^4 + M_A\frac{x^2}{2} + C_1 x + C_2$$

At $x = 0, y = 0$ gives $C_2 = 0$

At $x = 0, \dfrac{dy}{dx} = 0$ gives $C_1 = 0$

$$Ely = -R_A\frac{x^3}{6} - 10(x-2)^2 + \frac{2.5}{12}(x-35)^4 + M_A\frac{x^2}{2}$$

At $x = 7.5$ m, y = 0 and $\dfrac{dy}{dx} = 0$

$$0 = -70.3125\,R_A - 302.5 + 53.33 + 28.125\,M_A$$

$$70.3125\,R_A - 28.125\,M_A = -249.17 \qquad \text{...(1)}$$

$$0 = -28.125\,R_A - 110 + 53.33 + 7.5\,M_A$$

$$28125\,R_A - 7.5\,M_A = -56.67 \qquad \text{...(2)}$$

From Eqs. (1) and (2), we get

$$R_A = 1.0424 \text{ kN}, M_A = 11.465 \text{ kN.m}$$

$$7.5\,R_B = 5 \times 4 \times 5 \times 5 + 20 + 23.647 = 11.465 = 142.182$$

$$R_B = 18.9576 \text{ kN}$$

$$M_B = M_A - R_A \times l - 20 + 5 \times 4 \times 2$$

$$= 11.465 - 1.0424 \times 7.5 - 20 + 40$$

$$= 23.647 \text{ kN.m}$$

For y to be maximum, $\dfrac{dy}{dx} = 0$.

$$EI\,\frac{dy}{dx} = -0.5211\,x^2 - 20\,(x-2) + \frac{2.5}{3}\,(x-3.5)^3 + 11.465\,x = 0$$

$$-0.5211\,x^2 - 20x^2 + 40 + \frac{2.5}{3}\,(x^3 - 3.5^3 + 10.5x^2 - 3 \times 3.5^2\,x) + 11465x = 0$$

$$2.5\,x^3 + 24.6867\,x^2 - 117.48 \times + 12.8125 = 0$$

$$x^3 + 9.87468 - 46.992\,x + 5.125 = 0$$

$$x = 3.4 \text{ m}$$

$$y = 5.53 \text{ mm.}$$

Example 6.4 A *Fixed* bean of 12 m span *carries two point loads of* 20 *kN at* 3 *m from each end. Obtain the values of support moments and the central deflection. I* = *8520 cm*^{4}, E = 200 GPa. *Use moment area method.*

Solution:

The loaded beam is shown in Fig. 6.13.

Due to symmetry of loading, $M_A = M_B$

$$A_1 = -M_A l$$

$$A_2 = \frac{1}{2} \times 60 \times 3 \times 2 + 60 \times 6 = 540$$

$$-12\, M_A + 540 = 0$$

$$M_A = 45 \text{ kN.m}$$

$$R_A = R_B = 20 \text{ kN}$$

$$M_x = 20x - 45 - 20\,(x-3) - 20\,(x-9)$$

$$= 195 - 20\,x = 0$$

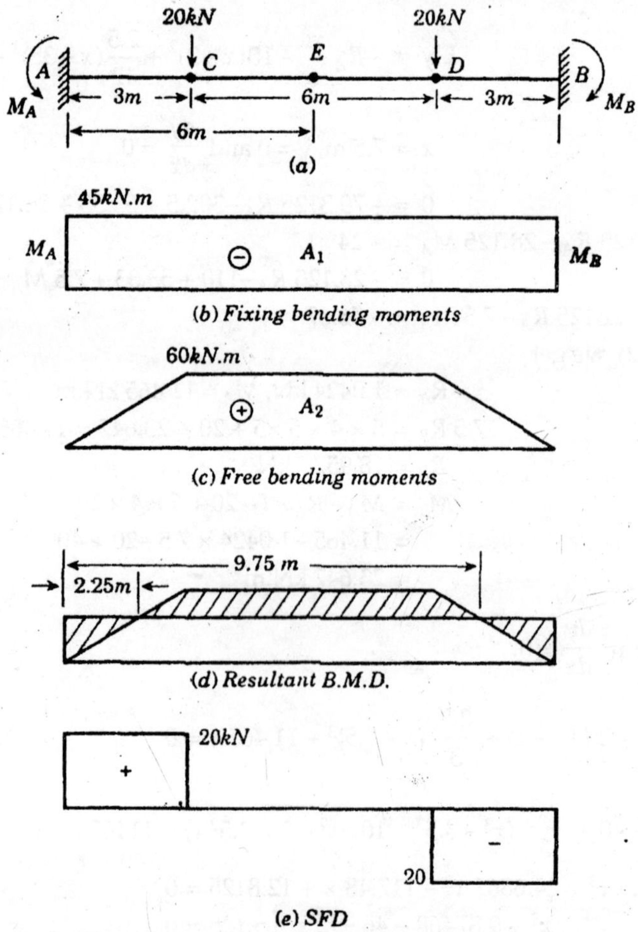

Fig. 6.13

$$x = 9.75 \text{ m}$$

and $$M_x = 20x - 45 = 0$$

$$x = 2.25 \text{ m}$$

$$y_E = \frac{A\bar{x}}{EI} = \frac{1}{EI}(A_1\bar{x}_1 + A_2\bar{x}_2)$$

Considering half the beam,

$$y_E = \frac{1}{EA}\left[-45\times6\times3-\left\{\frac{1}{2}\times60\times3\times\frac{2}{3}\times3+60\times3\times4.5\right\}\right]$$

$$= \frac{180\times10^3}{200\times10^9\times8520\times10^{-8}} = 10.56\text{mm}$$

Example 6.5 *A beam of length 10 m is fixed at both the ends and carries a uniformly distributed load of intensity 10 kNIm. Calculate the fixing moments and the central deflection. EI = 25 × 10^{-6} N-m^2. Use moment area method.*

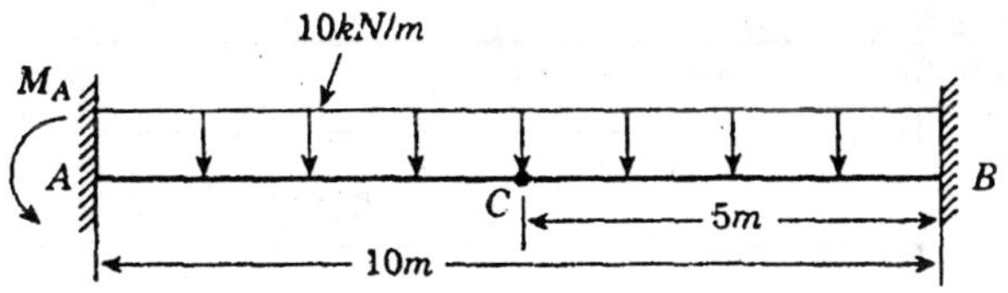

Fig. 6.14

Solution:

As shown in Fig. 6.14

$$M_A = \frac{wl^2}{12} = \frac{10\times100}{12} = 83.33\text{kN.m}$$

Central deflection, $$\delta_C = \frac{wl^4}{384EI} = \frac{10\times10^3\times10^4}{384\times25\times10^6} = 10.417 \text{ mm}$$

Example 6.6 A beam *of span* 8 *m has its ends fixed and carries a uniformly distributed load of* 15 *kN/m for the left hand half portion of the beam. Calculate the fixing moments and the support reactions. Also draw the B.M. and S.F. diagrams.*

Solution:

From Fig. 6.15, we have

$$A_1 = -\left(\frac{M_A+M_B}{2}\right)l$$

$$= -4(M_A+M_B)$$

$$R_A \times 8 = 15 \times 4 \times 6$$

$$R_A = 45 \text{ kN}$$

AC: $$M_x = 45\,x - 7.5\,x^2$$

$$M_c = 60 \text{ kN.m}$$

CB:
$$M_x = 45x - 15 \times 4\,(x-2)$$
$$= 120 - 15\,x$$

$$A_2 = \frac{2}{3}\times + 60\times 4 + \frac{1}{2}\times 60\times 4 = 280$$

$$A_1 + A_2 = 0$$
$$4(M_A + M_B) = 280$$
$$M_A + M_B = 70 \qquad \text{... (1)}$$
$$A_1\bar{x}_1 + A_2\bar{x}_2 = 0$$

$$-\left[M_B \times l \times \frac{l}{2} + \frac{1}{2}(M_A - M_B)\times l \times \frac{l}{3}\right]$$

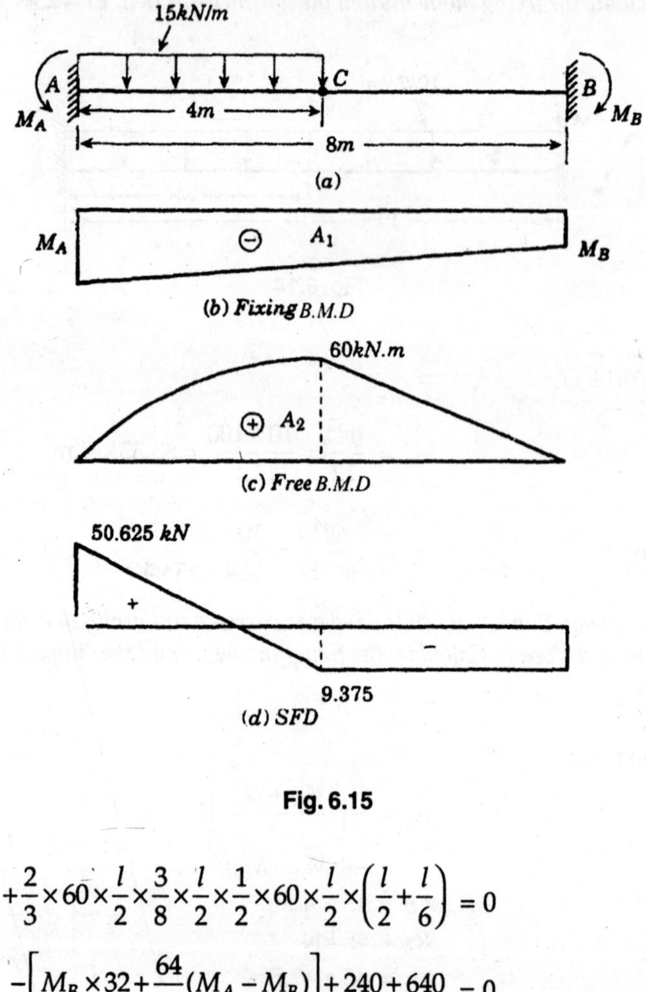

Fig. 6.15

$$+\frac{2}{3}\times 60\times\frac{l}{2}\times\frac{3}{8}\times\frac{l}{2}\times\frac{1}{2}\times 60\times\frac{l}{2}\times\left(\frac{l}{2}+\frac{l}{6}\right) = 0$$

$$-\left[M_B \times 32 + \frac{64}{6}(M_A - M_B)\right] + 240 + 640 = 0$$

$$-M_B - \frac{1}{3}(M_A - M_B) + 7.5 + 20 = 0$$

$$M_A + 2M_B = 82.5 \quad \text{...(2)}$$

Solving Eqs. (1) and (2), we get

$$M_A = 57.5 \text{ kN.m}, M_B = 12.5 \text{ kN.m}$$

$$8R_A - M_A + M_B - 15 \times 4 \times 6 = 0$$

$$8R_A - 57.5 + 12.5 - 360 = 0$$

$$R_A = 50.625 \text{ kN}, R_B = 9.375 \text{ kN}$$

Example 6.7 *A cantilever of effective length L carries a total load of wL uniformly distributed throughout the length. If the cantilever is propped at point L/4 from the free end and the level of the prop adjusted. so that there is no deflection at the free end. What is the reaction at the prop and the deflection of the beam at that point ? Sketch the S.F. and B.M. diagrams for the cantilever.*

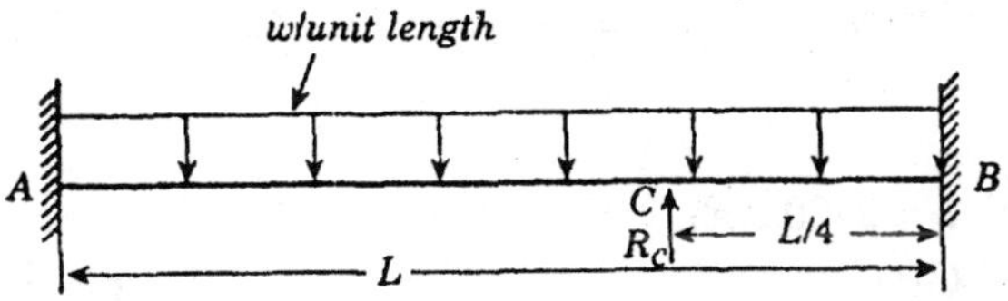

Fig. 6.16

Solution:

For Fig. 6.16, we have for *udl*,

$$\delta_B = \frac{wL^4}{8EI}$$

For prop,

$$\delta_B = \frac{R_C}{EI}\left[\frac{(l-a)^2(2l+a)}{6}\right]$$

Here

$$a = L/4$$

$$\delta_B = \frac{27R_CL^3}{128EI}$$

Now

$$\frac{27R_CL^3}{128EI} = \frac{wL^4}{8EI}$$

$$R_C = \frac{16}{27}wL$$

For *udl*,

$$Ely = \left(\frac{x^4}{24} - \frac{l^3x}{6} + \frac{l^4}{8}\right)$$

At

$$x = L/4$$

$$\delta_C = \frac{513 wL^4}{6144 EI}$$

For prop,
$$\delta_C = \frac{R_C (l-a)^3}{3EI} = \frac{16}{27} wL \times \frac{27L^3}{64 \times 3EI} = \frac{wL^4}{12 EI}$$

Net
$$\delta_C = \left(\frac{513}{6144} - \frac{1}{2}\right)\frac{wL^4}{EI} = \frac{wL^4}{6144 EI}$$

Example 6.8 *A beam AB of 6 m span is fixed at both ends and carries a load of 30 kN at C, 2m from A. For portion AC, I = 1200 cm⁴ and for portion BC, I = 2400 cm⁴. Find the fixed end moments and central deflection. Take E = 210 GPa.*

Solution:

Given: $I_1 = 1200 \text{ cm}^4, I_2 = 2400 \text{ cm}^4,$

$E = 210$ GPa

Refer to Fig. 6.17

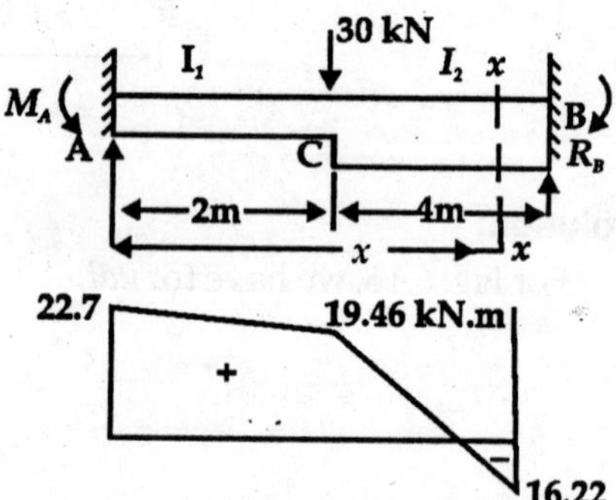

Fig. 6.17

$$M_x = R_A x - 30(x-2) - M_A$$

$$EI\frac{d^2y}{dx^2} = -M_x = -R_A x + 30(x-2) + M_A$$

$$EI\frac{dy}{dx} = -R_A\frac{x^2}{2} + 15(x-2)^2 + M_A x + C_1$$

$$EIy = -R_A\frac{x^3}{6} + 5(x-2)^3 + M_A\frac{x^2}{2} + C_1 x + C_2$$

For $0 \le x \le 2$m

$$\frac{dy}{dx} = \frac{1}{EI_1}\left[-R_A\frac{x^2}{2} + M_A x + C_1\right]$$

$$y = \frac{1}{EI_1}\left[-R_A\frac{x^3}{6} + M_A\frac{x^2}{2} + C_1 x + C_1\right]$$

At $x = 0, y = 0$ gives $C_2 = 0$

At $x = 0, \frac{dy}{dx} = 0$ gives $C_1 = 0$

At $x = 2$m

$$\frac{dy}{dx} = \frac{1}{EI_1}[-2R_A + 2M_A] \quad \text{... (1)}$$

$$y = \frac{1}{EI_1}\left[-\frac{4}{3}R_A + 2M_A\right] \quad \text{... (2)}$$

For $2\text{m} \le x \le 6\text{m}$,

$$\frac{dy}{dx} = \frac{1}{EI_2}\left[-R_A\frac{x^2}{2} + 15(x-2)^2 + M_A x + C_3\right]$$

$$y = \frac{1}{EI_2}\left[-R_A\frac{x^3}{6} + 5(x-2)^3 + M_A\frac{x^2}{2} + C_3 x + C_4\right]$$

At $x = 6\text{m}, y = 0$ and $\frac{dy}{dx} = 0$

$$0 = -18\,R_A + 240 + 6\,M_A + C_3$$

$$C_3 = 18\,R_A - 6\,M_A - 240 \quad \text{... (3)}$$

$$0 = -36\,R_A + 320 + 18\,M_A + 6\,C_3 + C_4$$

$$= -36\,R_A + 320 + 18\,M_A + 108\,R_A - 36\,M_A + 108\,\text{RA} - 36\,M_A - 1440 + C_4$$

$$C_4 = -72\,R_A + 18M_A + 1120 \quad \text{... (4)}$$

At $x = 2\text{m}$,

$$\frac{dy}{dx} = \frac{1}{EI_2}[-2R_A + 2\,M_A + 18\,R_A - 6\,M_A - 240]$$

$$= \frac{1}{EI_2}[+16\,R_A - 4M_A - 240] \quad \text{... (5)}$$

$$y = \frac{1}{EI_2}\left[-\frac{4}{3}R_A + 2M_A + 36 + 36R_a - 12M_A - 480 - 72R_A + 18M_A + 1120\right]$$

$$= \frac{1}{EI_2}[-37.33\,R_A + 8\,M_A + 640] \quad \text{... (6)}$$

From Eqs. (1) and (5) for $I_2 = 2I_1$, we have

$$-R_A + 2M_A = 8\,R_A - 2M_A - 120$$

$$10R_A - 4M_A = 120 \quad \text{... (7)}$$

From Eqs. (2) and (6), we have

$$\frac{-4}{3}R_A + 2M_A = -18.667\,R_A + 4M_A + 320$$

$$17.334\,R_A - 2M_A = 320$$

or

$$34.667\,R_A - 4M_A = 640 \quad \text{... (8)}$$

Substituting Eqs. (7) from (8), we have

$$24.667\ R_A = 520$$

$$R_A = 21.08 \text{ kN}, \quad R_B = 8.92 \text{ kN}$$

Substituting the value of R_A in Eq. (7), we get

$$210.8 - 4\ M_A = 120$$

$$M_A = 22.7 \text{ kN. m}$$

$$M_B = 22.7 + 30 \times 4 - 21.08 \times 6 = 16.22 \text{ kN m}$$

$$M_C = 21.08 \times 2 - 22.7 = 19.46 \text{ kN .m}$$

$$C_3 = 18 \times 21.08 - 6 \times 22.7 - 240 = 3.24$$

$$C_4 = -72 \times 21.08 + 18 \times 2.7 + 1120 = 10.84$$

At $x = 3$m

$$y = \frac{10^6}{210\times10^9\times2400\times10^{-8}}\left[-21.08\times\frac{27}{6}+5+22.7\times\frac{9}{2}+3.24\times3+10.84\right]$$

$$= \frac{32.85\times10^6}{210\times10\times2400} = 6.518 \text{ mm}$$

Example 6.9 *A built-in beam of 6 metre span carries a uniformly distributed load of 2 kN/m run from left end to the mid-span. In addition it carries a point load of 3 kN at 4 metre from the left hand support. Find the deflection at the mid-span and under the point load, if E = 12 × 10⁵ N/mm² and I = 10⁷ mm⁴.*

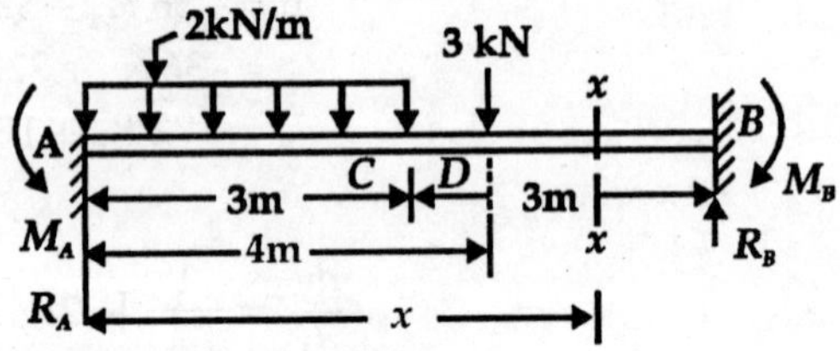

Fig. 6.18

Solution:

Given: $E = 12 \times 10^5$ N/mm², $I = 10^7$ mm⁴

Refer to Fig. 6.18

$$M_x = R_A\,x - 2\times3\times(x-1.5) - 3\,(x-4) - M_A$$

$$EI\frac{d^2y}{dx^2} = -M_x = -R_A\,x + 6\,(x-1.5) + 3\,(x-4) + M_A$$

$$EI\frac{dy}{dx} = -R_A\frac{x^2}{2} + 3\,(x-1.5)^2 + 1.5\,(x-4)^2 + M_A\,x + C_1$$

$$EIy = -R_A\frac{x^3}{6} + (x-1.5)^3 + 0.5\,(x-4)^3 + M_A\frac{x^2}{2} + C_1\,x + C_2$$

At $\quad x = 0, y = 0$ and $\frac{dy}{dx} = 0$

$\therefore \quad C_1 = 0$ and $C_2 = 0$

At $x = 6$m, $y = 0$ and $\frac{dy}{dx} = 0.$

$$-36R_A + 91.125 + 4 + 18M_A = 0$$

or $$-36\ R_A + 18\ M_A = -95.125 \quad \text{... (1)}$$

$$-18\ R_A + 60.75 + 6 + 6M_A = 0$$

$$-18\ R_A + 6M_A = -66.75$$

or $$-36\ R_A + 12\ M_A = -133.5 \quad \text{... (2)}$$

Substituting Eq. (2) from (1), we have

$$6M_A = 38.375$$
$$M_A = 6.396 \text{ kN m}$$

Substituting the value of M_A in Eq. (1), we get

$$-36 R_A + 18 \times 6.396 = -95.125$$
$$R_A = 5.84 \, kN$$

At $x = 3$ m,

$$EIy_c = -5.84 \times \frac{27}{6} + 3.375 + 6.396 \times \frac{9}{2} = 5.877$$

$$y_c = \frac{5.877 \times 10^{12}}{12 \times 10^5 \times 10^7} = 0.4 \text{ mm}$$

At $x = 4$m

$$EIy_D = -5.84 \times \frac{64}{6} + 15.625 + 6.396 \times 8 = 4.5$$

$$y_D = \frac{4.5 \times 10^{12}}{12 \times 10^5 \times 10^7} = 0.375 \text{ mm}$$

Example 6.10 *A fixed beam of span 6 metres carries point loads 200 kN and 150 kN at distances 2 m and 4 m from the left end. Find the fixed end moments and the reactions at the supports. Draw BM and SF diagrams.*

Solution:

Refer to Fig. 6.19

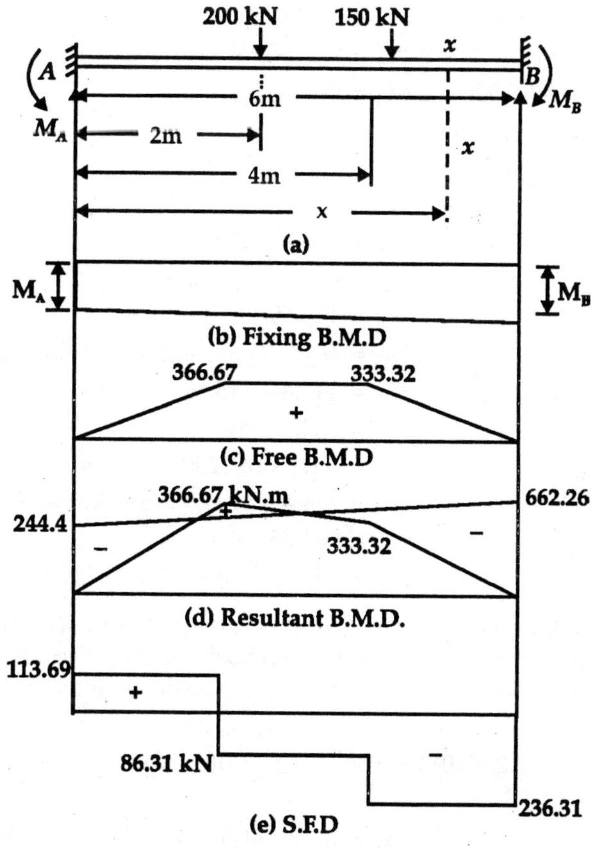

Fig. 6.19

$$M_x = R_A x - 200 (x-2) - 150 (x-4) - M_A$$

$$EI\frac{d^2y}{dx^2} = -M_x$$

$$= -R_A x + 200 (x-2) + 150 (x-4) + M_A$$

$$EI\frac{dy}{dx} = -R_A \frac{x^2}{2} + 100 (x-2)^2 + 75 (x-4)^2 + M_A x + C_1$$

$$EIy = -R_A \frac{x^3}{6} + \frac{100}{3} (x-2)^3 + 25 (x-4)^3$$

$$+ M_A \frac{x^2}{2} + C_1 x + C_2$$

At $x = 0$, $y = 0$ and $\frac{dy}{dx} = 0$, giving

$$C_1 = 0, C_2 = 0$$

At $x = 6\text{m}$, $y = 0$, $\frac{dy}{dx} = 0$

$$-36 R_A + 2133.3 + 200 + 18 M_A = 0$$

or $$-36 R_A + 18 M_A = -2333.3$$

or $$-18 R_A + 9 M_A = -1166.67 \quad \text{... (1)}$$

$$-18 R_A + 1600 + 300 + 6 M_A = 0$$

$$-18 R_A + 6M_A = -1900 \quad \text{... (2)}$$

Substituting Eq. (2) from (1), we have

$$3M_A = 733.33$$

$$M_A = 244.4 \text{ kN m}$$

$$-18 R_A = -1900 - 6 \times 244.4 = -2046.4$$

$$R_A = 113.69 \text{ kN}$$

$$M_B = M_A - R_A \times 6 + 200 \times 4 + 150 \times 2$$

$$= 244.4 - 113.69 \times 6 + 800 + 300$$

$$= 662.26 \; kN \text{ m}$$

$$M_B = -R_A \times 6 + 200 \times 4 + 150 \times 2$$

$$= 244.4 - 113.69 \times 6 + 800 + 300$$

$$= 662.26 \; kN \text{ m}$$

$$R_B = 350 - 113.69 = 236.31 \; kN$$

Example 6.11 *A girder of span 5 metres is built-in at each end and carries two point loads of 90 kN and 120 kN placed at 1 m and 2 metres from the left support respectively. Find the reactions and fixing moments at the ends, if E = 2 × 10⁵ /mm² and I = 0.8 × 10⁸* mm⁴. Also find the deflections at the load points.

Solution:

Given: $E = 2 \times 10^5$ N/mm,^{2} $I = 0.8 \times 10^8$ mm^4

Refer to Fig. 6.20.

$$M_x = R_A x - 90 (x-1) - 120 (x-2) - M_A$$

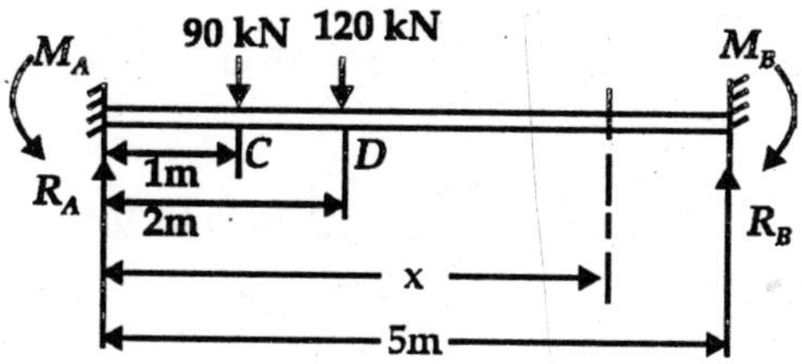

Fig. 6.20

$$EI\frac{d^2y}{dx^2} = -M_x$$

$$= -R_A x + 90(x-1) + 120(x-2) + M_A$$

$$EI\frac{dy}{dx} = -R_A\frac{x^2}{2} + 45(x-1)^2 + 60(x-2)^2 + M_A x + C_1$$

$$EIy = -R_A\frac{x^2}{2} + 15(x-1)^3 + 20(x-2)^3 + M_A\frac{x^2}{2} + C_1 x + C_2$$

At $x = 0, y = 0$ and $\frac{dy}{dx} = 0$ giving $C_1 = 0$, and $C_2 = 0$.

At $x = 5\text{m}, y = 0$ and $\frac{dy}{dx} = 0$

$$-\frac{125}{6}x^3 + R_A + 960 + 540 + 12.5 M_A = 0$$

$$-20.83 R_A + 12.5 M_A = -1500$$

or $$-1.67 R_A + M_A = -120 \quad \text{... (1)}$$

$$-12.5 R_A + 720 + 540 + 5 M_A = 0$$

$$-12.5 R_A + 5 M_A = -1260$$

or $$-2.5 R_A + M_A = -252 \quad \text{... (2)}$$

Substituting Eq. (2) from (1), we have

$$0.83 R_A = 132$$

$$R_A = 159\ kN$$

$$R_B = 210 - 159 = 51\text{kN}$$

$$M_A = -120 + 1.67 \times 159 = 145.53 \text{ kN m}$$

$$M_B = M_A - R_A \times 5 + 90 \times 4 + 120 \times 3$$

$$= 145.53 - 159 \times 5 + 360 + 360$$

$$= 70.53 \text{ kN m}$$

$$EIy = -\frac{159}{6}x^3 + 15(x-1)^3 + 20(x-2)^3 + \frac{145.53}{2}x^2$$

At $x = 1$m,

$$EIy_C = -26.5 + 72.76 = 46.26$$

$$y_C = \frac{46.26 \times 10^{12}}{2 \times 10^5 \times 0.8 \times 10^8} = 2.89 \text{ mm}$$

At $x = 2$m,

$$EI\,y_D = -159 \times \frac{8}{6} + 15 + 2 \times 145.53$$

$$= 94.06$$

$$y_D = \frac{94.06 \times 10^{12}}{2 \times 10^5 \times 0.8 \times 10^8} = 5.88 \text{ mm}$$

Example 6.12 *A clamped—ends bean. is subjected to a concentrated lode W at a distance 'a' from left end. Determine the reactions using Macaulay's singularity functions.*

Solution:

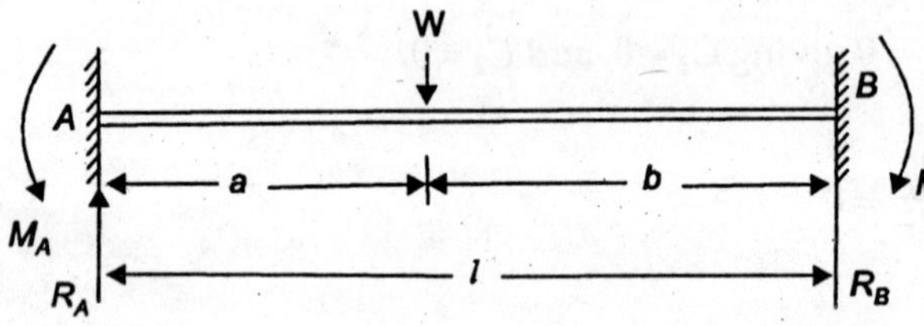

Fig. 6.21

Refer to Fig. 6.21

Equivalent distributive load,

$$w_c = -M_A < x >^{-2} + R_A < x >^{-1} - W(x-a)^{-1}$$
$$+ R_B < x - l >^{-1} + M_B < x - l >^{-2}$$

Integrating, we have

$$F(x) = -M_A < x >^{-1} + R_A < x >^{o} + W < x - l >^{o} + R_B < x - l >^{o} + M_B < x - l >^{-1}$$
$$M(x) = -M_A < x >^{o} + R_A < x >^{1} - W < x - a >^{1} + R_B < x - l >^{1} + M_B < x - l >^{o}$$
$$= EI\frac{d^2y}{dx^2}$$
$$= -M_A < x >^{o} + R_A < x > - W < x - a > + R_B < x - l > + M_B < x - l >^{o}$$

$$EI\frac{dy}{dx} = -M_A < x >^1 + \frac{R_A < x >^2}{2} - \frac{W < x - a >^2}{2} + \frac{R_n < x - l >^2}{2}$$
$$+ M_B < x - l >^1 + C_1 \qquad \text{...(a)}$$

When x = 0, $\frac{dy}{dx} = 0$ Hence $C_1 = 0$

Integrating again, we have

$$ELY = \frac{M_A <x>^2}{2} + \frac{R_A <x>^3}{6} - \frac{W <x-a>^3}{6} + \frac{R_B <x-l>^2}{2} + \frac{M_B <x-l>^2}{2} + C_2$$

when $x = 0, y = 0$ Hence $C_2 = 0$

when x = l, y = 0 and $\frac{dy}{dx} = 0$

$$0 = -M_A l + \frac{R_A l^2}{2} - \frac{Wb^2}{2} \quad ...(c)$$

$$0 = -\frac{M_A l^2}{2} + \frac{R_A l^3}{6} - \frac{Wb^3}{6} \quad ...(d)$$

Solving Eqs. (c) and (d) simultaneously, we get

$$M_A = \frac{Wab^2}{l^2}$$

$$R_A = \frac{Wb^2}{l^2} + \frac{2Wab^2}{l^3}$$

Now $$\frac{Wb^2}{l^2} + \frac{2Wab^2}{l^3} + R_B - W = 0$$

$\therefore$ $$R_B = \frac{Wa^2}{l^2} + \frac{2Wa^2b}{l^3}$$

Taking moments about A, we have,

$$M_B = M_A + R_B \times l - Wa$$

$$= \frac{Wab^2}{l^2} + \frac{Wa^2}{l} + \frac{2Wa^2b}{l^2} - Wa$$

$$= \frac{W}{l^2}[ab^2 + a^2l + 2a^2b - al^2]$$

$$= \frac{W}{l^2}[ab^2 + a^3 + a^2b + 2a^2b - a^3 - ab^2 - 2a^2b]$$

$$= \frac{Wa^2b}{l^2}$$

6.4 CONTINUOUS BEAMS

When a beam is supported on more than two supports, it is called continuous. Continuous beams are preferable over simply supported beams because the maximum bending moment in case of continuous beams is much lower in value, it is more over the supports than at mid span and hence they are more stiffer and lighter in weight. However, it is not possible to determine directly the reactions at the supports by the normal equations of static equilibrium since there are too many unknowns. The Clapeyron's theorem of three moments, which is an extension of the moment area method is used to analyse continuous beams.

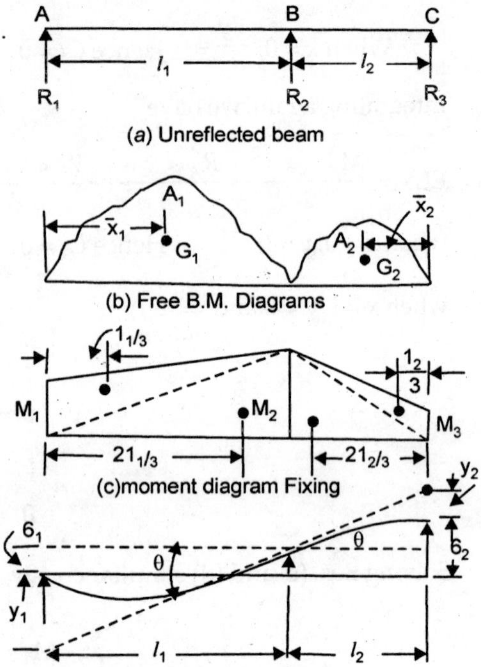

Fig. 6.22 Deriving Clapeyron's three-moment theorem

6.4.1 The Clapeyron's Three Moment Equation.

Consider a beam shown in Fig. 6.22. The areas of the free bending moment diagrams are A_1 and A_2 over two separate spans l_1 and l_2 respectively.

In general the bending moments at the three supports will not be zero but will have some values M_1, M_2 and M_3. Thus a fixing moment diagram must be introduced as shown in Fig. 6.22 (c), the actual bending moment diagram then being the algebraic sum of the two diagrams. Fig. 6.22 (d), shows the deflected position of the beam, the deflections δ_1 and δ_2 being relative to the left hand support. A tangent drawn at the central support has intercepts at the end of each span of y_1 and y_2 and θ is its slope, and hence the beam, at the same support.

Assuming small deflection,

$$\theta = \frac{y_1+\delta_1}{l_1} = \frac{y_2+\delta_2-\delta_1}{l_2}$$

$$\therefore \quad \frac{y_1}{l_1}+\frac{\delta_1}{l_1} = \frac{y_2}{l_2}+\frac{\delta_2-\delta_1}{l_2} \quad \text{...(a)}$$

From Mohr's moment area method

$$y = \frac{Ax}{EI}$$

$$\therefore \quad y_1 = \frac{1}{EI_1}\left[A_1\bar{x}_1+\frac{M_1l_1}{2}\times\frac{l_1}{3}+\frac{M_2l_1}{2}\times\frac{2l_1}{3}\right]$$

$$= \frac{1}{EI_1}\left[A_1\bar{x}_1+\frac{M_1l_1^2}{6}+\frac{M_2l_1^2}{3}\right]$$

and
$$y_2 = \frac{1}{EI_2}\left[A_2\bar{x}_2 + \frac{M_3 l_2}{6}\times\frac{l_2}{3} + \frac{M_2 l_2}{2}\times\frac{2l_2}{3}\right]$$

$$= \frac{1}{EI_2}\left[A_2\bar{x}_2 + \frac{M_3 l_2^2}{6} + \frac{M_2 l_2^2}{3}\right]$$

Substituting in Eq. (*a*), we get

$$= -\frac{1}{EI_1 l_1}\left[A_1\bar{x}_1 + \frac{M_1 l_1^2}{6} + \frac{M_2 l_1^2}{3}\right] + \frac{\delta_1}{l_1}$$

$$= -\frac{1}{EI_2 l_2}\left[A_2\bar{x}_2 + \frac{M_3 l_2^2}{6} + \frac{M_2 l_2^2}{3}\right] + \frac{\delta_2 - \delta_1}{l_2}$$

Simplifying, we get
$$= -\frac{M_1 l_1}{I_1} - 2M_2\left(\frac{l_1}{I_1} + \frac{l_2}{I_2}\right) - \frac{M_3 l_2}{I_2}$$

$$= 6\left(\frac{A_1\bar{x}_1}{I_1 l_1} + \frac{A_2\bar{x}_2}{I_2 l_2}\right) + 6E\left(\frac{\delta_2 - \delta_1}{l_2} - \frac{\delta_1}{l_1}\right) \qquad ...(6.31)$$

Eq. 6.31 is the three-moment equation. For a uniform beam,

$$l_1 = l_2 = I$$

Then Eq. (6.31) becomes,
$$= -M_1 l_1 - 2M_2(l_1 + l_2) - M_3 l_2$$

$$= 6\left(\frac{A_1\bar{x}_1}{l_1} + \frac{A_2\bar{x}_2}{l_2}\right) + 6EI\left(\frac{\delta_2 - \delta_1}{l_2} - \frac{\delta_1}{l_1}\right) \qquad ...(6.32)$$

If the supports are on the same level, i.e.,

$\delta_1 = \delta_2 = 0$, then we get

$$= -M_1 l_1 - 2M_2(l_1 + l_2) - M_3 l_2$$

$$= 6\left(\frac{A_1\bar{x}_1}{l_1} + \frac{A_2\bar{x}_2}{l_2}\right) \qquad ...(6.33)$$

The standard results for $\frac{6A\bar{x}}{l}$ are given below:

(*i*) *Concentrated Load.*

For a simply supported beam of span l carrying a concentrated load W at a distance '*a*' from left hand support, we have

$$\frac{6A\bar{x}}{l} = \frac{6}{l}\left[\frac{Wab}{l}\times\frac{a}{2}\times\frac{2a}{3} + \frac{Wab}{l}\times\frac{b}{2}\left(a + \frac{b}{3}\right)\right]$$

$$= \frac{Wab}{l^2}(2a+b)(a+b) = \frac{Wab}{l}(2a+b)$$

where $$b = l - a$$

$$\therefore \quad \frac{6A\bar{x}}{l} = \frac{Wa}{l}(l^2 - a^2) \qquad ...(6.34)$$

For $$a = \frac{l}{2}, \frac{6A\bar{x}}{l} = \frac{3}{8}Wl^2$$

For a simply supported beam of span l carrying a u.d.l. of intensity w per metre run, we have

$$\frac{6A\bar{x}}{l} = \frac{6}{l} \times \frac{2}{3} \times l \times \frac{wl^2}{8} \times \frac{l}{2} = \frac{wl^3}{4} \qquad ...(6.35)$$

Example 6.13 *A uniform beam is loaded as shown in Fig. 6.23. Determine the B.M. at the middle support and hence draw the B.M. and S.F. diagrams.*

Solution:

Applying the three-moment equation, we have

$$= -M_A l_1 - 2M_B(l_1 + l_2) - M_C l_2$$

$$= \frac{wl_1^3}{4} + \frac{Wa}{l_2}(l_2^2 - a^2)$$

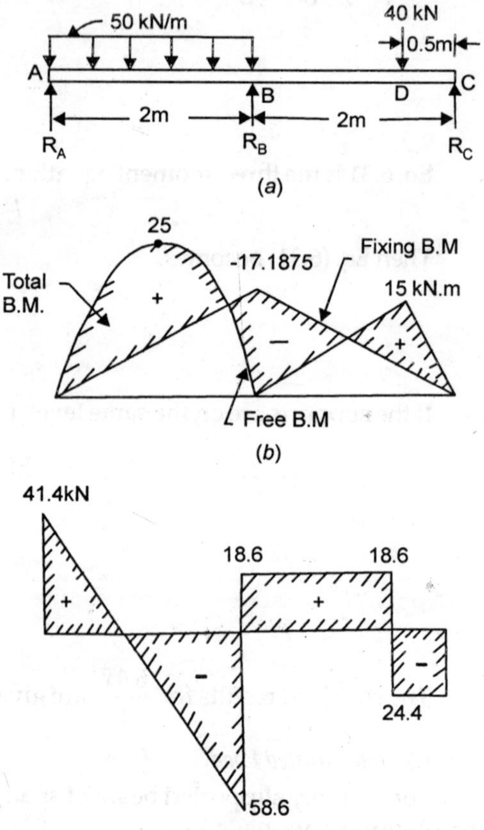

Fig. 6.23

With A and C simply supported

$$M_A = M_C = 0$$

$$\therefore \quad -2M_B(2+2) = \frac{50 \times 2^3}{4} + \frac{40 \times 0.5}{2}\left[2^2 - (0.5)^2\right]$$

$$-8M_B = 100 + 37.5 = 137.5$$

$$M_B = 17.1875 \text{ kN-m}$$

Taking moments about B from the left, we get

$$2R_A - 50 \times 2 \times 1 = -17.1875$$

$$R_A = 41.4 \text{ kN}$$

Taking moments about B, we have

$$2R_C - 40 \times 1.5 = -17.1875$$

$$R_C = 21.4 \text{ kN}$$

Now $$R_A + R_B + R_C = 50 \times 2 + 40 = 140 \text{ kN}$$

$$\therefore \quad R_B = 140 - 41.4 - 21.4 = 77.2 \text{ kN}$$

The B.M. and S.F. diagrams are shown in Fig. 8.49 (b) and (c) respectively.

Example 6.14 *A beam ABCDE is continuous over four supports and carries loads as shown in Fig. 6.24. Calculate the fixing moments at each support and hence draw the S.F. and B.M. diagrams.*

Solution.

Now $M_A = 0$

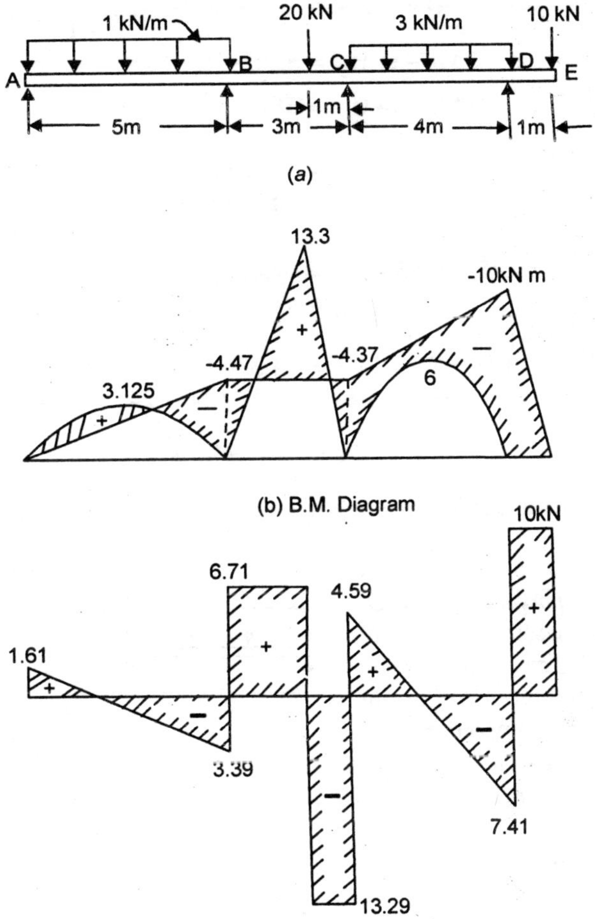

(a)

(b) B.M. Diagram

(c) S.F. Diagram

Fig. 6.24

$$M_D = -10 \times 1 = -10 \text{ kN-m}$$

Applying the three-moment equation for the spans AB and BC, we get

$$-M_A l_1 - 2M_B(l_1 + l_2) - M_C l_2 = \frac{wl_1^3}{4} + \frac{Wa}{l_2}\left(l_2^2 - a^2\right)$$

$$0 - 2M_B(5+3) - 3M_C = \frac{1 \times 5^3}{4} + \frac{20 \times 1}{3}\left(3^2 - 1^2\right).$$

$$-16\, M_B - 3M_C = 31.25 + 53.33$$

or $$-16\, M_B - 3M_C = 84.58 \qquad \text{...(a)}$$

For the spans BC and CD, we have

$$-3M_B l_2 - 2M_C(l_2 + l_3) - M_D l_3 = \frac{Wa}{l_2}\left(l_2^2 - a^2\right) + \frac{wl_3^3}{4}$$

$$-3M_B - 2M_C(3+4) + 10 \times 4 = \frac{20 \times 2}{3}\left(3^2 - 2^2\right) + \frac{3 \times 4^3}{4}$$

$$-3M_B - 14M_C = 66.67 + 48 - 40$$

or

$$-3M_B - 14M_C = 74.67 \quad \text{...(b)}$$

Solving Eqs. (a) and (b), we get

$$M_B = -4.47 \text{ kN-m}$$

$$M_C = -4.37 \text{ kN-m}$$

Taking moments about B, we get

$$R_A \times 5 - 1 \times 5 \times 2.5 = -4.47$$

$$5R_A = -4.47 + 12.5$$

Moments about C from the left, give

$$R_A \times 8 - (1 \times 5 \times 5.5) + R_B \times 3 - 20 \times 1 = -4.37$$

$$4R_D = -4.37 + 50 + 24$$

$$R_B = \frac{30.3}{3} = 10.1kN$$

The B.M. and S.F. diagrams have been shown in Fig. 6.24 (b) and (c) respectively.

Example 6.15 *A continuous beam ABCD is pinned at A and simply supported at B and D, these points being at the same level. AB = 3 m, BD = 4 m and DE = 2 m. It carries a point load of 100 kN at the mid point of BD and a uniformly distributed load of 15 kN/m run from A to E. Find the moments at B and D and the reaction on the pin A. Sketch the B.M. diagram for the Ibeam.*

Solution:

By inspection in Fig. 6.25

$$M_A = 0$$

$$M_D = $$

$$\frac{-15 \times 2^2}{2} = 30\text{kN - m}$$

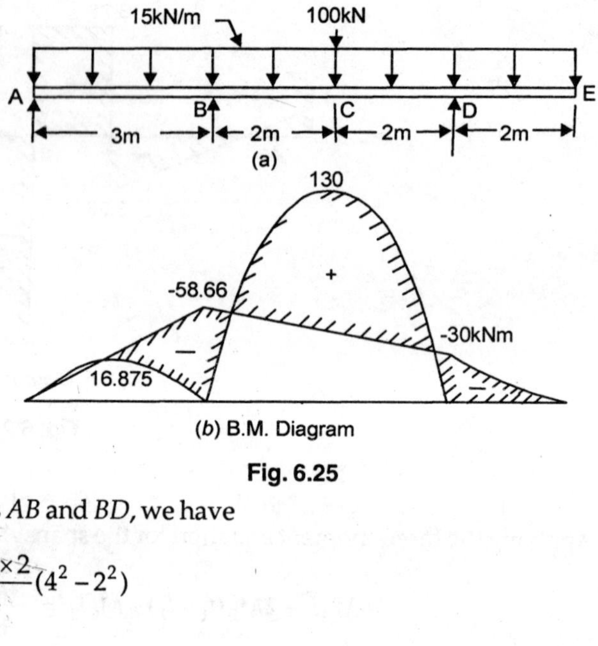

(b) B.M. Diagram

Fig. 6.25

Applying three-moment equation for the spans *AB* and *BD*, we have

$$0 - 2M_B\,(3+4) + 30 \times 4 = \frac{15 \times 3^3}{4} + \frac{15 \times 4^2}{4} + \frac{100 \times 2}{4}(4^2 - 2^2)$$

$$-14M_B + 120 = 101.25 + 240 + 600$$

$$M_B = -58.66 \text{ kN-m}$$

Taking moments about *B* from the left, we have

$$R_A \times 3 = 15 \times 3 \times 1.5 - 58.66$$

$$R_A = \frac{8.84}{3} = 2.947 \text{ kN}$$

The B.M. diagram is shown in Fig. 6.25 (b).

Example 6.16 *Fig. 6.26 (a) shows a continuous beam ABCD of uniform section (EI = 13.5 kN m2) encastre at A and freely supported at B, C and D, the supported at C is 1 cm below A, B, and D which are at same level. Determine the support moments and construct the bending moment diagram for beam.*

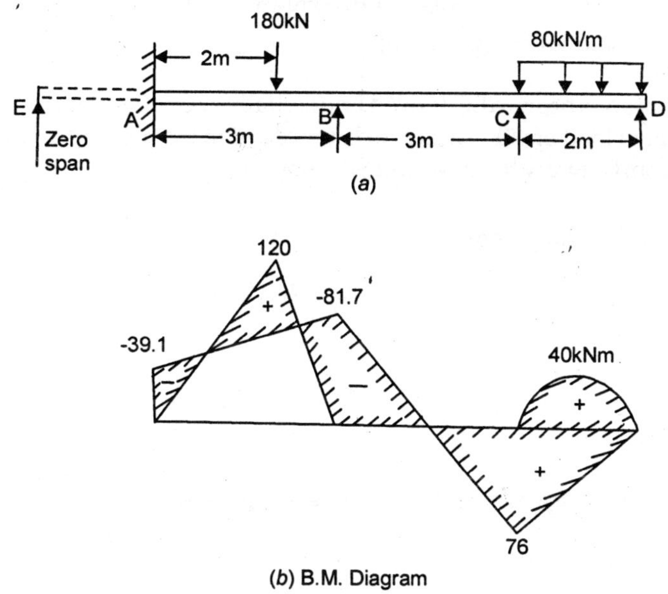

Fig. 6.26

Solution:

By inspection $M_D = 0$

For spans EA and AB

$$0 - 2M_A \times 3 - MB \times 3 = 3M_B + 10M_C$$

or

$$-6\,M_A - 3M_B = 480$$

$$-M_A - 0.5\,M_B = 80 \qquad \ldots(a)$$

For span AB and BC

$$-3M_A - 2M_B\,(3+3) - M_C \times 3 = \frac{180\times 2}{3}(9-4)+0+0-6\times 13.5\times 10^3\left(\frac{-1}{300}\right)$$

$$-3M_A - 12M_B - 3M_C = 600 + 270 = 870$$

$$-M_A - 4M_B - M_C = 290 \qquad \ldots(b)$$

For spans BC and CD

$$-3\,M_B - 2M_C(3+2)+0 = 0+\frac{80\times 2^3}{4}-6\times 13.5\times 10^3\left(\frac{1}{300}+\frac{1}{200}\right)$$

$$-3M_B - 10M_C = 160 - 675 = -515$$

or

$$3M_B + 10M_C = 515 \qquad \ldots(c)$$

Solving Eqs. (*a*), (*b*) and (*c*), we get

$$M_A = -39.1 \text{ kN-m}$$
$$M_B = -81.7 \text{ kN-m}$$
$$M_C = +76.0 \text{ kN-m}$$

The B.M. diagram is shown in Fig. 6.26 (*b*).

Example 6.17 A continuous beam ABC is fixed horizontally at C and supported at A and B. AB = 3m, BC = 4m. A concentrated load of 40kN acts at a distance of 2 m from A and another load of 60 kN acts at a distance of 2 m from C. Draw the B.M. and S.F. diagrams.

Solution:

Free bending moments, (Fig. 6.27)

$$M_D = \frac{Wab}{l} = \frac{40 \times 2 \times 1}{3} = 26.67 \text{ kN.m}$$

$$M_E = \frac{60 \times 2 \times 2}{4} = 60 \text{ kN.m}$$

Fixing moments:

The beam at end *C* may be assumed to be extended upto *F* so that *CF* is of zero length and carries no load.

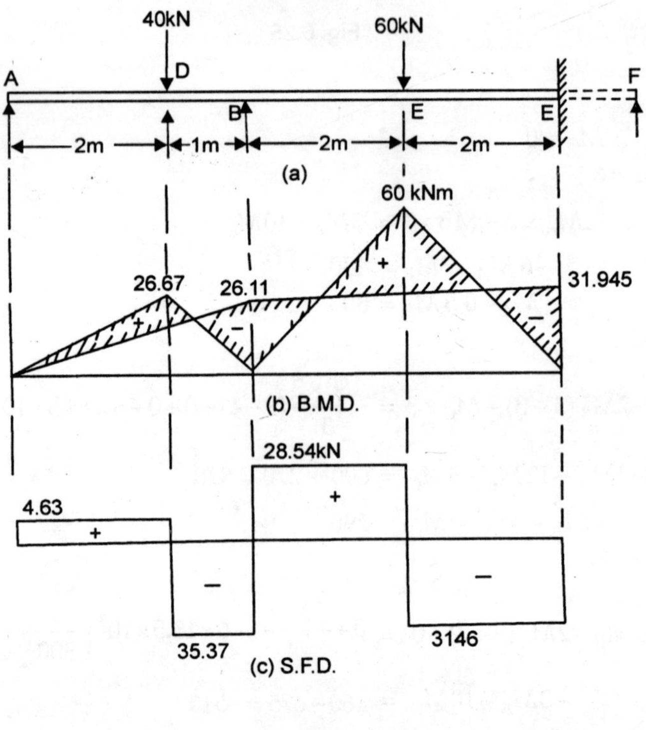

Fig. 6.27

Applying Clapeyron's equation for spans *AB* and *BC*,

$$3M_A + 2M_B(3+4) + 4M_C = -6\left[\frac{1}{3}\left\{\frac{1}{2}\times 2\times\frac{80}{3}\times\frac{2}{3}\times 2+\frac{1}{2}\times 1\times\frac{80}{3}\times\left(2+\frac{1}{3}\right)\right\}\right.$$

$$-6\left[\frac{1}{4}\left\{\frac{1}{2}\times 2\times 60\times\frac{2}{3}\times 2+\frac{1}{2}\times 2\times 60\times\left(2+\frac{2}{3}\right)\right\}+0\right]$$

$$M_A = 0, \text{being a simply supported end.}$$

$$14M_B + 4M_C = -493.32 \qquad ...(a)$$

For spans BC and CF,

$$4M_B + 2M_C(4+0)+0 = -6\left[\frac{1}{4}\left\{\frac{1}{2}\times 2\times 60\times\frac{2}{3}\times 2+\frac{1}{2}\times 2\times 60\times\left(2+\frac{2}{3}\right)\right\}+0\right]$$

$$4M_B + 8M_C = -360 \qquad ...(b)$$

Solving Eqs. (a) and (b), we get

$$M_B = -26.11 \text{ kN-m}$$

$$M_C = -31.945 \text{ kN-m}$$

Taking moments about B from left hand side,

$$R_A \times 3 - 40 \times 1 = M_B$$

$$3R_A - 40 = -26.11$$

$$R_A = 4.63 \text{ kN}$$

Taking moments about C,

$$4.63 \times 7 - 40 \times 5 + R_B \times 4 - 60 \times 2 = M_C$$

$$32.41 - 200 + 4R_B - 120 = -31.945$$

$$R_B = 63.91 \text{ kN}$$

$$R_A + R_B + R_C = 40 + 60$$

$$R_C = 100 - (4.63 + 63.91)$$

$$= 31.46 \text{ kN}$$

The B.M. and S.F. diagrams are shown in Fig. 6.27 (*b*) and (*c*) respectively.

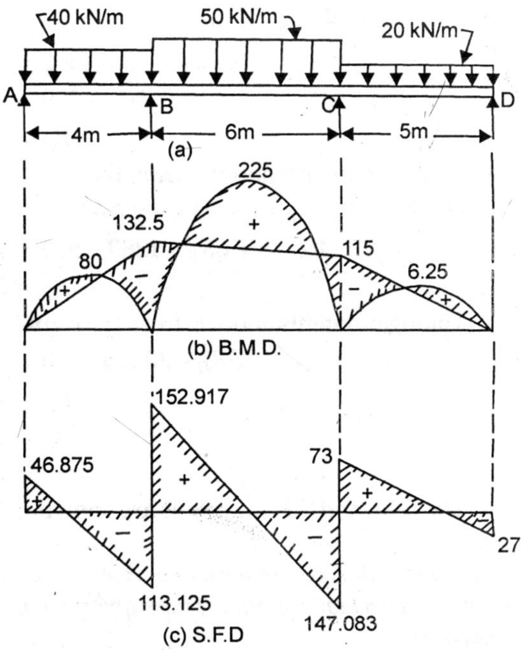

Fig. 6.28

Example 6.18 *Draw the B.M. and S.F. diagrams for the continuous beam shown in Fig. 6.28.*

Solution:

Free bending moments:

Span *AB*,

$$M_{max} = \frac{wl^2}{8}$$

$$= \frac{40\times 16}{8}$$

$$= 80 \text{ kN-m}$$

Span BC,

$$M_{max} = \frac{50 \times 36}{8}$$

$$= 225 \text{ kN-m}$$

Span CD,

$$M_{max} = \frac{20 \times 25}{8}$$

$$= 62.5 \text{ kN-m}$$

Using Clapeyron's equation for spans AB and BC,

$$4M_A + 2M_B(4+6) + 6M_C = \delta_1 = \delta_2 = 0$$

$$M_A = 0, \text{ being simply supported end.}$$

$$20M_B + 6M_C = -3340 \qquad ...(a)$$

Spans BC and CD,

$$6M_B + 22M_C + 5M_D = \frac{50 \times 6^3}{4} - \frac{20 \times 5^3}{4}$$

$$M_D = 0, \text{ being simply supported end.}$$

$$6M_B + 22M_C = -3325 \qquad ...(b)$$

Solving Eqs. (*a*) and (*b*), we get

$$M_B = -132.5 \text{ kN.m}$$

$$M_C = -115 \text{ kN.m}$$

Taking moments about B,

$$R_A \times 4 - 40 \times 4 \times 2 = M_B$$

$$4R_A - 320 = -132.5$$

$$R_A = 46.875 \text{ kN}$$

Taking moments about C from the left,

$$46.875 \times 10 + R_B \times 6 - 40 \times 4 \times 8 - 50 \times 6 \times 3 = M_C$$

$$468.75 + 6R_B - 1280 - 900 = -115$$

$$R_B = 226.042 \text{ kN}$$

Taking moments about C from the right,

$$5R_D - 20 \times 5 \times 2.5 = -115$$

$$R_D = 27 \text{ kN}$$

$$R_C = (40 \times 4 + 50 \times 6 + 20 \times 5) - 46.875 + 226.042 + 27)$$

$$= 220.083 \text{ kN}$$

The B.M. and S.F. diagrams are shown in Fig. 6.28 (*b*) and (*c*) respectively.

Example 6.19 *A continuous beam ABCD is loaded as shown in Fig. 6.29. The support B sinks by 5 mm and support C by 8 mm. Determine the support moments E= 200 GPa, I = 30 × 10⁻⁸ m⁴.*

Solution:

Spans AB and BC

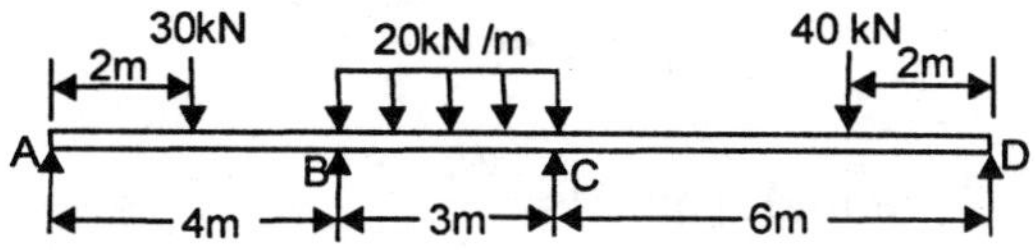

Fig. 6.29

Here

$$\delta_1 = 5\text{ mm}$$

$$\delta_2 = 5 - 8 = -3\text{ mm}$$

$$4\,MA + 2\,MB\,(4+3) + 3\,MC = -\left(\frac{3}{8}\times 30\times 16 + \frac{20\times 27}{4}\right)$$

$$+6\times 200\times 10^9\times 30\times 10^{-8}\times\left(\frac{5}{4}-\frac{3}{3}\right)\times 10^{-3}$$

$$4M_A + 14M_B + 3M_C = -(180+135)+90$$

$$M_A = 0,\text{ being simply supported end.}$$

$$14M_B + 3M_C = -225 \qquad \text{...}(a)$$

Spans BC and CD:

$$\delta_1 = 8 - 5 = 3\text{ mm}$$

$$\delta_2 = 8\text{ mm}$$

$$3M_B + 2M_C\,(3+6) + 6M_D = -\left[\frac{20\times 27}{4}+\frac{40\times 2}{6}(36-4)\right]$$

$$+6\times 200\times 10^9\times 30\times 10^{-8}\left(\frac{3}{3}+\frac{8}{6}\right)\times 10^{-3}$$

$$M_D = 0,\text{ being simply supported end.}$$

$$3M_B + 18\,M_C = 278.3 \qquad \text{...}(b)$$

Solving Eqs. (a) and (b), we get

$$M_B = -20.1\text{ kN.m}$$

$$M_C = 18.82\text{ kN.m}$$

SUMMARY

1. Fixed beams have zero deflection and slope at the ends. But there are reactions and fixing moments over there.
2. Fixed beams carry smaller bending moment, deflection and are more stronger and stiffer as compared to simply supported beams.
3. Fixed beams require high accuracy in aligning and fixing the supports during erection, which increases the cost.

4. Fixed beam carrying a concentrated load W at mind span.

$$M_A = M_B = \frac{Wl}{8}$$

$$\text{At mid span, } M = \frac{Wl}{8}$$

End reactions, $$R_A = R_B = \frac{W}{2}$$

Location of points of inflection from the ends $= \dfrac{l}{4}$

$$\text{Mid span deflection} = \frac{Wl^3}{192\,EI}$$

5. Fixed beam carrying concentrated load W off midspan.

$$M_A = \frac{Wab^2}{l^2}, \; M_B = \frac{Wa^2b}{l^2}$$

$$\text{Deflection under the load} = \frac{Wa^3b^3}{3EIl^3}$$

$$\text{Maximum deflection} = \frac{2Wa^3b^2}{3(3a+b)^2 EI} \text{ at } x = \frac{2al}{a+l}$$

6. Fixed beam carrying udl over whole span.

$$M_A = M_B = \frac{wl^2}{12}$$

$$R_A = R_B = \frac{wl}{2}$$

$$\text{Maximum deflection} = \frac{wl^4}{384\,EI}$$

Location of points of contraflexure = $0.211l$ and $0.789l$

7. Fixed beam carrying a clockwise couple M_0 at distance 'a' from left end.

$$R_A = \frac{6M_0\,ab}{l^3}, \; b = l - a$$

$$M_A = \frac{M_0\,b(l-3a)}{l^2}$$

$$\text{Deflection at the couple location} = \frac{M_0\,a^2b^2(l-2a)}{2\,EIl^3}$$

$$\text{Maximum deflection} = \frac{M_0\, a(2l-3a)^2}{54\, EIb^2} \text{ at } x = \frac{l^2}{3b}$$

8. Moment area method.

$$\text{Deflection,} = \frac{A\bar{x}}{EI} = \frac{1}{EI}(A_1\bar{x}_1 + A_2\bar{x}_2)$$

9. When a beam is supported on more than two supports, it is called a continuous beam.
10. Clapeyron's three moment equation:

$$\frac{-M_1 l_1}{I_1} - 2M_2\left(\frac{l_1}{I_1} + \frac{l_2}{I_2}\right) - \frac{M_3 l_2}{I_2} = 6\left[\frac{A_1\bar{x}_1}{I_1 l_1} + \frac{A_2\bar{x}_2}{I_2 l_2}\right] + 6E\left[\frac{\delta_2 - \delta_1}{l_2} - \frac{\delta_1}{l_1}\right]$$

For a uniform beam, $I_1 = I_2 = I$, then

$$-M_1 l_1 - 2M_2(l_1 + l_2) - M_3 l_2 = 6\left[\frac{A_1\bar{x}_1}{l_1} + \frac{A_2\bar{x}_2}{l_2}\right] + 6EI\left[\frac{\delta_2 - \delta_1}{l_2} - \frac{\delta_1}{l_1}\right]$$

For supports on the same level, $\delta_1 = \delta_2 = 0$, then

$$-M_1 l_1 - 2M_2(l_1 + l_2) - M_3 l_2 = 6\left[\frac{A_1\bar{x}_1}{l_1} + \frac{A_2\bar{x}_2}{l_2}\right]$$

Short Answers Type Questions

1. What is a fixed beam?
2. What are the advantages of fixed beams over simply supported beams?
3. What are the disadvantages of fixed beams?
4. How much is the maximum deflection at the centre of a fixed beam carrying concentrated load at the centre
5. How much is the maximum deflection at the centre of a fixed beam carrying udl over the whole span?
6. Explain the moment area method for fixed beams.
7. What is Clapeyron's three moment theorem?

Exercises

1. A fixed beam of span 5 m carries a concentrated load of 10 kN at the centre. calculate the deflection at the mid span. Take $E = 205$ GPa and $I = 2000$ cm^4. **[Ans.** 1.588 mm]
2. A fixed beam of span 8 m carries a concentrated load of 20 *kN* at 3m from the left hand support. Calculate the deflection under the load and maximum deflection. Take $E = 210$ GPa and $I = 2500$ cm^4. **[Ans.** 8.37 mm, 8.75 mm]
3. A fixed beam of span 10 m is subjected to a clochwise couple of 80 *kN*. m at a distance of 6 m from the left hand end. Calculate the deflection under the couple. Take $E = 205$ GPa and $I = 3000$ cm^4.

[**Ans.** 9.2 mm]

4. A fixed beam of span 12 m carries a udl of intensity 10kN/m. Calculate the maximum deflection of the beam. Take $E = 210$ GPa and $I = 28000$ cm^4. [**Ans.** 9.2 mm]
5. Solve Questions 1 to 4 by moment area method.

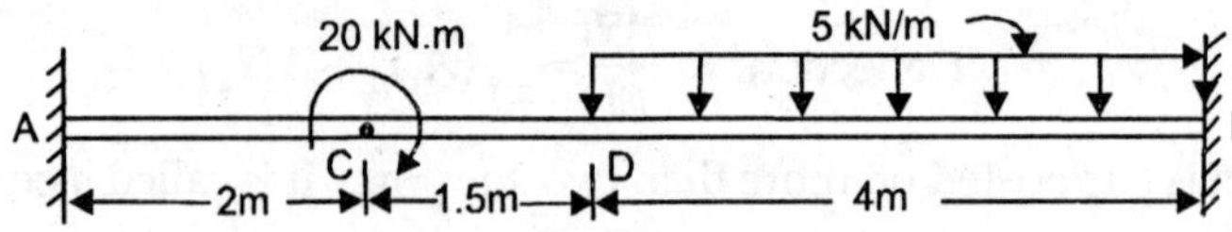

Fig. 6.30

6. A fixed beam is loaded as shown in Fig. 6.30. Determine the maximum deflection, end reactions and fixing moments. $E = 200$ GPa, $I = 3600$ cm^4.
[**Ans.** 5.53 mm at 3.4 m from A, 1.0424 kN, 18.9576 kN; 11.465 kN.m, 23.647 kN.m]
7. A fixed beam of 12 m span carries two point loads of 20 kN at 3 m from each end. Obtain the values of support moments and the central deflection. I = 8520 cm4, E = 200 GPa. Use moment area method. [**Ans.** 45 kN.m, 10.56 mm]
8. A beam of length 10m is fixed at both the ends and carries a uniformly distributed load of intensity 10 kN/m. Calculate the fixing moments and the central deflection. EI = 25×10^6 N-m^2. Use moment area method. [**Ans.** 83.33 kN.m, 10.417 mm]
9. A beam of span 8 m has its ends fixed and carries a uniformly distributed load of 15 kN/m for the left hand half portion of the beam. Calculate the fixing moments and the support reactions. Also draw the B.M. and S.F. diagrams. [**Ans.** 57.5 kN.m, 12.5 kN.m, 50.625 kN, 9.375 kN]
10. A cantilever of effective length L carries a total load of wL uniformly distributed throughout the length. If the cantilever is propped at point L/4 from the free end and the level of the prop adjuated so that there is no deflection at the free end, what is the reaction at the prop and the deflection of the beam at that point? Sketch the S.F. and B.M. diagrams for the cantilever.

$$\left[\textbf{Ans.} \frac{16wL}{27}, \frac{WL^4}{6144EI}\right]$$

11. Analyse the continuous beam shown in Fig. 6.31 if the support B sinks by 1 cm. Section is constant throughout. $E = 200$ GN/m^2, $I = 8500 \times 10^{-8}$ m^4.
[**Ans.** $M_B = 60.6$ kN.m, $M_C = -31.83$ kNm, $R_A = 9.9$ kN, $R_B = 62.8$ kN, $R_C = 52.6$ kN, $R_D = 24.7$ kN]

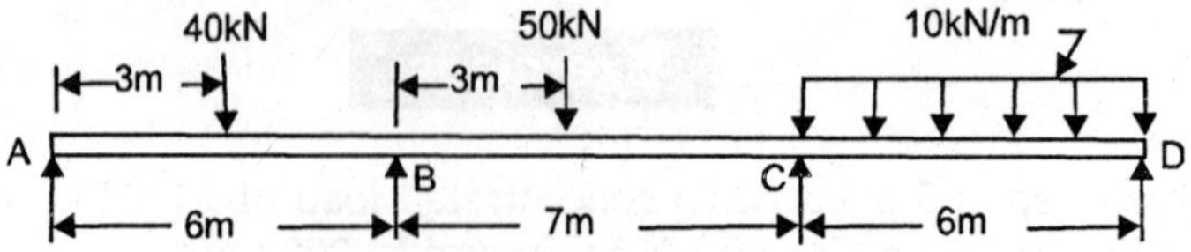

Fig. 6.31

12. A continuous beam of uniform cross-section is loaded as shown in Fig. 6.32. Draw the B.M. and S.F. diagrams.
[**Ans.** $M_A = -11.785$ kN.m, $M_C = -11.07$ kN.m,

$M_D = -20$kN.m, $R_C = 32.33$ kN, $R_D = 32.33$ kN, $R_A = -12.32$ kN]

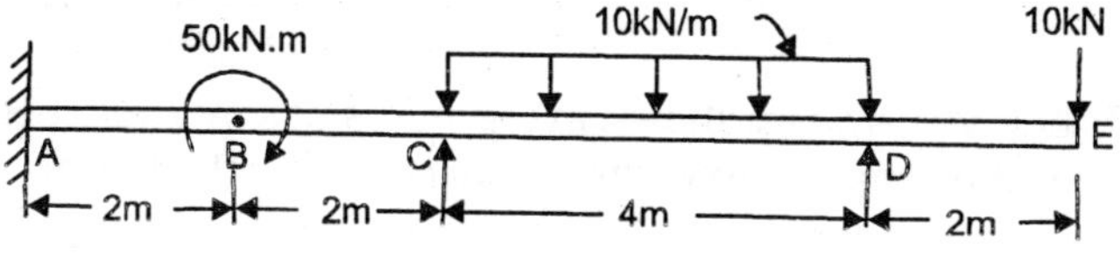

Fig. 6.32

13. Draw the B.M. diagram for the continuous beam shown in Fig. 6.33.

[**Ans.** $M_A = 9.75$ kN.m, $M_B = 70.44$ kN.m, $M_C = 68.13$ kN.m, $M_D = 32.60$ kN.m]

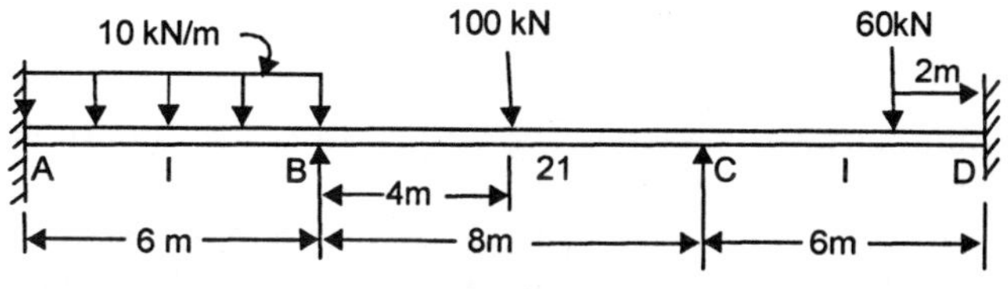

Fig. 6.33

14. A girder of 12 m span is fixed horizontally at the ends. A downward vertical load of 120 kN acts on the girder at a distance of 4 m from the left hand end and an upward vertical force of 80 kN acts at a distance of 6 m from the right hand end. Determine the end reactions and fixing moments and draw the bending moment and shearing force diagrams for the girder.

[**Ans.** 48.9 kN; –8.9kN; 93.3 kN.m; –13.3 kN-m]

15. A beam AB of 6 m span is fixed at both ends and carries a load of 30 kN at C, 2m from A. For portion AC, I = 1200 cm4 and for portion BC, I = 2400 cm4. Find the fixed end moments and central deflection. E = 210 GPa. [**Ans.** –22.7 kN.m, –16.22 kN.m; 3.52 mm]

16. A continuous horizontal beam *ABC* of uniform section is fixed at *A* and simply supported at *B* and C. The span *AB* is 5 m long and has a central load of 60 kN. The span *BC* is 4 m long and is loaded with uniformly distributed load of 40 kN/m. Draw the B.M. and S.F. diagrams.

[**Ans.** $M_A = -26.53$ kN.m, $M_B = -59.43$ kN.m, $R_A = 23.42$ kN, $R_B = 131.44$ kN, $R_C = 64.14$ kN]

17. A uniform girder is continuous over three spans, the end spans being 20 m each, and the centre span 5 m. Find the reactions at the supports when a load of 250 kN is placed at the centre of the first span. Draw S.F. and B.M. diagrams.

[**Ans.** $M_C = 75.76$ kN.m, $M_B = -757.6$ kN.m; $R_A =$ 87.12 kN, $R_B = 329.55$ kN, $R_C = -170.46$ kN, $R_D = 3.79$ kN]

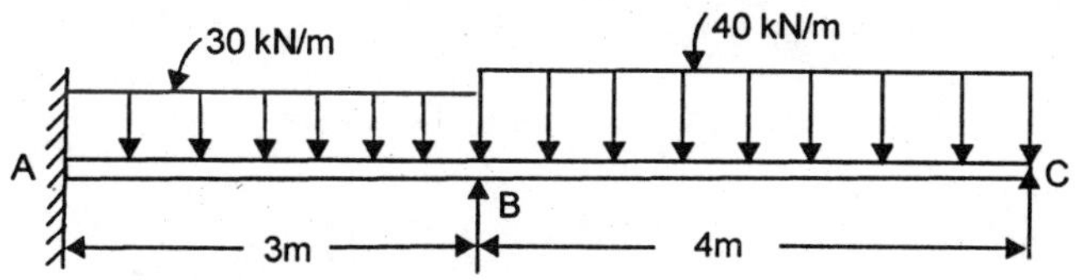

Fig. 6.34

18. Draw the B.M. and S.F diagrams for the beam shown in Fig. 6.34.

[**Ans.** MA = –4.102kN.m, MB = –59.295 kN.m; RA = 26.6 kN, RB = 158.22 kN, RC = 65.18 kN]

19. The support B of continuous beam shown in Fig. 6.35 sinks by 3 mm below the level of other supports. Determine the support moments. E = 200 GPa, I = 21 ×10^{-8} m^4.

[**Ans.** M_B = –40.64 kN.m, M_A = M_C = 0]

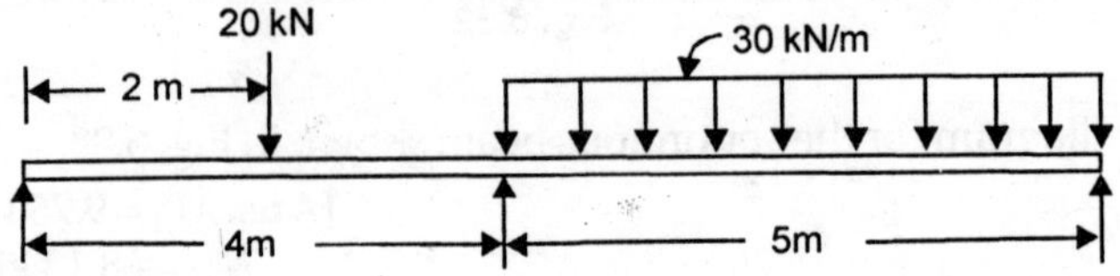

Fig. 6.35

20. Analyse the continuous beam shown in Fig. 6.36 by use of the three moment equation. Draw S.F. and B.M. diagrams.

[**Ans.** M_B = –30kN.m, M_C = –6.021 kN.m, M_D = –44.398 kN.m; R_B = 36 kN, R_C = 56.2 kN, R_D = 129.266 kN, R_E = 38.534 kN]

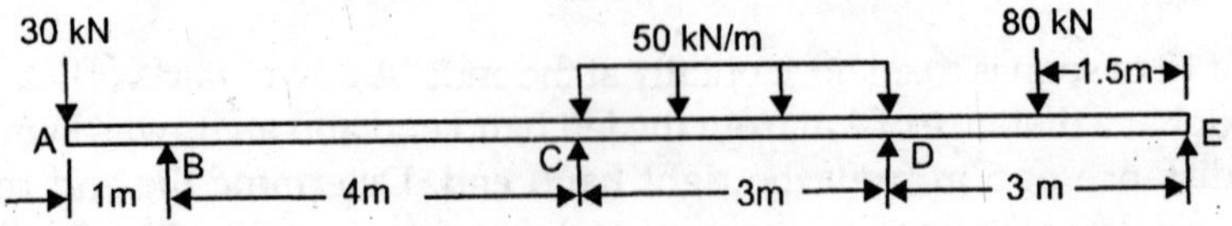

Fig. 6.36

21. Analyse the continuous beam shown in Fig. 6.37.

[**Ans.** M_A = –50.89 kN.m, M_B = –49.42 kNm; R_A = 36.294 kN, R_B = 116.061 kN, R_C = 67.645 kN]

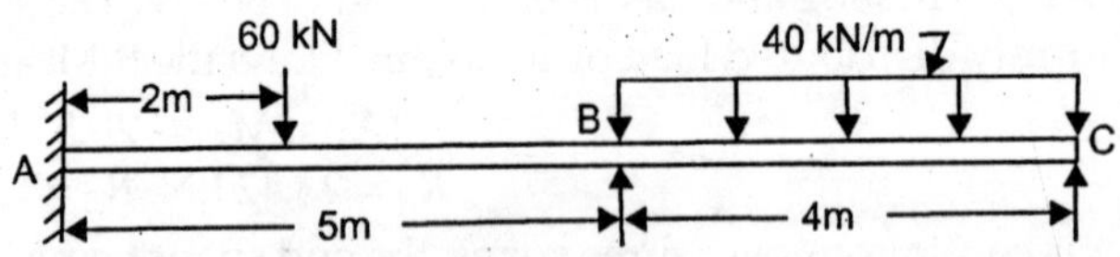

Fig. 6.37

Chapter-7

Torsion of Shafts

7.1 INTRODUCTION

A shaft is a rotating machine member which is required to transmit power in the industry from one unit to another. During this process it is subjected to twisting action, i.e. shearing of fibres of the shaft. The power is transmitted on account of a couple acting in a plane perpendicular to the longitudinal axis of the shaft. This couple is called the torque. The angular twist caused by the torque is called torsion.

7.2 TORSION OF A CIRCULAR SHAFT

To establish a, relation between the torque and the shear stress set up in members with circular solid and tubular cross-sections, it is necessary to make the following assumptions:

1. The material of the shaft is homogeneous, perfectly elastic, continuous, isotropic and obeys Hooke's law.
2. A plane section of material perpendicular to the axis of a circular shaft remains plane after the torque is applied, i.e. no warpage or distortion of parallel planes normal to the axis of the shaft takes place.
3. All cross-sections of the shaft rotate as if rigid, i.e. every diameter rotates thorough the same angle.
4. The shear strains γ vary linearly from the central axis, reaching γ_{max} at the periphery of the shaft.

7.2.1 Torsion Formula

Consider a circular shaft of length l and diameter d subjected to torque T, as shown in Fig. 7.1(a). One end of the shaft is rigidly fixed. A line AB on the surface of the shaft, which is parallel to the axis before twisting, takes up the form of a helix AC after twisting. Let ϕ be the angle of shear strain on the surface. Then

$$BC = l\phi$$

or

$$\phi = \frac{BC}{l}$$

But $$\phi = \frac{\tau}{G}$$

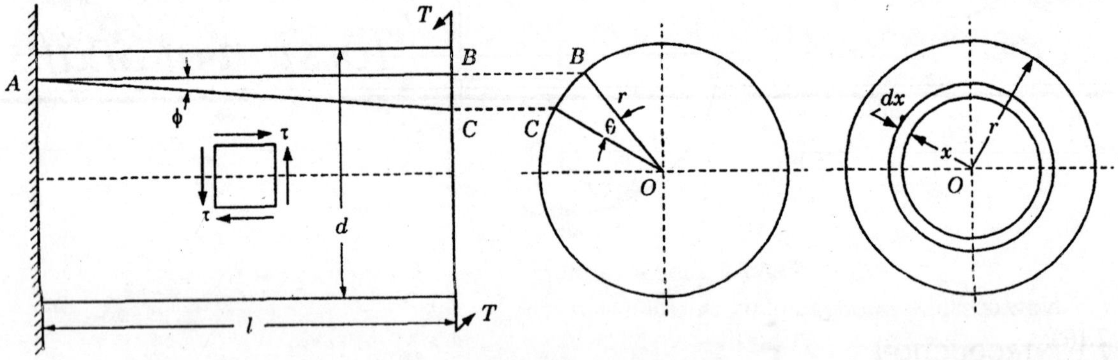

(a) Circular shaft under torsion (b) Twist of shaft. (c) Elementary ring of shaft.

Fig. 7.1 Torsion of a circular shaft.

where τ = shear stress in the shaft

G = modulus of rigidity of shaft material

$$\therefore \quad \tau = G\phi$$

Now $$BC = \frac{d}{2}\cdot\theta \text{ from Fig. 7.1(b)}$$

where θ = angle of twist of shaft

$$\therefore \quad \tau = G\times\frac{d}{2}\times\frac{\theta}{l} = \left(\frac{G\theta}{l}\right)\cdot\frac{d}{2} = k.r$$

where $k = \dfrac{G\theta}{l}$ is a constant for a given shaft

$r = \dfrac{d}{2}$ is the shaft radius.

Therefore, we find that the shear stress in the shaft is proportional to the radius of the shaft, as shown in Fig. 7.2. Hence

$$\frac{\tau}{r} = \frac{G\theta}{l}$$

At any radius x, we have

$$\frac{\tau_x}{\tau} = \frac{x}{r}$$

or $$\tau_x = \frac{\tau x}{r}$$

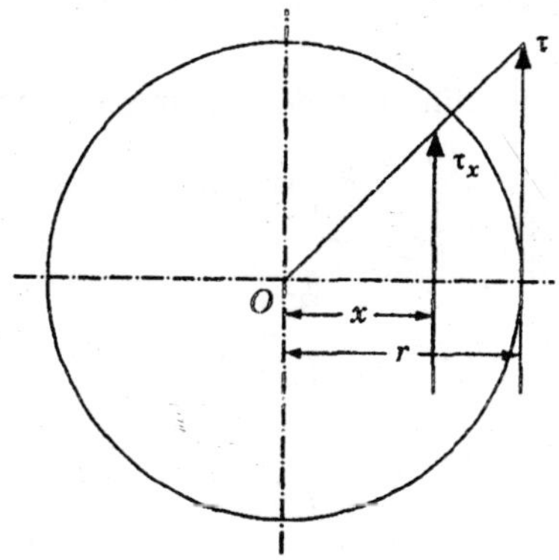

Fig. 7.2 Linear variation of torsional shear stress.

Now consider the elementary ring of the shaft at a radius x and of thickness dx, as shown in Fig. 7.1(c).

Total force on the ring, $F = \tau_x \times$ area of the ring $= \tau_x \times 2\pi x dx$

Moment of force F about the axis of the shaft,

$$dM = F \times x = \tau_x \times 2\pi x\, dx \times x = 2\pi x^2 \tau_x \cdot dx$$

$$= 2\pi x^2 \times \frac{\tau x}{r} \times dx = \frac{2\pi\tau}{r} \cdot x^3 dx$$

Total resisting moment of the shaft,

$$M = \frac{2\pi\tau}{r} \int_0^r x^3 dx = \frac{2\pi\tau}{r} \left| \frac{x^4}{4} \right|_0^4 = \frac{\pi r^3}{2} \tau$$

Total resisting moment, M = Applied torque T.

$$\therefore \quad T = \frac{\pi r^3}{2} \cdot \tau = \frac{\pi d^3}{16} \tau = \frac{\tau}{(d/2)} \times \frac{\pi d^4}{32} = \frac{2\tau}{d} \cdot J = \frac{\tau}{r} \cdot J$$

where $J = \dfrac{\pi d^4}{32}$ is the polar moment of inertia of shaft cross-section.

$$\therefore \quad \frac{T}{J} = \frac{\tau}{r} = \frac{G\theta}{l} \quad \text{...(7.1)}$$

This is the well known torsion formula for a circular shaft. The SI units of various quantities are :

T = torque, N.m.
J = polar moment of inertia, m^4
τ = shear stress, MN/m^2 (MPa)
r = radius, *m*
G = shear modulus, GPa
θ = angle of twist, radians
l = length, *m*

7.2.2 Hollow Circular Shaft

For a hollow circular shaft of outer diameter d_0 and inner diameter d_i., we have

$$J = \frac{\pi}{32}(d_0^4 - d_i^4)$$

$$\frac{T}{J} = \frac{32T}{\pi(d_0^4 - d_i^4)}$$

$$\frac{\tau}{r} = \frac{2\tau}{d_0}$$

$$\therefore \quad \frac{32T}{\pi(d_0^4 - d_i^4)} = \frac{2\tau}{d_0}$$

$$\tau = \frac{16T}{\pi d_0^3\left[1-\left(\frac{d_i}{d_0}\right)^4\right]} = \frac{16T}{\pi d_0^3(1-k^4)} \qquad \text{...(7.2)}$$

where $k = \frac{d_i}{d_0}$ is the ratio of inner diameter d_i to outer diameter d_0.

7.3 COMPARISON OF SOLID AND HOLLOW SHAFTS

Let d_0 = outer diameter of hollow shaft

d_i = inner diameter of hollow shaft

d = diameter of solid shaft

Let the density of the shaft material and length for the solid and hollow shafts be the same.

7.3.1 For the Same Weight

Weight of shaft is proportional to area of cross-section. For same weight and material of solid and hollow circular shafts, we have

$$\frac{\pi}{4}d^2 = \frac{\pi}{4}(d_0^2 - d_i^2)$$

or

$$d = \sqrt{d_0^2 - d_i^2}$$

For a solid shaft,

$$T_s = \frac{\pi d^3}{16}\tau_s$$

For a hollow shaft,

$$T_h = \frac{\pi(d_0^4 - d_i^4)\tau_h}{16 d_0}$$

$$\frac{T_h}{T_s} = \left(\frac{d_0^4 - d_i^4}{d_0}\right) \times \frac{1}{d^3} \times \frac{\tau_h}{\tau_s} = \left(\frac{d_0^2 - d_i^4}{d_0}\right) \times \frac{1}{(d_0^2 - d_i^2)d} \times \frac{\tau_h}{\tau_s}$$

$$= \left(\frac{d_0^2 + d_i^2}{d_0\, d}\right) \times \frac{\tau_h}{\tau_s} = \frac{d_0}{d}(1+k^2) \times \frac{\tau_h}{\tau_s}$$

where $$k = \frac{d_i}{d_0}$$

Now $$\left(\frac{d_0}{d}\right)^2 = \frac{d_0^2}{d_0^2 - d_i^2} = \frac{1}{1-k^2}$$

$$\frac{d_0}{d} = \frac{1}{\sqrt{1-k^2}}$$

$$\frac{T_h}{T_s} = \frac{1+k^2}{\sqrt{1-k^2}} \times \frac{\tau_h}{\tau_s} \qquad \text{... (7.3 a)}$$

If $\tau_h = \tau_s$, then

$$\frac{T_h}{T_s} = \frac{1+k^2}{\sqrt{1-k^2}} \qquad \text{... (7.3 b)}$$

7.3.2 For the same Strength

Here $\tau_h = \tau_s$,

$$\frac{T_h}{T_s} = \left(\frac{d_0^4 - d_i^4}{d_0}\right) \times \frac{1}{d^3}$$

For the same outer diameter, $d = d_0$.

$$\frac{T_h}{T_s} = \frac{d_0^4 - d_i^4}{d_0^4} = 1 - \left(\frac{d_i}{d_0}\right)^4 = 1 - k^4 \qquad \text{... (7.4)}$$

7.4 POWER TRANSMITTED

The power transmitted by a shaft is given by:

$$P = \frac{\omega T_{mean}}{10^3} - \frac{2\pi N T_{mean}}{60 \times 10^3} kW \qquad \text{... (7.5)}$$

where N = revolutions/min

$$\omega = \frac{2\pi N}{60} \text{ rad/s}$$

T_{mean} = mean torque transmitted, N.m.

$$T_{max} = \text{maximum torque transmitted, N.m.}$$
$$= (1 + \text{over load factor}) \times T_{mean}$$

7.5 TORSIONAL STIFFNESS

It is the torque per unit angle of twist.

From torsional formula, we have

$$\frac{T}{J} = \frac{G\theta}{l}$$

Torsional stiffness,
$$k_t = \frac{T}{\theta} = \frac{GJ}{l} \quad \text{... (7.6)}$$

7.6 TORSIONAL SECTION MODULUS

The torsional section modulus,

$$z_p = \frac{J}{r}$$

$$= \frac{\pi d^4}{32} \times \frac{2}{d} = \frac{\pi d^3}{16} \text{ for a solid shaft.} \quad \text{... (7.7a)}$$

$$= \frac{\pi}{16}\left(\frac{d_0^4 - d_i^4}{d_0}\right) = \frac{\pi}{16} d_0^3 (l - k^4) \text{ for a hollow shaft.} \quad \text{... (7.7b)}$$

where
$$k = \frac{d_i}{d_0}$$

Torque = Torsional section modulus x shear stress

$$T = z_p \tau \quad \text{... (7.8)}$$

Example 7.1 *A hollow shaft of diameter ratio 0.6 is required to transmit 500 kW at 100 rpm. The maximum torque being 15% greater than the mean. The shear stress is not to exceed 60 MPa and the angle of twist in a length of 3 m is limited to 1°. Calculate the minimum external diameter satisfying these conditions G = 80 GPa.*

Solution:

Given : $k = 0.6$, $P = 500$ kN, $N = 100$ rpm, $T_{max} = 1.15\,T_{mean}$, $t_{max} = 60$ MPa, $l = 3$M, $q = 1°$, $G = 80$ GPa.

$$P = \frac{2\pi N T_{mean}}{60 \times 10^3} \text{kW}$$

$$500 = \frac{2\pi \times 100 \times T_{mean}}{60 \times 10^3}$$

$$T_{mean} = \frac{500 \times 60 \times 10^3}{2\pi \times 100} = 47{,}746.5 \text{ N.m}$$

$$T_{max} = 1.15 \times 47{,}746.5 = 54{,}908.5 \text{ N.m}$$

$$d_i = 0.6\, d_0$$

$$J = \frac{\pi d_0^4}{32}[1-(0.6)^4] = 0.08545 d_0^4 \text{ mm}^4$$

$$\frac{T_{max}}{J} = \frac{\tau}{r}$$

$$\frac{54,908.5\times 10^3}{0.0854 d_0^4} = \frac{60\times 2}{d_0}$$

$$d_0^3 = \frac{54.908.5\times 10^3}{0.08545\times 120} = 5354.75\times 10^3$$

$$d_0 = 175.\text{ mm}$$

$$\frac{T_{max}}{J} = \frac{G\theta}{l}$$

$$\frac{54.908.5\times 10^3}{0.08545 d_0^4} = \frac{80\times 10^3\times \pi}{180\times 3\times 10^3}$$

$$d_0^4 = \frac{54,908.5\times 10^3\times 180\times 3}{0.08454\times 80\times \pi} = 1380643056$$

$$d_0 = 192.8 \text{ mm} \simeq 193 \text{ mm}$$

Hence $d_0 = 193$ mm will satisfy both the conditions.

Example 7.2 *Compare the weights of two solid shafts of identical length, one of steel and the other of aluminium alloy, both designed for the same angle of twist per unit length when subjected to the same torque.* $G_s = 3G_{al}$ *and* $\gamma_s = 2\gamma_{al}$, *where* γ = *specific weight.*

Solution:

Weight of steel shaft, $W_s = \frac{\pi}{4} d_s^2 l_s \gamma_s$

Weight of aluminium shaft, $W_{al} = \frac{\pi}{4} d_{al}^2 l_{al} \gamma_{al}$

Now $T_s = T_{al} = T$

and $l_s = l_{al} = l$

$$\frac{\theta}{l} = \frac{T}{JG}$$

$$J_s = \frac{\pi d_s^4}{32}, J_{al} = \frac{\pi d_{al}^4}{32}$$

$$\left(\frac{\theta}{l}\right)_s = \frac{32T}{\pi d_s^4 G_s}$$

$$\left(\frac{\theta}{l}\right)_{al} = \frac{32T}{\pi d_{al}^4 G_{al}}$$

$$\left(\frac{\theta}{l}\right)_s = \left(\frac{\theta}{l}\right)_{al}$$

$$\frac{1}{d_s^4\, G_s} = \frac{1}{d_{al}^4\, G_{al}}$$

$$\left(\frac{d_s}{d_{al}}\right)^4 = \frac{G_{al}}{G_s} = \frac{1}{3} \text{ or } \left(\frac{d_s}{d_{al}}\right)^2 = \frac{1}{\sqrt{3}}$$

$$\frac{W_s}{W_{al}} = \left(\frac{d_s}{d_{al}}\right)^2 \times \frac{\gamma_s}{\gamma_{al}} = \frac{1}{\sqrt{3}} \times 3 = \sqrt{3} = 1.732$$

Example 7.3 *A solid aluminium shaft 1 m long and 45 mm outside diameter is to be replaced by a tubular steel shaft of the same length and the same outside diameter so that either shaft could carry the same torque and have the same angle of twist over the whole length. What must be the inner radius of the tubular shaft ? $G_s = 3\,G_{al} = 84$ GPa.*

Solution:

Given:

$$l = 1\ m,\ d_0 = 45 \text{ mm},\ d_i = ?$$

$$l_{al} = l_s,\ T_{al} = T_s,\ \theta_{al} = \theta_s,\ d_{al} = d_0$$

$$J_{al} = \frac{\pi d_0^4}{32} = \frac{\pi}{32} \times 45^4 \text{ mm}^4$$

$$J_s = \frac{\pi}{32}(d_0^4 - d_i^4) = \frac{\pi}{32}(45^4 - d_i^4) \text{ mm}^4$$

$$\frac{T}{J} = \frac{G\theta}{l}, \frac{\theta}{l} = \frac{T}{GJ}$$

$$\left(\frac{T}{GJ}\right)_{al} = \left(\frac{T}{GJ}\right)_s$$

$$\frac{32\, T_{al}}{G_{al} \times \pi \times 45^4} = \frac{32 T_s}{G_s \times \pi \times (45^4 - d_i^4)}$$

$$\frac{G_{al}}{G_s} = \frac{45^4 - d_i^4}{45^4} = \frac{1}{3}$$

$$3(45^4 - d_i^4) = 45^4$$

$$3d_i^4 = 45^4\,(3 - 1) = 8201250$$

$$d_i^4 = 2733750$$

$$d_i = 40.66 \text{ mm}$$

$$r_i = 20.33 \text{ mm}$$

Example 7.4 *A solid shaft is to transmit 300 kW at 80 rpm. The shear stress is not to exceed 60 MPa. Find the shaft diameter. Calculate the percentage saving in weight if this shaft were replaced by a hollow one whose internal diameter equals 60 percent of external diameter. The length, material and maximum shear stress remain the same.*

Solution:

Given: $P = 300$ kW, $N = 80$ rpm, t = 60 MPa, $k = 0.6$

Power transmitted $$P = \frac{2\pi NT}{60\times10^3}\text{ kW}$$

$$T = \frac{300\times60\times10^3}{2\pi\times80} = 35809.86\text{ N.m}$$

For the hollow shaft, $$J_h = \frac{\pi}{32}\left(d_0^4 - d_i^4\right) = \frac{\pi}{32}d_0^4\left[1-(0.6)^4\right]$$

$$= 0.0272\times\pi d_0^4$$

For the solid shaft,

$$J_s = \frac{\pi}{32}d^4$$

Now $$T_s = T_h \text{ and } \tau_h = \tau_s$$

Now $$\frac{T}{J} = \frac{\tau}{r}$$

or $$\frac{T}{\tau} = \frac{J}{r}$$

or $$\frac{J_h}{d_0} = \frac{J_s}{d}$$

$$\frac{0.0272\times\pi d_0^4}{d_0} = \frac{\frac{\pi}{32}d^4}{d}$$

$$d^3 = 0.8704\ d_0^3$$

Also $$\tau = \frac{16T}{\pi d^3}$$

$$60\times10^6 = \frac{16\times35809.86}{\pi d^3}$$

$$d^3 = 3039.6\times10^{-6}$$

$$d = 14.486\times10^{-2}\text{ m}$$

$\therefore$ $$3039.6\times10^{-6} = 0.87504\ d_0^3$$

$$d_0 = 15.17\times10^{-2}\text{ m}$$

$$d_i = 9.10 \times 10^{-2}\ \text{m}$$

Area of solid shaft, $A_s = \frac{\pi}{4} \times (14.486 \times 10^{-2})^2 = 164.81 \times 10^{-4}\ \text{m}^2$

Area of hollow shaft, $A_h = \frac{\pi}{4} \times [(15.17)^2 - (9.10)^2 \times] \times 10^{-4} = 115.7 \times 10^{-4}\ \text{m}^2$

For same length and material of solid and hollow shafts, the weights are proportional to areas. Hence,

$$\text{Percentage saving in weight} = \frac{A_s - A_h}{A_s} \times 100$$

$$= \frac{164.81 - 115.7}{164.81} = 29.8\%$$

Example 7.5 *A hollow shaft of diameter ratio 0.6 is required to transmit 450 kW at 100 rpm, the maximum torque being 20 % greater than the mean value. The shear stress is not to exceed 56 MPa and the angle of twist in a length of 3m is not to exceed 1.5 degrees. Calculate the external diameter satisfying above conditions. Take* $G = 84$ GPa.

Solution:

Given: $k = 0.6$, $P = 450$ kW, N = 100 rpm, overload = 0.2, $t_{max} = 56$ MPa, $l = 3$m, q = 1.5°.

Power transmitted, $P = \frac{2\pi N T_{mean}}{60 \times 10^3}$

$$T_{mean} = \frac{4580 \times 60 \times 10^3}{2\pi \times 100} = 42972\ \text{N.m}$$

(*i*) For a hollow shaft, $T_{max} = 1.2 \times 42972 = 51566.2$ N.m

$$\tau_{max} = \frac{16 T_{max}}{\pi d_0^3 (1 - k^4)}$$

$$d_0^3 = \frac{16 \times 51566.2 \times 10^3}{\pi \times 56 \times [1 - (0.6)^4]} = 5388 \times 10^8$$

$$d_0 = 175.3\ \text{mm}$$

(*ii*) $\frac{T}{J} = \frac{G\theta}{l}$

$$\frac{T}{\frac{\pi}{32} d_0^4 (1 - k^4)} = \frac{G\theta}{l}$$

$$d_0^4 = \frac{51566.2 \times 3 \times 32 \times 180}{\pi[-(0.6)^4] \times 84 \times 10^9 \times \pi \times 1.5} = 82322.7 \times 10^{-8}$$

$$d_0 = 0.1694\ \text{m or } 169.4\ \text{mm}$$

Adopt, $d_0 = 175.3$ m

Example 7.6 *A solid circular shaft is to transmit* **250** *kW at* **120** *rpm. (a) If shear stress is not the exceed 63 MPa, find its diameter, (b) If this shaft is replaced by a hollow shaft whose diameter ratio is 0.6, what percentage saving in weight shall be obtained keeping the material and maximum shear stress same.*

Solution:

Given : $P = 250$ kW, $N = 120$ rpm, $t_{max} = 63$ MPa, $k = 0.6$

(a)
$$P = \frac{2\pi NT}{60\times10^3}$$

$$T = \frac{250\times60\times10^3}{2\pi\times120} = 19894 \text{ N.m}$$

Now
$$T = \frac{\pi}{16}d^3\tau$$

$$d = \sqrt[3]{\frac{16\times19894}{gp\times63\times10^6}} = 117 \text{ mm}$$

(b) For
$$\tau_s = \tau_h \text{ and } T_s = T_h$$

$$d^3 = d_0^3(1-k^4)$$

$$d_0^3 = \frac{117^3}{1-(0.6)^4} = 1840088$$

$$d_0 = 122.54 \text{ mm}$$

$$\frac{W_h}{W_s} = \frac{A_h}{A_s} = \frac{d_0^2-d_i^2}{d^2} = \frac{(122.54)^2[1-(0.6)^2]}{117^2} = 0.702$$

Percentage saving in weight
$$= \left(1-\frac{W_h}{W_s}\right)\times100$$

$$= (1\text{–}0.702)\times100 = 29.8\ \%$$

Example 7.7 *A hollow steel shaft is to be replaced by an aluminium shaft of the same external diameter. The material of steel shaft is 25% stronger than aluminium shaft in shear. Calculate the diameter ratios. Also find the percentage saving in weight assuming that steel is 3 times heavier than aluminum.*

Solution:

Given: $t_s = 1.25\ t_{al}$, $r_s = 3r_{al}$.

Let d_0, d_i = outside and inside diameters of steel shaft respectively

$d = d_0$ diameter of aluminium shaft

For same torque to be transmitted by both the shafts,

$$T_s = T_{al}$$

$$\frac{\pi}{16}d_0^3(1-k^4)\tau_s = \frac{\pi}{16}d_0^3\,\tau_{al}$$

$$(1-k^4)\tau_s = \tau_{al}$$

Now
$$\tau_s = 1.25\,\tau_{al}$$

$$\therefore \quad 1.25\,(1-k^4) = 1$$

$$1-k^4 = 0.8$$

$$k^4 = 0.2$$

$$k = 0.6687$$

Weight $\propto$ (area of cross-section × density)

$$W_s = \frac{\pi}{4} d_0^2 (1-k^2)\,\rho_s$$

$$W_{al} = \frac{\pi}{4} d_0^2\,\rho_{al}$$

$$\frac{W_s}{W_{al}} = (1-k^2)\frac{\rho_s}{\rho_{al}}$$

$$= [1-(0.6687)^2] \times 3 = 165852$$

or $$\frac{W_{al}}{W_s} = 0.603$$

Percentage saving in weight = (1 – 0.603) × 100 = 39.7 %

Example 7.8 *A solid shaft of medium carbon steel 100 mm diameter is to be replaced by a hollow shaft of alloy steel. The shear strength of alloy steel is 30% greater than medium carbon steel. The power to be transmitted is increased by 20 % and the speed of rotation decreased by 5%. Determine the maximum internal diameter of hollow shaft if its external diameter is to be 100 mm.*

Solution:

Given : $d_s = 100$ mm, $d_0 = 100$ mm $d_i = ?$, $\tau_h = 1.3\,\tau_s$, $P_h = 1.2\,P_s$, $N_h = 0.95\,N_s$.

$$P = \frac{2\pi NT}{60\times 10^3}$$

$$P_s = \frac{2\pi N_s T_s}{60\times 10^3}$$

$$P_h = \frac{2\pi N_h T_h}{60\times 10^3}$$

$$P_h = 1.2\,P_s$$

$$\therefore \quad 2\pi N_h T_h = 1.2 \times 2\pi N_s T_s$$

$$N_h T_h = 1.2\,N_s T_s$$

$$0.95\,T_h = 1.2\,T_s$$

or $$T_h = 1.263\,T_s \qquad \text{...(1)}$$

$$T_s = \frac{\pi}{16} d_s^3 \tau_s$$

$$T_h = \frac{\pi}{16} d_s^3 (1-k^4)\,\tau_h$$

$$\frac{T_h}{T_s} = \frac{(01-k^4)}{\tau_s} = 1.3(1-k^4) \quad \text{... (2)}$$

Substituting Eq. (1) in (2), we have

$$1.263 = 1.3\ (1-k^4)$$
$$0.9715 = 1-k^4$$
$$k^4 = 0.02846$$
$$k = 0.41$$
$$d_i = 0.41 \times 100 = 41 \text{ mm}$$

7.7 SHAFTS IN SERIES

Consider two shafts in series as shown in Fig. 7.3 and subjected to torque T.

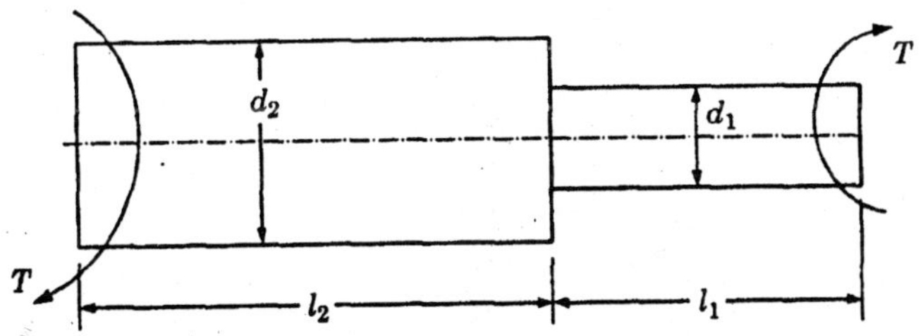

Fig. 7.3 Shafts in series.

$$T_1 = \frac{J_1 G_1 \theta_1}{l_1} = \frac{\tau_1 J_1}{r_1}$$

$$T_2 = \frac{J_2 G_2 \theta_2}{l_2} = \frac{\tau_2 J_2}{r_2}$$

Since $T_1 = T_2 = T$

$\therefore$ Ratio of angles of twist, $$\frac{\theta_1}{\theta_2} = \frac{J_2}{J_1} \times \frac{G_2}{G_1} \times \frac{l_2}{l_1} = \left(\frac{d_2}{d_1}\right)^4 \times \frac{G_2}{G_1} \times \frac{l_1}{l_2} \quad \text{... (7.9)}$$

and ratio of shear stress, $$\frac{\tau_1}{\tau_2} = \frac{J_2}{J_1} \times \frac{r_1}{r_2} = \frac{d_2^4}{d_1^4} \times \frac{d_1}{d_2} = \left(\frac{d_2}{d_1}\right)^3 \quad \text{... (7.10)}$$

Total angle of twist, $$\theta = \theta_1 + \theta_2$$

$$= T\left[\frac{l_1}{J_1 G_1} + \frac{l_2}{J_2 G_2}\right] \quad \text{... (7.11)}$$

7.8 SHAFTS IN PARALLEL

Consider two shafts in parallel as shown in Fig. 7.4. Let T be the torque applied to the composite shaft, then

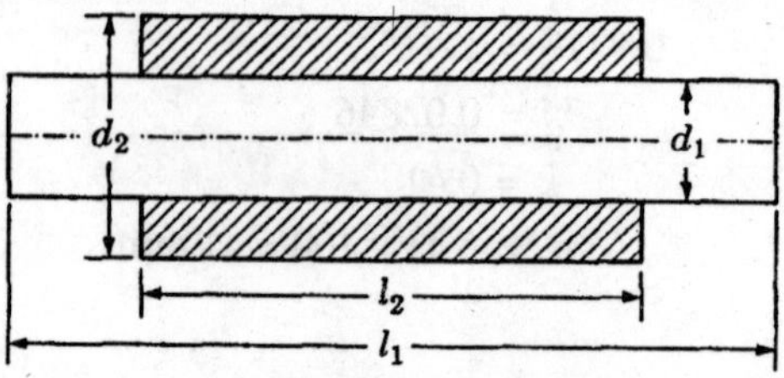

Fig. 7.4 Shafts in parallel

$$T = T_1 + T_2$$

$$= \frac{J_1 G_1 \theta_1}{l_1} + \frac{J_2 G_2 \theta_2}{l_2} \quad \text{... (7.12)}$$

But

$$\theta_1 = \theta_2 = \theta$$

$$\frac{T_1 l_1}{G_1 J_1} = \frac{T_2 l_2}{G_2 J_2}$$

$$T = \theta\left[\frac{J_1 G_1}{l_1} + \frac{J_2 G_2}{l_2}\right] \quad \text{... (7.13)}$$

and

$$\frac{T_1}{T_2} = \frac{G_1}{G_2} \times \frac{J_1}{J_2} \times \frac{l_2}{l_1} \quad \text{... (7.14)}$$

where

$$J_1 = \frac{\pi d_1^4}{32} \text{ and } J_2 = \frac{\pi}{32}(d_2^2 - d_1^4)$$

Example 7.9 *A stepped steel shaft shown in Fig. 7.5 is subjected to torsion. Calculate the total angle of twist at the free end, if the maximum shear stress in the shafts is limited to 60 MPa. Take G = 84 GPa.*

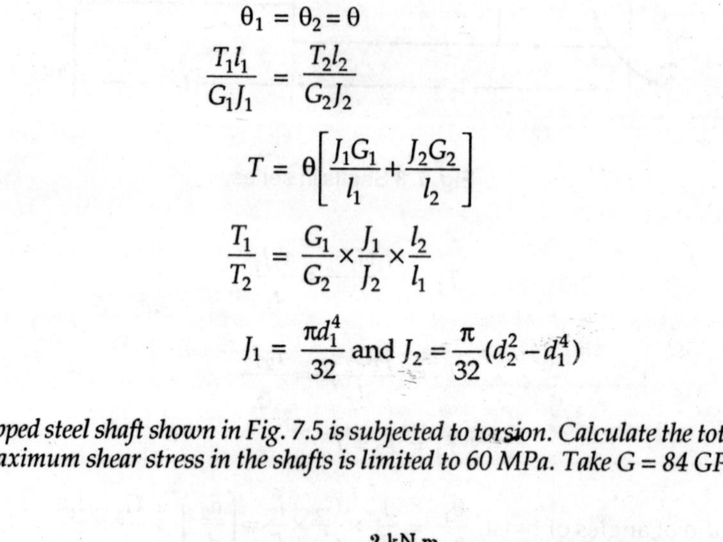

Fig. 7.5

Solution:

Given: $d_1 = 25$ mm, $l_1 = 1$m, $d_2 = 50$ mm, $l_2 = 1$m, $G = 84$ GPa.

Length AB:

$$T = T\ (ccw)$$

$$l_1 = 1\text{m}, d_1 = 25 \text{ mm}$$

$$J_1 = \frac{\pi d_1^4}{32} = \frac{\pi \times 25^4}{32} = 38349.5 \text{ mm}^4$$

$$\theta = \frac{Tl_1}{GJ_1} = \frac{T \times 1.0}{84 \times 10^9 \times 38349.5 \times 10^{-12}}$$

$$= 3.1043 \times 10^{-4}\ T \text{ rad } (ccw)$$

Length BC:

$$T = (1.5 - 1.0) = 0.5\,T\ (cw)$$

$$l_2 = 1.0 \text{ m}, d_2 = 50 \text{ mm}$$

$$J_2 = \frac{\pi d_2^4}{32} = \frac{\pi \times 50^4}{32} = 613592.3 \text{ mm}^4$$

$$\theta_2 = \frac{Tl_2}{GJ_2} = \frac{0.5T \times 1.0}{84 \times 10^9 \times 613592.3 \times 10^{-12}}$$

$$= 9.7 \times 10^{-6} \times T \text{ rad } (cw)$$

Total angle of twist,

$$\theta = \theta_1 - \theta_2$$

$$= (3.1043 - 0.097) \times 10^{-4} \times T$$

$$= 3.0073\ 5\ 10^{-4} \times T \text{ rad (ccw)}$$

Maximum shear stress shall occur in AB portion.

$$T = \frac{\pi}{16} d_1^3 \tau_1 = \frac{\pi}{16} \times 25^3 \times 60 \times 10^{-3} = 184.08 \text{ N.m}$$

$$\theta = 3.0073 \times 10^{-4} \times 184.08 = 0.05536 \text{ rad or } 3.17°$$

Example 7.10 *A steel shaft ABCD shown in Fig. 7.6 is subjected to equal opposite torques at the ends. Find the maximum permissible value of d_i if the maximum shearing stress in AB is not to exceed that in CD. Also calculate the total angle of twist if torque applied is 15 kN. m. Take G = 82 GPa.*

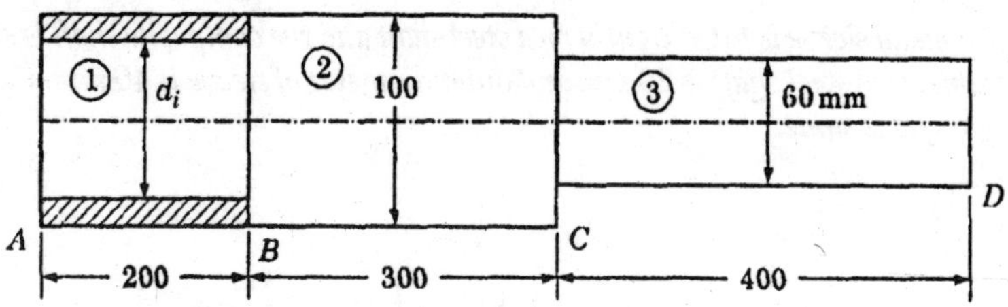

Fig. 7.6

Solution:

Length AB: $d_0 = 100$ mm, $l_1 = 200$ mm

$$T = \frac{\pi}{16} d_0^3 (1 - k^4) \tau_1$$

$$= \frac{\pi}{16} \times (0.1)^3 \times (1-k^4)\tau_1$$

Length CD: $d_3 = 60$ mm, $l_3 = 400$ mm

$$T = \frac{\pi}{16} \times (0.06)^3\ \tau_3$$

$$\tau_1 = \tau_3 \text{ (given)}$$

$$\frac{16T}{\pi \times (0.1)^3 \times (1-k^4)} = \frac{16T}{\pi \times (0.06)^3}$$

or

$$(0.1)^3 (1-k^4) = (0.06)^3$$

$$1 - k^4 = 0.216$$

$$k^4 = 0.784, \quad k = 0.941$$

$$\frac{d_i}{d_0} = 0.941$$

$$d_i = 0.941 \times 100 = 94.1 \text{ mm}$$

Total angle of twist,

$$\theta = \theta_1 + \theta_2 + \theta_3$$

$$= \frac{T}{G}\left[\frac{l_1}{J_1} + \frac{l_2}{J_2} + \frac{l_3}{J_3}\right]$$

$$= \frac{32T}{\pi G}\left[\frac{l_1}{d_0^4 - d_i^4} \frac{l_2}{d_2^4} + \frac{l_3}{d_3^4}\right]$$

$$= \frac{32 \times 15 \times 10^3}{\pi \times 82 \times 10^9}\left[\frac{200}{\{100^4 - (94.1)^4\}} + \frac{300}{100^4} + \frac{400}{60^4}\right] \times 10^9$$

$$= 0.08036 \text{ rad or } 4.6°$$

Example 7.11 *A gun metal sleeve is fitted tightly on a steel shaft and the composite shaft is subjected to a torque of 100 kN. m. The diameter of steel shaft is 20 mm and outer diameter of sleeve is 40 mm. Calculate the sharing of torque if length of shafts is same.*

Take $G_s = 2.5\ G_{gm}$

Solution:

Given

$$T = 100 \text{ kN.m}, l_s = l_{gm}, \frac{G_s}{G_{gm}} = 2.5$$

$$\frac{J_s}{J_{gm}} = \frac{20^4}{40^4 - 20^4} = 0.0667$$

$$\frac{T_s}{T_{gm}} = \frac{G_s}{G_{gm}} \times \frac{J_s}{J_{gm}} \times \frac{l_{gm}}{l_s}$$

$$= 2.5 \times 0.667 \times 1 = 0.167$$

$$T_s + T_{gm} = 100 \text{ kN.m}$$

$$T_{gm} = \frac{100}{1.167} = 85.69 \text{kN.m}$$

$$T_s = 14.31 \text{ kN.m}$$

Example 7.12 *Two solid shafts of different materials are fastened together and attached to rigid supports 4m apart as shown in Fig. 7.7. A torque of 2 kN.m is applied at the junction of two segments. Calculate the maximum shear stress developed in the assembly.*
Take $G_s = 3G_{al}$.

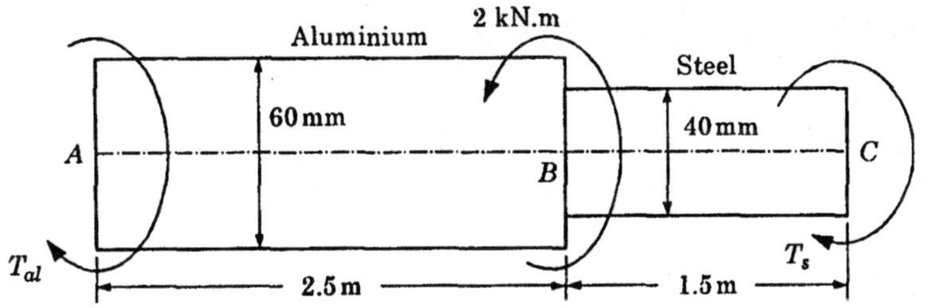

Fig. 7.7

Solution:

Given: $d_s = 40$ mm, $l_s = 1.5$ m, $d_{al} = 60$ mm, $l_{al} = 2.5$ m

$$T_s + T_{al} = \text{T} = 2 \text{ kN.m}$$

$$\theta_s = \theta_{al}$$

$$\frac{T_s l_s}{G_s J_s} = \frac{T_{al}\, l_{al}}{G_{al}\, J_{al}}$$

$$\frac{T_{al}}{T_s} = \frac{G_{al}}{G_s} \times \frac{J_{al}}{J_s} \times \frac{l_s}{l_{al}}$$

$$= \frac{1}{3} \times \left(\frac{60}{40}\right)^4 \times \frac{1.5}{2.5} = 1.0125$$

$$T_s = \frac{2 \times 10^3}{2.0125} = 993.8 \text{ N.m}$$

$$T_{al} = 1006.2 \text{ N.m}$$

Maximum shear stress in steel, $\tau_s = \dfrac{16T_s}{\pi d_s^3} = \dfrac{16 \times 993.8 \times 10^3}{\pi \times 40^3} = 79.08$ MPa

Example 7.13 *A composite shaft is subjected to a torque of 3 kN.m as shown in Fig. 7.8. Calculate the maximum shear stress developed in steel and copper.* $G_s = 2G_{cu}$

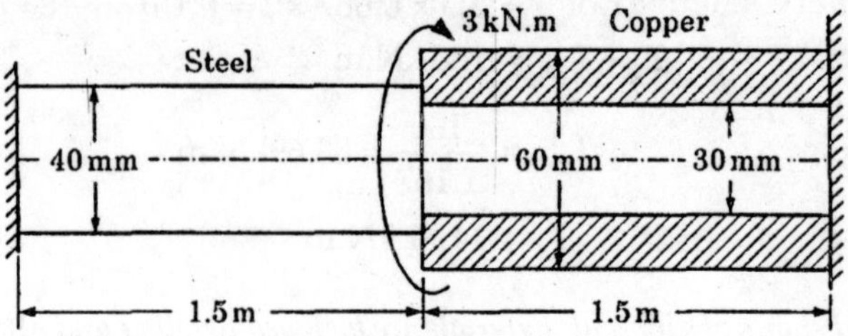

Fig. 7.8

Solution:

Given: $d_s = 40$ mm, $l_s = 1.5$ m, $d_0 = 60$ mm, $d_i = 30$ mm, $l_{cu} = 1.5$ m

$$J_s = \frac{\pi \times 40^4}{32}$$

$$J_{cu} = \frac{\pi}{32}(60^4 - 30^4)$$

$$\frac{J_s}{J_{cu}} = \frac{40^4}{60^4 - 30^4} = 0.2107$$

$$\theta_s = \theta_{cu} \quad \text{gives}$$

$$\frac{T_s}{T_{cu}} = \frac{G_s}{G_{cu}} \times \frac{J_s}{J_{au}} \times \frac{l_{al}}{l_s} = 2 \times 0.2107 \times 1 = 0.4214$$

$$T_s + T_{cu} = 3000 \text{ N.m}$$

$$T_{cu} = \frac{3000}{1.4214} = 2110.6 \text{ N.m}$$

$$T_s = 889.4 \text{ N.m}$$

$$\tau_s = \frac{T_s d_s}{2J_s} = \frac{88.9 \times 40 \times 32 \times 10^3}{2\pi \times 40^4} = 70.77 \text{ N/mm}^2$$

$$\tau_{cu} = \frac{2110.6 \times 10^3 \times 32}{\pi(60^4 - 30^4)} = 53.08 \text{ N/mm}^2$$

7.9 TAPERED CIRCULAR SHAFT

Consider a tapered circular shaft as shown in Fig. 7.9 subjected to torque T applied at the ends. Let τ be the shear stress at a distance x from the bigger end. Since torque is same throughout the shaft, therefore,

$$\frac{\pi}{16} d_1^3 \tau_1 = \frac{\pi}{16} d_2^3 \tau_2 = \frac{\pi}{16} d^2 \tau$$

or

$$d_1^3 \tau_1 = d_2^3 \tau_2 = d^3 \tau$$

Consider an elementary length dx of the shaft at a distance x from the larger end. Let d be the diameter of the shaft of this section.

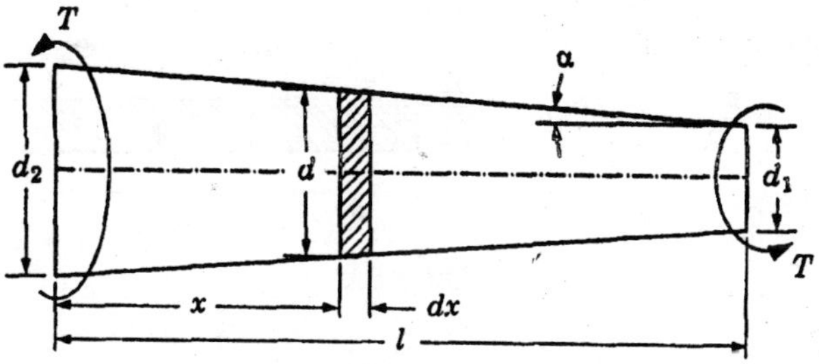

Fig. 7.9 Tapered shaft

Angle of twist of length dx,

$$d\theta = \frac{T \times dx}{GJ} = \frac{32T \times dx}{\pi d^4 G}$$

Now

$$\tan\alpha = \frac{d_2 - d_1}{2l}$$

$$d = d_2 - \left(\frac{d_2 - d_1}{l}\right)x$$

$$= d_2 - kx$$

where

$$k = \frac{d_2 - d_1}{l}$$

$$d\theta = \frac{32T \times dx}{\pi G (d_2 - kx)^4}$$

Total angle of twist of the whole shaft,

$$\theta = \frac{32T}{\pi G}\int_0^l \frac{dx}{(d_2 - kx)^4}$$

$$= \frac{32T}{3\pi Gk}\left|(d_2 - kx)^{-3}\right|_0^l$$

$$= \frac{32T}{3\pi Gk}\left[\frac{1}{(d_2 - kl)^3} - \frac{1}{d_2^3}\right]$$

$$= \frac{32Tl}{3\pi G(d_2 - d_1)}\left[\frac{1}{d_1^3} - \frac{1}{d_2^3}\right]$$

$$= \frac{32Tl}{3\pi G(d_2 - d_1)}\left[\frac{d_2^3 + d_1^3}{d_1^3 d_2^3}\right]$$

$$= \frac{32Tl}{3\pi G}\left(\frac{d_1^2 + d_2^2 + d_1 d_2}{d_1^3 d_2^3}\right) \quad \text{... (7.15)}$$

Maximum shear strees,

$$\tau_{max} = \frac{16T}{\pi d_1^3} \quad \text{... (7.16)}$$

7.10 STATICALLY INDETERMINATE SHAFTS

A shaft is said to be statically indeterminate if the internal torque at a given cross-section cannot be determined from moment equilibrium alone. For such a shaft the moment equilibrium equation needs to be augmented with an equation that involves the deformations of the shaft. If the axis of the shaft is along z-axis, then the equilibrium equation for moments about the axis of shaft gives,

$$\Sigma M_z = 0 \quad \text{... (7.17)}$$

The geometric compatibility imposes the condition that the ends of the shaft cannot rotate. If A and B are the fixed ends of the shaft, then

$$\theta_{B/A} = 0 \quad \text{... (7.18)}$$

The Eqs. (7.17) and (7.18) shall enable to solve statically indeterminate shafts.

Example 7.14 *A solid steel shaft, 6 m long, is fixed at both ends. A torque of 100 N.m is applied to the shafts at a section 2 m from one end. Determine the fixing torques set up at the ends of the shaft. Calculate the shear stress developed in the two portions if the shaft diameter is 30 mm. What is the angle of twist for the section where the torque is applied ? G = 84 GPa.*

Solution:

Let T_A and T_B be the torques set up at the ends of the shaft. Then from Fig. 7.10, we have

$$T_A + T_B = 100 \quad \text{... (a)}$$

$$\theta_{B/A} = \frac{T_B \times 6}{GJ} - \frac{100 \times 2}{GJ} = 0$$

or

$$T_B = \frac{100}{3} \text{ N.m}$$

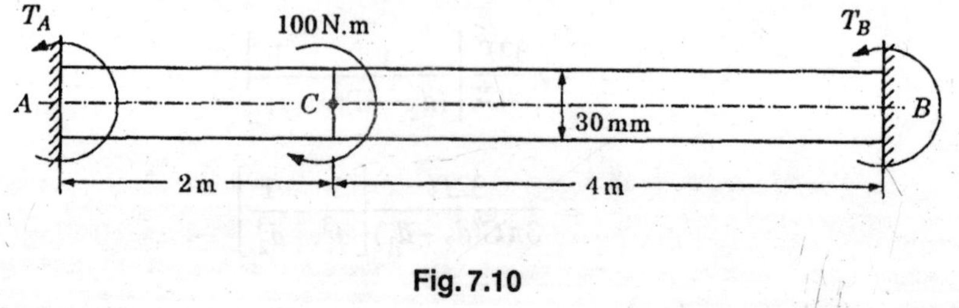

Fig. 7.10

From Eq. (a),

and $$T_A = 100 - \frac{100}{3} = \frac{200}{3} \text{ N.m}$$

Shear stress in each portion is,

$$\tau_{AC} = \frac{T_A r}{J} = \frac{200}{3} \times \frac{15 \times 10^{-3} \times 32}{\pi \times (30)^4} = 12.57 \text{ MPa}$$

$$\tau_{BC} = \frac{T_B r}{J} = \frac{100}{3} \times \frac{15 \times 10^{-3} \times 32}{\pi \times (30)^4 \times 10^{-12}} = 6.29 \text{ MPa}$$

Angle of twist, $$\theta = \frac{T_A l_1}{GJ}$$

$$= \frac{100}{3} \times \frac{2 \times 32}{84 \times 10^6 \times \pi \times (30)^4 - 10^{-12}} = 0.01996 \text{ rad}$$

Example 7.15 *A stepped steel shaft shown in Fig. 7.11 is subjected to a torque T and the free end and torque 2T in the opposite direction at the junction of two sizes. Calculate the total angle of twist at the free end if the maximum shear stress in the shaft is limited to 75 MPa. G = 84 GPa.*

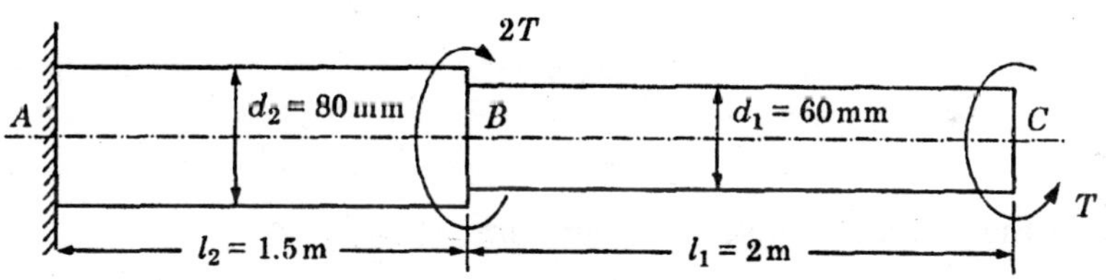

Fig. 7.11

Solution:

Torque in portion $BC = T$

Torque in portion $AB = T - 2T = -T$

$$T = \frac{\pi}{16} d_1^3 \tau$$

$$= \frac{\pi}{16} \times (60)^3 \times 10^{-9} \times 75 \times 10^6 = 3180.86 \text{ N.m}$$

Angle of twist: $$\theta_{BC} = \frac{T l_1}{G J_1}$$

$$= \frac{3180.86 \times 2 \times 32}{84 \times 10^9 \times \pi \times (60)^4 \times 10^{-12}} = 0.0595 \text{ rad.}$$

$$\theta_{AB} = \frac{-Tl_2}{GJ_2}$$

$$= \frac{3180.86 \times 1.5 \times 32}{84 \times 10^9 \times \pi \times (80)^4 \times 10^{-12}} = -0.0141 \text{ rad.}$$

$$\text{Total angle of twist} = \theta_{BC} + \theta_{AS}$$

$$= 0.0596 - 0.0141 = 0.0454 \text{ rad.} = 2° - 36'$$

Example 7.16 *A composite step shaft (Fig. 7.12), consists of an aluminium section 50 mm in diameter and a steel section 25 mm in diameter. The ends of the shaft are fixed so that rotation cannot occur there. Determine (a) the resisting torques exerted by the supports on the shaft, and (b) the maximum stress in aluminium and steel.* $G_s = 84$ *GPa,* $G_a = 28$ *GPa.*

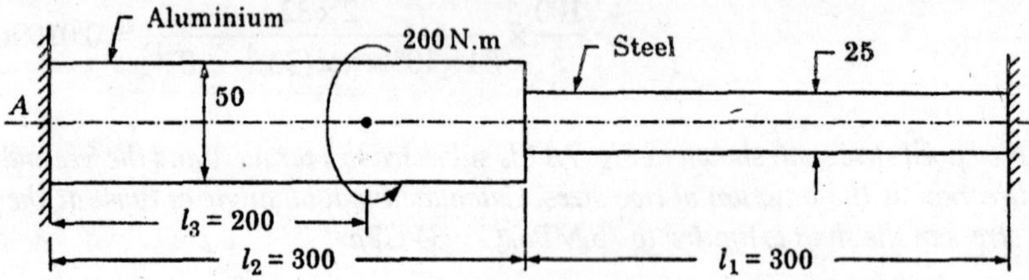

Fig. 7.12

Solution:

Let T_A and T_B be the torques exerted by the supports. Then

$$T_A + T_B = 200 \quad \text{...(a)}$$

Also

$$\theta_{B/A} = 0$$

or

$$\frac{T_B l_1}{G_1 J_1} + \frac{T_B l_2}{G_2 J_2} - \frac{200 l_3}{G_2 J_2} = 0$$

or

$$T_B \left(\frac{300 \times 10^{-3}}{84 \times 10^9 \times \frac{\pi}{32} \times (24)^4 \times 10^{-12}} + \frac{300 \times 10^{-3}}{28 \times 10^9 \times \frac{\pi}{32} \times (50)^4 \times 10^{-12}} \right)$$

$$= \frac{200 \times 200 \times 10^{-3}}{28 \times 10^9 \times \frac{\pi}{32} (54)^4 \times 10^{-12}}$$

$$T_B \left(1 + \frac{3}{16}\right) = 25$$

$$T_B = \frac{25 \times 16}{19} = 21.05 \text{ N.m}$$

From Eq. (*a*), are get

$$T_A = 178.95 \text{ N.m}$$

$$(\tau_a)_{max} = \frac{178.95 \times 0.025 \times 32}{\pi \times (0.05)^4} = 7.29 \text{ MPa}$$

$$(\tau_s)_{max} = \frac{21.05 \times 0.0125 \times 32}{\pi \times (0.025)^4} = 6.86 \text{ MPa}$$

7.11 COMBINED BENDING AND TORSION OF A SHAFT

Consider a shaft subjected to the combined action of bending and torsion. Let M be the bending moment and T the torque applied to the shaft. Then

Bending stress, $$\sigma_b = \frac{32M}{\pi d^3}$$

Torsional shear stress, $$\tau = \frac{16T}{\pi d^3}$$

The principal stress are, $$\sigma_{1,2} = \frac{1}{2}\sigma_b \pm \sqrt{\sigma_b^2 + 4\tau^2}$$

$$= \frac{16M}{\pi d^3} \pm \frac{1}{2}\sqrt{\left(\frac{32M}{\pi d^3}\right)^2 + 4\left(\frac{16T}{\pi d^3}\right)^2}$$

$$= \frac{16M}{\pi d^3} \pm \frac{16}{\pi d^3}\sqrt{M^2 + T^2}$$

$$= \frac{16}{\pi d^3}[M \pm \sqrt{M^2 + T^2}] = \frac{32}{\pi d^3}\left[\frac{1}{2}(M \pm \sqrt{M^2 + T^2})\right]$$

$$= \frac{32}{\pi d^3} \times M_e$$

Maximum shear stress, $$\tau_{max} = \frac{1}{2}(\sigma_1 - \sigma_2)$$

$$= \frac{16}{\pi d^3}\sqrt{M^2 + T^2} = \frac{16}{\pi d^3} \times T_e$$

where $$M_e = \frac{1}{2}[M \pm \sqrt{M^2 + T^2}] \quad \text{.... (7.20)}$$

= Equivalent bending moment

$$T_e = \sqrt{M^2 + T^2}$$

= Equivalent torque .... (7.21)

Equivalent bending moment: It is the bending moment which produces the same normal stress as produced by the combined effect of bending and torsion.

Equivalent torque: It is the torque which produces the same torsional shear stress as produced by the combined effect of bending and torsion.

Example 7.17 *A shaft is required to transmit 600 kW at 90 rpm. The maximum torque is 1.6 times the mean torque. The main bearings are 4m apart and the shaft carries a flywheel midway between the bearings weighing 300 kN. Calculate the diameter of the shaft so that the maximum tensile stress does not exceed 60 MPa and the shear stress is limited to 40 MPa.*

Solution:

Given: $P = 600$ kW, N = 90 rpm, $T_{max} = 1.6\ T_{mean}$, $L = 4$m, W = 300 kN, $s_b = 60$ MPa, t = 40 MPa.

Power transmitted, $$P = \frac{2\pi\ NT_{mean}}{60\times10^3}\text{kW}$$

$$T_{mean} = \frac{600\times60\times10^3}{2\pi\times90} = 63662\ \text{N.m}$$

$$T_{max} = 1.6\times63662 = 1011859\ \text{N.m}$$

$$M = \frac{WL}{4} = \frac{300\times10^3\times4}{4} = 300\times10^3\text{N.m}$$

Equivalent bending moment,

$$M_e = \frac{1}{2}[M+\sqrt{M^2+T^2}]$$

$$= \frac{1}{2}\left[300\times10^3+\sqrt{(300\times10^3)^2\times(101859)^2}\right] = 308410\ \text{N.m}$$

$$\sigma_b = \frac{32M_e}{\pi d^3}$$

$$d = \sqrt[3]{\frac{32\times308410}{\pi\times60\times10^6}} = 0.374\ \text{m}$$

Equivalent torque, $$T_e = \sqrt{M^2+T^2} = \sqrt{(300\times10^3)^2+(101859)^2} = 316820\ \text{N.m}$$

$$\tau_{max} = \frac{16T_e}{\pi d^3}$$

$$d = \sqrt[3]{\frac{16\times316820}{\pi\times40\times10^6}} = 0.343\text{m}$$

Safe diameter of shaft $= 0.374$ m

Example 7.18 *A hollow shaft is subjected to a torque of 40 kN.m and a bending moment of 30 kN.m. The internal diameter of shaft is 60% of the external diameter. If the maximum shear stress is not to exceed 63 MPa, find the diameter of the shaft.*

Solution:

Given: $M = 30$ kN.m, $T = 40$ kN.m, $k = \dfrac{d_i}{d_0} = 0.6$, $t_{max} = 63$ MPa

Equivalent torque,

$$T_e = \sqrt{M^2 + T^2}$$

$$= \sqrt{30^2 + 40^2} = 50 \text{ kN.m}$$

$$T_e = \frac{\pi}{16} d_0^3 (1 - k^4)\tau$$

$$50 \times 10^6 = \frac{\pi}{16} d_0^3 [1 - (0.6)^4] \times 63$$

$$d_0 = 166.8 \simeq 167 \text{ mm}$$

$$d_i = 100 \text{ mm}$$

Example 7.19 *A solid shaft is subjected to a bending moment of 2.5 kN.m and a torque of 3.5 kN.m. Calculate the diameter of the shaft if the permissible tensile and shear stresses for the shafts material are 90 MPa and 56 MPa respectively.*

Solution:

Given: $M = 2.5$ kN.m, $T = 3.5$ kN.m, $s_t = 90$ MPa, t = 56 MPa.

Equivalent torque,

$$T_e = \sqrt{M^2 + T^2}$$

$$= \sqrt{(2.5)^2 + (3.5)^2} = 4.30 \text{ kN.m}$$

$$T_e = \frac{\pi}{16} d^3 \tau$$

$$4.30 \times 10^6 = \frac{\pi}{16} d^3 \times 56$$

$$d = 73 \text{ mm}$$

Equivalent bending moment,

$$M_e = \frac{1}{2}[M + \sqrt{M^2 + T^2}]$$

$$= \frac{1}{2}[2.5 + \sqrt{(2.5)^2 + (3.5)^2}]$$

$$= 3.4 \text{ kN.m}$$

$$M_e = \frac{\pi}{32} d^3 \sigma_t$$

$$3.4 \times 10^6 = \frac{\pi}{32} d^3 \times 90$$

$$d = 72.7 \text{ mm}$$

Adopt $d = 73$ mm

Example 7.20 *A shaft transmits 300 kW at 240 rpm. The shaft is supported in bearings 1.6m apart and carries a concentrated load of 20 kN at midspan. The shaft in made of a material having ultimate tensile strength of 540 MPa and shear strength of 390 MPa. Calculate the diameter of shaft for a factor of safety of 6.*

Solution:

Given: $P = 300$ kW, $N =$ rpm, $l = 1.6$ m, $W = 20$ kN, $s_{ut} = 540$ MPa, t = 390 MPa, $n = 6$

$$P = \frac{2\pi NT}{60\times 10^3}$$

$$T = \frac{300\times 60\times 10^3}{2\pi\times 240} = 11936.6\text{N.m}$$

$$M = \frac{Wl}{4} = \frac{20\times 10^3\times 1.6}{4} = 8000\text{ N.m}$$

Equivalent torque,

$$T_e = \sqrt{M^2+T^2}$$

$$= \sqrt{8000^2+(11936.6)^2} = 14369.5\text{ N.m}$$

$$T_e = \frac{\pi}{16}d^3\tau\times\frac{1}{n}$$

$$14369.5\times 10^3 = \frac{\pi}{16}d^3\times\frac{390}{6}$$

$$d = 104\text{ mm}$$

Equivalent bending moment,

$$M_e = \frac{1}{2}[M+\sqrt{M^2+T^2}]$$

$$= \frac{1}{2}[8000+14369.5] = 11184.75\text{ N.m}$$

$$M_e = \frac{\pi}{32}d^3\frac{\sigma_{ut}}{n}$$

$$1184.75\times 10^3 = \frac{\pi}{32}d^3\frac{540}{6}$$

$$d = 108\text{ mm}$$

Adopt $d = 108$ mm

Example: 7.21 *Find the internal and external diameters required for a hollow shaft, which is to transmit 40 kW of power at 240 rev per minute. The shear stress is to be limited to 100 MN /m². Take outside diameter to be twice the inside diameter.*

Solution:

Given $P = 40$ kW, N = 240 rpm, $\tau = 100$ MN/m², $d_0 = 2d_i$

$$J = \frac{\pi}{32}(d_0^4 - d_i^4) = \frac{\pi}{32}d_0^4\left[-\left(\frac{1}{2}\right)^4\right] = \frac{15}{16}\times\frac{\pi}{32}d_{d_0}^4$$

$$P = \frac{2\pi NT}{60\times 10^3}$$

$$T = \frac{40\times 60\times 10^3}{2\pi\times 240}\ 1591.55\text{ N.m}$$

$$\tau = \frac{T_r}{J} = \frac{T_{d_0}}{2J}$$

$$100\times 10^6 = \frac{1591.55\times d_0}{2\times\frac{15}{16}\times\frac{\pi}{32}d_0^4}$$

$$d_0^3 = 86.46\times 10^{-6}$$

$$d_0 = 4.42\times 10^{-2} = 0.442\text{ m or } 44.2\text{ mm}$$

$$d_i = 22.1\text{ mm}$$

Example.7.22 *Find the diameter of a solid shaft to transmit 400 kW at 100 rpm. The maximum shear stress allowed in the material is 80* **MPa.** *If instead* ***of this solid shaft,*** *a* ***hollow*** *shaft with diameter ratio as 0.5 and* ***of*** *same length and material is* ***used, what percentage*** *saving in weight will be obtained ?*

Solution:

Given: P = 400 kW, N - 100 rpm, r_{max} = 80 MPa, $k = \frac{d_i}{d_0} = 0.5,\ l_s = l_h,\ \rho_s = \rho_h.$

$$P = \frac{2\pi NT}{60\times 10^3}$$

$$400\times 60\times 10^3 = 2\pi\times 100\times T$$

$$T = 38197\text{ N.m}$$

Solid Shaft:

$$T = \frac{\pi}{16}d_s^3\tau_{max}$$

$$d_s = \left(\frac{16\times 38197\times 10^3}{\pi\times 80}\right)^{1/3} = 134.5\text{ mm}$$

Hollow shaft:

$$T = \frac{\pi}{16}d_0^3[1-(0.5)^4]\times 80$$

$$d_0 = 137.4\text{ mm}$$

$$d_i = 0.5\,d_0 = 68.7 \text{ mm}$$

Weight of solid shaft, $W_s = \frac{\pi}{4} d_s^2 l\,\rho_s \propto d^2$

Weight of hollow shaft, $W_h = \frac{\pi}{4}(d_0^2 - d_i^2) l \rho_h \propto (d_0^2 - d_i^2) \propto d_0^2(1-k^2)$

$$\propto d_0^2[1-(0.5)^2] \propto 0.75 d_0^2$$

$$\text{Percentage saving in weight} = \left[1 - 0.75\left(\frac{137.4}{134.5}\right)^2\right] \times 100$$

$$= 21.73\%$$

Example 7.23 *Determine the outer diameter D for a hollow steel shaft (inner diameter 0.45 times the outer diameter) to transmit 200 HP at 3600 rpm, if the working stress in shear in the bar is 42 MPa. Also find the angle of twist per unit length of the shaft. Given that G =10⁵ MPa for the material of the shaft.*

Solution:

Given: $k = \frac{D_i}{D_0} = 0.45$, $P = 200$ HP, $N = 3600$ rpm, $\tau_w = 42$ MPa, G = 10GPa

$$P = \frac{2\pi NT}{45 \times 60} \text{HP}$$

$$200 = \frac{2\pi \times 3600 \times T}{4500}$$

$$T = 39.8 \text{ N.m}$$

$$J_h = \frac{\pi}{32}(D_0^4 - D_i^4) = \frac{\pi D_0^4}{32}(1-k^4)$$

$$T = \frac{\pi}{16} D_a^3 (1-k^4)\tau_w$$

$$39.8 \times 10^3 = \frac{\pi}{16} D_a^3 \,[1-(0.45)^4] \times 42$$

$$D_0 = 17.1 \text{ mm}$$

Example 7.24 *A compression member of 500 mm effective length consists of a solid 24 mm diameter aluminium rod. In order to reduce the weight of the member by 25%, the solid rod is replaced by a hollow rod with D_0 = 24 mm and D_i = 12 mm. Determine the percent reduction in the critical load.*

Solution.

Given: $l_s = 500$ mm, $d_s = 24$ mm, $D_0 = 24$ mm, $D_i = 12$ mm

Weight of solid rod, $W_s = \frac{\pi}{4} d_s^2\, l_s \rho = \frac{\pi}{4}(24)^2 \times 500 \times 10^{-9} \rho \text{kg}$

Weight of hollow rod, $W_h = \frac{\pi}{4}(D_0^2 - D_i^2)l\rho = \frac{\pi}{4}(24^2 - 12^2)\times l_h \times \rho \times 10^{-9}$ kg

$$W_h = 0.75\ W_s$$

$$\therefore \quad \frac{\pi}{4}(24^2 - 12^2)\times l_h \times 10^{-9} \times \rho = 0.75 \times \frac{\pi}{4} \times 24^2 \times 500 \times 10^{-9} \times \rho$$

$$432 \times l_h = 216000$$

$$l_h = 500 \text{ mm}$$

$$P_s = \frac{\pi^2 E_s I_s}{l_s^2},\ P_h = \frac{\pi^2 E_h I_h}{l_h^2}$$

$$E_s = E_h, l_s = l_h$$

$$\therefore \quad \frac{P_s}{P_h} = \frac{I_s}{I_h} = \frac{\frac{\pi}{64}\times 24^4}{\frac{\pi}{64}(24^4 - 12^4)} = \frac{331776}{311040} = 1.0667$$

$$\text{Reduction in the critical load} = \frac{P_s - P_h}{P_s} = 1 - \frac{P_h}{P_s} = 1 - \frac{1}{1.0667} = 0.0625 \text{ or } 6.25\%$$

Example 7.25 *Design a hollow steel shaft of length 4m to transmit 225 kW. at 75 rpm without exceeding a shear stress of 40 MPa. The ratio of outside diameter to the inside diameter is 1.2 :1. Also determine the diameter if a solid shaft is used instead. Calculate the angle of twist for hollow and solid shaft. Take G = 80 GPa.*

Solution:

Given : L = 4 m, P = 225 kW, N = 75 rpm, t = 40 MPa, $\frac{d_0}{d_i} = 1.2$, G = 80 GPa

Hollow shaft:

$$J = \frac{\pi}{32}d_0^4\left[1 - \left(\frac{d_i}{d_0}\right)^4\right] = \frac{\pi}{32}d_0^4\left[1 - \left(\frac{1}{1.2}\right)^4\right] = 0.05083\, d_0^4$$

$$P = \frac{2\pi NT}{60\times 10^3}$$

$$T = \frac{225\times 60\times 10^3}{2\pi\times 75} = 28648 \text{ N.m}$$

$$\frac{T}{J} = \frac{2\tau}{d_0}$$

$$\frac{28648\times 10^3}{0.05083\, d_0^4} = \frac{2\times 40}{d_0}$$

$$d_0^3 = 7045025$$

$$d_0 = 191.7 = 192 \text{ mm}$$

$$d_i = \frac{192}{1.2} = 160 \text{ mm}$$

Angle of twist

$$\frac{2\tau}{d_0} = \frac{G\theta}{L}$$

$$\theta = \frac{2\tau L}{d_0 G} = \frac{2 \times 40 \times 4 \times 10^3}{192Ú80 \times 10^3} = 0.02083 \text{ rad or } 1.194°$$

Solid shaft:

$$J = \frac{\pi d^4}{32}$$

$$\frac{T}{J} = \frac{2\tau}{d}$$

$$\frac{28648 \times 10^3 \times 32}{\pi d^4} = \frac{2 \times 40}{d}$$

$$d^3 = 3647576$$

$$d = 154 \text{ mm}$$

$$\theta = \frac{2 \times 40 \times 4 \times 10^3}{154 \times 80 \times 10^3} = 0.026 \text{ rad or } 1.49°$$

Example 7.26 *A torsion member is fabricated from two concentric thin tubes. At the ends, the two tubes are welded to rigid discs so that both the tubes are twisted as a unit. The radius of the outer tube is 2r and that of the inner tube is r. Find the ratio of th shear stress produced in the two tubes.*

Solution.

Given: $r_0 = 2r$, $r_i = r$

Refer to Fig. 7.13

Shear stress in a thin tube, $\quad r = \dfrac{T}{2\pi r^2 t}$

where $\quad T$ = torque applied

t = tube thickness

$$\therefore \quad \tau \propto \frac{1}{r^2}$$

Ratio of shear stress in tubes, $\dfrac{r_i}{\tau_0} = \left(\dfrac{r_0}{r_i}\right)^2 = \left(\dfrac{2r}{r}\right)^2 = 4.$

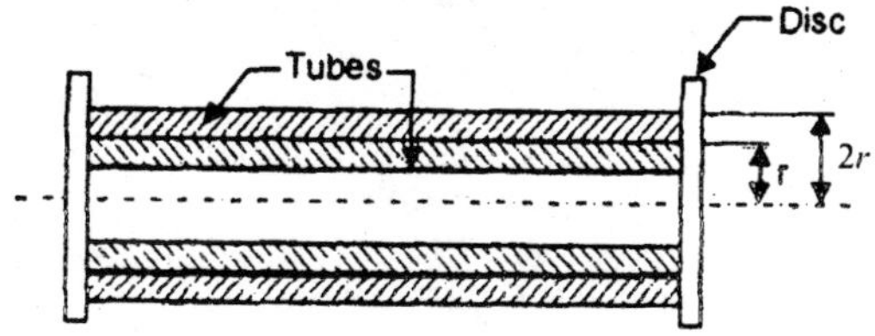

Fig. 7.13

Example 7.27 *Find the ratio of the torsional moments of resistance of a solid circular shaft of diameter 'D' and a hollow circular shaft having external diameter 'D' and internal diameter 'd'.*

Solution:

$$(z_p)_s = \frac{2J}{D} = \frac{2 \times \frac{\pi}{32} \times D^4}{D} = \frac{\pi}{16} D^3$$

$$(z_p)_h = \frac{2 \times \frac{\pi}{32}(D^4 - d^4)}{D} = \frac{\pi}{16} D^3 = \left[1 - \left(\frac{d}{D}\right)^4\right]$$

$$\frac{(z_p)_h}{(z_p)_s} = 1 - \left(\frac{d}{D}\right)^4$$

Example 7.28 *A hollow steel shaft 4 m long is to transmit 150 kW power at 150 rpm. The total angle of twist in this length is not to exceed 2.5° and the allowable shear stress is 60 MPa. Determine the inside and outside diameters, if G = 82 GPa.*

Solution:

Given: L = 4m, P = 150 kW, N = 150 rpm, θ = 2.5°, t = 60 MPa, G = 82 GPa

Let $$\frac{d_i}{d_0} = 0.6 \text{ [assumed]}$$

$$J = \frac{\pi}{32}(d_0^4 - d_i^4) = \frac{\pi}{32} d_0^4 \left[1 - \left(\frac{d_i}{d_0}\right)^4\right]$$

$$= \frac{\pi}{32} d_0^4 \,[1 - (0.6)^4] = 0.08545\, d_0^4$$

Power transmitted, $$P = \frac{2\pi NT}{60 \times 10^3} \text{kW}$$

$$150 = \frac{2\pi \times 150 \times T}{60 \times 10^3}$$

$$T = 9549.3 \text{ Nm.}$$

$$\frac{T}{J} = \frac{2\tau}{d_0}$$

$$\tau = \frac{Td_0}{2J}$$

$$60 \times 10^6 = \frac{9549.3 \times d_0}{2 \times 0.08545 \times d_0^4}$$

$$d_0^3 = 931.275 \times 10^{-6}$$

$$d_0 = 9.765 \times 10^{-2} \text{ m or } 97.65 \text{ mm}$$

Also
$$\frac{T}{J} = \frac{G\theta}{L}$$

$$\theta = \frac{TL}{GJ}$$

$$\frac{\pi}{180} \times 25 = \frac{9549.3 \times 4}{82 \times 10^9 \times 0.05845 \times d_0^4}$$

$$d_0^4 = 12493.62 \times 10^{-8}$$

$$d_0 = 10.52 \times 10^{-2} \text{ m or } 105.572 = 106 \text{ mm}$$

Adopt $d_0 = 106$ mm, $d_i = 63.6$ mm

Example 7.29 *A steel shaft ABCD having a total length of 4.8 m consists of three lengths having different sections as follows :*

AB	Hollow	d_0 = 12 cm	d_i = 10.2 cm
BC	Solid	d_2 = 12 cm	
CD	Solid	d_3 = 9.6 cm	

If the angle of twist is same for each section, determine the lengths of each section. Find the value of applied torque and total angle of twist if maximum shear stress in hollow section is limited to 50MPa. G = 84 GPa.

Solution:

Given: $\theta_1 = \theta_2 = \theta_3$, $t_1 = 50$ MPa, G= 84 GPa

Refer to Fig. 7.14.
$$\theta = \frac{Tl}{GJ}$$

Since T and G are same, therefore

$$\frac{l_1}{J_1} = \frac{l_2}{J_2} = \frac{l_3}{J_3}$$

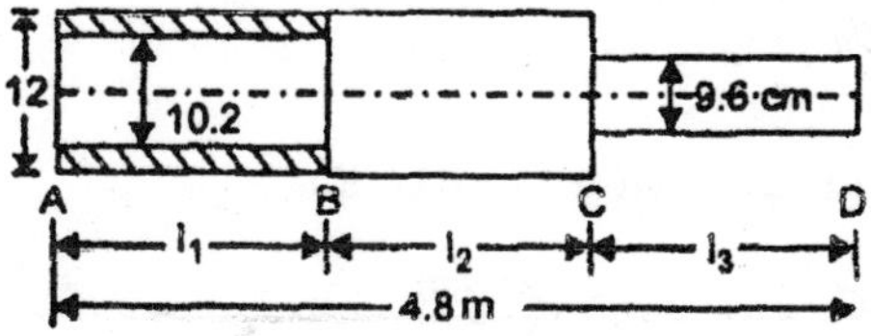

Fig. 7.14

$$l_3 = \frac{l_1 J_2}{J_1} = \frac{\frac{\pi}{32} \times 12^4 \times l_1}{\frac{\pi}{32}[12^4 - (10.2)^4]} = 0.857\, l_1$$

$$l_1 + l_2 + l_3 = 4.8 \text{ m}$$

$$l_1(1 + 2.092 + 0.857) = 4.8$$

$$l_1 = 1.215\text{m}, l_2 = 2.543 \text{ m}, l_3 = 1.042\text{m}$$

$$T = \frac{\tau J}{r} = \frac{50 \times \frac{\pi}{32} \times [12^4 - (10.2)^4] \times 10^{-6}}{6}$$

$$= 8.1089 \text{ kN.m}$$

$$\theta_{AB} = \frac{\tau_1 l_1}{Gr} = \frac{50 \times 1.215 \times 10^6}{84 \times 10^9 \times 6 \times 10^{-2}} = 0.6906°$$

$$\text{Total angle of twist} = 3 \times 0.6906 = 2.072°$$

Example 7.30 *A solid circular shaft is subjected to a bending moment M and torque T. What is the ratio of maximum bending stress to maximum shear stress in the shaft ?*

Solution:

Equivalent bending moment,

$$M_e = \frac{1}{2}[M + \sqrt{M^2 + T^2}]$$

Max. Bending stress,

$$\sigma = \frac{M_e}{z} = \frac{32 M_e}{\pi d^3}$$

Equivalent torque,

$$T_e = \sqrt{M^2 + T^2}$$

Maximum shear stress,

$$\tau = \frac{16 T_e}{\pi d^3}$$

$$\frac{\sigma}{\tau} = \frac{2M_e}{T_e}$$

Example 7.31 *A solid shaft 6.5 m long is securely fixed at both ends. A torque of 91 Nm is applied to the shaft at a section 2.5 m from one end. Find the fixing torques setup at the ends of the shaft. If the shaft is 35 mm in* *diameter, find the maximum shear stresses in the two portion. Also find the angle of twist for the section where torque is applied. Take G = 84 GPa.*

Solution:

Given: $l = 6.5$ m, T = 91 N.m, $d = 35$ mm, G = 84 GPa.

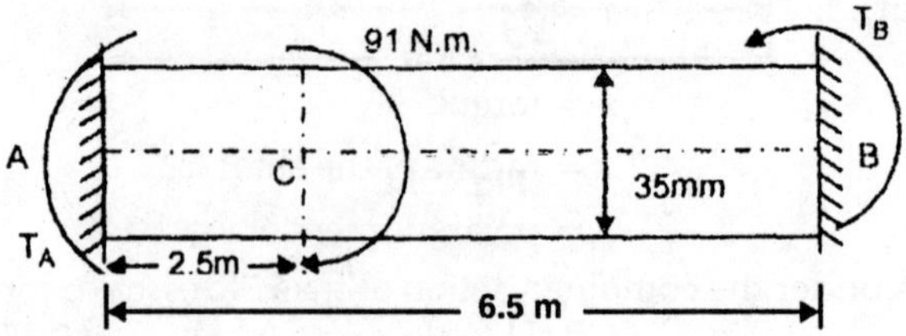

Fig.7.15

Refer to Fig. 7.15.

Let T_A and T_B be the torques set up at the ends of the shafts. Then

$$T_A + T_B = 91 \text{ N.m.}$$

$$\theta_{B/A} = \frac{T_B \times 6.5}{GJ} = \frac{91 \times .25}{GJ} = 0$$

$$T_B = \frac{91 \times 2.5}{6.5} = 35 \text{N.m}$$

$$T_A = 91 - 35 = 56 \text{ N.m}$$

Shear stress in each portion :

$$\tau_{AC} = \frac{16T_A}{\pi d^3} = \frac{16 \times 56 \times 10^3}{\pi \times 35^3} 6.652 \text{ MPQ}$$

$$\tau_{BC} = \frac{16T_B}{\pi d^3} = \frac{16 \times 35 \times 10^3}{\pi \times 35^3} 4.157 \text{ MPQ}$$

$$\theta = \frac{T_A l_i}{GJ} = \frac{5 \times 10^3 \times 2.5 \times 10^3 \times 32}{84 \times 10^3 \times \pi \times 34^4}$$

$$= 0.0113 \text{ rad or } 0.648°$$

7.12 COMBINED AXIAL, BENDING AND TORSIONAL STRESSES

Consider a body under the combined effect of axial, bending and torsional stresses. Then

Direct stress, $\sigma_d = \frac{P}{A}$

where P = axial load applied

A = area of cross-section

Bending stress, $\sigma_b = \frac{My}{I}$

where M = bending moment

y = distance of fibre from the neutral axis

I = moment of inertia

Torsional stress (shear), $\tau = \frac{Tr}{J}$

where T = torque

r = radius of the shaft

J = polar moment of inertia.

Thr principal stresses under the combined action of these stresses become,

$$\sigma_{1,2} = \frac{\sigma_d \pm \sigma_b}{2} \pm \frac{1}{2}\sqrt{(\sigma_d \pm \sigma_b)^2 + 4\tau^2}$$

and maximum shear stress, $\tau_{max} = \frac{1}{2}\sqrt{(\sigma_d \pm \sigma_b)^2 + 4\tau^2}$

Equivalent bending moment M_e may be defined as the bending moment which will produce the same maximum direct stress as produce by the bending moment and torque acting simultaneously. Similarly the equivalent torque T_e may be defined as the torque which will produce the same maximum shear stress as produced by the bending moment and torque acting simultaneously.

For a circular shaft of diameter d, $\sigma_\alpha = \frac{4P}{\pi d^2}$

$$\sigma_b = \frac{32M}{\pi d^3}$$

$$\tau = \frac{16T}{\pi d^3}$$

$$\therefore \quad \sigma_{1,2} = \frac{1}{2}\left(\frac{4P}{\pi d^2} \pm \frac{32M}{\pi d^3}\right) \pm \frac{1}{2}\sqrt{\left(\frac{4P}{\pi d^2} \pm \frac{32M}{\pi d^3}\right)^2 + \left(\frac{32T}{\pi d^3}\right)^2}$$

$$= \frac{16}{\pi d^3}\left(\frac{Pd}{8} \pm M\right) \pm \frac{16}{\pi d^3}\sqrt{\left(\frac{Pd}{8} \pm M\right)^2 + T^2}$$

$$= \frac{16}{\pi d^3}\left[\left(\frac{Pd}{8} \pm M\right) \pm \sqrt{\left(\frac{Pd}{8} \pm M\right)^2 + T^2}\right] \qquad \text{.... (7.22)}$$

$$\tau_{max} = \frac{16}{\pi d^3}\sqrt{\left(\frac{Pd}{8} \pm M\right)^2 + T^2} \qquad \text{... (7.23)}$$

Example 7.32 *A shaft 7.5 cm diameter is subjected to a bending moment of 400 N-m and a torque of 300 N-m. Determine (a) the maximum normal stress on a section perpendicular to the axis,*

Solution: *(a) Bending stress,*

$$\sigma_b = \frac{My}{I} = \frac{400\times7.5\times64}{2\times\pi\times(7.5)^4\times10^{-6}} = \frac{400\times64\times10^6}{2\times\pi\times(7.5)^3} = 9.658\text{ MPa}$$

(b) Shear Stress,

$$\tau = \frac{Tr}{J} = \frac{300\times7.5\times32}{2\times\pi\times(7.5)^4\times10^{-6}} = \frac{300\times32\times10^6}{2\pi\times(7.5)^3} = 3.621\text{ MPa}$$

(c) Principal stresses,

$$\sigma_{1,2} = \frac{1}{2}\sigma_b \pm \frac{1}{2}\sqrt{\sigma_b^2+4\tau^2} = 4.829 \pm \frac{1}{2}\sqrt{9.277+52.4464}$$

$$= 4.829 \pm 6.036 = 10.865, -1.207\text{ MPa}$$

(d) Maxmum shear stress,

$$\tau_{max} = \frac{1}{2}(\sigma_1-\sigma_2) = \frac{1}{2}(10.865+1.207) = 6.036\text{ MPa}$$

Example 7.33 *A shaft is required for an engine which indicates 750 kW at 80 r.p.m. The maximum twisting moment on the shaft is 1.8 times the mean twisting moment. The main bearings are 5 m apart, and the shaft carries a flywheel midway between bearings weighing 400 kN, the bending moment due to this weight is additional to that due to the steam pressure, which is 0.8 times the mean twisting moment. Find the diameter of shaft to satisfy the condition that the maximum tensile stress in the material is 56 MPa.*

Solution:

$$Power = \frac{2\pi N T_{mean}}{60\times10^3}\text{ kW}$$

$$T_{mean} = \frac{750\times60\times10^3}{2\pi\times80} = 89.5246\text{ kN.m}$$

$$T_{max} = 1.8\,T_{mean} = 1.8\times89.5246 = 161.1444\text{ kN.m}$$

$$M = \frac{Wl}{4}+0.8T_{mean} = \frac{400\times5}{4}+89.5246$$

$$= 500 + 71.6197 = 571.6197\text{ kN.m}$$

$$\sigma_1 = \frac{16}{\pi d^3}[M+\sqrt{M^2+T^2}]$$

$$= \frac{16}{\pi d^3}\{571.6197+\sqrt{(571.6197)^2+(161.1444)^2}\}\times10^3$$

$$= \frac{16}{\pi d^3}(571.6197+593.8994)\times10^3 = \frac{16}{\pi d^3}1165.5192\times10^3$$

$$d^3 = \frac{16}{\pi \times 56 \times 10^6} \times 1165.5192 \times 10^3 = 105.99894 \times 10^{-3}$$

$$d = 0.4733 \text{ m}$$

7.13 TORSION OF THIN CIRCULAR TUBES

Consider a thin circular tube as shown in Fig. 7.16, subjected to torque T

Let $\quad h$ = thickness of wall of the tube

$$= \frac{D-d}{2}$$

Now $\quad J = \frac{\pi}{32}(D^4 - d^4) = \frac{\pi}{32}[D^4 - (D-2h)^4$

$$= \frac{\pi}{32}\left[D^4 - D^4\left(1 - \frac{2h}{D}\right)^4\right]$$

$$= \frac{\pi D^4}{32}\left[1 - \left(1 - \frac{8h}{D} + \ldots\right)\right]$$

$$= \frac{\pi D^4}{32} \times \frac{8h}{D} = \frac{\pi}{4} D^3 h$$

$$T = \frac{\tau . J}{\frac{D}{2}} = \frac{2\tau}{D} \times \frac{\pi}{4} D^3 h$$

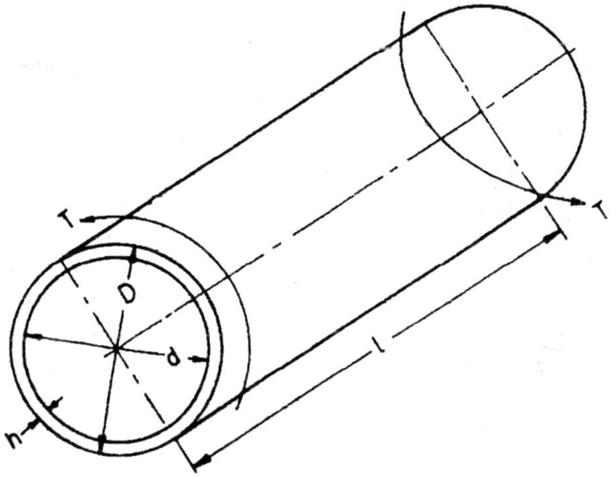

7.16 Torsion of a thin circular tube.

$$T = \frac{\pi}{2} D^2 h \tau \qquad \text{... (4.24)}$$

$$\theta = \frac{Tl}{JG} = \frac{Tl}{\frac{\pi}{4} D^3 hG} = \frac{4Tl}{\pi D^3 hG} \qquad \text{... (4.25)}$$

Example 7.34 *A shaft is 2 m long and tapers uniformly from10 cm diameter at one end to 5 cm diameter at the other end. The larger end is firmly fixed and a twisting moment of 5 kN.m is applied to the smaller end. Find the total angle of twist and the maximum shear stress induced Assume*

G = 84 GPa.

Soution:

Given: $\quad d_1 = 5$ cm, $d_2 = 10$ cm, $l = 2$ m, $T = 5$ kN.m

$$\theta = \frac{32Tl}{3\pi G}\left(\frac{d_1^2 + d_2^2 + d_1 d_2}{d_1^3 \, d_2^3}\right)$$

$$= \frac{32\times5\times10^3\times2}{3\pi\times84\times10^9}\left(\frac{25+100+50}{125\times1000}\right)\frac{1}{10^{-8}}$$

$$= \frac{32\times5\times10^3\times2\times175}{3\pi\times84\times10^9\times125\times1000}\times\frac{180}{\pi}\times\frac{1}{10^{-8}}$$

$$= 3.242°$$

Maximum shear stress, $$\tau = \frac{16T}{\pi d_1^3} = \frac{16\times5\times10^3}{\pi\times125\times10^{-6}}$$

$$= 203.718 \text{ MPa}$$

Example 7.35 *A thin tube is 5 cm mean diameter and 0.25 cm thick. If the allowable shear stress is 80 MPa, determine the twisting moment which can safely be applied to the tube. If the length of the tube is 50 cm and density 78000 kg/m³, determine the torsional strength to weight ratio.*

Solution:

Given: $D = 5$ cm, $h = 0.25$ cm, $\tau = 80$ MPa

$$T = \frac{\pi}{2}D^2 h\tau$$

$$= \frac{\pi}{2}\times25\times0.25\times10^{-6}\times80\times10^6$$

$$= 785.57 \text{ N.m}$$

Area of cross-section, $$A = \frac{\pi}{4}[D^2 - (D-2h)^2]$$

$$= \frac{\pi D^2}{4}\left[1-\left(1-\frac{2h}{D}\right)^2\right] = \frac{\pi D^2}{4}\left(2\cdot\frac{2h}{D}\right) = \pi Dh$$

Weight of the tube, $$W = Al\gamma = \pi Dhl\gamma$$

$$= \pi \times 5 \times 0.25 \times 50 \times 10^{-6} \times 7800 \times 9.81 = 15.02 \text{ N}$$

$$\frac{\text{Torsional strength}}{\text{Weight of tube}} = \frac{T}{W} = \frac{785.57}{15.02} = 52.3$$

Summary

1. Torsion formula:

$$\frac{T}{J} = \frac{\tau}{r} = \frac{G\theta}{l}$$

$$J = \frac{\pi d^4}{32} \text{ for a solid shaft.}$$

$$= \frac{\pi}{32}(d_0^4 - d_i^4) - \frac{\pi}{32}d_0^4(1-k^4) \text{ for a hollow shaft}$$

$$k = \frac{d_i}{d_0}$$

2. Maximum shear stress.

$$\tau_{max} = \frac{16T}{\pi d^3}, \text{ for a solid shaft}$$

$$= \frac{16T}{\pi d_0^3(1-k^4)}, \text{ for a hollow shaft.}$$

3. Comparison of solid and hollow shafts.

$$\frac{T_h}{T_s} = \frac{1+k^2}{\sqrt{1-k^2}} \times \frac{\tau_h}{\tau_s} \text{ for same weight}$$

$$= \left(\frac{d_0}{d_s}\right)^3 [1-k^4] \text{ for same strength}$$

4. Power transmitted.

$$P = \frac{2\pi NT_{mean}}{60 \times 10^3} kW$$

$$T_{max} = (1 + \text{over load factor}) \times T_{mean}$$

5. Shafts in series.

$$\theta = \theta_1 + \theta_2 = T\left[\frac{l}{J_1G_1} + \frac{l_2}{J_2G_2}\right]$$

$$\frac{\theta_1}{\theta_2} = \frac{J_2}{J_1} \times \frac{G_2}{G_1} \times \frac{l_1}{l_2}$$

$$\frac{\tau_1}{\tau_2} = \frac{J_2}{J_1} \times \frac{d_1}{d_2}$$

6. Shafts in parallel.

$$T = T_1 + T_2 = \theta\left[\frac{J_1G_1}{l_1} + \frac{J_2G_2}{l_2}\right]$$

$$\frac{T_1}{T_2} = \frac{G_1}{G_2} \times \frac{J_1}{J_2} \times \frac{l_2}{l_1}$$

7. Tapered circular shaft.

$$\theta = \frac{32Tl}{3\pi G}\left[\frac{d_1^2 + d_2^2 + d_1d_2}{d_1^3 d_2^3}\right]$$

$$\tau_{max} = \frac{16T}{\pi d_1^3}$$

8. Statically indeterminate shafts.

$$\Sigma M_z = 0$$

$$\theta_{B/A} = 0$$

9. Combined bending and torsion

Equivalent torque, $$T_e = \sqrt{M^2 + T^2}$$

Shear stress, $$\tau = \frac{16T_e}{\pi d^3}$$

Equivalent bending moment

$$M_e = \frac{1}{2}[M + \sqrt{M^2 + T^2}]$$

Bending stress, $$\sigma = \frac{32\,M_e}{\pi d^3}$$

Short Answer Type Questions

1. Define a shaft.
2. Write the tasion formula and give the units of all terms.
3. Why hollow shafts are preferred over solid shafts?
4. Define torsional stiffness.
5. What is the total angle of twist for two shafts in series?
6. How the torque is shared between two shafts of same length and material assembled in parallel?
7. What is a statically indeterminate shaft?
8. Define equivalent torque
9. What is polar moment of inertia?
10. What is the torsional section modulus of a circularshaft?

EXERCISES

1. A solid steel shaft is required to transmit 45 kW at 150 rpm. If the permissible shear stress is 75 MPa, find the diameter of the shaft.

2. A shaft is required to transmit 7.5 kW at 600 rpm. The maximum torque exceeds the mean torque by 40%. Calculate the diameter of the shaft if shear stress is not to exceed 70 MPa. Also calculate the angle of twist in a length of 2m. Take G = 84 GPa.

3. A hollow steel shaft of external diameter 400 mm and metal thickness 25 mm is required to transmit power at 240 rpm. If the shear stress is not to exceed 63 MPa and maximum torque exceeds the mean by 35%, find the power it can transmit.

4. A solid shaft of 100 mm diameter is to be replaced by a hollow shaft with internal diameter equal to 40% of external diameter. Find the size of hollow shaft and saving in material. Both the shafts are made of same material.

5. A composite shaft 1.2 m long is fixed at one end and is subjected to a twisting moment of 100 kN.m (cw) at the free end and 150 kN.m (cw) at a distance of 0.8 m form the fixed end. The shaft has a diameter of 100 mm for 0.8m length and 60 mm for the remaining length. Calculate (a) the maximum shear stress developed in each portion of shaft, and (b) the angle of twist at 0.8 m length and at the free end. Take G = 84 GPa.

6. A solid shaft 5m long is rigidly fixed at both ends. A torque of 1.5 kN.m is applied to the shaft at 2.0 m from one end. Calculate the fixing torques set up at the ends of the shaft. If the diameter of shaft is 45 mm, find the maximum shear stress in the two portions and the angle of twist at the location of applied torque. G = 84 GPa.

7. A propeller shaft of a ship is 400 mm in diameter and supports a propeller of 150 kN at the end of a cantilever of length 1.5 m. The propeller revolves at 100 rpm and the engine develops 15 MW. The propulsion efficiency is 85%. Calculate the principal stresses and maximum shear stress in the shaft.
8. A shaft 1.25 m long tapers uniformly from a diameter of 60 mm to 100 mm. The shaft transmits a torque of 10 kN.m. Find the angle of twist and maximum shear stress developed in the shaft. Take G = 84 GPa.
9. A composite shaft 3m long consists of a steel shaft of 25 mm diameter surrounded by an aluminium shaft of 50 mm diameter. The ends of the composite shaft are rigidly fixed and a torquc of 2 kN.m is applied at 1m from one end. Calculate the stresses in steel and aluminium. $G_s = 3G_a = 84$ GPa.
10. A shaft is subjected to a bending moment of 300 N.m and a torque of 400 N.m. If permissible stress in shear is 56 MPa, calculate the diameter of the shaft.

Chapter-8

Columns and Struts

8.1 INTRODUCTION

Strut: A bar or a member of a structure in any position and carrying an axial compressive load is called a strut.

Column: A bar or a member of a structure inclined at 90° to the horizontal and carrying an axial compressive load is called a column.

Slenderness ratio: The ratio of the length of the column to the least radius of gyration is called the slenderness ratio.

Buckling factor: It is the ratio between the equivalent length of the column to the minimum radius of gyration.

Buckling load: The minimum axial load at which the column tends to have lateral displacement or buckle is called the buckling or crippling or critical load.

Safe load: The safe load for the column is the load under which the column will not buckle.

Equivalent length: Equivalent length for a column is the length which gives the same buckling load as given by a both ends hinged column.

Short column: A column for which the slenderness ratio is less than 32 is called a short column.

Medium column: A column for which the slenderness ratio lies between 32 and 120 is called a medium column.

Long column: A column for which the slenderness ratio is more than 120 is called a long column.

8.2 CONCEPT OF ELASTIC STABILITY

There are three conditions of equihbrium, *i.e.* stable, unstable and neutral. A stable equilibrium is one is which the body returns to its original position on being displaced from its position of equihbrium, i.e. the resulting unbalance force works as a restoring force. In unstable equilibrium, the body does not come back to its original position, *i.e.* the resulting unbalance force works as an upsetting force. Under neutral equilibrium the body remains in equilibrium on being slightly disturbed.

When a small axial load is applied to a column, on removal of the load the column comes back to its original position and the column is said to be in stable equilibrium. If the axial load now is increased to such a value so that a slight lateral load produces a deflection in the column which remains even if the lateral load is removed, then the column is said to be in neutral equilibrium and the corresponing axial load is called the critical or buckling load. If the axial load is increased beyond this value then the column goes to unstable equilibrium.

8.3 END CONDITIONS

There can be three important conditions at the end of the column.

(*i*) *Fixed end.* In this case the end is fixed both in position and direction.
For such an end the deflection and slope is zero, i.e.

$$y = 0, \frac{dy}{dx} = 0$$

(*ii*) *Pinned end.* In this case the end is fixed in position only, i.e. $y = 0$.

(*iii*) *Free end.* At such an end the column is neither fixed in position nor in direction.

Thus, we can have four types of columns.

1. Both ends pinned
2. Both ends fixed.
3. One end fixed other end pinned.
4. One end fixed other end free.

8.4 EULER'S THEORY OF BUCKLING OF COLUMNS

The following assumptions are made in this theory:

(*i*) The column is initially straight and the applied load is truly axial.

(*ii*) The material of the column is homogeneous, linear and isotropic.

(*iii*) The length of the column is very large as compared to the cross-sectional dimensions of the column.

(*iv*) The cross-section of the column is uniform throughout.

(*v*) The shortening of the column due to axial compression is negligible.

(*vi*) The self weight of the column is neglected.

(*vii*) The ends are frictionless.

8.4.1 Column Hinged at Both Ends

Consider a column having both ends hinged and carrying an axial compressive load P as shown in Fig. 8.1. Taking origin at A, the bending moment at a distance x is

$$M_x = Py$$

$$\therefore \quad EI\frac{d^2y}{dx^2} = -M_x = -P \times y$$

or

$$EI\frac{d^2y}{dx^2} + Py = 0$$

$$\frac{d^2y}{dx^2} + \frac{P}{EI}y = 0$$

Let $k^2 = \frac{P}{EI}$, then

$$\frac{d^2y}{dx^2} + k^2 y = 0 \quad \text{... (a)}$$

or $$(D^2 + k^2) y = 0$$

where $$D^2 = \frac{d^2}{dx^2}$$

The auxiliary equation is:

$$D^2 + k^2 = 0 \quad \text{... (a)}$$

or $$D = \pm ik, i = \sqrt{-1}$$

The general solution of Eq. (a) is

$$y = A \cos kx + B \sin kx$$

where A and B are arbitrary constants.

The end condition are:

At $x = 0$, $y = 0$ and at $x = l$, $y = 0$

$$\therefore \quad A = 0$$

$$B \sin kl = 0$$

Now $B \neq 0$, because if $B = 0$ then $y = 0$, and the column will remain straight, which is not true.

$$\therefore \quad \sin kl = 0$$

or $$kl = n\pi, \quad n = 0, 1, 2, \ldots .$$

Taking the fundamental value, $n = 1$, we get

$$kl = \pi$$

or $$\sqrt{\frac{P}{EI}}\, l = \pi \quad \text{... (8.1a)}$$

$$P = \frac{\pi^2 EI}{l^2}$$

This load is called the Euler's critical load and is denoted by P_e.

$$P_e = \frac{\pi^2 EI}{l^2}$$

Now $$I = AK^2$$

where A = area of cross-section of the column

K = least radius of gyration

$$P_e = \frac{\pi^2 EA}{\left(\frac{l}{K}\right)^2} \quad \text{... (8.1b)}$$

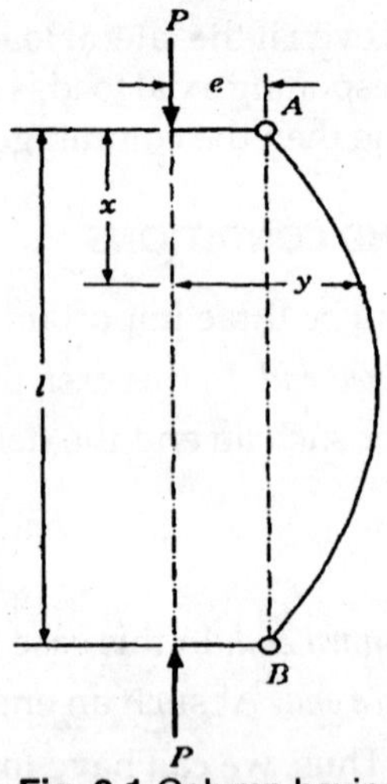

Fig. 8.1 Column having both ends hinged

Now $\dfrac{l}{K}$ = slenderness ratio of column = λ

$$P_e = \frac{\pi^2 EA}{\lambda^2}$$

Euler stress, $$\sigma_e = \frac{P_e}{A} = \frac{\pi^2 E}{\lambda^2} \qquad \text{...(8.2)}$$

8.4.2 Column Fixed at both ends

Consider a column fixed at both ends as shown in Fig. 8.2. Let M_A and M_B be the fixing moments at the ends. At a distance x from A,

$$M_x = P \times y - M_A$$

$$EI\frac{d^2y}{dx^2} = -M_x = M_A - Py$$

or $$EI\frac{d^2y}{dx^2} + Py = M_A$$

$$\frac{d^2y}{dx^2} + \frac{P}{EI}y = \frac{M_A}{EI} \qquad \text{...(a)}$$

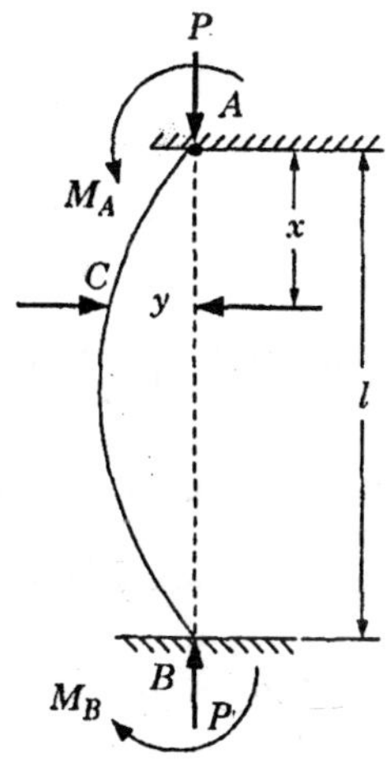

Fig. 8.2 Column having both ends fixed.

Let $$k^2 = \frac{P}{EI}$$

The general solution of Eq. (a) is:

$$y = A\cos kx + B\sin kx + \frac{M_A}{P}$$

The end conditions are

At $x = 0$, $y = 0$ and $\dfrac{dy}{dx} = 0$

$\therefore$ $$A + \frac{M_A}{P} = 0$$

$$A = -\frac{M_A}{P}$$

$$\frac{dy}{dx} = -Ak\sin kx + Bk\cos kx$$

$$0 = B$$

$$y = \frac{M_A}{P}(1 - \cos kx)$$

At $x = l, y = 0$ $\quad 0 = \frac{M_A}{P}(1-\cos kl)$

Now $\quad \frac{M_A}{P} \neq 0$

$\therefore \quad 1-\cos kl = 0$

$$\cos kl = 1$$

$$kl = 2n\pi, n = 0, 1, 2, ...$$

Taking the fundamental value, $n = 1$,

$$kl = 2\pi$$

$$\sqrt{\frac{P}{EI}}l = 2\pi$$

$$P = \frac{4\pi^2 EI}{l^2} = \frac{\pi^2 EI}{\left(\frac{l}{2}\right)^2} = \frac{\pi^2 EI}{l_e^2} \quad ...(8.3)$$

where $l_e = \frac{l}{2}$ is the equivalent length, of the column.

8.4.3 Column fixed at one end and hinged at the Other

Consider a column hinged at end A and fixed at B as shown in Fig. 8.3. There will be a fixing moment M_B at B and a horizontal force R_A will have to be applied at A to keep the column in equilibrium. At a distance x from A,

$$M_x = Py - R_A.x$$

$$EI\frac{d^2y}{dx^2} = -M_x = R_A x - Py$$

$$\frac{d^2y}{dx^2} + \frac{P}{EI}y = \frac{R_A.x}{EI} \quad ...(a)$$

The general solution, of Eq. (a) is:

$$y = A\cos kx + B\sin kx + \frac{R_A.x}{P}$$

where $k^2 = \frac{P}{EI}$

The end conditions are:

At $\quad x = 0, y = 0$

$\therefore \quad A = 0$

At $$x = l, y = 0$$

$$0 = B \sin kl + \frac{R_A \cdot l}{P} \quad \text{...(b)}$$

At $$x = l, \frac{dy}{dx} = 0$$

Now $$\frac{dy}{dx} = Bk \cos kx + \frac{R_A}{P}$$

$$0 = Bk \cos kl + \frac{R_A}{P}$$

or $$B = -\frac{R_A}{Pk} \times \frac{1}{\cos kl}$$

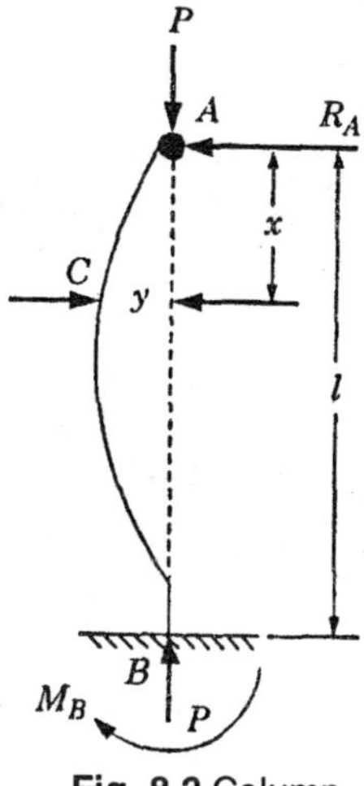

Fig. 8.3 Column having one end fixed and other hinged.

Substituting in Eq.(b), we get

$$0 = -\frac{R_A}{Pk} \cdot \tan kl + \frac{R_A \cdot l}{P} = \frac{R_A}{P}\left[l - \frac{\tan kl}{k}\right]$$

Now $$\frac{R_A}{P} \neq 0$$

∴ $$\tan kl - kl = 0$$

or $$kl = 4.49$$

$$\sqrt{\frac{P}{EI}}\, l = 4.49$$

$$P = \frac{20EI}{l^2} = \frac{2\pi^2 EI}{l^2} = \frac{\pi^2 EI}{\left(\frac{l}{\sqrt{2}}\right)^2} = \frac{\pi^2 EI}{l_e^2} \quad \text{...(8.4)}$$

∴ Equivalent length, $l_e = \frac{l}{\sqrt{2}}$

8.4.4 Column Fixed at One End and Free at the Other

Consider a column as shown in Fig. 8.4. Let the horizontal deflection of end A be e and fixing moment at end B be M_B.

At a distance x from A,

$$M_x = -P(e - y)$$

$$EI\frac{d^2y}{dx^2} = -M_x = P(e - y)$$

$$\frac{d^2y}{dx^2}+\frac{P}{EI}y = \frac{Pe}{EI} \quad ...(a)$$

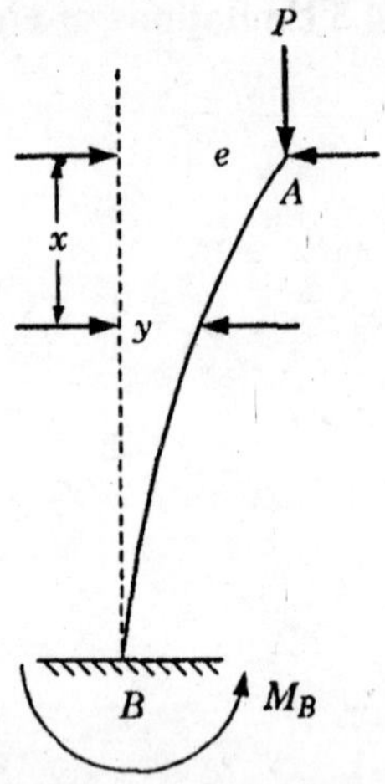

Fig. 8.4 Column having one end fixed and other free.

The general solution of Eq. (c) is :

$$y = A\cos kx + B\sin kx + e$$

where $k^2 = \dfrac{P}{EI}$

The end conditions are:

$$\text{At } x = l,\ y = 0 \text{ and } \frac{dy}{dx} = 0$$

$$0 = A\cos kl + B\sin kl + e \quad ...(b)$$

$$\frac{dy}{dx} = -Ak\sin kx + Bk\cos kx$$

$$0 = -Ak\sin kl + Bk\cos kl \quad ...(c)$$

$$\text{At } x = 0,\ y = e$$

$$e = A + e$$

or $$A = 0 \quad ...(d)$$

From Eq. (*b*), we get

$$B = \frac{-e}{\sin kl}$$

Substituting in Eq. (*c*), we get

$$\frac{-ek}{\sin kl}\cdot\cos kl = 0$$

or $$\cot kl = 0$$

$$kl = (2n+1)\frac{\pi}{2},\ n = 0, 1, 2, ...$$

For $n = 0$, $$kl = \frac{\pi}{2}$$

$$\sqrt{\frac{P}{EI}}l = \frac{\pi}{2}$$

$$P = \frac{\pi^2 EI}{4l^2} = \frac{\pi^2 EI}{(2l)^2} = \frac{\pi^2 EI}{l_e^2} \quad ...(8.5)$$

Where equivalent length, $l_e = 21$.

8.4.5 Limitations of Euler's Theory

1. The Euler's theory is applicable to columns which are initially exactly straight and the load is truly axial. However, there is always some crookedness in the column and the load may not be exactly axial.

2. This theory is applicable to long columns $\left(\frac{l_e}{K} > 120\right)$ only.

3. This theory does not take into account the axial compressive stress.

8.5 RANKINE-GORDON FORMULA

This formula takes into account the effect of direct compressive stress. Rankine proposed that crippling load,

$$\frac{1}{P} = \frac{1}{P_c} + \frac{1}{P_e}$$

where P_c = compressive load = σ_c x A

σ_c = ultimate compressive stress for the column material

A = area of cross-section of the column.

$$P_e = \text{Euler load} = \frac{\pi^2 EI}{l_e^2}$$

$$P = \frac{P_e P_c}{P_e + P_c} = \frac{P_c}{1 + \frac{P_c}{P_e}} = \frac{\sigma_c A}{1 + \frac{\sigma_c A l_e^2}{\pi^2 EI}}$$

$$= \frac{\sigma_c A}{1 + \frac{\sigma_c l_e^2}{\pi^2 EK^2}} = \frac{\sigma_c A}{1 + \frac{\sigma_c}{\pi^2 E}\left(\frac{l_e}{K}\right)^2}$$

$$= \frac{\sigma_c A}{1 + a\left(\frac{l_e}{K}\right)^2} \quad \text{...(8.6)}$$

where $a = \frac{\sigma_c}{\pi^2 E}$ is a constant for the column material.

8.6 EMPIRICAL FORMULAE

1. Straight line formula

$$P = \sigma_c A\left[1 - b\left(\frac{l_e}{K}\right)\right] \quad \text{...(8.7)}$$

where $$b = \sqrt{\frac{4E}{3\pi^2\sigma_c}}$$

2. Johnson's parabolic formula

$$P = \sigma_c A\left[1 - c\left(\frac{l_e}{K}\right)^2\right] \qquad ...(8.8)$$

where $$c = \frac{1}{4\pi^2 E}$$

Example 8.1 *A hallow steel column having both ends fixed is 50 mm external diameter and 25 mm internal diameter. If safe compressive stress is 320 MPa and E = 200 GPa, calculate the length.*

Solution:

Given : $d_0 = 50$ mm, $d_i = 25$ mm, $\sigma_e = 320$ MPa, E = 200 GPa

For both ends fixed,

$$P_e = \frac{4\pi^2 EI}{l^2} = \frac{4\pi^2 EAK^2}{l^2}$$

$$\sigma_e = \frac{P_e}{A} = \frac{4\pi^2 E}{\left(\frac{l}{K}\right)^2}$$

$$I = AK^2$$

$$K^2 = \frac{I}{A} = \frac{(d_o^4 - d_i^4)}{16\,(d_o^2 - d_i^2)} = \frac{(50^4 - 25^4)}{16\,(50^2 - 25^2)} = 195.3125 \text{ mm}^2$$

$$320 \times 10^6 = \frac{4\pi^2 \times 200 \times 10^9}{\left(\frac{l^2}{195.3125 \times 10^{-6}}\right)}$$

$$l^2 = \frac{4\pi^2 \times 200 \times 10^9 \times 195.3125 \times 10^{-6}}{320 \times 10^6} = 4.819$$

$$l = 2.195 \text{ m}$$

Example 8.2 *Using Rankine's formula, calculate the critical load for a column having both ends hinged when slenderness ratio = 60, E = 200 GPa, σ_c = 320 MPa and A = 100 mm².*

Solution:

Given : $\frac{l_e}{k} = 60, E = 200GPa, \sigma_e = 320MPa, A = 100mm^2$

$$P = \frac{\sigma_c A}{1 + a\left(\frac{l_e}{K}\right)^2}$$

$$a = \frac{\sigma_c}{\pi^2 E} = \frac{320}{\pi^2 \times 2 \times 10^5} = \frac{1}{6168.5}$$

$$P = \frac{320 \times 100}{1 + \frac{1}{6168.5}(60)^2} = 20.207 \text{ kN}$$

Example 8.3 *Compare the crippling load given by Euler's and Rankine's formulae for tubular strut 2 m long and 30 mm diameter loaded through pin joints at both ends.*

The yield stress is 320 MPa, $a = \frac{1}{7500}$ and E = 200 GPa.

Solution:

Given : $l = 2m, d = 30mm, \sigma_c = 320MPa, a = \frac{1}{7500}, E = 200GPa$

Euler's formula:

$$P_e = \frac{\pi^2 EI}{l_e^2} = \pi^2 \times 200 \times 10^9 \times \frac{\pi}{64} \times \frac{30^4 \times 10^{-12}}{4} = 19.621 \text{ kN}$$

Rankine's formula:

$$P = \frac{\sigma_c A}{1 + a\left(\frac{l_e}{K}\right)^2}$$

$$I = AK^2$$

$$K^2 = \frac{\pi d^4}{64} \times \frac{4}{\pi d^2} = \frac{d^2}{16} = \frac{30^2}{16} = 56.25 \text{ mm}^2$$

$$P = \frac{320 \times \frac{\pi}{4} \times 30^2}{1 + \frac{1}{7500} \times \frac{4}{56.25 \times 10^{-6}}} = 21.58 \text{ kN}$$

Example 8.4 *An I-section beam is shown in Fig. 8.5, simply supported at the ends. Compute its length, given that when it is subjected to a load of 20 kN per meter length, it deflects by 10 mm. Also find the safe load when used as a column with both ends fixed. Assume a factor of safety of 4. Take E = 210 GPa.*

Solution:

Given : $w = 20kN/m, \delta = 10mm, FOS = 4, E = 210GPa$

$$I_{xx} = \frac{1}{12}[0.3 \times (1.1)^3 - 0.28 \times 1^3] = 9.9417 \times 10^{-3} \text{ m}^4$$

$$I_{yy} = \frac{1}{12}[2 \times 0.05 \times 0.3^3 + 1 \times (0.02)^2] = 2.2567 \times 10^{-4} \text{ m}^4$$

$$\delta = \frac{5}{384}\frac{wl^4}{EI}$$

$$10 \times 10^{-3} = \frac{5}{384} \times \frac{20 \times 10^3 \times l^4}{210 \times 10^9 \times 9.9417 \times 10^{-3}}$$

$$l^4 = 80169.87$$

$$l = 16.82 \text{ m}$$

$I_{yy} < I_{xx}$, therefore, buckling will take place about y-axis

$$P_e = \frac{4\pi^2 EI_{yy}}{l^2}$$

$$= \frac{4\pi^2 \times 210 \times 10^9 \times 2.2567 \times 10^{-4}}{(16.82)^2} = 6613 \text{ kN}$$

$$\text{Safe load} = \frac{P_e}{FOS} = \frac{6613}{4} = 16.53 \text{ kN}$$

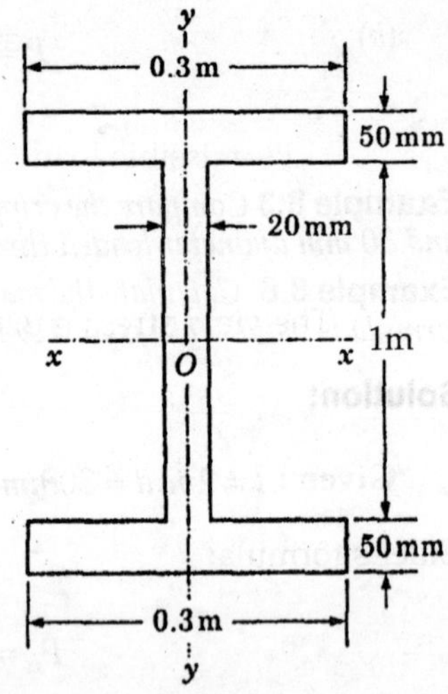

Fig. 8.5

Example 8.5 *A 2m long C.I. column has a circular cross-section of 100 mm diameter. One end of the column is fixed in direction and position and the other end is free. Taking factor of safety as 3, calculate the safe load using*

(a) Rankine-Gordon formula, taking yield stress = 540 MPa, and

$$a = \frac{1}{1600} \textit{ for pinned ends.}$$

(b) Euler's formula, taking E = 120 GPa.

Solution:

$$\text{Given: } d = 100mm, FOS = 3, \sigma_c = 540MPa, a = \frac{1}{1600}, E = 120GPa$$

(a)
$$P = \frac{\sigma_c A}{1 + a\left(\frac{l_e}{K}\right)^2}$$

For one end fixed and other free, $l_e = 2l = 4\text{m}$

$$K = \frac{d}{4} = \frac{100}{4} = 25\text{mm}, \; A = \frac{\pi}{4} \times 100^2 = 7854 \text{ mm}^2$$

$$P = \frac{540 \times 7854}{1 + \frac{1}{1600}\left(\frac{4 \times 10^3}{25}\right)^2} = 249480 \text{ N}$$

$$\text{Permissible load} = \frac{P}{FOS} = \frac{249480}{3} = 83160 \text{ N}$$

(*b*) $$P_e = \frac{\pi^2 EI}{l_e^2} = \frac{\pi^2 \times 120 \times 10^3 \times \pi \times 100^4}{64 \times (4 \times 10^3)^2} = 363354 \text{ N}$$

$$\text{Permissible load} = \frac{363354}{3} = 121118 \text{ N}$$

Example 8.6 *Calculate the maximum value of the slenderness ratio of a mild steel column for which Euler's formula is valid. Take* σ_c *= 320 MPa and E = 200 GPa.*

Solution:

$$P_e = \frac{\pi^2 EI}{l^2} = \frac{\pi^2 EAK^2}{l^2}$$

$$\sigma_e = \frac{P_e}{A} = \frac{\pi^2 E}{\left(\frac{l}{K}\right)^2}$$

$$\left(\frac{l}{K}\right)^2 = \frac{\pi^2 E}{\sigma_c}$$

$$= \frac{\pi^2 \times 200 \times 10^9}{320 \times 10^6} = 6168.5$$

$$\frac{l}{K} = 78.53$$

∴ For Euler's formula to be valid, $\frac{l}{K} > 78.53$ (say 80)

Example 8.7 *Compare the crippling loads given by Euler's and Rankine's formulae for a strut 2.5m long having 50 mm diameter and loaded through pin joints at both ends. Take ∴* σ_c *= 320 MPa,* $a = \frac{1}{7500}$*, and E = 200 GPa.*

Solution:

Given : $l = 2.5m, d = 50mm, \sigma_c = 320MPa, a = \frac{1}{7500}, E = 200GPa$

$$A = \frac{\pi}{4} \times 50^2 = 1963.5 \text{ mm}^2$$

$$K = \frac{d}{4} = \frac{50}{4} = 12.5 \text{ mm}$$

$$I = \frac{\pi \times 50^4}{64} = 306796 \text{ mm}^4$$

Euler's load, $$P_e = \frac{\pi^2 EI}{l^2} = \frac{\pi^2 \times 200 \times 10^9 \times 306796 \times 10^{-12}}{(2.5)^2} = 96894.6 \text{ N}$$

Rankine load, $$P_r = \frac{\sigma_c A}{1+a\left(\frac{l}{K}\right)^2}$$

$$= \frac{320 \times 1963.5}{1+\frac{1}{7500}\left(\frac{2.5\times 10^3}{12.5}\right)^2}$$

$$= 99208.4 \text{ N}$$

Example 8.8 *A column 2 m long is to support 225 kN of load. Calculate its diameter if $\sigma_c = 350$ MPa and $a = \frac{1}{7500}$. Assume both ends fixed.*

Solution:

Given : $l = 2m, W = 225kN, \sigma_c = 350MPa, a = \frac{1}{7500}$

Rankine load, $$P_r = \frac{\sigma_c A}{1+a\left(\frac{l_e}{K}\right)^2}$$

$$A = \frac{\pi d^2}{4}$$

$$K = \frac{d}{4}$$

$$l_e = \frac{l}{2} = \frac{2}{2} = 1\text{m}$$

$$225 \times 10^3 = \frac{350\times 10^6 \times \frac{\pi d^2}{4}}{1+\frac{1}{7500}\left(\frac{1\times 4}{d}\right)^2}$$

$$d^4 - 0.8185 \times 10^{-3} d^2 - 1.746 \times 10^{-6} = 0$$

$$d^2 = \frac{10^{-3}}{2}\left[0.8185 + \sqrt{(0.8185)^2 + 4\times 1.746}\right]$$

$$= 1.7925 \times 10^{-3}$$

$$d = 0.0423 \text{ m or } 42.3 \text{ mm}$$

Example 8.9 *A uniform round bar 2 m long is simply supported at the ends. It deflects 2.5 mm under a central load of 300 N. Calculate the Euler's buckling load for this bar when used as a strut with pinned ends.*

Solution:

Given : $l = 2m, \delta = 2.5mm, W = 300N$

Deflection, $$\delta = \frac{Wl^3}{48EI}$$

$$EI = \frac{300 \times 2^3}{48 \times 2.5 \times 10^{-3}} = 2000$$

Euler's load, $$P_e = \frac{\pi^2 EI}{l^2} = \frac{\pi^2 \times 2000}{2^2} = 4934.8 \text{ N}$$

Example 8.10 *Calculate the buckling load by Johnson's parabolic formula for a column having one end fixed and the other free having the following data:*

$\sigma_c = 380$ MPa, $A = 300$ mm^2, $E = 200$ GPa, $l = 3$ m, $I = 8 \times 10^4$ m^4

Solution:

Johnson's parabolic formula, $$P = \sigma_c A \left[1 - c\left(\frac{l_e}{K}\right)^2\right]$$

$$c = \frac{\sigma_c}{4\pi^2 E} = \frac{380 \times 10^6}{4\pi^2 \times 200 \times 10^9} = 20778$$

$$K^2 = \frac{I}{A} = \frac{8 \times 10^4}{800 \times 10^{-6}} = 1 \times 10^8 m^2, l_e = 2l = 6m$$

$$P = 380 \times 10^6 \times 300 \times 10^{-6} \left[1 - 20778 \times \frac{36}{1 \times 10^8}\right]$$

$$= 113{,}147 \text{ N}$$

8.7 ECCENTRIC LOADING

An eccentric load is that which does not pass through the centroid of the area of cross-section of a column or a machine member. Such a load produces bending in the member in addition to direct compressive stress. The eccentricity of load may be either about one axis or both axes. The resulant stress at any point of the member shall be the algebraic sum of direct compressive stress and bending stress.

8.7.1 Short Columns

1. Rectangular Cross-section

(a) No buckling. When there is no buckling of the column, then resultant stress.

$$\sigma_r = \sigma_d \pm \sigma_b$$

where direct stress, $\sigma_d = \frac{P}{A}$ (compressive)

If the load is at an eccentricity e_x, then bending moment,

$$M = P \times e_x$$

Bending stress, $$\sigma_b = \frac{My}{I} = \frac{P \times e_x \times y}{AK^2}$$

Resultant stress, $\sigma_r = \sigma_d \pm \sigma_b$

$$= \frac{P}{A}\left(1 \pm \frac{e_x \times y}{K^2}\right)$$

y = distance of the fibre of the column from the neutral axis.

For example for a rectangular column of width b and thickness d, we get

$$(\sigma_r)_{max} = \frac{P}{A}\left(-1 - \frac{6e_x}{b}\right) \text{ (compressive)}$$

$$= \frac{P}{A}\left(-1 + \frac{6e_x}{b}\right) \text{ (tensile)}$$

For tensile stress to be zero

$$-1 + \frac{6e_x}{b} = 0$$

or $$e_x = \frac{b}{6} = \frac{1}{2}\left(\frac{b}{3}\right) \quad \text{... (8.9)}$$

Thus if the value of eccentricity e_x on either side of y-axis does not exceed $bl6$, there will be no tensile stress in the column. Similarly by considering eccentricity e_y about x-axis, we can show that the eccentricity should not exceed $d/6$.

Thus for a rectangular section to avoid tension in the column the load must lie within the middle third of the cross-section of the structure. This is known as the *Middle Third Rule.* The permissible region has been shown in Fig. 8.6.

As a further explanation, the eccentric load may be considered as the sum of axial load P and a moment P. ex, as shown in Fig. 8.7(a). The stress distribution on faces AB and CD is shown in Fig.8.7(b).

If the eccentricity of the load is about both the axes, then (Fig. 8.8).

Resultant stress, $$\sigma_r = \left[\frac{P}{A} \pm \frac{Pe_x}{I_{yy}} x \pm \frac{Pe_y}{I_{xx}} y\right] \quad \text{...(8.10)}$$

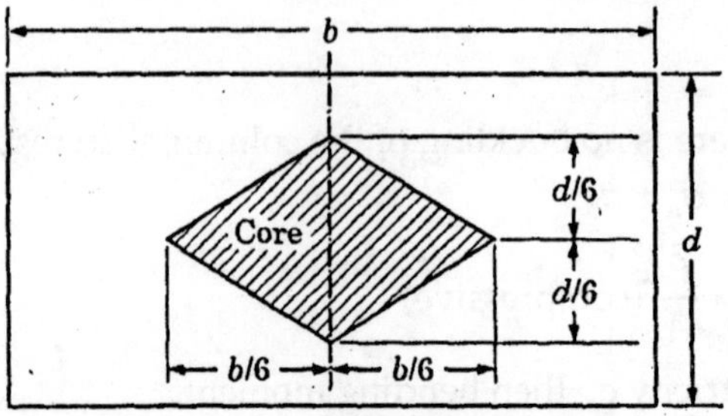

Fig. 8.6 Middle third rule

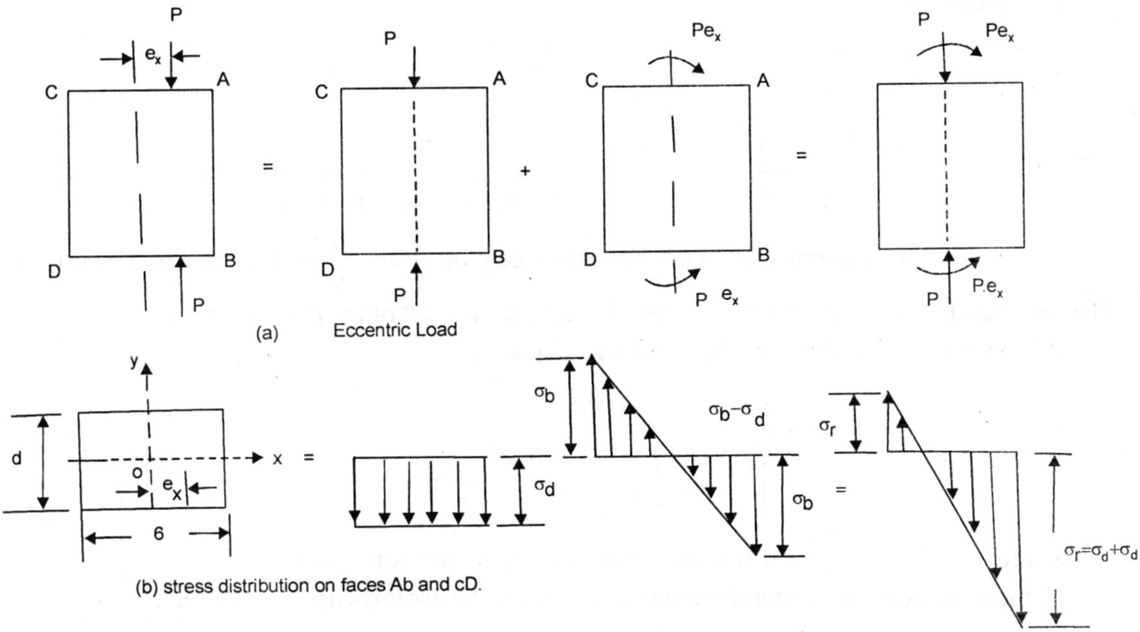

Fig. 8.7 Eccentric load on rectangular cross-section

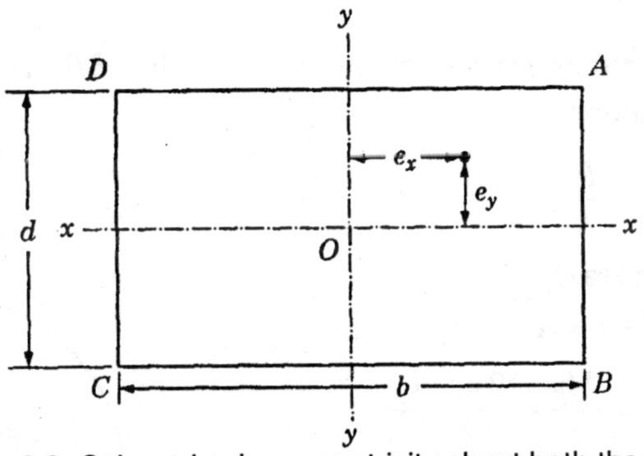

Fig. 8.8 Column having eccentricity about both the axes.

At A,
$$\sigma_A = \frac{P}{A}\left[1+\frac{6e_x}{b}+\frac{6e_y}{d}\right]$$

Similarly,
$$\sigma_B = \frac{P}{A}\left[1+\frac{6e_x}{b}-\frac{6e_y}{d}\right]$$

$$\sigma_C = \frac{P}{A}\left[1-\frac{6e_x}{b}-\frac{6e_y}{d}\right]$$

$$\sigma_D = \frac{P}{A}\left[1-\frac{6e_x}{b}+\frac{6e_y}{d}\right]$$

For no tension, $\sigma_c = 0$

or $$\frac{6e_x}{b}+\frac{6e_y}{d}=1$$

$$\frac{e_x}{b}+\frac{e_y}{d}=\frac{1}{6} \qquad \text{...(8.11)}$$

This represents a straight line, which makes intercepts of $\frac{b}{6}$ and $\frac{d}{6}$ on x-and y-axes respectively. The rhombus formed in the centre is called the core or 'kernel' of the column section.

(*b*) Considering buckling, the Rankine formula is,

$$P=\frac{\sigma_c \times A}{\left\{1+a\left(\frac{l_e}{K}\right)^2\right\}\left\{1+\frac{e_x \cdot x}{K_y^2}+\frac{e_y \cdot y}{K_x^2}\right\}} \qquad \text{... (8.12)}$$

where x, y = distances of the fibre from the neutral axis.

The second factor in the denominator accounts for the eccentricity effect of load.

2. Circular cross-section

The eccentric load on circular cross-section is shown in Fig. 8.9.

Let d = diameter of cylinder

Direct stress, $\sigma_d = \frac{P}{A}$

For eccentricity e_x along x-axis.

$$M = P\times e_x$$

Bending stress, $\sigma_b = \frac{My}{I} = \frac{P\times e_x \times y}{AK^2}$

Resultant stress, $\sigma_r = \frac{P}{A}\left[1 \pm \frac{e_x \times y}{K^2}\right]$

For σ_r to be maximum,

$$y = \frac{d}{2}$$

$$(\sigma_r)_{max} = \frac{P}{A}\left[1 \pm \frac{e_x \times \left(\frac{d}{2}\right)}{\left(\frac{d^2}{16}\right)}\right]$$

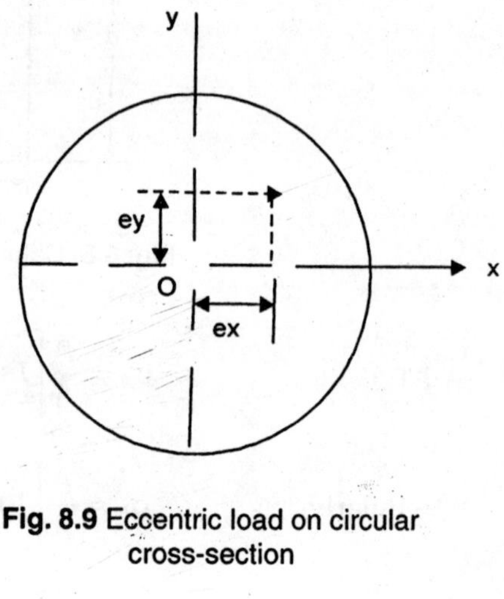

Fig. 8.9 Eccentric load on circular cross-section

$$= \frac{P}{A}\left[-1-\frac{8e_x}{d}\right] \text{ (compressive)}$$

$$= \frac{P}{A}\left[-1+\frac{8e_x}{d}\right] \text{ (tensile)}$$

For tensile stress to be zero,

$$-1+\frac{8e_x}{d} = 0$$

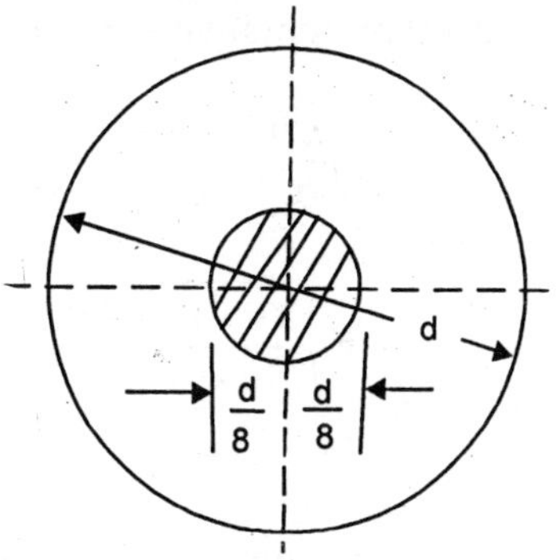

Fig. 8.10 Middle quarter rule

or $$e_x = \frac{d}{8} = \frac{1}{2}\left(\frac{d}{4}\right) \quad \text{...(8.13)}$$

Thus, if the value of eccentricity ex on either side of y-axis does not exceed , there shall be no tensile stress in the column. Due to symmetry of the circular cross-section this shall be true for eccentricity ey along y-axis also. Therefore to avoid tension in a circular cross-section the load must lie within the middle quarter of the circular cross-section. This is known as the middle quarter rule, as shown in Fig.8.10.

8.8 EULER'S THEORY AND EXPERIMENTAL RESULTS

According to Euler's theory for buckling of columns, the crippling load for a column having both ends hinged is given by,

$$P_r = \frac{\pi^2 EI}{L^2} = \frac{\pi^2 EAK^2}{L^2}$$

Crippling stress, $$\sigma_{cr} = \frac{P_{cr}}{A} = \frac{\pi^2 E}{(L/K)^2}$$

where $$\frac{L}{K} = \text{slenderness ratio.}$$

For a column material,

$$E = \text{const.}$$

$$\therefore \quad \sigma_{cr} \propto = \frac{1}{(L/K)^2}$$

The variation of crippling stress *v's* slenderness ratio is shown in Fig. 8.11. As L/K increases the crippling stress decreases. The crippling stress cannot exceed the crushing strength of the column material with a fairly good value for factor of safety. Let *OB* represent the crushing strength then slenderness ratio cannot exceed the value represented by *OA*.

As a representative estimate, if E = 200 GPa and σ_{cr} = 320 MPa, then

$$\frac{L}{K} = \sqrt{\frac{\pi^2 E}{\sigma_{cr}}}$$

$$= \pi\sqrt{\frac{200\times10^9}{320\times10^6}}$$

$$=78.5 \simeq 80$$

Therefore, Euler's formula shall be applicable for a steel column if $\frac{L}{K}$ is more than 80.

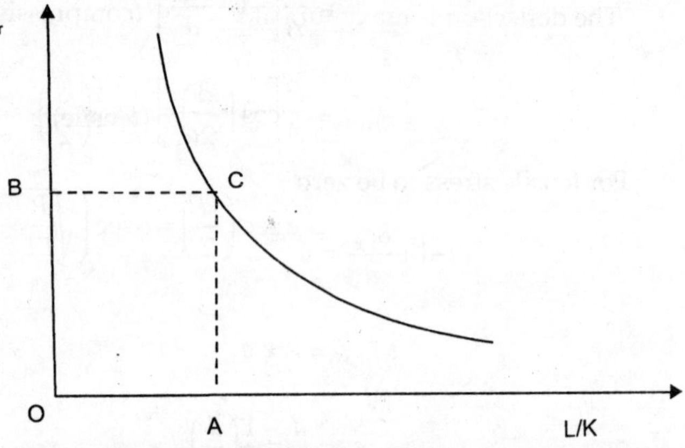

Fig. 8.11 Plot of crippling stress against slenderness ratio.

8.9 Long Columns-The Secant Formula

Consider a long column hinged at both ends and subjected to an eccentric load *P* at eccentricity 'e' as shown in Fig. 8.12. At a distance *x* from *A*, let *y* be the deflection of the column.

$$M_x = Py$$

$$\frac{EId^2y}{dx^2} = -M_x = -Py$$

$$\frac{EId^2y}{dx^2} + Py = 0$$

$$\frac{d^2y}{dx^2} + \frac{P}{EI}y = 0 \qquad \text{... (1)}$$

Let $\qquad \frac{P}{EI} = k^2$

The general solution of Eq.(l) is,

$$y = A\cos kx + B\sin kx \qquad \text{...(2)}$$

At $x = 0$, $\qquad y = e$

$\therefore \qquad A = e$

At $x = l$, $\qquad y = e$

$$e = e\cos kl + B\sin kl$$

$$B = \frac{e(1-\cos kl)}{\sin kl} = e\tan\left(\frac{kl}{2}\right)$$

$$y = e\cos kx + e\tan\left(\frac{kl}{2}\right)\sin kx \qquad \text{...(3)}$$

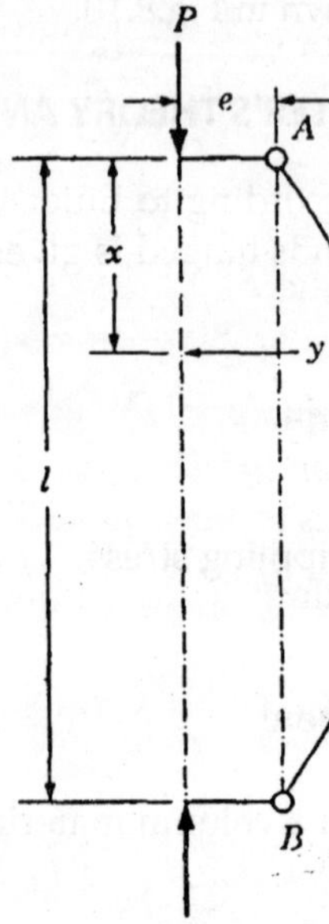

Fig. 8.12 Long Column under eccentric load

The deflection is maximum at $x = \frac{l}{2}$

$$y_{max} = e\cos\left(\frac{kl}{2}\right) + e\tan\left(\frac{kl}{2}\right)\cdot\sin\left(\frac{kl}{2}\right)$$

$$= e\sec\left(\frac{kl}{2}\right) = e\sec\left(\sqrt{\frac{P}{EI}}\frac{l}{2}\right) \quad \text{... (8.14)}$$

$$M_{max} = P \times y_{max} = Pe\sec\left(\sqrt{\frac{P}{EI}}\frac{l}{2}\right)$$

Resultant stress, $$\sigma_r = \frac{P}{A} \pm \frac{M}{z} = \frac{P}{A} \pm \frac{Pey_c}{AK^2}\sec\left(\sqrt{\frac{P}{EI}}\frac{l}{2}\right)$$

$$\sigma_{max} = \frac{P}{A}\left[1 + \frac{ey_c}{K^2}\sec\left(\sqrt{\frac{P}{EI}}\frac{l}{2}\right)\right]$$

where y_c = distance of the outer most fibre in compression from the neutral axis.

For other end conditions, we have

$$\sigma_{max} = \frac{P}{A}\left[1 + \frac{ey_c}{K^2}\sec\left(\sqrt{\frac{P}{EI}}\frac{l_e}{2}\right)\right] \quad \text{...(8.15)}$$

where l_e = equivalent length.

Example 8.11 *A concrete pier 1 m × 1 m in cross-section and 5 m long carries a compressive load P at a point on the top which is 0.3 m from each of the two axes. Calculate the maximum value of P, if no tensile stress is to occur in the base. Concrete weighs 20 kN/m³.*

Solution: Self weight of pier, $W = 1 \times 1 \times 5 \times 20 = 100$ kN

Total tensile stress in pier $= \frac{6P}{A}\left(\frac{e_x}{b} + \frac{e_y}{d}\right) - \frac{P}{A} - \frac{W}{A} = 0$

$$\frac{P}{1\times1}\left[6\left(\frac{0.3}{1} + \frac{0.3}{1}\right) - 1\right] - \frac{100}{1\times1} = 0$$

$$2.6\,P = 100$$

$$P = 38.4615 \text{ kN}$$

Example 8.12 *A steel tube of 50 mm external diameter and 40 mm internal diameter is 3 m long and carries a compressive load of 25kN acting parallel to the axis at an eccentricity of 5 mm. Calculate the maximum stress induced in the tube. Take E = 205 GPa.*

Solution: **Given :** $d_0 = 50$ mm, $d_i = 40$ mm, $l = 3$m, $P = 25$ kN, $e = 5$ mm, $E = 205$ GPa.

$$A = \frac{\pi}{4}(50^2 - 40^2) = 706.86 \text{ mm}^2$$

$$I = \frac{\pi}{64}(50^4 - 40^4) = 181132.5 \text{ mm}^2$$

$$K^2 = \frac{I}{A} = \frac{181132.5}{706.86} = 256.25 \text{ mm}^2$$

$$k^2 = \sqrt{\frac{P}{EI}} = \sqrt{\frac{25\times10^3}{205\times10^3\times1181132.5}} = 8.2\times10^{-4} \text{ mm}$$

$$\sigma_{max} = \frac{P}{A}\left[1+\frac{ey_c}{K^2}\sec\left(\frac{kl_e}{2}\right)\right]$$

$$= \frac{25\times10^3}{706.86}\left[1+\frac{5\times25}{256.25}\sec\left(\frac{8.2\times10^{-4}\times3\times10^3}{2}\right)\right]$$

$$\sigma_{max} = 52.62 \text{ N/mm}^2$$

Example 8.13 *Find the greatest length for which a mild steel strut of T-shaped cross-section of area 3000 mm² and least moment of inertia 24 × 10⁵ mm⁴, may be used with one end fixed and other free in order to limit the stress to 70 MPa. The safe load is one-fourth of the crippling load. Rankine constants for mild steel are :* $a = \frac{1}{7500}$ *and* $\sigma_c = 330$ *MPa.*

Solution:

Given: A = 3000 mm², I_{min} = 24 × 10⁵ mm⁴, σ = 70 MPa, fos = 4

$$K^2 = \frac{I}{A} = \frac{24\times10^5}{3000} = 800 \text{ mm}^2$$

Crippling load, $P_{cr} = \sigma\times A\times fos$

$$= 70\times3000\times4 = 840 \text{ kN}$$

$$P_{cr} = \frac{\sigma_c A}{1+a\left(\frac{l_e}{K}\right)^2}$$

$$840\times10^3 = \frac{330\times3000}{1+\frac{1}{7500}\left(\frac{l_e^2}{800}\right)} = \frac{99\times10^4}{1+\frac{l_e^2}{6\times10^6}}$$

$$1+\frac{l_e^2}{6\times10^6} = 1.1786$$

$$l_e^2 = 1071428$$
$$l_e = 1035 \text{ mm}$$
$$2l = 1035$$
$$l = 517.5 \text{ mm}$$

Example 8.14 *Determine the section of a cast iron hollow cylindrical column 5 metre long with ends firmly built-in if it carries an axial load of 300 kN. The ratio of internal to external diameter is 3/4. Use factor of safety of 8. Take $\sigma_c = 567$ N/mm^2 and Rankine's constant $a = 1/1600$.*

Solution:

Given : $L = 5\text{m}, P = 300 \text{ kN}, \frac{d_i}{d_o} = 0.75, fos = 8, \sigma_c = 567 \text{ N/mm}^2, a = \frac{1}{1600}$

According to Rankine formula,

$$P_{cr} = \frac{\sigma_c \times A / fos}{1+a\left(\frac{L_e}{K}\right)^2}$$

$$L_e = \frac{L}{2} = 2.5 \text{ m}$$

$$A = \frac{\pi}{4}(d_o^2 - d_i^2) = \frac{\pi}{4}d_o^2\,[1-(0.75)^2] = 0.3436\, d_o^2$$

$$I = \frac{\pi}{64}(d_o^4 - d_i^4) = AK^2$$

$$K^2 = \frac{I}{A} = \frac{\frac{\pi}{64}(d_o^4 - d_i^4)}{\left[\frac{\pi}{4}(d_o^2 - d_i^2)\right]} = \frac{1}{16}(d_o^2 + d_i^2)$$

$$= \frac{1}{16}d_o^2[1+(0.75)^2] = 0.0976\, d_o^2$$

$$300 \times 10^3 = \frac{\left(\frac{567}{8}\right)\times 0.3436 d_o^2}{1+\frac{1}{1600}\times\frac{(2500)^2}{0.0976 d_o^2}} = \frac{24.35 d_o^2}{1+\frac{40023}{d_o^2}}$$

$$= \frac{24.35 d_o^4}{d_o^2 + 40023}$$

$$24.35\, d_o^2 = 300 \times 10^3 d_o^2 - 120069 \times 10^5 = 0$$

$$d_o^4 = 12320\, d_o^2 - 4931 \times 10^5 = 0$$

$$d_o^2 = \frac{1}{2}\left[12320 \pm \sqrt{(12320)^2 + 4 \times 4931 \times 10^5}\right]$$

$$= \frac{1}{2}[12320 \pm 46089] = 29204$$

$$d_o = 171 \text{ mm}, d_i = 128.25 \text{ mm}$$

Example 8.15 *A 2m long pin ended column of square cross-section is to be made of wood. Assuming E = 13 GPA, σ_{allow} = 12 MPa, and using a factor of safety of 2.5 in computing the Euler's critical load for buckling, determine the size of the cross-section if the column is to safely support 100 kN load.*

Solution:

Given : l = 2m, E = 13 GPa, σ_a = 12 MPa, FOS = 2.5, P = 100 kN

Euler cripping load,

$$P_{cr} = \frac{\pi^2 EI}{l^2}$$

$$\text{Euler stress, } \sigma = \frac{P_{cr}}{A} = \frac{\pi^2 EI}{Al^2}$$

$$\text{Allowable stress, } \sigma_a = \frac{\sigma}{FOS} = \frac{\pi^2 EI}{Al^2 \times FOS}$$

Let a = side of square, m

$$I = \frac{a^4}{12}, A = a^2$$

$$\sigma_a = \frac{\pi^2 E a^4}{12 \times a^2 \times l^2 \times FOS}$$

$$= \frac{\pi^2 E a^2}{12 l^2 \times FOS}$$

$$12 \times 10^6 = \frac{\pi^2 \times 13 \times 10^9 \times a^2}{12 \times 4 \times 2.5}$$

$$a = 0.106 \text{ m or } 106 \text{ mm}$$

Example 8.16 *A solid round bar 3 m long and 5 cm in diameter is used as a column with both ends hinged. Determine the percentage change in the Euler's crippling load of the column if the end conditions are changed to both ends fixed. Take E = 200 GPa.*

Solution:

Given : $l = 3$ m, $d = 5$ cm, E = 200 GPa

$$(P_{cr})_{\text{hinged}} = \frac{\pi^2 E1}{l^2}$$

$$I = \frac{\pi d^4}{64} = \frac{\pi \times 5^4}{64} - 30.68 \text{ cm}^4$$

$$(P_{cr})_{\text{fixed}} = \frac{4\pi^2 EI}{l^2}$$

$$\frac{(P_{cr})_{\text{fixed}} - (P_{cr})_{\text{hinged}}}{(P_{cr})_{\text{hinged}}} = \frac{4-1}{1} = 3 \text{ } i.e. \text{ } 300\%$$

Example 8.17 *Find the Euler's crippling load for a hollow cylindrical cast iron (E = 80 GPa) column 150 mm external diameter and 20 mm thick. The column is 6 m long and hinged at both ends.*

Solution:

Given : $E = 80$ GPa, $d_0 = 150$ mm, $d_i = 110$ mm, $l = 6$ m

$$P_{cr} = \frac{\pi^2 EI}{l^2}$$

$$= \pi^2 \times 80 \times 10^9 \times \frac{\pi}{64}(150^4 - 110^4) \times 10^{-12} \times \frac{1}{36} = 387406 \text{ N}$$

or $$= 387.406 \text{ kN}$$

Example 8.18 *Calculate the safe compressive load on a hollow cast iron column (one end rigidly fixed and other end hinged) of 150 mm external diameter, 100 mm internal diameter and 10 m in length. Use Euler's formula with a factor of safety of 5 and E = 95 GPa.*

Solution:

Given: $d_o = 150$ mm, $d_i = 100$ mm, L = 10m, FOS = 5, E = 95 GPa

$$I = \frac{\pi}{64}(d_o^4 - d_i^4) = \frac{\pi}{64}(150^4 - 100^4) = 19941750 \text{ mm}^4$$

Euler's load for a column with one end fixed and other end hinged,

$$P_e = \frac{2\pi^2 EI}{L^2} = \frac{2\pi^2 \times 95 \times 10^3 \times 19941750}{(10 \times 10^3)^2} = 373.95 \text{ kN}$$

$$\text{Safe load} = \frac{P_e}{FOS} = \frac{373.95}{5} = 74.79 \text{ kN}$$

Example 8.19 *A steel tubular column of outer and inner diameters 40 mm and 30 mm respectively, is 2m long. Compare the crippling loads given by Rankine's and Euler's formula. They are loaded through pin joints at each end. Take yield stress = 330 N/mm², the Rankine's constant = 1/7500 and E = 200 GPa, also determine the length at which the Euler's formula ceases to apply (in the same cross section).*

Solution:

Given: $d_o = 40$ mm, $d_i = 30$ mm, L = 2m, $\sigma_c = 330$ N/mm², $a = \dfrac{1}{7500}$, E = 200 GPa

$$I = \frac{\pi}{64}(d_o^4 - d_i^4) = \frac{\pi}{64}(40^4 - 30^4) = 85902.92 \text{ mm}^4$$

Euler's formula:

$$P_e = \frac{\pi^2 EI}{L^2}$$

$$= \frac{\pi^2 \times 200 \times 10^9 \times 85902.92 \times 10^{-12}}{2^2} = 42391.4 \text{ N}$$

Rankine's formula:

$$P_r = \frac{\sigma_C \times A}{1 + a\left(\dfrac{L}{K}\right)^2}$$

$$A = \frac{\pi}{4}(d_o^2 - d_i^2) = \frac{\pi}{4}(40^2 - 30^2) = 549.78 \text{ mm}^2$$

$$K = \sqrt{\frac{I}{A}} = \sqrt{\frac{85902.92}{549.78}} = 12.5 \text{ mm}$$

$$P_r = \frac{330 \times 549.78}{1 + \dfrac{1}{7500}\left(\dfrac{2000}{12.5}\right)^2} = 41108.9 \text{ N}$$

$$\frac{P_e}{P_r} = \frac{42391.4}{41108.9} = 1.0312$$

Eiler's formula shall cease to apply, when

$$\frac{\pi^2 EI}{L^2} = \sigma_C \times A$$

or

$$\frac{\pi^2 EAK^2}{L^2} = \sigma_C \times A$$

$$\frac{\pi^2 EK^2}{L^2} = \sigma_C$$

$$L^2 = \frac{\pi^2 EK^2}{\sigma_C} = \frac{\pi^2 \times 200 \times 10^9 \times (12.5 \times 10^{-3})^2}{330 \times 10^6} = 934.62 \times 10^{-3}$$

$$L = 0.967 \text{ m}$$

Example 8.20 *Derive the formula for the Euler critical stress for a column with fixed-free end conditions as a function of slenderness ratio. Discuss the validity of curve formula from cruve between Euler critical stress and slenderness ration*

Solution:

Given : Refer to art. 8.4.4.

For a column with fixed-free ends,

$$P_{cr} = \frac{\pi^2 EI}{4l^2} = \frac{\pi^2 EAK^2}{4l^2}$$

Euler critical stress, $\sigma_{cr} = \frac{P_{cr}}{A} = \frac{\pi^2 E}{4} \times \frac{1}{\left(\frac{l}{K}\right)^2}$

$$\sigma_{cr} = f\left[\frac{1}{\left(\frac{l}{K}\right)^2}\right]$$

where $\frac{l}{K}$ = slenderness ratio

$\frac{l}{K} = \frac{\pi}{2}\sqrt{\frac{E}{\sigma_{cr}}}$ for a column having one and fixed and the other free.

For E = 210 GPa and σ_{cr} = 300 MPa, $\frac{l}{K} = \frac{\pi}{2}\sqrt{\frac{210 \times 10^9}{300 \times 10^6}} = 41.5$

For a medium column, $\frac{l}{k}$ = 32 to 120.

Therefore, for $\frac{l}{K} < 41.5$, Euler's formula shall not be applicable for a column having one end fixed and other free.

Example 8.21 *A cast iron column 200 mm external diameter is 20 mm thick and has a length of 4.5 m. Assuming it can be treated as rigidly fixed at each end, calculate the safe load by Rankine's formula, using the following empirical constants.*

σ_c = 550 MN/m², a (for hinged ends) = 1/1600, factor of safety = 4.

Solution:

Given: d_o = 200 mm, d_i = 200 – 2 × 20 = 160 mm, l = 4.5 m, $l_e = \frac{l}{2}$ (both ends fixed),

σ_c = 550 MN/m², $a = \frac{1}{1600}$, FOS = 4

$$A = \frac{\pi}{4}(d_o^2 - d_i^2) = \frac{\pi}{4}(200^2 - 160^2) = 11,309.7 \text{ mm}^2$$

$$I = \frac{\pi}{64}(d_o^4 - d_i^4) = \frac{\pi}{64}(200^4 - 160^4) = 46,369,907 \text{ mm}^4$$

$$K^2 = \frac{I}{A} = 4100 \text{ mm}^2$$

Rantine load,

$$P = \frac{\sigma_c \times \frac{A}{FOS}}{1 + \frac{1}{a}\left(\frac{l_e}{K}\right)^2}$$

$$= \frac{550 \times \frac{11,309.7}{4}}{1 + \frac{1}{1600}\left[\frac{(2.25 \times 10^3)^2}{4100}\right]} = 877.7 \text{ kN}$$

Example 8.22 *A column of circular section made of cost iron 200 mm external diameter and 20 mm thick is used as a column 4 metre long. Both ends of column are fixed. The column carries a load of 150 kN at an eccentricity of 25 mm from the axis of the column. Find the external stresses on the column section. Find also the maximum eccentricity in order there may be no tension anywhere on the section. Take E = 94000 N/mm².*

Solution:

Given: $d_o = 200$ mm, $d_i = 200 - 2 \times 20 = 160$ mm, $l = 4$ m, $l_e = \frac{l}{2} = 2$m, $P = 150$ kN, $e = 2$ mm, $E = 94000$ N/mm²

$$A = \frac{\pi}{4}(d_o^2 - d_i^2) = \frac{\pi}{4}(200^2 - 160^2) = 11,309.7 \text{ mm}^2$$

$$I = \frac{\pi}{64}(d_o^4 - d_i^4) = \frac{\pi}{64}(200^4 - 160^4) = 46,369,907 \text{ mm}^4$$

$$K^2 = \frac{I}{A} = 4100 \text{ mm}^2$$

Maximum deflection,

$$y_{max} = e \sec\left(\sqrt{\frac{P}{EI}} \times \frac{l_e}{2}\right)$$

$$= 25 \sec\left[\sqrt{\frac{150 \times 10^3}{94,000 \times 46,369,907}} \times 10^3\right]$$

$$= 25 \sec 0.1855 = 25 \sec 10.63°$$

$$= 25.43 \text{ mm}$$

$$M_{max} = P \times y_{max} = 150 \times 25.43 = 3815.5 \text{ kN.mm}$$

Resultant stress, $\sigma_r = \frac{P}{A}+\frac{M}{z}$

$$= \frac{150\times10^3}{11,309.7}+\frac{3815.5\times10^3\times100}{46,369,907}$$

$= 13.26 + 8.22 = 21.48$ N/mm^2 (compressive)

Tensile stress $= -\frac{P}{A}+\frac{M}{z}$

$$= -13.26+\frac{Pe\sec 10.63^\circ \times y_c}{I}$$

$$= -13.26+\frac{150\times10^3\times e\times 1.0175\times100}{46,369,907}=0$$

$e = 40.3$ mm

Example 8.23 *A straight cylindrical bar is 16 mm diameter and 1.2 m long. It is freely supported at its two ends in a horizontal position and loaded at the centre with concentrated load of 90 N. The central deflection is found to be 5 mm.*

If placed vertical and loaded along its axis, what load would it cause to buckle? (Use Euler's formula). What is the ratio of the maximum stresses in the two cases?

Solution:

Given: $d = 16$ mm, $l - 1.2$m, $W = 90$ N, $\delta - 5$ mm

$$\delta = \frac{Wl^3}{48EI}$$

$$5 = \frac{90\times(1.2\times10^3)^3}{48\times EI}$$

$EI = 648 \times 10^6$ N.mm^2

$$P_{cr} = \frac{\pi^2 EI}{l^2} = \frac{\pi^2\times648\times10^6}{(1.2\times10^3)^2} = 4441 \text{ N}$$

$$\sigma_1 = \frac{W}{A} = \frac{90\times4}{\pi\times16^2} = 0.4476 \text{ N/mm}^2$$

$$\sigma_2 = \frac{P_{cr}}{A} = \frac{4441\times4}{\pi\times16^2} = 22.088 \text{ N/mm}^2$$

$$\frac{\sigma_2}{\sigma_1} = \frac{22.088}{0.4476} = 49.35$$

SUMMARY

1. Euler's buckling load for a column,

$$P_e = \frac{\pi^2 EI}{l_e^2}$$

where l_e = equivalent length of a column

$= l$ for both ends hinged

$= \frac{l}{2}$ for both ends fixed

$= \frac{l}{\sqrt{2}}$ for one end fixed and other hinged

$= 2l$ for one end fixed and other free.

2. Slenderness ratio $= \dfrac{\text{length of column}}{\text{least radius of gyration}}$

3. Rankine-Gordon formula,

$$P = \frac{\sigma_c \times A}{1 + a\left(\frac{l_e}{K}\right)^2}, \; a = \frac{\sigma_c}{\pi^2 E}$$

4. Johnson's parabolic formula,

$$P = \sigma_c \times A\left[1 - c\left(\frac{l_e}{K}\right)^2\right]$$

where $c = \dfrac{1}{4\pi^2 E}$

5. Straight line formula,

$$P = \sigma_c \times A\left[1 - b\left(\frac{l_e}{K}\right)\right]$$

where $b = \sqrt{\dfrac{4E}{2\pi^2 \sigma_c}}$

6. Eccentric loading.

(a) Short column

Resultant stress,

$$\sigma_r = \frac{P}{A}\left[1 \pm \frac{6e_x}{b} \pm \frac{6e_y}{d}\right]$$

where b = width of column
d = depth of column

For no tension, $\frac{e_x}{b}+\frac{e_y}{d}=\frac{1}{6}$

(*b*) Long column

$$\sigma_r = \frac{P}{A}\left[1+\frac{ey_c}{K^2}\sec\left(\sqrt{\frac{P}{EI}}\times\frac{l_e}{2}\right)\right]$$

7. **Middle Third Rule:**
   To avoid tension in a rectangular cross-section column the load must lie within the middle third of the cross-section of the column.
8. **Middle Quarter Rule:**
   To avoid tension in a circular cross-section column the load must lie within the middle quarter of the cross-section of the column.

Short Answer Type Questions

1. Differentiate between a strut and a column.
2. Define equivalent length of a column.
3. Define slenderness ratio.
4. What is buckling load for a column?
5. Differentiate between a short and a long column.
6. Compare the equivalent length for a column having both ends fixed and another having one end fixed and the other free.
7. If P_c is the crushing load and P_e the Euler load then what is the Rankine load.
8. The Rankine consant depends on which column material properties.
9. State the middle third rule.

EXERCISES

1. Compare the crippling loads given by Euler's and Rankine's formulae for a column of tubular cross-section 3m long and having outer and inner diameters of 108 mm and 60 mm respectively. The one end of the column is fixed and other hinged. Take yield strength = 380 MPa, $a = \frac{1}{7500}$ and E = 210 GPa.
2. A built up beam is shown in Fig. 8.9. Calculate the length when it deflects 10 mm when subjected to a central load of 50 kN on simply supported ends. Find the safe load if this beam is used as a column with both ends fixed to give a factor of safety of 4. E = 200 GPa.
3. A column of channel cross-section fixed at both ends is shown in Fig. 6.10. Calculate the safe load if its length is 6 m. E = 210 GPa.

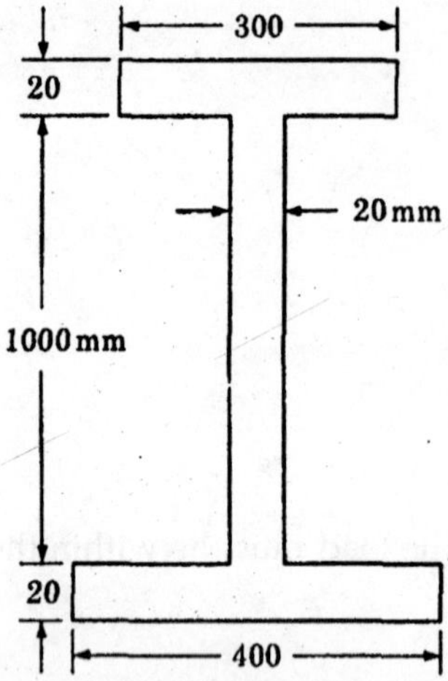

Fig. 8.9

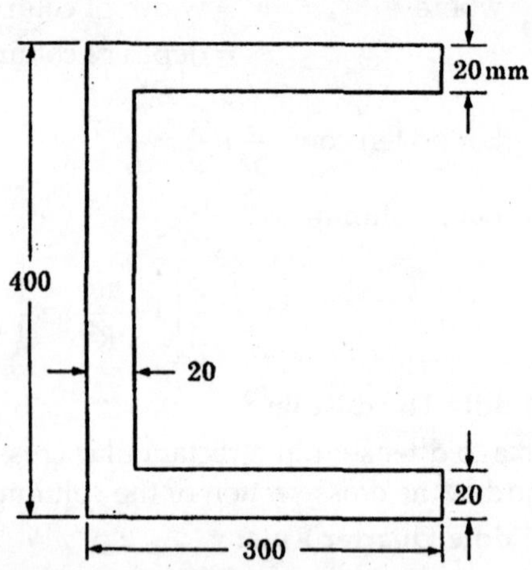

Fig. 8.10

4. A tubular steel strut 80 mm external diamter and 40 mm internal diameter is 4 m long and hinged at both ends. The load is parallel to the axis but eccentric. Find the maximum eccentricity for a crippling load of 80% of the Euler load. The yield strength is 320 MPa and $E = 200$ GPa.
5. A rectangular column is 200 mm × 150 mm. Determine the core of the column and show on a sketch.
6. A column is made of two RSJ of I-section 200 × 80 × 10 mm³ with plate 200 mm × 10 mm riveted with flanges one each on top and bottom. The edges of plates being flush with the outside edges of joist's flanges. Determine the safe load the column of 4 m length with both ends fixed can carry with factor of safety of 4. Take, $a = \dfrac{1}{7500}$ and $\sigma_c = 320$ MPa.
7. A steel rod having 100 mm diameter and 3 m long is used as a strut with both ends hinged. The load is parallel to the axis of the strut but is eccentric. Find the maximum value of eccentricity so that crippling load on strut is 60% of the Euler's crippling load. Take $E = 200$ GPa and $\sigma_c = 320$ MPa.

Chapter-9

Helical and Leaf Springs

9.1 INTRODUCTION

A helical spring is a piece of wire coiled in the form of helix. If the slope of the helix of the coil is so small that the bending effects can be neglected, then the spring is called a close-coiled spring. In such a spring only torsional shear stresses are introdcued. On the other hand, if the slope of the helix of the coil is quite appreciable, then both the bending as well as torsional shear stresses are introduced. in the spring and a spring of this type is called an open coiled spring. Generally if the hexlix angle is less than 5° then it is considered as a close-coiled helical spring.

A leaf spring is a beam of uniform strength supported at the centre and loaded at the ends. It consists of a number of over-lapping leaves, each of the same width and depth but varying in length. Each leaf is shorter than the one above it by a constant amout, called the overlap. Each plate is free to slide relative to its neighbour, *i.e.* each plate will act as a separate beam. Each plate has initially the same curvature, *i.e.* contact is only at the ends.

9.2 DEFINITIONS

Spring Index. It is the ratio of the mean coil diameter to the diameter of the spring wire. It is denoted by C.

$$C = D_m/d$$

Stiffness. It is defined as the load per unit deflection and is denoted by k.

Torisional rigidity. Torisional rigidity or torsional stiffness is defined as the torque per unit angular twist.

Helix angle. It is the angle which the axis of the spring wire makes with a horizontal line perpendicular to the axis of the spring.

9.3 CLOSE-COILED HELICAL SPRING

9.3.1 Axial Load

A close-coiled helical spring under the action of axial load is shown in Fig. 9.1

Let W = axial load

D_m = mean coil diameter

d = diameter of spring wire

δ = axial deflection

G = modulus of rigidity

θ = angular deflection

n = number of active coils

τ = maximum shearing stress induced.

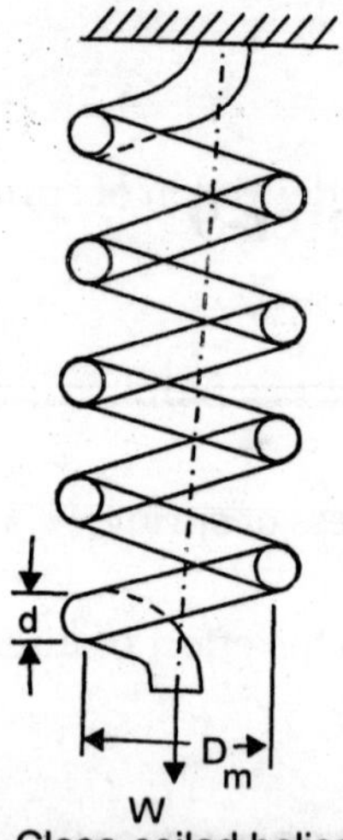

Fig. 9.1 Close-coiled helical spring under axial load

The following assumptions are made :

1. An element of the spring wire behaves essentially as a straight bar under pure torsion.
2. The planes perpendicular to the bar axis do not warp or distort during deformation. As a result of this the shearing stress will have a linear distribution along the radius.
3. The material of the wire is linear, homogeneous, isotropic and obeys Hooke's law.
4. Direct shear and curvature effects of wire are neglected.

Torque acting on the spring, $T = \frac{1}{2} W D_m$ ... (9.1)

Shear stess induced in spring wire,

$$\tau = \frac{Td}{2J} = \frac{1}{2} W D_m \times \frac{32d}{2 \times \pi d^4} = \frac{8 W D_m}{\pi d^3} \quad \text{... (9.2)}$$

Length of wire, $l = \pi D_m n$

Now $$\frac{T}{J} = \frac{G\theta}{l}$$

$\therefore$ $$\theta = \frac{Tl}{GJ} = \frac{1}{2} W D_m \times \frac{\pi D_m n \times 32}{G \times \pi d^4}$$

$$= \frac{16 W D_m^2 n}{G d^4} \quad \text{... (9.3)}$$

Strain energy stored in the spring,

$$U = \frac{1}{2} T\theta$$

$$= \frac{1}{2}\left(\frac{1}{2} W D_m\right) \times \frac{16 W D_m^2 n}{G D^4}$$

$$= \frac{4W^2 D_m^3 n}{Gd^4}$$

Axial deflection of spring by Castigliano's theorem is given by,

$$\delta = \frac{dU}{dW} = \frac{8WD_m^3 n}{Gd^4} \quad \text{... (9.4)}$$

Stiffness of spring, $$k = \frac{W}{\delta} = \frac{Gd^4}{8D_m^3 n} \quad \text{... (9.5)}$$

When the effect of direct shear stress and curvature of wire is taken into account, then

$$\tau_{max} = K_w \frac{8WD_m}{\pi d^3} \quad \text{... (9.6)}$$

where K_w = Wahl's correction factor

$$= \frac{4C-1}{4C-4} + \frac{0.615}{C} \quad \text{... (9.7)}$$

9.3.2 AXIAL TORQUE

A close-coiled helical spring under axial torque M_0 is shown in Fig. 9.2. The torque (or couple) produces bending stress in the coils of the spring.

Wind up angle, $$\phi = \frac{M_0 l}{EI}$$

Strain energy, $$U = \frac{1}{2} M_0 \phi$$

Now $$l = \pi D_m n$$

$$\therefore \quad \phi = \frac{64 M_0 \times \pi D_m n}{E \times \pi d^4} = \frac{64 M_0 D_m n}{Ed^4}$$

$$U = \frac{1}{2} M_0 \times \frac{64 M_0 D_m n}{Ed^4}$$

$$= \frac{32 M_0^2 D_m n}{Ed^4}$$

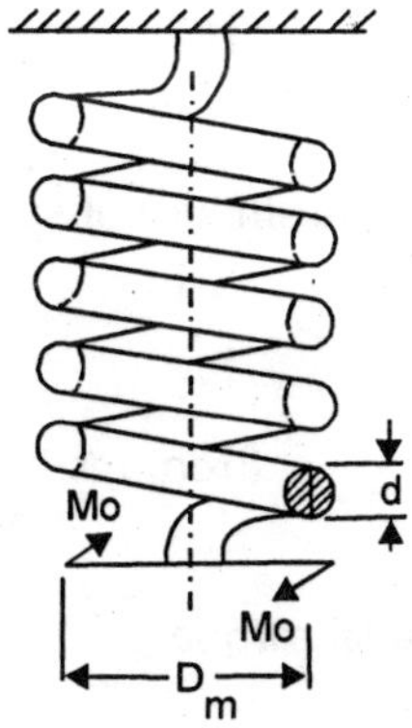

Fig. 9.2 Close-coiled helical spring under axial torque

Applying Castigliano's theorem, we have

Angle of twist of spring, $$\beta = \frac{\partial U}{\partial M_0} = \frac{64 M_0 D_m n}{Ed^4} \quad \text{... (9.8)}$$

Bending stress, $$\sigma_b = \frac{32 M_0}{\pi d^3}$$

9.3.3 Strain Energy Stored by the spring

(a) Axial Load

$$U = \frac{1}{2}T\theta = \frac{4W^2D_m^3n}{Gd^4}$$

$$\tau = \frac{8WD_m}{\pi d^3}$$

or
$$W = \frac{\pi d^3\tau}{8D_m}$$

$$\therefore \quad U = \left(\frac{4D_m^3n}{Gd^4}\right)\times\left(\frac{\pi d^3\tau}{8D_m}\right)^2$$

$$= \frac{\tau^2}{4G}\times(\pi D_m n)\times\left(\frac{\pi}{4}d^2\right)$$

$$= \frac{\tau^2}{4G}\times l\times A$$

$$= \frac{\tau^2}{4G}\times \text{Volume of spring wire}$$

$$\text{Resilience of spring} = \frac{\tau^2}{4G} \quad \text{... (9.9)}$$

$$\text{Proof resilience} = \frac{\tau_{max}^2}{4G} \quad \text{... (9.10)}$$

(b) Axial torque

$$U = \frac{1}{2}M_0\phi = \frac{32M_0^2D_mn}{Ed^4}$$

$$M_0 = \frac{\pi d^3}{32}\times\sigma_b$$

$$U = \frac{32D_mn}{Ed^4}\times\left(\frac{\pi d^3}{32}\times\sigma_b\right)^2$$

$$= \frac{\sigma_b^2}{8E} \times \pi D_m\, n \times \frac{\pi}{4} d^2$$

$$= \frac{\sigma_b^2}{8E} \times \pi D_m n \times \frac{\pi}{4} d^2$$

$$= \frac{\sigma_b^2}{8E} \times \text{Volume of spring wire}$$

$$\text{Resilience of spring} = \frac{\sigma_b^2}{8E} \qquad \text{... (9.11)}$$

$$\text{Proof resilience} = \frac{(\sigma_b)^2_{max}}{8E} \qquad \text{... (9.12)}$$

9.3.4 Axial Load and Twisting Moment acting Simultaneoulsy

Let W = axial load.

M_0 = axial twisting moment

Then twisting moment on the spring, $T = \frac{1}{2} WD_m$

Bending moment on the spring, $M = M_0$

Strain energy of the system,

$$U = \frac{T^2 l}{2GJ} + \frac{M^2 l}{2EI}$$

$$= \left(\frac{1}{2} WD_m\right)^2 \frac{l}{2GJ} + \frac{M_0^2 l}{2EI}$$

$$= (\pi D_m n)\left[\frac{1}{4} \times W^2 \times D_m^2 \times \frac{16}{G \times \pi d^4} + \frac{32 M_0^2}{E \times \pi d^4}\right]$$

$$= \frac{32 D_m n}{d^4}\left[\frac{W^2 D_m^2}{8G} + \frac{M_0^2}{E}\right]$$

Applying Castigliano's theorem, we have

$$\text{Axial deflection,} \quad \delta = \frac{\partial U}{\partial W} = \frac{32 D_m n}{d^4} \times \frac{W D_m^2}{4G}$$

$$= \frac{8 W D_m^3 n}{G d^4} \qquad \text{... (9.13)}$$

Axial twist, $$\beta = \frac{\partial U}{\partial M_0} = \frac{32D_m n}{d^4} \times \frac{2M_0}{E}$$

$$= \frac{64D_m n M_0}{Ed^4} \quad \text{... (9.14)}$$

Example 9.1 *A helical spring, in which the mean diameter of the coils is 8 times the wire diameter, is to be designed to absorb 200 N-m of energy with an exstension of 10 cm. The maximum shear stress is not to exceed 125 MPa. Determine the mean diameter of the helix, diameter of the wire and the number of turns. Also find the load with which an extension of 4 cm could be produced in the spring. G = 84 GPa.*

Solution.

Given $D_m = 8d$, $U = 200$ N.m, $\delta = 0.1$m, $\tau_{max} = 12.5$ MPa, $G = 84$ GPa

Work done on the spring = Energy abosorbed

$$\frac{1}{2} W\delta = 200$$

$$\frac{1}{2} W \times 0.1 = 200$$

$$W = 4 \text{ kN}$$

Maximum shear stress, $$\tau = \frac{8WD_m}{\pi d^3}$$

$$125 \times 10^6 = \frac{8 \times 4 \times 10^3 \times 8d}{\pi d^3}$$

$$d^2 = \frac{8 \times 4 \times 10^3 \times 8}{\pi \times 125 \times 10^6} = 6.519 \times 10^{-4}$$

$$d = 25.53 \times 10^{-3} \text{ m}$$

$$D_m = 8 \times 25.53 \times 10^{-3} = 204.26 \times 10^{-3} \text{ m}$$

Now deflection, $$\delta = \frac{8WD_m^3 n}{Gd^4}$$

$$0.10 = \frac{8 \times 4 \times 10^3 (204.26)^3 \times 10^{-9} n}{84 \times 10^9 (25.33)^4 \times 10^{-12}}$$

$$n = 13.08, \text{ say 13 turns}$$

When $\delta = 4$ cm

$$W = \frac{0.04 \times 84 \times 10^9 (25.53)^4 \times 10^{-12}}{8(204.26)^3 \times 10^{-9} \times 13} = 1610.5 \text{N}$$

Example 9.2 *A railway wagon weighing 6.5 kN and moving with a speed of 10 km/hr is to be stopped by 4 buffer springs in which the maximum compression allowed is 20 cm. Calculate the number of turns in each spring in which diameter of the wire is 2 cm and that of the coils 20 cm. G = 84 GPa.*

Solution.

Given : $W = 6.5\text{kN}, v = 10 \text{ km/h}, \delta = 0.2\text{m}, d = 2\text{cm}, D_m = 20\text{cm}, G = 84 \; GPa, i = 4$

Kinetic energy of the wagon,

$$\text{K.E.} = \frac{1}{2}\frac{W}{g}v^2$$

$$= \frac{1}{2} \times \frac{6.5 \times 10^3}{9.81}\left(\frac{10 \times 1000}{60 \times 60}\right)^2 = 2.5563 \,\text{kN.m}$$

$$\text{Energy to be abosorbed by each spring} = \frac{K.E.}{i}$$

$$= \frac{2.5563}{4} = 639.1 \,\text{N.m}$$

Energy absorbed = Work done on the spring

$$639.1 = \frac{1}{2} \times W_1 \times \delta$$

$$W_1 = \frac{2 \times 639.1 \times 10^3}{0.20} = 6391 \,\text{N}$$

Where W_1 is the equivalent static load.

$$\delta = \frac{8 W_1 D_m^3 n}{G d^4}$$

$$0.20 = \frac{8 \times 6391 \times 20^3 \times 10^{-6} \times n}{84 \times 10^9 \times 2^4 \times 10^{-8}}$$

$$n = 6.57 \simeq 7$$

Example 9.3 *A weight of 300 N is dropped through a height of 700 mm on to a closely coiled compression spring which instantaneously compresseses by 200 mm under the impact. If the diameter of the steel rod (of which the spring is made) is 25 mm and the mean radius of the coil is 100 mm, find the maximum instantaneous stress produced by the impact and the number of coils in the spring. Take shear modulus of spring material = 80 GPa.*

Solution.

Given : $W = 300 \text{ N}, h = 70 \text{ cm}, \delta = 20 \text{ cm}, d = 2.5 \text{ cm}, D_m = 20 \text{ cm}$

$G = 80 \text{ GPa}$

Let W_1 be the equivalent static load.

Energy lost by weight = Work done on the spring

$$W(h + \delta) = \frac{1}{2} W_1 \delta$$

$$300\,(70 + 20) \times 10^{-2} = \frac{1}{2} W_1 \times 20 \times 10^{-2}$$

$$W_1 = 2700 \text{ N}$$

$$\delta = \frac{8WD_m^3 n}{GD^4}$$

$$0.20 = \frac{8 \times 2700 \times 20^3 \times 10^{-6} \times n}{80 \times 10^9 (2.5) \times 10^{-8}}$$

$$n = 36.17$$

$$\tau = \frac{8WD_m}{\pi d^3} = \frac{8 \times 2700 \times 20 \times 10^{-2}}{\pi \times (2.5)^3 \times 10^{-6}} \; 87.971 \text{MPa}$$

Example 9.4 *A close-coiled helical spring is to have a stiffness of 1 kN/m of compression under a maximum load of 45 N and a maximum shearing stress of 126 MPa. The solid length of the spring (i.e. when the coils are touching) is to be 4.5 cm. Find the diameter of the wire, the mean diameter of the coils required. Modulus of rigidity, G = 42 GPa.*

Solution.

Given : $k = 1$ *kN/m*, $W = 45$ *N*, $\tau_{max} = 126$ *MPa*, $l_s = 4.5$ cm, $G = 42$ *GPa*

Stiffness,

$$k = \frac{W}{\delta} = \frac{Gd^4}{8D_m^3 n}$$

$\therefore$

$$10^3 = \frac{42 \times 10^9 \times d^4}{8D_m^3 n} \quad \text{...(1)}$$

Maxmum shear stress,

$$\tau_{max} = \frac{8WD_m}{\pi d^3}$$

$$126 \times 10^6 = \frac{8 \times 45 \times D_m}{\pi d^3}$$

or

$$D_m = 35 \times 10^4 \times \pi d^3 \quad \text{...(2)}$$

Solid length, $l_s = dn = 4.5 \times 10^{-2}$ m

$$n = \frac{4.5 \times 10^{-2}}{d} \quad \text{...(3)}$$

Substituting Eq. (3) in (1), we have

$$10^3 = \frac{42 \times 10^9 \times d^4}{8D_m^3 \times \left(\frac{4.5 \times 10^{-2}}{d}\right)} = 1.167 \times 10^{11} \times \frac{d^5}{D_m^3}$$

$$D_m^3 = 1.167 \times 10^8 \times d^5 \quad \text{...(4)}$$

Substituting Eq. (2) in (4),

$$(35 \times 10^4 \times \pi d^3) = 1.167 \times 10^8 \times d^5$$

$$d = 3.06 \times 10^{-3}\ \text{m} = 3.06\ \text{mm}$$

$$D_m = 35 \times 10^4 \times \pi \times (3.06)^3 = 31.5\ \text{mm}$$

$$n = \frac{4.5 \times 10^{-2}}{3.06 \times 10^{-3}} = 14.7 \simeq 15$$

Example 9.5 *(a) A close-coiled helical spring made of 12 mm round steel wire has 12 coils and the mean diameter of the coils is 16 cm. The spring is subjected to an axial load of 150 N. Determine the elongation, intensity of torsional stress and strain energy per cubic meter under the loaded condition. G = 84 GPa.*

(b) If the axial load is removed and an axial torque of 10 N.m is applied, determine the axial twist, intensity of bending stress, and work stored per cubic meter in the spring. E = 210 GPa.

Solution.

Given : $d = 12$ mm, $n = 12$, $D_m = 16$ cm, $W = 150$ N, $G = 84$ GPa, $M_0 = 10$ N.m, $E = 210$ GPa

(a) Axial load

Deflection, $$\delta = \frac{8WD_m^3 n}{Gd^4} = \frac{8 \times 150 \times 16^3 \times 10^{-6} \times 12}{84 \times 10^9 (1.2)^4 \times 10^{-8}} = 3.386 \times 10^{-2}\ \text{m}$$

Shear stress, $$\tau = \frac{8WD_m}{\pi d^3} = \frac{8 \times 150 \times 16 \times 10^{-2}}{\pi (1.2)^3 \times 10^{-6}} = 35.368\ \text{MPa}$$

Strain energy stored per unit volume,

$$u = \frac{\tau^2}{4G} = \frac{(35.368)^2 \times 10^{12} \times 12}{4 \times 84 \times 10^9} = 3.723\ \text{kN.m/m}^3$$

(b) Axial torque

Axial twist, $$\phi = \frac{64\, M_0 D_m n}{Ed^4} = \frac{64 \times 10 \times 16 \times 10^{-2}}{210 \times 10^9 (1.2)^4 \times 10^{-8}} = 0.2822\ \text{radians}$$

Bending stress, $$\sigma_b = \frac{32 M_0}{\pi d^3} = \frac{32 \times 10}{\pi (1.2)^3 \times 10^{-6}} = 58.946\ \text{MPa}$$

Strain energy stored per unit volume,

$$u = \frac{\sigma_b^2}{8E} = \frac{(58.946)^2 \times 10^{12}}{8 \times 210 \times 10^9} = 2.068\ \text{kNm/m}^3$$

Example 9.6 *A close-coiled helical spring is required to have an axial stiffness of 5 kN/m and an angular stiffness of 0.1 Nm per degree angle of twist. If the spring is made of steel wire 6 mm diameter, find the mean diameter of the coil and the number of turns required. Assuming E = 200 G Pa and G= 80 GPa.*

Solution.

Given : $k = 5$ kN/m, $k_\phi = 0.1$ N.m/deg, $d = 6$mm, $E = 200$ GPa, $G = 80$ GPa

$$k = \frac{W}{\delta} = \frac{Gd^4}{8D_m^3 n}$$

$$5\times10^3 = \frac{80\times10^9\times6^4\times10^{-12}}{8D_m^3 n}$$

$$D_m^3 n = 2592\times10^{-6} \quad \text{... (1)}$$

$$k_\phi = \frac{M_0}{\phi} = \frac{0.1\times180}{\pi\times1} = \frac{Ed^4}{64D_m n} = \frac{200\times10^9\times6^4\times10^{-12}}{64\times D_m n}$$

$$D_m n = 0.7068556 \quad \text{... (2)}$$

Solving Eqs. (1) and (2), we get

$$n = 11.67,\ D_m = 60.55 \text{ mm}$$

Example 9.7 *(a) A close-coiled helical spring is made of steel wire 6 mm diameter; it has 10 turns of mean diameter 7.5 cm and the pitch of the coils when the spring is unloaded is 2 cm. Determine the axial load which when gradually applied will cause the coils just to close up. Take G= 80 GPa.*

(b) If the spring stands on a rigid horizontal surface with the axis veritcal and a load of 150 N is allowed to fall freely on the top of the coil, find the height through which it must fall in order just to close to the coils together.

Solution.

Given : $d = 6$ mm, $n = 10$, $D_m = 7.5$ cm, $p = 2$ cm, G = 80 GPa

(a) Solid length = $nd = 10\times6 = 60$ mm

$$Gap = 20\times10 - 60 = 140 \text{ mm}$$

$$\delta = Gap = \frac{8WD_m^3 n}{Gd^4}$$

$$W = \frac{140\times10^{-3}\times80\times10^9\times6^4\times10^{-12}}{8\times(7.5)^3\times10^{-6}\times10} = 430\,\text{N}$$

(b)

$$W(h+\delta) = \frac{1}{2}k\delta^2,\ k = \frac{W}{\delta} = \frac{430}{140} = 3.071\,\text{N/mm}$$

$$150(h+140)\times10^{-3} = \frac{1}{2}\times3071\times(140\times10^{-3})^2$$

$$h + 140 = 200.638$$

$$h = 60.638 \text{ mm}$$

Example 9.8 *A helical spring of circular cross-section wire 18 mm in diameter is loaded by force of 500 N. The mean coil diameter of the spring is 125 mm. The modulus of rigidity is 80 GPa. Determine the maximum shear stress in the material of the spring. What number of coils must the spring have for its deflection to be 6 mm?*

Solution.

Given : $d = 18$ mm, $W = 500$ N, $D_m = 125$ mm, G = 80 GPa, $\delta = 6$ mm

$$\delta = \frac{8WD_m^3 n}{Gd^4}$$

$$6\times10^{-3} = \frac{8\times500\times125^3\times10^{-9}\times n}{80\times10^9\times18^4\times10^{-12}}$$

$$n = 6.45$$

$$\tau = \frac{8WD_m}{\pi d^3} = \frac{8\times500\times125\times10^{-3}}{\pi\times18^3\times10^{-9}} = 27.29\,\text{MPa}$$

Example 9.9 *A wire 6 mm in diameter is to be wound into a helical spring with a diameter of the generating cylinder equal to 5 cm. This spring is to deflect 2.5 cm under a load of 90N. Determine the required length of wire. What will then be the angle of twist between the ends of the wire, and the maximum shear stress in the spring? G = 80 GPa.*

Solution.

Given : $d = 6$ mm, $D_m = 5$ cm, $\delta = 2.5$ cm, $W = 90$ N, $G = 80$ GPa

$$\delta = \frac{8WD_m^3 n}{Gd^4}$$

$$2.5\times10^{-2} = \frac{8\times90\times56^3\times10^{-9}\times n}{80\times10^9\times6^4\times10^{12}}$$

$$n = 20.5 \simeq 21$$

$$l = \pi Dn = \pi\times56\times21 = 3.6945\text{ m}$$

$$\theta = \frac{16WD^2n}{Gd^4} = \frac{16\times90\times56^2\times10^{-9}\times21}{80\times10^9\times6^4\times10^{-12}} = 52.4°$$

$$\tau = \frac{8WD_m}{\pi d^3} = \frac{8\times90\times56\times10^{-3}}{\pi\times6^3\times10^{-9}} = 59.418\,\text{MPa}$$

$$\tau_d = \frac{4W}{\pi d^2} = \frac{4\times90}{\pi\times36\times10^{-6}} = 3.183\text{ MPa}$$

Total shear stress $= 59.418 + 3.183 = 62.601$ MPa

Example 9.10 *A safety valve 7.5 cm diameter, closed by a spring, must open at a pressure of 8 atmosphere. The valve opens only after it is lifted by 2 cm. The spring has a mean diameter of 6 cm, and is made of a wire 12 mm diameter. The clear distance between coils in the unloaded spring equals 5 mm. Determine the required number of coils, if the spring has a further capacity of deflection of 16 mm, when the valve is open. Determine also the initial compressive force, and the maximum shear stress in the material of the spring. G = 80 GPa and 1 atm = 1.013 × 10⁵ N/m².*

Solution.

Given : $d_p = 7.5$ cm, $p_v = 8$ atm, $\delta_1 = 2$ cm, $D_m = 6$ cm, $d = 12$ mm, $p = 5$ mm, $\delta_2 = 16$ mm, $G = 80$ GPa, $1\text{atm} = 1.013\times10^5\text{ N/m}^2$

Spring force to open the valve,

$$W = \frac{\pi}{4}d_p^2 p_v$$

$$= \frac{\pi}{4}\times(7.5)^2\times10^{-4}\times8\times1.013\times10^5 = 3580.24\,\text{N}$$

Let W_1 = initial compressive force on spring

Stiffness of spring, $k = \dfrac{W - W_1}{2\times10^{-2}} = \dfrac{3580.24 - W_1}{0.02}$

Also $k = \dfrac{Gd^4}{8D_m^3 n} = \dfrac{80\times10^3\times12^4\times10^{-12}}{8\times6^3\times10^{-6}\times n} = \dfrac{960\times10^3}{n}$

$$\therefore \quad \frac{3580.24 - W_1}{0.02} = \frac{960\times10^3}{n}$$

$$3580.24 - W_1 = \frac{19.2\times10^3}{n} \quad \text{... (1)}$$

$$\frac{np - (\delta_1 + \delta_2)}{10^3} = \frac{8W_1 D_m^3 n}{Gd^4}$$

$$\frac{5n - 36}{10^3} = W_1 n\left(\frac{8\times6^3\times10^{-6}}{80\times10^9\times12^4\times10^{-12}}\right)$$

$$5n - 36 = \frac{W_1 n}{960} \quad \text{... (2)}$$

From Eq. (1), we have

$$W_1 = 3580.24 - \frac{19.2\times10^3}{\pi} \quad \text{...(3)}$$

Substituting Eq. (3) in (2), we get

$$5\,n - 36 = \left(3580.24 - \frac{19.2\times10^3}{n}\right)\times\frac{n}{960}$$

$$= 3.729\,n - 20$$

$$1.327\,n = 16$$

$$n = 12$$

$$W_1 = 3580.24 - \frac{19.2\times10^3}{12} = 1980.24\text{ N}$$

$$\tau = \frac{8WD_m}{\pi d^3} = \frac{8\times3580.24\times6\times10^{-2}}{\pi\times12^3\times10^{-9}} = 316.56\text{ MPa}$$

$$\tau_d = \frac{4W}{\pi d^2} = \frac{4\times3580.24}{\pi\times144\times10^{-6}} = 31.66\text{ MPa}$$

$$\tau_{max} = \tau + \tau_d = 348.12\text{ MPa}$$

Example 9.11 *Two close-coiled helical springs are compressed between two parallel plates by a load of 1 kN. The springs have a wire diameter of 2 cm and the radii of the coils are 5 cm and 7 cm. Each spring has 12 coils and is of the same initial length. If the smaller spring is placed inside the larger one such that both springs deflect the same amount, determine the deflection of the springs and the maximum stress in each. G = 60 GPa.*

Solution.

Given : $W = 1$ kN, $d_1 = d_2 = 2$cm, $D_1 = 10$cm, $D_2 = 14$ cm, $n_1 = n_2 = 12$, $G = 60$ GPa

$\delta_1 = \delta_2$

$$\frac{8W_1 D_1^3 n_1}{Gd_1^4} = \frac{8W_2 D_2^3}{Gd_2^4};\ n_1 = n_2,\ d_1 = d_2$$

$$\frac{W_1}{W_2} = \left(\frac{D_2}{D_1}\right)^3 = \left(\frac{14}{10}\right)^3 = 2.744$$

$$W_1 + W_2 = 1000 \text{ N}$$

$$W_1 = 731 \text{ N},\ W_2 = 267 \text{ N}$$

$$\delta = \frac{8\times 731\times 10^3 \times 10^{-6} \times 12}{60\times 10^9 \times 2^4 \times 10^{-8}} = 7.31\times 10^{-3} \text{ m}$$

$$\tau_1 = \frac{8W_1 D_1}{\pi d_1^3} = \frac{8\times 731\times 10\times 10^{-2}}{\pi\times 8\times 10^{-6}} = 23.268 \text{ MPa}$$

$$\tau_2 = \frac{8\times 267\times 14\times 10^{-2}}{\pi\times 8\times 10^{-6}} = 11.898 \text{ MPa}$$

9.4 OPEN-COILED HELICAL SPRING

9.4.1 Axial Load

(a) Axial Deflection

Consider an open-coiled helical spring, as shown in Fig. 9.3, under the action of axial load.

Fig. 9.3 (*a*) shows the annulus cylinder which contains the axis of the spring wire. Obviously the radius of this cylinder will be equal to the mean coil radius of the spring. The axial load *W* can be consdered as a direct load Wacting on the spring at the mean radius *R*, together with a couple *WR* about *OA* shown in Fig. 9.3 (*b*),

Let α = constant helix angle which the coils make with planes perpendicular to the axis of the spring.

$D_m = 2R$ = mean diameter of the spring coil.

n = number of open coils.

p = pitch of coils.

W = axial load.

δ = axial deflection of spring.

Then from Fig. 9.3 (c), we have $\alpha = \dfrac{p}{2\pi R}$

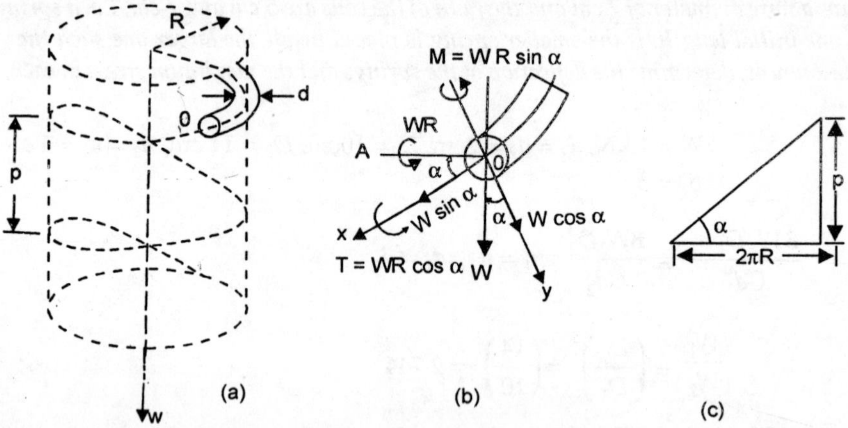

Fig. 9.3 Open coiled helical spring under axial load.

Length of spring wire, $l = \dfrac{2\pi Rn}{\cos\alpha} = 2\pi Rn\sec\alpha$

Twisting moment on the spring, $T = WR\cos\alpha$

Bending moment on the spring, $M = WR\sin\alpha$

Combined strain energy stored in the spring under bending and torsion,

$$U = \frac{1}{2}T\theta + \frac{1}{2}M\phi$$

$$= \frac{1}{2}\frac{T^2 l}{GJ} + \frac{1}{2}\frac{M^2 l}{EI}$$

$$= \frac{1}{2}\left[\frac{32(WR\cos\alpha)^2 l}{G\times\pi d^4} + \frac{64(WR\sin\alpha)^2 l}{E\times\pi d^4}\right]$$

$$= \frac{16l}{\pi d^4}\left[\frac{W^2R^2\cos^2\alpha}{G} + \frac{2W^2R^2\sin\alpha}{E}\right]$$

Applying Castigliano's theorem, we have

$$\delta = \frac{\partial U}{\partial W} = \frac{16lR^2}{\pi d^4}\left[\frac{2W\cos^2\alpha}{G} + \frac{4W\sin^2\alpha}{E}\right]$$

$$= \frac{32\,WlR^2}{\pi d^4}\left[\frac{\cos^2\alpha}{G} + \frac{2\sin^2\alpha}{E}\right]$$

$$= \frac{32WR^2}{\pi d^4}(2\pi Rn\sec\alpha)\left[\frac{\cos^2\alpha}{G}+\frac{2\sin^2\alpha}{E}\right]$$

$$= \frac{64WR^3n\sec\alpha}{d^4}\left[\frac{\cos^2\alpha}{G}+\frac{2\sin^2\alpha}{E}\right] \quad \text{... (9.15 a)}$$

$$= \frac{8WD_m^3n\sec\alpha}{d^4}\left[\frac{\cos^2\alpha}{G}+\frac{2\sin^2\alpha}{E}\right] \quad \text{... (9.15 b)}$$

For $\alpha = 0$, $\delta = \dfrac{8WD_m^3n}{Gd^4}$

(b) Angular Rotation

Let an imaginary axial torque T_0 be applied to the spring together with W. Then combined twisting moment on the spring,

$$T = WR\cos\alpha + T_0\sin\alpha$$

and combined bending moment,

$$M = T_0\cos\alpha - WR\sin\alpha$$

$$U = \frac{(WR\cos\alpha + T_0\sin)^2 l}{2GJ} + \frac{(T_0\cos\alpha - WR\sin\alpha)^2 l}{2EI}$$

Applying Castigliano's theorem, we have

Angular rotation, $\beta = \dfrac{\partial U}{\partial T_0}$, when $T_0 = 0$

$$= \frac{2WRl\cos\alpha\sin\alpha}{2GJ} - \frac{2WRl\ \sin\alpha\cos\alpha}{2EI}$$

$$= \frac{32WRl\sin\alpha\cos\alpha}{\pi d^4}\left[\frac{1}{G}-\frac{2}{E}\right]$$

Now $l = 2\pi Rn\sec\alpha$

$$\therefore \quad \beta = \frac{64WR^2n\sin\alpha}{d^4}\left[\frac{1}{G}-\frac{2}{E}\right] \quad \text{... (9.16 a)}$$

$$= \frac{16WD_m^2n\sin\alpha}{d^4}\left[\frac{1}{G}-\frac{2}{E}\right] \quad \text{... (9.16 b)}$$

9.4.2 Axial Torque

(a) Axial Rotation

Consider an open-coiled helical spring under the action of axial torque M_0, as shown in Fig. 9.4 (*a*). The axial torque can be split into two components.

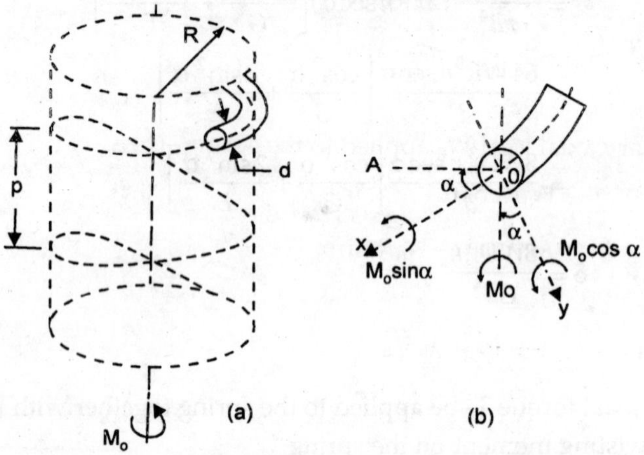

Fig. 9.4 Open coiled helical spring subjected to axial torque M_o.

Component along OX, T = = M_0 sin α, which at all sections produces torsion in the spring wire.

Component along OY, $M = M_0$ cos α, which at all sections produces bending moment in the spring wire.

Let β = angle through which the free end of the spring rotates.

Strain energy stored in the spring,

$$U = \frac{T^2 l}{2GJ} + \frac{M^2 l}{2EI}$$

$$= \frac{(M_0 \sin\alpha)^2 l}{2GJ} + \frac{(M_0 \cos\alpha)^2 l}{2EI}$$

Applying Castigliano's theorem, we have

$$\beta = \frac{\partial U}{\partial M_0}$$

$$= \frac{2M_0 l \sin^2\alpha}{2GJ} + \frac{2M_0 l \cos^2\alpha}{2EI}$$

$$= \frac{32 M_0 l \sin^2\alpha}{G \times \pi d^4} + \frac{64 M_0 l \cos^2\alpha}{E \times \pi d^4}$$

$$= \frac{32 M_0 l}{\pi d^4}\left[\frac{\sin^2\alpha}{G} + \frac{2\cos^2\alpha}{E}\right]$$

$$= \frac{32 M_0 \times 2\pi R n \sec\alpha}{\pi d^4}\left[\frac{\sin^2\alpha}{G} + \frac{2\cos^2\alpha}{E}\right]$$

$$\frac{64M_0Rn\sec\alpha}{d^4}\left[\frac{\sin^2\alpha}{G}+\frac{2\cos^2\alpha}{E}\right] \qquad ...(9.17)$$

(b) Axial Deflection

Assume an imaginary axial load W_0 applied to the spring. Then

$$T = W_0R\cos\alpha + M_0\sin\alpha$$

$$M = M_0\cos\alpha - W_0R\sin\alpha$$

Total strain energy stored,

$$U = \frac{1}{2}\frac{T^2l}{GJ}+\frac{1}{2}\frac{M^2l}{EI}$$

$$= \frac{[W_0R\cos\alpha + M_0\sin\alpha]^2 l}{2GJ}+\frac{[M_0\cos\alpha - W_0R\sin\alpha]^2 l}{2EI}$$

Applying Castigliano's theorem, we have

$$\delta = \frac{\partial U}{\partial W_0} \text{ when } W_0 = 0$$

$$= \frac{2[W_0R\cos\alpha + M_0\sin\alpha]lR\cos\alpha}{2GJ}+\frac{2(M_0\cos\alpha - W_0R\sin\alpha)\,l\times(-R\sin\alpha)}{2EI}$$

$$= \frac{M_0Rl\sin\alpha\cos\alpha}{GJ}-\frac{M_0Rl\sin\alpha\cos\alpha}{EI}$$

$$= \frac{32M_0Rl\sin\alpha\cos\alpha}{\pi d^4}\left[\frac{1}{G}-\frac{2}{E}\right]$$

$$= \frac{32\,M_0R\times 2\pi Rn\sec\alpha\times\sin\alpha\cos\alpha}{\pi d^4}\left[\frac{1}{G}-\frac{2}{E}\right]$$

$$= \frac{64\,M_0R^2n\sin\alpha}{d^4}\left[\frac{1}{G}-\frac{2}{E}\right] \qquad ...(9.18)$$

9.4.3 Stresses in the spring wire.

(a) Axial load

Torque, $T = WR\cos\alpha$

Bending moment, $M = WR\sin\alpha$

Shear stress, $\tau = \dfrac{16T}{\pi d^3} = \dfrac{16WR\cos\alpha}{\pi d^3}$

Bending stress, $\sigma = \dfrac{32M}{\pi d^3} = \dfrac{32WR\sin\alpha}{\pi d^3}$

Principal stresses are:

$$\sigma_{1,2} = \frac{\sigma}{2} \pm \frac{1}{2}\sqrt{\sigma^2 + 4\tau^2}$$

$$= \frac{16WR\sin\alpha}{\pi d^3} \pm \frac{1}{2}\sqrt{\left(\frac{32WR\sin\alpha}{\pi d^3}\right)^2 + 4\left(\frac{16WR\cos\alpha}{\pi d^3}\right)^2}$$

$$= \frac{16WR\sin\alpha}{\pi d^3} \pm \frac{16WR}{\pi d^3}\sqrt{\sin^2\alpha + \cos^2\alpha}$$

$$= \frac{16WR}{\pi d^3}(\sin\alpha \pm 1) \quad \text{... (9.19)}$$

Maximum shear stress,

$$\tau_{max} = \frac{1}{2}(\sigma_1 - \sigma_2)$$

$$= \frac{16WR}{\pi d^3}\left[\frac{(\sin\alpha + 1) - (\sin\alpha - 1)}{2}\right] = \frac{16WR}{\pi d^3} \quad \text{... (9.20)}$$

(b) Axial torque

Torque $T = M_0 \sin\alpha$

Bending moment $M = M_0 \cos\alpha$

$\therefore$ Shear Stress, $\tau = \dfrac{16T}{\pi d^3} = \dfrac{16M_0\sin\alpha}{\pi d^3}$

Bending stress, $\sigma = \dfrac{32M}{\pi d^3} = \dfrac{32M_0\cos\alpha}{\pi d^3}$

Principal stresses are :

$$\sigma_{1,2} = \frac{\sigma}{2} \pm \frac{1}{2}\sqrt{\sigma^2 + 4\tau^2}$$

$$= \frac{16M_0\cos\alpha}{\pi d^3} \pm \frac{1}{2}\sqrt{\left(\frac{32M_0\sin\alpha}{\pi d^3}\right)^2 + 4\left(\frac{16M_0\sin\alpha}{\pi d^3}\right)^2}$$

$$= \frac{16M_0\cos\alpha}{\pi d^3} \pm \frac{16M_0}{\pi d^3}\sqrt{\cos^2\alpha + \sin^2\alpha}$$

$$= \frac{16M_0}{\pi d^3}(\cos\alpha \pm 1) \quad \text{.... (9.21)}$$

Maximum shear stress,

$$\tau_{max} = \frac{16M_0}{\pi d^3} \quad \text{... (9.22)}$$

Example 9.12 *An open-coiled helical spring consists of 10 coils, each of mean diameter 5 cm; the wire forming the coils being 6 mm diameter, and making a constant angle of 30° with planes perpendicular to the axis of the spring. What load will cause the spring to elongate 1.25 cm and what will be the bending and shearing stresses due to this load? Calculate the value of axial twist which would cause a bending stress of 56 MPa in the coils. E = 210 GPa and G = 84 GPa.*

Solution.

Given : $n = 10, R = 2.5 \text{ cm}, d = 0.6 \text{ cm}, a = 30°, \; d = 1.25 \text{ cm}, W = ?, \sigma = ?, \tau = ?$

Under axial load, $$\delta = \frac{64 WR^3 n \sec\alpha}{d^4}\left[\frac{\cos^2\alpha}{G} + \frac{2\sin^2\alpha}{E}\right]$$

$$1.25 \times 10^{-2} = \frac{64 W (2.5)^3 \times 10 \times \sec 30°}{(0.6)^4 \times 10^{-2}}\left[\frac{\cos^2 30°}{84\times10^9} + \frac{2\times\sin^2 30°}{210\times10^9}\right]$$

$$= \frac{64(2.5)^3 \times 10 \times W}{(0.6)^4 \times 0.866 \times 10^{-2}}\left[\frac{(0.866)^2}{84\times10^9} + \frac{2\times(0.5)^2}{210\times10^9}\right]$$

$$= W \times 89099.875[8.9286 + 2.3809] \times 10^{-3} = 0.10076 \times 10^{-3}\, W$$

$$W = 124.05 \text{ N}$$

Torque, $$T = WR\cos\alpha = 124.05 \times 2.5 \times 10^{-2} \times 0.866 = 2.6857 \text{ N-m}$$

Shear stress, $$\tau = \frac{16T}{\pi d^3} = \frac{16\times 2.6857}{\pi(0.6)^3\times10^{-6}} = 63.325 \text{ MPa}$$

Bending moment, $$M = WR\sin\alpha = 124.05 \times 2.5 \times 10^{-2} \times 0.5 = 1.5506 \text{ N-m}$$

Bending stress, $$\sigma = \frac{32M}{\pi d^3} = \frac{32\times1.5506}{\pi(0.6)^3\times10^{-6}} = 73.122 \text{ MPa}$$

Let M_0 = axial torque

Bending moment, $M = M_0 \cos\alpha$

Bending stress, $$\sigma = \frac{32 M_0 \times \cos\alpha}{\pi d^3}$$

$$56 \times 10^6 = \frac{32 M_0 \times 0.866}{\pi(0.6)^3\times10^{-6}}$$

$$M_0 = \frac{56\times\pi(0.6)^3}{32\times0.866} = 1.371 \text{ N-m}$$

Example 9.13 *In an open coil helical spring having 10 coils, the stresses, due to bending and twisting are 98 MPa and 105 MPa respectively, when the spring is axially loaded. Assuming the mean diameter of the coils to be 8 times the diameter of wire, find the maximum permissible load and the diameter of wire for a maximum extension of 2 cm. E = 210 GPa and G = 82 GPa.*

Solution.

Given : $n = 10, \sigma = 98 \text{ MPa}, \tau = 105 \text{ MPa}, D = 8\,d, R = 4\,d, \delta = 2 \text{ cm}, W = ?, d = ?$

Under axial load, $$\delta = \frac{64 WR^3 n \sec\alpha}{d^4}\left[\frac{\cos^2\alpha}{G} + \frac{2\sin^2\alpha}{E}\right]$$

Also $$T = WR\cos\alpha$$

$$\tau = \frac{16T}{\pi d^3} = \frac{16 W R \cos\alpha}{\pi d^3}$$

$$105\times10^6 = \frac{16 W \times 4d \times \cos\alpha}{\pi d^3}$$

$$105\times10^6 = \frac{64 W \cos\alpha}{\pi d^2} \quad ...(1)$$

and

$$M = W R \sin\alpha$$

$$\sigma = \frac{32 M}{\pi d^3} = \frac{32 W R \sin\alpha}{\pi d^3}$$

$$98\times10^6 = \frac{32 \times W \times 4d \sin\alpha}{\pi d^3}$$

$$98\times10^6 = \frac{128 W \sin\alpha}{\pi d^2} \quad ...(2)$$

Dividing Eq. (2) by (1), we get

$$\frac{98}{105} = 2\tan\alpha$$

$$\tan\alpha = \frac{98}{210} = 0.466$$

$$\alpha = 25.017°$$

Substituting the value of α in Eq. (1), we get

$$105\times10^6 = \frac{64 W \cos 25.017°}{\pi d^2} = \frac{64\times0.9062 W}{\pi d^2}$$

$$W = 5.6878\times10^6\times d^2$$

$$0.02 = \frac{64\times5.6878\times10^6\times d^2(4d)^3\times10}{d^4\times0.9062}\times\left[\frac{(0.9062)^2}{82\times10^9}+\frac{2(0.4229)^2}{210\times10^9}\right]$$

$$= 0.02 = 25.708\, d[0.100146 + 0.017033]$$

$$0.02 = 3.0124\, d$$

$$d = 6.64\times10^{-3}\text{ m}$$

$$W = 5.6878\,(6.64)^2 = 250.8\text{ N}$$

Example 9.14 *An open-coiled helical spring of 5 cm mean diameter is made of steel of 6 mm diameter. Calculate the number of turns required in the spring to give a deflection of 1.2 cm for an axial load of 250 N if the angle of helix is 30°. Calculate also the rotation of one end of the spring relative to the other if it is subjected to an axial couple of 10 N-m. E= 210 GPa. G= 84 GPa.*

Solution.

Given : R=2.5 cm, d= 0.6 cm, δ =1.2 cm, W = 250 N, α = 30°, n = ?, M_0 = 10 N.m, β=?

$$\delta = \frac{64 W R^3 n \sec\alpha}{d^4}\left[\frac{\cos^2\alpha}{G}+\frac{2\sin^2\alpha}{E}\right]$$

$$1.2 \times 10^{-2} = \frac{64 \times 250(2.5)^3 n}{(0.6)^4 \times 0.866 \times 10^{-2}} \left[\frac{0.75}{84 \times 10^9} + \frac{2 \times 0.75}{210 \times 10^9} \right]$$

$$= 0.22275\, n\, [0.89286 + 0.23809] \times 10^{-2}$$

$$1.2 = 0.25192\, n$$

$$n = 4.76$$

$$\beta = \frac{64 M_0 R n \sec\alpha}{d^4} \left[\frac{\sin^2\alpha}{G} + \frac{2\cos^2\alpha}{E} \right]$$

$$= \frac{64 \times 10 \times 2.5 \times 5 \times 10^6}{(0.6)^4 \times 0.866} \left[\frac{0.25}{84 \times 10^9} + \frac{2 \times 0.75}{210 \times 10^9} \right]$$

$$= 0.71279\, [0.29762 + 0.71428] = 0.7216 \text{ rad}$$

Example 9.15 *The diameter of the wire of a helical spring with 6 coils is 1 cm, the mean coil diameter being 8 cm, and pitch of the coils 8 cm. Calculate the torque which acting coaxially with the spring, will produce a shearing stress of 300 MPa. Calculate also the angular twist of the spring produced by the torque. G =80 GPa, E = 200 GPa.*

Solution.

Given: $n = 6;\ d = 1$ cm, $R = 4$ cm, $p = 8$ cm; $M_0 = ?$, $\tau_{max} = 300$ MPa; $\beta = ?$

Helix angle, $$\alpha = \tan^{-1}\left[\frac{p}{2\pi R}\right] = \tan^{-1}\left[\frac{8}{2\pi \times 4}\right] = \tan^{-1} 0.3183$$

$$\alpha = 17^\circ\, 39'$$

Torque, $T = M_0 \sin\alpha$

Bending moment $M = M_0 \cos\alpha$

Shear stress, $$\tau = \frac{16 M_0 \sin\alpha}{\pi d^3} = \frac{16 \times M_0 \times 0.3032}{\pi \times 1}$$

Bending stress, $$\sigma = \frac{32\, M_0 \sin\alpha}{\pi d^3}$$

$$\sigma = \frac{32 M_0 \times 0.9529}{\pi \times 1}$$

Maximum shear stress,

$$\sigma_{max} = \frac{1}{2}\sqrt{\sigma^2 + 4\tau^2} = \frac{16\, M_0}{\pi d^3}$$

$$300 \times 10^6 = \frac{16 M_0}{\pi \times 10^{-6}}$$

$$M_0 = \frac{\pi \times 300}{16} = 58.9 \text{ N.m}$$

Angular twist, $\beta = \dfrac{64 M_0 R n \sec\alpha}{d^4}\left[\dfrac{\sin^2\alpha}{G} + \dfrac{2\cos^2\alpha}{E}\right]$

$$= \frac{64 \times 58.9 \times 4 \times 6}{1 \times 0.9529}\left[\frac{0.0919}{80 \times 10^9} + \frac{2 \times 0.9080}{200 \times 10^9}\right] \times 10^6$$

$$= 9.4942[0.0114875 + 0.09080] = 9.4942 \times 0.1022875 = 0.971 \text{ rad}$$

Example 9.16 *If the close coiled spring formula is used in finding the extension of an open-coiled spring under axial load, determine the maximum angle of helix for which the error in the value of the extension is not to exceed one per cent. Assume E = 2.5 G.*

Solution.

Given :For a close-coiled spring, deflection,

$$\delta_1 = \frac{64 W R^3 n}{G d^4}$$

and for an open-coiled spring, deflection,

$$\delta_2 = \frac{64 W R^3 n \sec\alpha}{d^4}\left[\frac{\cos^2\alpha}{G} + \frac{2\sin^2\alpha}{E}\right]$$

Percentage error $= \dfrac{\delta_2 - \delta_1}{\delta_2} \times 100$

$$\frac{\dfrac{1}{\cos\alpha}\left(\dfrac{\cos^2\alpha}{G} + \dfrac{2\sin^2\alpha}{E}\right) - \dfrac{1}{G}}{\dfrac{1}{\cos\alpha}\left(\dfrac{\cos^2\alpha}{G} + \dfrac{2\sin^2\alpha}{E}\right)} \times 100 = 1$$

$$\left[\frac{E\cos^2\alpha + 2G\sin^2\alpha - E\cos\alpha}{E\cos^2\alpha + 2G\sin^2\alpha}\right] \times 100 = 1$$

$$\left[\frac{2.5\cos^2\alpha + 2\sin^2\alpha - 2.5\cos\alpha}{2.5\cos^2\alpha + 2\sin^2\alpha}\right] \times 100 = 1$$

$$247.5\cos^2\alpha + 198\sin^2\alpha - 250\cos\alpha = 0$$

$$49.5\cos^2\alpha - 250\cos\alpha + 198 = 0$$

$$\cos^2\alpha - 5.05\cos\alpha + 4 = 0$$

$$\cos\alpha = \frac{5.05 \pm \sqrt{25.50 - 16}}{2}$$

$$= \frac{5.05 \pm 3.082}{2} = \frac{1.968}{2} = 0.984$$

$$\alpha = 10.263°$$

9.4.4 Combined Action of Axial Load and Moment

If an axial load W and a moment M_0 act simultaneously on the spring, then (Fig. 9.5)

$$T = M_0 \sin \alpha + WR \cos \alpha$$

$$M = M_0 \cos \alpha - WR \cos \alpha$$

Total strain energy stored in the spring becomes :

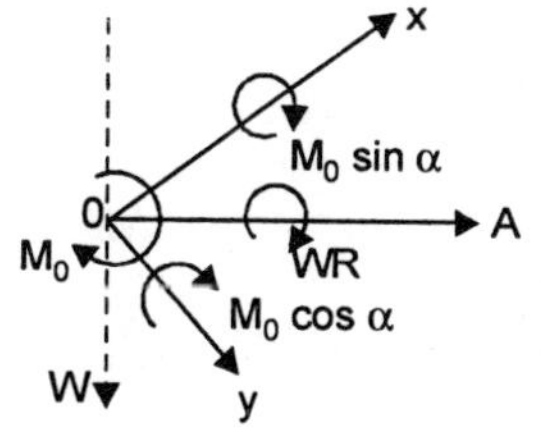

Fig. 9.5 Open coiled helical spring subjected to axial load and moment.

$$U = \frac{M^2 l}{2EI} + \frac{T^2 l}{2GJ}$$

$$= \frac{(M_0 \cos\alpha - W R \sin\alpha)^2 l}{2EI} + \frac{(M_0 \sin\alpha + W R \cos\alpha)^2 l}{2GJ}$$

Using Castigliano's theorem.

$$\delta = \frac{\partial U}{\partial W} = \frac{1}{EI}(M_0 \cos\alpha - W R \sin\alpha)(-R \sin\alpha) + \frac{1}{GJ}(M_0 \sin\alpha + W R \cos\alpha)(R \cos\alpha)$$

$$= \frac{lR}{GJ}(M_0 \cos\alpha \sin\alpha + W R \cos^2\alpha) - \frac{lR}{EI}(M_0 \cos\alpha \sin\alpha - W R \sin^2\alpha)$$

$$= l R M_0 \sin\alpha \cos\alpha \left(\frac{1}{GJ} - \frac{1}{EI}\right) + WR^2 l \left(\frac{\cos^2\alpha}{GJ} + \frac{\sin^2\alpha}{EI}\right)$$

$$= 2\pi R^2 n \sec\alpha M_0 \sin\alpha \cos\alpha \left[\frac{32}{\pi d^4 G} - \frac{64}{\pi d^4 E}\right] + 2\pi R^3 n \sec\alpha . W \left[\frac{32\cos^2\alpha}{\pi d^4 G} + \frac{64 \sin^2\alpha}{\pi d^4 E}\right]$$

$$= \frac{64 M_0 R^2 n \sin\alpha}{d^4}\left[\frac{1}{G} - \frac{2}{E}\right] + \frac{64 W R^3 n \sec\alpha}{d^4}\left[\frac{\cos^2\alpha}{G} + \frac{2\sin^2\alpha}{E}\right] \quad \text{... (9.23)}$$

$$\beta = \frac{\partial U}{\partial M_0} = \frac{1}{EI}(M_0 \cos\alpha - W R \sin\alpha)\cos + \frac{1}{GJ}(M_0 \sin\alpha + W R \cos\alpha)\sin\alpha$$

$$= l M_0 \left[\frac{\cos^2\alpha}{EI} + \frac{\sin^2\alpha}{GJ}\right] + lWR\left[\frac{\cos\alpha \sin\alpha}{GJ} - \frac{\sin\alpha \cos\alpha}{EI}\right]$$

$$= \frac{64 M_0 R n \sec\alpha}{d^4}\left[\frac{2\cos^2\alpha}{E}+\frac{\sin^2\alpha}{G}\right]+\frac{64 W R^2 n \sin\alpha}{d^4}\left[\frac{1}{G}-\frac{2}{E}\right] \quad \text{... (9.24)}$$

If the end is fixed against rotation, then

$$\beta = 0$$

and

$$M_0 = \frac{W R \sin\alpha\cos\alpha\left[\frac{2}{E}-\frac{1}{G}\right]}{\left[\frac{2\cos^2\alpha}{E}+\frac{\sin^2\alpha}{G}\right]} \quad \text{... (9.25)}$$

9.5 SEMI-ELLIPTICAL SPRING

Consider a semi-elliptical leaf spring as shown in Fig. 9.6 (a).

Let W = Load acting at mid-point'

n = Number of plates

a = Overlap at each end.

l = Length of the spring.

b = Width of plate

d = Depth of plate

Then $l = 2\,an$

$$a = \frac{l}{2n}$$

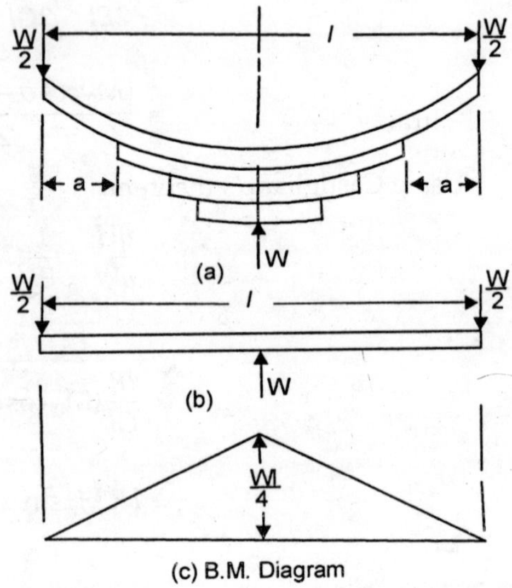

Fig. 9.6 Semi-elliptical leaf spring

Maximum bending moment as shown in Fig. 9.6 (*b*) and (*c*) is,

$$M_{max} = \frac{Wl}{4}$$

Bending moment for each plate,

$$M = \frac{Wl}{4n}$$

Now bending stress, $\sigma = \frac{M}{Z}$

$$\therefore \quad \sigma = \frac{Wl}{4n}\cdot\frac{6}{bd^2} = \frac{3}{2}\frac{Wl}{nbd^2} \quad \text{... (9.26)}$$

Change of curvature for each plate after loading,

$$\frac{1}{R} = \frac{M}{EI} = \frac{Wl}{4n}\times\frac{12}{Ebd^3} = \frac{3Wl}{nEbd^3}$$

Now $$EI\frac{d^2y}{dx^2} = M$$

$\therefore$ $$\frac{d^2y}{dx^2} = \frac{M}{EI} = \frac{3Wl}{nEbd^3}$$

$$y = \iint \frac{d^2y}{dx^2}\cdot dx.dx = \int_0^{l/2}\int_0^{l/2} \frac{3Wl}{nEbd^3}\cdot dx.dx$$

$\therefore$ $$= \frac{3Wl}{nEbd^3}\cdot\int_0^{l/2} xdx = \frac{3Wl}{nEbd^3}\times\frac{l^2}{8}$$

$\therefore$ Deflection at the centre,

$$\delta = \frac{3}{8}\frac{Wl^3}{nEbd^3} \quad \text{... (9.27)}$$

Strain-energy absorbed,

$$U = \frac{M^2}{2EI}\times \text{ Length of the beam}$$

$$= \frac{\left(\frac{bd^2}{6}\sigma\right)^2}{2E\cdot\frac{bd^3}{12}}\times \text{Total length of all the leaves}$$

$$= \frac{bd.\sigma^2}{6E}\times \text{ Total length of all the leaves}$$

$$= \frac{\sigma^2}{6E}\times \text{ Volume of spring}$$

$\therefore$ Resilience $$= \frac{\sigma^2}{6E} \quad \text{... (9.28)}$$

9.5.1 Quarter-Elliptic Leaf Spring

If we substitute $W = 2W$ and $l = 2l$ in Eqs. (9.26) and (9.27), then we get the expression for a quarter-elliptic spring shown in Fig. 9.7.

$\therefore$ $$\sigma = \frac{3}{2}\cdot\frac{2W\cdot 2l}{nbd^2} = \frac{6Wl}{nbd^2} \quad \text{... (9.29)}$$

$$\delta = \frac{3}{8}\cdot\frac{2W\cdot(2l)^3}{nEbd^3} = \frac{6Wl^3}{nEbd^3} \quad \text{... (9.30)}$$

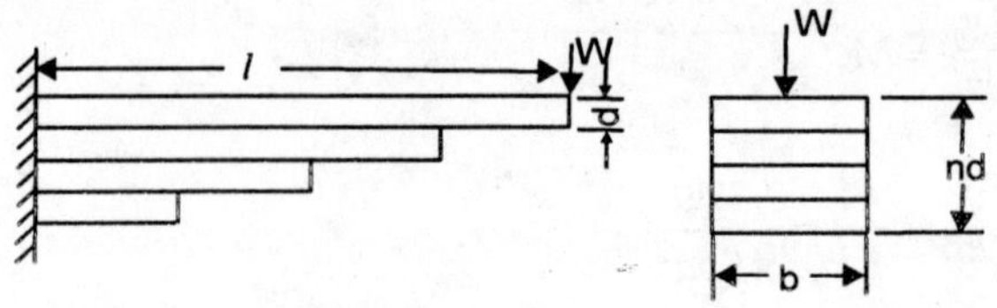

Fig. 9.7 Quarter-elliptic leaf spring.

Example 9.17 *A leaf spring of the semi-elliptical type has 10 plates each 8 cm wide and 10 mm thick. The length of the spring is 120 cm and the plates are of steel having a proof stress in bending of 630 MPa. To what radius should the plates be initially bent?*

From what height can a load of 450 N fall onto the centre of the spring if the maximum stress produced is to be one half of the proof stress. E = 210 GPa.

Solution.

Given : $b = 8$ cm, $d = 1$ cm, $l = 120$ cm, $\sigma_{max} = 630$ MPa

Now $$\frac{\sigma}{y} = \frac{E}{R}$$

$\therefore$ $$R = \frac{Ey}{\sigma} = \frac{Ed}{2\sigma} = \frac{210\times10^9\times10\times10^2}{2\times630\times10^6} = 1.67\text{ m}$$

Let W_e be the equivalent static load which would produce the same maximum stress and deflection as the impact load.

$\therefore$ $$\frac{W_e l}{4} = n\cdot\frac{bd^2}{6}\times\frac{\sigma_{max}}{2}$$

$$W_e = \frac{2}{3}\frac{nbd^2}{l}\cdot\frac{\sigma_{max}}{2} = \frac{2\times10\times8\times1\times630}{3\times1.2\times2} = 14\text{ kN}$$

Deflection, $$\delta = \frac{3}{8}\times\frac{W_e l^3}{nEbd^3} = \frac{3}{8}\times\frac{14\times10^3\times(1.20)^3}{10\times210\times10^9\times8\times1\times10^{-8}}$$

$$= 5.4\times10^{-3}\text{ m}$$

If h = height through which the weight W is allowed to fall

Work done by $W = W(h+\delta)$

Strain energy stored in the spring $= \frac{1}{2}W_e\cdot\delta$

$\therefore$ $$W(h+\delta) = \frac{1}{2}W_e\,\delta$$

$$450\,(h+0.054) = \frac{1}{2}\times14\times10^3\times0.054$$

$$h + 0.054 = \frac{14 \times 10^3 \times 0.054}{2 \times 450} = 0.84$$

$$\therefore \quad h = 78.6 \text{ cm}$$

Example 9.18 *A semi-elliptical laminated spring has the following specifications :*

Length of longest plate = 60 cm

Central load = 3 kN

Central deflection = 1.5 cm

Allowable bending stress = 180 MPa

Determine the size of plates, overlap, number of plates, and radius of curvature. Take width of plates to be 12 times the thickness and E = 206 GPa.

Solution.

Given : $l = 60$ cm, $W = 3$kN, $S = 1.5$ cm, $\sigma_b = 180$ MPa, $b = 12d$, $E = 206$ GPa

Deflection, $$d = \frac{3}{8}\frac{Wl^3}{nEbd^3}$$

Stress, $$\sigma = \frac{3}{2}\frac{Wl}{nbd^2}$$

$$\therefore \quad \frac{\delta}{\sigma} = \frac{l^2}{4Ed}$$

$$\therefore \quad d = \frac{\sigma}{\delta}\cdot\frac{l^2}{4E} = \frac{180 \times 10^6 \times 0.36}{1.5 \times 10^{-2} \times 4 \times 206 \times 10^9} = 5.25 \text{ mm}$$

$\therefore$ Width of plates, $b = 12 \times 5.25 = 63$ mm

$$\therefore \quad 180 \times 10^6 = \frac{3}{2} \times \frac{3 \times 10^3 \times 0.6}{n \times (0.525)^2 \times 6.3 \times 10^{-6}}$$

$$n = 8.6 \approx 9 \text{ (say)}$$

Now $$\frac{\sigma}{y} = \frac{E}{R}$$

$$\therefore \quad R = \frac{Ey}{\sigma} = \frac{Ed}{2\sigma} = \frac{206 \times 10^9 \times 0.525 \times 10^{-2} \times 9}{2 \times 180 \times 10^6 \times 8.6} = 3.1474 \text{ m}$$

Overlap, $$a = \frac{1}{2n} = \frac{0.6}{2 \times 9} = 33.3 \text{ mm}$$

Example 9.19 *A carriage spring 130 cm long has leaves of 10 cm width and 12 mm thickness. The maximum bending stress is 150 MPa and the spring must absorb 120 N.m of energy when straightened. Calculate the number of leaves and their initial radius of curvature. E = 200 GPa.*

Solution.

Given : $l = 130$ cm, $b = 10$ cm, $d = 12$ mm, $\sigma = 150$ MPa, $U = 120$ N.m, $E = 200$ GPa

Energy stored,
$$U = \frac{\sigma^2}{6E} \times \text{Volume}$$

$$\therefore \quad 120 = \frac{(150)^2 \times 10^{12}}{6 \times 200 \times 10^9} \times \frac{n \times 10 \times 1.2 \times 1.3 \times 10^{-4}}{2}$$

$$n = 8.21 \approx 9 \text{ (say)}$$

Actual bending stress,
$$120 = \frac{\sigma^2}{6 \times 200 \times 10^9} \times \frac{9 \times 10 \times 1.2 \times 1.3 \times 10^{-4}}{2}$$

$$\therefore \quad \sigma^2 = 205.13 \times 10^{14}$$

$$\sigma = 143.2 \text{ MPa}$$

Also
$$\frac{\sigma}{y} = \frac{E}{R}$$

$$\therefore \quad R = \frac{Ey}{\sigma} = \frac{Ed}{2\sigma} = \frac{200 \times 10^9 \times 1.2 \times 10^{-2}}{2 \times 143.2 \times 10^6} = 8.38 \text{ m}$$

Example 9.20 *A quarter-elliptic leaf spring has a length of 50 cm and consists of plates each 5 cm wide and 6 mm thick. Find the least number of plates which can be used if the defelction under a gradully applied load of 2 kN is not to exceed 7 cm.*

If instead of being gradually applied the load of 2 kN falls from a distance of 6 mm on to the undefelected spring, find the maximum defelection and stress produced, E = 200 GPa.

Solution.

Given : $l = 50$ cm, $b = 5$ cm, $d = 5$ mm, $W = 2$ kN, $\delta = 7$ cm, $h = 6$ mm, $E = 200$ GPa

Deflection,
$$\delta = \frac{6Wl^3}{nEbd^3}$$

$$0.07 = \frac{6 \times 2 \times 10^3 \times 50^3 \times 10^{-6}}{n \times 200 \times 10^9 \times 5 \times (0.6)^3 \times 10^{-8}}$$

$$n \approx 9.92 = 10 \text{ (say)}$$

Let W_e be equivalent gradually applied load which would produce the same deflection as is caused by the impact load.

$$\therefore \quad \delta_1 = \frac{6W_e l^3}{nEbd^3} = \frac{6W_e \times 50^3 \times 10^{-6}}{10 \times 200 \times 10^9 \times 5 \times (0.6)^3 \times 10^{-8}}$$

$$W_e = 28.8 \times 10^3 \times \delta_1 \text{ N}$$

Loss of potential energy
$$W(h + \delta_1) = 2 \times 10^3 (6 \times 10^{-3} + \delta_1)$$

Strain energy absorbed by spring = $\frac{1}{2}W_e\,\delta_1$

$$\therefore \quad 2 \times 10^3\,(6\times10^{-3} + \delta_1) = \frac{1}{2}W_e\delta_1 = \frac{1}{2}\times 28.8\times10^3\times\delta_1^2$$

$$\delta_1^2 - 0.1389\,\delta_1 - 0.833\times10^{-3} = 0$$

Solving for δ_1, we get $\delta_1 = 144.65$ mm

$$W_e = 28.8 \times 10^3 \times 144.65 \times10^{-3} = 4.166 \text{ kN}$$

Maximum stress, $\sigma_{max} = \dfrac{6W_e l}{nbd^2} = \dfrac{6\times4.166\times10^3\times0.5}{10\times5\times(0.6)^2\times10^{-6}} = 694.32$ MPa

Example 9.21 *A laminated steel spring, simply supported at the ends and centrally loaded with a span of 75 cm is required to carry a proof load of 7500 N and the central deflection is not to exceed 6 cm. The bending stress must be greater than 400 MPa. Plates are available in multiples of 1 mm for thickness and in multiples of 4 mm for width.*

Determine suitable values for the thickness, width and number of plates and the radius to which the plates should be formed. Assume the width to be twelve times the thickness. E = 200 GPa.

Solution.

Given : $l = 75$ cm, $W = 7500$ N, $\delta = 6$ cm, $\sigma = 400$ MPa, $b = 12d$, $E = 200$ GPa

Deflection $\delta = \dfrac{3}{8}\dfrac{Wl^3}{nEbd^3}$

$$0.05 = \frac{3\times7500\times75^3\times10^{-6}}{8\times200\times10^9\times12d\times d^3\times n}$$

$$nd^4 = 9.8877 \times 10^{-9}$$

Bending stress, $\sigma = \dfrac{3Wl}{2nbd^2}$

$$400 \times 10^6 = \frac{3\times7500\times75\times10^{-2}}{2\times n\times12d\times d^2}$$

$$nd^3 = 175.78 \times 10^{-8}$$

Dividing, we get $d = \dfrac{9.8877\times10^{-9}}{175.78\times10^{-8}} = 5.625$ mm

Adopt $d = 6$ mm

$b = 6$ mm

Now $n = \dfrac{9.8877\times10^{-9}}{d^4} = \dfrac{9.8877\times10^{-9}}{6^4\times10^{-12}} = 7.62 \simeq 8$

Also $$n = \frac{175.78\times10^{-8}}{(6)^3\times10^{-9}} = 8.13 \simeq 9$$

Hence adopt 9 plates

$$\therefore \text{ Actual deflection, } \delta = \frac{3}{8}\times\frac{7500\times75^3\times10^{-6}}{9\times200\times10^9\times(6)^3\times7.2\times10^{-11}} = 42.4\,\text{mm}$$

$$\text{Radius of curvature, } R = \frac{l^2}{8\delta} = \frac{(0.75)^2}{8\times42.4\times10^{-3}} = 1.66\text{ m}$$

9.6 SQUARE CROSS-SECTION WIRE HELICAL SPRINGS

Let b = side of the square.

Then $$\tau_{max} = \frac{T}{k_1 b^3}$$

where $k_1 = 0.208$

Angle of twist, $$\theta = \frac{Tl}{k_2 b^4 G}$$

where $k_2 = 0.141$

9.6.1 Axial Load.

$$T = \frac{1}{2}WD_m,\quad l = \pi D_m n$$

$$\tau_{max} = \frac{\frac{1}{2}WD_m}{k_1 b^3} = \frac{WD_m}{0.416 b^3} \qquad \text{.... (9.31)}$$

$$\theta = \frac{\frac{1}{2}WD_m\times\pi D_m n}{0.141 b^4 G} = \frac{11.14 WD_m^2 n}{Gb^4} \qquad \text{... (9.32)}$$

$$U = \frac{1}{2}T\theta = \frac{1}{2}WD_m\times\frac{11.14 WD_m^2 n}{Gb^4}$$

$$= \frac{5.57 W^2 D_m^3 n}{Gb^4}$$

Applying Castigliano's theorem, we have

$$\delta = \frac{\partial U}{\partial W} = \frac{11.14 WD_m^3 n}{Gb^4} \qquad \text{... (9.33)}$$

9.6.2 Axial Torque

$$\phi = \frac{M_0 l}{EI} = \frac{M_0 \times \pi D_m n}{EI} = \frac{12\, M_0 \times \pi D_m n}{Eb^4}$$

$$U = \frac{1}{2} M_0\, \phi$$

$$= \frac{1}{2} M_0 \times \frac{12\, M_0 \times \pi D_m n}{Eb^4} = \frac{6\pi\, M_0^2\, D_m n}{Eb^4}$$

Applying Castigliano's theorem, we have

$$\beta = \frac{\partial U}{\partial M_0} = \frac{12\pi\, M_0\, D_m n}{Eb^4} \qquad \text{... (9.34)}$$

$$\sigma_b = \frac{6M_0}{b^3} \qquad \text{... (9.35)}$$

SUMMARY

1. Close-coiled helical springs.
   (a) Axial load.

Deflection, $\delta = \dfrac{8WD_m^3\, n}{Gd^4}$

Stiffness, $k = \dfrac{Gd^4}{8D_m^3 n}$

Shear stress, $\tau = \dfrac{8WD_m}{\pi d^3}$

Strain energy stored $= \dfrac{\tau^2}{4G} \times$ spring wire volume.

   (b) Axial torque

Angular rotation, $\beta = \dfrac{64\, M_0 D_m n}{Ed^4}$

Strain energy stored $= \dfrac{\sigma_b^2}{8E}$

2. Open-coiled helical spring
   (a) Axial load.

Deflection, $\delta = \dfrac{8WD_m^3 n \sec\alpha}{d^4}\left[\dfrac{\cos^2\alpha}{G} + \dfrac{2\sin^2\alpha}{E}\right]$

Angular rotation, $\beta = \dfrac{16 W D_m^2 n \sin\alpha}{d^4}\left[\dfrac{1}{G} - \dfrac{2}{E}\right]$

(b) Axial torque.

Axial rotation, $\beta = \dfrac{64 M_0 R n \sec\alpha}{d^4}\left[\dfrac{\sin^2\alpha}{G} + \dfrac{2\cos^2\alpha}{E}\right]$

Axial deflection, $\delta = \dfrac{64 M_0 R^2 n \sin\alpha}{d^4}\left[\dfrac{1}{G} - \dfrac{2}{E}\right]$

3. Stress in spring wire

(a) Axial load

Shear stress, $\tau = \dfrac{16 WR \cos\alpha}{\pi d^3}$

Bending stress, $\sigma_b = \dfrac{32 WR \sin\alpha}{\pi d^3}$

Principal stress, $\sigma_{1,2} = \dfrac{16 WR}{\pi d^3}(\sin\alpha \pm 1)$

Maximum shear stress, $\tau_{max} = \dfrac{16 WR}{\pi d^3}$

(b) Axial torque

$$\tau = \frac{16 M_0 \sin\alpha}{\pi d^3}$$

$$\sigma_b = \frac{32 M_0 \cos\alpha}{\pi d^3}$$

$$\sigma_{1,2} = \frac{16 M_0}{\pi d^3}(\cos\alpha \pm 1)$$

$$\tau_{max} = \frac{16 M_0}{\pi d^3}$$

4. Laminated springs.

(a) Semi-elliptical spring

Bending stress, $\sigma_b = \dfrac{3 Wl}{2n\, bd^2}$

Deflection at the centre, $\delta = \dfrac{3}{8}\dfrac{Wl^3}{nE\, bd^3}$

Resilience $= \dfrac{\sigma_b^2}{6E}$

(b) Quarter -elliptic spring

$$\sigma_b = \frac{6Wl}{nbd^2}$$

$$\delta = \frac{6Wl^3}{nEbd^3}$$

Short Answer Type Questions

1. What is a helical spring?
2. Define spring index and stiffness.
3. What is Wahl's correction factor?
4. What is the effect of curvature of wire on shear stress?
5. Define torsional rigidity.
6. When a helical spring is called open-coiled?
7. What is a laminated spring? Where they are used?
8. What is cambre and overlap in leaf springs?

EXERCISES

1. A close-coiled helical spring having 20 turns is made of 6mm diameter wire.. The mean coil diameter is 60 mm. It carrier an axial load of 300 N. Calculate: (a) maximum shear stress developed in the spring wire, (b) axial deflection, and (c) spring stiffness.
   Take G = 83170 MPa. **[Ans.** 212.2 N/mm^2, 96.2 mm, 3.12 N/mm]
2. A spring loaded saftey valve of 100 mm diameter is designed to blow off at a gauge pressure of 1 N/mm^2. A close-coiled helical spring of 150 mm mean coil diameter is used to keep the valve in position. Calculate the wire diameter and number of turns required if the initial compression of the spring is 50 mm and allowable shear stress is 65 MPa. Take G = 84 GPa. **[Ans.** 35.8 mm, 32]
3. A close-coiled helical spring is required to carry an axial load of 50 N. The maximu shear stress induced is 100 MPa. The spring stiffness is 500 N/m. The solid length of the spring is 60 mm. Calculate: (a) wire diameter, (b) mean diameter, and (e) number of coils. Take G = 84 GPa

   [Ans. 5.2 mm, 110.4 mm, 12]
4. An open-coiled helical spring of 12.5 cm mean diameter has 10 coils of 12 mm diameter wire, at a slope of 30° to the horizontal when the coil axis is vertical. It is subjected to the combined action of axial load and torque. It is allowed to extend by 5 mm. If rotation is prevented, determine whether the torque tends to wind up or unwind the spring. E = 210 GPa and G = 84 GPa.

   [Ans. Unwind]
5. An open-coiled helical spring is made out of 1.5 cm diameter steel rod, the coils having 12 complete truns and a mean diameter of 10 cm, the angle of helix being 12°. Calculate the deflection under axial load of 300 N and the maximum intensities of direct and shear stresses induced in the section of the rod.

If the axial load of 300 N is replaced by an axial torque of 8 N.m, calculate the angle of rotation about the axis of the coil and the axial deflection, E = 210 GPa, G = 84 GPa.

[**Ans.** 6·86 mm, 27·34 MPa, 22·63 MPa, 3 42°, 0·15 mm]

6. Find by what percentage the axial extension is under estimated if the inclination of the coils is neglected for a spring in which $\propto$ = 25°, E = 210 GPa, G = 78·5 GPa, [Ans. 5·364%]
7. Find the mean radius of an open-coiled spring ($\propto$ = 30°) to give a vertical displacement of 2.25 cm and an angular rotation of the loaded end of 0·02 radians under an axial load of 40 N. The material available is steel rod of 6 mm diameter. E = 210 GPa, G = 84 GPa. [**Ans.** 10·26 cm]
8. An open-coiled helical spring of 5 cm mean diameter is made of steel of 6 mm diameter. Calculate the number of turns required in the spring to give a deflection of 12 mm for an axial load of 250 N if the angle of helix is 30°. Calculate also the rotation of one end of the spring relative to the other if it is subjected to an axial couple of 12·5 N.m. E = 210 GPa and G = 84 GPa.

   [**Ans.** 4·76, 4917°]
9. A semi-elliptic automobile spring 1·25 m long carries a load of 7500 N. The spring has 8 graduated leaves, all 5 cm wide. Determine the necessary thickness of leaves to give a deflection of 5 cm. E = 200 GPa. [**Ans.** 13.3 mm]
10. A carriage spring centrally loaded and simply supported at the ends has 10 steel plates each

    5 cm wide by 6 mm thick. If the longest plate is 75 cm long, determine the initial radius of curvature of the plates when the greatest bending stress is 800 MPa and the plates are finally straight. Neglecting the loss of energy at impact, find the greatest height from which a mass weighing 250 N may be dropped centrally on the spring without exceeding the limiting bending stress of 800 MPa. Take E = 210 GPa. [**Ans.** 78·75 cm, 219·6 cm]
11. Calculate the thickness and number of leaves of a semi-elliptic carriage spring which is required to support a central load of 2 kN on a span of 1 m if the maximum stress is limited to 225 MN/m^2 and the central deflection to 75 mm. Each leave is 100 mm wide. E = 210 GN/m^2.

    [**Ans.** 3·65 mm, 10]
12. A leaf spring 75 cm long is required to carry a central load of 8 kN. It the central defection is not to exceed 2 cm and bending stress no greater than 200 MPa, determine the thickness, width and number of plates.

    Also compute the radius to which plates should be curved. Assume width of plate to be 12 times its thickness and E = 200 GPa. [**Ans.** 0·703 cm, 8·436 cm, 11,351·5 cm]
13. A laminated spring 1 m long is made up of plates, each 5 cm wide and 1 cm thick. If the bending stress in plates is limited to 100 MPa, how many plates would be required to enable the spring to carry a central point load of 2 kN. If E = 210 GPa, what is the deflection under the given load of 2 kN? Derive the formula used for deflection. [**Ans.** n = 6, δ = 1·19 cm]
14. A leaf spring is to be made of seven steel plates 6·5 cm wide and 6·3 mm thick. Calculate the length of the spring so that it may carry a central load of 2·75 kN, the stress being limited to 160 MPa. Calculate also the deflection at the centre of the spring. E = 210 GPa. [**Ans.** 70 cm, 1·48 cm]

Chapter-10

Thin Cylinders and Spheres

10.1 STRESSES IN THIN CYLINDERS

Consider a thin cylinder of mean diameter d and thickness t subjected to internal pressure p, as shown in Fig. 10.1 (a). Consider a small element $ABCD$ of the circumference of the cylinder subtending an angle $d\theta$ at the centre O, as shown in Fig. 10.1 (*b*). This element has been shown separately in Fig. 10.1 (*c*).

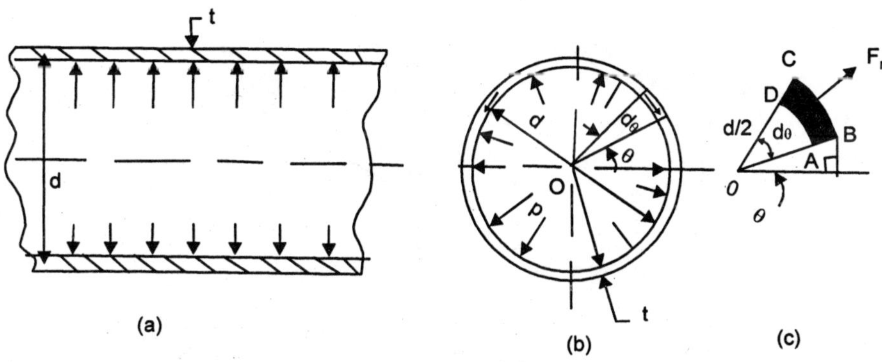

Fig. 10.1 Thin cylinder under internal pressure.

Radially outward force acting on the elment $ABCD$ per unit length of cylinder,

$$F_r = p \cdot \frac{d}{2} \cdot d\theta$$

Vertical component of force $F_r = F_r \sin\theta = p \cdot \frac{d}{2} \cdot d\theta \sin\theta$

Total vertical force on half the ring

$$= \int_0^{\pi} \frac{pd}{2} \sin\theta \cdot d\theta$$

Vertical component of force $F_r = F_r \sin\theta = p\cdot\frac{d}{2}\cdot d\theta \sin\theta$

Total vertical force on half the ring

$$= \int_0^{\pi} \frac{pd}{2} \sin\theta \cdot d\theta$$

$$= \frac{pd}{2}\int_0^{\pi} \sin\theta\, d\theta = \frac{pd}{2} \left| -\cos \right|_0^{\pi} = \frac{pd}{2}(1+1) = pd$$

This is the bursting force in the cylinder.
Let σ_θ = hoop stress set up in the cylinder wall.
Then the resisting force of the cylinder per unit length, as shown in Fig. 10.2 is,

$$= 2\sigma_\theta t$$

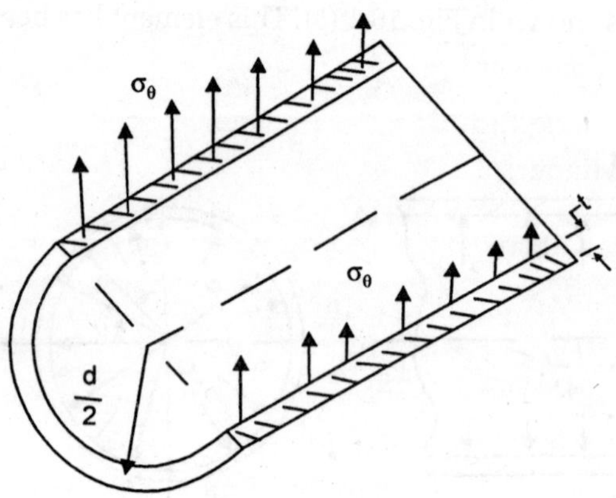

Fig. 10.2 Resisting force

For equilibrium of half the cylinder,

$$\text{Resisting force} = \text{Bursting force}$$

$$2\sigma_\theta t = pd$$

$$\sigma_\theta = \frac{pd}{2t} \qquad \text{... (10.1)}$$

Horizontal force acting on half the ring

$$= \frac{pd}{2}\int_0^{\pi} \cos\theta\, d\theta = \frac{pd}{2} \left| \sin\theta \right|_0^{\pi} = 0$$

Therefore, the horizotal forces in a long cylinder mutually balance.

If the cylinder is closed on both sides, then the presssure p acts in the longitudinal direction also perpendicular to the ends, as shown in Fig. 10.3

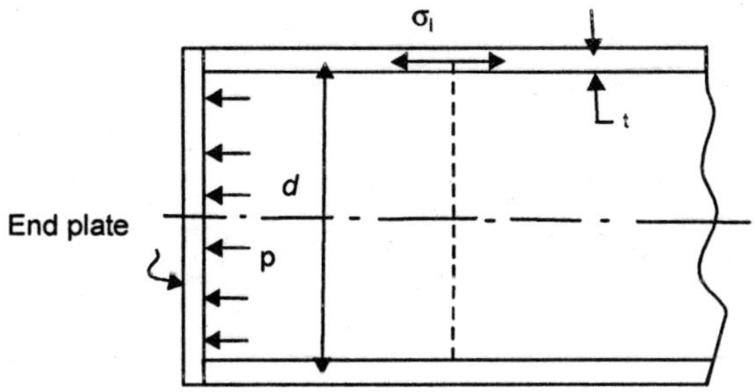

Fig. 10.3 Longitudinal stress in a closed cylinder.

$$\text{Force acting on the end plate} = \frac{\pi}{4}d^2p$$

$$\text{Resisting force} = \pi dt \times \sigma_l$$

where σ_l = longitudinal stress in the cylinder wall.

For equilibrium of the cylinder,

$$\pi dt.\sigma_l = \frac{\pi}{4}d^2p$$

$$\sigma_l = \frac{pd}{4t} \qquad \text{... (10.2)}$$

Maximum shear stress in the cylinder wall,

$$\tau_{max} = \frac{1}{2}(\sigma_\theta - \sigma_l)$$

$$= \frac{1}{2}\left[\frac{pd}{2t} - \frac{pd}{4t}\right]$$

$$= \frac{pd}{8t} \qquad \text{...(10.3)}$$

10.2 PRINCIPAL STRAINS IN THE CYLINDER WALL

Hoop strain,

$$\varepsilon_\theta = \frac{1}{E}[\sigma_\theta - \nu\sigma\ ]$$

$$= \frac{1}{E}\left[\frac{pd}{2t} - \nu\frac{pd}{4t}\right]$$

$$= \frac{pd}{4tE}[2-\nu] \quad \text{... (10.4)}$$

where ν = Poisson's ratio

Longitudinal strain, $\varepsilon_l = \frac{1}{E}[\sigma_l - \nu\sigma_\theta]$

$$= \frac{1}{E}\left[\frac{pd}{4t} - \nu\frac{pd}{2t}\right]$$

$$= \frac{pd}{4tE}[1-2\nu] \quad \text{... (10.5)}$$

10.3 CHANGE IN VOLUME OF CYLINDER

Volume of cylinder, $V = \frac{\pi}{4}d^2 l$

Taking log of both sides, we have

$$\log V = \log\frac{\pi}{4} + 2\log d + \log l$$

Differentiating, we get

$$\frac{\delta V}{V} = 0 + \frac{2\delta d}{d} + \frac{\delta l}{l}$$

$$= 2\varepsilon_\theta + \varepsilon_l$$

$$= \frac{pd}{4tE}[4-2\nu+1-2\nu]$$

$$= \frac{pd}{4tE}(5-4\nu)$$

$$= \frac{\sigma_\theta}{E}(2.5-2\nu)$$

Change in volume, $\delta V = \frac{V \times \sigma_\theta}{E}(2.5-2\nu)$ ...(10.6)

10.4 STRESSES IN A THIN SPHERE

Consider a thin spherical shell of mean diameter d and wall thickness t subjected to internal pressure p, as shown in Fig. 10.4. Consider the forces acting on an annular ring of sphere subtending an angle $d\theta$ at the centre as shown in Fig. 10.4.

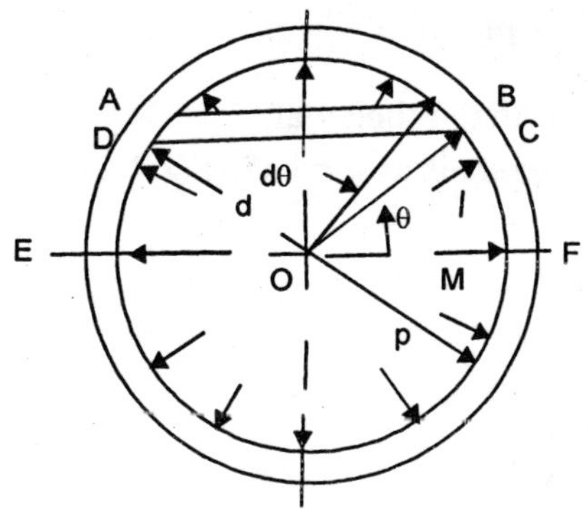

Fig. 10.4 Thin spherical shell under internal pressure

Diameter of the ring $ABCD = d \cos \theta$

Length of the ring $ABCD = \pi d \cos\theta$

Width of the ring $BC = \frac{d}{2} \cdot d\theta$

Surface area of ring $ABCD = \pi d \cos\theta . \frac{d}{2} \cdot d\theta$

$$= \frac{\pi d^2}{2} \cos\theta . d\theta$$

Raidal force acting on the ring ABCD

$$= \frac{\pi d^2}{2} \cos\theta \cdot d\theta . p$$

Vertical component of radial force

$$= \frac{\pi d^2}{2} \cos\theta . d\theta . p . \sin\theta$$

$$= \frac{\pi d^2}{4} \times p \sin 2\theta = d\theta$$

Total vertical force tending to burst the spherical shell along EF

$$= \frac{\pi d^2}{4} p \int_0^{\pi/2} \sin 2\theta . d\theta$$

$$= \frac{\pi d^2}{4} p \left| -\frac{\cos 2\theta}{2} \right|_0^{\pi/2}$$

$$= \frac{\pi d^2}{8} p(1+1) = \frac{\pi d^2}{4} \cdot p$$

If σ_θ = hoop stress developed in the shell wall.

Then total resisting force of the shell, as shown in Fig. 10.5 is,

$$= \pi dt\, \sigma_\theta$$

For equilibrium of the shell,

$$\pi\, dt\, \sigma_\theta = \frac{\pi d^2}{4} p$$

$$\sigma_\theta = \frac{pd}{4t} \qquad \text{... (10.7)}$$

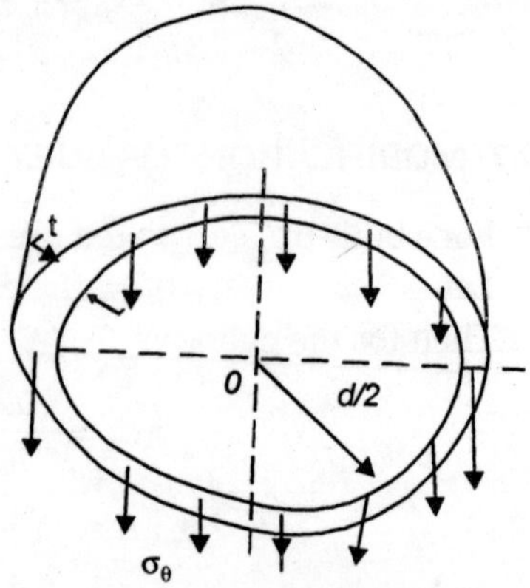

Fig. 10.5 Hoop stress in the wall of spherical shell

The principal stresses at any point in the shell are $\frac{pd}{4t}$.

10.5 PRINCIPAL STRAINS IN SPHERICAL SHELL

Hoof strain, $$\varepsilon_\theta = \frac{1}{E}[\sigma_\theta - \nu\sigma_\theta]$$

$$= \frac{\sigma_\theta}{E}(1-\nu)$$

$$= \frac{pd}{4tE}(1-\nu) \qquad \text{... (10.8)}$$

10.6 CHANGE IN VOLUME OF SPHERICAL SHELL

Volume, $$V = \frac{\pi}{6} d^3$$

Taking log of both sides, we have

$$\log V = \log\frac{\pi}{6} + 3\log d$$

Differentiating, we get

$$\frac{\delta V}{V} = 0 + 3\frac{\delta d}{d} = 3\varepsilon_\theta$$

$$= \frac{3pd}{4tE}(1-\nu)$$

$$\delta V = \frac{\pi d^3}{6} \times \frac{3pd}{4tE}(1-\nu)$$

$$= \frac{\pi}{8}\frac{d^4 p}{tE}(1-\nu) \quad \text{... (10.9)}$$

10.7 MODIFICATION FOR BUILT-UP CYLINDER AND SPHERE

For a built-up, i.e. riveted pressure vessel the efficienty of the joint should be taken into account.
Let η_θ, η_l =efficiency of circumferential and longitudinal joints respectively.
Then for the cylinder,

$$\sigma_\theta = \frac{pd}{2\eta_\theta t} \quad \text{... (10.10)}$$

$$\sigma_l = \frac{pd}{4\eta_l t} \quad \text{... (10.11)}$$

and for the sphere,

$$\sigma_\theta = \frac{pd}{4\eta_\theta t} \quad \text{... (10.12)}$$

10.8 EFFECT OF FLUID PRESSURE

When a fluid is used to fill the pressure vessel, the bulk modulus of the fluid should be taken into account in filling to raise the pressure by a speficied amount.

Bulk modulus fo fluid, $K = \dfrac{\text{Volumetric stress}}{\text{Volumetric strain}}$

$$= \frac{p}{(\delta V / V)} = \frac{pV}{\delta V}$$

Change in volume of fluid under pressure,

$$\delta V = \frac{pV}{K} \quad \text{... (10.13)}$$

Extra fluid required to raise the cylinder pressure by p

$$= \frac{pd}{2tE}(2.5-2\nu)V + \frac{pV}{K} \quad \text{... (10.14)}$$

Extra fluid required to raise the sphere pressure by p

$$= \frac{pd}{4tE}(1-\nu)V + \frac{pV}{K} \quad \text{... (10.15)}$$

Example 10.1 *A cast iron pipe 1 m in diameter is required to withstand 80 m head of water. Calculate its thickness of allowable tensile stress for the pipe material is 20 MPa.*

Solution.

Given : $d = 1\text{m},\ h = 80\text{m},\ \sigma_\theta = 20\text{ MPa},\ p = \rho g h = 10^3 \times 9.81 \times 80 = 784.8\text{ kPa}$

Hoop stress, $\sigma_\theta = \dfrac{pd}{2t}$

$$t = \frac{pd}{2\sigma_\theta} = \frac{784.8\times10^3\times1}{2\times20\times10^9} = 0.01962 \text{ m or } 19.62 \text{ mm}$$

Example 10.2 *A boilder shell 1.5 m diameter is built of steel plates 15 mm thick. The allwable tensile stress for the shell plates is 65 MPa and joint efficiency is 80 %. Calculate the safe pressure for the boiler.*

Solution.

Given: $d = 1.5$ m, $t = 15$ mm, $\sigma_\theta = 65$ MPa, $\eta_\theta = 0.8$

$$\sigma_\theta = \frac{pd}{2t\eta_\theta}$$

$$p = \frac{65\times2\times15\times0.8}{1.5\times10^3} = 1.04 \text{ MPa}$$

Example 10.3 *A cylindrical pressure vessel is 2m long, 0.6 m diameter and 10 mm thick at atmospheric pressure. Calculate its diamensions when subjected to an internal pressure of 1.4 MPa. Take E = 20.5 GPa and v = 0.3.*

Solution.

Given : $l = 2m$, $d = 0.6$ m, $t = 10$ mm, $p = 1.4$ MPa, $E = 205$ GPa, $n = 0.3$

Hoop stress,
$$\sigma_\theta = \frac{pd}{2t} = \frac{1.4\times0.6\times10^3}{2\times10} = 42 \text{ MPa}$$

Longitudinal stress,
$$\sigma_l = \frac{pd}{4t} = 21 \text{ MPa}$$

Hoop strain,
$$\varepsilon_\theta = \frac{1}{E}(\sigma_\theta - v\sigma_l)$$

$$= \frac{1}{205\times10^3}(42-0.3\times21) = 174.15\times10^{-6}$$

Longitudinal strain,
$$\varepsilon_l = \frac{1}{E}(\sigma_l - v\sigma_\theta)$$

$$= \frac{1}{205\times10^3}(21-0.3\times42) = 40.97\times10^{-6}$$

$$\varepsilon_\theta = \frac{\delta d}{d}$$

$$\delta d = 174.15\times10^{-6}\times0.6 = 10.4.49\times10^{-6} \text{ m}$$

$$\varepsilon_l = \frac{\delta l}{l}$$

$$\delta l = 40.97 \times 10^{-6} \times 2 = 81.94 \times 10^{-6} \text{ m}$$

Changed diamensions of shell are:

$$d' = d + \delta d = 0.6 + 104.49 \times 10^{-6} = 0.60010449 \text{ m}$$

$$l' = l + \delta l = 2 + 81.94 \times 10^{-6} = 2.00008194 \text{ m}$$

Example 10.4 *A steel cylinder 1 m long, 0.2 m internal diameter is made of steel plates 6 mm thick. It is subjected to internal pressure of 6 MPa. The increase in volume due to internal pressure is 30 cm³. Calculate the value of Poisson's ratio and shear modulus. Take E = 210 GPa.*

Solution.

Given: $l = 1$ m, $d = 0.2$ m, $t = 6$ mm, $p = 6$ MPa, $\delta V = 30$ cm^3, E = 210 GPa

$$\delta V = \frac{V\,pd}{2tE}(2.5 - 2\nu)$$

$$30 \times 10^{-6} = \frac{\pi (0.2)^2}{4} \times 1 \times \frac{6 \times 10^6 \times 0.2}{2 \times 6 \times 10^{-3} \times 210 \times 10^9}(2.5 - 2\nu)$$

$$2.005352 = 2.5 - 2\nu$$

$$\nu = 0.247$$

Shear modulus, $$G = \frac{E}{2(1+\nu)} = \frac{210}{2(1+0.247)} = 84.2\,GPa$$

Example 10.5 *A cylindrical tank is 2 m in diameter, 2.5 m long and 20 mm thick Calculate the increase in capacity when the inside pressure is 1.25 MPa (a) without axial compressive load on the tank, and (b)* with an axial compressive load of 300 kN. Take $E = 210$ GPa and $\nu = 0.3$.

Solution.

Given: $d = 2$m, $l = 2.5$ m, $t = 20$ mm, $p = 1.25$ MPa, $F_a = 300$ kN, $E = 210$ GPa, $\nu = 0.3$

(a) Change in volume, $$\delta V = \frac{V\sigma_\theta}{E}(2.5 - 2\nu)$$

$$V = \frac{\pi}{6}d^3 = \frac{\pi}{6} \times 2^3 = \frac{4\pi}{3}\text{m}^3$$

$$\sigma_\theta = \frac{pd}{2t} = \frac{1.25 \times 2 \times 10^3}{2 \times 20} = 62.5 \text{ MPa}$$

$$\delta V = \frac{4\pi}{3} \times \frac{62.5}{210 \times 10^3}(2.5 - 2 \times 0.3)$$

$$= 23.68.6 \times 10^{-6} \text{ m}^3$$

(b) Axial compressive stress, $$\sigma_a = \frac{F_a}{\pi dt} = \frac{300 \times 10^3}{\pi \times 2 \times 10^3 \times 20} = 2.387 \text{ MPa}$$

Longitudinal stress, $$\sigma_l = \frac{\sigma_\theta}{2} = 31.25 \text{ MPa}$$

Resultant longitudinal stress, $\bar{\sigma}_l = \sigma_l - \sigma_a = 31.25 - 2.387 = 28.863$ MPa

$$\frac{\partial V}{V} = 2\varepsilon_\theta + \varepsilon_l$$

$$\varepsilon_\theta = \frac{1}{E}(\sigma_\theta - \nu\sigma_l')$$

$$= \frac{1}{210\times10^3}(62.5 - 0.3\times28.863)$$

$$= 2.5638\times10^{-4}$$

$$\varepsilon_l = \frac{1}{E}(\sigma_l' - \nu\sigma_\theta)$$

$$= \frac{1}{210\times10^3}(28.863 - 0.3\times62.5)$$

$$= 4.8157\times10^{-5}$$

$$\delta V = \frac{4\pi}{3}(2\times2.5638\times10^{-4} + 4.8157\times10^{-5})$$

$$= 2349.6\times10^{-6}\ \text{m}^3$$

Example 10.6 *A thin spherical shell has a diameter of 0.5 m and wall thickness 3 mm. It is full of water at atmospheric pressure. Calculate the volume of water pumped in to raise the inside pressure by 2 MPa. Take E = GPa, K = 2.5 GPa and ν = 0.25.*

Solution.

Given: $d = 0.5$ m, $t = 3$ mm, $p = 2$ MPa, $E = 100$ GPa, $K = 2.5$ GPa, $\nu = 0.25$

$$\frac{\delta V}{V} = \frac{3pd}{4tE}(1-\nu) + \frac{p}{K}$$

$$V = \frac{\pi}{6}d^3 = \frac{\pi}{6}\times(0.5)^3 = 0.06545\ \text{m}^3$$

$$\delta V = 0.06545\times\left[\frac{3\times2\times0.5\times10^3}{4\times3\times100\times10^3}(1-0.25) + \frac{2}{2.5\times10^3}\right]$$

$$= 175.0.8\times10^{-6}\ \text{m}^3$$

Example 10.7 *A cylindrical vessel of 1 diameter and 2 m long is made of steel 10 mm thick plates. It is filled with water at 15° C. The temperature is then raised to 50°C. Calculate the stress induced in the vessel material if over this range of temperature water increases 0.005 per unit volume. Bulk modulus of water is 2.9 GPa. For steel, E = 210 GPa and ν = 0.3.*

Solution.

Given: $d = 1$m, $l = 2$m, $t = 10$ mm, $\delta V/V = 0.005$, $K = 2.9$ GPa, $E = 210$ GPa, $\nu = 0.3$

$$V = \frac{\pi d^2}{4}\times l = \frac{\pi}{4}\times1\times2 = \frac{\pi}{2}m^3$$

$$\delta V = 0.005 \times \frac{\pi}{2} = 7.854 \times 10^{-3}\ \text{m}^3$$

$$\frac{\delta V}{V} = \frac{pd}{tE}(1.25 - \nu) + \frac{p}{K}$$

$$0.005 = \frac{p \times 10 \times 10^3}{10 \times 210 \times 10^3}(1.25 \times 0.3) + \frac{p}{2.9 \times 10^3}$$

$$= 7.972 \times 10^{-4} \times p$$

$$p = 6.27\ \text{MPa}$$

$$\sigma_\theta = \frac{pd}{2t} = \frac{6.27 \times 1 \times 10^3}{2 \times 10} - 313.5\ \text{MPa}$$

$$\sigma_l = \frac{\sigma_\theta}{2} = 156.75\ \text{MPa}$$

Example 10.8 *A thin spherical shell of 1m diameter is made of 10 mm thick steel plate. It is filled with oil at a pressure of 3 MPa. Due to leakage, oil escapes until the pressure falls to 1.5 MPa. Calculate the volume of oil that has escaped and also the change in the internal diameter of shell due to this leakage. Bulk modulus of oil, K = 3GPa, E = 200 GPa, and ν = 0.3.*

Solution.

Given: $d = 1\text{m}, t = 10\ mm, p_1 = 3\ MPa, p_2 = 1.5\ MPa, K = 3GPa, E = 200\ GPa, \nu = 0.3$

$$\sigma_{\theta 1} = \frac{p_1 d}{4t} = \frac{3 \times 1 \times 10^3}{4 \times 10} = 75\ \text{MPa}$$

$$\varepsilon_{\theta 1} = \frac{dV}{V} = \frac{3\sigma_{\theta 1}}{E}(1 - \nu) = \frac{3 \times 75}{200 \times 10^3}(1 - 0.3) = 78.75 \times 10^{-5}$$

$$dV = \frac{\pi}{6} \times (1)^3 \times 78.75 \times 10^{-5} = 41.23 \times 10^{-5}\ \text{m}^3$$

Total volume of water at $p_1 = 3$ MPa $= V + dV$

$$= \frac{\pi}{6} \times 1^3 + 41.23 \times 10^{-5} = 0.524\ \text{m}^3$$

$$\text{Volumetric strain of oil} = \frac{p_1}{K} = \frac{3}{3 \times 10^3} = 1 \times 10^{-3}$$

Change in volume of oil = 0.524×10^{-3} m^3

Actual volume of oil at 3 MPa, $V_{t1} = 0.524 + 0.524 \times 10^{-3} = 0.524542$ m^3

Let V_1 = volume of oil leaked to reduce the pressure to 1.5 MPa.

$$\sigma_{\theta 2} = \frac{p_2 d}{4t} = \frac{1.5 \times 1 \times 10^3}{4 \times 10} 37.5\ \text{MPa}$$

$$\varepsilon_{\theta 2} = \frac{3\sigma_{\theta 2}}{E}(1 - \nu) = \frac{3 \times 37.5}{200 \times 10^3}(1 - 0.3) = 39.375 \times 10^{-5}$$

$$\text{Total volume, } V_t = V(1 + \varepsilon_{\theta 2}) = \frac{\pi}{6}(1 + 39.375 \times 10^{-5}) = 0.5238\ \text{m}^3$$

Change in volume due to compression of oil $= \frac{p_2}{K} \times V_t$

$$= \frac{1.5}{3 \times 10^3} \times 0.5238 = 26.19 \times 10^{-5}\,\text{m}^3$$

Actual volume of oil, $V_{t2} = 0.5238 + 26.19 \times 10^{-5} = 0.524062\ \text{m}^3$

Oil escaped $= V_{t1} - V_{t2} = 0.524524 - 0.524062 = 0.000\,426\ \text{m}^3$

$$\frac{\pi}{6}(\delta d)^3 = 0.000426$$

$$\delta d = 0.0933\ \text{m or } 93.3\ \text{mm}$$

SUMMARY

1. When the thickness to diameter ratio of a pressure vessel is less than 1/15, it is called a thin pressure vessel.
2. Stresses in a thin cylindrical pressure vessel.

(i) Hoop stress, $\sigma_\theta = \frac{pd}{2t}$

(ii) Longitudinal stress, $\sigma_l = \frac{pd}{4t}$

(iii) Maximum shear stress, $\tau_{max} = \frac{1}{2}(\sigma_\theta - \sigma_l) = \frac{pd}{8t}$

2. Strains in a thin cylindrical pressure vessel.

Hoop strain, $\varepsilon_\theta = \frac{1}{E}(\sigma_\theta - \nu\sigma_l)$

Longitudinal strain, $\varepsilon_l = \frac{1}{E}(\sigma_l - \nu\sigma_\theta)$

Volumetric strain, $\varepsilon_v = 2\varepsilon_\theta + \varepsilon_l$

4. Stress and strains in a spherical pressure vessel.

Hoop stress, $\sigma_\theta = \frac{pd}{4t}$

Hoop strain, $\varepsilon_\theta = \frac{\sigma_\theta}{E}(1-\nu)$

Volumetric strain, $\varepsilon_v = 3\,\varepsilon_\theta$

5. Effect of built-up ends.

(i) cyinder

$$\sigma_\theta = \frac{pd}{2\eta_\theta t}$$

$$\sigma_l = \frac{pd}{4\eta_l t}$$

(ii) sphere

$$\sigma_\theta = \frac{pd}{4\eta_\theta t}$$

6. Effect of fluid pressure.
Extra fluid required to raise the cylinder pressure by p,

$$= \frac{pd}{2tE}(2.5 - 2\nu)\,V + \frac{pV}{K}$$

and for the sphere

$$= \frac{pd}{4tE}(1 - \nu)\,V + \frac{pV}{K}$$

Short Answer Type Questions

1. Define a thin pressure vessel.
2. When the longitudinal stress develops in a thin cylindrical pressure vessel?
3. What is the ratio of hoop stress to longitudinal stress in a thin cylinder?
4. What is the ratio of hoop stress to maximum shear stress in a thin cylinder?
5. What is the effect of built-up pressure vessels on their wall thickness?
6. What is the ratio of hoop stresses in a thin cylinder and thin sphere of same diameter and wall thickness?
7. What is the effect of bulk modulus of fluid on the capacity of thin pressure vessels?

EXRCISES

1. A thin cylindrical shell of 1.5 m diameter is subjected to an internal pressure of 2.5 MPa. Calculate the wall thickness if allowable tensile stress is 150 MPa and joint efficiency is 85%.

   [**Ans.** 14.7 mm]
2. A boiler shell is made of 20 mm thick steel plate having tensile stress of 120 MPa. The efficiencies of longitudinal and circumferntial joints are 72% and 28% respectively. Determine the permissible diameter of the shell. [**Ans.** 1.344 m]
3. A cylindrical vessel whose ends are closed by means of rigid flange plates is made of steel plate 4 mm thick. The internal length and diameter of the vessel are 0.6 m and 0.3 m respectively. Calculate the longitudinal and circumferential stresses in the vessel due to an internal pressure of 2.5 MPa. [**Ans.** 93.75 MPa, 46.875 MPa]

4. A cylindrical shell 2.5 m long and closed at the ends has an internal diameter of 1.25 m and wall thickness of 20 mm. Calculate the change in dimensions when subjected to an internal pressure of 1.5 MPa. Take E = 200 GPa, ν = 0.3. **[Ans.** δd = 0.234 mm, δl = 0.117 mm]

5. A copper cylinder, 0.9 m long 0.4 m internal diameter and wall thickness of 6 mm has closed ends. It is initially full of oil at atmospheric pressure. Calculate the additional volume of oil which must be pumped into in order to raise the oil pressure 5 MPa above atmospheric pressure. Take E = 100 GPa and Poisson's ratio = $\frac{1}{3}$. **[Ans.** 550. 245 cm^3]

6. A thin spherical shell 1.5 m in diameter and wall thickness 10 mm is filled with a fluid at atmospheric pressure. 200 cm^3 of more fluid is pumped into the shell. Calculate the pressure developed. E = 200 GPa, ν = 0.3. **[Ans.** 0.287 MPa]

7. A thin cylindrical pressure vessle has a diameter of 2 m and its wall thickness is 10 mm. (a) Calculate the hoop and longitudinal stresses in the cylindrical wall caused by an internal pressure of 0.8 MPa. (*b*) Determine the change in diameter of the cylinder caused by pressurization. Take E = 200 GPa and ν = 0.25. **[Ans.** 80 MPa, 40 MPa, 0.35 mm]

8. A spherical pressure vessel of 2 m diameter has a wall thickness of 10 mm. (a) Determine the maximum membrane stresses caused by an internal pressure of 0.8 MPa. (b) Calculate the change in diameter of the sphere caused by pressurization. Assume E = 200 GPa and ν = 0.25. **[Ans.** 40 MPa, 0.15 mm]

Chapter-11

Thick Cylinders and Spheres

11.1 INTRODUCTION

A pressure vessel is said to be thick when the wall thickness to mean diameter ratio is more than $\frac{1}{15}$. The problem of thick cylinders under internal or external pressure is axisymmetric in nature. The solution to this problem was presented first of all by Lame and has become popular as Lame's theory. We Shall study this theory and then apply to solve compound and interference fitted thick cylinders.

11.2 LAME'S THEORY FOR THICK CYLINDERS

Assumptions

The following assmuptions are made in this theory:

1. The material of the cylinder is linear, homogeneous and isotropic.
2. Plane transverse sections of the cylinder remain plane under the action of pressure. This assumption is nearly true at a considerable distance from the ends of the cylinder. As a consequence of this assumption, the longitudinal strain remains constant at all points in the cylinder wall, i.e. it is independent of the cylinder radius.

Let σ_r, σ_θ, σ_z = radial, hoop and longitudinal stresses in the (r, θ, z) cylindrical coordinate system. The z-axis concides with the longitudinal axis of the cylinder.

All these stresses are assumed tensile in nature.

Longitudinal strain, $$\varepsilon_z = \frac{1}{E}[\sigma_z - \nu(\sigma_\theta + \sigma_r)]$$

Since ε_z, σ_z, E and ν are all constants, therefore

$$\sigma_\theta + \sigma_r = \text{constant} = 2A \text{ (say)} \quad \text{... (1)}$$

Consider a thick cylinder of inner radius r_1, outer radius r_2, and length l, subjected to uniformly distributed pressures p_1 and p_2 on the inner and outer surface respectively, as shown in Fig. 11.1 (a). Consider an annular ring of the cylinder between radii r and $r + dr$, as shown in Fig. 11.1 (*b*). The radial and hoop stresses on an element of the ring are shown there. Let the internal radial stress on this ring be σ_r and the external radial stress be $\sigma_r + d\sigma_r$.

Now consider the stresses acting on half the ring as shown in Fig. 11.1 (*c*).

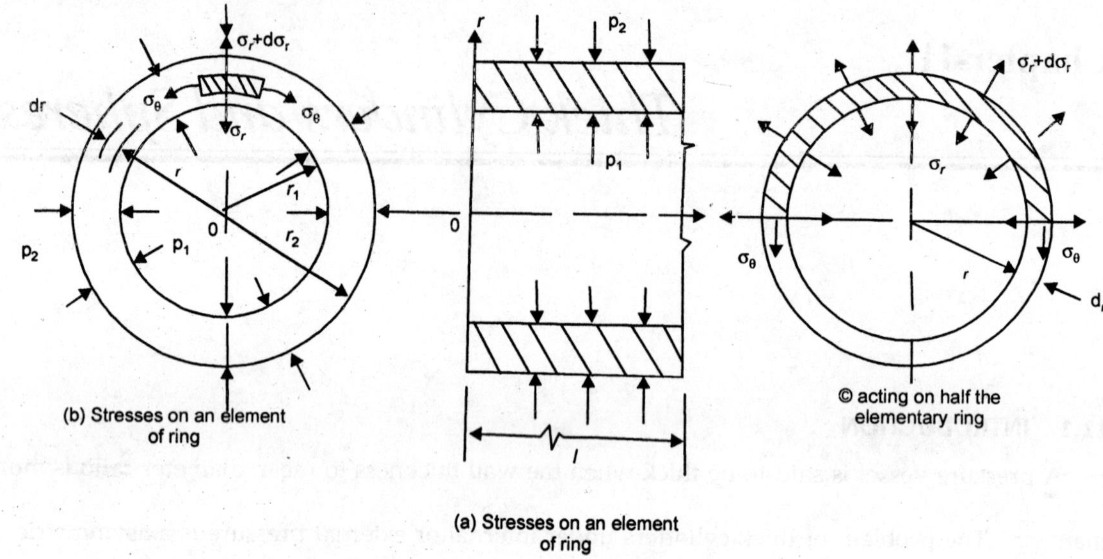

Fig. 11.1 Analysis of thick cylindrical pressure vessel.

Bursting force, $F_1 = (\sigma_r + d\sigma_r) \times 2(r + dr)l - \sigma_r \times 2rl$

Simplifying and neglecting small quantities, we get

$$F_1 = 2\sigma_r \cdot dr.l + 2d\sigma_r \times r.l$$

Resisting force, $\qquad F_2 = 2\sigma_\theta \cdot dr.l$

For equilibrium of the ring, $F_1 = F_2$

$$\therefore \qquad 2\sigma_\theta \cdot dr.l = 2\sigma_r \cdot dr\,l + 2d\sigma_r \cdot r.l$$

$$\sigma_\theta = \sigma_r + \frac{r\,d\sigma_r}{dr}$$

$$= \frac{d}{dr}(r\sigma_r) \qquad \text{... (2)}$$

Substituting Eq. (2) in Eq. (1), we get

$$\frac{d}{dr}(r\sigma_r) + \sigma_r = 2A$$

or

$$r\frac{d\sigma_r}{dr} + 2\sigma_r = 2A$$

or

$$r\frac{d\sigma_r}{dr} = 2(A - \sigma_r)$$

or $$\frac{d\sigma_r}{dr} = \frac{2(A-\sigma_r)}{r}$$

or $$\frac{d\sigma_r}{(A-\sigma_r)} = \frac{2dr}{r}$$

Intergrating, we get

$$\int \frac{d\sigma_r}{(A-\sigma_a)} = 2\int \frac{dr}{r}$$

$$-\ln (A - \sigma_r) = 2\ln r - \ln B$$

$$\ln(A-\sigma_r) = -2\ln r + \ln B$$

$$= \ln \frac{B}{r^2}$$

$$A-\sigma_r = \frac{B}{r^2}$$

or $$\sigma_r = A - \frac{B}{r^2} \quad (3)$$

where B is another constant

Substituting Eq. (3) in eq. (1), we get

$$\sigma_\theta = A + \frac{B}{r^2}$$

Thus the Lame's equations become:

$$\left.\begin{aligned} \sigma_r &= A - \frac{B}{r^2} \\ \sigma_\theta &= A + \frac{B}{r^2} \end{aligned}\right\} \quad \text{... (11.1)}$$

The constants A and B can be determined from the boundary conditions.

(*i*) At $r = r_1$, $\sigma_r = -p_1$

(ii) At $r = r_2$, $\sigma_r = -p_2$

Substiuting in first Lame's equation, wet have

$$-p_1 = A - \frac{B}{r_1^2}$$

$$-p_2 = A - \frac{B}{r_2^2}$$

$$p_1 - p_2 = B\left(\frac{1}{r_1^2} - \frac{1}{r_2^2}\right) = B\left(\frac{r_2^2 - r_2^2}{r_1^2 r_2^2}\right)$$

$$B = \frac{(p_1 - p_2) r_1^2 r_2^2}{(r_2^2 - r_2^2)}$$

$$A = \frac{B}{r_1^2} - p_1 = \frac{(p_1 - p_2) r_2^2}{(r_2^2 - r_1^2)} - p_1 = \frac{p_1 r_2^2 - p_2 r_2^2 - p_1 r_2^2 + p_1 r_1^2}{(r_2^2 - r_1^2)}$$

$$= \frac{p_1 r_1^2 - p_2 r_2^2}{(r_2^2 - r_1^2)}$$

$$\sigma_r = \frac{p_1 r_1^2 - p_2 r_2^2}{(r_2^2 - r_1^2)} - \frac{(p_1 - p_2) r_1^2 r_2^2}{r^2 (r_2^2 - r_1^2)} \quad \text{... (11.2)}$$

$$\sigma_\theta = \frac{p_1 r_1^2 - p_2 r_2^2}{(r_2^2 - r_1^2)} + \frac{(p_1 - p_2) r_1^2 r_2^2}{r^2 (r_2^2 - r_1^2)} \quad \text{... (11.3)}$$

Now we discuss some particular cases.

(a) Internal pressure only, $p_2 = 0$, then

$$\sigma_r = \frac{p_1 r_1^2}{(r_2^2 - r_1^2)} - \frac{p_1 r_1^2 r_2^2}{r^2 (r_2^2 - r_1^2)}$$

$$= \frac{p_1 r_1^2}{(r_2^2 - r_1^2)} \left[1 - \left(\left(\frac{r_2}{r} \right)^2 \right) \right] = \frac{p_1}{[(r_2 / r_1)^2 - 1]} \left[1 - \left(\frac{r_2}{r} \right)^2 \right] \quad \text{... (11.4 a)}$$

$$\sigma_\theta = \frac{p_1 r_1^2}{(r_1^2 - r_1^2)} \left[1 + \left(\frac{r_2}{r} \right)^2 \right] = \frac{p_1}{[(r_2 / r_1)^2 - 1]} \left[1 + \left(\frac{r_2}{r} \right)^2 \right] \quad \text{... (11.4 b)}$$

(b) External pressure only, $p_1 = 0$, then

$$\sigma_r = \frac{p_2}{\left[1 - \left(\frac{r_1}{r_2} \right)^2 \right]} \left[\left(\frac{r_1}{r} \right)^2 - 1 \right] \quad \text{... (11.5 a)}$$

$$\sigma_\theta = \frac{-p_2}{\left[1 - \left(\frac{r_1}{r_2} \right)^2 \right]} \left[1 + \left(\frac{r_1}{r} \right)^2 \right] \quad \text{...(11.5 b)}$$

(c) Solid circular shaft having external pressure only.

At $r_1 = 0, p_1 = 0$, then

$\sigma_r = -p_2$ and $\sigma_\theta = -\text{p}_2$

$\therefore \quad \sigma_r = \sigma_\theta = -p_2 = \text{constant}$ ... (11.6)

Thus the radial and hoop stresses are equal and constant througout the solid shaft and are of the same nature.

11.2.1 Longitudinal Stress

Consider a thick cylinder with closed ends subjected to an internal pressure p_1 and external pressure p_2. Let σ_z be the longitudinal (or axial) stress induced in the cylinder. Then for horizontal equlibrium, we have

$$\pi r_1^2 p_1 - \pi r_2^2 p_2 = \pi(r_2^2 - r_1^2)\sigma_z$$

$$\therefore \qquad \sigma_z = \frac{p_1 r_1^2 - p_2 r_2^2}{r_2^2 - r_1^2} = \text{const.} = A \qquad \text{...(11.7)}$$

11.2.2 Maximum Shear Stress

$$\tau_{max} = \frac{1}{2}(\sigma_\theta - \sigma_r)$$

$$= \frac{1}{2}\left(A + \frac{B}{r^2} - A + \frac{B}{r^2}\right)$$

$$= \frac{B}{r^2}$$

Shear stress will be maximum at $r = r_1$.

$$\tau_{max} = \frac{r_2^2(p_1 - p_2)}{r_2^2 - r_1^2} = \frac{(p_1 - p_2)}{\left[1 - \left(\frac{r_1}{r_2}\right)^2\right]} \qquad \text{... (11.8)}$$

11.2.3 Radial Deflection

Hoop Strain, $\varepsilon_\theta = \frac{1}{E}(\sigma_\theta - \nu\sigma_r)$

Change in circumference of cylinder at radius $r = 2\pi r \varepsilon_\theta$

Change in radius r, $\delta r = \frac{2\pi r \varepsilon_\theta}{2\pi} = r\varepsilon_\theta$

Radial deflection, $\delta r = \frac{r}{E}(\sigma_\theta - \nu\sigma_r)$

For the cylinder subjected to internal pressure only,

$$\delta r = \frac{r}{E}\left[\left\{1 + \left(\frac{r_2}{r}\right)^2\right\} - \nu\left\{1 - \left(\frac{r_2}{r}\right)^2\right\}\right] \times \frac{p_1}{\left[\left(\frac{r_2}{r_1}\right)^2 - 1\right]}$$

$$= \frac{r}{E}\left[(1-\nu)+(1+\nu)\left(\frac{r_2}{r}\right)^2\right]\times\frac{p_1}{[(r_2/r_1)^2-1]}$$

At $r = r_1$

$$\delta r_1 = \frac{p_1 r_1}{E}\left[\frac{(r_2/r_1)^2+1}{\{(r_2/r_1)^2-1\}}+\nu\right] \quad \text{... (11.9)}$$

For the cylinder subjected to external pressure only,

$$\delta r = \frac{r}{E}[\{1+(r_1/r)^2\}+\nu\,\{(r_1/r)^2-1\}]\times\left[\frac{-p_2}{\{1-(r_1/r_2)^2\}}\right]$$

$$= \frac{r}{E}[(1-\nu)+(1+\nu)(r_1/r)^2]\times\left[\frac{-p_2}{\{1-(r_1/r_2)^2\}}\right]$$

Ar $r = r_2$,

$$\delta r_2 = \frac{-p_2 r_2}{E}\left[\frac{\{(r_2/r_1)^2+1\}}{\{(r_2/r_1)^2-1\}}-\nu\right] \quad \text{... (11.10)}$$

Example 11.1 *A thick cylinder having internal and external radii of 8 cm and 16 cm respectively is subjected to an internal fluid pressure of 8 MPa. Draw the variation of radial and hoop stresses in the cylinder wall.*

Solution.

Given : $r_1 = 8$ cm, $r_2 = 16$ cm, $p_1 = 8$ MPa

Now
$$\sigma_r = \frac{p_1 r_1^2}{(r_2^2-r_1^2)}\left[1-\frac{r_2^2}{r^2}\right] = \frac{8\times 64}{(256-64)}\left[1-\frac{256}{r^2}\right]$$

$$= \frac{8\times 64}{192}\left[1-\frac{256}{r^2}\right] = 2.667\left[1-\frac{256}{r^2}\right]$$

r, cm	8	10	12	14	16
σ_r, MPa	−8	−4.16	−2.074	−0.816	0

$$\sigma_\theta = \frac{p_1 r_1^2}{(r_2^2-r_1^2)}\left[1+\frac{r_2^2}{r^2}\right] = 2.667\left[1+\frac{256}{r^2}\right]$$

r, cm	8	10	12	14	16
σ_θ, MPa	13.333	9.493	7.408	6.150	5.333

The variation of stresses has been plotted in Fig. 11.2.

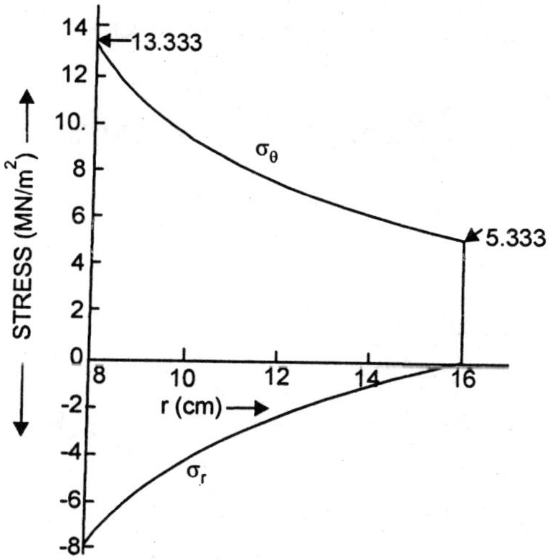

Fig. 11.2

Example 11.2 *The internal and external diameters of a thick hollow cylinder are 8 cm and 12 cm respectively. It is subjected to an external pressure of 40 MPa when, the internal pressure is 120 MPa. Calculate the circumferential stress at the external and internal surfaces and determine the radial and circumferential stresses at the mean radius. Plot the stresses on the thickness of the cylinder as base.*

Solution.

Given: $r_1 = 4$ cm, $r_2 = 6$ cm, $p_1 = 120$ MPa, $p_2 = 40$ MPa

Now
$$\sigma_\theta = \frac{p_1 r_1^2 - p_2 r_2^2}{(r_2^2 - r_1^2)} + \frac{r_1^2 r_2^2 (p_1 - p_2)}{r^2 (r_2^2 - r_1^2)}$$

$$= \frac{120 \times 16 - 40 \times 36}{(36 - 16)} + \frac{16 \times 36 (120 - 40)}{r^2 (36 - 16)} = \frac{1920 - 1440}{20} + \frac{16 \times 36 \times 80}{20 r^2}$$

$$= \frac{480}{20} + \frac{2304}{r^2} = 24 + \frac{2304}{r^2}$$

At $r = 4$ cm

$$\sigma_\theta = 24 + \frac{2304}{16} = 24 + 144 = 168 \text{ MPa}$$

At $r = 6$ cm

$$\sigma_\theta = 24 + \frac{2304}{36} = 24 + 64 = 88 \text{ MPa}$$

At the mean radius,

$r = 5$ cm

$$\sigma_\theta = 24+\frac{23.4}{25} = 24+92.16 = 116.16 \text{ MPa}$$

$$\sigma_r = 24-\frac{2304}{25} = 24-92.16 = -68.16 \text{ MPa}$$

The variation of stresses has been shown in Fig. 11.3.

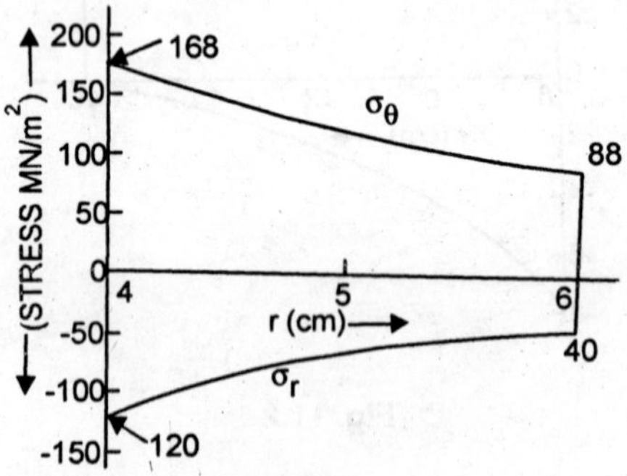

Fig. 11.3

Example 11.3 *The cylinder of a hydraulic press has an internal diameter of 30 cm and is to be designed to withstand a pressure of 10 MPa without the material being stressed over 20 MPa. Determine the thickness of the material and the stress on the outside of cylinder.*

Sketch a diagram showing the variation of radial and hoop stresses across the thickness of the wall of the cylinder.

Solution:

Given: $r_1 = 15$ cm, $p_1 = 10$ MPa $\sigma_\theta = 20$ MPa $r_2 = ?$

$$\sigma_\theta = \frac{p_1 r_1^2}{(r_2^2 - r_1^2)}\left[1+\frac{r_2^2}{r^2}\right]$$

The hoop stress is maximum at $r = 15$ cm.

$$\therefore \quad 20 = \frac{10\times 225}{(r_2^2-225)}\left[1+\frac{r_2^2}{225}\right]$$

$$r_2^2 - 225 = 112.5\left(1+\frac{r_2^2}{225}\right) = 112.5+\frac{r_2^2}{2}$$

$$\frac{r_2^2}{2} = 337.5$$

$$r_2^2 = 675$$

$$r_2 = 25.98 \text{ cm}$$

∴ Wall thickness

$$= 25.98 - 15 = 10.98 \text{ cm}$$

On the outside surface, the hoop stress is

$$\sigma_\theta = \frac{10 \times 225}{(675-225)}\left[1+\frac{675}{675}\right]$$

$$= \frac{10 \times 225}{450}(1+1)$$

$$= \frac{10 \times 225 \times 2}{450} = 10 \text{ MPa}$$

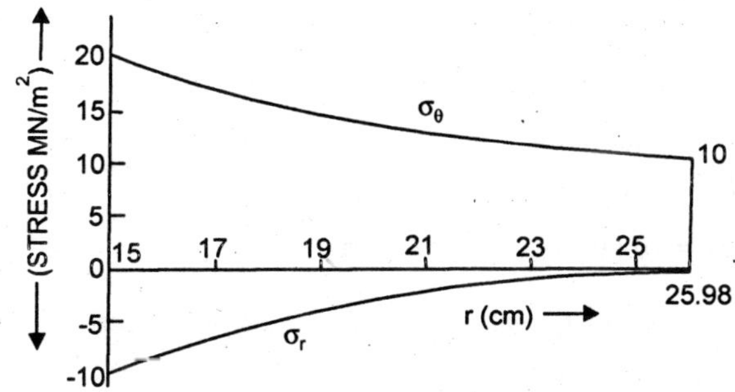

Fig. 11.4

r, cm	15	17	19	21	23	25.98
σ_r, MPa	–10	–6.678	–4.349	–2.653	–1.38	0
σ_θ MPa	20	16.678	14.349	12.653	11.38	10

The variatin of stressess has been shown in Fig. 11.4.

Example 11.4 *Find the ratio of thickness to the internal diameter of a tube subjected to internal pressure when the pressure is 75% of the value of the maximum permissible circumferential stress.*

Find the increase in internal diameter of such a tube, 20 cm internal diameter when the internal pressure is 42 MPa.

$$E = 200 \text{ GPa}. \ \nu = 0.28$$

Solution.

Given: Now
$$s_q = \frac{p_1 r_1^2}{(r_2^2 - r_1^2)}\left[1+\frac{r_2^2}{r^2}\right]$$

At $r = r_1$, $$\frac{\sigma_\theta}{p_1} = \frac{r_1^2}{(r_2^2 - r_1^2)}\left[1+\frac{r_2^2}{r_1^2}\right]$$

$$\frac{\sigma_\theta}{0.75\sigma_\theta} = \frac{r_1^2 + r_2^2}{r_2^2 - r_1^2}$$

$$\therefore \quad \frac{4}{3} = \frac{r_1^2 + r_2^2}{r_2^2 - r_1^2}$$

$$4r_2^2 - 4r_1^2 = 3r_1^2 + 3r_2^2$$

$$r_2^2 = 7r_1^2$$

$$r_2 = 2.64575\, r_1$$

$\therefore$ Thickness $$t = r_2 - r_1 = (2.64575 - 1)\, r_1$$
$$= 1.64575\, r_1 = 0.822875\, d_1$$

$$\therefore \quad \frac{t}{d_1} = 0.822875$$

For $$r_1 = 10 \text{ cm},\ r_2 = 26.4575 \text{ cm}$$

Increase in diameter: $$\delta d = \frac{p_1 d_1}{E}\left[\frac{r_1^2 + r_2^2}{r_2^2 - r_1^2} + \nu\right]$$

$$= \frac{42\times 20}{200\times 10^3}\left[\frac{100+700}{700-100} + 0.28\right] = \frac{42\times 20}{200\times 10^3}[1.333 + 0.28]$$

$$= \frac{42\times 20\times 1.613}{200\times 10^8} = 67.76 \times 10^{-3} \text{ mm}$$

Example 11.5 *A cylinder of 15 cm internal diameter and 20 cm external diameter is subjected to liquid pressure from inside. There is also a compressive load of 200 kN applied at the ends of the cylinder. Find the greatest pressure of the liquid so that the maximum stress in the material may not exceed 42 MPa.*

Solution.

Given: $$r_1 = 7.5 \text{ cm},\ r_2 = 10 \text{ cm},\ P = 200 \text{ kN (compressive)}$$
$$(\sigma_\theta)_{max} = 42 \text{ MPa}$$

At $$r = r_1$$

$$\sigma_\theta = p_1\left(\frac{r_1^2 + r_2^2}{r_2^2 - r_1^2}\right) = p_1\left(\frac{100+56.25}{100-56.25}\right) = \frac{156.25}{49.75}p_1 = 3.57p_1$$

Now $$\sigma_z.\pi(r_2^2 - r_1^2) = \pi r_1^2 . p_1$$

$$\therefore \qquad \sigma_z = \frac{r_1^2}{(r_2^2 - r_1^2)} p_1 = \frac{56.25}{43.75} \cdot p_1 = 1.2857\, p_1$$

Direct stress,
$$\sigma_z = \frac{P}{A} = \frac{200 \times 10^3}{\frac{\pi}{4}(400 - 225) \times 10^{-4}} = \frac{200 \times 10^3 \times 4}{\pi \times 175 \times 10^{-4}} = 14.551\,\text{MPa}$$

$$\therefore \qquad 3.57 p_1 = 42$$
$$p_1 = 11.765 \text{ MPa}$$
$$1.2857\, p_1 = 14.551$$
$$p_1 = 11.3176 \text{ MPa}$$

$\therefore$ Maximum pressure = 11.765 MPa

Example 11.6 (a) *Prove that the maximum hoop stess in a thick cylinder is given by*

$$(\sigma_\theta)_{max} = \left(\frac{k^2 + 1}{k^2 - 1}\right)$$

where p = *internal pressure*
k = *outside to inside diameter ratio.*

(b) If a cylinder of internal diameter d, wall thickness t and subjected to internal pressure only, is assumed to be a thin cylinder, what is the greatest value for the ratio t/d if the error in the estimated maximum hoop stress is not to exceed 5 per cent?

Solution.

Given : *(a)*
$$s_q = \frac{p_1 r_1^2}{(r_2^2 - r_1^2)} \left[1 + \frac{r_2^2}{r^2}\right]$$

At $r = r_1$, σ_θ is maximum

$\therefore$ For $p_1 = p, r = r_1$, we get

$$(\sigma_\theta)_{\max} = p\left(\frac{r_1^2 + r_2^2}{r_2^2 - r_1^2}\right) = p\left(\frac{k^2 + 1}{k^2 - 1}\right)$$

where $k = r_2/r_1$

(b) Hoop stress for a thin cylinder,

$$(\sigma_\theta)_1 = \frac{pd}{2t}$$

and for a thick cylinder, $(\sigma_\theta)_2 = p\left(\dfrac{r_1^2 + r_2^2}{r_2^2 + r_1^2}\right) = p\left[\dfrac{(d + 2t)^2 + d^2}{(d + 2t)^2 - d^2}\right]$

$$(\sigma_\theta)_2 = \left(\frac{d^2 + 2dt + 2t^2}{2dt + 2t^2}\right) p$$

$$\frac{(\sigma_\theta)_2-(\sigma_\theta)_1}{(\sigma_\theta)_2}=\frac{\dfrac{d^2+2dt+2t^2}{2dt+2t^2}-\dfrac{d}{2t}}{\dfrac{d^2+2dt+2t^2}{2dt+2t^2}}=\frac{2d^2t+4dt^2+4t^3-2d^3-2d^2t-2dt^2}{2t(d^2+2dt+2t^2)}$$

$$=\frac{2dt^2+4t^2}{2t(d^2+2dt+2t^2)}=\frac{dt+2t^2}{d^2+2dt+2t^2}=0.05$$

$$\therefore \quad dt+2t^2=0.05\,d^2+0.10\,dt+0.10\,t^2$$

$$0.05d^2-0.90\,dt-1.9t^2=0$$

$$d^2-18dt-38t^2=0$$

$$d=\frac{18t\sqrt{324t^2+152t^2}}{2}=9t\pm\frac{1}{2}\sqrt{476t^2}=9t\pm\frac{1}{2}\times 21.82\,t$$

$$=9\pm 10.91t=19.91t$$

$$\frac{t}{d}=0.0502=\frac{1}{19.92}\simeq\frac{1}{20}$$

11.3 COMPOUND CYLINDER

In a thick cylinder subjected to internal pressure, the maximum hoop stress occurs on the inner surface and diminishes to a minimum value on the outer surface. The intensity of the internal pressure which a cylinder of given thickness can carry is, therefore, limited by the maximum hoop stress which should not exceed the permissible tensile stress for the material. In order to reduce stresses inside a thick cylinder subjected to a high internal pressure, as in the case of guns two or more cylinders are assembled together by "shrink fitting". In the case of two cylinders, the outside cylinder has internal diameter slightly less than the external diameter of the inside cylinder. The outside cylinder is expanded by heating and the two cylinders are assembled together. After the cylinders attain the same temperature, the outside cylinder causes circumferential compressive stresses in the inner cylinder and the inner cylinder causes circumferential tensile stresses in the outside cylinder. When this compound cylinder is subjected to internal pressure, tensile hoop stresses shall be set up throughout the material of the cylinder. The final stresses will be algebraic sum of the initial shrinkage stresses and the stresses due to internal pressure.

Due to the initial compressive stress in the inner cylinder, the final stress (tensile) shall be less than if the cylinder was initially unstressed. For the outer cylinder it will be higher due to initial tension in the outer cylinder. Thus there shall be more or less uniform distribution of hoop stress throughout the wall of the cylinder and it will be able to carry greater internal pressure. Such a cylinder is also called a prestressed cylinder. Sometimes, instead of using only two cylinders, multi layered construction is also used.

(a) **Determination of Resultant stresses.** Consider a compound cylinder of inner radius r_1, outer radius r_2 and junction radius r_3 as shown in Fig. 11.5.

Let $\quad p_s$ = Shrinkage pressure at the common radius r_3.

(i) *Shrinkage stresses.* For the inner cylinder, (Fig. 11.6), let

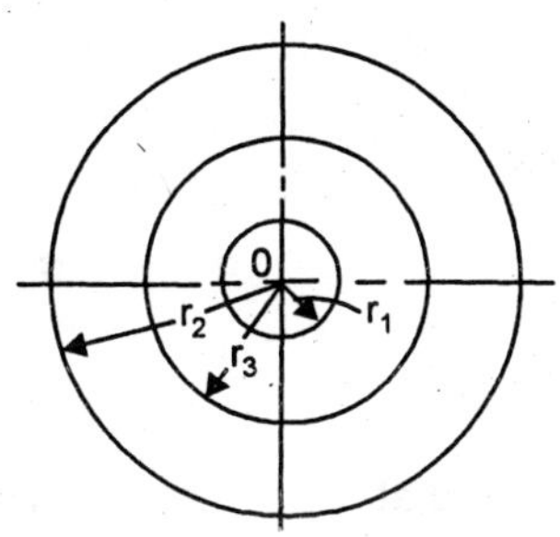

Fig. 11.5 Compound cylinder

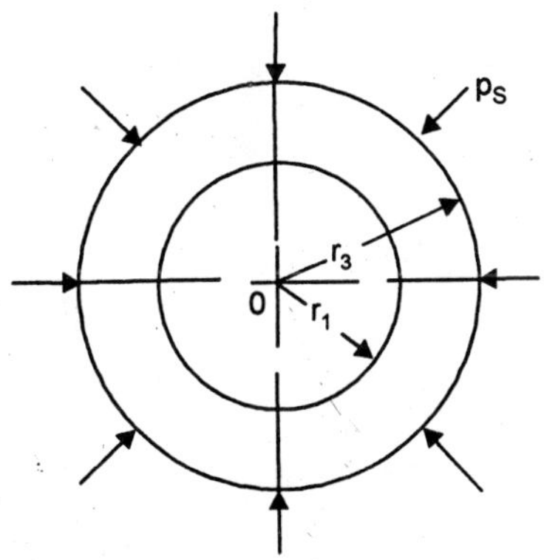

Fig. 11.6 Inner cylinder

$$\sigma_r = A - \frac{B}{r^2}$$

$$\sigma_\theta = A + \frac{B}{r^2}$$

At $\quad r = r_1, \sigma_r = 0$

and at $\quad r = r_3, \sigma_r = -p_s$

$$\therefore \quad 0 = A - \frac{B}{r_1^2}$$

$$-p_s = A - \frac{B}{r_3^2}$$

Hence $\quad A = \frac{B}{r_1^2}$

$$-p_s = B\left(\frac{1}{r_1^2} - \frac{1}{r_3^2}\right) = B\left(\frac{r_3^2 - r_1^2}{r_1^2 r_3^2}\right)$$

$$\therefore \quad B = -\frac{r_1^2 r_3^2 p_s}{(r_3^2 - r_1^2)}$$

$$A = -\frac{r_3^2 p_s}{(r_3^2 - r_1^2)}$$

$$\therefore \quad \sigma_r = \frac{r_3^2 p_s}{(r_3^2 - r_1^2)} + \frac{r_1^2 r_3^2 p_s}{r^2 (r_3^2 - r_1^2)} = \frac{r_3^2 p_s}{(r_3^2 - r_1^2)}\left[-1 + \frac{r_1^2}{r^2}\right]$$

$$\sigma_\theta = \frac{-r_3^2 p_s}{(r_3^2 - r_1^2)}\left[1 + \frac{r_1^2}{r^2}\right]$$

At $\quad r = r_3$

$$(\sigma_\theta)_3' = -p_s\left(\frac{r_3^2 + r_1^2}{r_3^2 - r_1^2}\right) \quad \text{... (11.11)}$$

At $\quad r = r_1$

$$(\sigma_\theta)_1' = \frac{-2 r_3^2 p_s}{(r_3^2 - r_1^2)} \quad \text{... (11.12)}$$

For the outer cylinder (Fig. 11.7), let

$$\sigma_r = a - \frac{b}{r^2}$$

$$\sigma_\theta = a + \frac{b}{r^2}$$

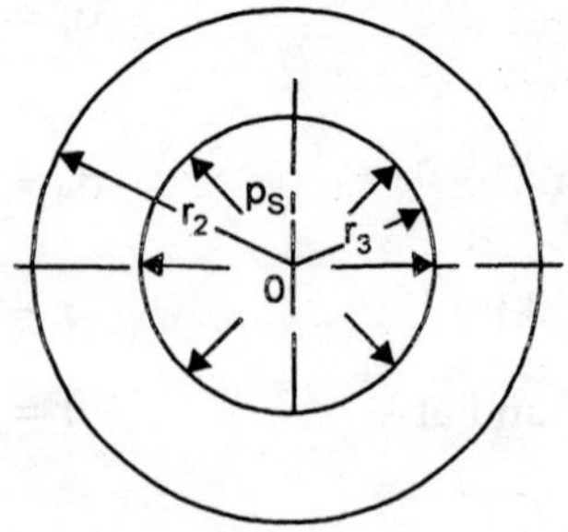

Fig. 11.7 Outer cylinder

At $\quad r = r_1, \sigma_3 = -p_s$

and at $\quad r = r_3, \sigma_r = 0$

$$\therefore \quad -p_s = a - \frac{b}{r_3^2}$$

$$0 = a - \frac{b}{r_2^2}$$

$$\therefore \quad a = \frac{b}{r_2^2}$$

$$-p_s = \frac{b}{r_2^2} - \frac{b}{r_3^2} = b\left[\frac{r_3^2 - r_2^2}{r_2^2 r_3^2}\right]$$

$$b = \frac{p_s r_2^2 r_3^2}{(r_2^2 - r_3^2)}$$

$$\therefore \quad \sigma_r = \frac{p_s r_3^2}{(r_2^2 - r_3^2)} - \frac{p_s r_2^2 r_3^2}{r^2(r_2^2 - r_3^2)} = \frac{p_s r_3^2}{(r_2^2 - r_3^2)}\left[1 - \frac{r_2^2}{r^2}\right]$$

$$\sigma_\theta = \frac{p_s r_3^2}{(r_2^2 - r_3^2)}\left[1 + \frac{r_2^2}{r^2}\right]$$

At $\quad r = r_3$

$$(\sigma_\theta)_3'' = p_s\left[\frac{r_3^2 + r_2^2}{r_2^2 - r_3^2}\right] \qquad \text{... (11.13)}$$

At $\quad r = r_2$

$$(\sigma_\theta)_2'' = \frac{2 p_s r_3^2}{(r_2^2 - r_3^2)} \qquad \text{... (11.14)}$$

(*ii*) Stresses due to fluid pressure.

Let $\quad p_i$ = internal fluid pressure.

Assuming that the compound cylinder behaves as one unit.

$$\sigma_r = \alpha - \frac{\beta}{r^2}$$

$$\sigma_\theta = \alpha + \frac{\beta}{r^2}$$

At $r = r_1, \sigma_r = -p_i$

and at $r = r_2, \sigma_r = 0$

$$\therefore \quad -p_i = \alpha - \frac{\beta}{r_2^2}$$

$$0 = \alpha - \frac{\beta}{r_2^2}$$

$$\therefore \quad \alpha = \frac{\beta}{r_2^2}$$

$$\therefore \quad -p_i = \beta\left(\frac{1}{r_2^2} - \frac{1}{r_1^2}\right)$$

$$\beta = \frac{p_i r_1^2 r_2^2}{r_2^2 - r_1^2}$$

$$\alpha = \frac{p_i r_1^2}{r_2^2 - r_1^2}$$

$$\therefore \quad \sigma_r = \frac{p_i r_1^2}{(r_2^2 - r_1^2)}\left[1 - \frac{r_2^2}{r^2}\right]$$

$$\sigma_\theta = \frac{p_i r_1^2}{(r_2^2 - r_1^2)}\left[1 + \frac{r_2^2}{r^2}\right]$$

At $r = r_1$

$$(\sigma_\theta)_1 = p_1\left(\frac{r_2^2 + r_1^2}{r_2^2 - r_1^2}\right) \qquad \text{... (11.15)}$$

At $r = r_3$

$$(\sigma_\theta)_3 = p_1 \frac{r_1^2}{r_3^2}\left(\frac{r_3^2 + r_2^2}{r_2^2 - r_1^2}\right) \qquad \text{... (11.16)}$$

At $r = r_2$

$$(\sigma_\theta)_2 = \frac{2p_i r_1^2}{(r_2^2 - r_1^2)} \qquad \text{... (11.17)}$$

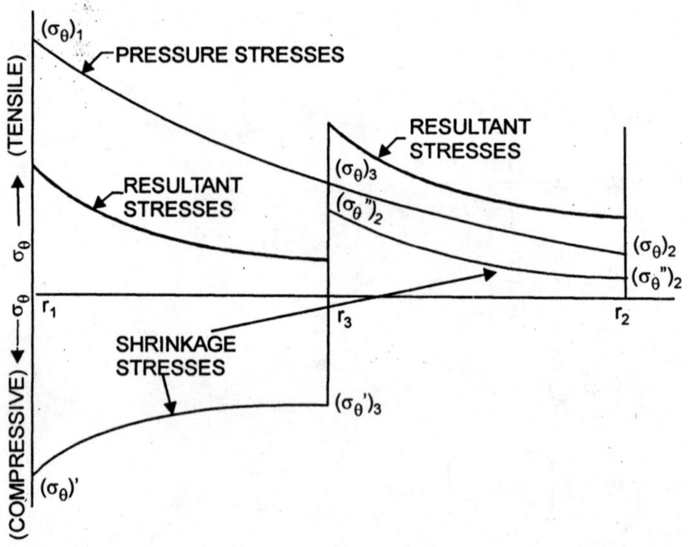

Fig.11. 8 Distribution of stresses in a compound cylinder.

(*iii*) *Resultant stresses.* The resultant stresses in the cylinder will be the algebraic sum of the stresses due to shrinkage pressure and fluid pressure, given by Eqs. (11.11) to (11.14) and (11.15) to (11.17) respectively. These have been shown in the following table and plotted in Fig. 11.8.

Radius	*Cylinder*	*Shrinkage stresses*	*Stresses due to fluid pressure*	*Reultant stresses*
r_1	inner	$(\sigma_\theta)_1'$	$(\sigma_\theta)_1$	$(\sigma_\theta)_1 + (\sigma_\theta)_1'$
r_3	inner	$(\sigma_\theta)_3'$	$(\sigma_\theta)_3$	$(\sigma_\theta)_3 + (\sigma_\theta)_3'$
r_3	outer	$(\sigma_\theta)_3''$	$(\sigma_\theta)_3$	$(\sigma_\theta)_3 + (\sigma_\theta)_3''$
r_2	outer	$(\sigma_\theta)_2''$	$(\sigma_\theta)_2$	$(\sigma_\theta)_2 + (\sigma_\theta)_2''$

(*b*) **Shrinkage allwance.** The inner radius of the outer cylinder shall be slightly smaller than the outer radius of inner cylinder, so that when the cylinder is shrunk on, the required radial pressure p_s is achieved at the common surface.

With reference to Fig. 11.9.

Let r_3 = Junction radius after shrinking on.

δr_1 = Difference between the outer radius of inner cylinder and r_3.

δr_2 = Difference between the inner radius of out cylinder and r_3.

= Shrinkage allowance, *i.e.* the original difference of radii of cylinder at the junction

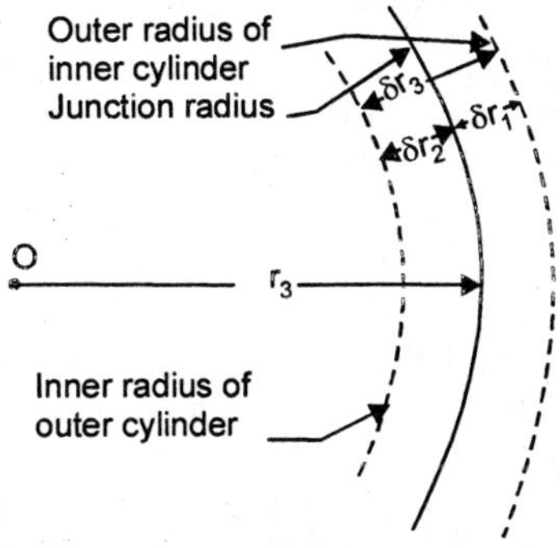

Fig 11.9 Determination of shrinkage allowance.

(*i*) *Cylinder made of same material*

(i.e. $E = E_1 = E_2$, $\nu = \nu_1 = \nu_2$)

For the inner cylinder, the circumferential strain at the common radius r_3 is given by:

$$\frac{\delta r_1}{r_3} = \frac{1}{E}[(\sigma_\theta)_3' - \nu(\sigma_r)_3']$$

where $(\sigma_\theta)_3'$ = hoop stress at the junction due to shrinking

$$= -p_s\left(\frac{r_3^2 + r_1^2}{r_3^2 - r_1^2}\right) \quad \text{from Eq. (11.11)}$$

$$(\sigma_r)_3' = -p_s$$

$$\therefore \quad \frac{\delta r_1}{r_3} = \frac{-p_s}{E}\left[\left(\frac{r_3^2 + r_1^2}{r_3^2 + r_1^2}\right) - \nu\right] \qquad \text{... (11.18)}$$

Similarly for the outer cylinder, the circumferential strain at the common radius r_3 is given by:

$$\frac{\delta r_2}{r_3} = \frac{1}{E}[(\sigma_\theta)_3' - \nu(\sigma_r)_3'] = \frac{p_s}{E}\left[\left(\frac{r_3^2 + r_2^2}{r_2^2 - r_3^2}\right) + \nu\right] \qquad \text{... (11.19)}$$

Taking the algerbraic sum, we get

$$\therefore \quad \frac{\delta r_1 + \delta r_2}{r_3} = \frac{p_s}{E}\left[\left(\frac{r_3^2 + r_1^2}{r_3^2 - r_1^2}\right) + \left(\frac{r_3^2 + r_2^2}{r_2^2 - r_3^2}\right)\right] \qquad \text{... (11.20)}$$

If $\Delta\theta$ = rise in temperature

α = coefficient of linear thermal expansion.

Then $\delta r_3 = r_3 \Delta\theta . \alpha$ ... (11.21

(*ii*) *Cylinder made of different materials. For this case, Eq. (11.18) may be written as,*

$$\frac{\delta r_1}{r_3} = -\frac{p_s}{E_1}\left[\frac{r_3^2+r_1^2}{r_3^2-r_1^2} - \nu_1\right] \quad \text{... (11.22)}$$

and Eq. (11.19) m ay be written as,

$$\frac{\delta r_2}{r_3} = \frac{p_s}{E_2}\left[\frac{r_2^2+r_3^2}{r_2^2-r_3^2} + \nu_2\right] \quad \text{... (11.23)}$$

Hence

$$\frac{\delta r_3}{r_3} = \frac{\delta r_1 + \delta r_2}{r_3} = p_s\left[\frac{1}{E_1}\left\{\frac{r_3^2+r_1^2}{r_3^2-r_1^2} - \nu_1\right\} + \frac{1}{E_2}\left\{\frac{r_2^2+r_3^2}{r_2^2-r_3^2} + \nu_2\right\}\right] \quad \text{... (11.24)}$$

and $\delta r_3 = r_3 . \Delta\theta . \alpha$ ... (11.25)

wher $\Delta\theta$ = rise of temperature

α = coefficient of linear thermal expansion

Example 11.7 (*a*) *A c ompoud cylinder is formed by shrinking one cylinder on the another. The final dimensions are: internal dia.= 15 cm, external dia.= 30 cm and dia. at junction = 25 cm. The shrinkage pressure is 10 MPa. Calculate the shrinkage allowance. What is the minimum temperature to which the outer cylinder must be heated so that it can be slipped down?*

Coefficient of linear expansion, $\alpha = 0.6 \times 10^{-5}$ *per °C.*

(*b*) *If the fluid at pressure 12 MPa is now introduced in the cylinder, what will be the resultant stresses in the cylinder wall?*

$E = 200$ *GPa.*

Solution.

(*a*) Given $r_1 = 7.5$ cm, $r_2 = 15$ cm, $r_3 = 12.5$ cm, $p_s = 10$ MPa

Shrinkage allowance,

$$\delta r_3 = \frac{p_s r_3}{E}\left[\frac{r_3^2+r_1^2}{r_3^2-r_1^2} + \frac{r_3^2+r_3^2}{r_2^2-r_3^2}\right]$$

$$= \frac{10\times12.5}{200\times10^3}\left[\frac{156.25+56.25}{156.25-56.25} + \frac{156.25+225}{225-156.25}\right] = \frac{10\times12.5}{200\times10^3}\left[\frac{212.5}{100} + \frac{381.25}{68.75}\right]$$

$$= \frac{10\times12.5}{200\times10^3}[2.12+5.545] = \frac{10\times12.5\times7.665}{200\times10^3} = 47.9\times10^{-3}\,\text{mm}$$

Let $\Delta\theta$ = rise of temperature

Then $\delta r_3 = r_3 . \Delta\theta . \alpha$

$$\therefore \quad \Delta\theta = \frac{4.79\times10^{-3}}{12.5\times0.6\times10^{-5}} = 63.87°\text{C}$$

(*b*) (*i*) *Shrinkage stresses:*

For the inner cylinder,

$$(\sigma_\theta)_1' = \frac{-2r_3^2 p_s}{(r_3^2 - r_1^2)} = \frac{-2\times(12.5)^2\times 10}{(12.5)^2-(7.5)^2}$$

$$= \frac{-2\times 156.25\times 10}{156.25-56.25} = \frac{-2\times 156.25\times 10}{100} = -31.25 \text{ MPa}$$

$$(\sigma_\theta)_3' = -p_s\left(\frac{r_3^2+r_1^2}{r_3^2 0 - r_1^2}\right)$$

$$= -10\left(\frac{156.25+56.25}{156.25-56.25}\right) = \frac{-10\times 212.5}{100} = -21.25 \text{ MPa}$$

For the outer cylinder,

$$(\sigma_\theta)_3' = p_s\left(\frac{r_3^2+r_2^2}{r_2^2-r_3^2}\right)$$

$$= 10\left(\frac{156.25+225}{225-156.25}\right) = \frac{10\times 381.25}{68.75} = 55.454 \text{ MPa}$$

$$(\sigma_\theta)_2'' = \frac{2p_s r_3^2}{r_2^2-r_3^2} = \frac{2\times 10\times 156.25}{225-156.25} = \frac{2\times 10\times 156.25}{68.75} = 45.454 \text{ MPa}$$

(ii) Stresses due to fluid pressure:

$$(\sigma_\theta)_1 = p_i\left(\frac{r_2^2+r_1^2}{r_2^2+r_1^2}\right) = \left(\frac{225+56.25}{225-56.25}\right) = \frac{12\times 281.25}{168.75} = 20 \text{ MPa}$$

$$(\sigma_\theta)_3 = p_i\frac{r_1^2}{r_3^2}\left(\frac{r_3^2+r_2^2}{r_2^2-r_1^2}\right)$$

$$= \frac{12\times 56.25}{156.25}\left(\frac{156.25+225}{225-56.25}\right) = \frac{12\times 56.25\times 381.25}{156.25\times 168.75} = 9.76 \text{ MPa}$$

$$(\sigma_\theta)_2 = \frac{2p_i r_1^2}{r_2^2-r_1^2} = \frac{2\times 12\times 56.25}{225-56.25} = \frac{2\times 12\times 56.25}{168.75} = 8 \text{ MPa}$$

Radius cm	*Cylinder*	*Shrinkage stresses MPa*	*Stresses due to fluid pressure MPa*	*Resultant stresses MPa*
7.5	inner	–31.25	20	–11.25
12.5	inner	–21.25	9.76	–11.49
12.5	outer	55.454	9.76	65.214
15	outer	45.454	8	53.454

Example 11.8 *A tube 8 cm inside and 12 cm outside diameter is to be reinforced by shrinking on second tube of 16 cm outside diameter. The compound tube is to withstand an internal pressure of 35 MPa and the shrinkage allowance is to be such that the final maximum stress in each tube is to be the same. Calculate this stress and show on a diagram the variation of hoop stress in the two tubes. What is the initial difference of diameters before shrinking on? E = 200 GPa.*

Solution.

Given: $r_1 = 4$ cm; $r_2 = 8$ cm; $r_3 = 6$ cm, $p_i = 35$ MPa

(i) Shrinkage stresses.

Let p_s = shrinkage pressure

For the inner tube:

$$\sigma_r = A - \frac{B}{r^2}$$

At $r = 4$ cm, $\sigma_r = 0$

and at $r = 6$ cm, $\sigma_r = -p_s$

$$\therefore \quad 0 = A - \frac{B}{16}$$

$$-p_s = A - \frac{B}{36}$$

$$\therefore \quad A = \frac{B}{16}$$

$$-p_s = B\left(\frac{1}{16} - \frac{1}{36}\right) = \frac{B \times 20}{16 \times 36}$$

$$\therefore \quad B = \frac{16 \times 36}{20} p_s = -28.8 p_s$$

$$A = 1.8 p_s$$

$$\therefore \quad \sigma_\theta = A + \frac{B}{r^2} = \left(-1.8 - \frac{28.8}{r^2}\right) p_s$$

At $r = 4$ cm, σ_θ is maximum

$$\therefore \quad (\sigma_\theta)_1 = \left(1 - 1.8\frac{28.8}{16}\right) p_s = (1.8 - 1.8) p_s = -3.6 p_s$$

For the outer tube, let

$$\sigma_r = a - \frac{b}{r^2}$$

At $r = 6$ cm, $\sigma_r = -p_s$

and at $r = 8$ cm, $\sigma_r = 0$

$$\therefore \quad -p_s = a - \frac{b}{36}$$

$$\therefore \quad 0 = a - \frac{b}{64}$$

$$\therefore \quad a = \frac{b}{64}$$

$$-p_s = b\left(\frac{1}{64} - \frac{1}{36}\right) = \frac{-b \times 28}{64 \times 36}$$

$$\therefore \quad b = \frac{64 \times 36}{28} p_s 82.286 p_s$$

$$\therefore \quad a = 1.286\, p_s$$

$$\therefore \quad \sigma_\theta = \left(1.286 + \frac{82.286}{r^2}\right) p_s$$

At $r = 6$ cm, σ_θ is maximum

$$\therefore \quad (\sigma_\theta)_3' = \left(1.286 + \frac{82.286}{36}\right) p_s = 3.572\, p_s$$

(ii) Stress due to fluid pressure.

Let $\quad \sigma_r = \alpha - \frac{\beta}{r^2}$

At $\quad r = 4$ cm, $\sigma_r = -35$ MPa

and at $\quad r = 8$ cm, $\sigma_r = 0$

$$\therefore \quad -35 = \alpha - \frac{\beta}{16}$$

$$0 = \alpha - \frac{\beta}{64}$$

$$\therefore \quad \alpha = \frac{\beta}{64}$$

$$-35 = \beta\left(\frac{1}{64} - \frac{1}{16}\right) = \frac{-\beta \times 48}{64 \times 16}$$

$$\therefore \quad \beta = \frac{35 \times 64 \times 16}{48} = 746.667$$

$$\alpha = 11.667$$

$$\sigma_\theta = 11.667 + \frac{746.667}{r^2}$$

At $\quad r = 4$ cm

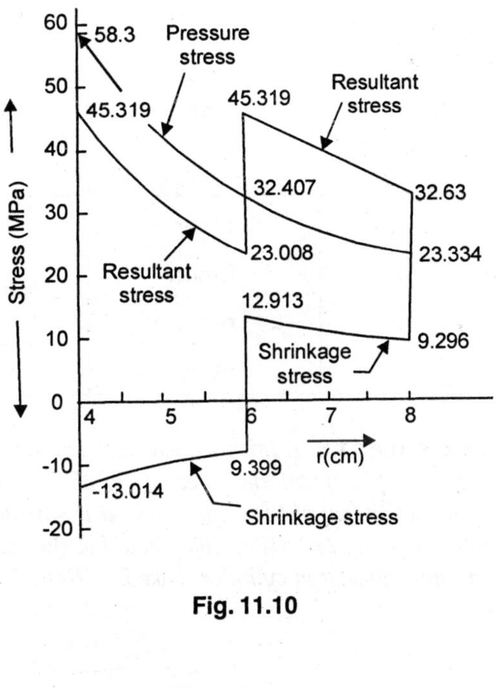

Fig. 11.10

$$(\sigma_\theta)_1 = 11.667 + \frac{746.667}{16} = 11.667 + 46.667 = 58.334 \text{ MPa}$$

Ar $\qquad r = 6$ cm

$$(\sigma_\theta)_3 = 11.667 + \frac{746.667}{36} = 11.667 + 20.741 = 32.408 \text{ MPa}$$

(*iii*) *Resultant stress.* In the inner tube,

$$\sigma_\theta = (\sigma_\theta)_1' + (\sigma_\theta)_1 = -3.6p_s + 58.334$$

In the outer tube, $\sigma_\theta = (\sigma_\theta)_3'' + (\sigma_\theta)_3 = 3.572\,p_s + 32.408$

$$\therefore \quad -3.6\,p_s + 58.334 = 3.5572\,p_s + 32.408$$

$$7.172\,p_s = 25.926$$

$$p_s = 3.615 \text{ MPa}$$

Maximum hoop stress = – 3.6 × 3.615 + 58.324 = – 13.014 + 58.324 = 45.31 MPa

Now
$$\delta r_3 = \frac{p_s . r_3}{E}\left[\frac{r_3^2 + r_1^2}{r_3^2 - r_1^2} + \frac{r_2^2 + r_3^2}{r_2^2 - r_3^2}\right]$$

$$= \frac{3.615 \times 6}{200 \times 10^3}\left[\frac{36+16}{36-16} + \frac{64+36}{64-36}\right] = \frac{3.615 \times 6}{2000 + 10^6}\left[\frac{52}{20} + \frac{100}{28}\right]$$

$$= \frac{3.615 \times 6}{200 \times 10^3} \times 6.17143 = 6.692914 \times 10^{-3} \text{ mm}$$

∴ Initial difference of diameters

$$= 2 \times 6.692914 \times 10^{-3} = 13.385828 \times 10^{-3} \text{ mm}$$

The stresses at various cross-sections have bee shown in the following table and plotted in Fig. 11.10.

Radius cm	*Tube*	*Shrinkage stresses MPa*	*Stresses due to fluid pressure MPa*	*Reultant stresses MPa*
4	inner	–13.014	58.333	45.389
6	inner	–9.399	32.407	23.008
6	outer	12.913	32.407	45.319
8	outer	9.296	23.334	32.630

Example 11.9 *A thick-walled steel cylinder has internal radius of 5 cm and external radius of 8 cm. This is subjected to an internal pressure of 250 MPa. The maximum hoop stress is not to exceed 280 MPa. If a cylindrical jacket of the same material is shrunk on the thick-walled cylinder in order to prevent the stress from exceeding 280 MPa, what should be the initial difference between the internal diameter of jacket and the external diameter of cylinder. Take E = 200 GPa, ν = 0.3.*

Solution:

Given: $r_1 = 5$ cm, $r_3 = 8$ cm $r_2 = ?$ $p_i = 250$ MPa

(*i*) Shrinkage stresses.

Let p_s = shrinkage pressure

For the cylinder, let $\sigma_r = A - \frac{B}{r^2}$

At $r = 5$ cm, $\sigma_r = 0$

and at $r = 8$ cm, $\sigma_r = -p_s$

$$\therefore \quad 0 = A - \frac{B}{25}$$

$$\therefore \quad -p_s = A - \frac{B}{64}$$

$$\therefore \quad A = \frac{B}{25}$$

$$-p_s = B\left(\frac{1}{25} - \frac{1}{64}\right)$$

$$-p_s = \frac{B \times 39}{25 \times 64}$$

$$\therefore \quad B = -\frac{25 \times 64}{39} p_s = -41.02\, p_s$$

$$A = -1.64\, p_s$$

$$\sigma_\theta = -\left(1.64 + \frac{41.02}{r^2}\right) p_s$$

At $r = 5$ cm

$$(\sigma_\theta)_1' = -\left(1.64 + \frac{41.02}{25}\right) p_s = -3.281\, p_s$$

(*ii*) Stresses due to fluid pressure.

Let $\sigma_r = \alpha - \frac{\beta}{r^2}$

At $r = 5$ cm, $\sigma_r = -\ 250$ MPa

and at $r = 8$ cm, $\sigma_r = 0$

$$\therefore \quad -250 = \alpha - \frac{\beta}{25}$$

$$0 = \alpha - \frac{\beta}{64}$$

$$\therefore \qquad \alpha = \frac{\beta}{64}$$

$$-250 = \beta\left(\frac{1}{64}-\frac{1}{25}\right) = -\frac{39\beta}{64\times 25}$$

$$\therefore \qquad \beta = \frac{250\times 64\times 25}{39} = 10256.41$$

$$\alpha = 160.256$$

$$\therefore \qquad \sigma_\theta = 160.256+\frac{10256.41}{r^2}$$

At $\qquad r = 5$ cm

$$(\sigma_\theta)_1 = 160.256+\frac{10256.41}{25} = 160.256+410 = 570.512 \text{ MPa}$$

Resultant stress $\qquad = 570.512 - 3.281\, p_s = 280$

$$p_s = \frac{290.512}{3.281} = 88.544 \text{ MPa}$$

Initial difference of diameter = $d_3 \times \varepsilon_\theta$

$$= 16\times\frac{1}{E}[\sigma_\theta - \nu\sigma_r]_{r=8\text{cm}}$$

$$= \frac{16}{200\times 10^3}\left[-\left(1.64+\frac{41.02}{64}\right)88.544 - 0.3(-88.544)\right]$$

$$= \frac{16\times 88.544}{200\times 10^3}[2.28+0.3] = \frac{-16\times 88.544\times 1.98}{200\times 10^3} = -0.140 \text{ mm}$$

11.4 PRESS AND SHRINK-FITTED ASSEMBLIES

When a hollow thick cylinder is pressed or force-fitted on to another soild cylinder, let

E_h = Modulus of elasti-city for the hub

E_s = Modulus of elastic-city for the shaft

ν_s = Poisson's ratio for the hub

ν_s = Poisson's ratio for the shaft.

Let r_1 be the radius of the shaft and r_2 the outer radius of the hub. Let u be the radial outward displacement of a point at radius r from the centre of the solid shaft as shown in Fig. 11.11 (actually u is negative here).

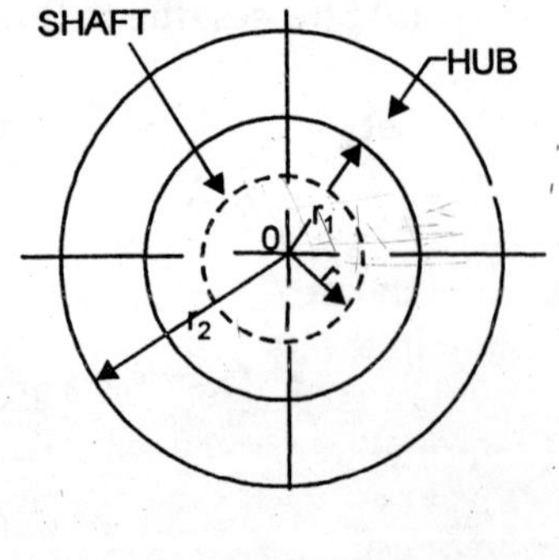

Fig. 11.11 Press-fitted shaft

Circumferential strain at radius r is,

$$\varepsilon_\theta = \frac{2\pi(r+u)-2\pi r}{2\pi r} = \frac{u}{r}$$

Hoop strain is also given by,

$$\varepsilon_\theta = \frac{1}{E_s}[\sigma_\theta - \nu_s\sigma_r] = \frac{u}{r}$$

$$\frac{u}{r} = \frac{1}{E_s}\left[\left(A+\frac{B}{r^2}\right)-\nu_s\left(A-\frac{B}{r^2}\right)\right] = \frac{1}{E_s}\left[A(1-\nu_s)+\frac{B}{r^2}(1+\nu_s)\right]$$

$$ur = \frac{1}{E_s}[Ar^2(1-\nu_s)+B(1+\nu_s)]$$

In a solid shaft at $r = 0, u = 0.$

$$\therefore \quad 0 = \frac{1}{E_s}[B(1+\nu_s)]$$

$$\therefore \quad B = 0$$

$$\therefore \quad \sigma_r = A \text{ and } \sigma_\theta = A$$

$$\therefore \quad \sigma_r = \sigma_\theta = \text{constant} = -p \qquad \text{... (11.26)}$$

where p = pressure between the shaft and the hub.

Therefore the radial and hoop stresses throughout a solid shaft are everywhere equal to the shrinkage (or interference) pressure and both are compressive in nature.

11.4.1 Force Fit or Tight Allowance

Let $(\sigma_\theta)_h$ = Circumferential stress in the hub at $r = r_1$.

δ = Excess of the original shaft diameter over that of the hub bore, called the tight allowance.

At $r = r_1$, the total circumferential strain

ε = tensile strain of hub + compressive strain of shaft.

$$\varepsilon = (\varepsilon_\theta)_h - (\varepsilon_\theta)_s$$

$$\frac{\delta}{2r_1} = \frac{1}{E_h}[(\sigma_\theta)_h = -\nu_h(\sigma_r)_h] - \frac{1}{E_s}[(\sigma_\theta)_s - \nu_s(\sigma_r)_s]$$

Now $(\sigma_\theta)_h = -p = (\sigma_r)_s$

$$\therefore \quad \frac{\delta}{2r_1} = \frac{1}{E_h}[(\sigma_\theta)_h + \nu_h . p] + \frac{1}{E_s}[1-\nu_s]p$$

$$\delta = 2r_1\left\{\frac{1}{E_h}[(\sigma_\theta)_h + \nu_h\, p] + \frac{p}{E_s}[1-\nu_s]\right\} \qquad \text{... (11.27)}$$

11.4.2 Axial Force Required

In the case of force-fitted cylinders the outer cylinder has to be pressed over the inner cylinder with a larg axial force. Let the interference pressure set up at the common surface be p then the normal force R between the mating cylinders is,

$$R = p \times 2\pi rl$$

where l is the axial length of the contact surfaces.

The force of friction between the cylinders which has to be overcome by the applied axial force is thus,

$$P = \mu R = \mu \times p \times 2\pi r l = 2\pi r l \mu p \qquad \text{... (11.28)}$$

where μ is coefficient of friction.

Example 11.10. *A bronze bush of 2 cm wall thickness is to be shrunk on to a steel shaft of 10 cm diameter. If the shrinkage pressure is to be 70 MPa, determine the tight allowance between the bush and shaft.*

For steel $E = 200$ GPa, $\nu = 0.28$

For bronze $E = 100$ GPa, $\nu = 0.32$.

Solution

Given $r_1 = 5$ cm, $r_2 = 7$ cm, $p = 70$ MPa

For the bush, let $\sigma_r = A - \dfrac{B}{r^2}$

At $r = 5$ cm, $\sigma_r = -p$

and at $r = 7$ cm, $\sigma_r = 0$

$$\therefore \quad -p = A - \frac{B}{25}$$

$$0 = A - \frac{B}{49}$$

$$A = 49$$

$$-p = B\left(\frac{1}{49} - \frac{1}{25}\right) = \frac{-B \times 24}{49 \times 25}$$

$$\therefore \quad B = \frac{49 \times 25}{24} p = 51.04p$$

$$A = 1.04\,p$$

$$\therefore \quad \sigma_r = \left(1.04 - \frac{51.04}{r^2}\right)p$$

$$\sigma_\theta = \left(1.04 + \frac{51.04}{r^2}\right)p$$

For the shaft, let $\sigma_r = a - \dfrac{b}{r^2}$

At $r = 0$, σ_r becomes infinite. Hence $b = 0$

$$\therefore \quad \sigma_r = a = \sigma_\theta = -p$$

Tensile strain in bush,

$$(\varepsilon_\theta)_b = \frac{1}{E_b}[(\sigma_\theta)_b - \nu_b(\sigma_r)_b]_{r=5\text{cm}}$$

Compressive strain in shaft,

$$(\varepsilon_\theta)_s = \frac{1}{E_s}[(\sigma_\theta)_s - \upsilon_s(\sigma_r)_s]$$

Total strain $= (\varepsilon_\theta)_b - (\varepsilon_\theta)_s$

$$= \frac{1}{10^5}\left[\left(1.04 + \frac{51.04}{25}\right)70 + 0.32 \times 70\right] - \frac{1}{2 \times 10^5}[-70 + 0.28 \times 70]$$

$$= \frac{1}{10^5}[72.8 + 143.92 + 22.4] - \frac{1}{2 \times 10^5}[-70 + 19.6]$$

$$= \frac{1}{10^5}(239.12 + 25.2) = \frac{264.32}{10^5} = 264.32 \times 10^{-5}$$

$$\therefore \quad \frac{\delta}{2r_1} = 0.0026432$$

$$\delta = 10 \times 0.0026432 = 0.026432 \text{ cm}$$

Example 11.11 *A steel plug 10 cm diameter is forced into a steel ring 15 cm external diameter and 6 cm wide. From a reading taken by fixing in a circumferntial directin an electric resistance strain-gauge on the external surface of the ring, the strain is found to be 1.55 × 10^{-4} cm/cm.*

Assuming coefficient of friction for the matching surface to be 0.25 find the force required to push the plug out of the ring. Also estimate the greatest hoop stress in the ring. E = 100 GPa.

Solution:

Given: $r_1 = 5$ cm, $r_2 = 7.5$ cm

For the steel ring, let $\sigma_r = A - \frac{B}{r^2}$

At $r = 5$ cm, $\sigma_r = -p$

At $r = 7.5$ cm, $\sigma_r = 0$

$$\therefore \quad -p = A - \frac{B}{25}$$

$$0 = A - \frac{B}{56.25}$$

$$\therefore \quad A = \frac{B}{56.25}$$

$$-p = B\left[\frac{1}{56.25} - \frac{1}{25}\right] = \frac{-31.25B}{56.25 \times 25}$$

$$\therefore \quad B = 45\,p$$

$$A = 0.8\,p$$

$$\therefore \quad \sigma_r = \left(0.8 - \frac{45}{r^2}\right)p$$

$$\sigma_\theta = \left(0.8 + \frac{45}{r^2}\right)p$$

At $\quad r = 7.5$ cm

$$\sigma_r = \left(0.8 + \frac{45}{56.25}\right)p = 1.6p$$

Hoop strain at the external surface of the ring,

$$\frac{\sigma_\theta}{E} = \frac{1.6p}{2\times10^5} = 1.55\times10^{-4}$$

$$\therefore \quad p = \frac{2\times10\times1.55}{1.6} = 19.375 \text{ MPa}$$

Area of friction surface = $\pi \times 10 \times 6 = 188.495$ cm^2

Total radial force $= 19.375 \times 106 \times 188.495 \times 10^{-4} = 364.7284$ kN

Tangential force $= 0.25 \times 364.7284 = 91.1821$ kN

Maximum hoop stress at $r = 5$ cm

$$= \left(0.8 + \frac{45}{25}\right)19.375 = 2.6\times19.375 = 50.375 \text{ MPa}$$

Example 11.12 *A thick cylindrical sleeve of inside diameter 20 cm and outside diameter 30 cm is shrunk on to a solid shaft. Due to force-fitting the inside diameter of the sleeve increases by 0.015 cm. Determine the force fit pressure between the sleeve and the shaft. E = 200 GPa, ν = 0.028.*

Solution.

Given : For the sleeve, let

$$\sigma_r = A - \frac{B}{r^2}$$

At $\quad r = 10$ cm, $\sigma_r = -p$

At $\quad r = 15$ cm, $\sigma_r = 0$

$$\therefore \quad -p = A - \frac{B}{100}$$

$$0 = A - \frac{B}{225}$$

$$\therefore \quad A = \frac{B}{225}$$

$$-p = B\left(\frac{1}{225}-\frac{1}{100}\right) = \frac{-125B}{225\times100}$$

$\therefore$ $$B = 180\,p$$

$$A = 0.8\,p$$

Stresses in the sleeve at $r = 10$ cm:

$$\sigma_r = \left(0.8-\frac{180}{100}\right)p = -p$$

$$\sigma_\theta = \left(0.8+\frac{180}{100}\right)p = 2.6p$$

Strain in the sleeve at $r = 10$ cm,

$$\varepsilon_\theta = \frac{1}{E}[\sigma_\theta - \nu\sigma_r] = \frac{1}{2\times10^5}[2.6p+0.28\,p]$$

$$= \frac{2.88\,p}{2\times10^5} = 1.44\times10^{-5}\,p = \frac{0.015}{20}$$

$\therefore$ $$p = \frac{0.015\times10^5}{20\times1.144} = 52.083 \text{ MPa}$$

Example 11.13 *A steel shaft 3 cm diameter is to be encased in bronze sleeve, 4.5 cm outside diameter, which is to be forced into position and before forcing on, the inside diameter of the sleeve is 0.004 cm smaller than the diameter of the shaft. Find (a) the radial pressure between the shaft and sleeve, (b) the maximum hoop stress in the sleeve, and (c) the change in outside diameter of the sleeve. E for steel = 200 GPa. E for bronze = 120 GPa. Poisson's ratio for steel = 0.29, and Poisson's ratio for bronze = 0.34.*

Solution.

Given : Let p = force fit pressure, MPa,

(*a*) For the sleeve, let $\sigma_r = A-\dfrac{B}{r^2}$

At $r = 1.5$ cm, $\sigma_r = -p$

At $r = 2.25$ cm, $\sigma_r = 0$

$\therefore$ $$-p = A-\frac{B}{2.25}$$

$$0 = A-\frac{B}{5.0625}$$

$\therefore$ $$A = \frac{B}{5.0625}$$

$\therefore$ $$-p = B\left[\frac{1}{5.0625}-\frac{1}{2.25}\right]-\frac{2.8125.B}{5.0625\times2.25}$$

$\therefore$ $$B = 4.05\,p$$

$$A = 0.8\,p$$

$$\therefore \quad \sigma_r = \left(0.8 - \frac{4.05}{r^2}\right)p$$

$$\sigma_\theta = \left(0.8 + \frac{4.05}{r^2}\right)p$$

Hoop strain in the sleeve at $r = 1.5$ cm is,

$$(\varepsilon_\theta)_b = \frac{1}{E_b}\left[\left(0.8 + \frac{4.05}{r^2}\right)p - \nu_b\left(0.8 - \frac{4.05}{r^2}\right)p\right]_{r=1.5\,\text{cm}}$$

$$= \frac{1}{1.2\times10^5}[2.6p + 0.34p] = \frac{2.94p}{1.2\times10^5}\ \text{(tensile)}$$

For the shaft, $\sigma_r = \sigma_\theta = -p$

∴ Hoop strain in the shaft,

$$(\varepsilon_\theta)_s = \frac{1}{E_s}[\sigma_\theta - \nu_s\sigma_r] = \frac{1}{2\times10^5}[-p + 0.29p] = \frac{-0.71p}{2\times10^5}\ \text{(compressive)}$$

Total strain $= (\varepsilon_\theta)_b - (\varepsilon_\theta)_s$

$$= \frac{2.94p}{1.2\times10^5} + \frac{0.71p}{2\times10^5} = 10^{-5}(2.45 + 0.355)$$

$$= 2.805\times10^{-5}\,p = \frac{0.004}{3}$$

$$\therefore \quad p = \frac{0.0004\times10^6}{3\times2.805} = 47.534\,\text{MPa}$$

(b) Maximum hoop stress in the sleeve occurs at r = 1.5 cm.

$$\sigma_\theta = \left(0.8 + \frac{4.05}{2.25}\right)\times47.534 = 2.6\times47.534 = 123.585\ \text{MPa}$$

(c) Change in outer diameter of sleeve. Hoop strain in the sleeve ar r = 2.25 cm.

$$(\varepsilon_\theta)_b = \frac{1}{E_b}\left[\left(0.8 + \frac{4.05}{5.0625}\right)p - 0.34\left(0.8 - \frac{4.05}{5.0625}\right)p\right]$$

$$= \frac{1}{1.2\times10^5}[1.6p - 0] = \frac{1.6\times47.534}{1.2\times10^5} = 633.787\times10^{-6}$$

∴ Change in diameter $= 4.5 \times 633.787 \times 10^{-6} = 28.52 \times 10^{-3}$ m

Example 11.14 *A short steel rod 4 cm diameter is subjected to an axial compressive load of 200 kN. It is surrounded by a steel sleeve 1 cm thick, slightly shorter than the rod, so that the compressive load is taken by*

the rod only. Assuming a close fit between the sleeve and the rod before the load is applied, and neglecting friction, find: (a) the pressure between the sleeve and the rod, and (b) the maxium tension in the sleeve. Take Poisson's ratio = 0.3.

Solution.

Given : Consider the rod and sleeve assembly as shown in Fig. 11.12.

$$\sigma_z = \frac{200\times10^3}{\frac{\pi}{4}\times16\times10^{-4}} = 159.155 \text{ MPa}$$

(*a*) Longitudinal stress in the rod = 159.155 MPa. (compressive)

Let p = Pressure between the sleeve and rod

Then in the rod $\sigma_r = \sigma_\theta = -p$

∴ Hoop strain in the steel rod at r = 2 cm,

$$\varepsilon_\theta = \frac{1}{E}[\sigma_\theta - \nu(\sigma_r+\sigma_z)] = \frac{1}{E}[-p-0.3(-p-159.155)]$$

$$= \frac{1}{E}[-p+0.3p+47.7465] = \frac{1}{E}[-0.7p+47.7465]$$

For the sleeve, let $\sigma_r = A - \frac{B}{r^2}$

At r = 2 cm, $\sigma_r = -p$

At r = 3 cm, $\sigma_r = 0$

∴ $$-p = A - \frac{B}{4}$$

$$0 = A - \frac{B}{9}$$

∴ $$A = \frac{B}{9}$$

$$-p = B\left[\frac{1}{9}-\frac{1}{4}\right] = \frac{-5B}{36}$$

∴ $$B = 7.2\,p$$

$$A = 0.8\,p$$

$$\sigma_r = \left(0.8-\frac{7.2}{r^2}\right)p$$

$$\sigma_r = \left(0.8+\frac{7.2}{r^2}\right)p$$

At r = 2 cm

$$\sigma_r = -p$$

$$\sigma_\theta = 2.6\,p$$

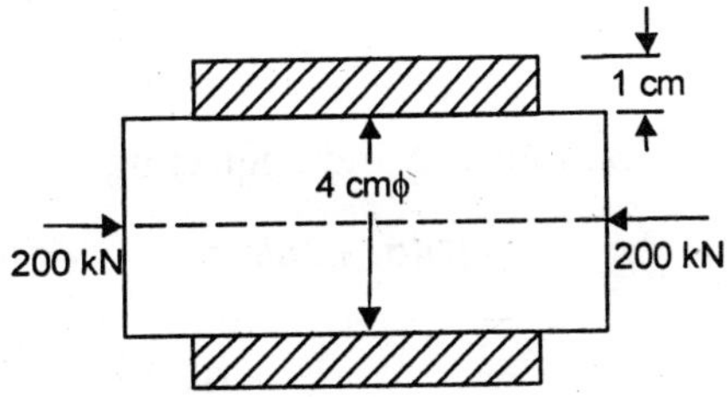

Fig. 11.12 Press-fitted shaft

∴ Hoop strain in the sleeve at $r = 2$ cm is,

$$\varepsilon_\theta = \frac{1}{E}[\sigma_\theta - \nu\sigma_r] = \frac{1}{E}[2.6p + 0.3p] = \frac{2.9p}{E}$$

Since hoop strain in the rod and sleeve will be equal at the common surface,

$$\therefore \quad \frac{1}{E}[-0.7p + 47.7465] = \frac{2.9p}{E}$$

$$\therefore \quad 3.6p = 47.7465$$

$$p = 13.263 \text{ MPa}$$

(b) *Maximum tension in sleeve,*

$$\sigma_\theta = 2.6p = 2.6 \times 13.263 = 34.484 \text{ MPa}$$

11.5 Thick Spherical Shell

Consider a thick spherical shell of internal radius r_1 and external radius r_2 subjected to internal pressure p_1 and external pressure p_2 as shown in Fig. 11.13. Let σ_r be the radial stress and the circumferential stress σ_θ is equal in the other two principal directions perpendicular to r. Consider an elementary ring of the spherical shell between radii r and $r + dr$. The forces acting on the elementary ring are:

Upward force due to radial stresses

$$= \pi(r+dr)^2(\sigma_r + d\sigma_r) - \pi r^2\sigma_r$$

Downward force due to circumferential stress

$$= 2\pi r dr.\sigma_\theta$$

For equilibrium of the elementary ring

$$= \pi(r+dr)^2(\sigma_r + d\sigma_r) - \pi r^2\sigma_r = 2\pi r.dr.\sigma_\theta$$

Simplifying and neglecting small quantities, we get

$$r^2 d\sigma_r + 2rdr.\sigma_r = 2r.dr.\sigma_\theta$$

$$\sigma_\theta = \sigma_r + \frac{r.d\sigma_r}{2dr} \quad ...(a)$$

Differentiating with respect to r, we get

$$\frac{d\sigma_\theta}{dr} = \frac{d\sigma_r}{dr} + \frac{r}{2}\frac{d^2\sigma_r}{dr^2} + \frac{1}{2}\frac{d\sigma_r}{dr} = \frac{3}{2}\frac{d\sigma_r}{dr} + \frac{r}{2}\frac{d^2\sigma_r}{dr^2} \quad ...(b)$$

Let increase in radius r be u, then hoop strain,

$$\varepsilon_\theta = \frac{2\pi(r+u) - 2\pi r}{2\pi r} = \frac{u}{r} \quad ...(c)$$

Radial strain, $$\varepsilon_r = \frac{\left(u+\frac{\partial u}{\partial r}dr\right)-u}{dr} = \frac{\partial u}{\partial r} \qquad ...(d)$$

Also $$\varepsilon_\theta = \frac{1}{E}\left[\sigma_\theta - \nu\left(\sigma_r + \sigma_\theta\right)\right] \qquad ...(e)$$

From Eqs. (*c*) and (*e*), we get

$$\frac{u}{r} = \frac{1}{E}\left[\sigma_\theta - \nu\left(\sigma_r + \sigma_\theta\right)\right]$$

$$u = \frac{r}{E}\left[\sigma_\theta - \nu\left(\sigma_r + \sigma_\theta\right)\right] \qquad ...(f)$$

Differentiating with respect to *r*, we get

$$\frac{du}{dr} = \frac{r}{E}\left[\frac{d\sigma_\theta}{dr} - \nu\left(\frac{d\sigma_r}{dr} + \frac{d\sigma_\theta}{dr}\right)\right] + \frac{1}{E}\left[\sigma_\theta - \nu\left(\sigma_r + \sigma_\theta\right)\right]$$

$$= \frac{r}{E}\left[(1-\nu)\frac{d\sigma_\theta}{dr} - \nu\frac{d\sigma_r}{dr}\right] + \frac{1}{E}\left[(1-\nu)\sigma_\theta - \nu\sigma_r\right] \qquad ...(g)$$

Similarly $$\varepsilon_r = \frac{1}{E}\left[\sigma_r - \nu\left(\sigma_\theta + \sigma_\theta\right)\right] = \frac{1}{E}\left[\sigma_r - 2\nu\sigma_\theta\right] = \frac{du}{dr} \qquad ...(h)$$

From Eqs. (*g*) and (*h*), we get

$$\frac{1}{E}\left[\sigma_r - 2\nu\sigma_\theta\right] = \frac{r}{E}\left[(1-\nu)\frac{d\sigma_\theta}{dr} - \nu\frac{d\sigma_r}{dr}\right] + \frac{1}{E}\left[(1-\nu)\sigma_\theta - \nu\sigma_r\right]$$

Substituting the values of σ_θ and $\frac{d\sigma_\theta}{dr}$ from Eqs. (*a*) and (*b*), we get

$$\frac{1}{E}\left[\sigma_r - 2\nu\left(\sigma_r + \frac{rd\sigma_r}{2dr}\right)\right] = \frac{r}{E}\left[(1-\nu)\left\{\frac{3}{2}\frac{d\sigma_r}{dr} + \frac{r}{2}\frac{d^2\sigma_r}{dr^2}\right\} - \nu\frac{d\sigma_r}{dr} + \frac{1}{E}(1-\nu)\left\{\sigma_r + r\frac{d\sigma_r}{2dr}\right\} - \nu\sigma_r\right]$$

$$\sigma_r - 2\nu\left(\sigma_r + \frac{rd\sigma_r}{2dr}\right) = (1-\nu)\left\{\frac{3}{2}\cdot\frac{rd\alpha_r}{dr} + \frac{r^2}{2}\frac{d^2\sigma_r}{dr^2}\right\}$$

$$= -r\nu\frac{d\sigma_r}{dr} + (1-\nu)\left\{\sigma_r + \frac{r}{2}\frac{d\sigma_r}{dr}\right\} - \nu\sigma_r$$

Simplifying, we get

$$4\frac{d\sigma_r}{dr} + r\frac{d^2\sigma_\rho}{dr^2} = 0 \qquad ...(i)$$

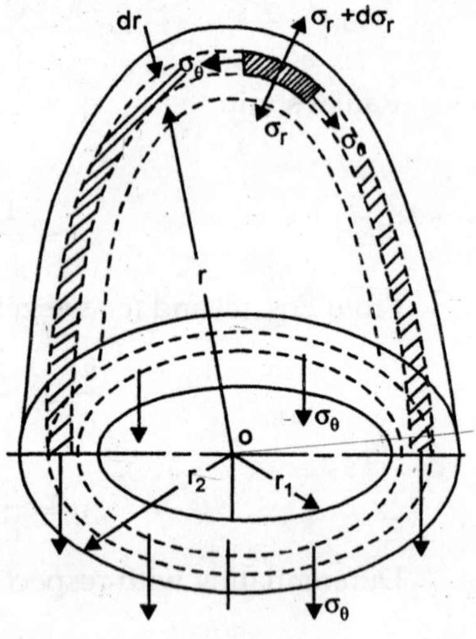

Fig. 11.13

Put $$x = \frac{d\sigma_r}{dr}$$

$$\therefore \quad \frac{dx}{x} = -4\frac{dr}{r}$$

Integrating, we get

$$ln\, x = -4\, ln\, r + ln\,(6B)$$

where $ln\,(6B) = a$ constant.

$$ln\,(xr^4) = ln\,(6B)$$

$$xr^4 = 6B$$

$$r^4 \frac{d\sigma_r}{dr} = 6B$$

$$d\sigma_r = \frac{6B}{r^4}dr$$

Integrating again, we get

$$\sigma_r = \frac{2B}{r^2} + A$$

or $$\sigma_r = A - \frac{2B}{r^3} \quad ...(j)$$

where A is another constant.

Substituting in Eq.(*a*), we get

$$\sigma_\theta = A - \frac{2B}{r^3} + \frac{r}{2}(-2B)\left(-3r^{-4}\right) = A - \frac{2B}{r^3} + \frac{3B}{r^3} = A + \frac{B}{r^3} \quad ...(k)$$

Hence the governing equations for a thick shell are

$$\sigma_r = A - \frac{2B}{r^3}$$

$$\sigma_\theta = A + \frac{B}{r^3}$$

The boundary conditions are

At $r = r_1, \sigma_r = -p_1$

$r = r_2, \sigma_r = -p_2$

$$-p_1 = A - \frac{2B}{r_1^3}$$

$$-p_2 = A - \frac{2B}{r_2^3}$$

$$p_1 - p_2 = 2B\left(\frac{1}{r_1^3} - \frac{1}{r_2^3}\right) = \left(\frac{r_2^3 - r_1^3}{r_1^3 r_2^3}\right) \cdot 2B$$

$$B = \frac{r_1^2 r_2^3 (p_1 - p_2)}{2\left(r_2^3 - r_1^3\right)}$$

$$A = \frac{2B}{r_1^3} - p_1 = \frac{r_2^3 (p_1 - p_2)}{\left(r_2^3 - r_1^3\right)} - p_1 = \frac{p_1 r_1^3 - p_2 r_2^3}{\left(r_2^3 - r_1^3\right)}$$

$$\therefore \qquad \sigma_r = \frac{p_1 r_1^3 - p_2 r_2^3}{r_2^3 - r_1^3} - \frac{r_1^3 r_2^3 (p_1 - p_2)}{r^3\left(r_2^3 - r_1^3\right)}$$

$$\sigma_\theta = \frac{p_1 r_1^3 - p_2 r_2^3}{r_2^3 - r_1^3} + \frac{r_1^3 r_2^3 (p_1 - p_2)}{2r^3\left(r_2^3 - r_1^3\right)}$$

(i) Interval pressure only, i.e. $p_2 = 0$.

$$\sigma_r = \frac{p_1 r_1^3}{\left(r_2^3 - r_1^3\right)} - \frac{p_1 r_1^3 r_2^3}{r^3\left(r_2^3 - r_1^3\right)} = \frac{p_1 r_1^3}{\left(r_2^3 - r_1^3\right)}\left[1 - \frac{r_2^3}{r^3}\right]$$

$$\sigma_\theta = \frac{p_1 r_1^3}{\left(r_2^3 - r_1^3\right)}\left[1 + \frac{r_2^3}{2r^3}\right]$$

(ii) External pressure only, i.e. $p_1 = 0$.

$$\sigma_r = \frac{p_1 r_2^3}{\left(r_2^3 - r_1^3\right)}\left[1 - \frac{r_1^3}{r^3}\right] \qquad \text{...(11.34)}$$

$$\sigma_\theta = \frac{p_2 r_2^3}{\left(r_2^3 - r_1^3\right)}\left[1 + \frac{r_1^3}{2r^3}\right] \qquad \text{...(11.35)}$$

(iii) Maximum shear stress,

$$\tau_{max}\big|_{r-r1} = \frac{1}{2}(\sigma_\theta - \sigma_r) = \frac{3}{4}\frac{r_2^3 (p_1 - p_2)}{\left(r_2^3 - r_1^3\right)}$$

Example 11.15 *A thick spherical shell 25 cm internal diameter and 5 cm thick is subjected to an integral pressure of 12 MPa. Determine the variation of radial and hoop stresses in the shell. Also find the increase in internal and external diameters. E = 200 GPa, v = 0.30.*

Solution.

Given : $\sigma_r = A - \frac{2B}{r^3}$

At $r = 12.5$ cm, $\sigma_r = -12$ MPa

$r = 17.5$ cm, $\sigma_r = 0$

$$-12 = A - \frac{2B}{1953 \cdot 125}$$

$$0 = A - \frac{2B}{5359 \cdot 375}$$

$$A = \delta d_2$$

$$-12 = \left[\frac{1}{2679 \cdot 6875} - \frac{1}{976 \cdot 5625}\right] B$$

$$-12 = \frac{-1703 \cdot 125B}{2679 \cdot 6875 \times 976 \cdot 5625}$$

$$B = 18438 \cdot 216 : A = 6.88$$

$$\sigma_r = 6.88 - \frac{36876 \cdot 432}{r^3}$$

$$\sigma_\theta = 6.88 + \frac{18438 \cdot 216}{r^3}$$

r, cm	12.5	13	14	15	16	17.5
σ_r, MPa	–12	–9	–6.559	–4.406	–2.123	0
s_q, MPa	16.32	15.272	13.599	12.343	11.381	10.320

The variation of stress is shown in Fig. 11.14. Hoop strain at $r = 12.5$ cm is,

$$e_q = \frac{1}{E}\left[\sigma_\theta - \nu(\sigma_r + \sigma_\theta)\right]$$

$$= \frac{1}{2 \times 10^5}\left[16.32 - 0.3(-12 - 16.32)\right]$$

$$= \frac{1}{2 \times 10^5}\left[16.32 - 0.3 \times 4.32\right]$$

$$= \frac{1}{2 \times 10^5}\left[16.42 - 1.296\right]$$

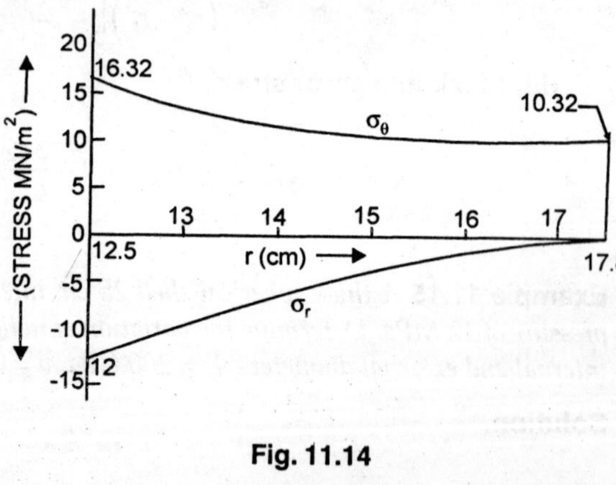

Fig. 11.14

$$= 75.12 \times 10^{-6} = \frac{\delta d_1}{d_1}$$

$\therefore$ $\delta d_1 = 25 \times 75.12 \times 10^{-6}$

$= 18.75 \times 10^{-3}$ mm

Hoop strain at

$r = 17.5$ cm is,

$$\varepsilon_\theta = \frac{1}{2\times10^5}\left[\sigma_\theta - \nu(\sigma_r + \sigma_\theta)\right] = \frac{1}{2\times10^5}\left[10.32 - 0.3(0+10.32)\right]$$

$$= \frac{1}{2\times10^5}[10.32 - 3.096] = 36.12\times10^{-6} = \frac{\delta d_2}{d_2}$$

$$\delta d_2 = 35\times36.12\times10^{-6} = 12.642 \times 10^{-3} \text{ mm}$$

SUMMARY

1. Lame's equations for thick cylinders.

$$\sigma_r = A - \frac{B}{r^2}$$

$$\sigma_\theta = A + \frac{B}{r^2}$$

2. Cylinder subjected to internal pressure only.

$$\sigma_r = \left(\frac{p_1}{k^2-1}\right)\left[1-\left(\frac{r_2}{r}\right)^2\right],\ k = \frac{r_2}{r_1}$$

$$\sigma_\theta = \left(\frac{p_1}{k^2-1}\right)\left[1+\left(\frac{r_2}{r}\right)^2\right]$$

3. Cylinder subjected to external pressure only.

$$\sigma_r = \frac{p_2}{\left(1-\frac{1}{k^2}\right)}\left[\left(\frac{r_1}{r}\right)^2 - 1\right]$$

$$\sigma_\theta = \frac{-p_2}{\left(1-\frac{1}{k^2}\right)}\left[\left(\frac{r_1}{r}\right)^2 + 1\right]$$

4. Solid circular shaft having external pressure only.

$$\sigma_r = \sigma_\theta = -p_2 = \text{const.}$$

5. **Longitudinal stress**

$$\sigma_z = \frac{p_1 r_1^2 - p_2 r_2^2}{(r_2^2 - r_1^2)} = \text{const.} = A$$

6. **Maximum shear stress,**

$$\tau_{max} = \frac{(p_1 - p_2)}{\left(1 - \frac{1}{k^2}\right)}$$

7. **Radial deflection**

$$\delta r = \frac{r}{E}(\sigma_\theta - \nu\sigma_r)$$

For $p_2 = 0$ ar $r = r_1$,

$$\delta r_1 = \frac{p_1 r_1}{E}\left[\frac{k^2+1}{k^2-1} + \nu\right]$$

For $p_1 = 0$ at $r = r_2$

$$\delta r_2 = \frac{-p_2 r_2}{E}\left[\frac{k^2+1}{k^2-1} - \nu\right]$$

8. **Shrinkage allowance.**

$$\frac{\delta r_3}{r_3} = p_s\left[\frac{1}{E_1}\left\{\frac{k_1^2+1}{k_1^2-1}\right\} - \nu_1\right\} + \frac{1}{E_2}\left\{\frac{k_2^2+1}{k_2^2-1} + \nu_2\right\}\right]$$

$$= \Delta\theta \,.\, \alpha$$

where $$k_1 = \frac{r_3}{r_1}, k_2 = \frac{r_2}{r_3}$$

9. **Press and shrink fitted assemblies.**
In a solid shaft force –fitted in a hub,

$$\sigma_r = \sigma_\theta = -p = \text{const.}$$

Force fit or tight allowance,

$$\delta = 2r_1\left[\frac{1}{E_h}\{(\sigma_\theta)_h + \nu_h p\} + \frac{p}{E_s}(1 - \nu_s)\right]$$

Axial force required, $P = 2\pi r l \mu p$

10. Thick spherical shell

$$\sigma_r = A - \frac{2B}{r^3}$$

$$\sigma_\theta = A + \frac{B}{r^3}$$

Short Answer Type Questions

1. When a pressure vessel is called thick?
2. What are the assumptions made in the Lame's theory for thick cylinders under fluid pressure?
3. Write Lame's equations for a thick cylinder.
4. What is the necessity for a thick cylinder.
5. What is the nature of stresses in a solid shaft under external pressure?
6. Give some examples of press-fitted assemblies

EXERCISES

1. A hydraulic pipe is 20 cm internal diameter and 5 cm thick. It is required to sustain an internal pressure of 8 MPa. Determine the intensities of radial and hoop stresses at various radii and plot their variation. **[Ans.** -8 MPa to zero and 20-8 to 12.8 MPa]
2. A field gun cylinder barrel is 120 mm internal diameter and 190 mm external diameter. The tensile and compressive proportional limit of the material is 300 MPa. Find.
   (i) the maximum internal pressure which the barrel may be subjected to assuming the external pressure to be negligible.
   (ii) the maximum external pressure which the barrel may be subjected to (by shrinking hoops), assuming then that the internal pressure is negligible. **[Ans.** 9 MPa, 9 MPa]
3. A compound cylinder is made by shrinking an outer tube on to an inner tube, the final dimensions being ; externa! diameter 25 cm, internal diameter 15 cm and diameter at the junction of tubes, 20 cm.

   If the common radial pressure between the two tubes due to shrinking be 17 MPa, determine the circumferential stresses at inner and outer diameters of both cylinders and show by means of a diagram how these stresses vary with the radius.**[Ans.** -77.71, –60.71, 77.44, 60.44 MPa]
4. A steel shaft, originally 15 cm diameter is subjected to an uniform radial compressive stress of 20 MPa. Assuming that the radial stress remains constant, find the uniform longitudinal stress required to reduce the initial diameter by 0.0015 cm and calculate the alteration of volume for a 15 cm length of shaft, E = 200 GPa, Poisson's ratio = 0.3. **[Ans.** 20 MPa, 0.106×10^{-6} m^3]
5. A steel tube is 3 cm internal diameter and 5 mm thick. One end is closed and the other end is screwed into a pressure vessel. The projecting length is 30 cm. Neglecting any constraints due to the ends, calculate the safe internal pressure if the allowable stress is not to exceed 140 MPa. Calculate the increase in volume under this pressure. E = 200 GPa. Poisson's ratio = 0.286.

   [Ans. 39.2 MPa, 0.313 cm^3]
6. A solid plug gauge of steel has a diameter of 2.0004 cm and is forced into a ring gauge of the same material, which measures 2 cm inside diameter and 4 cm outside diameter. Its axial length is 2 cm. What is the maximum stress in the ring, and what force is required to slide the plug, assuming the coefficient of friction to be 0.3 ? Take E = 200 GPa.**[Ans.** 25 MPa, 5655 N]

   A cylinder consists of an outer tube of internal and external diameters of 16 cm and 18 cm respectively shrunk on to an inner tube of internal and external diameters 8 cm and 16.01 cm respectively. The internal pressure developed in the cylinder is 300 MPa. *HE* - 200 GPa, find the maximum hoop stress developed in each tube. **[Ans.** 271.9 MPa, 415.3 MPa]
8. A steel rod, 6 cm diameter, is forced into a bronze casing having an outside diameter of 10 cm

and thereby produces a hoop tension at the outer circumference of the casing of 35 MPa. Determine (a) the radial pressure between the rod and the casing, and (b) the rise in temperature which would just eliminate the force fit. For steel, $E = 200$ GPa, $\nu = 0.28$, $\alpha = 12 \times 10^{-6}$ per °C. For bronze, $E = 110$ GPa $\nu = 0.33$, $\alpha = 19 \times 10^{-6}$ per °C]. **[Ans.** 31.1 MPa, 115° C]

9. A thick spherical shell having internal diameter of 200 mm is 50 mm thick. It is subjected to an internal pressure of 8 MPa. Calculate the maximum and minimum values of radial and hoop stresses. Also calculate the change in internal diameter. Take E = 200 GPa and $\nu = 0.3$.
[Ans. – 8MPa, 9.05 MPa; Zero, 5.05 MPa, 8.705×10^{-3} mm]

Chapter-12

Rotating Discs and Cylinders

12.1 INTRODUCTION

Stresses in rotating discs and cylinders are induced due to centrifugal action. These stresses are important in high speed machinery like turbines and compressors. While calculating these stresses, it is assumed that the density of the material is uniform throughout.

12.2 ROTATING DISC

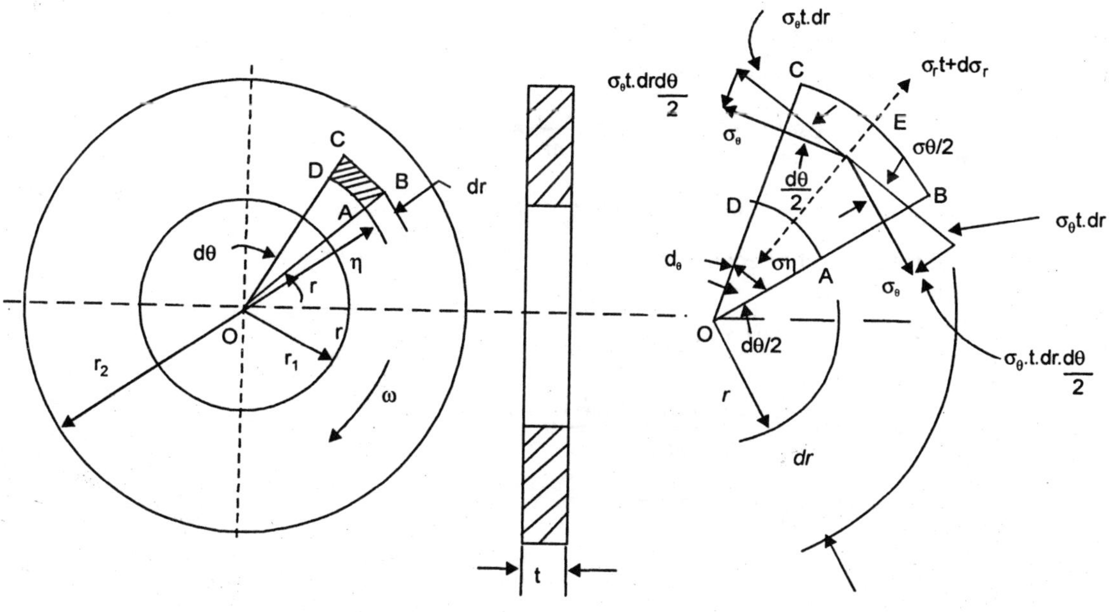

(a) Rotating disc of uniform thickness

(b) Forces acting on disc element ABCD

Fig. 12.1

Let t = uniform thickness of disc

r_1, r_2 = inner and outer radii of disc

ρ = uniform density of disc material

ω = angular speed of rotation of disc
ν = Poisson's ratio of disc material
σ_r, σ_θ = radial and hoop stresses in the disc

Consider a flat rotating disc of uniform thickness as shown in Fig. 12.1(*a*). An element of the disc ABCD between radii r and $r + dr$ is subjected to the following stresses:

Face AD : σ_r inwards
Face BC : $\sigma_r + d\sigma_r$ outwards
Face AB : σ_θ perpendicular to AB
Face CD : σ_θ perpendicular to CD

The forces acting on the element ABCD are shown in Fig. 12.1 (b).

Inward force on face AD = $\sigma_r . r d\theta t$
Outward force on face BC = $(\sigma_r + d\sigma_r) \cdot (r+dr) d\theta . t$
Force on face AB normal to AB = $\sigma_\theta . t . dr$
Force on face CD normal to CD = $\sigma_\theta . t . dr$

Resolving forces in radial outward direction, we have

$(\sigma_r + d\sigma_r) \cdot (r+dr) d\theta . t - \sigma_r . r d\theta . t - \sigma_\theta . t . dr . d\theta$

Simplifying and neglecting small quantities, we get

$(\sigma_r - \sigma_\theta)\, dr \times d\theta \times t + d\sigma_r \times r d\theta \times t$

For equilibrium of the element,

$\rho \times r d\theta \times dr \times t \times \omega^2 r + (\sigma_r - \sigma_\theta) dr \times d\theta \times t + d\sigma_r \times r d\theta \times t = 0$

$$\rho . \omega^2 r^2 + (\sigma_r - \sigma_\theta) + r \times \frac{d\sigma_r}{dr} = 0$$

$$\sigma_r - \sigma_\theta = -r\frac{d\sigma_r}{dr} - \rho . \omega^2 r^2 \quad ...(a)$$

On account of rotation let r becomes $r + u$ and $r + dr$ becomes $r + dr + du$.

Then circumferential strain,

$$\varepsilon_\theta = \frac{2\pi(r+u) - 2\pi r}{2\pi r} = \frac{u}{r}$$

and radial strain,
$$\varepsilon_r = \frac{(r + dr + du) - (r + dr)}{dr} = \frac{du}{dr}$$

Also
$$\varepsilon_\theta = \frac{1}{E}[\sigma_\theta - \nu\sigma_r] = \frac{u}{r}$$

$\therefore$
$$u = \frac{r}{E}[\sigma_\theta - \nu\sigma_r] \quad ...(b)$$

and
$$\varepsilon_r = \frac{1}{E}[\sigma_r - \nu\sigma_\theta] = \frac{du}{dr} \quad ...(c)$$

Differentiating Eq. (*b*) w.r.t. r, we get

$$\frac{du}{dr} = \frac{r}{E}\left[\frac{d\sigma_\theta}{dr} - \nu\frac{d\sigma_r}{dr}\right] + \frac{1}{E}[\sigma_\theta - \nu\sigma_r] \quad ...(d)$$

Comparing Eqs. (*c*) and (*d*), we get

$$\frac{1}{E}[\sigma_r - \nu\sigma_\theta] = \frac{r}{E}\left[\frac{d\sigma_\theta}{dr} - \nu\frac{d\sigma_r}{dr}\right] + \frac{1}{E}[\sigma_\theta - \nu\sigma_r]$$

$$\frac{1}{E}[\sigma_r - \sigma_\theta - \nu\sigma_\theta + \nu\sigma_r] = \frac{r}{E}\left[\frac{d\sigma_\theta}{dr} - \nu\frac{d\sigma_r}{dr}\right]$$

$$(1+\nu)(\sigma_r - \sigma_\theta) = r\left(\frac{d\sigma_\theta}{dr} - \nu\frac{d\sigma_r}{dr}\right) \quad \text{...(e)}$$

Substituting Eq. (*a*) in (*e*), we get

$$(1+\nu)\left[-r\frac{d\sigma_r}{dr} - \rho.\omega^2 r^2\right] = r\left(\frac{d\sigma_\theta}{dr} - \nu\frac{d\sigma_r}{dr}\right)$$

Simplifying, we get

$$\frac{d}{dr}(\sigma_r + \sigma_\theta) + (1+\nu)\rho.\omega^2 r = 0 \quad \text{...(f)}$$

Integrating, we get

$$\sigma_r + \sigma_\theta + \frac{1}{2}(1+\nu)\rho\omega^2 r^2 = C_1 \quad \text{...(g)}$$

where C_1 = Constant of integration

$$\sigma_\theta = C_1 - \sigma_r - \frac{1}{2}(1+\nu)\rho\omega^2 r^2$$

Substituting in Eq. (a), we get

$$2r\sigma_r + r^2\frac{d\sigma_r}{dr} = C_1 r - \left(\frac{3+\nu}{2}\right)\rho\omega^2 r^3$$

or

$$\frac{d}{dr}(r^2\sigma_r) = C_1 r - \left(\frac{3+\nu}{2}\right)\rho\omega^2 r^3$$

Integrating, we get

$$r^2\sigma_r = \frac{C_1 r^2}{2} - \left(\frac{3+\nu}{2}\right)\rho\omega^2\frac{r^4}{4} + C_2$$

$$\sigma_r = \frac{C_1}{2} - \left(\frac{3+\nu}{8}\right)\rho.\omega^2 r^2 + \frac{C_2}{r^2} \quad \text{...(12.1)}$$

where C_2 is another constant of integration.
Substituting Eq. (12.1) in (g), we get

$$\sigma_\theta = \frac{C_1}{2} - \frac{C_2}{r^2} - \left(\frac{1+3v}{8}\right)\rho\omega^2 r^2 \qquad ...(12.2)$$

The values of the constant C_1 and C_2 can be determined from the boundary conditions. Eq. (b) becomes,

$$u = \frac{r}{E}\left[(1-v)\frac{C_1}{2} - (1+v)\frac{C_2}{r^2} - \left(\frac{1-v^2}{8}\right)\rho\omega^2 r^2\right] \qquad ...(12.3)$$

Eq. (12.1) and (12.2) are the equations which give the stresses in the disc.

12.2.1 Solid Disc.

From Eq. (12.1) and (12.2), we have respectively

$$\sigma_r = \frac{C_1}{2} + \frac{C_2}{r^2} - \left(\frac{3+v}{8}\right)\rho\omega^2 r^2$$

$$\sigma_\theta = \frac{C_1}{2} + \frac{C_2}{r^2} - \left(\frac{1+3v}{8}\right)\rho\omega^2 r^2$$

At r = 0, u = 0 and σ_r and σ_θ become infinite from above equations, which is not possible. Hence $C_2 = 0$

$$\therefore \qquad \sigma_r = \frac{C_1}{2} - \left(\frac{3+v}{8}\right)\rho\omega^2 r^2 \qquad \sigma_\theta = \frac{C_1}{2} - \left(\frac{1+3v}{8}\right)\rho\omega^2 r^2$$

At $r = r_2$, $\quad \sigma_r = 0$

$$\therefore \qquad C_1 = \left(\frac{3+v}{4}\right)\rho\omega^2 r_2^2$$

$$\therefore \qquad \sigma_r = \left(\frac{3+v}{8}\right)\rho\omega^2 (r_2^2 - r^2)$$

$$\sigma_\theta = \left(\frac{3+v}{8}\right)\rho\omega^2 r_2^2 - \left(\frac{1+3v}{8}\right)\rho\omega^2 r^2$$

or

$$\sigma_\theta = \frac{\rho}{8}\omega^2\left[(3+v)r_2^2 - (1+3v)r^2\right]$$

At

$$r = r_2; \; \sigma_\theta = \frac{(1-v)\rho}{4}\omega^2 r_2^2$$

The values of σ_r and σ_θ are maximum at $r = 0$

$$(\sigma_r)_{max} = \left(\frac{3+v}{8}\right)\rho\omega^2 r_2^2$$

$$(\sigma_\theta)_{max} = \left(\frac{3+\nu}{8}\right)\rho\omega^2 r_2^2$$

Radial displacement of the disc becomes

$$u = \frac{r}{E}[\sigma_\theta - \nu\sigma_r]$$

$$u = \frac{r}{E}\left[\left(\frac{3+\nu}{8}\right)\rho\omega^2 r_2^2 - \left(\frac{1+3\nu}{8}\right)\rho\omega^2 r^2\right] - \nu\left[\left(\frac{3+\nu}{8}\right)\rho\omega^2\left(r_2^2 - r^2\right)\right]$$

$$= \frac{r(1-\nu)}{SE}\cdot\rho\omega^2\left[(3+\nu)r_2^2 - (1+\nu)r^2\right] \quad ...(12.9)$$

At $\quad r = r_2$

$$u = \left(\frac{1-\nu}{4E}\right)\rho\omega^2 r_2^3 \quad ...(12.10)$$

12.2.2 Hollow Disc

From Eqs. (12.1) and (12.2), we have respectively

$$\sigma_r = \frac{C_1}{2} + \frac{C_2}{r^2} - \left(\frac{3+\nu}{8}\right)\rho\omega^2 r^2$$

$$\sigma_\theta = \frac{C_1}{2} - \frac{C_2}{r^2} - \left(\frac{1+3\nu}{8}\right)\rho\omega^2 r^2$$

At $\quad r = r_1,\ \sigma_r = 0$ and at $r = r_2,\ \sigma_r = 0$

$$\therefore \quad 0 = \frac{C_1}{2} + \frac{C_2}{r_1^2} - \left(\frac{3+\nu}{8}\right)\rho\omega^2 r_1^2$$

$$0 = \frac{C_1}{2} + \frac{C_2}{r_2^2} - \left(\frac{3+v}{8}\right)\rho\omega^2 r_2^2$$

Solving for C_1 and C_2, we get

$$C_1 = \left(\frac{3+\nu}{4}\right)\rho\omega^2\left(r_1^2 + r_2^2\right);\ C_2 = -\left(\frac{3+\nu}{4}\right)\rho\omega^2 r_1^2 r_2^2$$

$$\therefore \quad \sigma_r = \left(\frac{3+\nu}{8}\right)\rho\omega^2\left[r_1^2 + r_2^2 - r^2 - \frac{r_1^2 r_2^2}{r^2}\right] \quad ...(12.11)$$

$$\sigma_\theta = \left(\frac{3+\nu}{8}\right)\rho\omega^2\left[r_1^2 + r_2^2 + \frac{r_1^2 r_2^2}{r^2} - \left(\frac{1+3\nu}{3+\nu}\right)r^2\right] \quad ...(12.12)$$

σ_θ is maximum when $r = r_1$

$$(\sigma_\theta)_{max} = \left(\frac{3+\nu}{4}\right)\rho\omega^2\left[r_2^2 + \left(\frac{1-\nu}{3+\nu}\right)r_1^2\right] \quad ...(12.13)$$

For σ, to be maximum,

$$\frac{d\sigma_r}{dr} = 0$$

Using Eq. 12.11, we get

$$2r - \frac{2r_1^2 r_2^2}{r^3} = 0$$

$$\therefore \quad r = \sqrt{r_1 r_2} \quad ...(12.14)$$

$$\therefore \quad (\sigma_r)_{max} = \left(\frac{3+\nu}{8}\right)\rho\omega^2(r_2 - r_1)^2 \quad ...(12.15)$$

When $r_1 \to 0$, i.e. a very small hole is drilled in the disc at the centre, then Eq. (12.12) gives,

$$(\sigma_r)_{max} = \left(\frac{3+\nu}{4}\right)\rho\omega^2 \cdot r_2^2 \quad ...(12.16)$$

Comparing with Eq. (12.8), it can be concluded that the maximum hoop stress in a rotating disc is twice as large, when there is a small hole at its axis of rotation, as that when the disc is solid.

Example 12.1 *A disc of 50 cm diameter and uniform thickness is rotating at 2400 r.p.m. Determine the maximum stress induced in the disc.*

If a hole of 10 cm diameter is drilled at the centre of the disc. determine the maximum intensities of radial and hoop stresses induced. Take $\nu = 0.28$, density of disc = 7800 kg/m³.

Solution.

Given: $r_2 = 25$ cm, $r_1 = 5$cm, $\nu = 0.28$, $\rho = 7800$ kg/m³, N = 2400 rpm

For the solid disc,

$$(\sigma_r)_{max} = (\sigma_\theta)_{max} = \left(\frac{3+\nu}{8}\right)\rho\omega^2 r_2^2$$

$$= \left(\frac{3+0.28}{8}\right)\times 7800 \times \left(\frac{2\pi\times 2400}{60}\right)^2 \times 25^2 \times 10^{-4}$$

$$= \frac{3.28}{8}\times 7800 \times (80\pi)^2 \times 625 \times 10^{-4} = 12.625\,MPa$$

For the hollow disc

$$(\sigma_r)_{max} = \left(\frac{3+\nu}{8}\right)\rho\omega^2(r_2 - r_1)^2 = \left(\frac{3+0.28}{8}\right)\times 7800 \times (80\pi)^2(25-5)^2 \times 10^{-4}$$

$$= \frac{3.28}{8}\times 7800 \times (80\pi)^2 \times 400 \times 10^{-4} = 8.080\,MPa$$

$$(\sigma_r)_{max} = \left(\frac{3+\nu}{4}\right)\rho\omega^2\left[\left(\frac{1-\nu}{3+\nu}\right)r_1^2 + r_2^2\right]$$

$$= \left(\frac{3+0.28}{4}\right)\times 7800\times(80\pi)^2\times\left[\left(\frac{1-0.28}{3+0.28}\right)25+625\right]\times 10^{-4}$$

$$= \frac{3.28}{4}\times 7800\times(80\pi)^2\times 630\cdot 488\times 10^{-4} = 25.472\,MPa$$

Example 12.2 *A circular thin disc of outer radius 20cm and inner radius 5 cm is rotating at 3000 r.p.m.. Determine the variation of radial and hoop stresses in the disc.* ρ = *7800 kg/m³* ν = *0.3.*

Solution.

Given: r_2 = 20cm. r_1 = 5cm, N=3000rpm, ρ = 7800 kg/m³, ν = 0.3

$$\sigma_r = \left(\frac{3+\nu}{8}\right)\rho\omega^2\left[r_1^2 + r_2^2 - r^2 - \frac{r_1^2 r_2^2}{r^2}\right]$$

$$= \left(\frac{3+0.3}{8}\right)\times 7800\times\left(\frac{2\pi\times 3000}{60}\right)^2\times\left[25+400-r^2-\frac{25\times 400}{r^2}\right]\times 10^{-4}$$

$$= 0.3175\times 10^{-4}\left[425-r^2-\frac{10000}{r^2}\right]$$

Also, σ_r is maximum at

$$r = \sqrt{r_1 r_2} = \sqrt{5\times 20} = 10 \text{ cm}$$

r, cm	5	10	15	20
σ_r, MPa	0	7.145	4.939	0

$$\sigma_\theta = \left(\frac{3+\nu}{8}\right)\rho\omega^2\left[r_1^2 + r_2^2 + \frac{r_1^2 r_2^2}{r^2} - \left(\frac{1+3\nu}{3+\nu}\right)r^2\right]$$

$$\times\left[25+400+\frac{25\times 400}{r^2}-\frac{1.9}{3.3}r^2\right]\times 10^{-4}$$

$$= \left(\frac{3+0.3}{8}\right)\times 7800\times\left(\frac{2\pi\times 3000}{60}\right)^2$$

$$= \left(\frac{3.3}{8}\right)\times 7800\times\left(\frac{2\pi\times 3000}{60}\right)^2$$

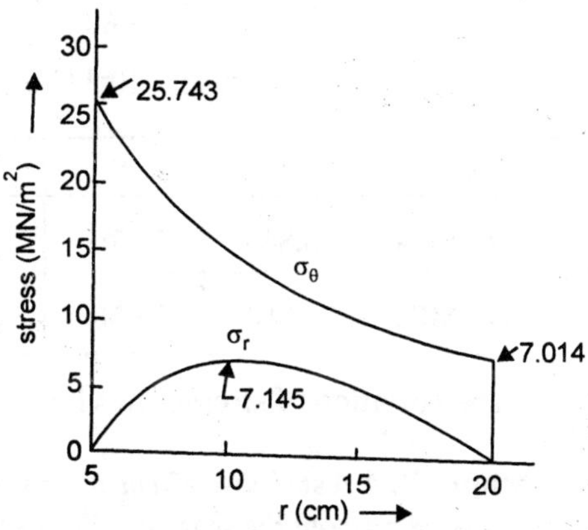

Fig. 12.2

$$= \left[425 + \frac{10{,}000}{r^2} - 0.57576r^2\right] \times 10^{-4}$$

$$= 0.3175 \times 10^{-4}\left[425 + \frac{10{,}000}{r^2} - 0.57576r^2\right]$$

r, cm	5	10	15	20
σ_θ, MPa	25.743	14.85	10.81	7.014

The variation of stress distribution is shown in Fig. 12.2.

Example 12.3 *A solid disc of diameter 60 cm is rotating at a speed of 2750 r.p.m. Determine the distribution of radial and hoop stresses in the disc. Density of disc material is 7800 kg/m³ and ν = 0.28.*

Solution.

Given: $r_2 = 30$cm. N=2750 rpm, $\rho = 7800$ kg/m³, $\nu = 0.28$.

$$\sigma_r = \left(\frac{3+\nu}{8}\right)\rho\omega^2\left(r_2^2 - r^2\right)$$

$$= \left(\frac{3+0.28}{8}\right)\times 7800$$

$$\times\left(\frac{2\pi\times 2750}{60}\right)^2\left(900 - r^2\right)\times 10^{-4}$$

$$= 0.2652 \times 10^{-4}(900 - r^2)$$

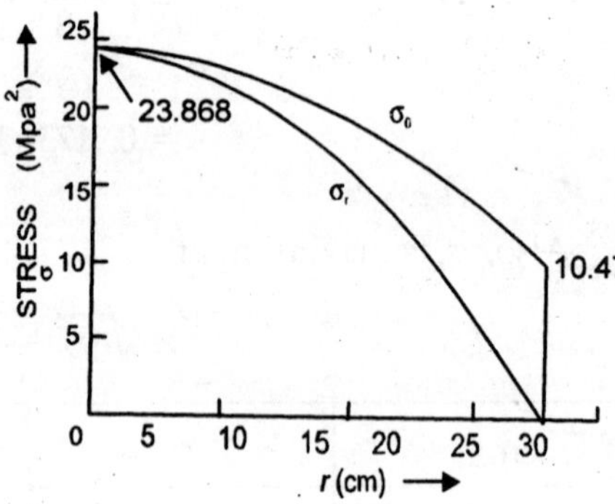

Fig. 12.3

$$\sigma_\theta = \frac{\rho}{8}\cdot\omega^2\left[(3+\nu)r_2^2 - (1+3\nu)r^2\right]$$

$$= \frac{7800}{8}\times\left(\frac{2\pi\times 2750}{60}\right)^2\times\left[3.28\times 900 - 1.84r^2\right]\times 10^{-4}$$

$$= 0.08086 \times 10^{-4}\,(2952 - 1.84r^2)$$

r, cm	0	5	10	15	20	25	30
σ_θ, MPa	23.868	23.20	21.22	17.90	13.26	7.29	0
σ_θ, MPa	23.868	23.50	22.38	20.52	17.92	14.57	10.479

The variation of stresses has been plotted in Fig. 12.3

Example 12.4 *A steel disc 25 cm outside diameter and 5 cm inside diameter is shrunk on a steel shaft so that the pressure between the shaft and the disc at stand still is 45 MPa.*

Assuming that the change in dimensions of the shaft are negligible, find at which speed the disc will loosen from the shaft. Density = 7800 kg/m³ and ν = 0.3.

Solution.

Given: $r_2 = 12.5$ cm. $r_1 = 2.5$ cm, $p_r = 45$MPa, $\rho = 7800\text{kg/m}^3$, $\nu = 0.3$.

For hollow disc

$$\sigma_r = \frac{C_1}{2}+\frac{C_2}{r^2}-\left(\frac{3+\nu}{8}\right)\rho\omega^2 r^2$$

At stand-still, $\omega = 0$

$$\therefore \quad \sigma_r = \frac{C_1}{2}+\frac{C_2}{r^2}$$

Similarly $\sigma_\theta = \frac{C_1}{2}+\frac{C_2}{r^2}$

Let p = Shrinkage pressure between the disc and the shaft.

For the disc,

At $r = 2.5$ cm, $\sigma_r = -p$

and at r = 12.5 cm, $\sigma_r = 0$

$$-p = \frac{C_1}{2}+\frac{C_2}{6.25}$$

$$0 = \frac{C_1}{2}+\frac{C_2}{156.25}$$

$$\therefore \quad \frac{C_1}{2} = -\frac{C_2}{156.25}$$

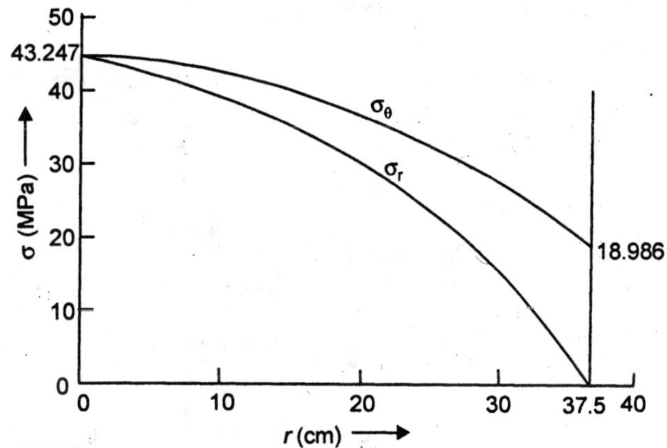

Fig. 12.4

$$-p = C_2\left[-\frac{1}{156.25}+\frac{1}{6.25}\right] = \frac{150C_2}{156.25\times 6.25}$$

$$C_2 = \frac{-156.25\times 6.25}{150}p = -6.51p$$

$$\frac{C_1}{2} = 0.0417\,p$$

$$\therefore \quad \sigma_r = 0.0417\,p - \frac{6.51p}{r^2}$$

$$\sigma_\theta = 0.0417\,p + \frac{6.51p}{r^2}$$

At $r = 2.5$ cm, the hoop stress becomes

$$\sigma_\theta = 0.0417p + \frac{6.51}{6.25}p$$

$= 0.0417$ p $+ 1.0416$ p $= 1.0833$ p $= 1.0833 \times 45 = 48.7485$ MPa

Hoop strain at inside radius of disc becomes,

$$\varepsilon_\theta = \frac{1}{E}[\sigma_\theta - \nu\sigma_r]$$

$$= \frac{1}{E}[48.7485 + 0.3 \times 45] = \frac{1}{E}[48.7485 + 13.5] = \frac{62.2485}{E}$$

Due to rotation, hoop stress in the disc at inside radius is given by,

$$\sigma_\theta = \left(\frac{3+\nu}{4}\right)\rho\omega^2\left[r_3^2 + \left(\frac{1-\nu}{3+\nu}\right)r_1^2\right]$$

$$= \left(\frac{3+0.3}{4}\right) \times 7800 \times \omega^2\left[156.25 + \frac{0.7}{3.3} \times 6.25\right] \times 10^{-4}$$

$$= \frac{3.3}{4} \times 7800 \times \omega^2\left[156.25 + 1.3257 \times 10^{-4}\right]$$

$$= \frac{3.3 \times 7800 \times 157.5758 \times 10^{-4}}{4} \times \omega^2 = 101.4\omega^2$$

$$\sigma_r = 0$$

$\therefore$ $$\text{Hoop strain} = \frac{\sigma_\theta}{E} = \frac{101.4\omega^2}{E}$$

$\therefore$ $$\omega^2 = \frac{62.2485 \times 10^6}{101.4}$$

$$\omega = 783.5 \text{ rad/s}$$

$$\frac{2\pi N}{60} = 783.5$$

$$N = \frac{783.5 \times 60}{2\pi} = 7482 \text{ r.p.m.}$$

Example 12.5 *A thin solid disc of 75 cm diameter is to rotate at 3000 r.p.m. The material density is 7600 kg/m³ and Poisson's ratio is 0.28. Plot the variation of radial and hoop stresses in the disc.*

Solution.

Given: $r_2 = 37.5$ cm. $N = 3000$ rpm, $\rho = 7600 \text{kg/m}^3$, $\nu = 0.28$.

$$\sigma_r = \left(\frac{3+\nu}{8}\right)\rho\omega^2\left(r_2^2 - r^2\right)$$

$$= \left(\frac{3.28}{8}\right) \times 7600 \times \left(\frac{2\pi \times 3000}{60}\right)^2$$

$$\left[(37.5)^2 - r^2\right] \times 10^{-4}$$

$$= 30753.68\ (1406.25 - r^2)$$

$$\sigma_\theta = \frac{1}{8}\rho\omega^2\left[(3+\nu)r_2^2 - (1+3\nu)r^2\right]$$

$$= \frac{1}{8} \times 7600 \times \left(\frac{2p \times 3000}{60}\right)^2$$

$$\left[3.28 \times (37.5)^2 - 1.84r^2\right] \times 10^{-4}$$

$$= 17252.06\ (2506.79 - r^2)$$

r, cm	0	10	20	30	37.5
σ_θ, MPa	43.247	40.172	30.946	15.569	0
σ_θ, MPa	43.247	41.522	36.346	27.720	18.986

The variation of stresses have been plotted in Fig. 12.4

Example 12.6 *A hollow thin disc of 60 cm outer diameter and 40 cm inner diameter is running at 1440 r.p.m Calculate the maximum intensities of radial and hoop stresses. At what radius the radial stress is maximum? Density of disc material = 7800 kg/m³. Poisson's ratio = 0.30.*

Solution.

Given: $r_1 = 20$ cm. $r_2 = 30$ cm, $N = 1440$ rpm, $\rho = 7800\text{kg/m}^3$, $\nu = 0.30$.

$$r = \sqrt{r_1 r_2} = \sqrt{20 \times 30} = 24.495 \text{ cm}$$

$$(\sigma_r)_{max} = \left(\frac{3+\nu}{8}\right)\rho\,\omega^2\,(r_2 - r_1)^2$$

$$= \left(\frac{3.3}{8}\right) \times 7800 \times \left(\frac{2\pi \times 1440}{60}\right)^2 (30-20)^2 \times 10^{-4} = 0.73 \text{ MPa}$$

$$(\sigma_\theta)_{max} = \left(\frac{3+\nu}{4}\right)\rho\,\omega^2\left(r_2^2 + \left(\frac{1-\nu}{3+\nu}\right)r_1^2\right)$$

$$= \left(\frac{3.3}{4}\right) \times 7800 \times \left(\frac{2\pi \times 1440}{60}\right)^2 \left[900 + \frac{0.7}{3.3} \times 400\right] \times 10^{-4} = 14.41 \text{ MPa}$$

Example 12.7 *A thin uniform disc of 25 cm diameter with a central hole of 5 cm diameter runs at 10,000 r.p.m. Calculate the maximum principal stresses and the maximum shearing stress in the disc. Poisson's ratio = 0.3 and density of material = 7470 kg/m³.*

Solution.

Given: $r_1 = 2.5$ cm. $r_2 = 12.5$ cm, $N = 10{,}000$ rpm, $\nu = 0.3$, $\rho = 7470\ \text{kg/m}^3$.

$$(\sigma_r)_{max} = \left(\frac{3.3}{4}\right) \times 7470 \times \left(\frac{2\pi \times 10{,}000}{60}\right)^2 \times (12.5 - 2.5)^2 \times 10^{-4} = 33.791 \text{ MPa}$$

$$(\sigma_\theta)_{max} = \left(\frac{3.3}{4}\right) \times 7470 \times \left(\frac{2\pi \times 10{,}000}{60}\right)^2 \left[156.25 + \frac{0.7}{3.3} \times 6.25\right] \times 10^{-4} = 106.493 \text{ MPa}$$

$$\tau_{max} = \frac{\sigma_\theta - \sigma_r}{2} = 36.351 \text{ MPa}$$

12.3 ROTATING CYLINDER

Consider a circular cylinder of inside radius r_1 and outside radius r_2 rotating at speed ω as shown in Fig. 12.5. Assume that plane sections of the cylinder remain plane during rotation, then the axial strain along the z-axis will be independent of the radius r of the cylinder and will be constant.

Radial strain, $$\varepsilon_r = \frac{1}{E}\left[\sigma_r - \nu(\sigma_\theta + \sigma_z)\right] = \frac{du}{dr} \quad ...(a)$$

Hoop strain, $$\varepsilon_\theta = \frac{1}{E}\left[\sigma_\theta - \nu(\sigma_r + \sigma_z)\right] = \frac{u}{r} \quad ...(b)$$

Axial strain, $$\varepsilon_z = \frac{1}{E}\left[\sigma_z - \nu(\sigma_r + \sigma_\theta)\right] \quad ...(c)$$

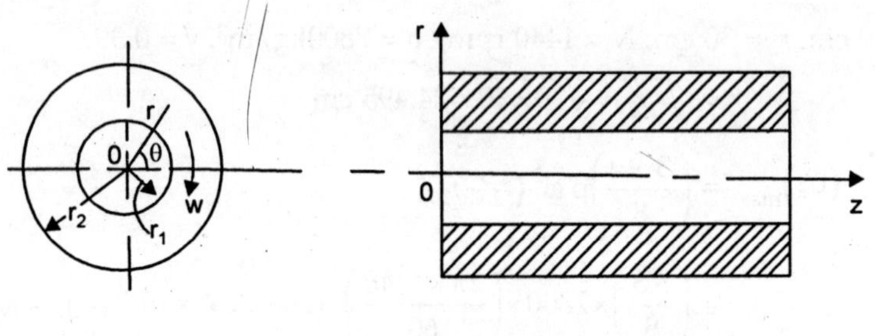

Fig. 12.5 Rotating Cylinder

From Eq. (*b*), we have

$$Eu = r\left[\sigma_\theta - \nu(\sigma_r + \sigma_z)\right]$$

Differentiating w.r.t.r, we get

$$E\frac{du}{dr} = \sigma_\theta - \nu(\sigma_r + \sigma_z) + r\left[\frac{d\sigma_\theta}{dr} - \nu\left(\frac{d\sigma_r}{dr} + \frac{d\sigma_z}{dr}\right)\right]$$

By using Eq.(*a*), we have

$$E\frac{du}{dr} = \sigma_r - \nu(\sigma_\theta + \sigma_z)$$

$$\therefore \quad (\sigma_r - \sigma_\theta)(1+\nu) = r\left[\frac{d\sigma_\theta}{dr} - \nu\left(\frac{d\sigma_r}{dr} + \frac{d\sigma_z}{dr}\right)\right] \qquad ...(d)$$

From Eq. (*c*), we get

$$E\varepsilon_z = \sigma_z - \nu(\sigma_r + \sigma_\theta) = \text{constant} = C_1$$

$$\therefore \quad \sigma_z = C_1 + \nu(\sigma_r + \sigma_\theta)$$

Differentiating w.r.t. r, we get

$$\frac{d\sigma_z}{dr} = \nu\left(\frac{d\sigma_r}{dr} + \frac{d\sigma_\theta}{dr}\right)$$

Substituting in Eq. (*d*), we get

$$(\sigma_r - \sigma_\theta)(1+\nu) = r\left[\frac{d\sigma_\theta}{dr} - \nu\left\{\frac{d\sigma_r}{dr} + \nu\left(\frac{d\sigma_r}{dr} + \frac{d\sigma_\theta}{dr}\right)\right\}\right]$$

$$= r\left[(1-\nu^2)\frac{d\sigma_\theta}{dr} - \nu(1+\nu)\frac{d\sigma_r}{dr}\right]$$

$$\sigma_r - \sigma_\theta = r\left[(1-\nu)\frac{d\sigma_\theta}{dr} - \nu\frac{d\sigma_r}{dr}\right] \qquad ...(e)$$

Also considering the equilibrium of an element of the cylinder between angular positions θ and $\theta + d\theta$ and radii r and $r + dr$, we can get as in the case of a rotating disc,

$$\sigma_r - \sigma_\theta = -\left(r\frac{d\sigma_r}{dr} + \rho\omega^2 r^2\right) \qquad ...(f)$$

Comparing Eqs. (*e*) and (*f*), we get

$$r\left[(1-\nu)\frac{d\sigma_\theta}{dr} - \nu\frac{d\sigma_r}{dr}\right] = -r\left(\frac{d\sigma_r}{dr} + \rho\omega^2 r\right)$$

$$(1-\nu)\frac{d\sigma_r}{dr} - \nu\frac{d\sigma_r}{dr} = -\left(\frac{d\sigma_r}{dr} + \rho\omega^2 r\right)$$

$$(1-\nu)\frac{d\sigma_r}{dr} + (1-\nu)\frac{d\sigma_\theta}{dr} = -\rho\omega^2 r$$

$$(1-\nu)\left[\frac{d\sigma_r}{dr} + \frac{d\sigma_\theta}{dr}\right] = -\rho\omega^2 r$$

$$\frac{d\sigma_r}{dr} + \frac{d\sigma_\theta}{dr} = -\frac{\rho}{(1-\nu)}\cdot\omega^2 r$$

$$\frac{d}{dr}(\sigma_r + \sigma_\theta) = -\frac{\rho}{(1-\nu)}\cdot\omega^2 r$$

Integrating ,we get $$\sigma_r + \sigma_\theta = \frac{-\rho}{(1-\nu)} \cdot \frac{\omega^2 r^2}{2} + C_2 \quad \text{... (g)}$$

where C_2 is a constant of integration

Adding Eqs. (*f*) and (*g*), we get, $$2\sigma_r = -\left[\frac{d\sigma_r}{dr} + \rho\omega^2 r^2\right] - \frac{\rho}{(1-\nu)} \cdot \frac{\omega^2 r^2}{2} + C_2$$

$$2\sigma_r + r\frac{d\sigma_r}{dr} = -\rho\omega^2 r^2 \left(\frac{3-2\nu}{2(1-\nu)}\right) + C_2$$

Multiplying both sides by r, we get

$$2r\sigma_r + r^2 \frac{d\sigma_r}{dr} = -\rho \frac{\omega^2 r^3}{2} \left(\frac{3-2}{1-\nu}\right) + rC_2$$

or $$\frac{d}{dr}(r^2 \sigma_r) = \rho \cdot \frac{\omega^2 r^3}{2} \left[\frac{3-2\nu}{1-\nu}\right] + rC_2$$

Integrating, we get $$r^2 \sigma_r = -\rho \cdot \frac{\omega^2 r^4}{8} \left(\frac{3-2\nu}{1-\nu}\right) + \frac{r^2}{2} C_2 + C_3$$

where C_3 is another constant of integration

$$\sigma_r = -\rho \cdot \frac{\omega^2 r^2}{8} \left(\frac{3-2\nu}{1-\nu}\right) + \frac{C_2}{2} + \frac{C_3}{r^2} \quad \text{... (12.17)}$$

Substituting in Eq. (*g*), we get $$\sigma_\theta = -\rho \cdot \frac{\omega^2 r^2}{8} \left(\frac{1+2\nu}{1-\nu}\right) + \frac{C_2}{2} - \frac{C_3}{r^2} \quad \text{... (12.18)}$$

Eqs. (12.17) and (12.18) are the governing equation for a rotating cylinder.

12.3.1 Solid Cylinder

From eqs. 12.17 and 12.18, we have respectively

$$\sigma_r = \frac{C_2}{2} + \frac{C_3}{r^2} - \rho \cdot \frac{\omega^2 r^2}{8} \left(\frac{3-2\nu}{1-\nu}\right)$$

$$\sigma_\theta = \frac{C_2}{2} - \frac{C_3}{r^2} - \rho \cdot \frac{\omega^2 r^2}{8} \left(\frac{1-2\nu}{1-\nu}\right)$$

Constant C_3 must be zero, since the stress remains finite at $r = 0$

$$\therefore \quad \sigma_r = \frac{C_2}{2} - \frac{1}{8}\left(\frac{3-2\nu}{1-\nu}\right)\rho\omega^2 r^2$$

$$\sigma_\theta = \frac{C_2}{2} - \frac{1}{8}\left(\frac{1+2\nu}{1-\upsilon}\right)\rho\omega^2 r^2$$

For a solid cylinder with a free surface, at $r = r_2, \sigma_r = 0$

$$\therefore \quad \frac{C_2}{2} = \frac{1}{8}\left(\frac{3-2\nu}{1-\nu}\right)\rho\omega^2 r_2^2$$

$$\therefore \quad \sigma_r = \frac{1}{8}\left(\frac{3-2\nu}{1-\nu}\right)\rho\omega^2 (r_2^2 - r^2) \qquad \text{... (12.19)}$$

$$\sigma_\theta = \frac{1}{8}\left(\frac{3-2\nu}{1-\nu}\right)\rho\omega^2\left[r_2^2 - \left(\frac{1+2\nu}{3-2\nu}\right)r^2\right] \qquad \text{... (12.20)}$$

The maximum stresses occur at the centre of the cylinder, where $r = 0$.

$$\therefore \quad (\sigma_r)_{\max} = (\sigma_\theta)_{\max} = \frac{1}{8}\left(\frac{3-2\nu}{1-\nu}\right)\rho\omega^2 r_2^2 \qquad \text{... (12.21)}$$

12.3.2 Hollow Cylinder.

From Eq. 12.17, we have

$$\sigma_r = \frac{C_2}{2} + \frac{C_3}{r^2} - \frac{1}{8}\left(\frac{3-2\nu}{1-\nu}\right)\rho\omega^2 r^2$$

$$\sigma_r = 0 \text{ at } r = r_1 \text{ and } r = r_2$$

$$\therefore \quad 0 = \frac{C_2}{2} + \frac{C_3}{r_1^2} - \frac{1}{8}\left(\frac{3-2\nu}{1-\nu}\right)\rho\omega^2 r_1^2$$

$$0 = \frac{C_2}{2} + \frac{C_3}{r_2^2} - \frac{1}{8}\left(\frac{3-2\nu}{1-\nu}\right)\rho\omega^2 r_2^2$$

Solving for C_2 and C_3, we get

$$\frac{C_2}{2} = \frac{1}{8}\left(\frac{3-2\nu}{1-\nu}\right)\rho\omega^2 (r_1^2 + r_2^2)$$

$$C_3 = -\frac{1}{2}\left(\frac{3-2\nu}{1-\nu}\right)\rho\omega^2 r_1^2 r_2^2$$

$$\therefore \quad \sigma_r = \frac{1}{8}\left(\frac{3-2\nu}{1-\nu}\right)\rho\omega^2\left(r_1^2 + r_2^2 - \frac{r_1^2 r_2^2}{r^2} - r^2\right) \qquad \text{... (12.22)}$$

$$\sigma_\theta = \frac{1}{8}\left(\frac{3-2\nu}{1-\nu}\right)\rho\omega^2\left[r_1^2 + r_2^2 ++ \frac{r_1^2 r_2^2}{r^2} - \left(\frac{1+2\nu}{3-2\nu}\right)r^2\right] \qquad \text{... (12.23)}$$

σ_θ is maximum at $\quad r = r_1$

$$\therefore \quad (\sigma_\theta)_{\max} = \frac{1}{4}\left(\frac{3-2\nu}{1-\nu}\right)\rho\omega^2 r_2^2\left[1 + \left(\frac{1-2\nu}{3-2\nu}\right)\frac{r_1^2}{r^2}\right] \qquad \text{... (12.24)}$$

If $\quad \dfrac{r_1}{r_2} \approx 0,\;$ then

$$(\sigma_\theta)_{max} = \frac{1}{4}\left(\frac{3-2\nu}{1-\nu}\right)\rho\omega^2 r_2^2 \qquad \text{... (12.25)}$$

Comparing Eqs. (12.21) and (12.25) we find that the maximum hoop stress in a cylinder with a small hole at the centre is twice that of an a solid cylinder.

For σ_r to be maximum, $$\frac{d\sigma_r}{dr} = 0$$

$$-2r + \frac{2r_1^2 r_2^2}{r^2} = 0$$

$$r = \sqrt{r_1 r_2}$$

$$(\sigma_r)_{max} = \frac{1}{8}\left(\frac{3-2\nu}{1-\nu}\right)\rho\omega^2 (r_2 - r_1)^2$$

Example 12.8 *A solid cylinder of 25 cm diameter is rotating at 1500 r.p.m. Determine the maximum hoop stress induced in the cylinder of its material density is 7800 kg/m³. Poisson's ratio is 0.28. Also draw the variation of radial and hoop stresses in the cylinder.*

Solution.

Given: $r_2 = 12.5$ cm, $N = 1500$ rpm, $\rho = 7800$ kg/m², $\nu = 0.28$

$$\sigma_r = \frac{1}{8}\left(\frac{3-2\nu}{1-\nu}\right)\rho\omega^2 (r_2^2 - r^2)$$

$$= \frac{1}{8}\left(\frac{3-0.56}{1-0.28}\right)\times 7800 \times \left(\frac{2\pi\times 1500}{60}\right)^2 \times (156.25 - r^2)\times 10^{-4}$$

$$= 0.08152 \times 10^{-4}\,(156.25 - r^2)$$

r, cm	0	2	4	6	8	10	12.5
σ_r, MPa	1.274	1.241	1.143	0.98	0.751	0.458	0

$$\sigma_\theta = \frac{1}{8}\left(\frac{3-\nu}{1-\nu}\right)\rho\omega^2 \left[r_2^2 - \left(\frac{1-2\nu}{3-2\nu}\right)r^2\right]$$

$$= \frac{1}{8}\left(\frac{3-0.56}{1-0.28}\right)\times 7800 \times \left(\frac{2\pi\times 1500}{60}\right)^2 \left[156.25 - \frac{1.56}{2.44}r^2\right]$$

$$= \frac{1}{8}\times\frac{2.44}{0.72}\times 7800 \times (50\pi)^2 [156.25 - 0.63934 r^2]$$

$$= 0.08152 \times 10{-}4\,(156.25 - 0.6393\, r^2).$$

r, cm	0	2	4	6	8	10	12.5
σ_θ, MPa	1.274	1.253	12.13	1.190	0.940	0.752	0.459

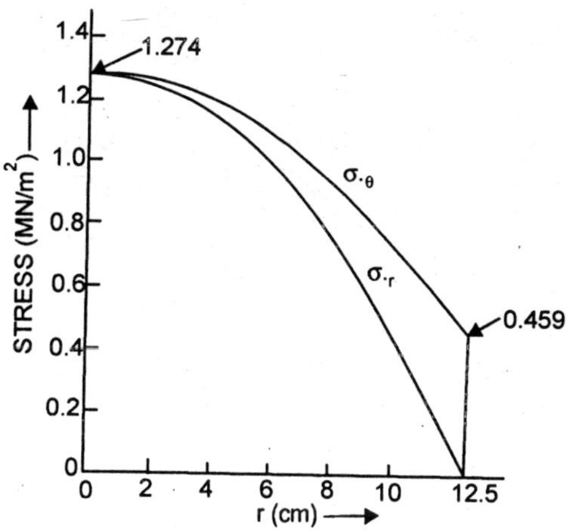

Fig. 12.6

The variation of stresses is shown in Fig. 12.6

Example 12.9 *A hollow cylinder of 40 cm external diameter and 20 cm internal diameter is rotating at 3000 r.p.m. Determine the distribution of radial and hoop stresses in the cylinder. Density of cylinder material is 7800 kg/m³, ν = 0.3.*

Solution.

Given: $r_2 = 20$ cm, $r_1 = 10$ cm, N = 3000 rpm, $\rho = 7800$ kg/m³, $\nu = 0.3$

$$\sigma_r = \frac{1}{8}\left(\frac{3-2\nu}{1-\nu}\right)\rho\omega^2\left(r_1^2 + r_2^2 - \frac{r_1^2 r_2^2}{r^2} - r^2\right)$$

$$= \frac{1}{8}\left(\frac{3-0.6}{1-0.3}\right)\times 7800\times\left(\frac{2\pi\times 3000}{60}\right)^2\left[10^2 + 20^2 - \frac{100\times 400}{r^2} - r^2\right]\times 10^{-4}$$

$$= 0.329927\times 10^{-4}\left(500 - \frac{40000}{r^2} - r^2\right)$$

r, cm	10	14.142	15	20
σ_r, MPa	0	3.299	3.207	0

σ_r is maximum at,

$$r = \sqrt{r_1 r_2} = \sqrt{10\times 20} = 14.142\,\text{cm}$$

$$\sigma_\theta = \frac{1}{8}\left(\frac{3-2\nu}{1-\nu}\right)\rho\omega^2\left[r_1^2 + r_2^2 + \frac{r_1^2 r_2^2}{r^2} - \left(\frac{1+2\nu}{3-2\nu}\right)r^2\right]$$

$$= \frac{1}{8}\times\frac{2.4}{0.7}\times 7800\times(100\pi)^2\left[100 + 400 + \frac{40,000}{r^2} - \frac{2}{3}r^2\right]\times 10^{-4}$$

$$= 0.329927 \times 10^{-4}\left[500 + \frac{40,000}{r^2} - 0.6667\, r^2\right]$$

r, cm	10	15	20
σ_r, MPa	27.493	17.41	10.997

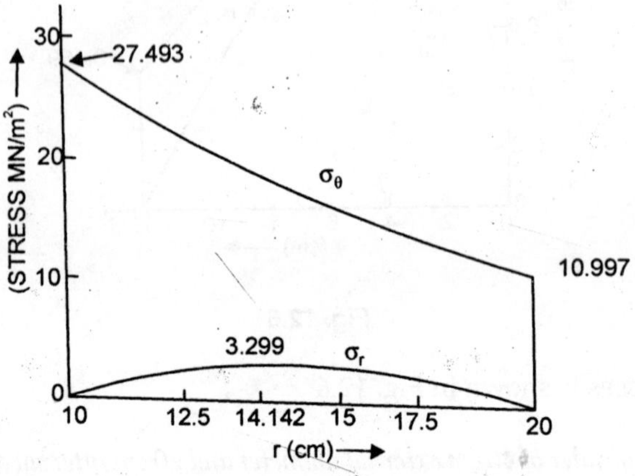

Fig. 12.7

The variation of stresses is shown in Fig. 12.7

Example 12.10 *A long hollow cylinder is of 20 cm external diameter and is 5 cm thick. It is revolving at a constant speed of 2400 r.p.m. Calcualte the maximum radial and hoop stresses induced in cylinder. The density of cylinder material is 7600 kg/m³ and Poisson's ratio = 0.30.*

Solution.

Given: $r_2 = 10$ cm, $r_1 = 5$ cm, $N = 2400$ rpm, $\rho = 7600$ kg/m3, $\nu = 0.30$.

$$(\sigma_r)_{max} = \frac{1}{8}\left(\frac{3-2\nu}{1-\nu}\right)\left(\frac{3-2\nu}{1-\nu}\right)\rho w^2 (r_2 - r_1)^2$$

$$= \frac{1}{8}\left(\frac{2.4}{0.7}\right)\times 7600 \times \left(\frac{2\pi \times 2400}{60}\right)^2 (10-5)^2 \times 10^{-4} 0.514 \text{ MPa}$$

$$(\sigma_r)_{max} = \frac{1}{4}\left(\frac{3-2\nu}{1-\nu}\right)\rho w^2 r_2^2\left[1 + \left(\frac{1-2\nu}{3-2\nu}\right)\left(\frac{r_1}{r_{02}}\right)^2\right]$$

$$= \frac{1}{4}\times\frac{2.4}{0.7}\times 7600 \times \left(\frac{2\pi\times 2400}{60}\right)^2 \times 100 \times 10^{-4} \times \left[1 - \frac{0.4}{2.4}\times 0.25\right]$$

$$= 4.286 \text{ MPa}$$

SUMMARY

1. Rotation causes centrifugal stresses in discs and cylinders.
2. Rotating disc.
   (a) Solid disc.

(*i*) $\sigma_r = \left(\frac{3+\nu}{8}\right)\rho\omega^2(r_2^2 - r^2)$

(*ii*) $\sigma_\theta = \frac{\rho}{8}\omega^2\,[(3+\nu)r_2^2 - (1+3\nu)\,r^2]$

(iii) At $r = r_2, \sigma_r = 0$

and $\sigma_\theta = \left(\frac{1-\nu}{4}\right)\rho\omega^2 r_2^2$

(iv) At $r = 0, (\sigma_r)_{max} = (\sigma_\theta)_{max} = \left(\frac{3+\nu}{8}\right)\rho\omega^2 r_2^2$

(*v*) Radial dispalcement,

$$u = \left(\frac{1-\nu}{8E}\right)\rho\omega^2 r[(3+\nu)r_2^2 - (1+\nu)r^2]$$

$$\text{At r } = r_2,\ u = \left(\frac{1-\nu}{+E}\right)\rho\omega^2 r_2^3$$

(b) Hollow disc.

(*i*) $\sigma_r = \left(\frac{3+\nu}{8}\right)\rho\omega^2\left[r_1^2 + r_2^2 - r^2 - \frac{(r_1 r_2)_2}{r^2}\right]$

(ii) $\sigma_\theta = \left(\frac{3+\nu}{8}\right)\rho\omega^2\left[r_1^2 + r_2^2 + \left(\frac{r_1 r_2}{r}\right)^2 - \left(\frac{1+3\nu}{3+\nu}\right)r^2\right]$

(iii) $(\sigma_\theta)_{max}\Big|_{r=r_1} = \left(\frac{3+\nu}{4}\right)\rho\omega^2\left[r_2^2 + \left(\frac{1-\nu}{3+\nu}\right)r_1^2\right]$

(iv) σ_r to be maximum, $r = \sqrt{r_1 r_2}$

$$(\sigma_r)_{max} = \left(\frac{3+\nu}{8}\right)\rho\omega^2(r_2 - r_1)^2$$

(v) When $r_1 = 0$ then $(\sigma_\theta)_{max} = \left(\frac{3+\nu}{4}\right)\rho\omega^2 r_3^2$

3. Rotating cylinder.

(a) Solid cylinder

(*i*) $\sigma_r = \frac{1}{8}\left(\frac{3-\nu}{1-\nu}\right)\rho\omega^2 (r_2^2 - r^2)$

(ii) $\sigma_\theta = \frac{1}{8}\left(\frac{3-2\nu}{1-2\nu}\right)\rho\omega^2\left[r_2^2 - \left(\frac{1+2\nu}{3-2\nu}\right)r^2\right]$

(iii) At $r = 0$, $(\sigma_r)_{max} = (\sigma_\theta)_{max} = \frac{1}{8}\left(\frac{3-2\nu}{1-\nu}\right)\rho\omega^2 r_2^2$

(b) Hollow cylinder.

(i) $\sigma_r = \frac{1}{8}\left(\frac{3-2\nu}{1-\nu}\right)\rho\omega^2\left[r_1^2 + r_2^2 - \left(\frac{r_1 r_2}{r}\right)^2 - r^2\right]$

(ii) $\sigma_\theta = \frac{1}{8}\left(\frac{3-2\nu}{1-\nu}\right)\rho\omega^2\left[r_1^2 + + r_2^2 + \left(\frac{r_1 r_2}{r}\right)^2 - \left(\frac{1+2\nu}{2-2\nu}\right)r^2\right]$

(iii) At $r = r_1$, $(\sigma_\theta)_{max} = \frac{1}{4}\left(\frac{3-2\nu}{1-\nu}\right)\rho\omega^2 r_2^2\left[1 + \left(\frac{1-2\nu}{3-2\nu}\right)\left(\frac{r_1}{r_2}\right)^2\right]$

(iv) $r_1 \to 0$, thebn

$$(\sigma_\theta)_{max} = \frac{1}{4}\left(\frac{3-2\nu}{1-\nu}\right)\rho\omega^2 r_2^2$$

(*v*) For σ_r to be maximum, $r = \sqrt{r_1 r_2}$

$$(\sigma_r)_{max} = \frac{1}{8}\left(\frac{3-2\nu}{1-\nu}\right)\rho\omega^2 (r_2 - r_1)^2$$

Short Answer Type Questions

1. How stresses due to rotation are induced in discs and cylinders?
2. Where the radial and hoop stresses are maximum in a solid disc under rotation?
3. Where the radial centrifugal stresses is maximum in a hollow disc?
4. What is the ratio of maximum hoop stress in a rotating solid disc to when a needle hole is drilled at its centre?
5. What is the location of maximum radial and hoop stresses in a rotating solid cylinder?
6. Where the maximum radial stress occurs in a rotating hollow cylinder?

EXERCISES

1. A thin circular disc of 600 mm outside diameter having a central hole, rotates at a uniform speed about its axis through its centre. The diameter of hole is such that the maximum stress due to rotation is 85% of that in a thin ring whose mean diameter is 600 mm. If both are of the same material and rotates at the same speed, find out the diameter of the centre hole and the speed of rotation if the maximum allowable stress in the disc is 100 MPa. Take $\rho = 7850$ kg/m^3 and $\nu = 0.3$. [Ans. 226 mm, 3896 rpm]
2. A hollow disc of uniform thickness has an external diameter of 300 mm and internal diameter of 60 mm. Find the radial and hoop stresses at 90 mm if the disc rotates at 1500 rpm. Find also the change in external diameter of the disc. Take $\rho = 7500$ kg/m^3, $\nu = 0.3$ and $E = 200$ GPa.
   [Ans. 0.977 MPa, 1.62 MPa, 1.3×10^{-3} mm]
3. A long hollow cylinder 150 mm external diameter and 30 mm thick is rotating at a constant speed of 2000 rpm. Calculate the maximum radial and hoop stresses developed in the cylinder Take $\rho = 7470$ kg/m^3 and$\nu = 0.30$.
4. A solid steel shaft 250 mm in diameter is rotating at a speed of 280 rpm. If the shaft is constrained at its ends so that it cannot expand or contract longitudinally, calculate the total longitudinal thrust over a cross-section due to rotational stresses. Take $\rho = 7800$ kg/m^3 and $\nu = 0.30$.
5. A long solid cylinder of 500 mm diameter is rotating at a speed of 3600 rpm. Uniform density of the cylinder material is 7800 kg/m^3 and Poisson's ratio is 0.30. Calculate the maximum hoop stress developed in the cylinder. **[Ans.** 29.69 MPa]
6. A long hollow cylinder of 0.2 m inside diameter and 0.5 m outside diameter is rotating at 3600 rpm. Calculate the maximum radial and hoop stresses and their locations, if density of material is 7800 kg/m^3 and Poisson's ratio is 0.30.
   [Ans. 10.68 MPa at 0.158 m radius; 60.93 MPa at r = 0.1 m]

Chapter-13

Curved Beams

13.1 INTRODUCTION

Machine members and structures subjected to bending loads are not always initially straight. Common examples of such members are: crane hook, circular and oval chain links, machine frames, etc. Simple theory of bending cannot be applied to such like members to determine the bending stresses. For deeply curved beams, the neutral axis does not coincide with the centroidal axis, as is the case in simple bending of straight members. Extensive research was carried out by Winkler-Bach to analyse this problem. This has become popular by the name Winkler-Bach theory, which we shall study in this chapter.

13.2 THE WINKLER-BACH THEORY

The following assumptions are made in this analysis :

1. Plane transverse sections before bending remain plane after bending.
2. Limit of proportionality is not exceeded.
3. Radial strain is negligible.
4. The material considered is isotropic and obeys Hooke's law.

Consider a portion of a beam *ABCD* initially curved in its unstrained state as shown in Fig. 13.1. Let,

R = Radius of curvature of the centroidal axis *GH*.

y = Distance of a fibre *EF* from *GH*.

Let *ABC'D'* be the strained position of the beam.

R_1 = radius of curvature of *GH'*.

y_1 = Distance between *EF* and *GH'* after straining,

M = Uniform bending moment applied to the beam, assumed positive when tending to increase the curvature.

ϕ = Original angle subtended by the centroidal axis *GH* at its centre of curvature *O*.

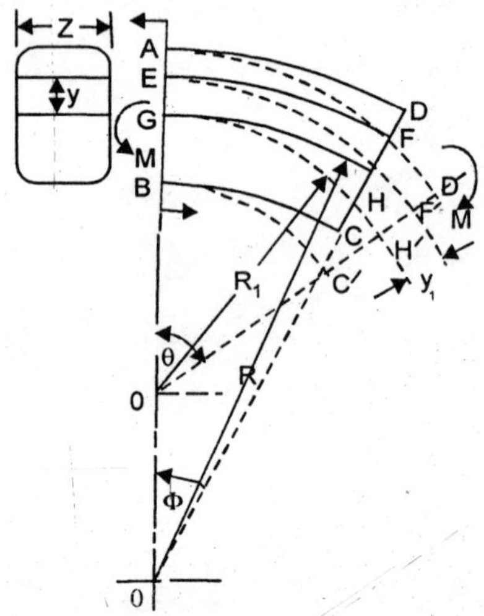

Fig.13.1 Bending of a curved beam

θ = Angle subtended by GH' (after bending) at the centre of curvature O'.

Let breadth of the beam section perpendicular to y be z, and let A be the constant area of cross-section, *i.e.* $\Sigma\delta A = \Sigma z dy$, where δA is an element of area.

Now
$$GH = R\phi$$
$$EF = (R + y)\phi$$
$$EF' = (R_1 + y_1)\theta$$

Circumferential strain in EF,

$$\varepsilon = \frac{EF' - EF}{EF} = \frac{(R_1 + y_1)\theta}{(R+y)\phi} - 1 = \frac{\sigma}{E} \quad \text{... (1)}$$

where σ = bending stress in EF.

Let ε_0 = Strain in GH when it increases to GH'.

$$\varepsilon_0 = \frac{GH' - GH}{GH} = \frac{R_1\theta}{R\phi} - 1$$

$$\therefore \quad (1+\varepsilon_0)R\phi = R_1\theta$$

or
$$\frac{\theta}{\phi} = \frac{R(1+\varepsilon_0)}{R_1} \quad \text{... (2)}$$

Substituting Eq. (2) in (1), we get

$$\varepsilon = \frac{(R_1 + y_1)}{R+y} \times \frac{R(1+\varepsilon_0)}{R_1} - 1 = \frac{\left(1+\frac{y_1}{R_1}\right)}{\left(1+\frac{y}{R}\right)} \times (1+\varepsilon_0) - 1$$

Accoriding to assumption (3), radial strain is zero.

$$\therefore \quad y = y_1$$

$$\therefore \quad \varepsilon = \frac{\left(1+\frac{y}{R_1}\right)}{\left(1+\frac{y}{R}\right)}(1+\varepsilon_0) - 1 = \frac{1+\frac{y}{R_1}+\varepsilon_0+\varepsilon_0\frac{y}{R_1}-1-\frac{y}{R}}{1+\frac{y}{R}}$$

Adding and subtracting $\varepsilon_0 \frac{y}{R}$ in the numerator, we get

$$\varepsilon = \varepsilon_0 + \frac{(1+\varepsilon_0)\left(\frac{1}{R_1} - \frac{1}{R}\right)y}{1+\frac{y}{R}} \quad \text{... (3)}$$

The tensile stress in EF′ becomes,

$$\sigma = E\varepsilon = E\left[\varepsilon_0 + \frac{(1+\varepsilon_0)\left(\frac{1}{R_1}-\frac{1}{R}\right)y}{1+\frac{y}{R}}\right] \quad \text{... (4)}$$

The total longitudinal force on the cros-section at *EF'* becomes,

$$F = \int \sigma \cdot dA = E\int \varepsilon dA = E\left[\int \varepsilon_0\, dA + \int \frac{(1+\varepsilon_0)\left(\frac{1}{R_1}-\frac{1}{R}\right)y}{1+\frac{y}{R}}\, dA\right]$$

$$= E\varepsilon_0 A + E(1+\varepsilon_0)\left(\frac{1}{R_1}-\frac{1}{R}\right)\int \frac{ydA}{1+\frac{y}{R}} \quad \text{... (5)}$$

The resisting moment about an axis through the centroid is,

$$\text{M} = \int \sigma . dA . y = \int \text{E}\varepsilon y dA = \int \text{E}\varepsilon_0 y dA + \int \frac{E(1+\varepsilon_0)\left(\frac{1}{R_1}-\frac{1}{R}\right)}{1+\frac{y}{R}} \cdot y^2 dA$$

Now $\int ydA = 0$, since y is measured from an axis through the centroid.

$$\therefore \quad M = E(1+\varepsilon_0)\left(\frac{1}{R_1}-\frac{1}{R}\right)\int \frac{y^2\, dA}{1+\frac{y}{R}} \quad \text{... (6)}$$

Let

$$\int \frac{y^2 dA}{1+\frac{y}{R}} = Ah^2 \quad \text{... (7)}$$

where h^2 = a constant for the cross-section of the beam.

$$\therefore \quad M = E(1+\varepsilon_0)\left(\frac{1}{R_1}-\frac{1}{R}\right)Ah^2 \quad \text{... (8)}$$

Now consider

$$\int \frac{ydA}{1+\frac{y}{R}} = \int \frac{RydA}{R+y} = \int\left(y - \frac{y^2}{R+y}\right)dA = \int ydA - \int \frac{y^2 dA}{R+y} = -\frac{1}{R}\int \frac{y^2\, dA}{1+\frac{y}{R}}$$

$$\therefore \quad \int \frac{ydA}{1+\frac{y}{R}} = -\frac{1}{R}\int \frac{y^2\, dA}{1+\frac{y}{R}} = -\frac{1}{R}\cdot Ah^2 \quad \text{... (9)}$$

Hence Eq. (5) becomes,

$$F = E\varepsilon_0 A - E(1+\varepsilon_0)\left(\frac{1}{R_1}-\frac{1}{R}\right)\frac{Ah^2}{R} \quad \text{... (10)}$$

Since transverse plane sections before bending remains plane after bending. hence

$$F = 0$$

or

$$0 = E_{\varepsilon_0} A - E(1+\varepsilon_0)\left(\frac{1}{R_1}-\frac{1}{R}\right)\frac{Ah^2}{R}$$

$$\therefore \quad \varepsilon_0 = (1+\varepsilon_0)\left(\frac{1}{R_1}-\frac{1}{R}\right)\frac{h^2}{R} \quad \text{... (11)}$$

Also from Eq. (8), we have

$$(1+\varepsilon_0)\left(\frac{1}{R_1}-\frac{1}{R}\right)h^2 = \frac{M}{AE}$$

Substituting in Eq. (11), we get

$$\varepsilon_0 = \frac{M}{EAR} \quad \text{... (12)}$$

Thus

$$\sigma = E\left[\varepsilon_0 + \frac{(1+\varepsilon_0)\left(\frac{1}{R_1}-\frac{1}{R}\right)y}{1+\frac{y}{R}}\right] = E\left[\frac{M}{EAR}+\frac{M}{EAh^2}\cdot\left(\frac{y}{1+\frac{y}{R}}\right)\right]$$

$$= \frac{M}{AR}+\frac{M}{Ah^2}\left(\frac{y}{1+\frac{y}{R}}\right) = \frac{M}{AR}\left[1+\frac{R^2}{h^2}\cdot\left(\frac{y}{R+y}\right)\right] \quad \text{... (13.1(a))}$$

On the other side of *GH*, *y* will be negative, and stress will be compressive.

$$\therefore \quad \sigma = \frac{M}{AR}\left[1-\frac{R^2}{h^2}\left(\frac{y}{R-y}\right)\right] \quad \text{... (13.1 (b))}$$

If the bending moment is applied in such a manner that it tends to decrease the curvature of the beam, then Eq. (13.1 (a)) will give compressive stress and Eq. 13.1 (b), tensile stress.

13.2.1 POSITION OF NEUTRAL AXIS

At the neutral axis $\sigma = 0$

$$\frac{M}{AR}\left[1+\frac{R^2}{h^2}\cdot\left(\frac{y}{R+y}\right)\right]=0$$

$$\frac{R^2}{h^2}\left(\frac{y}{R+y}\right)=-1$$

$$R^2y = -Rh^2 - h^2y$$

$$(R^2+h^2)y = -Rh^2$$

$$y = -\left(\frac{Rh^2}{R^2+h^2}\right) \quad \text{... (13.2)}$$

Hence the neutral axis is located below centroidal axis.

13.3 VALUES OF h^2

Now $$h^2 = \frac{1}{A}\cdot\int\frac{y^2 dA}{1+\frac{y}{R}}$$

$$h^2 = \frac{R}{A}\int\frac{y^2\,dA}{R+y} = \frac{R}{A}\left[\int ydA - \int RdA + \int\frac{R^2\,dA}{y+R}\right]$$

$$= \frac{R}{A}\left[0-RA+\int\frac{R^2dA}{y+R}\right] = \frac{R^3}{A}\int\frac{dA}{y+R}-R^2 \quad \text{... (13.3)}$$

13.3.1 Rectangular Cross-section

Consider an elementary strip of width B and depth dy at a distance y from the centrodial axis of a rectangular beam as shown in Fig. 13.2.

$$A = BD, dA = B\,dy$$

$$h^2 = \frac{R^3}{BD}\int_{-D2}^{+D2}\frac{B\,dy}{(R+y)} - R^2 = \frac{R^3}{BD}\,|\,ln(R+y)\,|_{-D2}^{+D2} - R^2$$

$$= \frac{R^3}{D}ln\left(\frac{2R+D}{2R-D}\right)-R^2 \quad \text{... (13.4)}$$

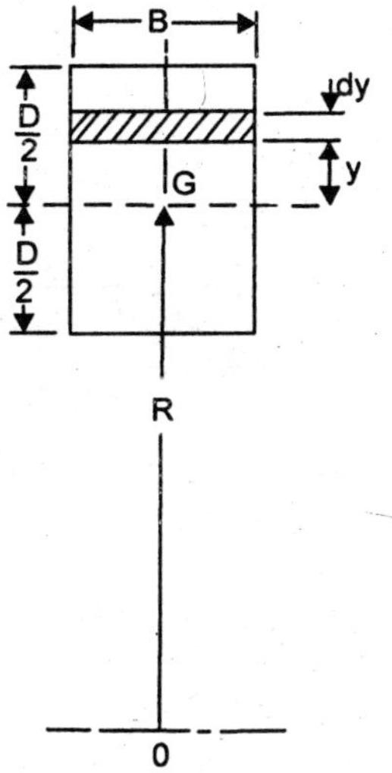

Fig. 13.2 Rectangular cross-section.

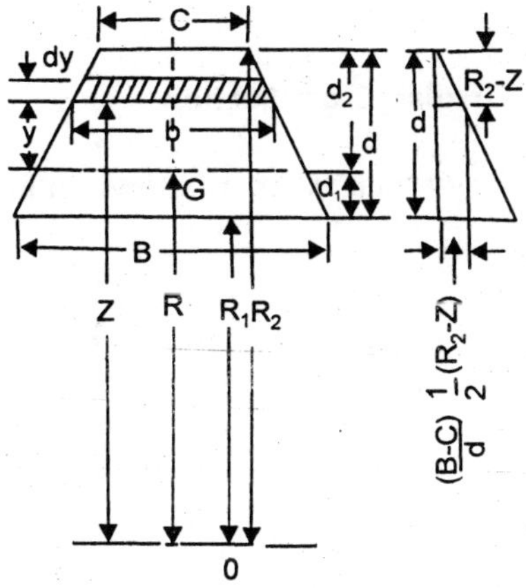

Fig. 13.3 Trapezoidal cross-section

13.3.2 Trapezoidal Cross-Section

For the trapezoidal cross-section shown in Fig. 13.3, let

$$z = R + y$$

$$\therefore \quad dz = dy$$

$$dy = b\,dy = \left[C + \frac{B-C}{d_1+d_2}(R_z - z)\right]dz$$

$$h^2 = \frac{R^3}{A}\int_{R_1}^{R_2}\left[C + \left(\frac{B-C}{d_1+d_2}\right)(R_2 - z)\right]\frac{dz}{z} - R^2$$

$$= \frac{R^3}{A}\left[\int_{R_1}^{R_2}\frac{C\,dz}{z} + \left(\frac{B-C}{d_1+d_2}\right)\int_{R_1}^{R_2}\left(\frac{R_2 - z}{z}\right)dz\right] - R^2$$

$$= \frac{R^3}{A}\left[C \mid \ln z\,|_{R_1}^{R_2} + \left(\frac{B-C}{d_1+d_2}\right) \mid R_2 \ln z - z\,|_{R_1}^{R_2}\right] - R^2$$

$$= \frac{R^3}{A}\left[C\ln\frac{R_2}{R_1} + \left(\frac{B-C}{d_1+d_2}\right)\left\{R_2 ln\frac{R_2}{R_1} - (R_2 - R_1)\right\}\right] - R^2$$

$$= \frac{R^3}{A}\left[C\,ln\left(\frac{R+d_2}{R-d_1}\right) + \left(\frac{B-C}{d}\right)(R+d_2)ln\left(\frac{R+d_2}{R-d_1}\right) - (B-C)\right] - R^2 \quad \ldots (13.5)$$

where

$$A = \left(\frac{B+C}{2}\right)d, d_1 = \frac{d}{3}\left(\frac{B+2C}{B+C}\right), d_2 = d - d_1$$

13.3.3 Circular Cross-Section

For the circular beam as shown in Fig. 13.4

$$b = 2\sqrt{\frac{D^2}{4} - y^2}$$

$$dA = b\,dy = 2\sqrt{\frac{D^2}{4} - y^2}\cdot dy$$

$$h^2 = \frac{R^3}{A}\int_{-D2}^{+D2} \frac{2\sqrt{\frac{D^2}{4} - y^2}}{R+y}.dy - R^2$$

$$= \frac{8R^3}{\pi D^2}\int_{-D2}^{+D2} \frac{\sqrt{\frac{D^2}{4} - y^2}}{R+y}\cdot dy - R^2$$

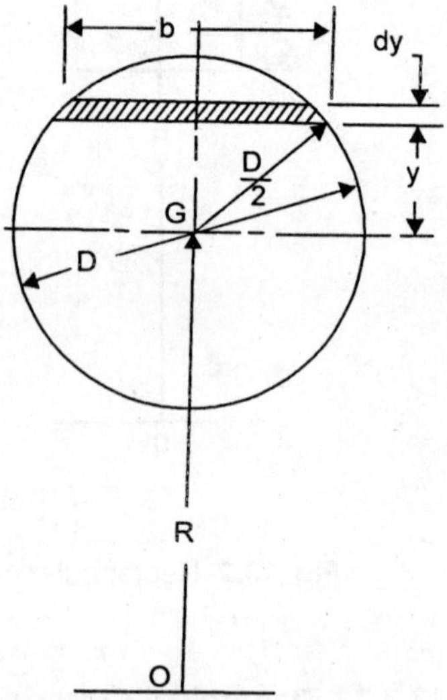

Fig. 13.4 Circular cross-section

Expanding the integrand by binomial expansion and then integrating, we get

$$h^2 = \frac{D^2}{16} + \frac{1}{128}\frac{D^4}{R^2} + \ldots \qquad \ldots (13.6)$$

13.3.4 T-section

Consider a *T*-section as shown in Fig. 13.5

Let $z = R + y$

$\therefore$ $dz = dy$

$$h^2 = \frac{R^3}{A}\left[\int_{r_1}^{r_2} \frac{dA}{R+y} + \int_{r_2}^{r_3} \frac{dA}{R+y}\right] - R^2 = \frac{R^3}{A}\left[\int_{r_1}^{r_2} \frac{b_2 dz}{z} + \int_{r_2}^{r_3} \frac{t_1 dz}{z}\right] - R^2$$

$$= \frac{R^3}{A}[b_2 \,|\, ln\, z\,|_{r_1}^{r_2} + t_1 \,|\, ln\, z\,|_{r_2}^{r_3}] - R^2 = \frac{R^3}{A}\left[b_2\, ln\frac{r_2}{r_1} + t_1 ln\frac{r_3}{r_2}\right] - R^2 \qquad \ldots (13.7)$$

where $A = b_1 t_1 + b_2 t_2$

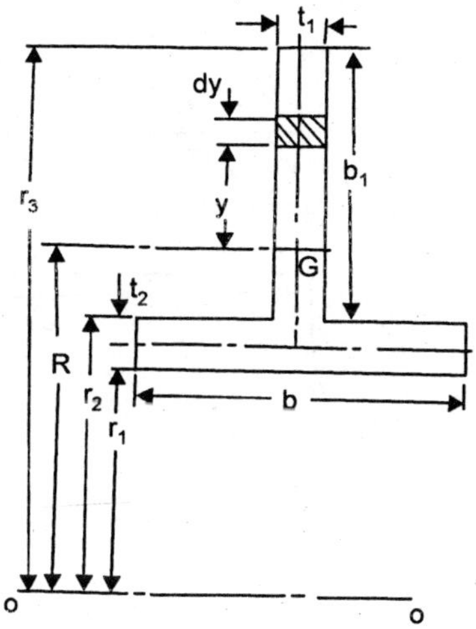

Fig. 13.5 T.Section

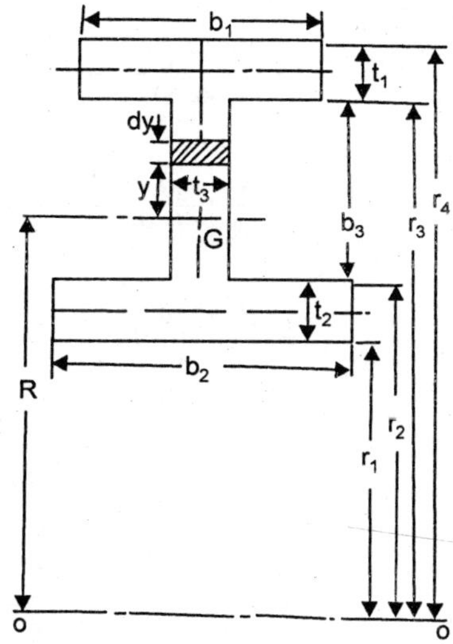

Fig. 13.6 I.Section

13.3.5 I-Section

Consider the I-section as shown in Fig. 13.6.

Let $\qquad R + y = z, \quad dy = dz$

$$h^2 = \frac{R^3}{A}\left[\int_{r_1}^{r_2} \frac{dA}{R+y} + \int_{r_2}^{r_3} \frac{dA}{R+y} + \int_{r_3}^{r_4} \frac{dA}{R+y}\right] - R^2$$

$$= \frac{R^3}{A}\left[\int_{r_1}^{r_2} \frac{b_2 d_z}{z} + \int_{r_2}^{r_3} \frac{t_3\,dz}{z} + \int_{r_3}^{r_4} \frac{b_1\,dz}{z}\right] - R^2$$

$$= \frac{R^3}{A}\left[b_2\, ln\frac{r_2}{r_1} + t_3\, ln\frac{r_3}{r_2} + b_2\, ln\,\frac{r_4}{r_3}\right] - R^2 \qquad \text{... (13.8)}$$

where $\qquad A = b_1 t_1 + b_2 t_2 + b_3 t_3$

13.3.6 Triangular Section

Consider a beam of triangular section as shown in Fig. 13.7

Let $\qquad R + y = z$

$\therefore \qquad dy = dz$

$$b = \left\{\frac{r_2 - z}{h}\right\} d$$

$$h^2 = \frac{R^3}{A}\int_{r_1}^{r_2} \frac{b\,dz}{z} - R^2$$

$$= \frac{R^3}{A}\int_{r_1}^{r_2}\left(\frac{r_2 - z}{h}\right)d.\frac{dz}{z} - R^2$$

$$= \frac{R^3}{A}\left[\frac{r_2 d}{h}\int_{r_1}^{r_2}\frac{dz}{z} - \frac{d}{h}\int_{r_1}^{r_2} dz\right] - R^2$$

$$= \frac{R^3}{A}\left[\frac{r_2 d}{h}\cdot ln\frac{r_2}{r_1} - \frac{d}{h}(r_2 - r_1)\right] - R^2$$

$$= \frac{R^3}{A}\left[\left(R + \frac{2h}{3}\right)\frac{d}{h}\cdot ln\left(\frac{2R+2h}{3R-h}\right) - d\right] - R^2$$

$$= \frac{2R^3}{dh}\left[(3R+2h)\frac{d}{h}\ ln\left(\frac{3R+2h}{3R-h}\right) - d\right] - R^2$$

$$= \frac{2R^3}{h}\left[\left(\frac{3R+2h}{3h}\right) ln\left(\frac{3R+2h}{3R-h}\right) - 1\right] - R^2 \qquad \dots (13.9)$$

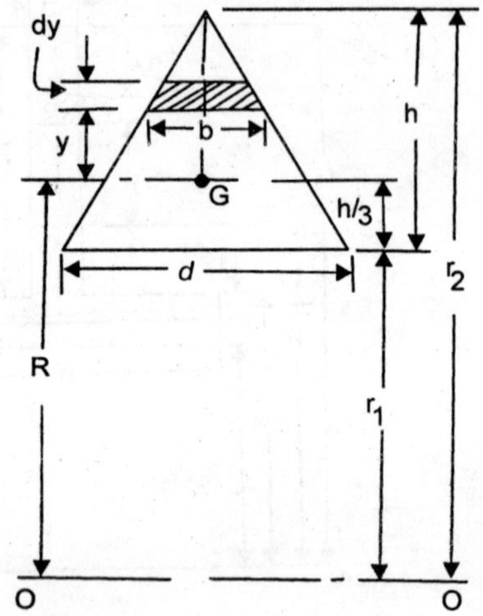

Fig. 13.7 Triangular section

13.4 STRESSES IN A CRANE HOOK

For the crane book shown n Fig 13.8, O is centre of curvature and the load line passes through O_1. The radius of curvature of the centroid is R.

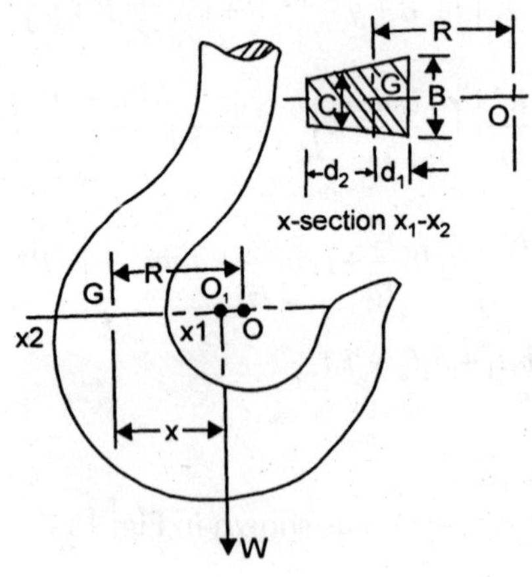

Fig. 13.8 Crane hook

Bending moment about the centroid G is

$$M = Wx$$

This bending moment is such that it is tending to decrease the curvature, i.e. this is a negative bending moment. Therefore, bending stresses at point x_1 and x_2 are respectively,

$$\sigma_1 = \frac{Wx}{AR}\left\{\frac{R^2}{h^2}\left(\frac{d_1}{R-d_1}\right)-1\right\} \text{ (tensile)} \qquad \text{... (3.10)}$$

$$\sigma_2 = \frac{Wx}{AR}\left[1+\frac{R^2}{h^2}\left(\frac{d_2}{R+d_2}\right)\right] \text{(compressive)} \qquad \text{... (13.11)}$$

Direct stress, $\sigma_d = \dfrac{W}{A}$ (tensile) ... (13.12)

$\therefore$ Resultant stress at $x_1 = \sigma_1 + \sigma_d$ (tensile) ... (13.13)

and resultant stress at $x_2 = \sigma_2 - \sigma_d$ (compressive) ... (13.14)

Example 13.1 *The diamensions of 100 kN crane hook are shown Fig. 13.9. Determine the stresses at the inside and outside fibres of the hook on a horizontal diameter passing through the centre of curvature.*

Solution.

Given : Here $B = 9$ cm, $C = 3$ cm, $d = 12$ cm, $W = 100$ kN

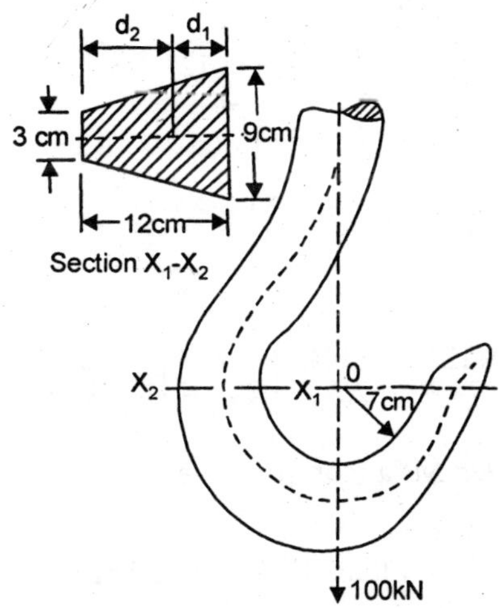

Fig. 13.9

Area of the cross-section

$$A = \left(\frac{9+3}{2}\right)\times 12 = 72 \text{ cm}^2$$

$$d_1 = \frac{d}{3}\left(\frac{B+2C}{B+C}\right) = \frac{12}{3}\left(\frac{9+6}{9+3}\right) = \frac{12}{3}\times\frac{15}{12} = 5 \text{ cm}$$

$\therefore \quad d_2 = 12 - 5 = 7 \text{ cm}$

$\therefore \quad R = 7 + 5 = 12 \text{ cm}$

$$h^2 = \frac{R^3}{A}\left[C\,ln\left(\frac{R+d_2}{R-d_1}\right)+\left(\frac{B-C}{d}\right)\times(R+d_2)ln\left(\frac{R+d_2}{R-d_1}\right)-(B-C)\right]-R^2$$

$$= \frac{12^3}{72}\left[3\,ln\left(\frac{12.+7}{12-5}\right)+\left(\frac{9-3}{12}\right)(12+7)ln\left(\frac{12+7}{12-5}\right)-(9-3)\right]-12^2$$

$$= 24\left[3\,ln\frac{19}{7}+\frac{19}{2}ln\frac{19}{7}-6\right]-144$$

$$= 24[3 \times 0.99853 + 9.5 \times 0.99853 - 6] - 144 = 24\,[2.9956 + 9.4860 - 6] - 144$$

$$= 24 \times 6.4816 - 144 = 155.584 - 144 = 11.584 \text{ cm}^2$$

Bendign stresses are:

$$\sigma_1 = \frac{Wx}{AR}\left[\frac{R^2}{h^2}\left(\frac{d_1}{R-d_1}\right)-1\right]$$

$$= \frac{10^5 \times 12}{720+12\times10^{-4}}\left[\frac{144}{11.584}\left(\frac{5}{12-5}\right)-1\right] = \frac{10^9}{72}[8.8792-1] = \frac{10^9 \times 7.8792}{72}$$

$$= 109.434 \text{ MPa (tensile)}$$

$$\sigma_2 = \frac{Wx}{AR}\left[1+\frac{R^2}{h^2}\left(\frac{d_2}{R+d_2}\right)\right]$$

$$= \frac{10^9 \times 12}{72\times12}\left[1+\frac{144}{11.584}\left(\frac{7}{12+7}\right)\right] = \frac{10^9}{72}[1+4.57982] = \frac{10^9 \times 5.57982}{72}$$

$$= 77.4975 \text{ MPa (compressive)}$$

Direct stress, $\quad \sigma_d = \dfrac{W}{A} = \dfrac{10^9}{72} = 13.889 \text{ MPa} \quad \text{(tensile)}$

Resultant stress on the inside fibres

$$= \sigma_1 + \sigma_d = 109.434 + 13.889 = 123.323 \text{ MPa (tensile)}$$

Resultant stress on the outside fibres

$$= \sigma_2 - \sigma_d = 77.495 - 13.889 = 63.6085 \text{ MPa (compressive)}$$

Example 13.2 *An open ring having T-section as shown in Fig. 13.10 is subjected to a compressive load of 100 kN. Detrmine the stresses at A and B.*

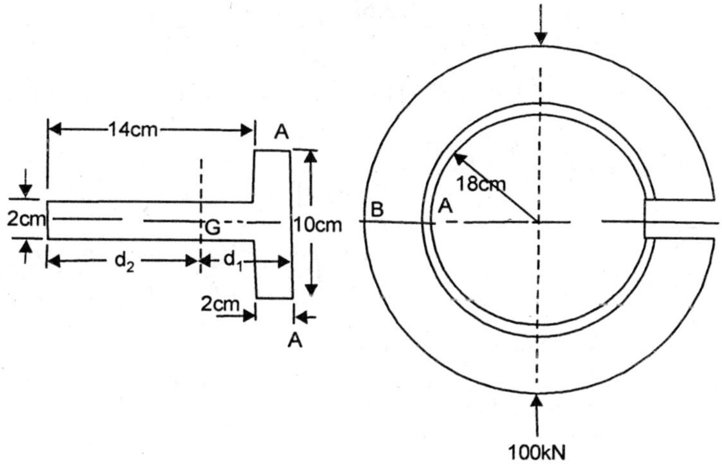

Fig. 13.10

Solution.

Given : Area of *T*-cross-section

$$A = 10 \times 2 + 14 \times 2 = 20 + 28 = 48 \text{ cm}^2$$

Taking moment about line *A-A*, we have

$$48 \times d_1 = 10 \times 2 \times 1 + 2 \times 14 \times 9 = 20 + 252 = 272$$

$$d_1 = 5.67 \text{ cm}; d = 16 - 5.67 = 10.33 \text{ cm}$$

$$r_1 = 18 \text{ cm}; r_2 = 18 + 2 = 20 \text{ cm}$$

$$r_3 = 18 + 16 = 34 \text{ cm}$$

$$R = 18 + 5.67 = 23.67 \text{ cm}$$

$$h^2 = \frac{R^3}{A}\left[b_2 ln\frac{r_2}{r_1} + t_1 ln\frac{r_3}{r_2}\right] - R^2$$

Here $b_2 = 10$ cm; $t_1 = 2$ cm

$\therefore$
$$h^2 = \frac{(23.67)^3}{48}\left[10\,ln\frac{20}{18} + 2\,ln\frac{34}{20}\right] - (23.67)^2$$

$$= \frac{13261.57}{48}[10 \times 0.10536 + 2 \times 0.53063] - 560.269$$

$$= 276.283\,[1.0536 + 1.06126] - 560.269 = 276.283 \times 2.11486 - 560.269$$

$$= 584.299 - 560.269 = 24.03 \text{ cm}^2$$

Here the bending moment is such that it is tending to increase the curvature of the ring. Therefore, it is positive in nature. Hence resultant stress at *A* becomes,

$$\sigma_A = -\frac{W}{A} + \frac{WR}{AR}\left[1 - \frac{R^2}{h^2}\cdot\left(\frac{d_1}{R - d_1}\right)\right]$$

$$= -\frac{10^9}{48} + \frac{10^9}{48}\left[1 - \frac{(23.67)^2}{24.03}\left(\frac{5.67}{23.67 - 5.67}\right)\right]$$

$$= -20.833 + 20.833\,[1 - 7.34] = 152.914 \text{ MPa (compressive)}$$

$$\sigma_B = -\frac{W}{A} + \frac{W}{A}\left[1 + \frac{R^2}{h^2}\left(\frac{d_2}{R + d_2}\right)\right]$$

$$= -\frac{10^9}{48} + \frac{10^9}{48}\left[1 + \frac{(23.67)^2}{24.03}\left(\frac{10.33}{23.67 + 10.33}\right)\right] = \frac{10^9}{48} \times \frac{(23.67)^2}{24.03} \times \frac{10.33}{34}$$

$$= 147.578 \text{ MPa (tensile)}$$

Example 13.3 *The load P applied to the frame shown in Fig. 13.11 is 2 kN. Calculate the resultant stresses at A and B. Also locate the position of the neutral axis.*

Solution.

Given : Area of cross-section

$$A = 4 \times 1.5 = 6 \text{ cm}^2$$

$$h^2 = 147.5\frac{R^3}{D} ln\left(\frac{2R + D}{2R - D}\right) - R^2$$

Here $R = 4$ cm; $D = 4$ cm

$$h^2 = \frac{4^3}{4} ln\left(\frac{2 \times 4 + 4}{2 \times 4 - 4}\right) - 4^2 = 16\, ln\frac{12}{4} - 16$$

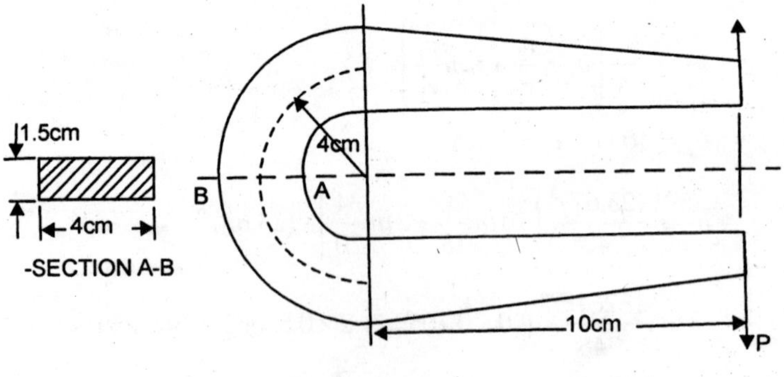

Fig. 13.11

$$= 16 \ln 3 - 16 = 16 \times 1.0986 - 16 = 17.5778 - 16 = 1.5775 \text{ cm}^2$$

$$M = -2 \times 10^3 \times 14 \times 10^{-2} = -280 \text{ N-m}$$

$$\sigma_A = \frac{200}{6 \times 10^{-4}} + \frac{280}{6 \times 4 \times 10^{-6}}\left[-1 + \frac{16}{1.5778}\left(\frac{2}{4 - 2}\right)\right]$$

$= 3.333 + 11.667 [-1 + 10.141] = 3.333 + 11.667 \times 9.141 = 3.333 + 106.645$
$= 109.978$ MPa (tensile)

$$\sigma_B = \frac{2000}{6\times10^{-4}} - \frac{280}{6\times4\times10^{-6}}\left[1+\frac{16}{1.5778}\left(\frac{2}{4.2}\right)\right]$$

$= 3.333 - 11.667 [1 + 3.3802] = 3.333 - 11.667 \times 4.3802 = 3.333 - 51.1042$
$= - 47.7712$ MPa (compressive)

Position of neutral axis is given by,

$$y = -\frac{Rh^2}{(R^2+h^2)} = \frac{-4\times1.5778}{(16+1.5778)} = -0.359 \text{ cm}$$

Hence neutral axis is at a radius of 4 – 0.359 = 3.641 cm.

13.4 STRESSES IN A RING

Consider a circular ring loaded as shown in Fig. 13.12. Let M_2 be the bending moment at any section $x_1 - x_2$ inclined at angle θ with the line of action of the applied load W. The portion x_1 DF x_2 of the ring is in equilibrium under the action of M_1 at DF, pull $\frac{W}{2}$ at DF and the moment M_2 at $x_1 - x_2$ along with pull T at $x_1 - x_2$, we get

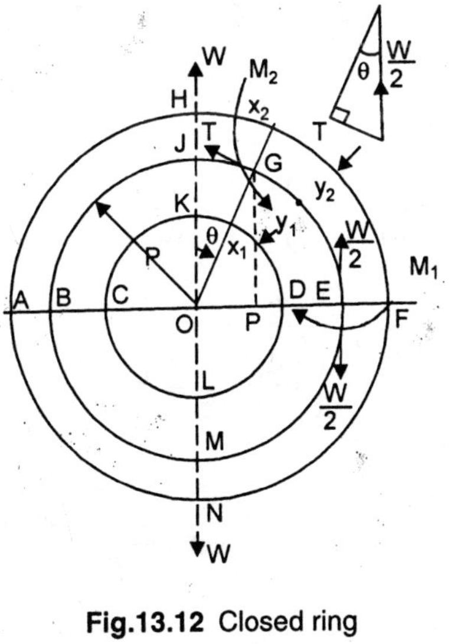

Fig.13.12 Closed ring

$$M_2 = M_1 + \frac{W}{2}R\times(1-\sin\theta) \quad \dots (a)$$

Also

$$M_2 = E(1+\varepsilon_0)\times\left(\frac{1}{R_1}-\frac{1}{R}\right)Ah^2 \quad \dots (b)$$

Comparign Eqs. (*a*) and (*b*), we get

$$E(1+\varepsilon_0)\left(\frac{1}{R_1}-\frac{1}{R}\right)Ah^2 = M_1+\frac{W}{2}R\times(1-\sin\theta) \quad \text{... (c)}$$

Multiplying both sides by $Rd\theta$ and integrating from 0 to $\pi/2$, we have

$$E\int_0^{\pi/2}(1+\varepsilon_0)\left(\frac{1}{R_1}-\frac{1}{R}\right)Ah^2.Rd\theta = \int_0^{\pi/2}M_1Rd\theta+\int_0^{\pi/2}\frac{W}{2}R^2\times(1-\sin\theta)d\theta$$

$$\text{or } \int_0^{\pi/2}\frac{R(1+\varepsilon_0)}{R_1}Ah^2d\theta-E\int_0^{\pi/2}(1+\varepsilon_0)Ah^2d\theta = \int_0^{\pi/2}M_1Rd\theta+\int_0^{\pi/2}\frac{W}{2}R^2\times(1-\sin\theta)d\theta$$

Now $$R_1 = R(1+\varepsilon_0)$$

$$\therefore \quad \int_0^{\pi/2}\frac{R(1+\varepsilon_0)}{R_1}d\theta = \frac{\pi}{2}$$

$$\therefore \quad \frac{\pi}{2}EAh^2-\frac{\pi}{2}\cdot EAh^2(1+\varepsilon_0) = \frac{\pi}{2}M_1R+\frac{WR^2}{2}\left(\frac{\pi}{2}-1\right)$$

$$-\frac{\pi}{2}EAh^2\varepsilon_0 = \frac{\pi}{2}M_1R+\frac{WR^2}{2}\left(\frac{\pi}{2}-1\right) \quad \text{... (d)}$$

Now $$T = \frac{W}{2}\sin\theta \quad \text{... (e)}$$

$$\therefore \quad EA\left[\varepsilon_0-(1+\varepsilon_0)\left(\frac{1}{R_1}-\frac{1}{R}\right)\frac{h^2}{R}\right] = \frac{W}{2}\sin\theta \quad \text{... (f)}$$

From Eq. (*c*), we have

$$E(1+\varepsilon_0)\left(\frac{1}{R_1}-\frac{1}{R}\right)\frac{Ah^2}{R} = \frac{M_1}{R}+\frac{W}{2}(1-\sin\theta)$$

Substituting in Eq. (*f*), we get

$$EA\varepsilon_0 = \frac{W}{2}\sin\theta+\frac{M_1}{R}+\frac{W}{2}(1-\sin\theta) = \frac{W}{2}+\frac{M_1}{R}$$

$$\therefore \quad \varepsilon_0 = \frac{W}{2EA}+\frac{M_1}{EAR} \quad \text{... (g)}$$

Substituting in Eq. (*d*), we get

$$M_1 = \frac{WR}{2}\left[\frac{2}{\pi}\left(\frac{R^2}{R^2+h^2}\right)-1\right] \quad \text{... (h)}$$

Substituting in Eq. (*a*), we get

$$M_2 = \frac{WR}{2}\left[\frac{2}{\pi}\left(\frac{R^2}{R^2+h^2}\right)-\sin\theta\right] \quad \text{... (i)}$$

M_2 will be maximum at $\theta = 0°$ and $180°$

$$\therefore \quad M_{max} = \frac{WR^3}{\pi(R^2+h^2)} \quad \text{... (j)}$$

M_2 will be zero, when

$$\sin\theta = \frac{2R^2}{\pi(R^2+h^2)} \quad \text{... (k)}$$

Thus bending moment and bending stress is zero at four points, one in each quadrant. Substituting the value of M_1 in Eq. (*g*) from (*h*), we get

$$\varepsilon_0 = \frac{W}{AE}\left[\frac{R^2}{\pi(R^2+h^2)}\right] \quad \text{... (l)}$$

$$\therefore \quad \varepsilon = \varepsilon_0 + (1+\varepsilon_0)\left(\frac{1}{R_1}-\frac{1}{R}\right)\cdot\left(\frac{y}{1+\frac{y}{R}}\right) \text{ and } \sigma = E\varepsilon$$

Also from Eq. (*b*), we have $(1+\varepsilon_0)\left(\frac{1}{R_1}-\frac{1}{R}\right) = \frac{M_1}{EAh^2}+\frac{WR}{2Ah^2E}(1-\sin\theta)$

$$\varepsilon = \varepsilon_0 + \left[\frac{M_1}{EAh^2}+\frac{WR}{2Ah^2E}(1-\sin\theta)\right]\left[\frac{Ry}{R+y}\right]$$

$$= \frac{W}{AE}\cdot\left\{\frac{R^2}{\pi(R^2+h^2)}\right\}+\left\{\frac{WR}{2}\left[\frac{2}{\pi}\left(\frac{R^2}{R^2+h^2}\right)-1\right]\times\frac{1}{EAh^2}+\frac{WR}{2Ah^2E}(1-\sin\theta)\right\}\left(\frac{Ry}{R+y}\right)$$

$$\therefore \quad \sigma = E\varepsilon = \frac{W}{A}\left[\frac{R^2}{\pi(R^2+h^2)}+\frac{R^2}{2h^2}\left\{\frac{2R^2}{\pi(R^2+h^2)}-\sin\theta\right\}\times\left(\frac{Ry}{R+y}\right)\right] \quad \text{... (m)}$$

Direct stress, $\quad \sigma_d = \dfrac{W\sin\theta}{2A}$

Resultant stress, $\quad \sigma_r = \sigma_d \pm \sigma$

(i) On a section taken along the line of action W, $\theta = 0°$ and the stresses become:

(*a*) At outside of ring,

$$\sigma_r = \frac{W}{\pi A}\left(\frac{R^2}{R^2+h^2}\right)\left[1+\frac{R^2}{h^2}\cdot\left(\frac{y_2}{R+y_2}\right)\right] \quad \text{... (13.15)}$$

and is tensile in nature.

(*b*) At inside of ring,

$$\sigma_r = \frac{W}{\pi A}\left(\frac{R^2}{R^2+h^2}\right)\left[\frac{R^2}{h^2}\left(\frac{y_1}{R-y_1}\right)-1\right] \quad \text{... (13.16)}$$

and is compressive in nature.

(*ii*) On a section perpendicular to the line of action of W, $\theta = 90°$, and the stresses becomes:

(*a*) At outside of ring,

$$\sigma_r = -\frac{W}{A}\left[\frac{R^2}{\pi(R^2+h^2)}+\frac{R^2}{2h^2}\left\{\frac{2R^2}{\pi(R^2+h^2)}-1\right\}\times\left(\frac{y_2}{R+y_2}\right)\right]+\frac{W}{2A} \quad \text{... (13.17)}$$

and is compressive in nature.

(*b*) At inside of ring,

$$\sigma_r = \frac{W}{A}\left[\frac{R^2}{2h^2}\left\{\frac{2R^2}{\pi(R^2+h^2)}-1\right\}\left(\frac{-y_1}{R-y_1}\right)+\frac{R^2}{\pi(R^2+h^2)}\right]+\frac{W}{2A} \quad \text{... (13.18)}$$

and is tensile in nature.

Example 13.4 *A ring made of 2.5 cm diameter steel bar carries a pull of 10 kN. Calculate the maximum tensile and compressive stresses in the material of the ring. The mean radius of the ring is 15 cm.*

Solution.

Given : Refer to Fig. 13.13

Area of cross-section, $A = \frac{\pi}{4}\cdot(2.5)^2 = 4.91\,\text{cm}^2$

$$h^2 = \frac{D^2}{16}+\frac{1}{128}\frac{D^4}{R^2} = \frac{(2.5)^2}{16}+\frac{1}{128}\times\frac{(2.5)^4}{15^2}$$

$$= 0.3906 + 1.3563\times10^{-3} = 0.392\ \text{cm}^2$$

$$\sigma_D = \frac{W}{\pi A}\left(\frac{R^2}{R^2+h^2}\right)\times\left[1+\frac{R^2}{h^2}\left(\frac{y_2}{R+y_2}\right)\right]$$

$$= \frac{10^8}{\pi\times4.91}\times\left(\frac{225}{225+0.392}\right)\left[1+\frac{225}{0.392}\left(\frac{1.25}{15+1.25}\right)\right]$$

$$= \frac{10^8\times225}{\pi\times4.91\times225.392}[1.44.1523]$$

$$= \frac{10^8\times225\times45.1523}{\pi\times4.91\times225.392} = 292.2\ \text{MPa (tensile)}$$

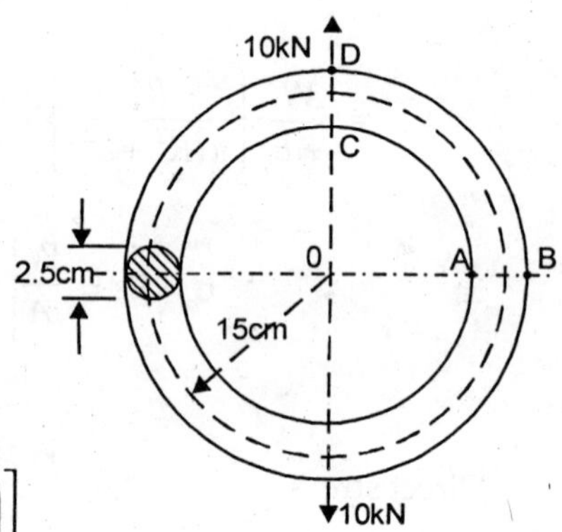

Fig. 13.13

$$\sigma_C = \frac{W}{\pi A}\left\{\frac{R^2}{R^2+h^2}\right\}\times\left[\frac{R^2}{h^2}\left(\frac{y_1}{R-y_1}\right)-1\right]$$

$$= \frac{10^8\times 225}{\pi\times 4.91\times 225.392}\left[\frac{225}{0.392}\times\frac{1.25}{13.75}-1\right]$$

$$= \frac{10^8\times 225}{\pi\times 4.91\times 225.392}\times 51.18 = 331.2\text{ MPa} \quad \text{(compressive)}$$

$$\sigma_A = \frac{W}{A}\left[\frac{R^2}{2h^2}\left\{\frac{2R^2}{\pi(R^2+h^2)}-1\right\}\left(\frac{y_1}{R-y_1}\right)-\frac{R^2}{\pi(R^2+h^2)}\right]+\frac{W}{2A}$$

$$= -\frac{10^8}{4.91}\left[\frac{225}{2\times 0.392}\left\{\frac{2\times 225}{\pi(225+0.392)}-1\right\}\left(\frac{1.25}{13.75}\right)-\frac{225}{\pi(225.392)}\right]+\frac{10^8}{2\times 4.91}$$

$$= \frac{10^8}{4.91}[-9.51-0.318]+10.183 = 200.163+10.183 = 210.346\text{ MPa (tensile)}$$

$$\sigma_B = -\frac{W}{A}\left[\frac{R^2}{\pi(R^2+h^2)}+\frac{R^2}{2h^2}\left(\frac{2R^2}{\pi(R^2+h^2)}-1\right)\left(\frac{y_2}{R+y_2}\right)\right]+\frac{W}{2A}$$

$$= \frac{10^8}{4.91}\left[\frac{225}{\pi\times 225.392}+\frac{225}{2\times 0.392}\left(\frac{2\times 225}{\pi\times 225.392}-1\right)\times\left(\frac{1.25}{16.25}\right)\right]+\frac{10^8}{2\times 4.91}$$

$$= \frac{10^8}{4.91}[0.31776-8.0465]+10.183 = -157.407+10.183 = -147.225\text{ MPa (compressive)}$$

∴ Maximum tensile stress = 292.2 MPa

Maximum com pressive stress = 331.2 MPa

Example 13.5 *A ring of mean diameter 100 mm is made of round steel 25 mm diameter. It is subjected to four equal pulls in two directions at right angles and possion through the centre. Determine the maximum value of the pulls if the maximum tensile stress is not to exceed 80 N/mm².*

Solution.

Given: $R = 5$ cm, $D = 2.5$ cm

$$A = \frac{\pi\times(2.5)^2}{4} = 4.91\text{ cm}^2$$

$$h^2 = \frac{D^2}{16}+\frac{1}{128}\frac{D^4}{R^2} = \frac{6.25}{16}+\frac{1}{128}\times\frac{(2.5)^4}{25} = 0.403\text{ cm}^2$$

$$\sigma_A = \frac{W}{\pi A}\left(\frac{R^2}{R^2+h^2}\right)\left[1+\frac{R^2}{h^2}\times\frac{y_2}{R+y_2}\right] = \frac{W}{\pi\times 4.91\times 10^{-4}}\left(\frac{25}{25.403}\right)\left[1+\frac{25}{0.403}\left(\frac{1.25}{6.25}\right)\right]$$

$= 8.5537 \times 10^3$ WN/m^2

$$\sigma_B = \frac{W}{\pi A}\left(\frac{R^2}{R^2+h^2}\right)\left[\frac{R^2}{h^2}\times\frac{y_1}{R-y_1}-1\right] = \frac{W}{\pi\times 4.91\times 10^{-4}}\left(\frac{25}{25.403}\right)\left[\frac{25}{0.403}\times\frac{1.25}{3.75}-1\right]$$

$= -12.554 \times 10^3$ W N/m^2

$$\sigma_C = \frac{W}{A}\left(\frac{R^2}{h^2}\right)\left\{\frac{2R^2}{\pi(R^2+h^2)}-1\right\}\left(-\frac{y_1}{R-y_1}\right)+\frac{R^2}{\pi(R^2+h^2)}+\frac{W}{2A} = 9.5208\times 10^3\ W\,N/m^2$$

$$\sigma_D = \frac{W}{A}\left[\frac{2R^2}{\pi(R^2+h^2)}+\frac{R^2}{2h^2}\left\{\frac{R^2}{\pi(R^2+h^2)}-1\right\}\left(\frac{y_2}{R+y_2}\right)\right]+\frac{W}{2A} = -5.099\times 10^3\ W\,N/m^2$$

Lo cation	*Vertical Load*	*Horizontal load*	*Resultant stress*
A	8.5537×10^3 W	-5.0990×10^3 W	3.4547×10^3 W
B	-12.5548×10^3 W	9.5208×10^3 W	-3.0340×10^3 W
C	9.5208×10^3 W	-12.5548×10^3 W	-3.0340×10^3 W
D	-5.0990×10^3 W	8.5537×10^3 W	3.4547×10^3 W

Maximum stress $= 3.4547 \times 10^3$ W N/m^2 = 80 × 106

W = 23.157 kN

Example 13.6 *A curved beam of mean radius of curvature 100 mm has recatngular cross-section 30 mm wide and 50 wide and 50 mm deep. Determine the location of neutral axis and resultant stresses on the top and bottom fibres if a positive bending moment of 1 kN.m is applied to the beam.*

Solution.

Given: B = 30 mm, D = 50 mm, R = 100 mm, M = + 1 kN.m

Refer to Fig. 13.14

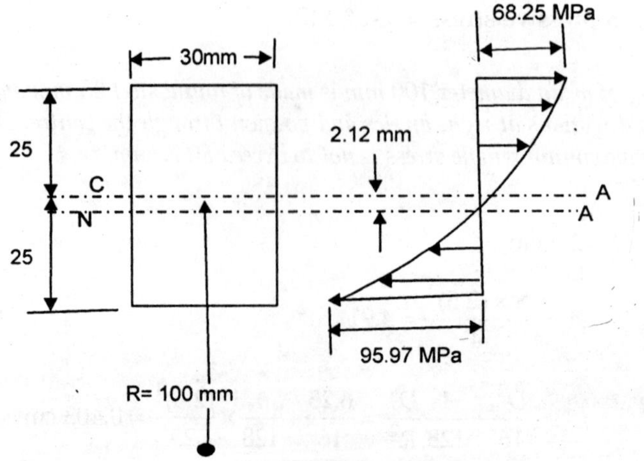

Fig. 13.14

$$A = B \times D = 30 \times 50 = 1500 \text{ mm2}$$

$$h^2 = \frac{R^3}{D} ln\left(\frac{2R+D}{2R-D}\right) - R^2$$

$$= \frac{(100)^3}{50} ln\left(\frac{2\times100+50}{2\times100-50}\right) - (100)^2$$

$$= 216.5 \quad \text{mm}^2$$

Position of neutral axis,

$$y = -\left(\frac{Rh}{R^2+h^2}\right) = \frac{160\times216.5}{[(100)^2+216.5]} = -2.12 \text{ mm}$$

–ve sign indicates that neutral axis is below the centroidal axis.

$$\sigma = \frac{M}{AR}\left[1+\frac{R^2}{h^2}\left(\frac{\pm y}{R\pm y}\right)\right]$$

At the top fibres, y is +ve. $y_t = y_c = 25$ mm

$$\therefore \qquad \sigma_{top} = \frac{1\times10^6}{1500\times100}\left[1+\frac{(100)^2}{216.6}\left(\frac{25}{100+25}\right)\right]$$

$$= 68.25 \text{ MPa (tensile)}$$

At bottom fibres, y is –ve.

$$\sigma_{bottom} = \frac{M}{AR}\left[1+\frac{R^2}{h^2}\left(\frac{-y}{R-y}\right)\right]$$

$$= \frac{1\times10^6}{1500\times100}\left[1-\frac{(100)^2}{216.5}\times\left(\frac{25}{100-25}\right)\right]$$

$$= -95.97 \text{ MPa (compressive)}$$

The stress distribution is hyperbolic in nature.

SUMMARY

1. In a deeply curved beam, the natural axis does not coincide with the centroidal axis.
2. The stress distribution in a curved beam is hyperbolic in nature.
3. Position of neutral axis.

$$y = -\left[\frac{Rh^2}{R^2+h^2}\right]$$

4. Value of h^2

$$h^2 = \frac{R^3}{A}\int\frac{dA}{y+R} - R^2$$

(i) Rectangular cross-section.

$$h^2 = \frac{R^3}{D} ln\left[\frac{2R+D}{2R-D}\right] - R^2$$

(ii) Trapezoidal cross-section

$$h^2 = \frac{R^3}{A}\left[C\,ln\left(\frac{R+d_2}{R-d_1}\right) + \left(\frac{B-C}{d}\right)(R+d_2)\,ln\left(\frac{R+d_2}{R-d_1}\right) - (B-C)\right] - R^2$$

where $A = \left(\frac{B+C}{2}\right)d, d_1 = \frac{d}{3}\left(\frac{B+2C}{B+C}\right), d_2 = d - d_1$

(iii) Circular cross-section

$$h^2 = \frac{D^2}{16} + \frac{1}{128}\frac{D^4}{R^2} + \cdots$$

(iv) Triangular cross-section

$$h^2 = \frac{2R^3}{h}\left[\left\{\frac{3R+2h}{3h}\right\}\ln\left\{\frac{3R+2h}{3R-h}\right\} - 1\right] - R^2$$

5. Bending Stresses

$$\sigma = \frac{M}{AR}\left[1 + \frac{R^2}{h^2}\left(\frac{\pm y}{R \pm y}\right)\right]$$

M is taken +ve if it increases the curvature, i.e. R decreases. In that case take +ve sign before y and σ shall be tensile in nature.

6. Crane hook having trapezoidal cross-section

Bending stress on the inner surface of hook,

$$\sigma_1 = \frac{Wx}{AR}\left[\frac{R^2}{h^2}\left(\frac{d_1}{R-d_1}\right) - 1\right] \text{(tensile)}$$

Bending stress on the outer surface of hook,

$$\sigma_2 = \frac{Wx}{AR}\left[1 + \frac{R^2}{h^2}\left(\frac{d_2}{R+d_2}\right)\right] \text{(compressive)}$$

Direct stress, $\sigma_d = \frac{W}{A}$ (tensile)

Resultant stresses are:

$$\sigma_{r1} = \sigma_1 + \sigma_d$$

$$\sigma_{r2} = \sigma_2 - \sigma_d$$

7. Stresses in a ring.

$$\sigma = \frac{W}{A}\left[\frac{R^2}{\pi\left(R^2+h^2\right)}+\frac{R^2}{2h^2}\left\{\frac{2R^2}{\pi\left(R^2+h^2\right)}-\sin\theta\right\}\times\left(\frac{Ry}{R+y}\right)\right]$$

$$\sigma_d = \frac{W\sin\theta}{2A}$$

$$\sigma_r = \sigma_d \pm \sigma$$

(i) At $\theta = 0°$ i.e. along the line of action of W

(a) Outside of ring

$$\sigma_r = \frac{W}{\pi A}\left[\frac{R^2}{R^2+h^2}\right]\left[1+\frac{R^2}{h^2}\left(\frac{y_2}{R+y_2}\right)\right] \text{(tensile)}$$

(b) Inside of ring

$$\sigma_r = \frac{W}{\pi A}\left[\frac{R^2}{R^2+h^2}\right]\left[\frac{R^2}{h^2}\left(\frac{y_1}{R-y_1}\right)-1\right] \text{(compressive)}$$

(ii) At $\theta = 90°$

(a) outside of ring

$$\sigma_r = \frac{W}{A}\left[\frac{R^2}{\pi\left(R^2+h^2\right)}+\frac{R^2}{2h^2}\left\{\frac{2R^2}{\pi\left(R^2+R^2\right)}-1\right\}\times\left(\frac{y_2}{R+y_2}\right)\right]+\frac{W}{2A} \text{(compressive)}$$

(b) Inside of ring

$$\sigma_r = \frac{W}{A}\left[\frac{R^2}{2h^2}\left\{\frac{2R^2}{\pi\left(R^2+h^2\right)}-1\right\}\times\left(\frac{-y_1}{R-y_1}\right)+\frac{R^2}{\pi\left(R^2+h^2\right)}\right]+\frac{W}{2A} \text{(tensile)}$$

Short answer type Questions

1. Define a curved beam.
2. What is the location of neutral axis with respect to the centroidal axis for a curved beam?
3. What is the nature of stress in a curved beam?
4. Why trapezoidal cross-section is preferred for a crane hook?
5. What are the assumptions made in the Winkler-Back theory?
6. What is the significance of h^2.

EXERCISES

1. A curved bar 40 mm × 60 mm has a mean radius of curvature of 50 mm. The bar is subjected to a bending moment of 400 N.m. tending to straighten it. Calculate the stresses at the inner and outer faces of the bar.
2. A crane hook having trapezoidal cross-section is 40 mm wide at inside and 30 mm wide at outside. The thickness of section is 36 mm. The crane hook carries a load of 10kN whose line of action is 30 mm from the inside edge of the section. The centre of curvature is 40 mm from the inside edge. Calculate the maximum tensile and compressive stresses in the section.

3. A frame of horse shoe type is shown in Fig. 13.15. Determine the stresses at the inside and outside of the curved surfaces.

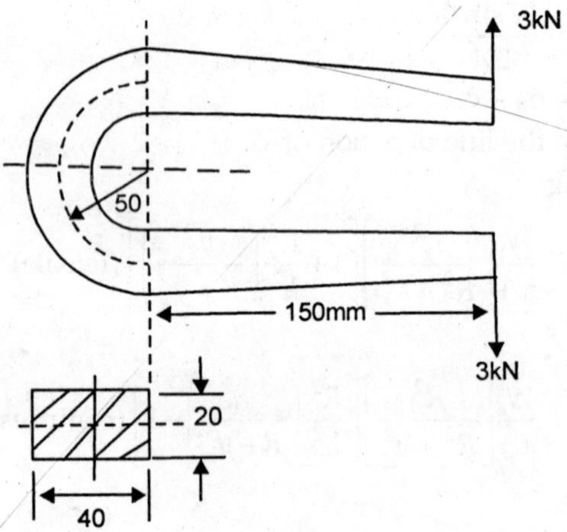

Fig. 13.15

4. A curved bar is formed of a tube 50 mm outside radius and 6 mm thick. The centre line of the bar has radius of curvature of 160 mm. A bending moment of magnitude 3 kN.m. tending to increases the curvature of the bar is applied. Calculate the maximum tensile and compressive stresses developed in the bar.
5. A ring is made of round steel bar 30 mm diameter with mean diameter 150 mm. Calculate the maximum stresses in the ring perpendicular to the load line if the ring is pulled by a force of 10 kN.

Chapter-14

Unsymmetrical Bending

14.1 INTRODUCTION

Bending caused by loads that do not lie in (or parallel to) a plane that contains the principal centroidal axes of ineritia of the cross-section is usually called unsymmetrical bending. In this chapter, we shall study the stresses caused by unsymmetrical bending and then determine the method to locate the shear centre of beams.

14.2 DEFINITIONS AND BASIC FORMULAE

1. Centroid

$$\bar{x} = \frac{\Sigma A_i x_i}{\Sigma A_i}, \; \bar{y} = \frac{\Sigma A_i y_i}{\Sigma A_i} \quad \text{... (14.1)}$$

2. Moment of inertia

$$I_x = \int y^2 \cdot dA, \; I_y = \int x^2 \cdot dA \quad \text{... (14.2)}$$

$I = AK^2$, where K = radius of gyration.

3. Parallel axes theorem

$$I = I_{\bar{x}} + A\bar{y}^2 \; I_y = I_{\bar{y}} + A\bar{x}^2 \quad \text{... (14.3)}$$

4. Product of inertia

It may be defined as the sum of the products obtained by multiplying each element of the area by the product of the two coordinates of the centroid of the element with respect to two rectangular axes. Thus

$$I_{xy} = \int xy.dA$$

The product of inertia may be positive, negative or zero.

5. Parallel axis theorem for product of inertia.

If $x'x'$ and $y'y'$ are the axes through the centroid C of the area and xx and yy are paralle axes through any other point O, then by definition.

$$I_{xy} = \int (x'+\bar{x})(y'+\bar{y})dA$$

$$= \int x'y'dA + \bar{x}\,\bar{y}\int dA + \bar{y}\int x'dA + \bar{x}\int y'dA$$

$$= \bar{I}_{xy} + A\bar{x}\,\bar{y} + 0 + 0$$

$$= \bar{I}_{xy} + A\bar{x}\,\bar{y} \quad \text{... (14.5)}$$

6. Tananansformation laws for moment and product of inertias

Let I_x, I_y, I_{xy} with respect to x–y coordinate system be known and we are interseted to known $I_{x'}$, $I_{y'}$, $I_{x'y'}$ with respect to x'– y' axes which have been obtained after rotation of x-y axes through an angles θ in the anti-clockwise direction, then

$$\left.\begin{aligned} I_{x'} &= I_x \cos^2\theta + I_y \sin^2\theta - I_{xy}\sin 2\theta \\ I_{y'} &= I_x \sin^2\theta + I_y \cos^2\theta + I_{xy}\sin 2\theta \\ I_{x'y'} &= I_{xy}\cos^2 2\theta + \frac{1}{2}(I_x - I_y)\sin 2\theta \end{aligned}\right\} \quad \text{... (14.6)}$$

7. Principal axes

The axes for which product of inertia is zero are called principal axes.

$$\therefore \quad I_{xy}\cos 2\theta + \frac{1}{2}(I_x - I_y)\sin 2\theta = 0$$

$$\tan 2\theta = \frac{2I_{xy}}{I_y - I_x} \quad \text{... (14.7)}$$

If $I_{xy} = 0$, then $\theta = 0$. Hence axes of symmetry are pricipal axes.
If $I_x = I_y$, then $\theta = 45°$

8. Flexure or bending axis

Bending axis of a beam is the longitudinal axis through which the transverse bending loads must pass in order that the bending of the beam shall not be accompained by twisting of the beam.

14.3 STRESSES DUE TO UNSYMMETRICAL BENDING

The following assunpitions are made:

1. The beam is origianally strrainght and of unifrom cross-section.
2. Loads are assumed to act through the bending axis of the beam so that the beam is not subjected to torsion in addition to bending.

3. Loads act in a direction perpendicular to the bending axis so that bending stresses are not accompanied by direct axial tensile or compressive stresses.

Let x', y' = Perpendicular axes lying in a transverse section of the beam and passing through any arbitrary chosen directions.

x, y = Perpendicular axes lying in a transverse section of the beam and passing through the centroid of the section, but having any arbitrarily chosen directions.

U, V = Centroidal principal axes of inertia of a transverse sectin of beam.

ϕ = Angle between the plane of loads and the vertical longitudinal plane.

0 = Shear centre.

C = *Centrroid.*

α = Angle between the principal axis of inertia CU and the neutralaxis CN and

N-N = Neutral axis.

u, v = Perpendicular distances of a point from principal axes U and V respectively.

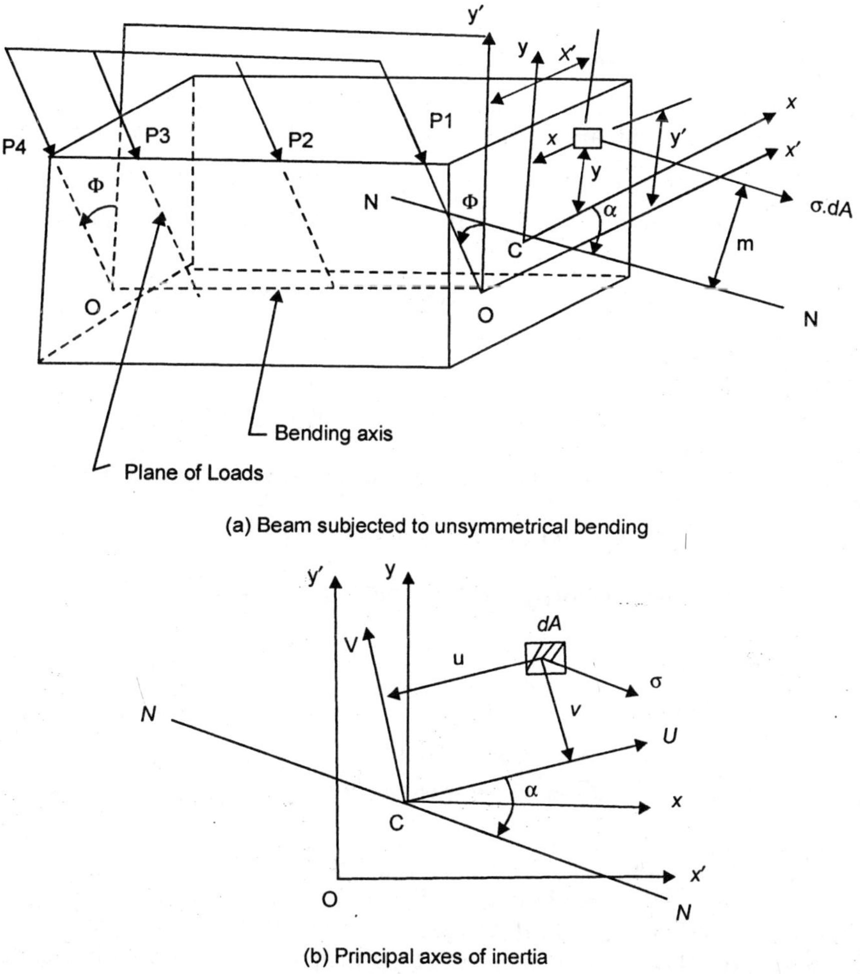

Fig. 14.1 Analysis of stress in unsymmetrical bending

Fig.14.1 (a) shows abeam subjected to unsymmetrical bending. Axes x' and y' are chosen here as horizontal and vertical through the shear centre O. The axes x and y pass through the centroid C and are parallel to the axes Ox' and Oy'. Neglecting shear stresses, the three equations for equilibrium of forces give,

$$\int \sigma dA = 0$$

$$M_{x'} = \int y' \sigma dA$$

$$M_{y'} = \int x' \sigma dA$$

The moments $M_{x'}$ and $M_{y'}$ are the components of the bending moment M that lie in the vertical and horizontal planes respectively.

The stress at any point is proportional to the distance m from the neutral axis as long as the elastic strength of the material is not exceeded.

$$\sigma = km$$

where k = a constant

$$\int \sigma dA = k \int m dA = 0$$

Therefore, the axis from which m is measured, *i.e.* the neutral axis, must pass through the centroid C of the cross-sectional area and not through the shear centre O. Now

$$m = (y' - \bar{y}) \cos\alpha + (x' - \bar{x}) \sin\alpha = y \cos\alpha + x \sin\alpha$$

where $(y'-y)$ and $(x'-x)$ are the distances from Ox' and Oy' respectively of the centroid of the section.

$$\therefore \quad \sigma = km = k\,[(y' - \bar{y})\cos\alpha + (x' - \bar{x})\sin\alpha] = k\,(y \cos\alpha + x \sin\alpha)$$

$$M_{x'} = \int y' \sigma dA$$

$$= k \int [(y'^2 - y'\bar{y})\cos\alpha + (x'y' - \bar{x}y')\sin\alpha] dA$$

$$= k \cos\alpha \left[\int y'^2 dA - \bar{y}\int x' dA\right] + k \sin\alpha \left[\int x'y' dA - \bar{x}\int y' dA\right]$$

$$= k \cos\alpha\,[I_{x'} - \bar{y}\,.\bar{y}\,A] + k \sin\alpha\,[I_{x'y'} - \bar{x}\bar{y}\,A]$$

$$= k \cos\alpha\,[I_{x'} - A\bar{y}^2] + k \sin\alpha\,[I_{x'y'} - A\bar{x}\bar{y}]$$

But $I_{x'} - A\bar{y}^2 = I_x$

and $I_{x'y'} - A\bar{x}\bar{y} = I_{xy}$

$$\therefore \quad M_{x'} = kI_x \cos\alpha + kI_{xy} \sin\alpha$$

where I_x = Moment of inertia of the cross-sectional area with respect to a centroidal axis Cx parallel to ox'.

I_{xy} = Product of inertia with respect to a set of rectangular centroidal axes Cx and Cy parallel to ox' and oy' respectively.

Since the axes Cx and Cy are any arbitrary set of perpendicular axes passing through the centroid of the section, they may be chosen to be principal axes of inertia designated as U and V as shown in Fig. 14.1 (b).

$\therefore \quad M_{x}' = M_x = M_u = k\, I_u \cos \alpha = M \cos \phi$

Since $\quad I_{uv} = 0$ for principal axes.

Similarly $\quad M_{y'} = M_y = \int x'\sigma dA = k\, I_{xy} \cos\alpha + kI_y \sin\alpha$

$\therefore \quad M_v = k\, I_v \sin \alpha = M \sin \phi$

and $\quad \sigma = k\,(v \cos \alpha + u \sin \alpha)$

Now $$\frac{M_v}{M_u} = \tan \phi = \frac{I_v}{I_u} \tan\alpha$$

$$\therefore \quad \tan \alpha = \frac{I_u}{I_v} \tan\phi \qquad ...(14.8)$$

Which gives the direction of the neutral axis. Angle ϕ is measured from V principal axis and α is measured from U principal axis in clockwise direction.

Also $$\cos \alpha = \frac{M_u}{k\, I_u}$$

$$\sin \alpha = \frac{M_v}{k\, I_v}$$

$$\therefore \quad \sigma = \frac{M_u . v}{I_u} + \frac{M_v . u}{I_v}$$

$$= \frac{(M\cos\phi)v}{I_u} + \frac{(M\sin\phi)u}{I_v} \qquad ...(14.9)$$

Sign *convention.* M_u is considered positive if it produces tensile stress at points that have a positive v co-ordinate. Similarly M_v is positive if it produces a tensile stress at points that have a positive u co-ordinate. Generally it is more preferable to assign positive sign to the term $M_u.v/I_u$ or $\frac{M_v u}{I_v}$ if it produces tension at a point and negative sign if it produces compression.

14.3.1 Combined Bending and Axial Loads

If the beam is subjected to loads that are not perpendicular to the bending axis of the beam, then these loads can be resolved into components along the bending axis and perpendicular to it. The component along the bending axis will produce direct stress and component perpendicular to it will produce bending stress. The resultant stress will be,

$$\sigma = \frac{P}{A} + \frac{M_u . v}{I_u} + \frac{M_v . u}{I_v} \qquad ...(14.10)$$

Example 14.1 *A 20 cm × 20 cm angle is loaded as shown in Fig. 14.2. The total load P being 50 kN. Find the direction of the neutral axis between the loads. Also calculate the values of bending stresses at A, B and C.*

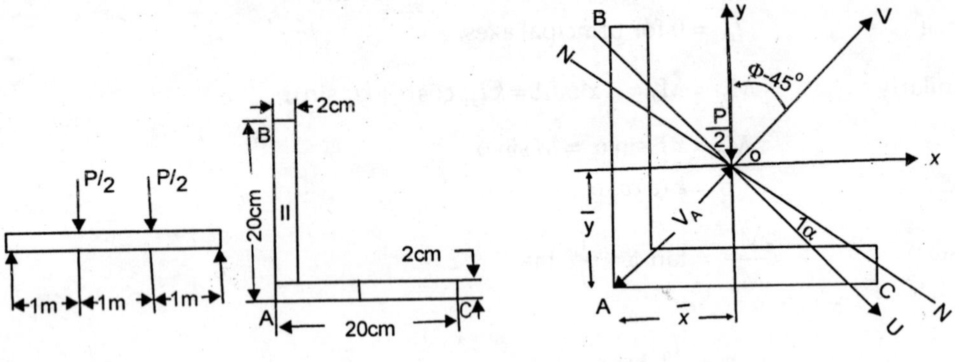

Fig. 14.2

Solution. $A_1 = 20 \times 2 = 40 \text{ cm}^2$; $A_2 = 18 \times 2 = 36 \text{ cm}^2$; $A = 40 + 36 = 76 \text{ cm}^2$

$$\bar{x} = \frac{40 \times 10 + 36 \times 1}{76} = \frac{436}{76} = 5.737 \text{ cm};$$

$$\bar{y} = \frac{40 \times 1 + 36 \times 11}{76} = \frac{436}{76} = 5.737 \text{ cm}$$

$$I_x = \frac{20 \times 2^3}{12} + 40 + (4.737)^2 + \frac{2 \times 18^3}{12} + 36 \times (11 - 5.737)^2$$
$$= 1333.333 + 725.685 + 12 + 805.810 = 2880.828 \text{ m}^4$$

$$I_y = \frac{2 \times 20^3}{12} + 40 \times (10 - 5.737)^2 + \frac{18 \times 2^3}{12} + 36 \times (5.737 - 1)^2 = 2880.828 \text{ cm}^4$$

$$I_{xy} = 40 \times 4.263 \times (-4.737) + 36 \times (-4.737)(5.263)$$
$$= -807.753 - 897.510 = -17050263 \text{ cm}^4$$

$$\tan 2\theta = \frac{2I_{xy}}{I_y - I_x} = -\infty$$

$$\theta = -45°$$

$$I_u = I_x \cos^2\theta + I_y \sin^2\theta - I_{xy} \sin 2\theta = 2880.828 - 1705.263 = 1175.565 \text{ cm}^4$$

$$I_v + I_u = I_x + I_y = 5761.656$$

$$I_v = 5761.656 - 1175.565 = 4586.091 \text{ cm}^4$$

$$\tan\alpha = \frac{I_u}{I_v}\tan\phi = \frac{1175.565}{4856.091} \times \tan(-45°) = -\frac{1175.565}{4586} = -0.2563$$

$$\alpha = -(14° \, 22')$$

$$M = 25 \times 10^3 \times 1 = 25 \text{ kN-m}$$

$$M_u = M_v = M \cos 45° = 25 \times \frac{1}{\sqrt{2}} = 17.67 = 17.67 \text{ kN-m}$$

With reference to Fig. 14.3.

Co-ordinates of point *A* are

$$u_A = 0$$

$$v_A = -\sqrt{(5.737)^2 + (5.737)^2} = -8.113 \text{ cm.}$$

M_u produces compression everywhere and M_v produces tension.

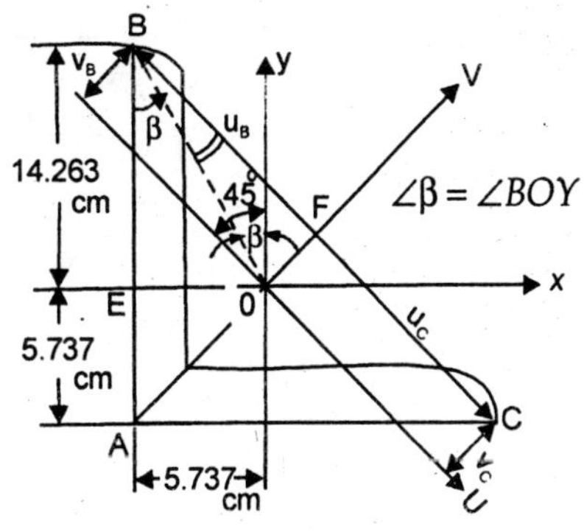

Fig. 14.3

$$\therefore \quad M_u = -17.67 \text{ kN-m}$$

$$M_v = +17.67 \text{ kN-m}$$

$$\sigma_A = \frac{M_u . v_A}{I_u} + \frac{M_v . u_A}{I_v}$$

$$= \frac{-17.67 \times 10^3 \times (-8.113) \times 10^{-2}}{1175.565 \times 10^{-8}}$$

$$= 122 \text{ MPa (tension)}$$

$$\tan \beta = \frac{OE}{BE} + \frac{5.737}{14.264} = 0.40223$$

Now $\beta = 21.91°$

$$\angle OBF = 46° - 21.91° = 23.09°$$

$$u_B = -OB \cos 23.09 = -\sqrt{OE^2 + BE^2} \times \cos 23.09°$$

$$= \sqrt{(5.737)^2 + (14.263)^2} \times \cos 23.09° = -\sqrt{32.913 + 203.433} \times \cos 23.09°$$

$$= -\sqrt{236.346} \times 0.91989 = -15.3735 \times 0.91989 = -14.142 \text{ cm.}$$

$$v_B = OB \sin 23.09° = 15.3735 \times 0.3222 = 6.03 \text{ cm}$$

$$\sigma_B = \frac{M_u . v_B}{I_u} + \frac{M_v . u_B}{I_v}$$

$$= \frac{-17.67 \times 10^3 \times 6.03 \times 10^{-2}}{1175.565 \times 10^{-8}} + \frac{17.67 \times 10^{-3} \times (-14.142) \times 10^{-2}}{4581.091 \times 10^{-8}}$$

$$= -90.67 + 54.77 = -35.90 \text{ MPa (compression)}$$

Example 14.2 *A standard I- beam as shown in Fig. 14.4 is bent by equal and opposite couples M acting at the ends of the beam in the plane m.m Find the maximmum stress*

$$I_u = 2400 \text{ cm}^4, I_v = 150 \text{ cm}^4, M = 5 \text{ kN-}$$
$$l = 3 \text{ m}, \phi = 30°, E = 200 \text{ GPa.}$$

Solution.

Given : Maximmum stress occurs at point A, where

$$u_A = 5 \text{ cm}$$

$$v_A = 10 \text{ cm}$$

$$\sigma_{max} = \frac{M_u \times v_A}{I_u} + \frac{M_v \times u_A}{I_v} = \frac{M \cos\phi \times v_A}{I_u} + \frac{M \sin\phi \times u_A}{I_v}$$

$$= \frac{5\times10^3\times0.866\times10\times10^6}{2400}+\frac{5\times10^{-3}\times0.5\times5\times10^6}{150}$$
$$= 18.0417 + 83.3333 = 101.375 \text{ MPa}$$

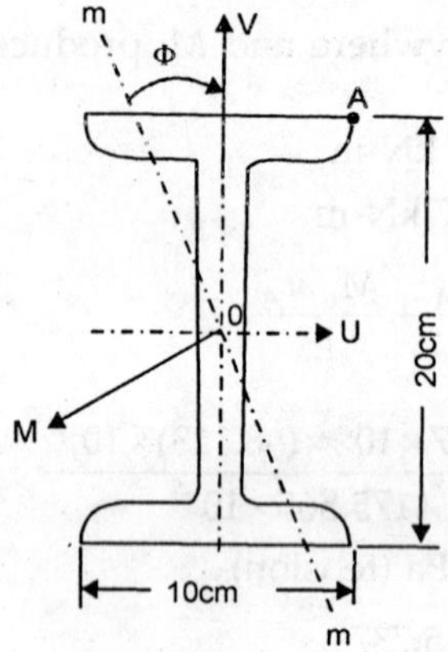

Fig. 14.4

Example 14.3 *A cantilever beam of I-section is used to support the loads inclined to the V-axis as shown in Fig. 14.5. Calculate the stresses at the corners A, B, C and D. Also locate the netural axis.*

Solution.

Given : Resolving loads vertically and calculating M_u, we have

$$M_u = 10^4 \cos 30° \times 1.5 + 5 \times 10^3 \cos 45° \times 2.5$$
$$= 10^4 \times 0.866 \times 1.5 + 5 \times 10^3 \times 0.707 \times 2.5$$
$$= 12.99 + 8.8375 = 21.8275 \text{ kN.m}$$

M_u is positive for points A and B and negative for C and D.

Resolving loads horizontally and calculating M_v, we have

$$M_v = -10^4 \sin 30° \times 1.5 + 10^3 \sin 45° \times 2.5$$
$$= -10^4 \times 0.5 \times 1.5 + 5 \times 10^3 \times 0.707 \times 2.5$$
$$= -7.5 + 8.8375 = 1.3375 \text{ kN-m}$$

M_v is positive for points A and C and negative for B and D.

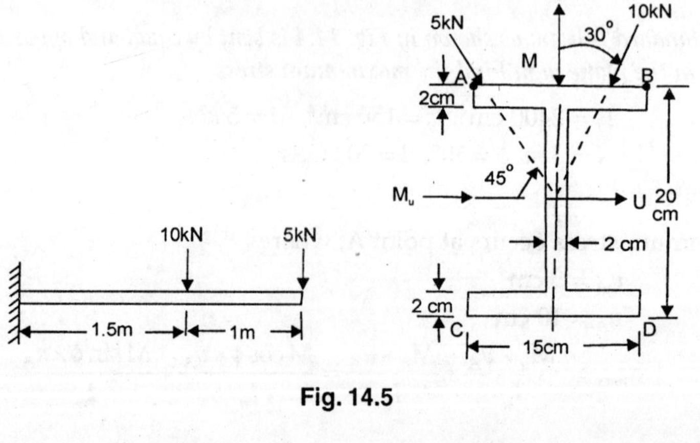

Fig. 14.5

$$I_u = \frac{15 \times 20^3}{12} - \frac{13 \times 16^3}{12} = 1000 - 4437.33 = 5562.67 \text{ cm}^4$$

$$I_v = \frac{2 \times 2 \times 15^3}{12} + \frac{16 \times 2^3}{12} = 1125 + 10.67 = 1135.67 \text{ cm}^4$$

Now $$\sigma = \frac{M_u . v}{I_u} + \frac{M_v . u}{I_v}$$

For point A, $u_A = 7.5$ cm, $v_A = 10$ cm

$M_u = + 21.8275$ kN-m

$M_v = + 1.3375$ kN.m

$$\therefore \quad \sigma_A = \frac{21.8275 \times 10 \times 10^9}{5562.67} + \frac{1.3375 \times 7.5 \times 10^9}{1135.67} = 39.24 + 8.33 = 48.07 \text{ MPa}$$

For point B, $u_B = 7.5$ cm, $v_B = 10$ cm

$M_u = + 21.8275$ kN-m, $M_v = - 1.3375$ kN.m

$$\sigma_B = \frac{21.8275 \times 10^9 \times 10}{5562.67} + \frac{1.3375 \times 7.5 \times 10^9}{1135.75} = 39.24 - 8.83 = 30.41 \text{ MPa}$$

For point C, $u_C = 7.5$ cm, $v_C = 10$ cm

$M_u = 21.8275$ kN-m

$M_v - +1.3375$ kN-m

$$\sigma_C = \frac{-21.8275 \times 10 \times 10^9}{5562.67} + \frac{1.3375 \times 7.5 \times 10^9}{1135.75} = -39.24 + 8.83 = -30.41 \text{ MPa}$$

For point D, $u_D = 7.5$ cm, $v_D = 10$ cm

$M_u = -21.8275$ *kN.m*, $M_v = - 1.3375$ kN-m

$$\sigma_D = \frac{-21.8275 \times 10^9 \times 10}{5562.67} + \frac{-1.3375 \times 7.5 \times 10^9}{1135.67} = -39.24 - 8.33 = -48.07 \text{ MPa}$$

$$\tan\alpha = \frac{I_u}{I_v} \cdot \tan\phi = \frac{5562.67}{1135.67} \times \frac{M_u}{M_v} = \frac{5562.67}{1135.67} \times \frac{1.3375}{21.8275} = 0.30014$$

$$\alpha = 16.70°$$

Example 14.3 *A simply supported beam of T-section, 2.5 m long carries a central concentrated load inclined at 30° the y-axis as shown in Fig. 14.6. If the maximum compressive and tensile stresses in bending are not to exceed 75 MPa and 35 MPa respectively, find the maximum load the beam can carry.*

Solution.

Given : $A_1 = 10 \times 2 = 20$ cm^2

$A_2 = 13 \times 2 = 26$ cm^3

$A = 20 + 26 = 46$ cm^2

$$\bar{y} = \frac{20\times1+26\times8.5}{46} = \frac{20+221}{46} = \frac{241}{46} = 5.239 \text{ cm}$$

$$I_x = \frac{10\times2^3}{12} + 20\times(4.239)^2 + \frac{2\times13^3}{12} + 26(8.5-5.239)^2$$

$$= 6.667 + 359.392 + 366.167 + 276.487 = 978.703 \text{ cm}^4$$

$$I_y = \frac{2\times10^3}{12} + \frac{13\times2^3}{12} = 166.667 + 8.667 = 175.334 \text{ cm}^4$$

$$I_{xy} = 0$$

$$\therefore \quad I_x = I_u$$

and $\quad I_y = I_v$

$$\tan\alpha = \frac{I_u}{I_v}\cdot\tan\phi = \frac{9780.703}{175.334}\times\tan 30°$$

$$= \frac{978.703}{175.334}\times\frac{1}{\sqrt{3}} = 3.2228$$

$$\alpha = 72.76°$$

Let P be the load in N that the beam can carry safely.

$$\therefore \quad M_u = \frac{P\cos 30°\times2.5}{4} = \frac{P\times0.866\times2.5}{4} = 54.125\times10^{-2}\times P \text{ N-m}$$

$$M_v = \frac{P\sin 30°\times2.5}{4} = \frac{P\times0.5\times2.5}{4} = 31.25\times10^{-2}\times P \text{N-m}$$

Maximum compressive stress will occur at point A.

$$\therefore \quad \sigma_A = \frac{M_u\cdot v_A}{I_u} + \frac{M_v\cdot u_A}{I_v}$$

$$75 = \frac{54.12\times10^{-2}\times P\times5.239}{978.703} + \frac{31.25\times10^{-2}\times P\times5}{175.334}$$

$$= (0.28973+0.89116)P\times10^{-2}$$

$$75 = 1.1808\ P\times10^{-2}$$

$$P = 6351.14 \text{ N}$$

Maximum tensile stress will occur at point B.

$$\sigma_B = \frac{M_u\times v_B}{I_u} + \frac{M_u\times u_B}{I_u}$$

$$35 = (-0.2893 + 8.89116)\ P\times10^{-2}$$

$$35 = 3.60143\ P\times10^{-2}$$

$$P = 5819.46\ N$$

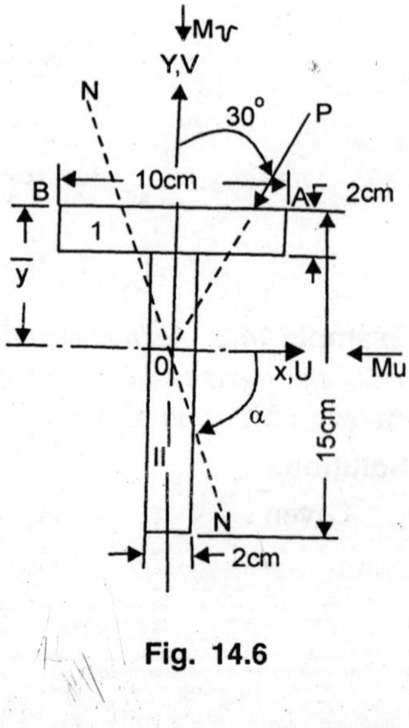

Fig. 14.6

Example 14.5 *If the maximum bending stress allowed in the cross-section of the beam shown in Fig. 14.7 is 15 MPa, determine the value of P.*

Solution.

$$\tan\theta = \frac{7.5}{5} = 1.5$$

$$\theta = 56.31°$$

$$M = 0.6\,P$$

$$M_u = M\cos(90° - \theta) = 0.6\,P\cos 33.69°$$

$$= 0.5\text{ P N.m}$$

$$M_u = M\sin(90° - \theta) = 0.6\,P\sin 33.69°$$

$$= 0.3328\,P\text{ N.m}$$

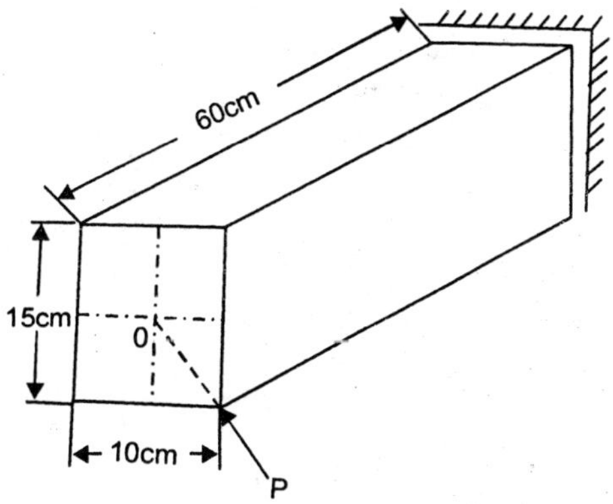

Fig. 14.7

$$I_u = \frac{10\times 15^3}{12} = 2812.5\text{ cm}^4$$

$$I_v = \frac{15\times 10^3}{12} = 1250\text{ cm}^4$$

$$\sigma_A = \frac{M_u v_A}{I_u} + \frac{M_v u_A}{I_v}$$

$$u_A = 5\text{ cm},\ v_A = 7.5\text{ cm}$$

$$\sigma_A = \left(\frac{0.5\,P\times 7.5\times 10^{-2}}{28120.5\times 10^{-8}} + \frac{0.3325\,P\times 5\times 10^{-2}}{1250\times 10^{-8}}\right) = -2664.5\,P = 15\times 10^6$$

$$P = 5.63\text{ kN}$$

Example 14.5 *For the beam loaded as shown in Fi. 14.8 (a), determine the stress at A and locate the neutral axis.*

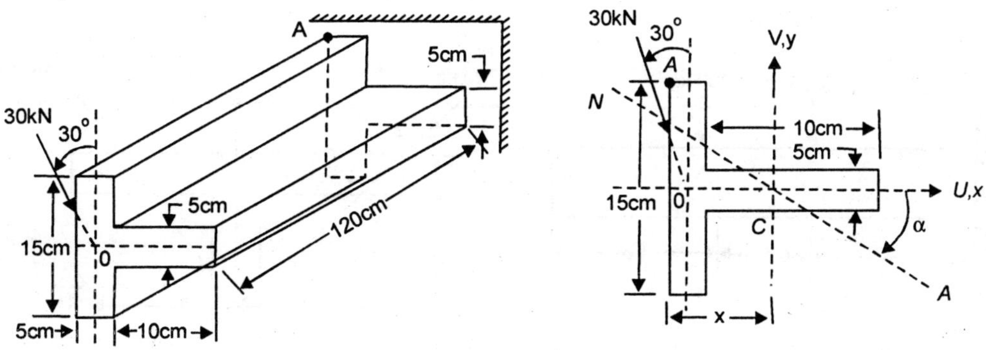

(a) (b)

Fig. 14.8

Solution.

Given : $M = 30 \times 1.2 = 36$ kN.m

$$A_1 = 75 \text{ cm}^2, \quad A_2 = 50 \text{ cm}^2, \quad A = 125 \text{ cm}^2$$

$$125\bar{x} = 75 \times 2.5 + 50 \times 10 = 687.5$$

$$\bar{x} = 5.5 \text{ cm}$$

$$I_{xx} = \frac{5 \times 15^3}{12} + \frac{10 \times 5^3}{12} = 1510.4 \text{ cm}^4$$

$$I_{yy} = \frac{15 \times 5^3}{12} + 75(5.5 - 2.5)^2 + \frac{5 \times 10^3}{12} + 50(10 - 5.5)^2 = 2260.42 \text{ cm}^4$$

$$I_{xy} = 0$$

$$\tan 2\theta = \frac{2 I_{xy}}{I_y - I_x} = 0$$

$$\theta = 0^\circ$$

Hence $I_u = I_x, I_v = I_y.$

$$\tan \alpha = \frac{I_u}{I_v} \tan \phi = \frac{1510.4}{2260.42} \times \tan 30^\circ = 0.38578$$

$$\alpha = 21.1^\circ$$

$$\sigma_A = M\left(\frac{v_A \cos \phi}{I_u} + \frac{u_A \sin \phi}{I_v}\right)$$

$$u_A = 5.5 \text{ cm}, v_A = 7.5 \text{ cm}$$

$$\sigma_A = 36 \times 10^3 \left[\frac{7.5 \cos 30^\circ}{1510.1} + \frac{5.5 \sin 30^\circ}{2260.42}\right] \times 10^6 = 198.608 \text{ MPa}$$

Example 14.7 *A certain beam of I-section as shown in Fig. 14.9 is used to support a load inclined as shown in figure. Determine the stress at corners A and B. What is the inclination of the neutral axis at the section? Take $I_u = 5000$ cm⁴, $I_v = 1500$ cm⁴.*

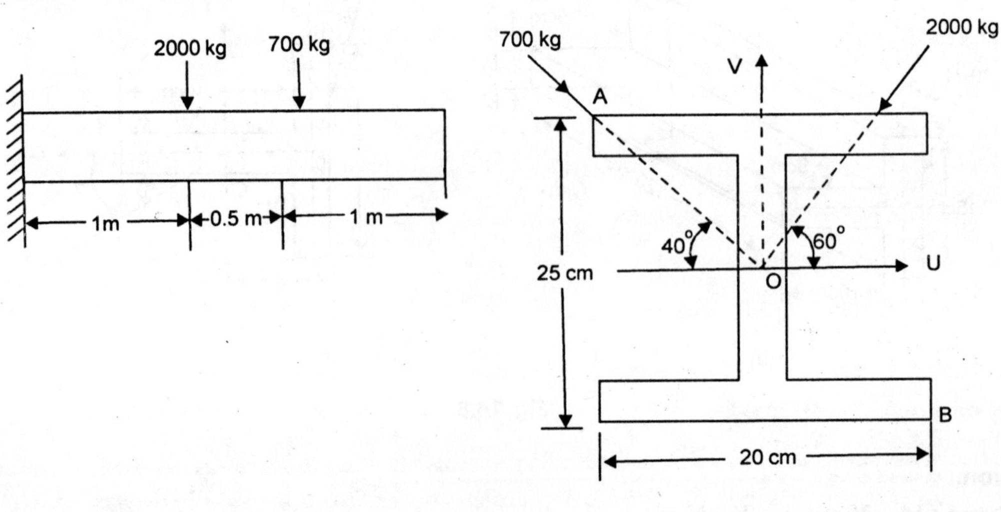

Fig. 14.9

Solution.

Given : $I_u = 5000 \text{ cm}^4, I_v = 1500 \text{ cm}^4$

$$M_u = (2000 \times \cos 30° \times 1 + 700 \cos 50° \times 1.5) \times 9.81$$
$$= 23612.45 \text{ N.m}$$

M_u is –ve for point A and +ve for point B.

$$M_v = (2000 \times \cos 60° \times 1 - 700 \times \cos 40° \times 1.5) \times 9.81$$
$$= 1919.36 \text{ N.m.}$$

M_v is +ve for point A and +ve for Point B.

Bending stress, $$\sigma = \frac{M_u v}{I_u} + \frac{M_v u}{I_v}$$

Point A: $u_A = +10 \text{ cm}, v_A = +12.5 \text{ cm}$

$$M_u = -23612.45 \text{ N.m.}, M_v = 1919.36 \text{ N.m}$$

$$\sigma_A = \frac{23612.45 \times 12.5 \times 10^{-2}}{5000 \times 10^{-8}} + \frac{1919.36 \times 10 \times 10^{-2}}{1500 \times 10^{-8}}$$
$$= (-59.03 + 12.79) \times 10^6$$
$$= -46.24 \text{ MPa}$$

Point B: $u_B = 10 \text{ cm},$

$$v_B = 12.5 \text{ cm}$$
$$M_u = +23612.45 \text{ N.m},$$
$$M_v = -1919.36 \text{ N.m}$$
$$\sigma_B = (59.03 - 12.79)\ 5\ 10^6$$
$$= 46.24 \text{ MPa}$$

14.4 SHEAR CENTRE

The shear centre for any transverse section of a beam is the point of intersection of the bending axis and the plane of the transverse section. Shear centre is also called the centre of twist. If a beam has two axes of symmetry, then shear centre coincides with the centroid. For section having one axis of symmetry, the shear centre does not coincide with the centroid but lies on the axis of symmetry. If a load passes through the shear centre then there will be only bending in the cross-section and no twisting.

14.4.1 Shear Centre for Symmetrical Sections

The principle involved in locating the shear centre for a cross-section of a beam is that the loads acting on the beam must lie in a plane which contains the resultant shear force on each cross-section of the beam as computed from the shearing stresses produced in the beam when it is loaded so that it does not twist as it bends. For, if this condition is not satisfied, the beam at any cross-section will be subjected to a twisting moment in addition to a bending moment. Thus the main problem in locating the shear centre for a cross-section of a beam is that of locating the line of action of the resultant shearing force on the section. Shear force has no meaning for pure bending as no resultant shear force acts in this case.

14.4.2 Shear Centre of Equal-leg Angle Section

Consider an angle section subjected to vertical forces P_1, P_2 etc, and simply supported at the ends as shown in Fig.14.10. Let F be the resultant of all these forces. F can be resolved into a force F' at O and a couple composed of a force F and F'' such that $F' = F'' = F$. Let F_1 and F_2 be the components of F_R. The action line of resultant shearing force F_R in any section must pass through the point of intersection of its components F_1 and F_2 which lie along the centre lines of the upper and lower legs. Thus the point O is the shear centre.

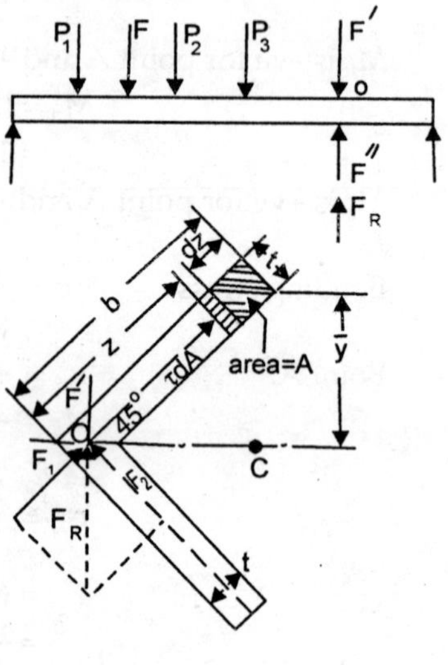

Fig.14.10

Now $\quad F_1 = \int \tau . dA$

where $\quad \tau$ = Shear stress at any point.

$$= \frac{FA\bar{y}}{It}$$

F = Vertical shear force.

A = Area of cross-section between the point at which τ occurs and the outermost edge of the leg.

$\bar{y}$ = Distance of the centroid of the area from the neutral axis.

I = Moment of Inertia of the entire cross-section with respect to the neutral axis.

t = Thickness of the area.

$$\therefore \quad F_1 = \int \tau dA = \int_0^b \frac{F}{It} A\bar{y}\, dA = \frac{F}{It}\int_0^b A\bar{y} d\, A$$

Now $\quad A = t(b-z)$

$$\bar{y} = \left(z + \frac{b-z}{2}\right)\sin 45° = \frac{1}{2\sqrt{2}}(b+z)$$

$$F_1 = \frac{F}{2\sqrt{2}It}\int_0^b t(b-z)(b+z)dA = \frac{F}{2\sqrt{2}I}\int_0^b \left(b^2 - z^2\right)t\, dz$$

$$F_1 = \frac{F}{2\sqrt{2}I}\left|b^2 z - \frac{z^3}{3}\right|_0^b = \frac{Ft}{2\sqrt{2}I}\cdot\frac{2}{3}b^3 = \frac{Ftb^3}{3\sqrt{2}I} = \frac{Ftb^3}{3\sqrt{2}\times\frac{tb^3}{3}} = \frac{F}{\sqrt{2}}$$

But in this case, $\quad F_1 = F_2 = \dfrac{F}{\sqrt{2}}$

Resultant shear force, $F_R = \sqrt{F_1^2 + F_2^2} = \sqrt{\dfrac{F^2}{2} + \dfrac{F^2}{2}} = F$

But for the equilibrium of the beam.

$$F = F' = F_R$$

$\therefore$ F_1 and F_2 are the components of the resisting force F_R and O is the shear centre.

14.4.3 Shear centre for channel section

For the channel sectin shown in Fig. 14.11, let F_1, F_2 and F_3 be the shear forces in the three legs of the channel

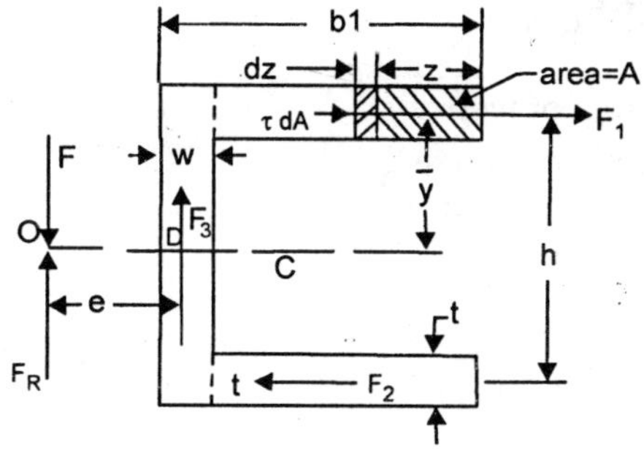

Fig. 14.11

$$F_1 = \int \tau dA = \int \frac{FA\bar{y}}{It} dA$$

$$A = tz$$

$$dA = tdz$$

$$\therefore \quad \bar{y} = \frac{h}{2}$$

$$F_1 = \frac{F}{It}\int_0^{b_1} tz\cdot\frac{h}{2}\cdot tdz = \frac{Fth}{4I}b_1^2$$

$$I = 2b_1t\left(\frac{h}{2}\right)^2 + \frac{1}{12}wh^3$$

$$= \frac{1}{2}b_1th^2\left[1+\frac{1}{6}\frac{wh}{b_1t}\right]$$

$$\therefore \quad F_1 = \frac{Fb_1}{2h\left[1+\frac{1}{6}\frac{wh}{b_1t}\right]} = F_2$$

Taking moments about D, we get

$$F_R \times e = F_1 \times \frac{h}{2} + F_2 \times \frac{h}{2} = F_1 h = \frac{Fb_1}{2\left(1+\frac{1}{6}\frac{wh}{b_1t}\right)}$$

Since $\qquad F_R = F$

$$\therefore \qquad e = \frac{b_1}{2\left(1+\frac{1}{6}\frac{wh}{b_1 t}\right)}$$

$$e = \frac{\frac{1}{2}b_1}{1+\frac{1}{6}\frac{A_w}{A_f}}$$

where $\qquad A_w$ = area of web

A_f = area of flange

14.4.4 Shear centre for unequal I-section

$$F_1 = \int \tau dA = \int \frac{F A\bar{y}}{I t_1} dA$$

$$= \int_0^{b_1/2} \frac{F\left(\frac{b_1}{2}-t\right)t_1 \cdot \left(\frac{b_1}{2}+y\right)}{b_{1/2}} dA = \frac{F}{2I}\int_0^{b_1/2}\left(\frac{b_1^2}{4}-y^2\right)\cdot t_1\, dy$$

$$= \frac{Ft_1}{2I}\left|\frac{b_1^2}{4}y-\frac{y^3}{3}\right| = \frac{Ft_1}{2I}\left[\frac{b_1^3}{8}-\frac{b_1^3}{24}\right] = \frac{Ft_1 b_1^3}{242} = \frac{FI_1}{I}$$

where $\qquad I_1 = \frac{t_1 b_1^3}{12}$

Similarly $\qquad F_2 = \frac{F\times I_2}{2I}, I_2 = \frac{t_2 b_2^3}{12}$

Let the shear centre be at a distance e_1 from left flange (Fig. 14.12) Then

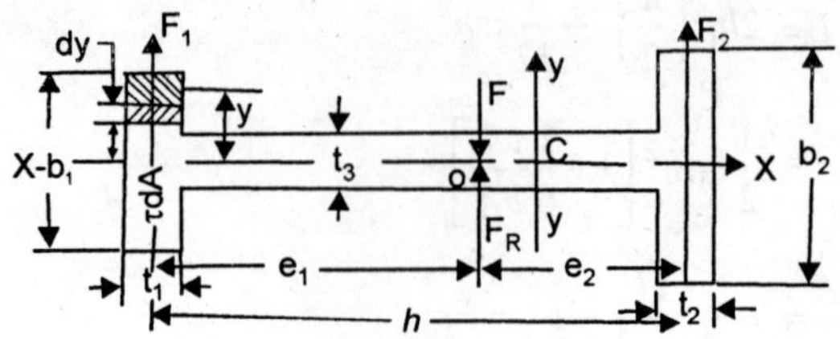

Fig. 14.12

$$F_1 \times e_1 = F_2 \times e_2$$

$$\therefore \qquad \frac{F\times I_1}{2I}\times e_1 = \frac{F\times I_2}{2I}\times e_2$$

$$\therefore \quad \frac{e_1}{e_2} = \frac{I_2}{I_1} = \left(\frac{t_2}{t_1}\right)\left(\frac{b_2}{b_1}\right)^3.$$

Example 14.8 *Determine the shear centre of the channel section shown in Fig. 14.13.*

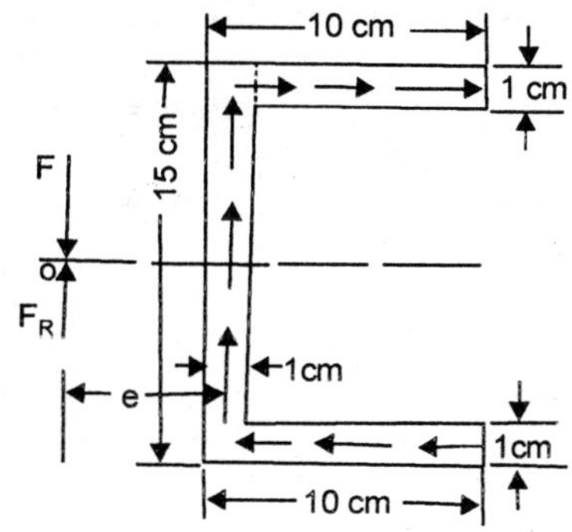

Fig. 14.13

Solution.

Given : Following the procedure of art. we have

$$e = \frac{\frac{1}{2}b_1}{1+\frac{1}{6}\frac{wh}{b_1 t}}$$

Here

$$b_1 = 10 - 1 = 9 \text{ cm}$$
$$h = 15 - 1 = 14 \text{ cm}$$
$$w = 1 \text{ cm}$$
$$t = 1 \text{ cm}$$

$$e = \frac{\frac{1}{2}\times 9}{1+\frac{1}{6}\times\frac{1\times 14}{9\times 1}}$$

$$= \frac{4.5}{1+0.25926} = \frac{4.5}{1.25926} = 3.57\text{cm}$$

Example 14.9 *Locate the shear centre of the cross-section shown in Fig. 14.14.*

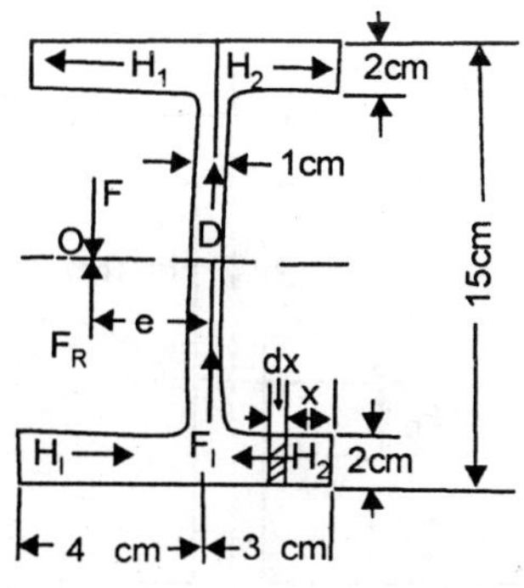

Fig. 14.14

Solution.

$$H_2 = \int \tau dA = \int \frac{FA\bar{y}}{It}\cdot dA = \frac{F}{It}\int A\bar{y}dA$$

$$= \frac{F}{It}\int_0^3 2(3-x)\times(6.5)\times 2dx$$

$$= \frac{F}{2I}\int_0^3 26(3-x)dx$$

$$= \frac{13F}{I}\left|3x-\frac{x^2}{2}\right|_0^3 = \frac{13F}{I}\left[9-\frac{9}{2}\right] = 58.5\frac{F}{I}$$

$$H_1 = \frac{F}{2I}\int_0^4 2(4-x)\times 6.5\times 2dx$$

$$= \frac{13F}{I}\left|4x - \frac{x^2}{2}\right|_0^4 = \frac{13F}{I}[16\times 8] = \frac{104F}{I}$$

Taking moments about point D, we get

$$F_R \times e = 2(H_1 - H_2)\,6.5 = 2(104 - 58.5)6.5 \times \frac{F}{I}$$

Now $$F_R = F$$

$$e = \frac{2\times 45.5\times 6.5}{I} = \frac{591.5}{I}$$

$$I = 2\left[\frac{7\times 2^3}{12} + 14\times(6.5)^2\right] + \frac{1\times 11^3}{12}$$

$$= 2\,[4.667+591.5] + 110.917 = 1192.334 + 110.917 = 1303.251 \text{ cm}^4$$

$$e = \frac{591.5}{1303.251} = 0.454\,\text{cm}$$

Example 14.10 *Determine the shear centre of the cross-section shown in Fig. 14.15.*

Solution.

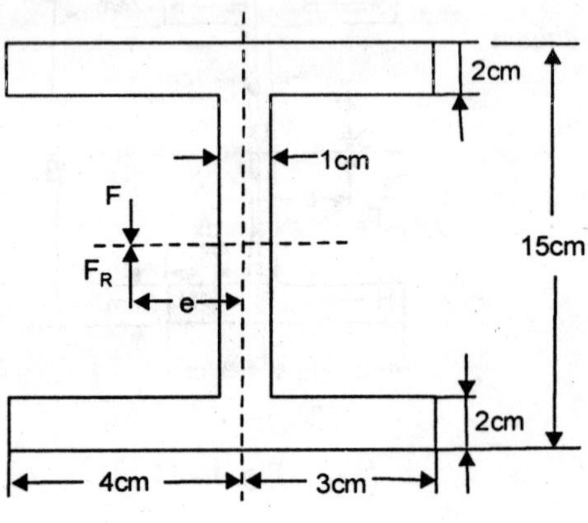

Fig. 14.15

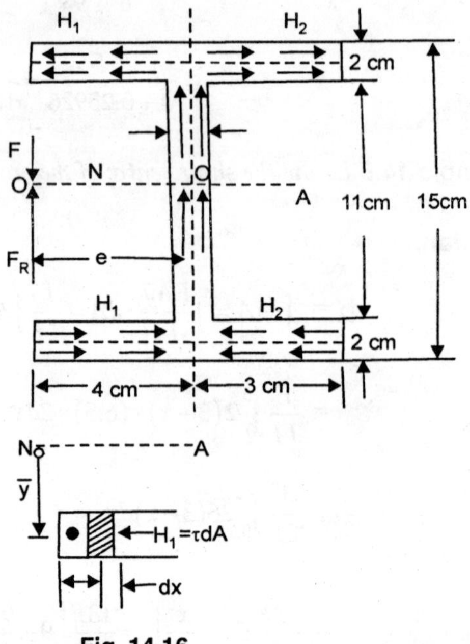

Fig. 14.16

$$I_{NA} = \frac{7\times 15^3}{12} - \frac{6\times 11^3}{12} = 1303.25 \text{ cm}^4$$

$$H_1 = \frac{FA\bar{y}}{1b}$$

$$A = 2x, dA = 2dx$$

$$\bar{y} = 7.5 - 1 = 6.5 \text{ cm}$$

$$b = 2 \text{ cm}$$

From Fig. 14.16, we have

$$H_1 = \frac{F}{2I}\int_0^4 2x \times 6.5 \times 2dx$$

$$= \frac{13F}{I}\int_0^4 xdx = \frac{13F}{I}\left|\frac{x^2}{2}\right|_0^4 = \frac{13F}{I}\times 8 = \frac{104F}{I}$$

Similarly, $$H_2 = \frac{F}{21}\int_0^3 26xdx = \frac{13F}{I}\left|\frac{x^2}{2}\right|_0^3 = 58.5\frac{F}{I}$$

Taking moments about point D, we have

$$F_R \times e = 2(H_1 - H_2) \times 6.5$$

$$= \frac{2F}{I}(104 - 58.5)\times 6.5$$

$$= \frac{591.5F}{I} = \frac{591.5F}{1303.25} = 0.454\ F$$

Now $$F_R = F$$

∴ $$e = 0.454 \text{ cm}$$

SUMMARY

1. Bending caused by loads that do not lie in (or parallel to) a plane that contains the principal centroidal axes of inertia of the cross-section is usually called unsymmetrical bending.
2. The axes for which product of inertia is zero are called principal axes.
3. Principal axes of inertia are given by,

$$\tan 2\theta = \frac{2I_{xy}}{I_y - I_x}$$

4. Bending stress under unsymmetrical bending is,

$$\sigma = \frac{M_u.v}{I_u} + \frac{M_v.u}{I_v}$$

$$= \frac{(M\cos\phi)v}{I_u} + \frac{(M\sin\phi)u}{I_v}$$

5. Location of neutral axis

$$\tan\alpha = \frac{I_u}{I_v}\tan\phi$$

6. The shear centre for any transverse section of a beam is the point of intersection of the bending axis and the plane of the transverse section.

Short Answer Type Questions

1. Define unsymmetrical bending.
2. Define product of moment of inertia.
3. What is the condition on product of moment of inertia to locate the principal axes.
4. Define bending axis.
5. Write the formula for bending stress under unsymmetrical bending.
6. Define shear centre.
7. What is the advantage of locating the shear centre of a beam?

EXERCISES

1. An unequal leg angle iron 60 mm × 40 mm × 6 mm is used as a cantilever beam of length 0.6 m, with 40 mm leg horizontal. The canti lever carries a load of 1 kN at the free end. Calculate:
   (a) maximum bending stress developed, and
   (b) position of the neutral axis.
2. A simply supported beam of span 4 m carries a load of 500 N at its centre. The cross-section of the beam is an equal leg angle iron of size 100 mm × 100 mm × 12 mm. The load line passes through the centroid of the section and lies in the vertical plane. Calculate the stresses at the outer corner points of the angle and position of the neutral axis.
3. An I-section beam shown in Fig. 14.17 is simply imported over a span of 3m. It carries a load of 5kN at mid span. The load line is inclined at an angle of 30° with the vertical and passes through the centroid of the beam cross-section. Calculate the maximum tensile stress in the beam.
4. A beam of rectangular cross-section 60 mm × 80 mm is simply supported over a span of 4 m and carries a mid span load of 3 kN. The load line is inclined at an angle of 30° with the vertical axis of the beam and passes through the centroid. Calculate the stresses at the corners of the beam.

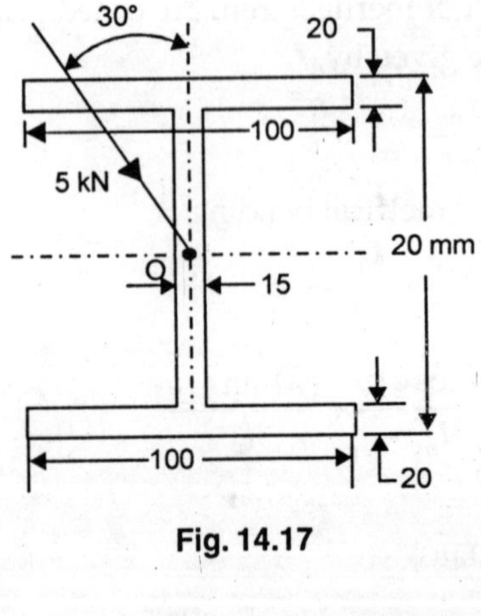

Fig. 14.17

5. Locate the shear centre of the unsymmetrical I-section beam shown in Fig.14.18. Take shear force equal to 1 kN.

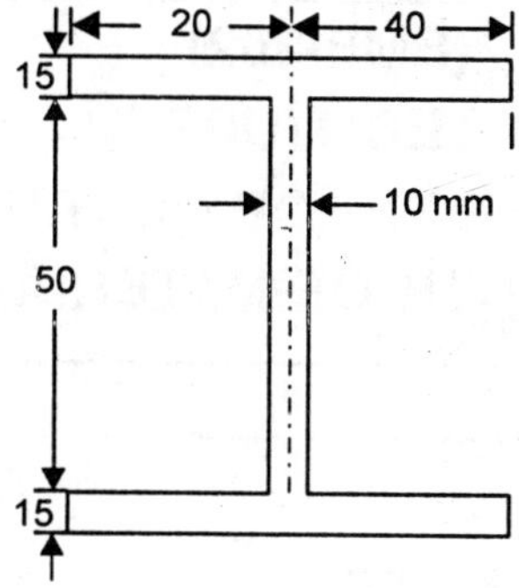

Fig. 14.18

6. Locate the shear centre for the channel section beam shown in Fig. 14.19. Take F = 1.5 kN.

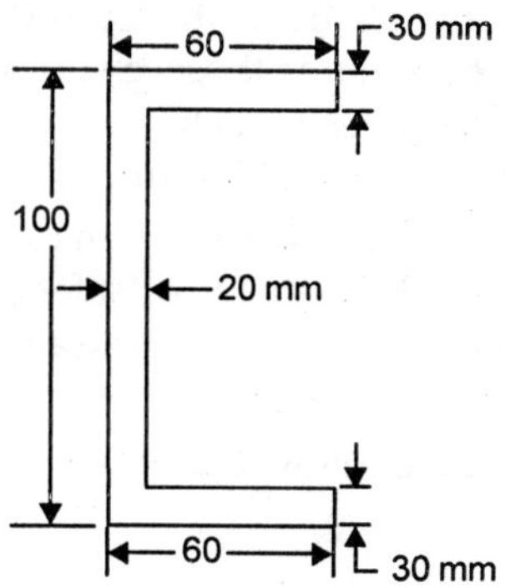

Fig. 14.19

Appendix–I

SOLVED UPTU/UKTU PREVIOUS QUESTION PAPERS

(EME-302)

B. TECH (UPTU)

(SEM. III) ODD SEMESTER THEORY EXAMINATION 2009-2010

STRENGTH OF MATERIALS

Time : 3 Hours *Total Marks : 100*

Note : (1) *Attempt* **all** *questions:*

(2) *Assume any* **data** *required, but not given.*

1. Attempt any **two** parts of the following : $10 \times 2 = 20$

(a) At a point in a strained material, there are normal stresses of 30 N/mm^2, tension and 20 N/mm^2, compression on two planes at right angles to one another, together with shearing stresses of 15 N/mm^2 on the same planes. If the loading on the material is increased so that the stresses reach values of K times those given, find the maximum permissible value of K if the maximum direct stress in the material is not to exceed 80 N/mm^2, and maximum shear stress is not to exceed 50 N/mm^2.

(b) A shaft of 15 cm diameter is subjected to a maximum torque of 20 kNm and a maximum bending moment of 18 kNm. Find the factor of safety (i) according to the maximum shear stress theory; (ii) according to the maximum strain energy theory if the elastic limit in simple tension is 240 MN/m^2. Take $\mu = 0.3$.

(c) Write short notes on any two of the following:

(i) Castigliano's theorem

(ii) Compatibility equations

(iii) Three-dimensional stresses.

2. Attempt any **two** parts of the following : **$10 \times 2 = 20$**

(a) A timber joist of 6 metre span has to carry a load of 15 kN/metre. Find the dimensions of the joist if the maximum permissible stress is limited to 8 N/mm^2. The depth of the joist has to be twice the width.

(b) A beam, simply supported at ends A and B is loaded with two point loads of 60 kN and 50 kN at distance 1 metre and 3 metre respectively from end A. Determine the position and magnitude of maximum deflection. Take $E = 2 \times 10^5$ 'N/mm^2 and $I = 8500$ cm^4.

(c) Find the internal and external diameters required for a hollow shaft, which is to transmit 40 kW of power at 240 rev/minute. The shear stress is to be limited to 100 MN/m^2. Take outside diameter to be twice the inside diameter.

3. Attempt any **two** parts of the following : **$10 \times 2 = 20$**

(a) A leaf spring has 12 plates each 50 mm wide and 5 mm thick, the longest plate being 600 mm long. The greatest bending stress is not to exceed 180 N/mm^2 and the central deflection is 15 mm. Estimate the magnitude of the greatest central load that can be applied to the spring. $E = 0.206 \times 10^6$ N/mm^2.

(b) Determine the section of a cast iron hollow cylindrical column 5 metre long with ends firmly built-in if it carries an axial load of 300 kN. The ratio of internal to external diameter is 3/4. Use factor of safety of 8. Take σ_c = 567 N/mm² and Rankine's constant a = 1/1600.

(c) From the first principles drive the expression for the critical buckling for a column having one end fixed and one end hinged.

4. Attempt any **two** parts of the following: **10 × 2 = 20**

(a) A cylindrical vessel 1.5 metre in diameter, 2 metre long and 1.5 cm thick is closed at both the ends by rigid plates and this cylinder is filled with water at atmospheric pressure. Find how much additional amount of water is required to be pumped so as to make the final pressure in the cylinder as 70 bar. Take E = 210 GN/m² and μ = 0.3 for the material of the cylinder. Bulk modulus of the water is 2.4 GN/m².

(b) Calculate the thickness of metal necessary for a cylindrical shell of internal diameter of 80 mm to withstand an internal pressure of 25 N/mm², if the maximum permissible tensile stress is 125 N/mm².

(c) Write short notes on any **two** of the following:

(i) Lame's theory of thick cylinders.

(ii) Compound Cylinders

(iii) Radial axial and circumferential stresses in thick cylinders.

5. Attempt any **two** parts of the following: **10 × 2 = 20**

(a) A curved bar of square section, 3 cm sides and mean radius of curvature 4.5 cm is initially unstressed. If a bending moment of 300 Nm is applied to the bar tending to straighten it, find the stresses at the inner and outer faces.

(b) A 60 mm × 40 mm × 6 mm unequal angle is placed with the longer leg vertical, and is used as a beam. It is subjected to a bending moment of 12 kN-cm acting in the vertical plane through the centroid of the section. Determine the maximum bending stress induced in the section.

(c) Write short notes on any **two** of the following:

(i) Centroidal Principal Axes

(ii) Assumptions for the theory of curved beams

(iii) Application of curved beams with large initial curvature.

Q. 1. (*a*) Given: σ_x = + 3`0 N/mm², σ_y = –20 N/mm², τ_{xy} = ± 15 N/mm², σ_1 = 80 N/mm², τ_{max}= 50 N/mm²

The compound stress system is shown in Fig. 1

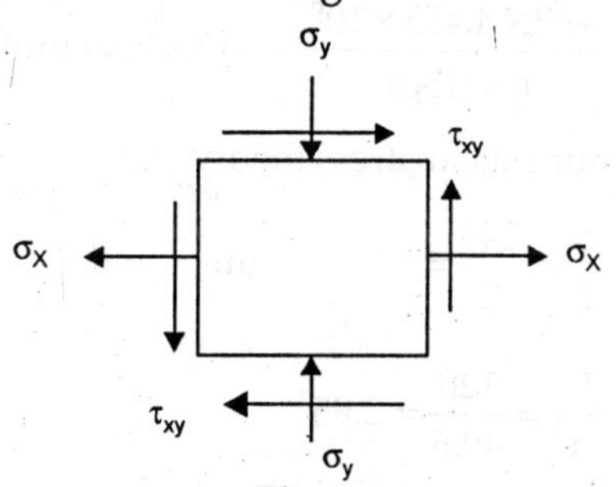

Fig. 1

Principal stress are given by,

$$\sigma_{1,2} = \frac{1}{2}(\sigma_x+\sigma_y)\pm\frac{1}{2}\sqrt{(\sigma_x-\sigma_y)^2+4\tau_{xy}^2}$$

$$= \frac{1}{2}(30-20)\pm\frac{1}{2}\sqrt{(30+20)^2+4\times15^2}$$

$$= 5 \pm 29.15$$

$$\sigma_1 = 34.15\text{ N/mm}^2, \sigma_2 = -24.15\text{ N/mm}^2$$

Maximum shear stress, $\tau_{max} = \frac{1}{2}(\sigma_1-\sigma_2) = \frac{1}{2}(34.15+24.15)$

$$= 29.15\text{ N/mm}^2$$

When the stress σ_x, σ_y and τ_{xy} are raised by K times, then maximum principal stress,

$$\sigma_1 = K\times 34.15\text{ N/mm}^2 = 80$$

$$\therefore \quad K = 2.342$$

and maximum shear stress becomes,

$$\tau_{max} = K\times 29.15 = 50$$

$$K = 1.715$$

Hence permissible value of $K = 1.715$

Q.1. (*b*) Given: $d = 15$ cm, $T = 20$ kN.m, $M = 18$ kN.m, $\sigma = 240$ MPa, $\mu = 0.3$

Eqivalenet torque, $T_e = \sqrt{M^2+T^2} = \sqrt{18^2+20^2} = 26.907$ kN.m

Shear stress, $\tau = \frac{16T_e}{\pi d^3} = \frac{16\times26.907\times10^6}{\pi\times(150)^3} = 40.6\text{ N/mm}^2$

Equivalent bending moment, $M_e = \frac{1}{2}[M\pm\sqrt{M^2+T^2}]$

$$= \frac{1}{2}[18\pm\sqrt{18^2+20^2}]$$

$$= 22.453\text{ kN.m}, -4.453\text{ kN.m}$$

Bending stress, $\sigma_1 = \frac{32M_e}{\pi d^3} = \frac{32\times22.453\times10^6}{\pi\times(150)^3} = 67.76\text{ N/mm}^2$

$$\sigma_2 = \frac{-32\times4.453\times10^6}{\pi\times(150)^3} = -13.44\text{ N/mm}^2$$

(i) According to the maximum shear stress theory,

$$\tau_y = \frac{\sigma}{2} = \frac{240}{2} = 120\text{ N/mm}^2$$

$$\text{Factor of safety} = \frac{\tau_y}{\tau} = \frac{120}{40.6} = 2.95$$

(ii) According to maximum strain energy theory, for two-dimensional case,

Resilience, $u = \frac{1}{2E}[\sigma_1^2 + \sigma_2^2 - 2\mu\,\sigma_1\sigma_2]$

$= \frac{2}{2E}[(67.76)^2 + (-13\cdot 44)^2 - 2\times 0.3\times 67.76\times(-13.44)]$

$= \frac{1}{2E}[5318.47]$

Resilience under simple tension $= \frac{\sigma^2}{2E} = \frac{(240)^2}{2E} = \frac{57600}{2E}$

Factor of safety $= \frac{57600}{5318.47} = 10.83$

Q.1. (*c*) (*i*) Refer to art. 2.5.

(*ii*) Refer to art. 3.15.

(iii) Refer to art. 3.4.

Q.2. (*a*) Given: $l = 6$ m, $w = 15$ kN/m, $\sigma_{max} = 8\text{N/mm}^2$, $d = 2b$

Maxmium bending moment, $M = \frac{wl^2}{8}$

$= \frac{15\times 6^2}{8} = 67.5$ kN.m

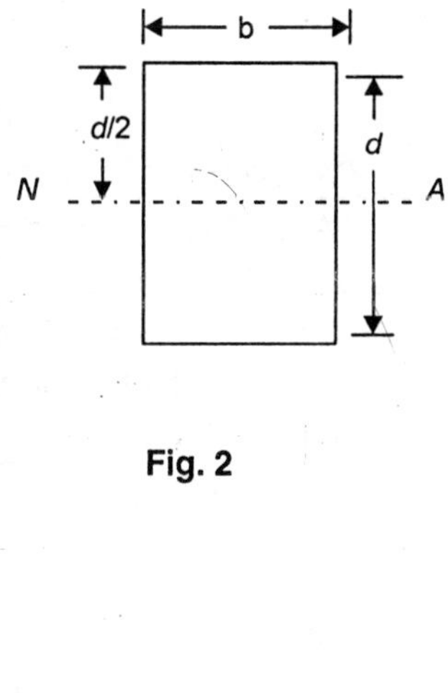

Fig. 2

Section modulus, $z = \frac{bd^2}{6} = \frac{b\times 4b^2}{6} = \frac{2}{3}b^3$

Bending stress, $\sigma = \frac{M}{z} = \frac{67.5\times 10^3}{\left(\frac{2}{3}b^3\right)} = \frac{101.25\times 10^3}{b^3}\text{N/m}^2$

$= 8 \times 10^6$

$b^3 = 12.65625 \times 10^{-3}$

$b = 0.233$ m or 233 mm

$d = 2b = 466$ mm

Q. 2. (*b*) Let span of beam $AB = 4$ m (assumed)

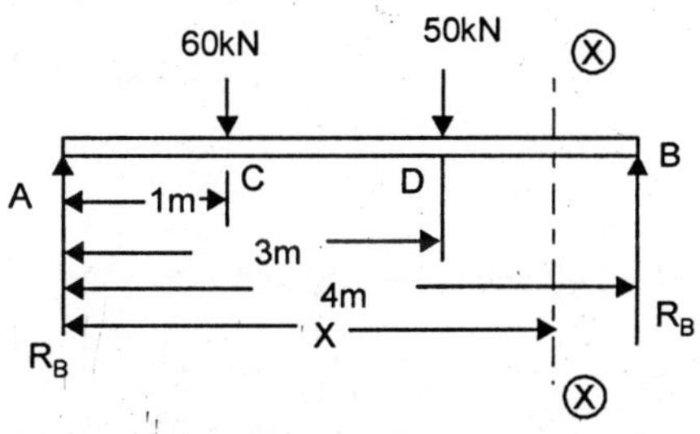

Fig. 3

Given: $E = 2\times10^5$ N/mm², $I = 8500$ cm⁴

$$\Sigma M_B = 0 \text{ gives,}$$

$$4R_A = 60 \times 3 + 50 \times 1 = 230$$

$$R_A = 57.5 \text{ kN}$$

Take a section at a distance x from A.

$$M_x = 57.5 \times x - 60\ (x-1) - 50\ (x-3)$$

$$EI\frac{d^2y}{dx^2} = -M_x = -57.5\ x + 60\ (x-1) + 50\ (x-3)$$

Integrating, we get

$$EI\frac{dy}{dx} = -28.75x^2 + 30(x-1)^2 + 25(x-3)^2 + C_1$$

Integrating again, we have

$$EIy = -9.583\ x^3 + 10\ (x-1)^3 + 8.333\ (x-3)^3 + C_1x + C_2$$

At $x = 0, y = 0$

$\therefore$ $$C_2 = 0$$

At $x = 4$ m, $y = 0$

$$0 = -9.583 \times 4^3 + 10 \times 3^3 + 8.333 + 4C_1$$

$$C_1 = 83.745$$

$\therefore$ $$EIy = -9.583\ x^3 + 10\ (x-1)^3 + 8.333\ (x-3)^3 + 83.745\ x$$

At $x = 1$m, $$EIy = 74.162$$

At $x = 3$m, $$EIy = 72.494$$

$\therefore$ Maximum deflection occurs between C and D.

$\therefore$ $$-28.75\ x^2 + 30\ (x-1)^2 + 83.745 = 0$$

$$-2.875\ x^2 + 30x^2 + 30 - 60x + 83.745 = 0$$

$$1.25\ x^2 - 60x + 113.745 = 0$$

$$x^2 - 48x + 91 = 0$$

$$x = \frac{1}{2}[48 \pm \sqrt{48^2 - 4\times91}]$$

$$= 1.977 \text{ m}$$

$$EIy_{max} = -9.583 \times (1.977)^3 \times 10 \times (0.977)^3 + 83.745 \times 1.977$$

$$= 100.84 \text{ kN.m}^3$$

$$y_{max} = \frac{100.84\times10^{12}}{2\times10^5\times8500\times10^4} = 5.93 \text{ mm}$$

Q.2. (*c*) Given: $P = 40$ kW $N = 240$ rpm, $\tau = 100$ MPa, $d_0 = 2d_i$

$$P = \frac{2\pi NT}{60\times10^3}$$

$$T = \frac{40 \times 60 \times 10^3}{2\pi \times 240} = 1591.55 \text{ N.m}$$

$$k = \frac{d_i}{d_0} = 0.5$$

$$\frac{T}{J} = \frac{2\tau}{d_0}$$

$$\tau = \frac{Td_0}{2J} = \frac{Td_0}{2} \times \frac{32}{\pi(d_0^4 - d_i^4)} = \frac{16T}{\pi d_0^3(1-k^4)}$$

$$d_0^3 = \frac{16T}{\pi\tau(1-k^4)} = \frac{16 \times 1591.55 \times 10^3}{\pi \times 100[1-(0.5)^4]} = 86460.8$$

$$d_0 = 44.22 \text{ mm}$$
$$d_i = 22.11 \text{ mm}$$

Q.3. (*a*) Given: $n = 12$, $b = 50$ mm, $t = 5$ mm, $l = 600$ mm, $\sigma_b = 180$ N/mm², $\delta = 15$ mm, $E = 0.206 \times 10^6$ N/mm²

Bending stress, $\sigma_b = \dfrac{3}{2}\dfrac{Wl}{nbt^2}$

$$180 = \frac{3}{2} \times \frac{W \times 600}{12 \times 50 \times 25}$$

$$W = 3000 \text{ N}$$

Central deflection, $\delta = \dfrac{3}{8}\dfrac{Wl^3}{nEbt^3}$

$$15 = \frac{3}{8} \times \frac{W \times (600)^3}{12 \times 0.206 \times 10^6 \times 50 \times 125}$$

$$W = 2861.11 \text{ N}$$

Safe load = 2861.11N

Q.3. (*b*) Given: $l = 5$m, $P = 300$ kN, $\dfrac{d_i}{d_0} = \dfrac{3}{4}$, $fos = 8$, $\sigma_c = 567$ N/mm², $a = \dfrac{1}{1600}$

For both ends fixed column, equilvalent length, $l_e = \dfrac{l}{2} = \dfrac{5}{2} = 2.5$ m

According to Rankine's formula,

$$P_{cr} = \frac{\sigma_c \times A / fos}{1 + \frac{1}{a}\left(\frac{l_e}{K}\right)^2}$$

For a hollow column, $A = \frac{\pi}{4}(d_0^2 - d_i^2) = \frac{\pi}{4}d_0^2\left[1-\left(\frac{d_i}{d_0}\right)^2\right]$

$$= \frac{\pi}{4}d_0^2[1-(0.75)^2] = 0.3436\,d_0^2$$

Moment of inertia, $I = \frac{\pi}{64}(d_0^4 - d_i^4) = \frac{\pi d_0^4}{64}[1-(0.75)^4] = 0.033556\,d_0^4$

$$I = AK^2$$

$$K^2 = \frac{I}{A} = \frac{0.033556\,d_0^4}{0.3436\,d_0^2} = 0.09766\,d_0^2$$

$$300 \times 10^3 = \frac{567 \times 0.3436\,d_0^2/8}{1+\frac{1}{1600}\times\frac{(2.5\times10^3)2}{0.09766\,d_0^2}}$$

$$= \frac{24.352\,d_0^2}{1+\frac{40,000}{d_0^2}}$$

$$3\times10^5 + \frac{12\times10^9}{d_0^2} = 24.352\,d_0^2$$

$$24.352\,d_0^4 - 3\times10^5 d_0^2 - 12\times10^9 = 0$$

$$d_0^4 - 1.232\times10^4 d_0^2 - 4.927\times10^8 = 0$$

$$d_0^2 = \frac{1}{2}[1.232\times10^4 \pm \sqrt{(1.232\times10^4)^2 + 4\times4.927\times10^8}]$$

$$= \frac{1}{2}[1.232\times10^4 \pm 4.607\times10^4]$$

$$= 2.9195 \times 104$$

$$d_0 = 1.708 \times 10^2 = 170.8 \text{ mm}$$

$$d_i = 0.75 d_0 = 0.75 \times 170.8 = 128.1 \text{ mm}$$

Q.3. (*c*) Refer to art. 8.4.3

Q.4. (*a*) Given: $d = 1.5$ m, $l = 2$ m, $t = 15$ mm, $p_1 = 1$ atm, $p_2 = 70$ bar, $E = 210$ GN/m², $\mu = 0.3$, $K = 2.4$ GN/m²

$$\frac{\delta V}{V} = \frac{pd}{2tE}(2.5-2\mu) + \frac{p}{K}$$

$$V = \frac{\pi}{4}d^2l = \frac{\pi}{4}\times(1.5)^2\times 2 = 3.534 \text{ m}^3$$

$$p = p_2 = 70 \text{ bar} = 70 \times 10^5 \text{ N/m}^2$$

$$\delta V = \frac{3.534\times 70\times 10^5\times 1.5}{2\times 15\times 10^{-3}\times 210\times 10^9}2.5 - 2\times 0.3) + \frac{70\times 10^5}{2.4\times 10^9}$$

$$= 0.0141 \text{ m}^3$$

Q.4. (*a*) **Given:** $d = 80$ mm, $p = 25$ N/mm^2, $\sigma_\theta = 125$ N/mm^2

Considering it to be a thin cylinder,

$$\sigma_\theta = \frac{pd}{2t}$$

$$t = \frac{25\times 80}{2\times 125} = 8 \text{ mm}$$

$$\frac{t}{d} = \frac{8}{80} = \frac{1}{10} > \frac{1}{15}$$

Therfore, this may be considered as a thick cylinder.

$$\sigma_r = A - \frac{B}{r^2}$$

At $r = 40$ mm, $\sigma_r = -p = -25$ N/mm^2

$r = 40 + t$, $\sigma_r = 0$

$$-25 = A - \frac{B}{(40)^2} = A - \frac{B}{1600}$$

$$0 = A - \frac{B}{(40+t)^2}$$

$$\therefore \quad A = \frac{B}{(40+t)^2}$$

$$-25 = \frac{B}{(40+t)^2} - \frac{B}{1600}$$

$$= B\left[\frac{1}{(40+t)^2} - \frac{1}{1600}\right]$$

$$= B\left[\frac{1600 - 1600 - t^2 - 80t}{1600(40+t)^2}\right]$$

$$= -B\left[\frac{t^2 + 80t}{1600(40+t)^2}\right]$$

$$B = \frac{40,000(40+t)^2}{(t^2+80t)}$$

$$A = \frac{40,000}{(t^2+80t)}$$

$$\sigma_\theta = A + \frac{B}{r^2}$$

σ_θ is maximum at r = 40 mm

$$125 = \frac{40,000}{(t^2+80t)} + \frac{40,000(40+t^2)}{1600(t^2+80t)}$$

$$= \frac{40,000}{(t^2+80t)} + \frac{25(40+t)^2}{(t^2+80t)}$$

$$125(t^2+80t) = 40,000 + 25(40+t)^2$$

$$125t^2 + 1000t = 40,000\ 25\ (1600 + t^2 + 80\ t)$$

$$= 40,000 + 40,000 + 25\ t^2 + 2000\ t$$

$$= 80,000 + 40000 + 25t^2 + 2000\ t$$

$$100\ t^2 + 8000\ t - 80,000 = 0$$

$$t^2 + 80t - 800 = 0$$

$$t = \frac{1}{2}[-80 \pm \sqrt{64000+3200}]$$

$$= 8.989 \simeq 9 \text{ mm}$$

Q.4. (*c*) (i) Refer to art.11.2.

(*ii*) Refer to art. 11.3.

(iii) Refer to art. 11.2 and art. 11.2.1.

Q.5. (*a*) Given: $B = D = 3$ cm, $R = 4.5$ cm, $M = -300$ N.m

$$A = 3 \times 3 = 9 \text{ cm}^2$$

$$h^2 = \frac{R^3}{D} ln\left(\frac{2R+D}{2R-D}\right) - R^2$$

$$= \frac{(4.5)^3}{3} ln\left(\frac{2\times4.5+3}{2\times4.5-3}\right) - (4.5)^2$$

$$= 0.8043 \text{ cm}^2$$

Compressive stress on top fibres of bar.

$$\sigma = \frac{M}{AR}\left(1 + \frac{R^2}{h^2}\left(\frac{y}{R+y}\right)\right)$$

Here $y = 1.5$ cm

$$\sigma_{top} = \frac{-300\times10^2}{9\times4.5}\left[1+\frac{(4.5)^2}{0.8043}\left(\frac{1.5}{4.5+1.5}\right)\right]\times10^{-2}$$

$$= -54 \text{ N/mm}^2 \text{ (compressive)}$$

$$\sigma_{bottom} = \frac{-M}{AR}\left[1-\frac{R^2}{h^2}\left(\frac{y}{R-y}\right)\right]$$

$$= \frac{-300\times10^2}{9\times4.5}1-\frac{(4.5)^2}{0.8043}\left(\frac{1.5}{4.5-1.5}\right)\times10^{-2}$$

$$= +85.84 \text{ N/mm}^2$$

Q.5. (*b*) Given: $M = 12$ kN.m or 120 N.m

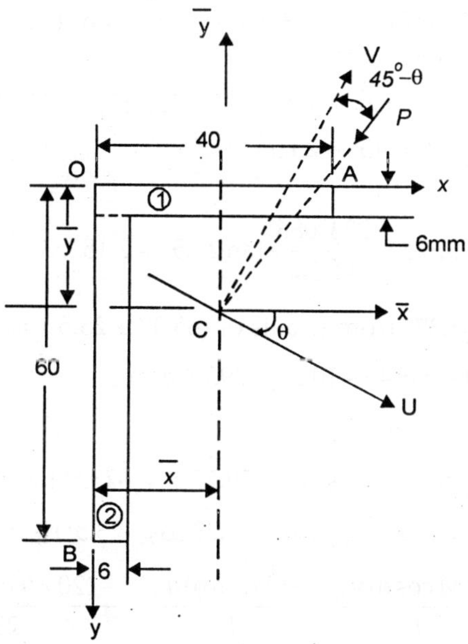

Fig. 4

$A_1 = 40 \times 6 = 240 \text{ mm}^2$

$A_2 = (60-6) \times 6 = 324 \text{ mm}^2$

$A = A_1 + A_2 = 240 + 324 = 564 \text{ mm}^2$

$$\bar{x} = \frac{A_1\bar{x}_1 + A_2\bar{x}_2}{A} = \frac{240\times20+324\times3}{584} = 10.23 \text{ mm}$$

$$\bar{y} = \frac{A_1\bar{y}_1 + A_2\bar{y}_2}{A} = \frac{240\times3+324\times(27+6)}{564} = 20.23 \text{ mm}$$

$$I_{\overline{xx}} = \frac{40\times6^3}{12} + 240(20.23-3)^2 + \frac{6\times54^3}{12} + 324(33-20.23)^2$$

$$= 203{,}537 \text{ mm}^4$$

$$I_{\bar{y}\bar{y}} = \frac{6\times 40^3}{12} + 240(20-10.23)^2 + \frac{54\times 6^3}{12} + 324(10.23-3)^2 = 72{,}817 \text{ mm}^4$$

$$I_{\bar{x}\bar{y}} = 240 \times (20 - 10.23) \times (20.23 - 3) + 324 \times (10.23 - 3) \times (33 - 20.23)$$

$$= 70{,}190 \text{ mm}^4$$

Principal axes are given by,

$$\tan 2\theta = \frac{2I_{\bar{x}\bar{y}}}{I_{\bar{y}\bar{y}} - I_{\bar{x}\bar{x}}} = \frac{2\times 70{,}190}{72{,}817 - 203{,}537} = -1.0739$$

$$2\theta = -47°$$

$$\theta = -23.5°$$

$$I_u = I_{\bar{x}} \cos^2\theta + I_{\bar{y}} \sin^2\theta - I_{\bar{x}\bar{y}} \sin 2\theta$$

$$= 203{,}537 \cos^2(-23.5°) + 72{,}817 \sin^2(-23.5°) - 70{,}190 \sin(-47°)$$

$$= 234{,}086 \text{ mm}^4$$

$$I_v = (I_{\bar{x}} + I_{\bar{y}}) - I_u = (203{,}537 + 72{,}817) - 234{,}0.86 = 42{,}268 \text{ mm}^4$$

$$\phi = 45° - \theta = 45° - 23.5 = 21.5°$$

The neutral axis is given by,

$$\tan\alpha = \frac{I_u}{I_v}\tan\phi = \frac{234{,}086}{42{,}268}\tan 21.5° = 2.18153$$

$$\alpha = -65.37° \text{ from U-axis or } (65.37 + 23.5) = 88.87° \text{ from } \bar{x} \text{ axis}$$

For point A, $\bar{x}_A = 40 - \bar{x} = 40 - 10.23 = 29.77$ mm

$$\bar{y}_A = 20.23 \text{ mm}$$

$$u_A = \bar{x}_A \cos\theta + \bar{y}_A \sin\theta = 29.77\cos(-23.5°) + 20.23\sin(-23.5°) = 19.23 \text{ mm}$$

$$v_A = \bar{y}_A \cos\theta - \bar{x}_A \sin\theta = 20.23\cos(-23.5°) - 29.77\sin(-23.5°) = 30.42\text{mm}$$

$$\sigma_A = \frac{-(M\cos\phi)v_A}{I_u} + \frac{-(M\sin\phi)u_A}{I_v} = \frac{-120\times(\cos 21.5°)\times 30.42}{234{,}086} +$$

$$= \frac{-120\times(\cos 21.5°)\times 30.42}{234{,}086} + \frac{-120\times 10^3 \sin 21.5^0 \times 19.23}{42{,}268}$$

$$= -34.51 \text{ N/mm}^2 \text{ (compressive)}$$

For point B, $\bar{x}_B = -10.23\text{mm}$, $\bar{y}_B = -(33 - 20.23) = -12.77\text{mm}$

$$u_B = -10.23 \cos(-23.5°) - 12.77 \sin(-23.5°) = -4.29 \text{ mm}$$

$$v_B = -12.77 \cos(-23.5°) + 10.23 \sin(-23.5°) = -15.79 \text{ mm}$$

$$\sigma_B = -120\times 10^3\left[\frac{-15.79}{234{,}086}\times\cos 21.5° - \frac{4.29\sin 21.5°}{42{,}268}\right]$$

$$= -12 \text{ N/mm}^2 \text{ (compressive)}$$

Q.5. (*c*) (i) Refer to art. 14.1.
(*ii*) Refer to art. 13.2.
(iii) Refer to art. 13.1.

(EME-302)
B. TECH (UPTU)
(SEM. III) ODD SEMESTER THEORY EXAMINATION 2010-2011
STRENGTH OF MATERIALS

Time : 3 Hours *Total Marks : 100*

Note : (1) *Attempt* **all** *questions:*
(2) *Assume any* **data** *required, but not given.*

1. Attempt any two of the following: **(10 × 2)**
   (a) A two dimensional state of stress is given by σ_{xx} = 10 MPa, σ_{yy} = 5 σ_{xy} MPa and σ_{xy} = 2.5MPa.
   Determine the following on a plane inclined at an angle of 30° from x-plane in anticlockwise direction.
   (i) Normal stress (ii) Shear stress
   (iii) Resultant stress and (iv) Principal stresses at the point.
   (*b*) One end of a spring is fixed on horizontal with its axis in vertical direction. When it supports a mass of 10 kg, the spring is compressed by 2.0 cm. The maximum principal stress in the spring wire is 20 MPa and maximum shear stress 10 MPa. Find the maximum value of the deflection and the stresses when the same mass is allowed to freely fall and strike the spring.
   (*c*) A bar of length 1.5 m, cross sectional area 20 cm^2 and mass 30 kg is hanging in vertical direction. A force of 200 N is applied at the free end of bar. Using Castigliano's theorem, determine the elongation in the bar due to the force and self weight of the bar.
2. Attempt any two of the following : 10 × 2
   (a) A 20 mm diameter shaft of length 500 mm is fixed at one end. A torque, T is applied at its free end. The linear strains at surface of shaft at an angle of 45° from the axis is 4.0×10^{-3} Determine :
   (i) Torque
   (ii) Angle of twist and
   (iii) Shear stress in the shaft.
   Take E=2.0×10^{11} and G = 8.0×10^{10} N/m^2
   (b) A fixed beam of length 2.0 m carries a concentrated load of 10 kN at its mid span. Determine the reaction and bending moment at the support and deflection at the mid span. Fiexural stiffness of the beam is 2.0×10^7 N/m^2.
   (c) A 100 mm x 150 mm wooden bar is to be symmetrically loaded with two equal forces P as shown in Fig. 1. Determine the position of loads and their magnitude when a bending stress of 10 MPa and shearing stress of 2.5 MPa are just reached. Neglect the weight of the beam.

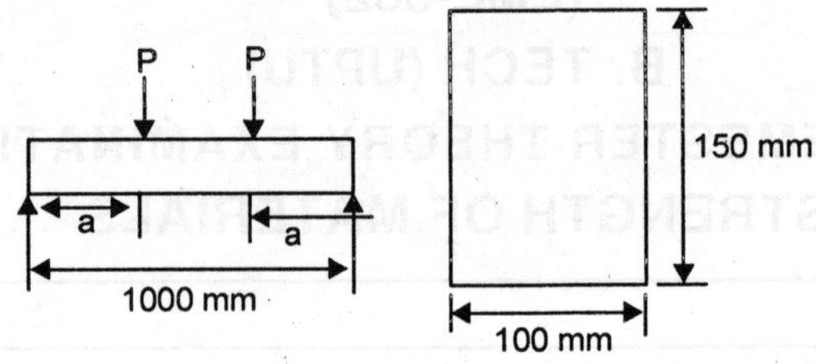

Fig. 1

3. Attempt any two of the following: **(10 × 2)**
   (a) A closed coiled helical spring made of 10 mm diameter steel bar has 8 coils of 150 mm mean diameter. Calculate the elongation, torsional stress and strain energy/unit volume when the spring is subjected to an axial load of 130 N. Take G = 80 GPa. If instead of axial load, an axial torque of 9 N.m is applied, find the axial twist, bending stress and strain energy/volume. Take E = 200 GPa.
   (b) A 5 m long hollow column with fixed ends supports an axial load of 800 kN. The external diameter of the column is 240 mm. Determine the thickness of the column using Rankine's formula. Given that a = 1/6400 and working stress of 80 MPa.
   (c) Explain the term core of a section with reference to short columns. Derive an expression for finding out the core of a rectangular section.
4. Attempt any two of the following: **(10 × 2)**
   (a) A thin spherical vessel having diameter of 1.50 m is of uniform thickness. It is filled with water at a gauge pressure of 2.0 MPa. A relief valve attached to the vessel is opened and water is allowed to escape until the pressure falls to atmospheric. If the volume of the water escaped is 4 litre find thickness of the plate of the vessel. Bulk modulus of water is 2 GPa and Young's modulus of vessel material is 200 GPa and Poisson's ratio is 0.30.
   (b) A thick cylinder of 160 mm internal, 240 mm external diameter is subjected to an external pressure 12 MPa. Determine the maximum internal pressure that can be applied if the maximum allowable normal stress is 36 MPa. Plot the variation of radial and hoop stresses.
   (c) Derive expressions for radial and hoop stresses in a thick cylinder with internal and external radii of a and b subjected to an internal pressure of p
5. Attempt any two of the following: **(10 × 2)**
   (a) A simply supported I-section beam of span 2 m carries a concentrated load of 4.0 kN at an angle of 20° from vertical as shown in Fig. 2. The load passes through CG of the section. Determine the maximum bending stress in the beam.

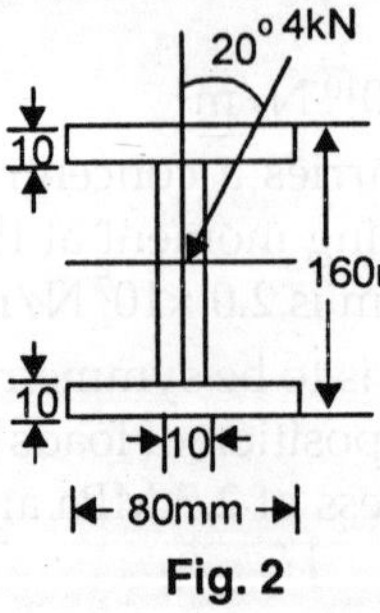

Fig. 2

(b) A steel ring made of a rod of circular cross-section is pulled by a force of 8.0 kN. The mean diameter of the ring is 160.0 mm. Determine the increase in diameter along the load. Take E = 200 GPa.

(c) An open ring has T-section as shown in the Fig. 3. Determine the stress at points P and Q.

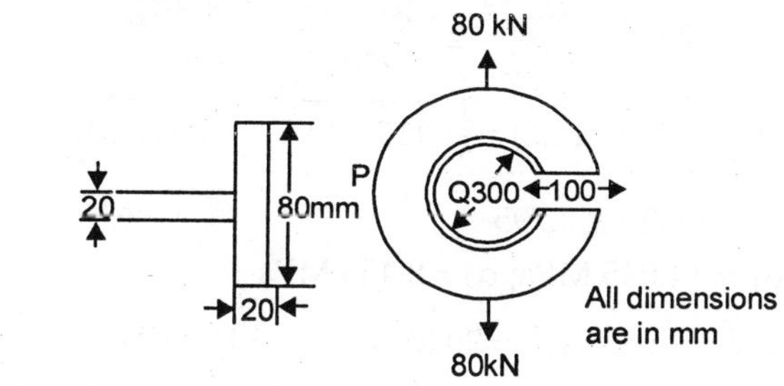

Fig. 3

Q.1. (a) Given: $\sigma_{xx} = 10$ MPa, $\sigma_{xy} = 2.5$ MPa, $\sigma_{yy} = 5\sigma_{xy} = 5 \times 2.5 = 12.5$ MPa, $\theta = 30°$

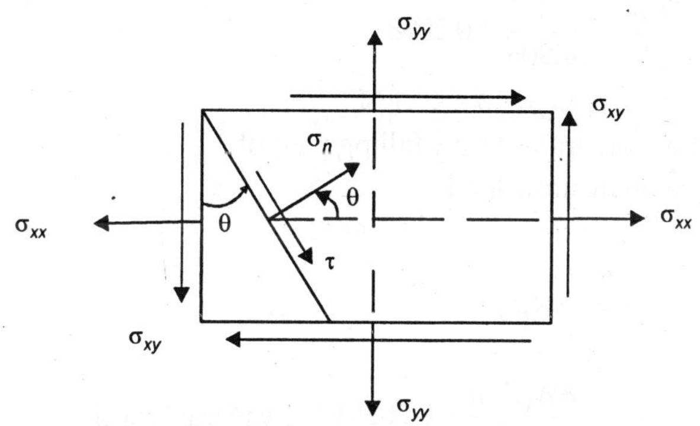

Fig. 4

(i) Normal stress, $\sigma_n = \frac{1}{2}(\sigma_{xx} + \sigma_{yy}) + \frac{1}{2}(\sigma_{xx} - \sigma_{yy})\cos 2\theta + \sigma_{xy}\sin 2\theta$

$$= \frac{1}{2}(10 + 12.5) + \frac{1}{2}(10 - 12.5)\cos 60° + 2.5\sin 60°$$

$$= 12.79 \text{ MPa}$$

(ii) Shear stress, $\tau = \frac{1}{2}(\sigma_{xx} - \sigma_{yy})\sin 2\theta - \sigma_{xy}\cos 2\theta$

$$= \frac{1}{2}(10 - 12.5)\sin 60° - 2.5\cos 60°$$

$= -2.33$ MPa

(iii) Resultant stress, $\sigma_r = \sqrt{\sigma_n^2 + \tau^2} = \sqrt{(12.79)^2 + (-2.33)^2} = 13\text{ MPa}$

(iv) Principal stress.

$$\sigma_{1,2} = \frac{1}{2}(\sigma_{xx} + \sigma_{yy}) \pm \frac{1}{2}\sqrt{(\sigma_{xx} - \sigma_{yy})^2 + 4\sigma_{xy}^2}$$

$$= \frac{1}{2}(10 + 12.5) \pm \frac{1}{2}\sqrt{(10 - 12.5)^2 + 4 \times (2.5)^2}$$

$$= 11.25 \pm 2.795$$

$$\sigma_1 = 14.045\text{ MPa}, \sigma_2 = 8.455\text{ MPa}$$

Q.1. (*b*) Given $m = 10$ kg, $\delta = 2$ cm, $\sigma_{max} = 20$ MPa, $\tau_{max} = 10$ MPa

$$\text{Stiffness of spring, } k = \frac{mg}{\delta} = \frac{10 \times 9.81}{20} = 4.905\text{ N/mm} \quad \text{... (1)}$$

Now
$$k = \frac{Gd^4}{8D^3n} = 4.905$$

$$\frac{8D^3n}{Gd^4} = \frac{1}{4.905} = 0.2038$$

and
$$W = \text{mg} = 10 \times 9.81\text{ N}$$

Let δ_1 = deflection under freely falling weight

W_1 = equivalent static load

h = height of fall = 10 cm (assumed)

Then
$$W(h + \delta_1) = \frac{1}{2} \times w_1 \delta_1$$

where
$$\delta_1 = \frac{8W_1D^3n}{Gd^4} = 0.2038\,W_1, \text{ by using Eq.(2).}$$

$$98.1\,(100 + 0.2038\,W_1) = \frac{1}{2}W_1 \times 0.2038\,W_1$$

$$0.1019W_1^2 - 20W_1 - 9810 = 0$$

$$W_1^2 - 196.27W_1 - 96270.8 = 0$$

$$W_1 = \frac{1}{2}[196.27 \pm \sqrt{(196.27)^2 + 4 \times 96270.8}]$$

$$= 423.56\text{ N}$$

$$\delta_1 = 0.2038 \times 423.56 = 86.3\text{ mm}$$

$$\tau_{max} = \frac{10}{98.1} \times 423.56 = 43.17\text{ N/mm}^2$$

Maximum Principal stress, $\sigma_{max} = 86.34\text{ N/mm}^2$

Q.1. (c) Given: $l = 1.5$ m, $A = 20$ cm^2, $m = 30$ kg, $P = 200$ N

Weight of the bar, $W = mg = 30 \times 9.81 = 294.3$ N

W can be considered acting as $\frac{W}{2}$ at the free end of the bar

Total force at the free end of the bar $= P + \frac{W}{2}$

Volume of the bar, $V = Al = 20 \times 10^{-4} \times 1.5 = 30 \times 10^{-4}\,\text{m}^3$

Stress in the bar, $\sigma = \frac{1}{A}\left(P + \frac{W}{2}\right) = \left(P + \frac{W}{2}\right) \times \frac{10^4}{20} = 500\left(P + \frac{W}{2}\right) N/m^2$

Strain energy stored in the bar,

$$U = \frac{1}{2}\frac{\sigma^2}{E} \times V = \frac{1}{2} \times (500)^2 \left(P + \frac{W}{2}\right)^2 \times \frac{30 \times 10^{-4}}{E}$$

$$\text{Deflection of the bar, } \delta = \frac{\partial U}{\partial\left(P + \frac{W}{2}\right)} = (500)^2 \times \left(P + \frac{W}{2}\right) \times \frac{30 \times 10^{-4}}{E}$$

Let $E = 200$ GPa.

$$\delta = (500)^2 \times \left(200 + \frac{294.3}{2}\right) \times \frac{30 \times 10^{-4}}{200 \times 10^{-9}}$$

$$= 1.3 \times 10^{-3}\text{ mm}$$

Q.2. (a) Given: $d = 20$ mm, $l = 500$ mm, $\varepsilon = 4 \times 10^{-3}$, $E = 2 \times 10^{11}$ N/m^2, $G = 8 \times 10^{10}$ N/m^2

At 45° planes in a shaft under torsion there occur tensile and compressive stresses whose magnitude is equal to the shear stress.

(iii) $\sigma_1 = -\sigma_2 = \tau$

$$G = \frac{E}{2(1+\nu)},\ 1 + \nu = \frac{E}{2G},\ \nu = \frac{E}{2G} - 1$$

$$\therefore \quad \nu = \frac{2 \times 10^{11}}{2 \times 8 \times 10^{10}} - 1 = 0.25$$

$$\varepsilon = \frac{1}{E}(\sigma_1 - \nu\sigma_2) = \frac{\tau}{E}(1 + \nu)$$

$$4 \times 10^{-3} = \frac{\tau}{2 \times 10^{11}}(1 + 0.25)$$

$$\tau = 640\text{ MPa}$$

(i) Torque, $T = \frac{\pi}{16}d^3\tau = \frac{\pi}{16} \times (20)^3 \times 640 = 1005.3$ N.m

(ii) Angle of twist, $\theta = \frac{Tl}{GJ} = \frac{1005.3 \times 0.5 \times 32}{8 \times 10^{11} \times \pi(0.02)^4} = 0.4$ rad or 22.92°

Q.2. (*b*) Given: $l = 2$ m, $W = 10$ kN, EI $= 2 \times 10^7$ N.m^2

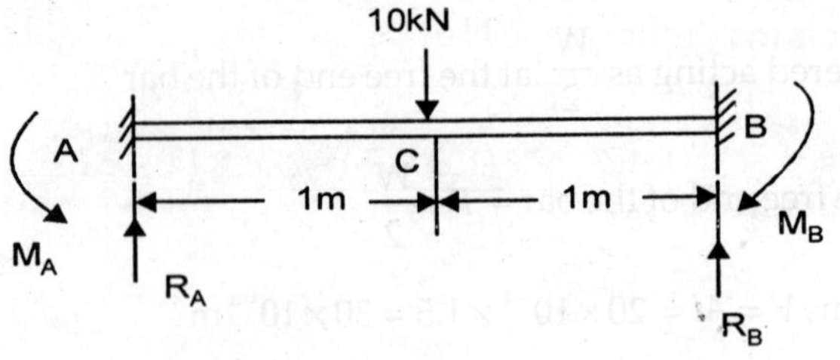

Fig. 5

$$R_A = R_B = \frac{W}{2} = \frac{10}{2} = 5 \text{ kN}$$

$$M_A = M_B = \frac{Wl}{8} = \frac{10 \times 2}{8} = 2.5 \text{ kN.m}$$

Deflection at midspan, $\delta_C = \dfrac{Wl^3}{192\,EI}$

$$= \frac{10 \times 10^3 \times 2^3}{192 \times 2 \times 10^7} = 0.0208 \text{ mm}$$

Q.2. (*c*) Given: $\sigma = 10$ MPa, $\tau = 2.5$ MPa, $b = 100$ mm, $h = 150$ mm

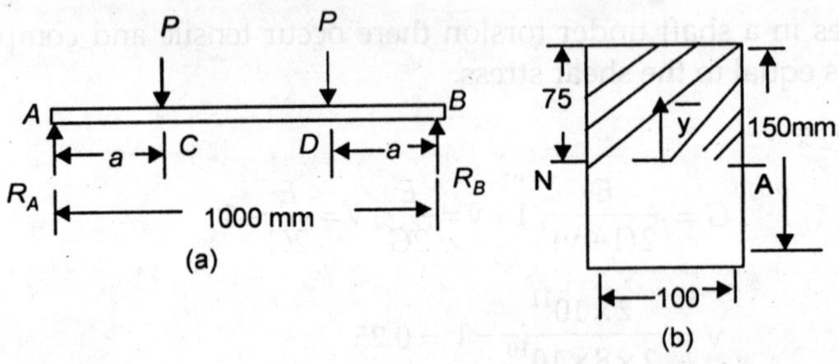

Fig. 6

Due to symmetry, $R_A = R_B = P$

$$M_{max} = P \times a$$

$$z = \frac{bh^2}{6} = \frac{100 \times 150^2}{6} = 375 \times 10^3 \text{ mm}^3$$

$$\sigma = \frac{M_{max}}{z}$$

$$10 = \frac{Pa}{375\times10^3},\ Pa = 375\times10^4 \text{ N.mm}$$

Horizontaql shear is maximum at the N-A

$$F = P$$

$$\tau_{max} = \frac{FA\bar{y}}{Ib} = \frac{P\times(100\times75)\times37.5\times12}{100\times150^3\times100} = 2.5$$

$$P = 25 \text{ kN}$$

$$a = \frac{375\times10^4}{25\times10^3} = 150 \text{ mm}$$

Q.3. (*a*) Given: $d = 10$ mm, $n = 8$, $D_m = 150$ mm, $W = 130$ N, $G = 80$ GPa, $M_0 = 9$ N.m, $E = 200$ GPa

(*i*) Axial load.

Elongation of spring, $\delta = \frac{8WD_m^3 n}{Gd^4} = \frac{8\times130\times150^3\times8}{80\times10^3\times10^4} = 35.1 \text{ mm}$

Torsional stress, $\tau = \frac{8WD_m}{\pi d^3} = \frac{8\times130\times150}{\pi\times10^3} = 49.66 \text{ N/mm}^2$

Strain energy/unit volume, $u = \frac{\tau^2}{4G} = \frac{(49.66)^2}{4\times80\times10^3} = 7.7054\times10^{-3} \text{ N.mm/mm}^3$

(ii) Axial torque.

Axial twist, $\phi = \frac{64\,M_0\,D_m n}{Ed^4} = \frac{64\times9\times10^3\times150\times8}{200\times10^3\times10^4} = 0.3456 \text{ rad or } 19.8°$

Bending stress, $\sigma_b = \frac{32\,M_0}{\pi d^3} = \frac{32\times9\times10^3}{\pi\times10^3} = 91.67 \text{ N/mm}^2$

$$u = \frac{\sigma_b^2}{8E} = \frac{(91.67)^2}{8\times200\times10^3} = 5.2525\times10^{-3} \text{ N.mm./mm}^3$$

Q.3. (*b*) Given: $L = 5$m, $P = 800$ kN, $d_0 = 240$ mm, $d_i = ?$, $a = \frac{1}{6400}$, $\sigma_c = 80$ MPa

According to Rankine's formula,

$$P = \frac{\sigma_c\times A}{1+a\left(\frac{L_e}{K}\right)^2}$$

For fixed ends column, $$L_e = \frac{L}{2} = \frac{5}{2} = 2.5 \text{ m}$$

$$800\times 10^3 = \frac{80\times10^6\times A}{1+\frac{1}{6400}\left(\frac{2.5}{K}\right)^2}$$

$$0.01+1.5625\times10^{-6}\times\left(\frac{2.5}{K}\right)^2 = A$$

$$0.01\, K^2 + 9.7656\times 10^{-6} = AK^2$$

$$K^2(A-0.01) = 9.7656\times 10^{-6}$$

$$I = AK^2,\ K^2 = \frac{I}{A} = \frac{\frac{\pi}{64}(d_0^4-d_i^4)}{\frac{\pi}{4}(d_0^2-d_i^2)} = \frac{d_0^2+d_i^2}{16}$$

$$\left(\frac{d_0^2+d_i^2}{16}\right)\left[\frac{\pi}{4}(d_0^2-d_i^2)-0.01\right] = 9.7656\times 10^{-6}$$

$$\frac{\pi}{64}(d_0^4-d_i^4)-0.01\left(\frac{d_0^2+d_i^2}{16}\right) = 9.7656\times10^{-6}$$

$$\frac{\pi}{64}\left[(0.24)^4-d_i^4\right]-6.25\times10^{-4}[(0.24)^2+d_i^2] = 9.7656\times10^{-6}$$

$$7.4940\times 10^{-3} = \pi d_i^4+0.04\,d_i^2$$

or $$d_i^4+0.01273d_i^2-0.0023854 = 0$$

$$d_i^2 = \frac{1}{2}[-0.01273\pm\sqrt{(0.01273)^3+4\times0.0023854}]$$

$$= 0.0406$$

$$d_i = 0.2015 \text{ m} \quad \text{or} \quad 201.5 \text{ mm}$$

Thickness of column, $$t = \frac{1}{2}(240-201.5) = 19.25 \text{ mm}$$

Q.3. (*c*) Refer to art. 8.7.1

Q.4. **(a)** Given: $d = 1.5$ m, $p = 2$ MPa, $V_e = 4$ litre or 0.004 m^3, $K = 2$GPa, $E = 200$ GPa, $\nu = 0.30$

Hoop stress, $$\sigma_\theta = \frac{pd}{4t} = \frac{2\times1500}{4t} = \frac{750}{t}\text{N/mm}^2$$

$$\varepsilon_\theta = \frac{\delta V}{V} = \frac{3\sigma_\theta}{E}(1-\nu) = \frac{3\times 750}{t\times 200\times 10^3}(1-0.3) = \frac{7.875\times 10^{-3}}{t}$$

$$V = \frac{\pi}{6}d^3 = \frac{\pi}{6}\times(1.5)^3 = 1.76714\text{ m}^3$$

$$\delta V = \varepsilon_\theta \times V = \frac{7.875\times 10^{-3}}{t}\times 1.76714 = \frac{0.013916}{t}m^3$$

Total volume of water at 2 MPa = $V + \delta V = \left(1.76714 + \frac{0.013916}{t}\right)\text{m}^3$

Volumetric strain of water = $\frac{p}{K} = \frac{2\times 10^6}{2\times 10^9} = 0.001$

Change in volume of water = $0.001\left(1.76714 + \frac{0.031916}{t}\right)\text{m}^3$

Actual volume of water at 2 MPa = $1.001\left(1.76714 + \frac{0.013916}{t}\right)\text{m}^3$

$$= \left(1.7689 + \frac{0.01393}{t}\right)\text{m}^3 \qquad \text{... (1)}$$

At atmospheric presure of 1.01325 bar,

$$\sigma_\theta = \frac{1.01325\times 10^{-1}\times 1500}{4t} = \frac{37.997}{t}\text{N/mm}^2$$

$$\varepsilon_\theta = \frac{3\sigma_\theta}{E}(1-\nu) = \frac{3\times 37.997}{E\times 200\times 10^3}(1-0.3) = \frac{0.39897\times 10^{-3}}{t}$$

Change in volume of spherical shell at atmospheric pressure,

$$\delta V = 1.76714\times\frac{0.39897\times 10^{-3}}{t} = \frac{0.70503\times 10^{-3}}{t}\text{m}^3$$

Total volume, $\quad V_t = V + \delta V = \left(1.76714 + \frac{0.70503\times 10^{-3}}{t}\right)\text{m}^3$

Change in volume due to compression of water, = $\frac{p}{K}V_t$

$$= \frac{1.01325\times 10^5}{2\times 10^9}\times\left(1.76714 + \frac{0.7.503\times 10^{-3}}{t}\right)$$

$$= \left(8.95277\times10^{-5} + \frac{3.57186\times10^{-8}}{t}\right)\text{m}^3$$

$$\text{Actual volume of water} = \left(1.76714 + \frac{0.70503\times10^{-3}}{t}\right) + \left(8.95277\times10^{-5} + \frac{3.57186\times10^{-8}}{t}\right)$$

$$= 1.76723 + \frac{0.70506\times10^{-3}}{t} \quad \text{... (2)}$$

$$\text{Water escaped} = \text{Eq. (1)} - \text{Eq. (2)}$$

$$= \left(1.7669 + \frac{0.01393}{t}\right) - \left(1.76723 + \frac{0.70506\times10^{-3}}{t}\right)$$

$$= 1.67\times10^{-3} + \frac{0.01322}{t}$$

$$= 0.004$$

$$t = \mathbf{5.67\ mm}$$

Q.4. (*b*) Given: $r_1 = 80$ mm, $r_2 = 120$ mm, $p_2 = 12$ MPa, $p_1 = ?$, $\sigma_\theta = 36$ MPa

Maximum hoop stress occurs at the inner radius.

$$\sigma_\theta = \frac{p_1 r_1^2 - p_2 r_2^2}{(r_2^2 - r_1^2)} + \frac{r_2^2 (p_1 - p_2)}{(r_2^2 - r_1^2)}$$

$$= \frac{1}{(r_2^2 - r_1^2)}[p_1 r_1^2 - p_2 r_2^2 + p_1 r_2^2 - p_2 r_2^2]$$

$$= \frac{1}{(r_2^2 - r_1^2)}[p_1 (r_1^2 + r_2^2) - 2p_2 r_2^2]$$

$$36 = \frac{1}{(120^2 - 80^2)}[p_1(80^2 + 120^2) - 2\times12\times120^2]$$

$$= \frac{1}{8000}[20800\times p_1 - 345600]$$

$$p_1 = 30.46 \text{ MPa}$$

Q.4. (c) Refer to art. 11.2.

Q.5. (*a*) Given: $l = 2$ m, $W = 4$ kN,, $\phi = 20°$

As the I-beam has symmetry about x - and y-axes, therefore, these are also the principal axes of inertia.

$$M_{max} = \frac{Wl}{4} = \frac{4\times2}{4} = 2\text{kN.m}$$

$$I_u = \frac{80 \times 160^3}{12} - \frac{70 \times 140^3}{12} = 113 \times 10^5 \text{ mm}^4$$

$$I_v = \frac{2 \times 10 \times 80^3}{12} + \frac{140 \times 10^3}{12} = 865 \times 10^3 \text{ mm}^4$$

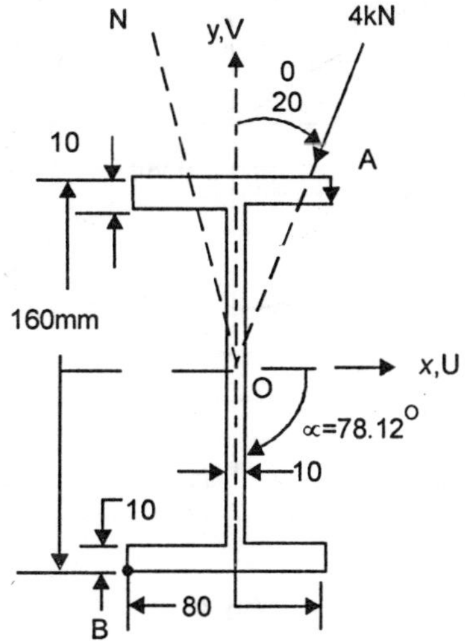

Fig. 7

Position of neutral axis,

$$\tan \alpha = \frac{I_u}{I_v} \tan \phi = \frac{113 \times 10^5}{865 \times 10^3} \times \tan 20° = 4.75476$$

$$\alpha = 78.12°$$

$$\sigma = \frac{M_u . v}{I_u} + \frac{M_v . u}{I_v},$$

For points A,

$$u_A = 40 \text{ mm}, v_A = 80 \text{ mm}.$$

Both M_u and M_u produce compressive stress at A and tensile stress at B.

$$\sigma_A = 2 \times 10^6 \left[\frac{-80 \times \cos 20°}{113 \times 10^5} + \frac{-40 \times \sin 20°}{865 \times 10^3} \right] = -44.937 N/\text{mm}^2$$

For point B, both M_u amd M_v produce tensile stress.

$\therefore$ $$\sigma_B = 44.937 \text{ N/mm}^2$$

Q.5. **(b)** Given: $W = 8$ kN, $R = 80$ mm, $E = 200$ GPa

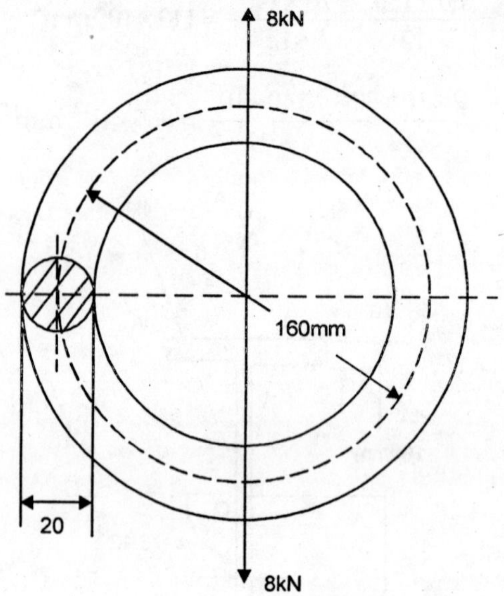

Fig. 9

Let diameter of rod, $D = 20$ mm

$$h^2 = \frac{D^2}{16} + \frac{1}{128}\frac{D^4}{R^2} + \ldots$$

$$= \frac{20^2}{16} + \frac{1}{128} \times \frac{20^4}{80^2} = 25.195 \text{ mm}^2$$

Deflection of ring along the load line,

$$\delta = \frac{WR^3}{EI}\left[\frac{\pi}{4} - \frac{2}{\pi}\left(\frac{R^2}{R^2 + h^2}\right)\right]$$

$$= \frac{8 \times 10^3 \times 80^3}{200 \times 10^3 \times \pi \times 20^4}\left[\frac{\pi}{4} - \frac{2}{\pi}\left(\frac{80^2}{80^2 + 25.195}\right)\right]$$

$$= 0.3945 \text{ mm}$$

Q.5. (*c*) Given : $W = 80$ kN

For the T-section,

$$A_1 = 80 \times 20 = 1600 \text{ mm}^2$$
$$A_2 = 80 \times 20 = 1600 \text{ mm}^2$$
$$A = A_1 + A_2 = 2 \times 1600 = 3200 \text{ mm}^2$$

$$\bar{x} = \frac{1600 \times 10 + 1600 \times 60}{3200} = 35 \text{ mm}$$

$\therefore$ $$R = 150 + 35 = 185 \text{ mm}$$

$$h^2 = \frac{R^3}{A}\left[b_2 ln\frac{r_2}{r_1} + t_1 ln\frac{r_3}{r_2}\right] - R^2$$

where $b_2 = 80$ mm, $t_1 = 20$ mm, $r_1 = 150$ mm

$r_2 = 150 + 20 = 170$ mm, $r_3 = 150 + 100 = 250$ mm

$$h^2 = \frac{185^3}{3200}\left[80\,ln\frac{170}{250} + 20\,ln\frac{250}{170}\right] - 185^2$$

$$= 848.84 \text{ mm}^2$$

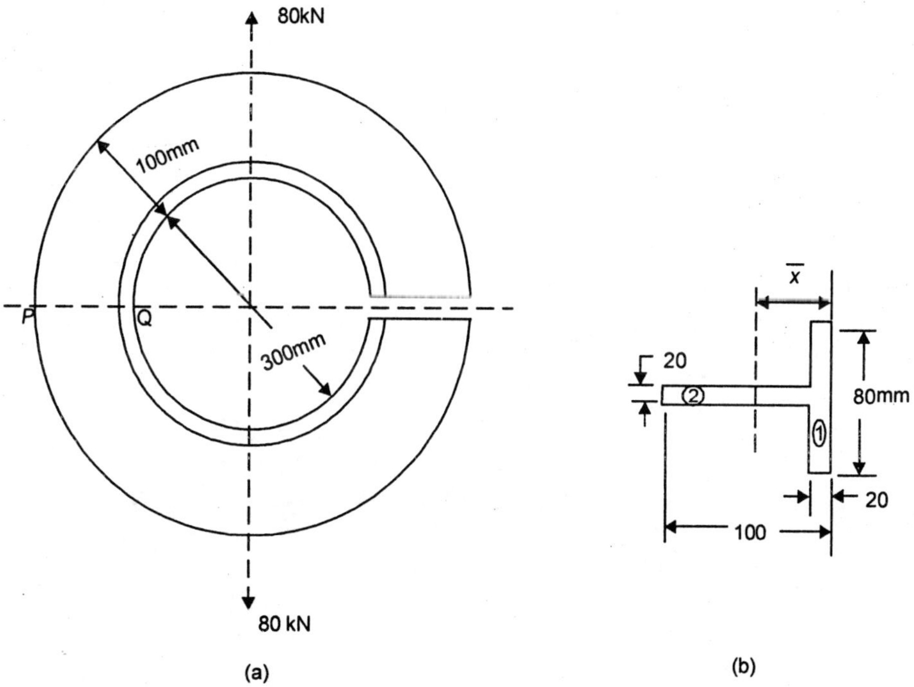

Fig. 8

Here the bending moment is such that it is tending to decrease the curvature of the ring. Therefore, it is -ve in nature.

$$\sigma_p = \frac{W}{A} - \frac{W}{A}\left[1 + \frac{R^2}{h^2}\left(\frac{d_2}{R + d_2}\right)\right]$$

Here $d_1 = \bar{x} = 35\text{mm}, d_2 = 100 - d_1 = 100 - 35 = 65\text{mm}$

$$\sigma_P = \frac{80\times10^3}{3200} - \frac{80\times10^3}{3200}\left[1+\frac{185^2}{848.84}\left(\frac{65}{185+65}\right)\right]$$

$$= 25 - 287.08 = -262.08 \text{ N/mm}^2$$

$$\sigma_Q = \frac{W}{A} + \frac{W}{A}\left[1-\frac{R^2}{h^2}\left(\frac{d_1}{R-d_1}\right)\right]$$

$$= \frac{80\times10^3}{3200} + \frac{-80\times10^3}{3200}\left[1-\frac{185^2}{848.84}\left(\frac{35}{185-35}\right)\right]$$

$$= 25 + 210.2 = 235.2 \text{ N/mm}^2$$

(TME-303)

B. TECH (UKTU)

(SEM. III) ODD SEMESTER THEORY EXAMINATION 2009-10

STRENGTH OF MATERIAL

Time : 3 Hours *Total Marks : 100*

Note:(i) Use of calculator is permitted.

(ii) Attempt all questions.

(iii) Assume any data required, but not given.

1. Attempt any **four** of the following : **5 × 4=20**

(a) In a steel member, at a point the major principal stress is 180 MPa and minor principal stress is compressive. If the tensile yield point of the steel is 225 MPa, find the value of minor principal stress using (i) Maximum shear stress criteria and (ii) maximum strain energy criteria. Take Poisson's ratio = 0.26.

(b) In a triaxial stress system, the six components of the stress at a point are given below :

$$\sigma_x = 6 \text{ MN/m}^2 \ \tau_{xy} = \tau_{yx} = 1 \text{ MN/m}^2$$
$$\sigma_y = 5 \text{ MN/m}^2, \tau_{yz} = \tau_{zy} = 3 \text{ MN/m}^2$$
$$\sigma_z = 4 \text{ MN/m}^2, \tau_{zx} = \tau_{xz} = 2 \text{ MN/m}^2$$

Find the magnitude of three principal stresses.

(c) A cylindrical shaft made of steel of yield strength 700 MPa is subjected to static load consisting of bending moment 10 kNm and a torsional moment 30 kNm. Determine the diameter of shaft using maximum shear stress theory and assuming a factor of safety 2. Take E = 210 GPa, Poisson's ratio = 0.25.

(d) A rod of 20 mm diameter and 500 mm length is first subjected to a steady load of 20 kg. What total stress would be produced in the bar if a weight of 5 kg falls through a height of 50 mm on the collar attached at the bottom end of the rod. The rod is hanging freely from the ceiling. Take $E = 1 \times 10^5$ N/mm^2,

(e) Discuss Castigliano's theorem.

(f) Explain Airy's stress function.

2. Attempt any **four** of the following : **5 × 4 =20**

(a) A simply supported cast iron rectangular beam of 800 mm length and 15 ×15 mm in section fails on applying a load of 360 N at the midspan. Find the maximum uniformly distributed load that can be applied safely to a cantilever made of the same material 75 mm, deep and 1.6m long.

(b) A 320 mm × 160 mm I-section joist has 20 mm thick flanges and a 15 mm thick web. At a certain cross-section it is acted upon by a bending moment of 100 kNm and a shear force of 200 kN. Determine the principal stresses in the flanges at 140 mm from neutral axis.

(c) An 80 mm wide and 180 mm deep cantilever is of 3 m span. It carries a uniformly distributed load of intensity 6 kN/m on a 2 m length of the span starting from the free end. Determine the slope and the deflection at the free end. E = 205 GPa.

(d) A beam AB simply supported at the ends is 6 m long. It carries a uniformly distributed load of intensity 15 kN/m over the length of 3m starting at point A. Calculate the deflection at the centre. Take E=200 GPa, I = 9000 cm^4. Also find the slope at A.

(e) In a hollow circular shaft, the external diameter is 100 mm and internal diameter is 60 mm. The allowable shear stress in the shaft material is 55 N/mm^2. Determine the angle of twist in a length of twenty times the external diameter of the shaft.

(f) A solid shaft transmits 1000 kW at 100 rpm, the maximum torque being 45% greater than the mean. The shaft is supported in ball bearings 1.5 m apart with a 7500 kg fly-wheel midway between bearings. Find the shaft diameters if the maximum tensile stress is 60 MN/m^2.

3. Attempt any **two** of the following : **10 × 2 = 20**

(a) The stiffness of a close-coiled helical spring is 1.5 N/mm of compression under a maximum load of 60 N. The maximum shearing stress produced in the wire of the spring is 130 N/mm^2. The solid length of the spring (when the coils are touching) is given as 5 cm. Find :

(i) Diameter of wire

(ii) Mean diameter of the coils, and

(iii) Number of coils required.

Take $G = 4.5 \times 10^4$ N/mm^2.

(b) A 4 m long fixed-end hollow cast-iron column supports an axial load of 1 MN. The external diameter of the column is 200 mm. Determine the thickness of the column by using Rankine formula taking a constant of 1/6400 and working stress as 78 MN/m^2.

(c) From the first principles derive the expression for the critical buckling for a column having one end fixed and one end hinged.

4. Attempt any **two** of the following : **10 × 2=20**

(a) A thin cylinder of 200 mm inside diameter is 4 mm thick. The ends of the cylinder are closed by rigid plates and then it is filled with water under pressure.
If an external axial pull of 50 kN is applied to the ends, the water pressure falls by 0.12 MPa. Find the value of the Poisson's ratio. K = 2100 MPa and E = 150 GPa.

(b) A thick spherical shell of 40 mm inside diameter is subjected to an internal pressure of 20 MPa. Determine the necessary thickness of the shell, if permissible stress of the shell material is 40 MPa.

(c) Write short notes on :

(i) Compound cylinders

(ii) Radial, axial and circumferential stresses in thick cylinders.

5. Attempt any **two** of the following : **10 × 2=20**

(a) A curved bar of rectangular section of 30 mm width and 40 mm depth and mean radius of curvature of 60 mm is initially unstressed. If a bending moment of 400 Nm is applied to the bar which tends to straighten it, determine the stresses at the inner and outer surfaces and sketch a diagram showing stress variation across the section.

(b) A ring made of 3.0 cm diameter steel bar carries a pull of 10 kN. Calculate the maximum tensile and compressive stresses in the material of the ring. The mean radius of the ring is 20 cm.

(c) In crane hooks trapezoidal section is very commonly used. Explain.

Q.1.(a) **Given:** $\sigma_1 = +180$ MPa, $\sigma_2 = ?$ (–ve), $\sigma_{yt} = 225$ MPa

(i) According to maximum shear stress theory,

$$\sigma_1 - \sigma_2 = \sigma_{yt}$$
$$180 - (-\sigma_2) = 225$$
$$\sigma_2 = 225 - 180 = 45 \text{ MPa (compressive)}$$

(ii) According to maximum strain energy theory,

$$\sigma_1^2 + \sigma_2^2 - 2\nu\sigma_1\sigma_2 = \sigma_{yt}^2$$

$$(180)^2 + \sigma_2^2 - 2 \times 0.26 \times 180 \times (-\sigma_2) = (225)^2$$

$$32400 + \sigma_2^2 + 93.6\sigma_2 - 50625 = 0$$

$$\sigma_2^2 + 93.6\sigma_2 - 18225 = 0$$

$$\sigma_2 = \frac{1}{2}\left[-93.6 \pm \sqrt{(93.6)^2 + 4 \times 18225}\right]$$

$$= \frac{1}{2}[-93.6 \pm 285.8]$$

$$= 96.1 \text{ MPa}, -189.7 \text{ MPa}$$

$\therefore$ $\sigma_2 = 189.7$ MPa (compressive)

Q.1. (b) Given:

$$\sigma_x = 6\text{MPa}, \sigma_y = 5\text{MPa}, \sigma_z = 4\text{MPa}, \tau_{xy} = 1\text{MPa}, \tau_{yz} = 3 \text{ MPa}, \tau_{xz} = 2 \text{ MPa}$$

The stress invariants are,

$$I_1 = \sigma_x + \sigma_y + \sigma_z = 6 + 5 + 4 = 15$$
$$I_2 = \sigma_x\,\sigma_y + \sigma_y\,\sigma_z + \sigma_z\,\sigma_x - \tau^2_{xy} - \tau^2_{yz} - \tau^2_{xz}$$
$$= 6 \times 5 + 5 \times 4 + 4 \times 6 - 1 - 3^2 - 2^2$$
$$= 60$$
$$I_3 = \sigma_x\,\sigma_y\,\sigma_z - \sigma_x\,\tau^2_{yz} - \sigma_y\,\tau^2_{xz} - \sigma_z\,\tau^2_{xy} + 2\tau_{xy}\,\tau_{yz}\,\tau_{xz}$$
$$= 6 \times 5 \times 4 - 6 \times 9 - 5 \times 4 - 4 \times 1 + 2 \times 1 \times 3 \times 2$$
$$= 54$$

Principal stresses are the roots of the cubic equation,

$$\sigma^3 - I_1\sigma^2 + I_2\sigma - I_3 = 0$$

$$\sigma^3 - 15\sigma^2 + 60\,\sigma - 54 = 0 \qquad ...(1)$$

Now $\cos^{3\theta} = 4\cos^3\theta - 3\cos\theta$

$\therefore$
$$\cos^3\theta - \frac{3}{4}\cos\theta - \frac{1}{4}\cos 3\theta = 0 \qquad ...(2)$$

Put
$$\sigma = r\cos\theta + \frac{I_1}{3} = r\cos\theta + \frac{15}{3} = r\cos\theta + 5$$

Eq. (1) becomes,

$$(r\cos\theta+5)^3-15(r\cos\theta+5)^2+60(r\cos\theta+5)-54=0$$

$$r^3\cos^3\theta+125+75r\cos\theta+15r^2\cos^2\theta-15r^2\cos^2\theta-375-150r\cos\theta+60r\cos\theta+300-54=0$$

$$r^3\cos^3\theta-15r\cos\theta-4=0$$

or
$$\cos^3\theta-\frac{15}{r^2}\cos\theta-\frac{4}{r^3}=0 \quad ...(3)$$

Comparing Eqs. (2) and (3), we have

$$\frac{15}{r^2}=\frac{3}{4}$$

or
$$r=\sqrt{20}=4.472$$

$$\frac{4}{r^3}=\frac{1}{4}\cos 3\theta$$

$$\cos 3\theta=\frac{16}{(4.472)^3}=0.17888$$

$$3\theta=79.69$$

or
$$\theta_1=26.56°,\ \theta_2=120°-\theta_1=93.44°,\ \theta_3=120°+\theta_1=146.56°$$

$$\sigma_1=r\cos\theta_1+5=4.472\cos 26.36°+5=9\text{ MPa}$$

$$\sigma_2=r\cos\theta_2+5=4.472\cos 93.44°+5=4.73\text{ MPa}$$

$$\sigma_3=r\cos\theta_3+5=4.472\cos 146.56°+5=1.27\text{ MPa}$$

Check: $\sigma_1+\sigma_2+\sigma_3=\sigma_x+\sigma_y+\sigma_z=15$

Hence, the solution is correct.

Q.1. (c) **Given:** $\sigma_y=700$ MPa, M=10 kN.m, $T=30$ kN.m, FOS = 2, $E=210$ GPa, $\nu=0.25$

Equivalent torque,

$$T_e=\sqrt{M^2+T^2}=\sqrt{10^2+30^2}=31.623\text{ kN.m}$$

Maximum shear stress,

$$\tau_{max}=\frac{16T_e}{\pi d^3}=\frac{16\times31.623\times10^6}{\pi d^3}\text{ N/mm}^2$$

$$=\frac{\sigma_y}{2\times FOS}=\frac{700}{2\times2}=175\text{ N/mm}^3$$

$$d^3=0.9203\times10^6$$

$$d=0.9727\times10^2=97.27\text{ mm}$$

∴ Shaft diameter = 97.27 mm

Q.1. (d) **Given:** $d=20$mm, $l=500$ mm, $P=20\times9.81=196.2$ N

Static stress, $\sigma_{st} = \frac{P}{A} = \frac{4P}{\pi d^2} = \frac{4 \times 196.2}{\pi \times 20^2} = 0.6245 \text{ N/mm}^2$

$$W = 5 \times 9.81 = 49.05N, h = 50mm, E = 1 \times 10^5 \text{ N/mm}^2$$

$$W_1 = W\left[1+\sqrt{1+\frac{2AEh}{Wl}}\right]$$

$$= 49.05\left[1+\sqrt{1+\frac{2\times\pi\times400\times1\times10^5\times50}{4\times49.05\times500}}\right]$$

$$= 17604.5 \text{ N}$$

Impact stress, $\sigma_1 = \frac{W_1}{A} = \frac{4 \times 17604.5}{\pi \times 400} = 56.04 \text{ N/mm}^2$

Total stress $\sigma_{st} + \sigma_1 = 0.6245 + 56.04 = 56.66 \text{ N/mm}^2$

Q.1. (e) Refer to art. 2.5.

Q.1. (f) Refer to art. 3.17.

Q.2 (a)Simply supported beam :

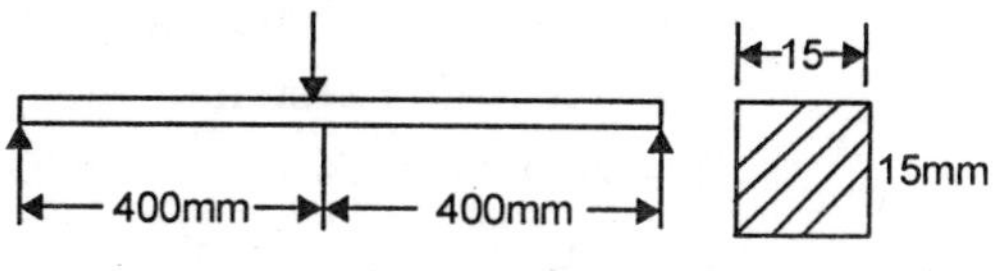

Fig.1

$$b = h = 15 \text{ mm}$$

$$M_{max} = \frac{Wl}{4} = \frac{360 \times 800}{4} = 72 \times 10^3 N.mm$$

Section modulus, $z = \frac{b^3}{6} = \frac{15}{6} = 562.5 \text{ mm}^3$

Maximum bending stress,

$$\sigma_b = \frac{M}{z} = \frac{72 \times 10^3}{562.5} = 128 \text{ N/mm}^2$$

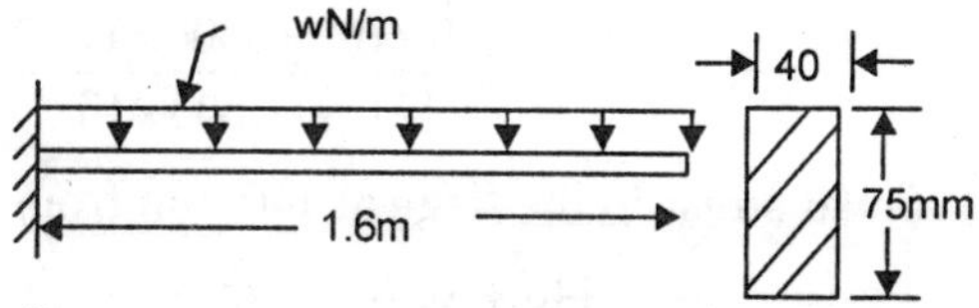

Fig.2

Cantilever beam:

$$l = 1.6 \text{ m}, b = 40 \text{ mm}, h = 75 \text{ mm}$$

$$M_{max} = \frac{wl^2}{2} = \frac{w \times (1600)^2}{2} = 128 \times 10^4 \times w \text{ N.mm}$$

$$z = \frac{bh^2}{6} = \frac{40 \times 75^2}{6} = 37500 \text{ mm}^3$$

$$\sigma_b = \frac{M_{max}}{z} = \frac{128 \times 10^4 \times w}{37500} = 34.13 \times w \text{ N.mm} = 128$$

$$w = 3.75 \text{ N/mm or } 3750 \text{ N/m}$$

Q.2 (b) Given: M = 100 kN.m, F = 200 kN

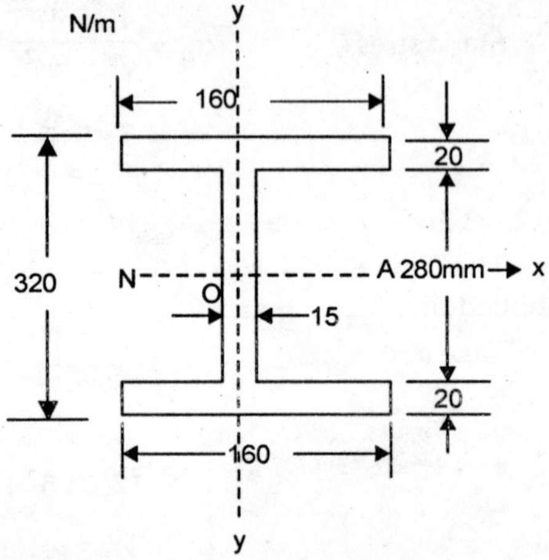

Fig. 3

$$I_{x-x} = \frac{160 \times 320^3}{12} - \frac{(160-15)(320-40^3)}{12}$$

$$= 171,653,333 \text{ mm}^4$$

Maximum bending stress occurs at the outer most fibres of the beam.

$$\frac{M}{I_{x-x}} = \frac{\sigma}{y}$$

$$\sigma = \frac{100 \times 10^6 \times 160}{171,653,333} = 93.2 \text{ N/mm}^2$$

Maximum shear stress occurs at the neutral axis.

Let $\bar{y}$ = distance of the centroid of the area above the N – A.

$$A_1 = 160 \times 20 = 3200 \text{ mm}^2$$

$$A_2 = 140 \times 15 = 2100 \text{ mm}^2$$

$$A = A_1 + A_2 = 3200 + 2100 = 5300 \text{ mm}^2$$

$$\bar{y} = \frac{A_1y_1 + A_2y_2}{A} = \frac{3200 \times 150 + 2100 \times 70}{5300}$$

$$= 118.3 \text{ mm}$$

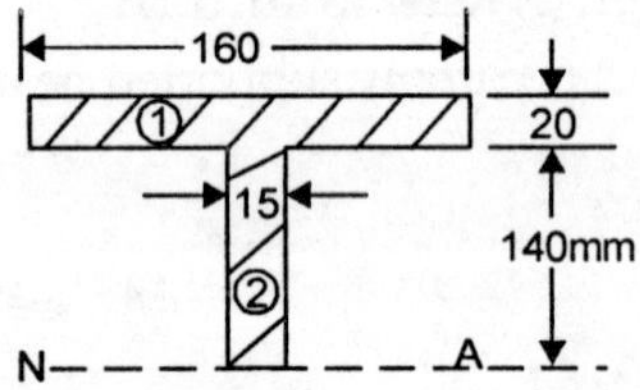

Fig. 4

Shear stress,

$$\tau = \frac{FA\bar{y}}{Ib} = \frac{200 \times 10^3 \times 5300 \times 118.3}{171,653,333 \times 15} = 48.7 \text{ N/mm}^2$$

Shear stress in the flage at 140 mm from the N.A.

$$\bar{y} = 140 + 10 = 150 \text{ mm}$$

$$A = 160 \times 20 = 3200 \text{ mm}^2$$

$$b = 160 \text{ mm}$$

$$\tau_{flange} = \frac{FA\bar{y}}{Ib} = \frac{200 \times 10^3 \times 3200 \times 150}{171,653,333 \times 160} = 3.495 \text{ N/mm}^2$$

Bending stress in the flange at the junction of web and flange,

$$\sigma_b = \frac{93.2}{160} \times 140 = 81.55 \text{ N/mm2}$$

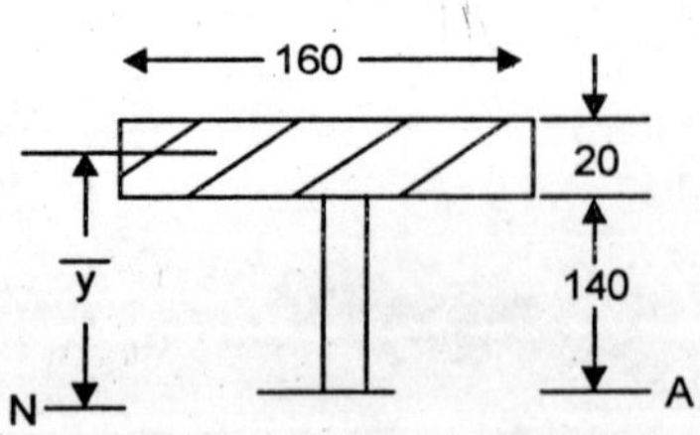

Fig. 5

Principal stresses are given by,

$$\sigma_{1,2} = \frac{\sigma_b}{2} \pm \frac{1}{2}\sqrt{\sigma_b^2 + 4\tau^2}$$

$$= \frac{81.55}{2} \pm \frac{1}{2}\sqrt{(81.55)^2 + 4\times(3.495)^2}$$

$$= 40.75 \pm 40.92$$

$$= 81.68 \text{ N/mm}^2, -0.16 \text{ N/mm}^2$$

Q.2(c) Given: E = 205 GPa

The Udl can be treated on the whole span of cantilever and an upward udl of same intensity applied over the span CB.

Take a section at a distance x from free end A.

$$M_x = -6x \times \frac{x}{2} + 6\times(x-2)\frac{(x-2)}{2}$$

$$= -3x^2 + 3(x-2)^2 \text{ kN.m}$$

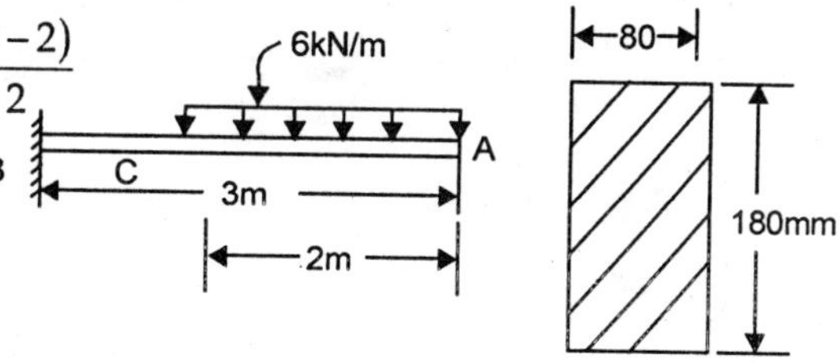

Fig. 6

$$EI\frac{d^2y}{dx^2} = -M_x = 3x^2 - 3(x-2)^2$$

Integrating twice, we have

$$EI\frac{dy}{dx} = x^3 - (x-2)^3 + C_1$$

$$EIy = \frac{x^4}{4} - \frac{(x-2)^4}{4} + C_1 x + C_2$$

At $x = 3$ m, $y = 0$ and $\frac{dy}{dx} = 0$

$$0 = 27 - 1 + C_1$$

$$C_1 = -26$$

$$EI\frac{dy}{dx} = x^3 - (x-2)^3 - 26$$

$$0 = 20.25 - 0.26 - 78 + C_2$$

$$C_2 = 58$$

At $x = 0$

$$EI\frac{dy}{dx} = -26$$

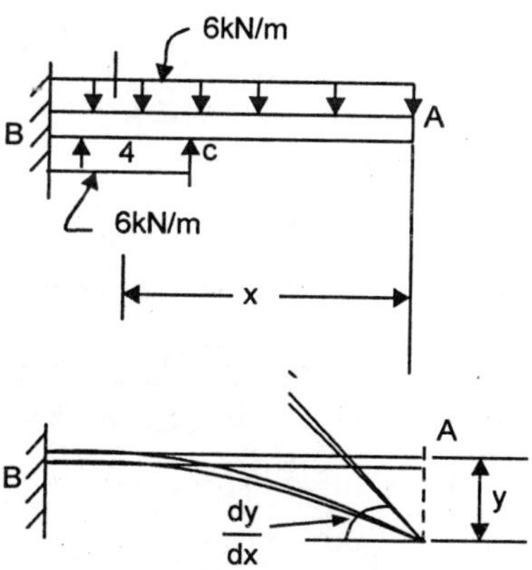

Fig. 7

$$I = \frac{80\times 180^3}{12} = 3888\times10^4 \text{ mm}^4$$

$$\frac{dy}{dx} = \frac{-26\times10^9}{205\times10^3\times3888\times10^4} = -3.262\times10^{-3} \text{ rad or } 0.187 \text{ deg}$$

$$y = \frac{58\times10^{12}}{205\times10^3\times3888\times10^4} = 7.27 \text{ mm}$$

Q.2. (d) **Given:** E = 200 GPa, I = 9000 cm^4

$$\Sigma M_B = 0 \text{ gives}$$
$$6R_A = 15 \times 3 \times 4.5$$
$$R_A = 33.75 \text{ kN}$$

The *udl* may be extended over the whole span and an upward *udl* of same intensity applied on the beam length *CB*.

Take a section at the distance x from support A and closer to B.

$$M_x = 33.75\times x - 15\times x\times\frac{x}{2} - 15(x-3)\frac{(x-3)}{2}$$

$$= 33.75x - 7.5x^2 + 7.5(x-3)^2 \text{ kN.m}$$

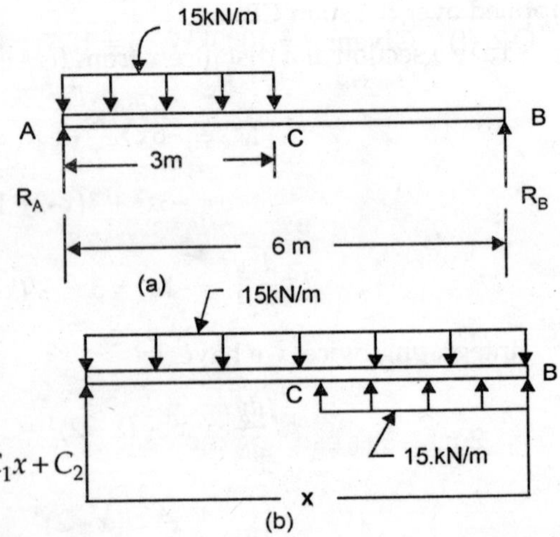

Fig. 8

$$EI\frac{d^2y}{dx^2} = -M_x = -33.75x + 7.5x^2 - 7.5(x-3)^2$$

Integrating twice, we have

$$EI\frac{dy}{dx} = -16.875x^2 + 2.5x^3 - 2.5\,(x-3)^3 + C_1$$

$$EIy = -5.625x^3 + 0.625x^4 - 0.625(x-3)^4 + C_1x + C_2$$

At $x = 0$ and $x = 6$m, $y = 0$

∴ $C_2 = 0$

$$0 = -1215 + 810 - 50.625 + 6C_1$$
$$C_1 = 75.9375$$

At $x = 3$m

$$EIy_C = -5.265 \times 3^3 + 0.625 \times 3^4 + 0 + 3 \times 75.9375 = 126.5625$$

$$y_c = \frac{126.5625\times 10^{12}}{200\times 10^3 \times 9000\times 10^4} = 7 \text{ mm}$$

At $x = 0$,

$$EI\frac{dy}{dx} = C_1 = 75.9375$$

$$\left(\frac{dy}{dx}\right)_A = \frac{75.9375\times 10^9}{200\times 10^3\times 9000\times 10^4} = 4.218 \times 10^{-3} \text{ rad or } 0.24°$$

Q.2. (e) **Given:** $d_0 = 100$ mm, $d_i = 60$ mm, $\tau_{allow} = 55$ N/mm^2, $l = 20d_0$.

$$J = \frac{\pi}{32}\left(d_0^4 - d_i^4\right) = \frac{\pi}{32}\left(100^4 - 60^4\right) = 8{,}545{,}132 \text{ mm}^4$$

$$\frac{2\tau}{d_0} = \frac{G\theta}{l}$$

$$\theta = \frac{2\tau l}{d_0 G} = \frac{40\tau}{G} \qquad \left(\because \frac{l}{d_0} = 20\right)$$

G is not given. Let G = 84 GPa

Angle of twist, $$\theta = \frac{40 \times 55}{84 \times 10^3}$$

$$= 0.02619 \text{ rad or } 1.5°$$

Q.2. (f) **Given:** P = 1000kW, N = 100 rpm, $T_{max} = 1.45T_m$, $l = 1.5$m, m = 7500 kg, $(\sigma_t)_{max} = 60$ MPa

$$M_{max} = \frac{mg \times l}{4} = \frac{7500 \times 9.81 \times 1.5}{4} = 27,5906 \text{ N.m}$$

$$P = \frac{2\pi N T_m}{60 \times 10^3}$$

$$T_m = \frac{1000 \times 60 \times 10^3}{2\pi \times 100} = 95,493 \text{ N.m}$$

$$T_{max} = 1.45 \times 95,493 = 138,465 \text{ N.m}$$

Equivalent bending moment,

$$M_e = \frac{1}{2}\left[M_{max} + \sqrt{M_{max}^2 + T_{max}^2}\right]$$

$$= \frac{1}{2}\left[27,590.6 + \sqrt{(27,590.6)^2 + (138,465)^2}\right]$$

$$= 84,388.7 \text{ N.m}$$

For a solid shaft, section modules,

$$z = \frac{\pi d^3}{32}$$

$$(\sigma_t)_{allow} = \frac{M_e}{z}$$

$$60 = \frac{84,388.7 \times 10^3 \times 32}{\pi d^3}$$

$$d = 242.87 \text{ mm}$$

Q.3(a) Given: $k = 1.5$N/mm, $W = 60$ N, $\tau = 130$N/mm^2, $l_s = 5$ cm, $G = 4.5 \times 10^4$ N/mm^2

Spring stiffness, $$k = \frac{W}{\delta} = \frac{Gd^4}{8D_m^3 n}$$

Shear stress, $$\tau = \frac{8WD_m}{\pi d^3}$$

Solid length of spring,

$$l_s = nd$$

$$\therefore \quad 1.5 = \frac{4.5\times10^4\times d^4}{8D_m^3 n} \quad \text{...(1)}$$

$$130 = \frac{8\times60\times Dm}{\pi d^3} \quad \text{...(2)}$$

$$50 = nd \quad \text{...(3)}$$

From Eq. (3), we have

$$n = \frac{50}{d} \quad \text{...(4)}$$

Substituting Eq.(4) in (1), we have

$$1.5 = \frac{4.5\times10^4\times d^4\times d}{8D_m^3\times50} = \frac{112.5d^5}{D_m^3}$$

$$\frac{D_m^3}{d^5} = 75 \quad \text{...(5)}$$

From Eq. (2), we have

$$\frac{D_m}{d^3} = 0.85085$$

or $$D_m = 0.85085\, d^3 \quad \text{...(6)}$$

Substituting Eq. (6) in (5), we have

$$(0.85085\, d^3)^3 = 75d^5$$

$$0.61596\, d^9 = 75d^5$$

$$d^4 = 121.76$$

$$d = 3.32 \text{ mm}$$

$$n = \frac{50}{3.32} = 15$$

$$D_m = 0.85085 \times (3.32)^3 = 31.14 \text{ mm}$$

Q.3. (b) **Given:** $l = 4\text{m}$. $P = 1\text{MN}$, $d_0 = 200$ mm, $a = \frac{1}{6400}$, $\sigma_C = 78\ \text{MN/m}^2$

Area of cross-section of hollow column,

$$A = \frac{\pi}{4}\left(d_0^2 - d_i^2\right)$$

Moment of inertia,

$$I = \frac{\pi}{64}\left(d_0^4 - d_i^4\right)$$

$$I = AK^2$$

$$K^2 = \frac{I}{A} = \frac{d_0^2 + d_i^2}{16}$$

According to Rankine's formula,

$$P_{cr} = \frac{\sigma_c \times A}{1 + a\left(\frac{l_e}{K}\right)^2}$$

For a fixed ends column,

$$l_e = \frac{l}{2} = \frac{4}{2} = 2 \text{ m}$$

$$\therefore \quad 1 \times 10^6 = \frac{78 \times \frac{\pi}{4}\left(200^2 - d_i^2\right)}{1 + \frac{1}{6400} \times \frac{2^2 \times 10^5 \times 16}{\left(200^2 + d_i^2\right)}}$$

$$= \frac{61.261\left(200^2 - d_i^2\right)}{1 + \frac{10,000}{\left(200^2 + d_i^2\right)}}$$

$$1 + \frac{10,000}{\left(200^2 + d_i^2\right)} = 61.261 \times 10^{-6}\left(200^2 - d_i^2\right)$$

$$200^2 + d_i^2 + 10,000 = 61.261 \times 10^{-6} \times \left(200^4 - d_i^4\right)$$

$$= 98,017.6 - 61.261 \times 10^{-6} \times d_i^4$$

$$61.261 \times 10^{-6} \times d_i^4 + d_i^2 - 48017.6 = 0$$

$$d_i^4 + 16323.6 d_i^2 - 783.82 \times 10^6 = 0$$

$$d_i^2 = \frac{1}{2}\left[-16323.6 \pm \sqrt{(16323.6)^2 + 4 \times 783.82 \times 10^6}\right]$$

$$= \frac{1}{2}[-16323.6 \pm 58324.4]$$

$$= 21,000$$

$$d_i = 145 \text{ mm}$$

$$\text{Thickness of column} = \frac{1}{2}(d_0 - d_i) = \frac{1}{2}(200 - 145) = 27.5 \text{ mm}$$

Q.3. (c) Refer to art. 8.4.3.

Q.4.(a)**Given:** d = 200 mm, t = 4 mm, P = 50 kN, Δp = 0.12 MPa, K = 2100 MPa, E = 150 GPa

Resultant longitudinal stress in cylinder wall when axial pull is applied,

$$\sigma_l = \frac{P}{\pi dt} + \frac{\Delta p \times d}{4t}$$

$$= \frac{50 \times 10^3}{\pi \times 200 \times 4} + \frac{0.12 \times 200}{4 \times 4} = 21.39 \text{ N/mm}^2 \text{ (tensile)}$$

$$\sigma_\theta = \frac{\Delta p \times d}{2t} = \frac{0.12 \times 200}{2 \times 4} = 3 \text{N/mm}^3 \text{ (compressive)}$$

Longitudinal strain,

$$\varepsilon_l = \frac{1}{\varepsilon}[\sigma_l - \nu\sigma_\theta]$$

$$= \frac{1}{150 \times 10^3}[21.39 + \nu \times 3]$$

Hoop strain, $$\varepsilon_\theta = \frac{1}{E}[\sigma_\theta - \nu\sigma_l]$$

$$= \frac{1}{150 \times 10^3}[-3 - \nu \times 21.39]$$

Volumentric strain of cylinder,

$$\frac{\delta V}{V} = \varepsilon_l + 2\varepsilon_\theta = \frac{1}{150 \times 10^3}[21.39 + 3\nu - 6 - 42.78 \times \nu]$$

$$= \frac{1}{150 \times 10^3}[15.39 - 39.78 \times \nu]$$

Also tensile volumetric strain of water

$$= \frac{\Delta p}{K} = \frac{0.12}{2100} = 5.714 \times 10^{-5}$$

$$\therefore \frac{1}{150 \times 10^3}[15.39 - 39.78 \times \nu] = 5.714 \times 10^{-5}$$

$$15.39 - 39.78 \times \nu = 8.571$$

$$\nu = 0.1714$$

Q.4. (b) **Given:** d = 40 mm, p = 20 MPa, σ_θ = 40 MPa

For a thick spherical shell,

$$\sigma_r = A - \frac{2B}{r^3} \text{ and } \sigma_\theta = A + \frac{B}{r^3}$$

For internal pressure only,

$$\sigma_\theta = \frac{p_1 r_1^3}{\left(r_2^3 - r_1^3\right)}\left[1 + \frac{r_2^3}{2r^3}\right]$$

σ_θ is maximum at $r = r_1$

$$\sigma_\theta = \frac{p_1 r_1^3}{\left(r_2^3 - r_1^3\right)}\left[1 + \frac{r_2^3}{2r_1^3}\right] = \frac{p_1}{\left(k^3 - 1\right)}\left(1 + \frac{k^3}{2}\right)$$

where $\quad k = \dfrac{r_2}{r_1}$

$$40 = \frac{20}{\left(k^3 - 1\right)}\left[1 + \frac{k^3}{2}\right]$$

$$2\left(k^3 - 1\right) = 1 + \frac{k^3}{2}$$

$$2k^3 - 2 = 1 + 0.5k^3$$

$$1.5k^3 = 3$$

$$k^3 = 2$$

$$k = 1.26$$

$$r_2 = 20 \times 1.26 = 25.2 \text{ mm}$$

Thickness of spherical shell,

$$t = r_2 - r_1 = 25.2 - 20 = 5.2 \text{ mm}$$

Q.4. (c) (*i*) Refer to art. 11.3.

(*ii*) Refer to art. 11.2.

Q.5(a) **Given:** $B = 30\text{mm}$, $D = 40$ mm, $R = 60$ mm, $M = -400$ N.m,

$$h^2 = \frac{R^3}{D} ln\left(\frac{2R + D}{2R - D}\right) - R^2 = \frac{60^3}{40} ln\left(\frac{2\times 60 + 40}{2\times 60 - 40}\right) - 60^2 = 143 \text{ mm}^2$$

Bending stress at the inside face,

$$\sigma_1 = \frac{-M}{AR}\left[1 - \frac{R^2}{h^2}\left(\frac{d_1}{R - d_1}\right)\right]$$

$$= \frac{-400\times 10^3}{30\times 40\times 60}\left[1 - \frac{60^2}{143}\left(\frac{20}{60 - 20}\right)\right]$$

$$= 64.37 \text{ N/mm}^2 \text{ (tensile)}$$

Bending stress at the outer face,

$$\sigma_1 = \frac{-M}{AR}\left[1 + \frac{R^2}{h^2}\left(\frac{d_2}{R + d_2}\right)\right]$$

$$= \frac{-400\times10^3}{30\times40\times60}\left[1+\frac{60^2}{143}\left(\frac{20}{60+20}\right)\right]$$

$= -40.52\ N/mm^2$ (compressive)

Position of neutral axis.

$$y = -\frac{Rh^2}{R^2+h^2} = -\frac{60\times143}{(3600+143)} = -2.29\,\text{mm}$$

Bending stress at any point,

$$\sigma = \frac{-M}{AR}\left[1+\frac{R^2}{h^2}\left(\frac{\pm y}{R\pm y}\right)\right]$$

$$= \frac{-400\times10^3}{30\times40\times60}\left[1+\frac{60^2}{14^3}\left(\frac{\pm y}{60\pm y}\right)\right]$$

$$= -5.555 - 139.86\,\frac{\pm y}{60\pm y}$$

y, mm	20	15	10	5	0	–2.29	–5	–10	–15	–20
$\sigma\ \frac{N}{mm^2}$	–40.52	–33.53	–25.53	–16.31	–5.55	0	7.16	22.42	41.06	64.37

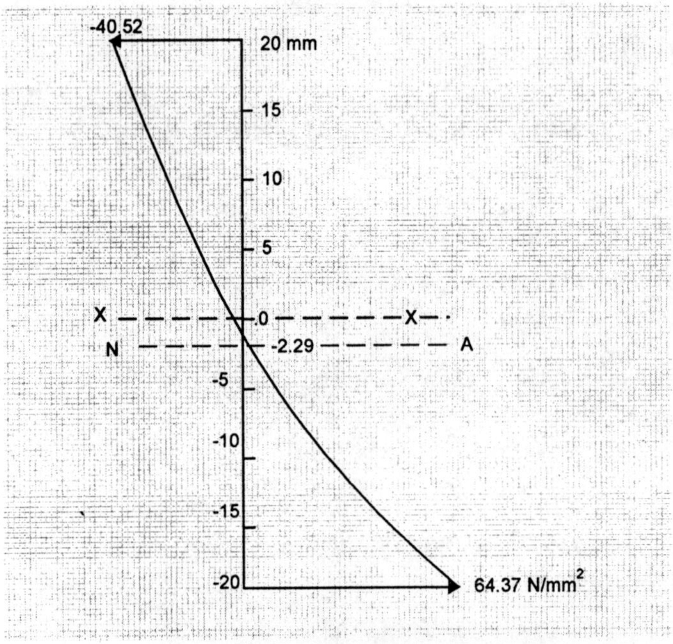

Fig. 9 Bending stress distribution.

(b) Given: D = 3 cm, W = 10 kN, R = 20 cm
Area of cross-section,

$$A = \frac{\pi}{4}D^2 = \frac{\pi}{4}\times 3^2 = 7.0686 \text{ cm}^2$$

$$h^2 = \frac{D^2}{16} + \frac{1}{128}\frac{D^4}{R^2}$$

$$= \frac{9}{16} + \frac{1}{128}\times\frac{81}{400} = 0.564 \text{cm}^2$$

The maximum tensile stress in the ring occurs at A and the maximum compressive stress occurs at B.

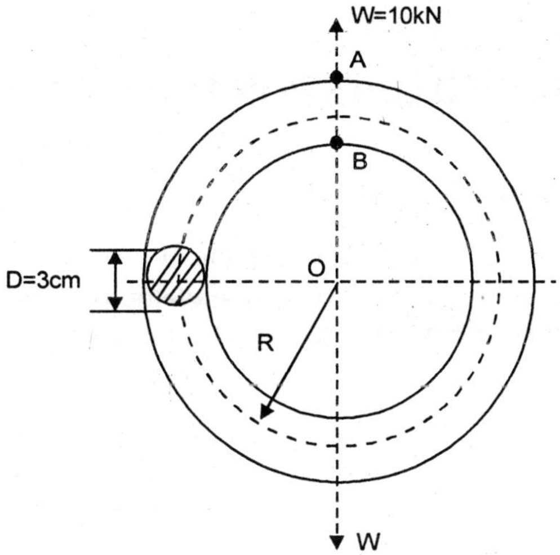

Fig. 10

$$\sigma_A = \frac{W}{\pi A}\left(\frac{R^2}{R^2+h^2}\right)\left[1+\frac{R^2}{h^2}\left(\frac{y_2}{R+y_2}\right)\right]$$

$$= \frac{10^4}{\pi\times 7.0686}\left(\frac{400}{400+0.564}\right)$$

$$\left[1+\frac{400}{0.564}\left(\frac{1.5}{20+1.5}\right)\right]$$

$= 22{,}700$ M/cm^2
$= 227$ MPa (tensile)

$$\sigma_B = \frac{W}{\pi A}\left(\frac{R^2}{R^2+h^2}\right)\left[\frac{R^2}{h^2}\left(\frac{y_1}{R-y_1}\right)-1\right]$$

$$= \frac{10^4}{\pi\times 7.0686}\left(\frac{400}{400+0.564}\right)\left[\frac{400}{0.564}\left(\frac{1.5}{20-1.5}\right)-1\right]$$

$= 25{,}408.9$ N/cm^2 or 254.09 MPa (compressive)

Q.5. (c) The stress distribution in a crane hook is hyperbolic in nature. The neutral axis N-A does not coincide with the centroidal axis x – x and is pulled towards the centre of curvature. Therefore, the tensile stress on the inner face of the hook, which is nearer to the centre of curvature, is larger than the compressive stress on the outer face. This requires larger area on the inner side of the hook as compared to the outer side. The trapezoidal cross-section, is therefore, most suitable to meet these requirements with larger side nearer to the centre of curvature.

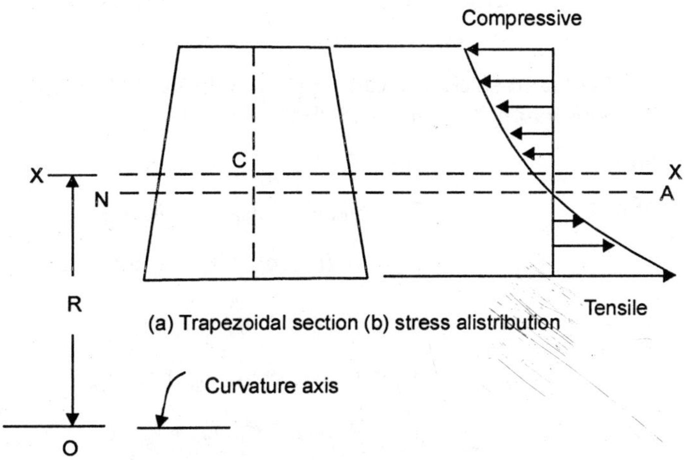

Fig. 11

Appendix–II

SOLVED QUESTION FROM PREVIOUS UNIVERSITY EXAMINATIONS

Example 1 *Derive expressions for normal and shear stress on an oblique plane.*

Solution. Normal and Shearing Stresses

Consider a rectangular parallelopiped, as shown in Fig.1. Let σ_r be the resultant stresses at a point O and $\sigma_{rx}, \sigma_{ry}, \sigma_{rz}$, are its components along the three axes of reference.

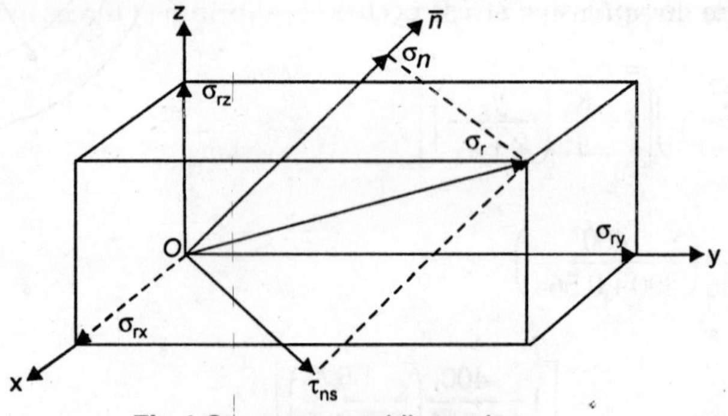

Fig.1 Stress on an oblique plane

Let a_{rx}, a_{ry}, a_{rz} be the direction cosines of the resultant stress σ_r, so that,

$$a_{rx} = \frac{\sigma_{rx}}{\sigma_r}$$

$$a_{ry} = \frac{\sigma_{ry}}{\sigma_r}$$

$$a_{rz} = \frac{\sigma_{rz}}{\sigma_r}$$

The resultant stress σ_r, can be resolved into normal stress σ_n along the unit normal $\bar{n}$ having direction cosines a_{nx}, a_{ny}, a_{nz} and shear stress τ_{ns}.

Now $\quad \sigma_n = \sigma_r \cdot a_{nr}$

where $\quad a_{nr} = a_{nx} . a_{rx} + a_{ny} . a_{ry} + a_{nz} . a_{rz}$

$\therefore \quad \sigma_n = \sigma_r (a_{nx} . a_{rx} + a_{ny} . a_{ry} + a_{nz} . a_{rz})$

$$\sigma_n = \sigma_{rx} . a_{nx} + \sigma_{ry} . a_{ny} + \sigma_{rz} . a_{nz}$$

Also $\quad \sigma_{rx} = \sigma_x . \sigma_{nx} + \tau_{xy} . a_{ny} + \tau_{xz} . a_{nz}$

$$\sigma_{ry} = \tau_{xy} . a_{nx} + \sigma_y . a_{ny} + \tau_{yz} . a_{nz}$$

$$\sigma_{rz} = \tau_{xz} . a_{nx} + \tau_{yz} . a_{ny} + \sigma_z . a_{nz}$$

These are called the *stress vectors.*

Hence the normal stress σ_n becomes,

$$\sigma_n = \sigma_x a_{nx}^2 + \sigma_y \cdot a_{ny}^2 + \sigma_z . a_{nz}^2 + (\tau_{xy} . a_{nx} a_{ny} + \tau_{yz} . a_{ny} a_{nz} + \tau_{xz} . a_{nx} a_{nz})$$

Also $$\sigma_n^2 + \tau_{ns}^2 = \sigma_r^2$$

$\therefore$ $$\tau_{ns} = \sqrt{\sigma_r^2 - \sigma_n^2}$$

where $$\sigma_r = \sqrt{\sigma_{rx}^2 + \sigma_{ry}^2 + \sigma_{rz}^2}$$

Let a_{sx}, a_{sy}, a_{sz} be the direction cosines of τ_{ns}. For equilibrium of forces in the x-direction, we have

$$\sigma_n . a_{nx} + \tau_{ns} . a_{sx} = \sigma_{rx}$$

$\therefore$ $$a_{sx} = \frac{1}{\tau_{ns}}[\sigma_{rx} - \sigma_n . a_{nx}]$$

$$= \frac{1}{\tau_{ns}}[(\sigma_x - \sigma_n) a_{nx} + \tau_{xy} a_{ny} + \tau_{xz} a_{nz}]$$

Similarly $$a_{sy} = \frac{1}{\tau_{ns}}[\tau_{xy} a_{nx} + (\sigma_y - \sigma_n) a_{ny} + \tau_{yz} a_{nz}]$$

$$a_{sz} = \frac{1}{\tau_{nz}}[\tau_{xy} a_{nx} + \tau_{yz} a_{ny} + (\sigma_z - \sigma_n) a_{nz}]$$

Example 2 *The stress component at a point are: $\sigma_x = 20$, $\sigma_y = 10$, $\sigma_z = 5$, $\tau_{xy}, = 10$, $\tau_{yz} = 4$ and $\tau_{xz} = 5$ MPa. Deteremine the normal and shearing stresses on the plane whose direction cosines are $\frac{1}{\sqrt{3}}, \frac{1}{\sqrt{3}}, \frac{1}{\sqrt{3}}$. Also calcualte the direction of the shear stress.*

Solution.

Resolving the stresses along the three co-ordinate axes, we have

$$\sigma_{rx} = \sigma_x a_{nx} + \tau_{xy} a_{ny} + \tau_{xz} \cdot a_{nz}$$

$$= \frac{1}{\sqrt{3}}(20 + 10 + 5) = \frac{35}{\sqrt{3}} \text{ MPa}$$

$$\sigma_{ry} = \tau_{xy} . a_{nx} + \sigma_y . a_{ny} + \tau_{yz} . a_{nz}$$

$$= \frac{1}{\sqrt{3}}(10 + 10 + 4) = \frac{24}{\sqrt{3}} \text{ MPa}$$

$$\sigma_{rz} = \tau_{xz} . a_{nx} + \tau_{yz} . a_{ny} + \sigma_z . a_{nz}$$

$$= \frac{1}{\sqrt{3}}(5 + 4 + 5) = \frac{14}{\sqrt{3}} \text{ MPa}$$

Resultant stress, $\sigma_r = \sqrt{\sigma_{rx}^2 + \sigma_{ry}^2 + \sigma_{rz}^2}$

$$= \sqrt{\frac{1}{3}[(35)^2] + (24)^2 + (14)^2]}$$

$$= \sqrt{\frac{1}{3}(1225 + 576 + 196)}$$

$$= \sqrt{\frac{1997}{3}} = 25.8 \text{ MPa}$$

Normal stress, $\sigma_n = \sigma_{rx} \cdot a_{nx} + \sigma_{ry}.a_{ny} + \sigma_{rz}.a_{nz}$

$$= \frac{1}{3}(35 + 24 + 14)$$

$$= \frac{73}{3} = 24.3 \text{ MPa}$$

∴ Shear stress, $\tau_{ns} = \sqrt{\sigma_r^2 - \sigma_n^2}$

$$= \sqrt{665.67 - 592.11}$$

$$= \sqrt{73.56} = 8.576 \text{ MPa}$$

The direction of the shear stress is obtained as follows:

$$a_{sx} = \frac{1}{\tau_{ns}}[\sigma_{rx} - \sigma_n.a_{nx}]$$

$$= \frac{1}{8.576}\left[\frac{35}{\sqrt{3}} - \frac{24.3}{\sqrt{3}}\right]$$

$$= \frac{10.7}{\sqrt{3} \times 8.576} = 0.7203$$

$$a_{sy} = \frac{1}{\tau_{nz}}[\sigma_{ry} - \sigma_n .a_{ny}]$$

$$= \frac{-0.3}{\sqrt{3} \times 8.576} = -0.0202$$

$$a_{sz} = \frac{1}{\tau_{ns}}[\sigma_{rz} - \sigma_n.a_{nz}]$$

$$= \frac{1}{8.576}\left[\frac{14}{\sqrt{3}} - \frac{24.3}{\sqrt{3}}\right]$$

$$= \frac{-10.3}{\sqrt{3} \times 8.576} = -0.6934$$

Example 3 *At a point P in a body the stresses are:*

$$\sigma_x = 10000 \text{ N/cm}^2, \sigma_y = -5000 \text{ N/cm}^2, \tau_{xy} = \tau_{yz} = \tau_{xz} = 10000 \text{ N/cm}^2.$$

Determine normal stress on a plane that is equally inclined to all three axes. *[UPTU, 2008-2009]*

Solution. The direction cosines of the plane equally inclined to the three axes are: $\frac{1}{\sqrt{3}}, \frac{1}{\sqrt{3}}, \frac{1}{\sqrt{3}}$.

The resultant stresses along the three coordinate axes are:

$$\sigma_{rx} = \sigma_x \cdot a_{nx} + \tau_{xy} \cdot a_{ny} + \tau_{xz} \cdot a_{nz}$$

$$= \frac{1}{\sqrt{3}}(1000 + 2 \times 1000) = 17320.5 \text{ N/cm}^2$$

$$\sigma_{ry} = \tau_{xy} \cdot a_{nx} + \sigma_y \cdot a_{ny} + \tau_{yz} \cdot a_{nz}$$

$$= \frac{1}{\sqrt{3}}(10000 - 5000 + 10000) = 8660.25 \text{ N/cm}^2$$

$$\sigma_{rz} = \tau_{xz} \cdot a_{nx} + \tau_{yz} \cdot a_{ny} + \sigma_z \cdot a_{nz}$$

$$= \frac{1}{\sqrt{3}}(1000 + 1000 - 5000) = 8660.25 \text{ N/cm}^2$$

Normal stress,

$$\sigma_n = \sigma_{rx} \cdot a_{nx} + \tau_{ry} \cdot a_{ny} + \sigma_{rz} \cdot a_{nz}$$

$$= \frac{1}{\sqrt{3}}(17320.5 + 8660.25 + 8660.25)$$

$$= 20{,}000 \text{ N/cm}^2$$

Example 4. *A point in a strained material is subjected to a tensile stress of 65 N/mm² and a compressive stress of 40 N/mm², acting on two mutually perpendicular planes and a shear stress of 10 N/mm² are acting on these planes. Find the normal stress, tangential stress and resultant stress on a plane inclined at 30° with the plane of the compressive stress.* *[UPTU, 2006-07]*

Solution.

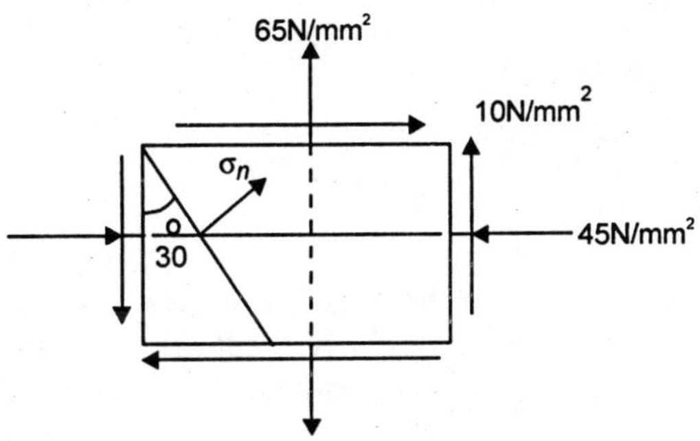

Fig. 2

Given: $\sigma_x = -45\ \text{N/mm}^2$, $\sigma_y = 65\ \text{N/mm}^2$, $\tau_{xy} = 10\ \text{N/mm}^2$

$$\sigma_n = \frac{1}{2}(\sigma_x + \sigma_y) + \frac{1}{2}(\sigma_x - \sigma_y)\cos 2\theta + \tau_{xy}\sin 2\theta$$

$$= \frac{1}{2}(-45+65) + \frac{1}{2}(-45j - 65)\cos 60° + 10\sin 60°$$

$$= 10 - 27.5 + 8.66 = -8.84\ \text{N/mm}^2$$

$$\tau = \frac{1}{2}(\sigma_x - \sigma_y)\sin 2\theta - \tau_{xy}\cos 2\theta$$

$$= \frac{1}{2}(-45-65)\sin 60° - 10\cos 60°$$

$$= -52.63\ \text{N/mm}^2$$

$$\sigma_r = \sqrt{\sigma_n^2 + \tau^2} = \sqrt{(-8.84)^2 + (-52.63)^2} = 53.37\ \text{N/mm}^2$$

Example 5. *At a point in a body, the displacement field is linear and is given by the following expressions. Find all the strains.*

$u = 0.07x + 0.05y + 0.01z$, $v = 0.01y - 0.04x$, $w = 0.02x + 0.022$. *[UPTU, 2008 - 09]*

Solution.

$$\varepsilon_x = \frac{\partial u}{\partial x} = 0.07$$

$$\varepsilon_y = \frac{\partial v}{\partial y} = 0.01$$

$$\varepsilon_z = \frac{\partial w}{\partial z} = 0.02$$

$$\gamma_{xy} = \frac{1}{2}\left(\frac{\partial u}{\partial y} + \frac{\partial v}{\partial x}\right) = \frac{1}{2}(0.05 - 0.04) = 0.005$$

$$\gamma_{yz} = \frac{1}{2}\left(\frac{\partial v}{\partial z} + \frac{\partial w}{\partial y}\right) = \frac{1}{2}(0 + 0) = 0$$

$$\gamma_{zx} = \frac{1}{2}\left(\frac{\partial w}{\partial x} + \frac{\partial u}{\partial z}\right) = \frac{1}{2}(0.02 + 0.01) = 0.015$$

Example 6. *At a point in a steel member, the major principal stress is 250 N / mm² and the minor principal stress is compressive. If the tensile yield stress is 250 N / mm², find the minor principal stress at which yielding will commence using the following theories of failure:*

(i) *Maximum total strain energy theory, and*

(ii) *Maximum shear strain energy theory.*

Take Poisson's ratio = 0.25 *[UPTU, 2001 - 02]*

Solution.

Given: $\sigma_1 = 250\ \text{N/mm}^2$, $\sigma_2 = -ve$, $\sigma_{yt} = 250\ \text{N/mm}^2$

(*i*) According to total strain energy theory,

$$\sigma_1^2 + \sigma_2^2 - 2\nu\sigma_1\sigma_2 = \sigma_{yt}^2$$

$$(250)^2 + (-\sigma_2)^2 - 2 \times 0.25 \times 250 \times (-\sigma_2) = (250)^2$$

$$\sigma_2 = 125\ \text{N/mm}^2$$

(*ii*) According to maximum shear strain energy theory,

$$(\sigma_1 - \sigma_2)^2 + (\sigma_2 - \sigma_3)^2 + (\sigma_3 - \sigma_1)^2 = 2\sigma_{yt}^2$$

$$(\sigma_1 + \sigma_2)^2 + \sigma_2^2 + \sigma_1^2 = 2\sigma_{yt}^2$$

$$\sigma_1^2 + \sigma_2^2 + \sigma_1\sigma_2 = \sigma_{yt}^2$$

$$(250)^2 + \sigma_2^2 - 250\sigma_2 = 250^2$$

$$\sigma_2 = 250\ \text{N/mm}^2$$

Example 7 In a *metallic body, the principal stresses are + 35 MN / m² and – 95 MN / m², the third principal stress being zero. The elastic limit stress in simple tension as well as in simple compression is same and is 220 MN / m². Find the factor of safety based on the elastic limit if the criterion of failure for the material is the maximum principal stress theory.* *[UPTU, 2003 – 04]*

Solution.

Given: $\sigma_1 = +35$ MPa, $\sigma_2 = 0$, $\sigma_3 = -95$ MPa, $\sigma_e = 220$ MPa

According to maximum principal stress theory,

$$FOS = \frac{\sigma_e}{\sigma_1} = \frac{220}{35} = 6.286$$

Example 8 *The three principal stresses at a point in a body subjected to a system of loading are 2.5 σ (tensile), 1.2 σ (tensile) and 0.8 σ (compressive). If yield stress for the material is 280 MN / m², using factor of safety = 2, Poisson's ratio = 0.3, determine the maximum value of σ so that failure may not occur according to maximum principal stress theory and maximum shear stress theory.* *[UPTU, 2008 - 09]*

Solution.

Given. $\sigma_1 = +2.5\sigma$, $\sigma_2 = +1.2\sigma$, $\sigma_3 = -0.8\sigma$, $\sigma_y = 280\ \text{MN/m}^2$, FOS = 2, $\nu = 0.3$

(*i*) According to maximum principal stress theory,

$$2.5\,\sigma = \frac{280}{2}$$

$$\sigma = 56\ \text{MN/m}^2$$

(ii) According to maximum shear stress theory,

$$\frac{\sigma_1 - \sigma_3}{2} = \frac{\sigma_y}{2 \times FOS}$$

$$2.5\,\sigma + 0.8\,\sigma = \frac{280}{2}$$

$$\sigma = \frac{140}{3.3} = 42.42 \text{ MN/m}^2$$

Example 9. *A mild steel shaft 100 mm diameter is subjected to a maximum torque of 15 kN.m and a maximum bending moment of 10 kN.m at a particular section. Find the factor of safety accord.ng to the maximum shear stress theory of failure if the elastic limit in simple tension is 240 MN/ m².* *[UPTU, 2006–07]*

Solution.

Given: $d = 100$ mm, $T = 15$ kN.m, $M = 10$ kN.m, $\sigma_{et} = 240$ MN/m^2

Equivalent torque, $T_e = \sqrt{M^2 + T^2} = \sqrt{100 + 225} = 18.028\ kN.m$

$$\text{Maximum shear stress, } \tau = \frac{16T_e}{\pi d^3} = \frac{16 \times 18.0.28 \times 10^6}{\pi \times 100^3} = 91.81 \text{ MN/m}^2$$

$$\text{FOS} = \frac{\sigma_{et}}{2} \times \frac{1}{\tau} = \frac{240}{2} \times \frac{1}{91.81} = 1.3$$

Example 10 *(i) The stresses in three principal directions are : + 65 MN / m², + 20 MN / m² and – 85 MN/ m². Find the principal strain. Take ν = 0.3 and E = 200 GN/m².*

(ii) A mild steel shaft 100 mm diameter is subjected to a maximum torque of 15 kN.m and a maximum bending moment of 10 kN.m at a particular section. Find the factor of safety according to the maximum shear stress theory of failure if the elastic limit in simple tension is 240 MN / m². [UPTU, 2006 – 07]

Solution.

Given: $\sigma_1 = +65$ MN/m2, $\sigma_2 = +20$ MN/m^2, $\sigma_3 = -85$ MN/m^2, $\nu = 0.3$, $E = 200$ GN/m^2

$$\varepsilon_1 = \frac{1}{E}[\sigma_1 - \nu(\sigma_2 + \sigma_3)]$$

$$= \frac{1}{200 \times 10^9}[65 - 0.3(20 - 85)] \times 10^6$$

$$= 422.5 \times 10^{-6}$$

$$\varepsilon_2 = \frac{1}{E}[\sigma_2 - \nu(\sigma_3 + \sigma_1)]$$

$$= \frac{1}{200 \times 10^9}[20 - 0.3(-85 + 65)] \times 10^6$$

$$= 130 \times 10^{-6}$$

$$\varepsilon_3 = \frac{1}{E}[\sigma_3 - \nu(\sigma_1 + \sigma_2)]$$

$$= \frac{1}{200 \times 10^9}[-85 - 0.3\,(65 + 20)] \times 10^6$$

$$= -552.5 \times 10^{-6}$$

(*ii*) Given d = 100 mm, T = 15 kN.m, M = 10 kN.m, σ_t = 240 MN.m²

Equivalent, torque, $T_e = \sqrt{M^2 + T^2} = \sqrt{100 + 225} = 18.025$ kN.m

Shear stress, $\tau = \frac{16T_e}{\pi d^3} = \frac{16 \times 18.028 \times 10^6}{\pi \times 100^3} = 91.81\ \text{MN/m}^2$

$$\text{FOS} = \frac{\sigma_t}{2} \times \frac{1}{\tau} = \frac{240}{2} \times \frac{1}{91.81} = 1.3$$

Example 11 *A circular bar ABC, 3 m long, is rigidly fixed at its ends A and C. The portion AB is 1.8 m long and 50 mm diameter and BC is 1.2 m long and 25 mm diameter. If the twisting moment of 680 N.m is applied at B, determine the values of the resisting moments at A and C and the maximum stress in each section of the shaft. What will be the angle of twist of each portion? For the material of the shaft, G = 80 GPa.*

[UPTU, 2005 - 06]

Solution.

Given: $l_1 = 1.8$ m, $d_1 = 50$ mm, $l_2 = 1.2$ m, $d_2 = 25$ mm, $T_B = 680$ N.m, $G = 80$ GPa

Refer to Fig. 3 $T_A + T_C = T_B = 680$ N.m

Also $\theta_{C/A} = 0$

$$\frac{T_C l_2}{GJ_2} + \frac{T_C l_1}{GJ_1} - \frac{T_B l_1}{GJ_1} = 0$$

$$T_C\left[\frac{l_2}{J_2} + \frac{l_1}{J_1}\right] = \frac{T_B l_1}{J_1}$$

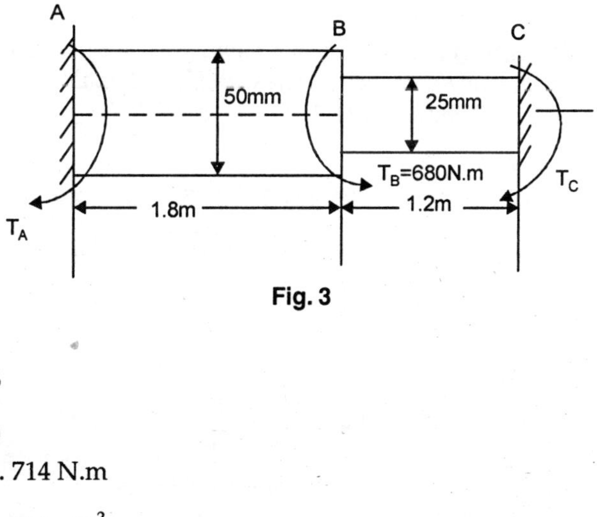

Fig. 3

$$T_c\left[\frac{1.2 \times 32}{\pi \times 25^4} + \frac{1.8 \times 32}{\pi \times 50^4}\right] = T_B\left[\frac{1.8 \times 32}{\pi \times 50^4}\right]$$

$$T_c\,[3.36 \times 10^{-6}] = T_B \times 0.288 \times 10^{-6}$$

$$T_B = 11.667\, T_C = 680$$

$$T_A = 680 - 286 = 621.714 \text{ N.m}$$

$$\tau_{AB} = \frac{16T_A}{\pi d_1^3} = \frac{16 \times 621.714 \times 10^3}{\pi \times 50^3} = 25.33\ \text{N/mm}^2$$

$$\tau_{BC} = \frac{16T_C}{\pi d_2^3} = \frac{16 \times 58.286 \times 10^3}{\pi \times 25^3} = 19\ \text{N/mm}^2$$

$$\theta_{BC} = \frac{T_C l_2}{GJ_2} = \frac{58.286 \times 10^3 \times 1.2 \times 10^3 \times 32}{80 \times 10^3 \times \pi \times 25^4} = 2.28 \times 10^{-3} \text{ rad or } 0.13°$$

$$\theta_{AC} = \frac{T_A l_1}{G J_1} = \frac{621.714\times10^3\times8\times10^3\times32}{80\times10^3\times\pi\times50^4} = 0.0228 \text{ rad or } 1.31°$$

Example 12 *In a hollow shaft, the external diameter is 100 mm and internal diameter is 50 mm. The allowable shear stress in the shaft material is 55 N / mm². Determine the angle of twist in a length of twenty times the external diameter of the shaft. G = 84 GPa.* [UPTU, 2006 - 07]

Solution.

$$d_0 = 100 \text{ mm}, d_i = 50 \text{ mm}, \tau = 55 \text{ N/mm}^2, l = 20\, d_0, G = 84 \text{ GPa}$$

$$\theta = \frac{Tl}{GJ}, \tau = \frac{T d_0}{2J}$$

$$T = \frac{2J\tau}{d_0} = \frac{2\times\pi(d_0^4 - d_i^4)\tau}{32\, d_0} = \frac{2\times\pi(100^4 - 50^4)\times55}{32\times100}$$

$$= 10{,}124,273 \text{ N.mm}$$

$$\theta = \frac{10{,}124{,}273\times20\times100\times32}{84\times10^3\times\pi(100^4-50^4)} = 0.0262 \text{ rad or } 1.5°$$

Example 13 *A solid shaft is subjected to a bending moment of 2.3 kN.m and a twisting moment of 3.45 kN.m. Find the diameter of the shaft if the allowable tensile and shear stresses for the shaft material are limited to 703 MN/m² and 421.8 MN/m² respectively.* [UPTU, 2008-09]

Solution.

Given: $M = 2.3$ kN.m , $T = 3.45$ kN.m , $\sigma_t = 703$ MN/m², $\tau = 421.8$ N/m²

Equivalent torque, $T_e = \sqrt{M^2 + T^2} = \sqrt{(2.3)^2 + (3.45)^2} = 4.146$ kN.m

$$T_e = \frac{\pi}{16} d^3 \tau$$

$$4.146\times10^6 = \frac{\pi}{16}\times d^3\times421.8$$

$$d = 36.85 \text{ mm}$$

Equivalent bending moment, $M_e = \frac{1}{2}[M + \sqrt{M^2 + T^2}]$

$$= \frac{1}{2}[2.3 + 4.146] = 3.223 \text{ kN.m}$$

$$M_e = \frac{\pi}{32} d^3 \sigma_t$$

$$3.223 \times 10^6 = \frac{\pi}{32} \times d^3 \times 703$$

$$d = 36.01 \text{ mm}$$

Adopt $$d = 36.85 \simeq 37 \text{ mm}$$

Example 14 *A simply supported beam has a flexural rigidity of 24 MN.m² and is loaded as shown in Fig. (a) Determine the deflection at mid - span.* *[UPTU, 2008-09]*

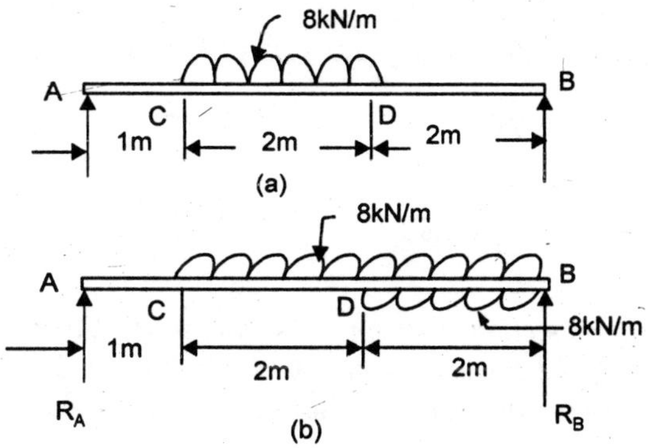

Fig. 4

Solution.

Given: $EI = 24$ MN.m², $w = 8$ kN/m

The *udl* can be extended upto the end of the beam and upwards *udl* of same intensity applied on the last 2 m of the span (see Fig. 4).

Taking moments about *B*, we have

$$5R_A = 8 \times 2 \times 3$$

$$R_A = 9.6 \text{ kN}$$

Taking a section nearer to the end *B* of the beam, the bending moment there is,

$$M_x = 9.6\,x - 4\,(x-1)^2 + 4\,(x-3)^2$$

$$EI\frac{d^2y}{dx^2} = -M_x = -9.6x + 4(x-1)^2 - 4(x-3)^2$$

Integrating twice, we have

$$EI\frac{dy}{dx} = -4.8\,x^2 + \frac{4}{3}(x-1)^3 - \frac{4}{3}(x-3)^3 + C_1$$

$$EIy = -1.6\,x^3 + \frac{1}{3}(x-1)^4 - \frac{1}{3}(x-3)^4 + C_1x + C_2$$

$$y = 0 \text{ at } x = 0 \text{ and } x = 5 \text{ m}$$

$\therefore \quad C_2 = 0$

$$0 = -1.6\times 5^3 + \frac{4^4}{3} - \frac{2^4}{3} + 5C_1$$

$$C_1 = 24$$

$$\therefore \quad EIy = -1.6x^3 + \frac{1}{3}(x-1)^4 - \frac{1}{3}(x-3)^4 + 24x$$

At $\quad x = 2.5$ m

$$EIy = -1.6(2.5)^3 + \frac{1}{3}(2.5-1)^4 + 24\times 2.5$$

$$= 36.6875$$

$$y = \frac{36.6875\times 10^3}{24\times 10^6} = 1.5286\ 10^{-3}\text{ m} \quad \text{or} \quad 1.5286 \text{ mm}$$

Example 15 *A beam of uniform section,10 m long, is simply supported at the ends. It carries point loads of 150 kN and 65 kN at a distance of 2.5 m and 5.5 m respectively from the left end. Calculate : (i) the deflection under each load, and (ii) the maximum deflection. Take; E = 200 GN/m² and I = 118 × 10⁻⁴ m⁴.*

[UPTU, 2006-07]

Solution

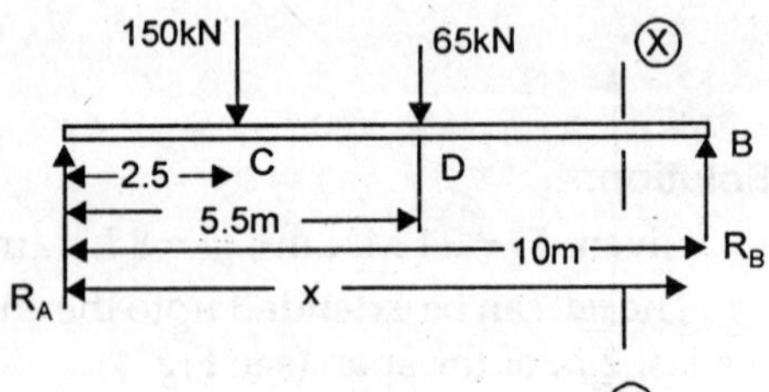

Fig. 5

Given: $E = 200$ GN/m², $I = 118 \times 10^{-4}$ m⁴

Taking moments about B, we have(Fig. 5)

$$10\, R_A = 150 \times 7.5 + 65 \times 4.5 = 1417.5$$

$$R_A = 141.75 \text{ kN}$$

$$M_x = 141.75 \times x - 150\,(x - 2.5) - 65\,(x - 5.5)$$

$$EI\frac{d^2y}{dx^2} = -M_x = -141.75\,x + 150\,(x - 2.5) + 65\,(x - 5.5)$$

Integrating twice, we have

$$EI\frac{dy}{dx} = -70.875\,x^2 + 75\,(x - 2.5)^2 + 32.5\,(x - 5.5)^2 + C_1$$

$$EIy = -23.625\,x^3 + 25\,(x - 2.5)^3 + 10.833\,(x - 5.5)^3 + C_1x + C_2$$

$$y = 0 \text{ at } x = 0 \quad \text{and} \quad x = 10 \text{ m}$$

$\therefore \quad C_2 = 0$

$$0 = -23.625 \times 10^3 + 25 \times (7.5)^3 + 10.833 \times (4.5)^3 + 10\,C_1$$

$$C_1 = 1209.1$$

$$\therefore \quad EIy = -23.625\,x^3 + 25(x-2.5)^3 + 10.833\,(x - 5.5)^3 + 1209.1\,x$$

(*i*) At $x = 2.5$ m

$$EIy_C = 2653.61$$

$$y_C = \frac{2653.61\times 10^3}{200\times 10^9\times 118\times 10^{-4}} = 1.12 \text{ mm}$$

At $x = 5.5$ m

$$EIy_D = 3394.44$$

$$y_D = \frac{3394 \times 44 \times 10^3}{200 \times 10^9 \times 118 \times 10^{-4}} = 1.438 \text{ mm}$$

(*ii*) For maximum deflection, $\frac{dy}{dx} = 0.$

$$-70.875\,x^2 + 75\,(x - 2.5)^2 + 1209.1 = 0$$

$$-70.875\,x^2 + 75x^2 + 75x^2 + 468.75 - 375 + 1209.1 = 0$$

$$4.125x^2 - 375x + 1677.6 = 0$$

$$x^2 - 90.91x + 406.69 = 0$$

$$x^2 - 90.91\,x + 406.69 = 0$$

$$x^2 - 90.91x + 406.69 = 0$$

$$x = \frac{1}{2}[90.91 \pm \sqrt{(90.91)^2 - 4 \times 406.69}]$$

$$= \frac{1}{2}[90.91 \pm 81.47]$$

$$= 4.72 \text{ m}$$

At $x = 4.72$ m

$$EIy = -23.625\,(4.72)^3 + 25\,(4.72 - 2.5)^3 + 1209.1 \times 4.72$$

$$= 3496.2$$

$$y = \frac{3496.2 \times 10^3}{200 \times 10^9 \times 118 \times 10^4} = 1.48 \text{ mm}$$

Example 16. *A timber beam of 3 m span carries a uniformly distributed toad of 5 kN / m and a point load 1 kN at the centre of the span. If the permissible bending stress be 100 N / mm², find the section taking depth as twice the breadth,* *[UPTU, 2006 - 07]*

Solution.

Given: $P = 1$ kN, w = 5 kN/m, $l = 3$ m, $h = 2b$, $\sigma = 100$ N/mm²

Refer to Fig. 6.

$$M_{max} = \frac{Pl}{4} + \frac{wl^2}{8} = \frac{1 \times 3}{4} + \frac{5 \times 9}{8} = 6.375 \text{ kN.m}$$

$$z = \frac{bh^2}{6} = \frac{b \times (2b)^2}{6} = \frac{2}{3}b^3$$

$$\sigma = \frac{M_{max}}{z}$$

$$100 = \frac{6.375 \times 10^6 \times 3}{2b^3}$$

$$b^3 = 95625$$

$$b = 45.72 \text{ mm}, h = 91.44 \text{ mm}$$

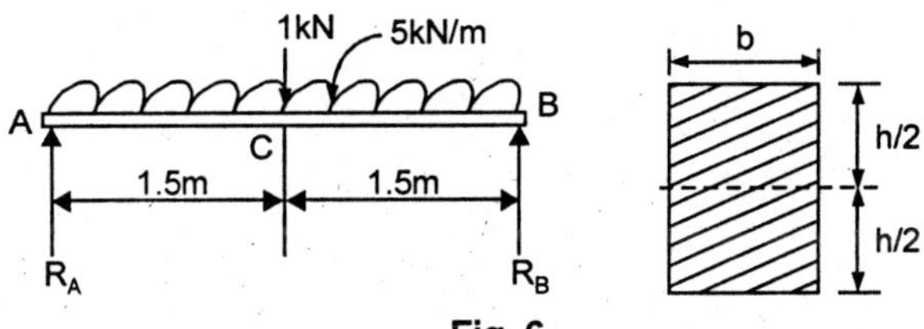

Fig. 6

Example 17. *Determine the numerical, value of the ratio $\sigma_{max}/\sigma_{min}$ for the case of plane bending of a curved beam having 2.5 x 2.5 cm square cross - section if the radius of curvature of the centroidal axis is 3.75 cm.*

[UPTU, 2007- 08]

Solution.

Given: $B = D = 2.5$ cm, $R = 3.75$ cm (Refer to Fig. 7)

Area of cross section, $A = B \times D = 2.5 \times 2.5 = 6.25\ \text{cm}^3$

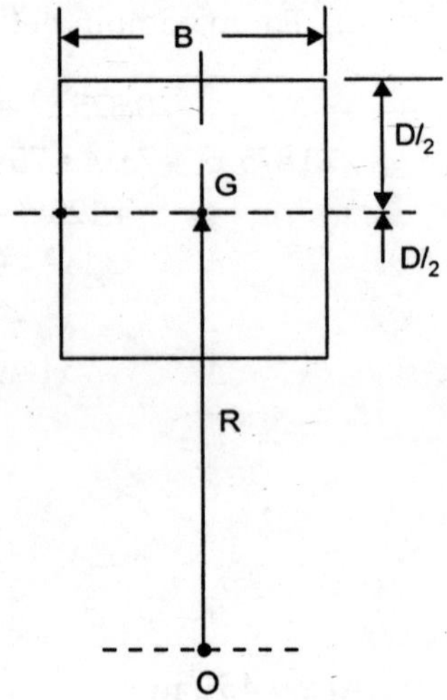

Fig. 7

$$h^2 = \frac{R^3}{D} ln\left[\frac{2R+D}{2R-D}\right] - R^2$$

$$= \frac{(3.75)^3}{(2.5)} ln\left[\frac{2\times3.75+2.5}{2\times3.75-2.5}\right] - (3.75)^2$$

$$= 0.5586\ \text{cm}^2$$

$$\sigma = \frac{M}{AR}\left[1+\frac{R^2}{h^2}\left(\frac{\pm y}{R\pm y}\right)\right]$$

$$\sigma_{max} = \frac{M}{AR}\left[1+\frac{R^2}{h^2}\left(\frac{D/2}{R+D/2}\right)\right]$$

$$\sigma_{min} = \frac{M}{AR}\left[1-\frac{R^2}{h^2}\left(\frac{D/2}{R-D/2}\right)\right]$$

$$\frac{\sigma_{max}}{\sigma_{min}} = \frac{1+\frac{R^2}{h^2}\left(\frac{D}{2R+D}\right)}{1-\frac{R^2}{h^2}\left(\frac{D}{2R-D}\right)} = \frac{1+\frac{(3.75)^2}{0.5586}\left[\frac{2.5}{2\times3.75+2.5}\right]}{1-\frac{(3.75)^2}{0.5586}\left(\frac{2.5}{2\times3.75-2.5}\right)}$$

$$= \left|\frac{7.29363}{-11.58721}\right| = -0.629$$

SUBJECT INDEX